IFAC

International Federation of Automatic Control

COMPUTER AIDED DESIGN IN CONTROL AND ENGINEERING SYSTEMS

IFAC Proceedings Series, 1986. Number 5

1985

No. 1 *PERKINS, MACFARLANE & LEININGER: Analysis and Synthesis of Control Systems

No. 2 *LJUNG & ÅSTRÖM: Identification, Adaptive and Stochastic Control

No. 3 *TITLI, SINGH, PACHTER, HAUTUS, KOKOTOVIC & RAUCH: Large-scale Systems, Decision-making, Mathematics of Control

No. 4 *NISENFELD, LEFKOWITZ, PAI & DY LIACCO: Process Industries, Power Systems

No. 5 *REMBOLD, KEMPF, TOWILL, JOHANNSEN, PAUL, HARRISON, EHRENBERGER, TABAK, LESKIEWICZ, FENTON & DEBRA: Manufacturing, Man–Machine Systems, Computers, Components, Traffic Control, Space Applications

No. 6 *VOSSIUS, WEED, RESWICK, COBELLI, HAIMES, TAKAMATSU, KAYA, NAJIM, MARTENSSON, CHESTNUT & LARSEN: Biomedical Applications, Water Resources, Environment, Energy Systems, Development, Social Effects, SWIIS, Education

(**A Bridge between Control Science and Technology, Ninth Triennial World Congress, General Editors: J. Gertler and L. Keviczky*)

No. 7 BARKER & YOUNG: Identification and System Parameter Estimation (1985)

1986

No. 1 NORRIE & TURNER: Automation for Mineral Resource Development

No. 2 CHRETIEN: Automatic Control in Space

No. 3 DA CUNHA: Planning and Operation of Electric Energy Systems

No. 4 VALADARES TAVARES & EVARISTO DA SILVA: Systems Analysis Applied to Water and Related Land Resources

No. 5 LARSEN & HANSEN: Computer Aided Design in Control and Engineering Systems

No. 6 PAUL: Digital Computer Applications to Process Control

No. 7 YANG JIACHI: Control Science & Technology for Development

No. 8 MANCINI, JOHANNSEN & MARTENSSON: Analysis, Design and Evaluation of Man–machine Systems

No. 9 GELLIE, FERRATE & BASANEZ: Robot Control "Syroco '85"

No. 10 JOHNSON: Modelling and Control of Biotechnological Processes

IFAC Related Titles

BROADBENT & MASUBUCHI: Multilingual Glossary of Automatic Control Technology

EYKHOFF: Trends and Progress in System Identification

ISERMANN: System Identification Tutorials (*Automatica Special Issue*)

COMPUTER AIDED DESIGN IN CONTROL AND ENGINEERING SYSTEMS

ADVANCED TOOLS FOR MODERN TECHNOLOGY

Proceedings of the 3rd IFAC/IFIP Symposium
Lyngby, Denmark, 31 July–2 August 1985

Edited by

P. MARTIN LARSEN

Technical University of Denmark

and

N. E. HANSEN

NEH Engineering, Denmark

Published for the

INTERNATIONAL FEDERATION OF AUTOMATIC CONTROL

by

PERGAMON PRESS

OXFORD · NEW YORK · TORONTO · SYDNEY · FRANKFURT
TOKYO · SAO PAULO · BEIJING

U.K.	Pergamon Press, Headington Hill Hall, Oxford OX3 0BW, England
U.S.A.	Pergamon Press Inc., Maxwell House, Fairview Park, Elmsford, New York 10523, U.S.A.
CANADA	Pergamon Press Canada, Suite 104, 150 Consumers Road, Willowdale, Ontario M2J 1P9, Canada
AUSTRALIA	Pergamon Press (Aust.) Pty. Ltd., P.O. Box 544, Potts Point, N.S.W. 2011, Australia
FEDERAL REPUBLIC OF GERMANY	Pergamon Press GmbH, Hammerweg 6, D-6242 Kronberg, Federal Republic of Germany
JAPAN	Pergamon Press, 8th Floor, Matsuoka Central Building, 1-7-1 Nishishinjuku, Shinjuku-ku, Tokyo 160, Japan
BRAZIL	Pergamon Editora Ltda., Rua Eça de Queiros, 346, CEP 04011, São Paulo, Brazil
PEOPLE'S REPUBLIC OF CHINA	Pergamon Press, Qianmen Hotel, Beijing, People's Republic of China

First edition 1986

British Library Cataloguing in Publication Data

Computer aided design in control and engineering systems :
advanced tools for modern technology : proceedings of the 3rd
IFAC/IFIP symposium, Lyngby, Denmark, July 31 – August 2,
1985. — (IFAC proceedings series; 1986, no. 5)
1. Automatic control — Data processing
I. Larsen, P. Martin II. Hansen, N. E. III. International
Federation of Automatic Control IV. International Federation for
Information Processing V. Series
629.8'312 TJ213

ISBN 0-08-032557-2

In order to make this volume available as economically and as rapidly as possible the author's typescript has been reproduced in its original form. This method unfortunately has its typographical limitations but it is hoped that they in no way distract the reader.

Printed in Great Britain by A. Wheaton & Co. Ltd., Exeter

COMPUTER AIDED DESIGN IN CONTROL AND ENGINEERING SYSTEMS

Sponsored by:
International Federation of Automatic Control (IFAC)
IFAC Committee on Systems Engineering (SECOM)

Co-sponsored by:
IFAC Committee on Application (APCOM)
IFAC Committee on Computers (COMPUT)
IFAC Committee on Education (EDCOM)
IFAC Committee on Economic & Management Systems (ENSCOM)
IFAC Committee on Social Effects (SOC:EFF)
IFAC Committee on Theory (THEORY)
International Federation for Information Processing (IFIP)
Institute of Electrical and Electronics Engineers (IEEE)
Control System Society
Denmark Section

Organized by:
The Danish Automation Society

National Organizing Committee
Ole Feddersen
Niels Erik Hansen
Mogens Kümmel (Chairman)
Peer Martin Larsen
Jørgen Pontoppidan
Christoffer Svendsen
Holger B. Thomasen (Secretary)

International Programme Committee
H. Akashi, Japan
S. P. Bingulac, Yugoslavia
A. van Cauwenberghe, Belgium
Chen Zhen-Yu, China
J. B. Cruz, USA
M. Cuenod, Switzerland
A. R. Curran, USA
C. Foulard, France
L. Fransson (IFIP), Sweden
R. W. Gellie, Australia
G. Guardabassi, Italy
N. E. Hansen, Denmark
R. Isermann, Germany
Z. Kozar, Czechoslovakia
P. Martin Larsen (Chairman), Denmark
R. Lauber, Germany
G. G. Leininger, USA
L. Ljung, Sweden
M. Mansour, Switzerland
D. Q. Mayne, UK
L. F. Pau, Switzerland
M. Paul, Austria
W. Schaufelberger, Switzerland
H. Austin Spang, III Junior, USA
B. Tamm, France
A. Titli, France
M. Wozny, USA
K. J. Åström, Sweden

PREFACE

The object of the 3rd International IFAC Symposium on **Computer aided Design in Control and Engineering Systems** was to present and review the state of the art and the trends in development of Computer Aided Design of Control Systems (CADCS), techniques, procedures and concepts.

The two preceeding symposia were held in Zürich, Switzerland, 1979, and in West Lafayette, USA, 1982.

The 3rd symposium was held at the Technical University in Lyngby, Denmark, from July 31 to August 2, 1985. 264 participants from 33 countries attended the symposium, where totally 77 papers were presented in 4 plenary sessions, 3 special sessions, and 15 technical sessions. Addiditionally the program included a panel discussion on standardization of software.

These Proceedings contain the text of 73 papers selected for publication of the Symposium Editors and a summary of the panel discussion. The papers are ordered alphabetically with respect to the first listed author within 15 sections corresponding to the headings of the symposium sessions. An author index is included for the convenience of the reader.

The next symposium on the same subject is planned to take place in 1988 in Beijing, PRC.

P. Martin Larsen
N. E. Hansen

CONTENTS

EXPERT SYSTEMS TECHNIQUES IN A COMPUTER-BASED CONTROL SYSTEM ANALYSIS AND DESIGN ENVIRONMENT

J. D. Birdwell*, J. R. B. Cockett*, R. Heller*, R. W. Rochelle*, A. J. Laub, M. Athans*** and L. Hatfield******

**Departments of Electrical Engineering and Computer Science, The University of Tennessee, Knoxville, TN 37996-2100, USA*
***Department of Electrical and Computer Engineering, The University of California, Santa Barbara, CA 93106, USA*
****Laboratory of Information and Decision Systems, Massachusetts Institute of Technology, 35-406, Cambridge, MA 02139, USA*
*****Lawrence Livermore National Laboratory, PO Box 5504, L-156, Livermore, CA 94550, USA*

Abstract. We explore the fusion of expert control system analysis and design tools into a prototype computer environment. The use of expert systems technology allows the transfer of recent developments in control system design to users who may not be experts in either theoretical developments or computer technology, in a vehicle which provides considerable design flexibility. The technology does not provide as general a framework as, for example, a command language based environment; however, in specific applications, such as system modeling and linear control systems, it is useful and appropriate. Criteria for such applications is that they have a mature theoretical basis, and that the user interaction influence the sequence of the design process.

Keywords. Expert systems; computer-aided design; artificial intelligence; linear systems; computer-aided system design; natural languages.

INTRODUCTION

The interaction of two areas is examined: the development of knowledge-based systems, and the analysis and design of control systems. Our viewpoint is the result of work performed on the CASCADE (Computer-Aided Systems and Control Analysis and Design Environment) project, sponsored by the U. S. Department of Energy. The project attempts to generate new concepts in computer environments for the systems community by bringing together an interdisciplinary team of researchers from computer science, numerical analysis, and systems science. The emphasis of this project is the examination of ideas relevant to the development of computer-based aides for the systems community, rather than implementation of a specific design environment. The first phase of this project studied the issues which affect the design of a computer-based environment; this phase is summarized in Birdwell (1983a, 1983b, 1984). In the past year, the majority of this effort has focused on the feasibility of using expert system technology to construct the user interface portion of a computer environment relevant to system engineers. Some of this effort was reported in Birdwell (1985).

An expert system-based environment for aiding the analysis and design of control systems has two primary features: The environment must have solid numerical routines for the necessary calculations; it must also have knowledge of the design methods available, and be able to judge when and which to apply to the particular problem. It was clear early in the project that the area of system engineering to which the environment was to be applicable would be sharply limited; the area also had to be mature and well-defined to fit within the expert system framework. Linear multivariable control system design was chosen as the application area. In particular, the environment would emphasize the very recent results using the robust properties of the linear quadratic regulator or the Kalman filter, and loop transfer recovery techniques. This methodology allows synthesis of multivariable control design which satisfy frequency domain constraints on robustness to unmodeled plant uncertainty, command following error limits or disturbance rejection, and closed loop bandwidth.

To address this problem, it was proper to use an expert system which would sit over numerical and graphical routines and provide a knowledgeable interface to the application of these routines. A system written over Prolog (Clocksin (1981)) is being used as the knowledge interpreter; this paper describes how it is intended to work and the progress to date.

There is a clear distinction in the design of a system which is concerned with highly numerical problems between the numerical calculations that the computer carries out and what the human wants to see. The human is concerned with general descriptions (is this greater than that; have I met my design specifications?), while often the major concern in performing the numerical calculations is to maintain the precision of the computation. Humans can quickly digest graphical and conceptual data; however, the precise numerical values which gave rise to these are often difficult to digest. Thus, particularly for numerically-based tasks, there is a serious requirement to provide a mechanism for translating the numerical calculations into human terms.

This distinction gives rise to an important observation: If a human cannot handle large numbers of high precision numbers, there is no need for the machinery at the interface to particularly good at that task. The machinery at the interface needs to be good at the sorts of things humans are good at; for example, providing illustrative sketches and discussing concepts. This sort of task is matched by the abilities of workstations and the current wave of symbolic machines.

This distinction implies the software for handling the interface could be on a different type of machine than that for handling the numerical calculations. There can be a physical division of labor: work-station versus work-horse. The types of task which the workstation needs to be capable at include:

1. symbolic manipulations,
2. graphical displays, and
3. interactive computing,

while the task that the work-horse needs to be good at is numerical calculations.

In this environment, the role of the workstation is as a facilitator where the needs of a human can be handled adequately, rather than a place for major computation. The expert system task provides interaction with the user and arranges that the necessary programs be run and the results be appropriately presented to the user. Most important, the expert system must guide the user through the steps of a design. If possible, this should be done in a way which does not aggravate the experienced user, while allowing the inexperienced user to have every step explained and thus learn from the process.

Finally, the expert system must not be so complicated to use that it requires a hefty manual; it must be able to explain itself. A user should be able to do something almost immediately, so that there is no lengthy learning cycle, and so the program can be self contained.

The CASCADE prototype expert system was developed in DECIDE, which is an interface implemented in Prolog. It allows a problem solving method to be programmed in such a way that the computer can lead a user through a problem, while being tolerant of user typing errors and misunderstandings. At the same time, it is capable of explaining its actions and any terms it uses.

The DECIDE system is based on a decision expression methodology, which allows a highly modular design of a knowledge base (Cockett (1983a, 1983b)). The knowledge is clustered into self-contained subtopics, each of which can be independently designed and implemented. The clusters are direct analogs of subroutines.

GENERIC ENVIRONMENT

The designs we investigated in the CASCADE project can be generalized into a structure for a generic computer environment for systems analysis and design. This environment is primarily a partition of the software into three distinct parts: The expert system; a command interpreter, which serves as an interface between the expert system and numeric or graphics software, and the numeric and graphics software. With the exception of the command interpreter, these functions will be available in any expert system-based environment; here, the emphasis is on a clean, well-defined partition. A diagram of this generic environment and its interaction with the user and the computer resources is shown in Fig. 1.

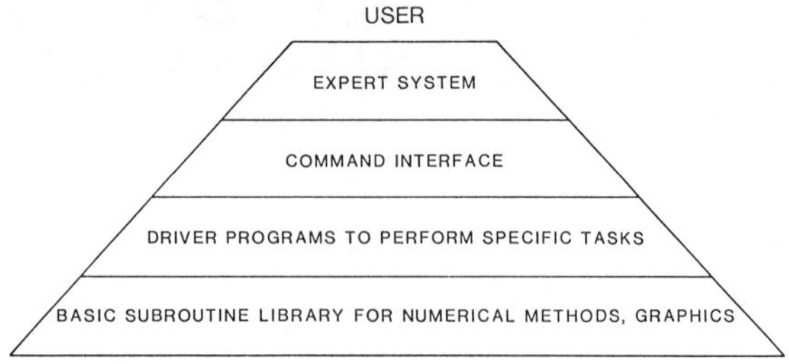

Fig. 1. Generic environment.

There are valid reasons for separating the expert system from the remainder of the environment. First, the command interpreter and numeric/graphics software could be an existing software package. The expert system then assumes the primary functions of data organization and repository for a knowledge base which specifies how the software package is applied to specific problem classes. If the expert system generates commands to the package, over a defined interface, then little or no change is required in the existing code.

Second, many artificial intelligence/expert system languages are not designed to efficiently manipulate numerical information in the quantity or accuracy required in a systems application. CProlog on the VAX is an example of this, where the floating point implementation is single precision. This requires the numeric, and possibly graphics, processing to be implemented in another language; the designer is then forced to provide an interface between the expert system environment and the computation environment. There are exceptions to this generalization, specifically, the Lisp Machine; however, one is still faced with the large amount of existing code written in other languages which is well-tested, and applicable to systems analysis and design.

The command interpreter is required as an interface between the implementation language of the expert system (Prolog in our application) and numeric/graphics software (Fortran 77 and C). The alternative is to provide a direct linkage between the implementation language, which is usually an interpreter, and the remaining software. The feasibility of this is governed not only by the constraints of the machine environment (for example, VMS or UNIX), but also by the details of the expert system implementation language. We used CProlog, and avoided almost all modification of the prolog system. The single exception was the addition of a "system" function to CProlog, which allowed the expert system to spawn tasks to perform specific computations.

The expert system/command interpreter interface can be implemented in a variety of ways. We experimented with two approaches, and some variants. The interface must provide two functions: commands to perform a specific task, and locations of any data which is required or will be generated. Thus, one interface is for the expert system to run a specific program, with data in named files. The program is expected to place results in a file which is mutually agreeable. This is the first interface we implemented, and is found in other packages. There are some reasons that this is not an ideal implementation;

notably, it is not efficient. For example, on a VAX using the VMS operating system, this requires spawning a process in which to run the program, and VMS has a significant overhead associated with task creation and termination. We experimented with a multi-tasking interface between the subsystems, with the expert system passing commands to the command interpreter through VMS mailboxes (the VMS analog of UNIX pipes). This is rather difficult to set up under VMS, due in part to the large number of file or device attributes, and rather easy under UNIX.

We did learn a valuable rule in designing this interface; regardless of the manner in which the expert system/command interpreter interface is done, the expert system should be in complete control of the location of information. This is required if the expert system provides a mechanism for backtracking, or going back to a previous point in the design and allowing the user to choose a different path of decisions. When backtracking is performed, the expert system should avoid recomputation of information which has been previously obtained. The expert system associates the location of computational results with the information those results depend upon. If the command interpreter is forcing an arbitrary (to the expert system) naming convention for results, then it becomes very hard for the expert system to maintain these associations. For example, if the name for a particular result is derived from that of one input parameter, then if the expert system recomputes the results using a different set of input data which contains the same parameter value, the expert system must provide an ad hoc resolution of the duplicate output names. This could be done either by renaming an existing file, or destroying existing information and recomputing the information later if required. Either method is more difficult that forcing the command interface to rely on the expert system for all names.

DEVELOPMENT TEAM

The most difficult part of expert system development is the integration of expertise in the application area (systems analysis and design) with expertise in knowledge base design. This expertise can not generally reside in an individual due to the depth of knowledge required in both areas. The structure of the decision process must be designed by the systems analysis and design expert; however, that structure is influenced by the characteristics of the inference engine.

Our solution was the formulation of a "meta-language" which was used for communications between the implementor of the inference engine and the systems engineer. The knowledge base was written in this meta-language, and its form was modified by the knowledge engineer.

This was not completely satisfactory for a number of reasons. First, the inference engine DECIDE was so new that there was no mature knowledge engineer. This eventually forced the designer of the engine to devote most of his efforts to knowledge base development. Second, the understanding that the expert system was based on a type of decision tree methodology let the experts to produce a rather lengthy (and difficult to digest) manuscript. It was not until the knowledge engineer and the experts got together that it was discovered that there were fundamental symmetries (of which the expert was well aware) in the design technique. These could be heavily exploited to significantly reduce the knowledge base, and the knowledge engineering effort.

Perhaps the major lesson from that experience is the importance of face-to-face discussions between the knowledge engineer and the expert. As with any computing language there will be different ways of approaching problems. The task of the knowledge engineer is to implement in the chosen language the subtleties of the knowledge. The eventual form may not be quite what the expert imagined nor indeed what the knowledge engineer expected.

DETAILS

In the following sections, we present an overview of some of the implementation details of the CASCADE prototype software.

Purpose and Area of Application

The prototype CASCADE provides a methodology for the design of multivariable linear control systems, using linear quadratic regulator (LQR) or Kalman-Bucy filter (KBF) designs to meet frequency domain measures of performance, and loop transfer recovery (LTR) techniques to preserve the performance characteristics in the full design. An overview of this design method is available in Doyle (1981). The primary criticism of the linear quadratic gaussian (LQG) methodology is that the cost functional which is being optimized and the noise characteristics which determine the filter design are unrealistic. This criticism is not valid for LQR/LTR or KBF/LTR designs. The selection of a controller is not based on an optimization problem; rather, the LQR or KBF algorithm is used as a vehicle to construct a design which satisfies frequency-based performance constraints. These constraints are intuitively satisfying, and are direct extensions of standard measures of performance used for single-input/single-output (SISO) compensator design (gain margin, closed loop bandwidth, disturbance rejection, command following error, robustness to plant perturbations).

Theoretical Details

We assume the plant for which we wish to construct a controller is a linear time-invariant (LTI) multi-input/multi-output system represented by the unknown transfer function $\underline{G}(s)$. The design problem is to find a controller of the structure given in Fig. 2 which satisfies a set of design constraints. We will discuss the characteristics of the design problem in three parts: The plant and plant model, the controller structure, and the performance criteria.

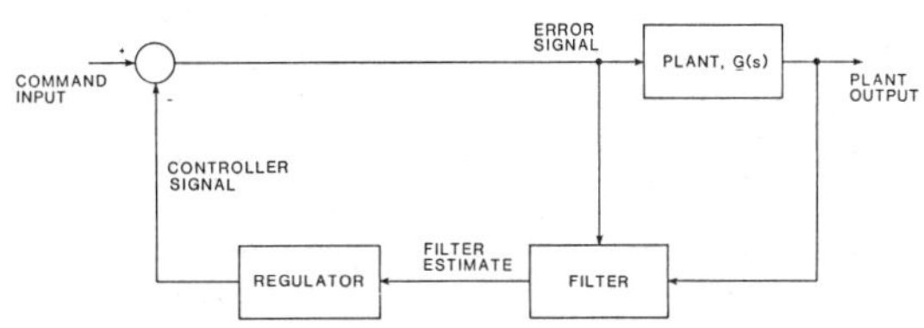

Fig. 2. The controller structure.

The plant and plant model. We assume the designer has a model, $\underline{G0}$, of the plant, which can be described in state-space form as a LTI system:

$$\underline{x}(t) = \underline{A}\underline{x}(t) + \underline{B}\underline{u}(t) \qquad (1)$$
$$\underline{y}(t) = \underline{C}\underline{x}(t) \qquad (2)$$

where $\underline{x}$ is the state vector and has dimension n, $\underline{u}$ is the input vector and has dimension m, and $\underline{y}$ is the output vector and has dimension p. This is a general representation of a linear time-invariant (LTI) multi-input/multi-output (MIMO) system which has finite bandwidth.

The design methodology assumes the accuracy with which the designer can represent the plant is high for low frequencies, and decreases monotonically with frequency. The error in the model's representation of the plant is modeled by a transfer function $\underline{D}(s)$:

$$\underline{G}(s) = \underline{G0}(s)(\underline{I} + \underline{D}(s)) \text{ or } (\underline{I} + \underline{D}(s))\underline{G0}(s) \quad (3)$$

The methodology assumes the designer knows some information about the uncertainties of his model, at either the input or the output of the plant, in the form of a frequency-dependent bound.

$$|| \underline{D}(jw) || < h(w) \text{ for } w > 0 \quad (4)$$

<u>The controller structure</u>. The feedback controller can be decomposed into two parts, a filter, which is based on a nominal plant transfer function, and a regulator gain. This structure is sometimes referred to as a model-based controller (MBC). There are two free parameters of the controller design, as shown in Fig. 3: The regulator gain matrix G, and the filter gain matrix F. The design objective is to select these matrices so several frequency-domain performance criteria are satisfied; these will be developed in the next section.

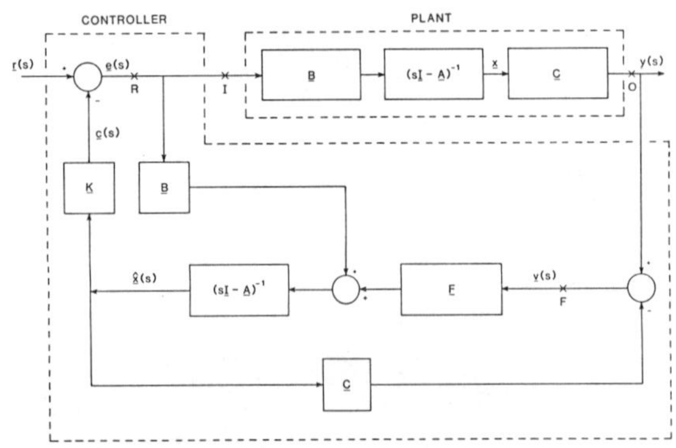

Fig. 3. Closed-loop structure.

<u>Performance criteria</u>. Three measures of performance are used to evaluate a MBC design: Robustness to unstructured perturbations in the plant dynamics, command following performance, and rejection of disturbance signals. Robustness is defined as the maximum bound h(w), such that for all perturbations in the plant dynamics $\underline{D}(s)$ which satisfy Eqn. (4), the closed-loop system is stable; it is developed in detail in Lehtomaki (1981) and Doyle (1981).

Command following error is defined as the signal at node R in Fig. 3. Disturbance rejection is measured as the minimum attenuation along the path from $\underline{d}(s)$ to $\underline{y}(s)$. These concepts are developed fully in Safonov (1981).

For this design methodology, the engineer must decide which performance measure is more important: command following error or disturbance rejection. Either performance measure is then traded off against the robustness measure in the design.

These measures of performance are determined by the characteristics of the system return ratio, $\underline{L}(s)$, the return difference, $\underline{I} + \underline{L}(s)$, or the inverse return difference, $\underline{I} + \underline{L}(s)$**(-1), as in Safonov (1981).

Since the system is MIMO, the location of the break at which $\underline{L}(s)$ is calculated is significant. The design must be based on performance at either the input or the output of the plant. The choice of the point of analysis is made by the expert system, as described in the next section. The return ratio at the plant input, $\underline{LI}(s)$, is:

$$\underline{LI}(s) = \underline{K}(\underline{Z}(s)^{-1} + \underline{BK} + \underline{FC})^{-1}\underline{FCZ}(s)\underline{B} \quad (5)$$

where

$$\underline{Z}(s) = (s\underline{I} - \underline{A})^{-1} \quad (6)$$

The return ratio at the plant output, $\underline{LO}(s)$, is:

$$\underline{LO}(s) = \underline{CZ}(s)\underline{BK}(Z(s)^{-1} + \underline{BK} + \underline{FC})^{-1}\underline{F} \quad (7)$$

The three performance measures are calculated from either $\underline{LO}$ or $\underline{LI}$. The robustness performance measure is the minimum singular value of the inverse return difference transfer function. Either the input node or output node return ratio is used, depending on whether the plant uncertainty bound is known at the plant input or plant output, respectively. The command following error performance measure is the minimum singular value of the return difference transfer function. The input node return ratio must be used. The disturbance rejection performance measure is the minimum singular value of the return difference transfer function. The output node return ratio must be used. All MOP's are evaluated on the imaginary axis; details are in Laub (1979, 1981), and Klema (1980).

<u>Construction of performance constraints</u>. For each design, two performance measures are used. We will assume the design is based on command following error and robustness measures in the following discussion; thus, the design is evaluated at the input node of the plant. The development using the disturbance rejection performance measure is similar.

In CASCADE, constraints are divided into two classes: low and high frequency constraints. Low frequency constraints are specified as minimum values for the minimum singular value of the return difference for all frequencies below a specified frequency. These constraints correspond to a minimum allowable loop gain for the closed-loop system. High frequency constraints are specified as minimum values for the minimum singular value of the inverse return difference for all frequencies above a specified frequency. Each candidate design is evaluated against these constraints, and the expert system advises the user to make appropriate adjustments in design parameters if the design does not meet certain constraints.

The design proceeds in two stages. The first stage creates either a LQR or a KBF design, depending on the node selected for analysis. This design is developed to satisfy the performance constraints, and may require some iteration. The second stage (LTR) uses the remaining component of the controller (filter or regulator) to force the return ratio of the closed-loop system to approximate the return ratio of the initial design (LQR or KBF).

<u>LQR/KBF design properties</u>. The expert system uses the performance constraints to calculate a maximum desirable crossover frequency for the return ratio of the closed-loop design. It then performs either the regulator or the filter design based on information provided by the user about what type of performance measure is more important, whether or not there is a preference for either design,

whether sufficient information exists to construct either design, and the number of plant inputs and outputs. The regulator requires the design evaluation to be at the plant input node; the filter requires it to be at the plant output node. The regulator and filter design methods are symmetric; thus, the regulator-based design is developed here. A detailed development is given in Lehtomaki (1981).

The regulator constructed in the first stage is evaluated using its return ratio. The return ratio of the regulator is equivalent to the return ratio of a closed-loop system consisting of the plant with full state observation and the regulator implementing full state feedback. The regulator's return ratio is the return ratio of the closed-loop system in Fig. 3, calculated at node R, and given by

$$\underline{LR}(s) = \underline{KZ}(s)\underline{B} \tag{8}$$

The first step in constructing the regulator is to "balance" the singular values of the return ratio of the LQR design at either low or high frequencies. This balancing operation causes the singular values in the selected frequency range to be approximately equal (equal in the limiting case as frequency tends toward either zero or infinity). The singular values are balanced by constructing a state weighting matrix with the following structure for the LQR cost functional:

$$\underline{Q} = \underline{P}^T\underline{P} \tag{9}$$

where for high frequency balancing $\underline{P}$ is the least squares solution to the equation $\underline{PB} = \underline{I}$ and for low frequency balancing it is the least squares solution to the equation $\underline{BP} = \underline{A}$.

The choice of high or low frequency balancing is decided by the expert system, based on input from the user and an estimate of the width of the window of frequencies within which the crossover frequencies of the closed-loop system return ratio must lie. The expert system prefers to use low frequency balancing, since by equalizing the singular values at low frequencies, the command following error performance measure is maximized. If, however, the window for the crossover frequencies is narrow, then it may be easier to satisfy all the performance criteria by using high frequency balancing, since the singular values near the crossover frequency are grouped more closely. In CASCADE, the threshold for using high frequency balancing is a window which spans less than a factor of four in frequency. This is a heuristic test.

Given the matrix from the balancing procedure, if the control weighting matrix parameter required by the LQR design is set to R = tI, then the maximum crossover frequency of a singular value of the return ratio of the regulator system, $\underline{LR}(s)$, is approximately

$$\text{wcmax} = \sigma_{max}(\underline{PB})/\sqrt{t} \tag{10}$$

An important property of this LQR design is that near the crossover frequencies of the singular values a plot of either the maximum or minimum singular value versus frequency has a slope of approximately -20dB/decade. This property allows a rough approximation of the minimum crossover frequency required to satisfy the low frequency constraints and the maximum crossover frequency which will allow the high frequency constraints to be satisfied. The expert system uses this to set a desired value for the maximum crossover frequency, and solves for the parameter t.

At this point, there is sufficient information to develop an initial LQR design. This design is evaluated by the expert system using the algorithm in Laub (1981), and calculating the gain margin by which the design satisfies the performance constraints. The minimum gain margin and the corresponding frequency for each measure of performance is calculated and provided to the user for evaluation.

If a constraint is violated, the corresponding gain margin is negative. If all the constraints are satisfied, the expert system recommends proceeding to the next stage of the design, loop transfer recovery. If a constraint is violated, the expert system recommends an adjusted value for the parameter t and allows the user to iterate the design. Under certain conditions, the expert system recommends the system be balanced at high frequencies if it had been balanced at low frequencies.

The singular value and gain margin information is presented to the user both numerically and graphically. The user can select views of the minimum and maximum singular values of the return ratio, return difference, and inverse return difference versus frequency. The constraints which are appropriate for each graph are displayed with the singular value curves. A sample plot of the return ratio singular values is shown in Fig. 4.

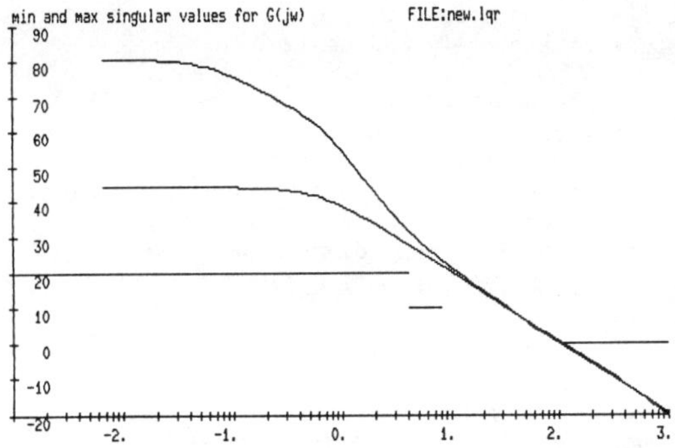

Fig. 4. Singular value plot of return ratio.

Recovery of the LQR/KBF design properties. The objective of stage one of the design process is to satisfy the performance criteria using only a regulator or filter design. These properties exist at either node R or node F in the system. The objective of stage two of the design process is to construct the remainder of the controller loop such that the properties of the system evaluated at node I (O) approximate those at node R (F). If this approximation is sufficiently close, then the final design will satisfy the performance criteria.

The loop transfer recovery (LTR) procedure only works if the plant has no non-minimum phase zeros "near" the bandwidth of the initial design. Note that non-minimum phase zeros represent a fundamental limitation on system performance, and are not a limitation of this design methodology. Neglecting this limitation sacrifices the robustness properties of the controller.

The regulator return ratio can be recovered at the plant output node, O, given that the plant is minimum phase over a suitable frequency range and that the plant has at least as many outputs as inputs, by constructing a Kalman filter using specific noise covariance matrices parametized by a positive number, q, As q goes to infinity, the return ratio at the plant output node approaches the return ratio of the regulator design pointwise. Thus, for a sufficiently large q, the closed-loop system satisfies the performance constraints.

The loop transfer recovery stage in the expert system is implemented as an iterative design, where q is increased until the performance constraints are satisfied. Tests are made against the performance constraints at each stage, rather than against the original design, since the objective is to satisfy the constraints rather than to approximate the regulator design.

Each design is evaluated using the same procedure as is used for the regulator and filter designs. When the user and expert system agree that the design is satisfactory, the expert system concludes that the design process is finished and exits. The information describing the controller is available in files specified by the expert system.

Developing the Knowledge Base

A decision which the system must make early in the design is whether to design the regulator or the filter first. An expert in the area can determine the answer to this question by knowledge of which performance measures have the highest priority and what information he has about the model. There are two questions which an expert system must evaluate: Which performance measures are most important, and does sufficient information exist to perform the preferred design?

As an example of CASCADE, we will describe how this knowledge was encoded. The first step for the knowledge engineer was to discover that this was the intent of the expert through discussion. Next, we produced a diagram of the decision process (Fig. 5). This diagram was translated into the code for the DECIDE system, given in Fig. 6.

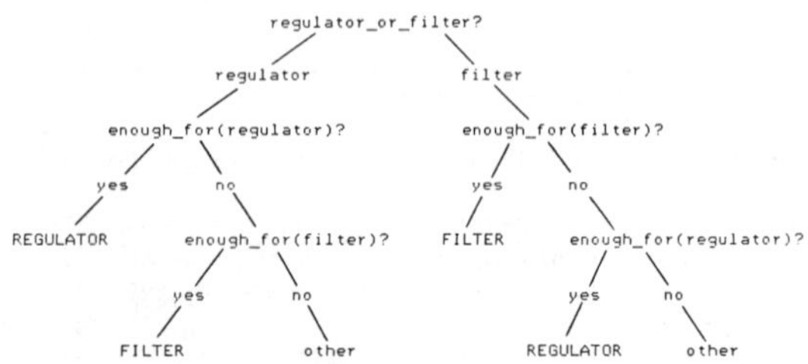

Fig. 5. Top level of regulator or filter decision.

```
choose_reg_or_filter(equal):>
   reg_or_filter:[choose1<<regulator,
                  choose2<<filter,
                  choose1<<conflict].

choose1:=enough_for(regulator):
   [<:regulator<<yes,choose11<<no].

choose11:=enough_for(filter):
   [<:filter<<yes,other<<no].

choose2:= enough_for(filter):
   [<:filter<<yes,choose22<<no].

choose22:= enough_for(regulator):
   [<:regulator<<yes,other<<no].
```

Fig. 6. Top level DECIDE knowledge base.

The code in Fig. 6 breaks down the decision to do a regulator or filter design into deciding which design is wanted, followed by the pragmatic decision about whether the user has enough information to do this design. If the user does not have sufficient information for either design, the system replies it cannot help; this is the significance of the "other" answer in Figs. 5 and 6.

Deciding whether there is enough information to do a regulator or a filter design can in turn be broken down. There is a symmetry between having enough to do a regulator or a filter design; thus, it is possible to parametize the decision process according to which design method the user wishes to follow.

The user has enough information to do a regulator design if he knows either a minimum permissible bandwidth or bounds on the command following error, and if he knows either a maximum permissible bandwidth or bounds on the plant uncertainties. It is important to note that if he has bounds, he also may have calculated his permissible bandwidth. The explicit bounds, however, are much more useful to the system not only to determine the quality of the eventual design, but also as if only the bandwidth limits are known, the system takes some liberties with theory. Thus, if the user knows bounds, the expert system should elicit them from him; it must thus ask first about whether he knows bounds. The form of the decision logic is given in Fig. 7. The DECIDE input for this logic is given in Fig. 8.

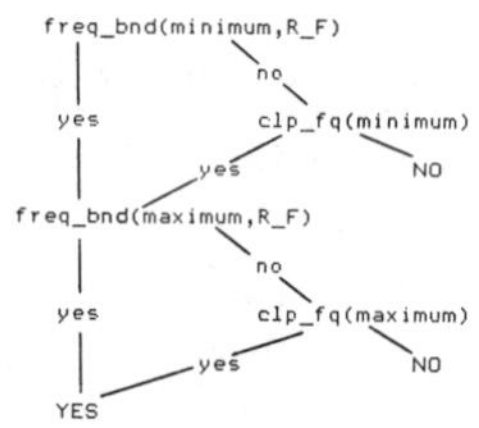

Fig. 7. Decision if enough information exists.

```
enough_for(R_H):>
   freq_bnds(minimum,R_H):
     [enough_1(R_H)<<yes,enough_2(R_H)<<no].

enough_1(R_H):=freq_bnds(maximum,R_H):
   [<:yes<<yes,enough_3(R_H)<<no].

enough_2(R_H):=clp_fq(minimum):
   [enough_1(R_H)<<yes,<:no<<no].

enough_3(R_H):=clp_fq(maximum):
   [<:yes<<yes,<:no<<no].
```

Fig. 8. Enough_for DECIDE knowledge base.

The DECIDE language allows one to explicitly determine the order in which the interaction proceeds. This distinguishes it from the rule base approach where the control is implicit. This has both advantages and disadvantages: The principal disadvantage is that the designer must explicitly program the order. The advantage of having total control over decision ordering can be considerable in some applications.

Each of the decisions which Figs. 6 and 8 are making is a knowledge cluster. It is determined by its "governing decision"; for example, in Fig. 8, this is whether one has enough for a regulator or a filter design. In the DECIDE system each cluster may be held as a separate file, which is called into memory only for as long as it is needed. In a large expert system this means that one can focus the expert system on the current problems very effectively. This also has the significant advantage that the knowledge base is extremely modular. A knowledge cluster can be altered independently of others provided the interface, determined by the governing question, is not altered. In rule based approaches modularity is often difficult to achieve.

Just as subroutines can be parametized, so can knowledge clusters. This means they can be called recursively provided their context changes. The DECIDE system automatically responds to decisions which have already been made with the same response as before. This means if the recursive call is to work, some parameter in the call must be altered. This facility can introduce complex behavior, and is extensively used in the latter part of the CASCADE knowledge base to implement the iterative design steps.

Initially, to debug the system the text is omitted from the decision processes. Once the form of the interaction is satisfactory, the text is added. Text can be associated with each node in the decision tree; in this case, each node will result in an interaction with the user. Alternatively, one can associate a process with a node. For example, an external program could be run, or a file could be read. In this example, each node will have associated text.

The DECIDE system has a text generation system which knows about simple English language constructions and questions. It can change the tense and sense of a sentence so that it can be used in an explanation or as a question. The facility relies heavily on the definite clause grammar rules of Prolog (Pereira (1980)), which provide a convenient method of text generation. The basic structure of the sentence is specified to DECIDE using a slots grammar system (McCord (1982)), which then manipulates it into the forms it needs. For example, the following variants:

"What do you want to design first regulator or filter?"

"You want to design the regulator first."

"You want to design the filter first."

are all generated from the specification:

```
tx_gen(reg_or_filter,T,q([regulator,filter],
   w(obj,X,cl(subj,design,[subj:you,obj:d X,
     adv:first],i(want,T))))).
```

while the following variants

"Do you have enough information to do a regulator design?"

"You have enough information to do a filter design."

"You did not have enough information to do a regulator design."

are all generated from the specification:

```
tx_gen(enough_for(R_F),q(yesno,cl(subj,have,
  [subj:you,obj:d np(information,
  [adj enough,'to do':a np(design,
    [adj R_F])])],T))).
```

Notice in the last case how the decision node parametization can be used to parametize the text itself. This is one of the advantages of text generation, as complex changes in text can be effected by the parameters. This can dramatically reduce the size of the knowledge base. It also closely couples the text and the underlying logic, which means that subsequent alterations to the logic will not result in inappropriate text. The disadvantage is that the knowledge engineer must carefully structure the knowledge, right down to the wording of the questions.

The following text is an interaction from CASCADE which covers the interaction discussed above. The DECIDE system can explain its actions. The prompt to CASCADE reflects the state of the expert system and the availability of detail or default values. A "(?)-" prompt signifies that a more detailed interaction is available, while the prompt "(d)-" indicates there is a default. The response "<ESC>?" (which is shown as "$?") forces the DECIDE system to either take a default or go into detail. The "<ESC>" character actually signals that the user is responding to the DECIDE system as opposed to the CASCADE knowledge base. DECIDE has a number of other facilities available including glossaries, explanations, and backing up. The "(s)-" prompt means that one is talking to the DECIDE system. It occurs in particular when a conclusion is reached and the system wants to know whether the user agrees. It is also a change to find out how the system has reached its conclusion; note the use of how in this interaction. Often it is a point where a user will backtrack in the interaction to correct a mistake.

```
We are going to attempt to apply the LTR design
procedure to your control problem.
What is the name of the file containing the system
description?
(?)-new.sys
Your control problem has an equal number of inputs
and outputs.
(s)-how
I read the value of the number of states, number
of inputs and number of outputs from new.sys and I
found that the number of inputs equals the number
of outputs which led me to conclude:
Your control problem has an equal number of inputs
and outputs.
(s)-
What do you want to design first the regulator or
filter?
(?)-$?
In the design either the command following or the
noise performance characteristics are traded
against robustness to plant perturbations. If the
filter is designed first then the regulator design
will be used to approximate the filter
characteristics of the original design. In this
case the regulator poles may be faster than
desired. If the regulator is designed first the
converse problem arises..
Is command following error performance more
important than closedloop system noise
performance?
(d)-yes
In your application is it important to minimize
the effect of system noise on the plant output?
( )-yes
( yes )
You want to design the both first as you have a
conflicting requirement! I suggest you design the
regulator first.
(s)-
Do you have enough information to do a regulator
design?
(?)-$?
Can you specify bounds on the command following
error response as a function of frequency?
( )-y
Can you specify bounds on the multiplicative plant
uncertainties at the plant input as a function of
frequency?
```

( )-n
(no)
Can you specify a maximum permissible bandwidth for the closed loop system?
( )-yes
You have enough information to do a regulator design.
(s)-
The regulator will be designed first.
(s)-

CONCLUSIONS

A successful application of expert systems technology to user interface design requires maturity of the knowledge base. This removes many of the research-oriented application areas of computer aided design environments from consideration. The existence of a knowledge base which drives the analysis/design process requires that the knowledge base be applicable to all potential problems without modification. This introduces the second requirement: The capabilities of the expert system-based environment must be sharply defined and limited to those applications to which the knowledge base is clearly applicable.

At the present time, a design structure which provides a division of labor between the expert system/inference engine and the computational component is desirable. The existence of this interface allows the expert system implementation to concentrate on the symbolic computation without adding the complexity, and the possibility of errors, which a low-level linkage represents. We found this division also allowed a clearer perception of the numeric data required for decision-making, as opposed to the much larger set of data which was associated with the numeric computations. Last, the division is a natural framework for the future exploration of distributed computation, with hardware specialized to each task.

ACKNOWLEDGMENTS

This research was funded by the U. S. Department of Energy, Division of Electric Energy Systems, as part of the Systems Engineering for Power research program. The project was managed at Oak Ridge National Laboratory, operated by Martin Marietta Energy Systems, Inc. for the U. S. Department of Energy under contract no. DE-AC05-84OR21400. Work was performed at Oak Ridge National Laboratory and under subcontract 7685 S13 at the University of Tennessee, under subcontract 11X-39024V to M. Athans, under subcontract 11X-39060V to A. J. Laub, and under subcontract 12X-43107V at Lawrence Livermore National Laboratory.

REFERENCES

Birdwell, J. D. (1983a). Future directions in computer-aided control system design software development. IEEE Control System Magazine, 3, 11-14.

Birdwell, J. D., S. A. Bly, and A. J. Laub (1983b). The DOE/EES/ORNL CACSD development effort: an overview. (abstract) CACSD '83: IEEE Control Systems Society Symposium on Computer-Aided Control System Design.

Birdwell, J. D., M. Athans, S. A. Bly, J. R. B. Cockett, M. T. Heath, R. W. Heller, C. J. Herget, A. J. Laub, R. W. Rochelle, J. P. Stovall, R. Strunce (1984). Issues in the Design of a Computer-Aided Systems and Control Analysis and Design Environment, ORNL/TM-9038. Oak Ridge National Laboratory, Oak Ridge, TN.

Birdwell, J. D., M. Athans, S. A. Bly, J. R. B. Cockett, L. Hatfield, M. T. Heath, R. W. Heller, A. J. Laub, R. W. Rochelle, J. P. Stovall, B. Cabral, G. Chen, K. Fung, S. Heard, C. McCullough, A. Mak (1985). CASCADE: experiments in the development of knowledge-based computer-aided systems and control analysis and design environments. (abstract) IEEE Control Systems Society 2nd Symposium on Computer-Aided Control System Design (CACSD).

Clocksin, W. F., and C. S. Mellish (1981). Programming in Prolog. Springer-Verlag, New York.

Cockett, J. R. B. (1983a). File Handling for Detail, Extent and Sub-Tasks in the Implementation of Decision Processes, Tech. Rpt. CS-83-56. Dept. of Comp. Sci., Univ. Tennessee, Knoxville, TN.

Cockett, J. R. B. (1983b). The Theory of Discrete Decision Processes, Tech. Rpt. CS-83-57. Dept. of Comp. Sci., Univ. Tennessee, Knoxville, TN.

Davis, R., B. Buchanan, and E. Shortcliffe (1977). Production rules as a respresentation for a knowledge based consultation program. Artificial Intelligence, 8, 15-45.

Doyle, J. C., and G. Stein (1981). Multivariable feedback design: Concepts for a classical/modern synthesis. IEEE Trans. Auto. Control, 26, 4-16.

Hayes-Roth, F., D. A. Waterman, and D. B. Lenat, (Eds.) (1983). Building Expert Systems. Addison-Wesley, Reading, MA.

Klema V., and A. J. Laub (1980). The singular value decomposition: Its computation and some applications. IEEE Trans. Auto. Control, 25, 164-176.

Laub, A. J. (1979). An inequality and some computations related to robust stability of linear dynamic systems. IEEE Trans. Auto. Control, 24, 318-320.

Laub, A. J. (1981). Efficient multivariable frequency response computations. IEEE Trans. Auto. Control, 26, 407-408.

Lehtomaki, N., N. Sandell, and M. Athans (1981). Robustness results in LQG based multivariable control designs. IEEE Trans. Auto. Control, 26, 75-93.

McCord, M. C. (1982). Using slots and modifiers in logic grammars for natural language. Artificial Intelligence, 18, 327-367.

Michie, D. (ed.) (1982). Introductory Readings in Expert Systems. Gordon and Breach, New York.

Pereira, F., and D. H. D. Warren (1980). Definite clause grammars for language analysis--a survey of the formalism and a comparison with augmented transition networks. Artificial Intelligence, 13, 237-278.

Safonov, M. G., A. J. Laub, and G. L. Hartmann (1981). Feedback properties of multivariable systems: The role and use of the return difference matrix. IEEE Trans. Auto. Control, 26, 47-65.

Weiss, S. M., and C. A. Kulikowaski (1984). A Practical Guide to Designing Expert Systems. Rowman and Allanheld.

DECISION AIDS FOR THE MANAGEMENT OF LARGE NON-JOB SHOP TYPE MANUFACTURING SYSTEMS

M. G. Singh

Control Systems Centre, UMIST, PO Box 88, Manchester M60 1QD, UK

Abstract. In this paper we examine the problems of management and control of large scale multi-product multi-line batch manufacturing outside the mechanical engineering industries. The main examples of such non-job shop type manufacturing arises in the making of frequently purchased products e.g. small consumer goods, processed foods, cigarettes, candy, etc. The decision making problems which arise in such systems are extremely complex and the decision support tools for helping management at various levels could be of significant advantage in improving the productivity of such businesses. This paper describes a number of decision support tools which could be used at the different levels of the production hierarchy.

The production hierarchy in virtually all manufacturing can be conveniently divided into 4 levels each of which is characterised by a significantly different time scale from that of the other levels. At the highest level, the time horizon of interest is 1 to 5 years and the decision making is of a strategic nature. At the next level down where the time horizon of interest is from 1-2 months to 1 year, the decision making is operational/tactical. The third level down concentrates on issues of production planning and scheduling and its horizon of interest varies from 1 hour to about 20 weeks. The lowest level deals with disturbances occurring over a very short horizon (a few seconds to about 1 hour) and at this level, most of the decision making is automatic (PID controllers, optimal control, self tuning regulators, etc.). Each higher level provides performance targets for the next level down.

In this paper we examine the main issues which arise at the top three levels in the hierarchy and describe some decision support tools which have been developed by the author and others. At the top two levels the tools are designed to cope with the issues of competition whilst at the lower level (production planning and scheduling) the problems of allocating tasks to limited resources in the most cost effective way are addressed. Overall, we show how improved efficiency at each level can lead to significant improvements in productivity.

Keywords. Manufacturing systems for frequently purchased products, strategic decision making, operational and tactical decision making, production planning and scheduling.

INTRODUCTION

In the first part of this century, production systems were rationalised through the use of dedicated production lines. Immense economies of scale were achieved in the mechanical engineering industries (particularly in manufacture of automobiles) through the installation of high volume dedicated production lines. Unfortunately, these systems become uneconomic for lower volumes and this led the drive towards standardisation of products. The Information age is in the process of shifting the push towards standardisation since with the advent of flexible manufacturing systems, low unit costs can be achieved with medium volume production and in principle, even with low volume production. This is being achieved by the introduction of several nondedicated machines which can perform different functions on different parts and which are interconnected together through an automated transportation system.

At the present time, flexible manufacturing systems are virtually exclusively confined to the mechanical engineering industries. However, even outside these industries, there is much scope for improving the productivity of manufacturing by more non-dedicated usage of plants. Thus even in the manufacture of small consumer goods like detergents, chocolates, cornflakes, toothpaste, etc, it should be possible to build non-dedicated units for production or packing which can be cleaned and/or re-set as the unit is used for making or packing a product different from the one which is currently being made. From a straight design point of view, it certainly makes sense to have a substantial number of non-dedicated manufacturing units since a typical consumer goods plant outside the mechanical engineering industry (i.e. a non-job shop type plant) will have a designed life of about 30 years and in that period market forces would have ensured that the plant manufactures a different range of products from the ones that it was originally designed for. The desire of the managers of consumer goods companies to go down this path enthusiastically has been tempered to a large extent by the preceived difficulty in managing and controlling such large non-job type plants since very little attention has so far been paid to development of the appropriate management and control hardware and software tools. It has also become clear that the problems of management and control of such multi-product multi-line batch manufacturing plants are different from those of job shop type plants where most of the current research effort is going. Thus, it is not possible to await the resolution of the problems in flexible manufacturing and a subsequent transfer of the know-how and tools to non-job shop type plants.

In this paper we describe the problems which arise in such systems and give an overview of the computerised decision support tools which have been developed or are being developed for the management and control of such plants.

HIERARCHIES IN THE MANAGEMENT OF NON-JOB SHOP TYPE PLANTS

The essential problem for consumer goods manufacture as for any other business is to maximise profits. In the present context, this implies the manufacture and sale of the range of products being manufactured by the non-job shop type plant. There are many times scales over which planning scheduling and control decisions need to be taken and computerised decision support tools are desirable for each level of the hierarchy.

The highest level decisions are long term or strategic decisions. Typical examples are those of capacity expansion or contraction, achievement of a larger market share through improvements of brand image, pricing policies, etc. Strategic decision making is invariably done over a long time horizon (several years) and usually requires capital expenditure or redeployment.

The next level down is concerned with operational decision making in a competitive environment. In the small consumer goods sector, in most countries of the world there are only 4-5 major companies which dominate the market. Since most of the goods being manufactured can only be sold through intermediaries (supermarkets, hypermarkets and other retail outlets) who are also very concentrated, the decision making at this level is on pricing, advertising, etc. in order to obtain competitively substantial shelf-space from the retailers. The time horizon of interest here varies from about one month to about one year.

At the next level down, the problem is one of production planning and scheduling where one attempts to operate and schedule the manufacturing plant hour by hour with a view to meeting the market demand in the most cost effective way. The time horizon of interest here is between one hour and about 10 weeks.

At the lowest level, we do not consider the whole factory but rather the individual manufacturing units and the objective is to run each unit as efficiently as possible. The horizon of interest here is therefore a few seconds to about an hour and most of the units at this level have automatic control (P.I.D. controllers, self-tuning regulators, multivariable control, etc.).

Figure 1 shows the various levels of the manufacturing hierarchy. There is a significant amount of interaction between the levels of this hierarchy. In general each higher level decision provides a goal for the next level down.

In what follows, we will describe the main decision problems faced at the top three of the levels of the hierarchy in Fig. 1 and give an overview of the decision tools which have been developed or need to be developed.

DECISION SUPPORT SYSTEMS FOR STRATEGIC DECISION MAKING [1]

Strategic decision making is usually done at a very senior level within a company (Board or near Board level). At this level, most of the factors which need to be weighed up are qualitative and very few decision support tools exist.

Strategic decision making for a firm is essentially concerned with the selection of business strategies which will enable management to achieve its long term objectives based on limited information about

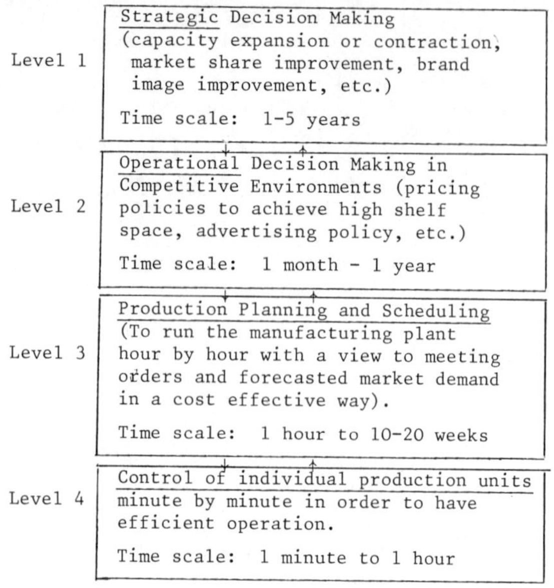

Fig. 1 Manufacturing System Hierarchy

the company's macroeconomic, competitive, technological and factor supply environments.

In the last three decades, a number of tools have been developed for strategic planning in companies. The principal tools are:

- Experience Curves
- Growth/Share Matrix
- Shell Screen/General Electric/McKinsey approach
- Product life cycle
- the PIMS data base
- Competitive Analysis

Of these, the Competitive Analysis framework of M. Porter of Harvard Business School [2] provides the most attractive approach to strategy formulation since it recognises explicitly that the strategy of a firm must take into account the potential reactions of competitors. Competition here is not limited to the direct competitors who provide the same goods or services as the subject firm but it also includes the Supplier Groups, Buyer Groups, new entrants and potential substitutes as shown in Fig. 2. Thus the competitive position of the subject firm is governed at any point in time by the balance of the forces shown in Fig. 2, together with the company's own internal skills at satisfying the market.

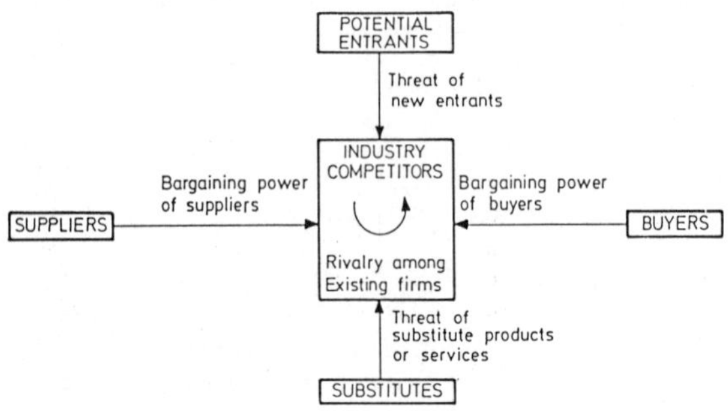

Fig. 2 Fundamental Determinants of Industry Competition

Although the framework developed by Porter is an extremely attractive one, it has many limitations. The main difficulty lies in the fact that only descriptive checklists emerge as a result of the analysis. Since the strategy of an enterprise involves taking costly decisions which often affect, in a negative way the bottom line in any year, and may in many cases affect the very survival of the firm, it is important to be able to quantify the effects of the different options. This is particularly difficult to do in competitive situations since one's competitors will inevitably respond to one's strategic moves, thus, in many cases negating one's effort or making it more costly.

We see that competition is one of the key dimensions which marketing managers of manufacturing companies must take into account when they make decisions on pricing, advertising, shelf space allocation, etc. in the area of frequently purchased products which are manufactured in non-job shop type plants. The competitive aspects become even more important when strategic decisions are taken on, for example, capacity expansion or contraction, vertical integration, etc. Up to now, most decision making of this kind has relied on the skill and experience of the managers concerned. Given the complexity of the environment within which such decision making takes place, it is not surprising that there is considerable scepticism amongst senior managers about the use of mathematical models as an aid since it is generally believed that there are many judgemental factors where managerial "gut-feel" which has not been modelled far outweighs the components which are amenable to modelling. Clearly, what is required, is a framework which allows the manager to "paint" his own view of the world which incorporates both the "Hard" information wbich is available to the manager as well as the "Gut feel" information. The framework could then be manipulated in order to act as a decision aid.

In the area of competitive decision making, the "hard" information is the information available about one's own company e.g. the level of efficiency of production, brand image, sales, costs, etc. The "soft" or "gut feel" information is the information about similar parameters for the competing companies i.e. whether a competitor has equivalent manufacturing efficiency, better marketing ability and similar considerations of this kind affecting his overall performance and profits.

In order to incorporate both the "Hard" and the "Gut feel" information, it is necessary to have [3,4]

(1) A Shell which constitutes the knowledge base

(2) A mechanism for knowledge acquisition in order to parametrise the Shell, and

(3) A mechanism (Inference Engine) for helping in decision making after the Shell has been suitably parameterised.

The "Shell" must be sufficiently general to enable us to organise a wide range of "hard" and "soft" information for the class of applications for which the system is to be used.

In Fig. 3 we show a competitive marketing situation where a number of competing manufacturers make a product which they sell through a number of competing retail chains. Each of the managers of the manufacturing companies needs not only to take into account the direct competition from other manufacturers but also how to play off the Retailers amongst themselves in order to gain an advantage for their direct competition with other manufacturers. Similarly, buyers from Retail companies need to take into account their own direct competition when they negotiate prices, shelf spaces, etc. for the products of the manufacturers. This is an immensely complicated situation, subsets of which have been studied in various guises for many years. For example, Helmy et al [5] considered the subset which lumped the manufacturer level into one and the retailer level into one and analysed the resulting vertical marketing structures. Balderston [6] examined the optimal entry of new members in marketing

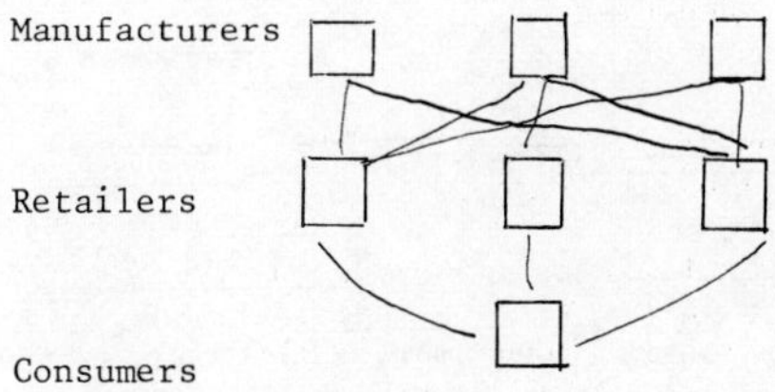

Fig. 3

channel networks whilst Corstjens et al [7] and Zusman et al [8] considered strategic decision making aspects in marketing channels. In all this work and in the more recent papers of Arnold et al [9] and Singh et al [10], the basic "Shell" which was used was "static" i.e. all decision making was assumed to take place at a particular point in time so that longer term decisions could only be represented as "snap shots" at different points in time.

Although such static shells are useful in that they do lead to efficient decision support tools as shown by "Mark-opt" [10] which has been made into a commercial product and is being used by a number of organisations. However, since the shell within which competition is studied is purely static, it is unable to take into account a number of important real world aspects of this problem which the managers taking the ultimate decisions cannot afford to ignore. The main factors which a more advanced system than "Mark-opt" should try and take into account are:

(1) <u>The Dynamics of Stocks:</u>

Since the "Shell" in previous systems [5-10] is static, it is not possible to incorporate the dynamic behaviour of the stocks kept by each manufacturer and each retailer since these systems can only examine a particular point in time and not a contimum of time points.

2. <u>Reactions Over Time:</u>

It is important to keep in mind that negotiations between manufacturers and retailers are not one shot deals but rather a part of an on-going process. Thus systems which only consider decision making at a particular point in time are unable to take into account the importance of maintaining good channel relationships over a time horizon.

3. <u>Strategic Decision Making and Scenario Analysis</u>

Strategic decision making by definition considers a time horizon as opposed to a particular point in time and here again, even though one could go some way towards taking sensible strategic decisions by considering different time "snap shots", it is more attractive to consider the time contimum and to examine decisions over a time horizon.

In [1] we examined a dynamic Shell and appropriate dynamic decision making mechanisms for the class of marketing channel systems shown in Fig. 3. Since the dynamic decision problem in marketing channel systems is at least an order of magnitude more complicated than the corresponding static problem, we considered only a system which comprised 2 manufacturers in competition who sell to 2 retailers who are in competition as shown in Fig. 4.

It is conceptually quite easy to extend the channel structure to more than two manufacturers and more than two retailers although at the expense of significant additional computation.

In [1], the dynamic decision support tool for

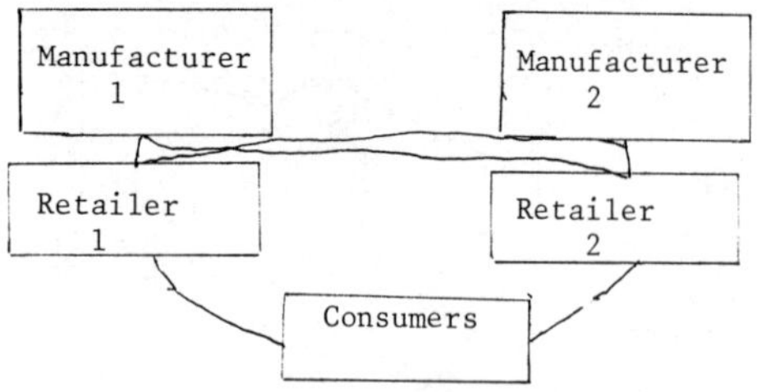

Fig. 4

strategic decision making which is described has the following structure:

(a) A "Shell" or a knowledge base with Causal models. Each of the models represents the profit maximising behaviour of a manufacturer/retailer in the form of an objective function with a summation over time subject to difference equation constraints.

(b) A knowledge acquisition mechanism which marries "hard" information with managerial gut feel information and uses parameter estimation techniques to parameterise the "Shell".

(c) An Inference Engine which computes the Nash equilibrium of the game between the profit maximising players described in (a) for the particular situation described in (b).

In [1], a number of interesting strategic decision making scenarios have been analysed with this tool.

Next, we describe the operational decision making level.

DECISION SUPPORT SYSTEMS FOR OPERATIONAL DECISION MAKING

At the operational decision making level, the time horizon of interest is much smaller (1 month to 1 year). Here the issues of competition are still quite important since in successful companies, this level needs to be closely tuned to the needs of the market place. In the area of frequently purchased products which are manufactured in non-job shop type plants, it is important to be able to market them through the 4-5 major retail chains which exist in each country since this is the only real outlet for such manufacturing. This problem is similar to the one discussed in section 3 except that in this case, the issues of importance are the shorter term ones i.e. negotiations with retailers, allocating of marketing budgets, etc.

For the negotiation problem, the system "Mark-opt" which was mentioned in section 3 may prove quite useful as a decision support tool. We will therefore start by describing the intelligent knowledge based system "Mark-opt".

Mark-opt [9,10]

"Mark-opt" is a software suite which provides a decision support capability in a key area of marketing i.e. the negotiation process between manufacturers and retailers. In any negotiation, it is of great importance to understand the imperatives of the person you are negotiating with. "Mark-opt" allows the negotiator to feed in all the information that he has about his own company, that about the company he is negotiating with, information about his own company's competitors as well as that on the competitors of the company he is negotiating with as well as information about the market. From all this information, Mark-opt allows the user to calculate various scenarios e.g. where the Retailers respond quickly to each other's competitive price moves, or where one Manufacturer optimises whilst the others don't or where all others optimise as well.

"Mark-opt" can also be used as a short horizon strategic marketing tool. Some of the more interesting strategic scenarios where "Mark-opt" can be used include:

- pricing policies for white label and own label products from the point of view of a manufacturer as well as from that of a retailer
- marketing strategies for manufacturers with poor brand image: trade-offs between brand image improvement (through advertising, ReD, etc.) against improvements in manufacturing efficiency.
- use of optimisation where the competitors are not doing so.
- the impact on the profitability of all manufacturers and retailers if a single manufacturer raises or lowers his prices by a small amount.

For manufacturers, it is possible to calculate safe negotiating ranges which might not invite retaliation from competitors but which nevertheless yield reasonable profits. For Retailers, Mark-opt enables them to obtain better margins and calculate their best shelf space allocation for the brands of different manufacturers so as to maximise total profits.

This description is not only a valid one for products sold through supermarkets and hypermarkets but it could also be used to describe the selling of any product which is sold through intermediaries before reaching the end user.

The problem of competition between manufacturers has been considered in the mathematical economics literature as has the problem of vertical integration. However, the whole system above has never been considered mathematically since it was thought to be an unsolvable problem. Arnold. Singh and Corstjens [9 10] recently provided the first solution ever of this hitherto unsolved problem in marketing and "Mark-opt" incorporates their basic method as extended to the case of multiple manufacturers and multiple retailers.

"Mark-opt" takes into account the manufacturers costs (including economies of scale), the retailers costs (fixed plus variable) as well as the perception of the consumers about the brands of different manufacturers, etc.

Figure 5 shows the structure of the system being considered.

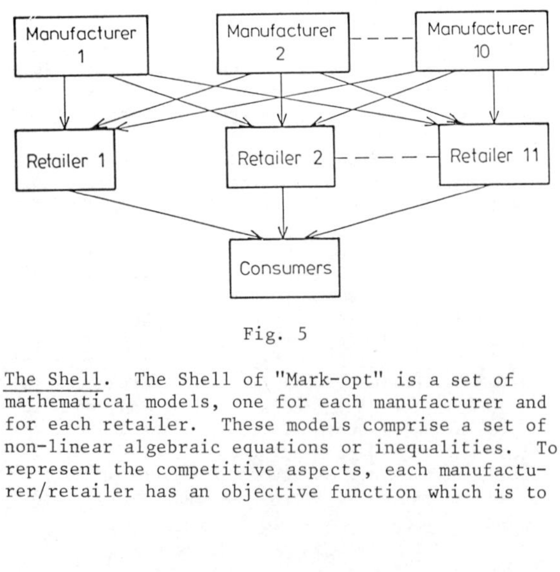

Fig. 5

The Shell. The Shell of "Mark-opt" is a set of mathematical models, one for each manufacturer and for each retailer. These models comprise a set of non-linear algebraic equations or inequalities. To represent the competitive aspects, each manufacturer/retailer has an objective function which is to maximise profits. Thus the Shell comprises in this case a set of interacting optimisation problems.

Knowledge Acquisition. Knowledge Acquisition in "Mark-opt" involves the fitting of parametric values into the mathematical models of each manufacturer/retailer in the Shell. Since the system "Mark-opt" will most likely be used by a manufacturer, the process of knowledge acquisition starts by the manufacturer feeding in "hard" information about his own production costs, economies of scale, brand image (as represented by price elasticity, shelf space elasticity, etc.). Next, the user feeds in "Gut feel" information about the competing manufacturers (Is their manufacturing as efficient as ours? Is their brand considered better or worse than ours by the consumers? etc.) and the retailers (Are they all equally efficient, etc.). These two kinds of information are integrated into the Shell in order to provide the framework for Scenario Analysis.

Inference Engine. The Inference Engine in "Mark-opt" allows one to examine various scenarios where the manufacturers/retailers respond to each other's moves by maximising their profits in each situation. The mathematical framework of interest here is Game Theory" and essentially, the Inference Engine in "Mark-opt" has, at its heart, a mechanism for finding Nash Equilibria of various Gaming situations amongst different players.

Other decision making problems at this level are essentially concerned with the allocation of limited resources in the most profitable way e.g. sales force allocation, advertising budget allocation, etc.

For such problems, an Intelligent Knowledge Based System called "Retail-opt" [11] has been developed. This system uses, in the same way as the systems described above, and in the section entitled "Mark-opt [9,10], a combination of "hard" information and "gut feel" information. The Shell or knowledge base here has also causal maximising models represented by a set of non-linear polynomial equations and linear profit functions. The Inference Engine is an appropriate optimisation algorithm.

Both the systems "Mark-opt" and "Retail-opt" are being used by a number of industrial companies.

Finally, we describe the production planning and scheduling level.

PRODUCTION PLANNING AND SCHEDULING

Production planning and scheduling needs to cope with events over a much smaller time horizon than the ones considered in sections 3 and 4. Here the basic system is shown in Fig. 6 and the objective is to provide help to plant managers to run the plants so as to meet the market demand over a medium term horizon (up to about 20 weeks) whilst using the flexibility inherent in the non-dedicated nature of some or all of the production and packing units. At the same time, since perturbations occur over fairly short periods of time (a worker sick, a machine out of order, etc.), it is important for the scheduler to be able to provide "work to" lists going down possibly to hour by hour.

For many such plants, orders may exist for the next 2-3 weeks and a reasonable forecast may be available of the demand over the longer horizon of 3 to 20 weeks. In these plants, demand is typically "lumpy" i.e. there may be periods when demand (in terms of orders to be met) is below capacity and others when it is far in excess of capacity. The plants need to cope with this lumpy demand in the

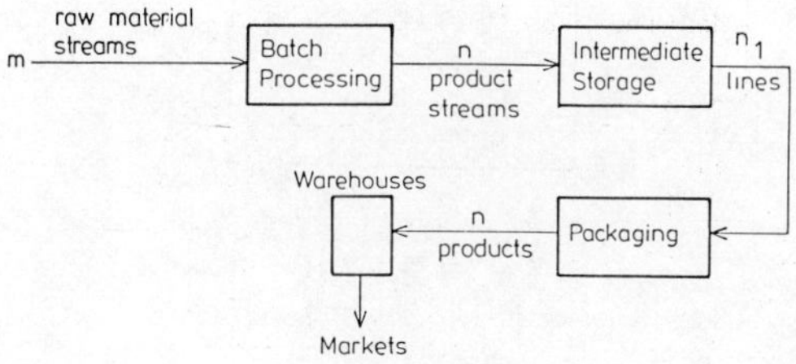

Fig. 6

most cost effective way by weighing up the cost of work in progress and inventories against the lost production due to changeovers.

At this level, the amount of judgemental aspects as opposed to modelisable aspects is quite different from the two upper levels i.e. here one can use much better models. The difficulty is that with a sampling interval of 1 hour and a time horizon of interest of say 10-20 weeks, the optimisation problem which one can formulate for operating these plants in the most cost effective way often ends up with several million constraints, and several hundred thousand variables, many of which are integer valued (i.e. production unit/packing unit on/off, ready to come on/ready to go off, etc.) There are no techniques available at the present time for solving mixed integer nonlinear programs of such high dimension.

One way of getting around this has been pioneered by ICI PLC and commercialised in the decision support system "Barnett" [12]. Essentially, the horizon of interest is broken down into a Short Horizon and a Long Horizon. The Short Horizon and the Long Horizon schedulers are partly rule based and they do not take into account longer term demand than that specified by the existing orders. No optimisation is done.

Another approach which is more novel has been developed by the author and his collaborators whilst working with a major consumer goods manufacturer. This approach involves the use of a 3 level structure as shown in Fig. 7. Here levels 1 and 2 form a distributed knowledge based system comprising an expert system each for production, intermediate storage and packing with a coordinator at level 2. These 2 levels solve the Short Horizon (2 weeks) scheduling problem using artificial intelligence techniques whilst at the 3rd level, a mathematical optimisation problem is solved with an aggregated set of constraints and from which the optimal lot sizes for each item are fed to level 2. The resulting structure provides an extremely attractive and efficient approach to tackling these very difficult and important problems. The approach has been successfully used on the data bases of several plants to provide attractive near optimal schedules.

Thus, for production planning and scheduling for such non-job shop type plants, useful computerised decision tools are beginning to emerge although much still remains to be done.

CONCLUSIONS

In this paper we have described the major decision making problems which arise within manufacturing hierarchies where frequently purchased products are being produced. Since such products need to be sold via a small number of retail chains, the problems at the top two levels of the hierarchy concentrate on competitive analysis within marketing channels. At the highest level where strategic decision making takes place, we have described a decision support system which uses a differential game framework. For the operational/tactical level, a static game framework which seeks Nash Equilibria as a part of

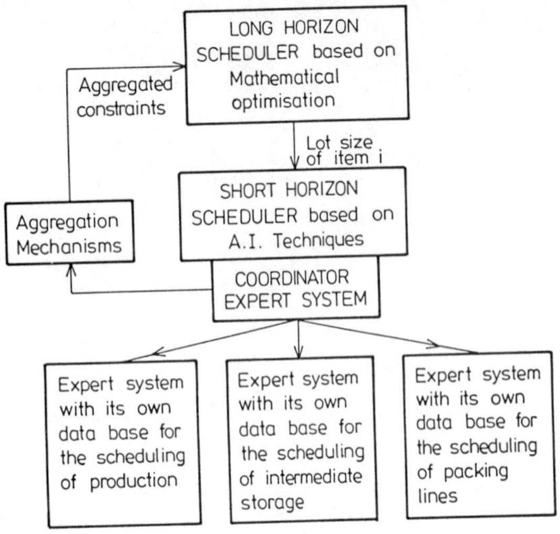

Fig. 7

Scenario Analysis has been outlined. A tool for allocation decisions was also mentioned. Finally, the problems of production planning and scheduling for non-job shop type plants have been described and some of the work going on in industry in this area has been highlighted.

REFERENCES

1. Ambrosiadou, V. and M.G. Singh (1985). A decision support system for strategic decision making in marketing channels using a differential games framework. CSC Report No. 631. Submitted for publication.

2. Porter, M. (1980). Competitive strategy: techniques for analysing industries and competitors, Free Press, NY.

3. Singh, M. G. and R. Cook (1985). A new class of intelligent knowledge based systems with an optimisation based inference engine. CSC Report No. 629. Submitted for publication.

4. Singh, M. G. (1985). Decision support systems for marketing and retailing. Invited plenary paper for the US PD Conference on Expert Systems and Decision Support Systems, Washington, DC. April.

5. Helmy, H., Baligh L. and L.E. Richards (1984). An Analysis of Vertical Marketing Structures. Managemet Science, Vol. 10, No. 4, July.

6. F.E. Balderston (1958). Communication Networks in Intermediate Markets. Management Science, Vol. 4, January.

7. Corstjens M. and P. Doyle (1979). Channel Optimisation in Complex Marketing Systems. Management Science, Vol. 25, No. 10, October.

8. Zushan, P. and M. Etgar (1981). The Marketing Channel in an Equilibrium Set of Contracts. Management Science, Vol. 27, No. 3, March.

9. Arnold, J., Singh M.G. and M. Corstjens (1984). A Decision Support System for Marketing. LSS, Vol. 4, 2.

10. Singh, M. G., Bhondi J. and M. Corstjens. "Mark-opt": a decision support system for marketing. IEEE Trans SMC, 1985 (in the press)

11. "Retail-opt" User Manual, Control Sciences Ltd. Manchester, U.K., 1984.

12. "Barnet", ICI PLC, Runcorn, U.K., 1984.

STANDARDIZATION OF SOFTWARE FOR CACSD

Panel Discussion

Chairman: Mohamed Mansour
Swiss Federal Institute of Technology (ETH),
Zurich, Switzerland

Members : Derek P. Atherton
University of Sussex, Brighton, U.K.

Michel A. Floyd
Integrated Systems Inc., Palo Alto, California USA

Hartmut Hensel
Technische Hochschule Darmstadt, FRG

Magnus Rimvall
Swiss Federal Institute of Technology (ETH),
Zurich, Switzerland

Keywords: CAD, Software, Standardization

INTRODUCTION

The aim of this discussion is to point out roughly what kind of standardizations can be made to CACSD software and how can this be realized. Hopefully at the end of the discussion we can find a way inside IFAC to begin work in this direction.

During the last decade, a large number of CACSD packages have been developed at different universities and in the industry. These programs/packages are often very specific, unflexible, unhandy in their man-machine interfaces, or are constructed to meet special requirements of the hardware or the operatinhg system software (Hensel). In the following, standardization in the following areas are adressed.

1. Programming languages
2. Simulation languages
3. General purpose operating systems
4. Graphical software
5. Modelling and design data bases
6. User interfaces
7. Data format standardization
8. Interfaces to numerical algorithms

1. PROGRAMMING LANGUAGES

The introduction of new powerful programming languages like PASCAL, MODULA-2 and ADA may facilitate the wide use of some program packages due to the transparent program-structures, the high level portability and the computer independent user-surfaces. However the acceptance of these new languages is rather poor due to the excessive effort spent on programs written in the old languages (Hensel). To counteract the increase in language confusion (Fortran, Ada, Modula, Pascal, Prolog, Lisp ... etc.) a command language standard and a data formal standard for the exchange of data between different control packages is necessary (Rimvall).

2. SIMULATION LANGUAGES

Atherton points out that simulation is the core of a CACSD package. The system description software should be closely linked to a simulation language. Some simulation languages allow linearized models to be produced, although in general one can see the need for other forms of reduction of the simulation model. Floyd also considers simulation as the core of a CACSD package and therefore there is a need for a standardization of dynamic data representation.

Hensel sees that there is a large number of simulation languages for continuous as well as discrete systems. Standardization in this area should reduce their number and support the development of a few more general simulation languages.

Rimvall considers the CSSL'67 standard to be a step in this direction. The development of simulation languages after this standard is typical and should encourage standardization in the area of CACSD. However, to avoid duplication of work, any standardization group in CADSD software should work in close conjunction with the standardization group within the simulation community (IMACS and SCS).

3. GENERAL PURPOSE OPERATING SYSTEMS

The most promising candidate for standardization of general purpose operating systems is the UNIX-system which offers a computer-family independent operating system surface and programming environment. Major drawbacks of this operating system are the rather slow system reaction performance, the large variety of extensions to the basic standard and the absence of a good real time capability (Hensel). Floyd gave UNIX as an example how standardization has failed in the area of opeating systems, which would likely be the case also in CACSD.

4. GRAPHICAL SOFTWARE

One very popular representative of graphical software which already became ISO-norm is the graphical kernel system (GKS). It offers a uniform user data interface which is almost completely independent from the installed hardware (achieved by writing special drivers for each connected device). The GKS-system offers different capability levels so that the system can be tuned with respect to the used device (Hensel). Atherton states that the problem is greatly compounded when moving software to the new generation of powerful workstations with colour graphics, multiple windows etc.

5. MODELING AND DESIGN DATA BASES

In control engineering the engineer's design is eventually transformed into real-time control software or hardware. The key to the long term success of CACSD lies in the automation of the real-time programming process, the critical software component will be the design data base.(Floyd) A fully evolved database should be able to describe any dynamic model. The component library should include linear, nonlinear and logical elements. Discrete-time attributes such as sampling rates (including multiple rates) and delays must be clearly defined. A complete data base can provide accurate and reliable communication among control engineer and real time developers. Floyd gives as example "System Build" used by MATRIX-x.

Atherton also states that entering the description of the system is the starting point for standardization. Any package must be capable of accepting the complete system description, which can presumably include nonlinear and possibly time varying elements, using several possible techniques and be able to convert between them (state space, transfer function, connection statements, block diagrams, signal flow graphs). The files holding the system description data of the appropriate form should then be readable directly into analytical design routines. Obviously it should also be easy to make changes to the system description, either because errors have been made during the entry or for the purpose of studying the behaviour for different parameter sets.

6. USERS INTERFACES

This is the part of CACSD package which is visible to the user. Floyd states that the design of user interface for CACSD software is fairly subjective and can not be subject to standardization. Standardization of the user interface would only stifle creativity in one of the most rapidly evolving areas in computer science. On the other hand Rimvall sees that a standard interactive command-language is needed to enable the exchange of macros and to unify the user interfaces of different packages. Such a standard can and should be described consistently using a set of "BNF"-productions for the syntax with textual additions for the semantic details. Any standard on this level will be hard to enforce on already existing packages, but is likely to be followed by any new packages. Hensel sees that the dialog interface (at its program end) must be standardized in order to achieve simple structuring of dialogs during the program development phase. The dialog itself should be uniform and easy to learn. In the case of command or menue-technique an automatic adaptation of help features dependent on the user's advancement can be very convenient. A future aspect to be considered might be how to change the man-machine to a man-model-communication, which allows the user to work directly on a model-base without the need of using related system data. This would also reduce the required level of abstraction. Athertons considers the standardization of user interface mainly in view of the standardization of system description as mentioned before under 5.

7. DATA FORMAT STANDARDIZATION

A standard data format (supporting all common control structures) for the exchange of data between different control packages over external files is very much needed. Such a standard is easy to enforce: each program can have several input/output modes, one such mode can support the standard while the other modes are left as they are (Rimvall).

8. INTERFACES TO NUMERICAL ALGORITHMS

Floyd states that standardization of numerical algorithms would be difficult to achieve in a non commercial environment. However, Rimvall sees the possibility to specify how the algorithm is to be accessed interactively. This facilitates the exchange of algorithms between different CACSD-packages, but also couples the subprogram standardization with the interactive command-language standardization, and is not totally free from implementation restrictions (e.g. the supported data structures of the implementation language and target machine limitations).

CONCLUSIONS

The discussion in the session has shown that there is a necessity to standardize some aspects of the software for CACSD, mainly in the man-machine and program-program interface area. It was felt that IFAC should establish a working group in this area. The organizer of the session has already taken steps in this direction and it is hoped that soon there will be a result of the efforts in this direction.

M. Mansour

CAD FOR EDUCATION OF DIGITAL CONTROL SYSTEMS

K. Furuta

Tokyo Institute of Technology, Oh-Okayama, Meguro-ku, Tokyo, Japan

Abstract This paper is concerned with the use of CAD for the eduction of the modern control theory and of the digital control. It is pointed out that the theory can only be understood through its use in practice, and the control system design of the experimental apparatus is found one of such practices, where the analysis and design of control system should be done by the help of CAD. CAD developed by the authors and the control of experimental apparatus are briefly described.

Keywords CAD, multivariable control, digital control, education

INTRODUCTION

The computer usage has become inevitable not only as the digital controller but also as the means to assist the design and analysis of multivariable control systems. In the controller synthesis based on the classical control theory, the identification of the plant by the frequency response and its representation by Bode deagram can derive the optimum parameter of commercially available PID controller. However if one wishes to design the control system based on the modern control theory, he has to know the wide spectrum of the complex multivariable control theory from the identification theory to the optimal control including the observer. And in order to get all of these knowledge, one has to read through many text books or papers. Therefore leymen can not learn the theory by themselves through text books in spite that it looks effective, and they murmur that the classical control theory is more useful than the modern one, since modern one is too theoretical. But from the theoretical point of view, the so-called classical control theory is using as well complex theory as the modern one. The difference is that in the classical theory, they use theory practically without paying attention to the proofs, but in the modern theory, they are lost themselves in the maze of the theorems and proofs. To overcome this problem the wide spectrum of modern multivariable control theory should be educated from the view point of how to be used in design of control system including its realization by the digital computer.

For this purpose, instead of text books, the computer aided design program package is much useful to be used to educate the control system design, where the computer is used to provide appropriate information such as computational algorithms and to store the data. Thus by the help of CAD, the control system design from the parameter identification to the control system design can be done easily. But the structure of the designed control system is not simple and it can be only realized by the digital computer. The last step has been most difficult and time consuming. But control engineers should be educated from the design of control systems to its implementation. Therefore at suthor's laboratory, as thesis projects, the design of controllers and its realization have been done using CAD developed by author's group[1)] for the experimental apparatus, such as

i) double inverted pendulum[2)]
ii) triple inverted pendulum[3)]
iii) swing up of single pendulum[4)]
iv) magnetic levitated iron plate[5)]
v) radio control helicopter[6)]
vi) industrial furnace[7)]
vii) linear motion mechanism[8)] and
viii) manipulator[9)].

They will be described briefly. The temperature control of the industrial furnace has been done for the manufacturing of magnet wire and it is now used practically. Through these projects it is concluded that not only the design and analysis but also the implementation should also be assisted by the computer and this kind of system should be called the computer aided control (CAD).

BRIEF DESCRIPTION OF DPACS

The computer aided program package named DPACS has been developed by the author's group, which can assist to design and to analyze the multivariable control system by providing the programs of algorithms and by handling the systems data. The systems treated in this CAD are linear multivariable systems in the forms of state space equation, transfer function, and input-output data, can be either continuous or discrete. Systems data are stored in the database with the system's name, and can be utilized by its name. DPACS is used interactively through command from the console. The command consists of the program name and the system name to be treated such as

OPT SYS 1,

which is interpreted by the supervisor that the optimal control law for the system named SYS 1 should be calculated. Sometimes another system name is added so that the resultant system data given by the pragram is stored by the name. Parameters and specifications in the algorithm for the pragram are set in the question & answer way. The conversation example using DPACS is shown in Fig.1, where the system named EXAM is stored in the transfer function form and minimally realized in the state space. The list of commands which provide algorithms is given Table 1. DPACS consists of program package of several levels such as matrix calculation (MAT), time series processing (TS) and system data processing (SYS). They are also used interactively and the matrix data are handelled in DPACS. So even new algorithm which is not included in DPACS is to be executed.

When control system is designed and its data are

```
    FURUTA LAB. DPACS (REV.2)

/DPACS/ 19: 0:33
SYSIN EXAM
SPECIFY TYPE OF SYSTEMDATA :  T
TO INPUT SYSTEMDATA
CONTINUOUS OR DISCRETE TYPE ? [C OR D] :  C
SPECIFY NO. OF INPUTS & OUTPUTS :  1.1
(1) DENOMINATOR & NUMERATOR COEFFICIENT MATRICIES
(2) DENOMINATOR POLINOMIAL & NUMERATOR POLINOMIAL MATRIX
(3) RATIONAL POLINOMIAL MATRIX
SPECIFY INPUT-MODE :  3
G(1.1) =  S2+2S+3/(S+1)(2S2+3S+1)

TYPE THE RATIONAL POLYNOMIAL MATRIX ?  N

REVISE THE RATIONAL POLYNOMIALS ?  N

EPSILON FOR COMPILATION :  1D-10

GIVE COMMENT FOR SYSTEMDATA :  EXAMPLE FOR IFAC

/DPACS/ 19: 2:35
CMR/T EXAM
TO OBTAIN CANONICAL MINIMAL REALIZATION

WHICH PAIR IN MINIMAL REALIZATION IS TRANSFORMED
INTO CANONICAL FORM. (1) (A.B) OR (2) (A.C) ? :  1
SPECIFY TYPE OF CANONICAL FORM [1 OR 2] :  1

GIVE EPSILON. TO DETERMINE RANK OF MATRIX U0
EPS =  1D-10
COLUMN INDEPENDENCY OF MATRIX U0 IS INVESTIGATED
( 1)       1
( 2)       1
( 3)       1
( 4)       0
CONDITION NUMBERS ARE AS FOLLOWS
--- ( 4. 1) MATRIX CNUM  ---
           ( 1)
( 1)  1.00000000D  0
( 2)  6.123724360D -1
( 3)  5.701261120D -1
( 4)  6.600614190D-23
CHANGE THE EPSILON ?  N

GIVE COMMENT FOR SYSTEMDATA :  EXAMPLE FOR IFAC  -2

/DPACS/ 19: 4: 2
LIST EXAM
------------------------------------------------------------------
CONTINUOUS SYSTEM EXAM     ( 1. 1)
------------------------------------------------------------------
 1) EXAM.T0      (    3) :SYSIN   /T/C/ /   ( 1/ 9/85) :EXAMPLE
 2) EXAM.S0      (    3) :CMR     /T/C/1/   ( 1/ 9/85) :EXAMPLE
```

Fig.1 Conversation example of the DPACS

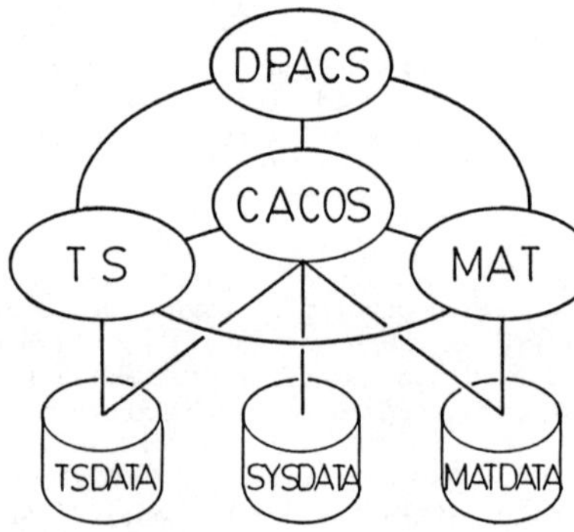

Fig.2 Relation between DPACS and CACOS

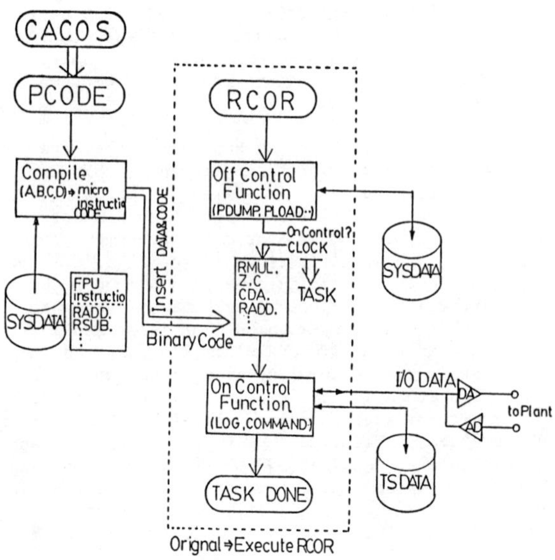

Fig.3 Flow chart of CACOS to generate real time control code

[1] System and Systemdata Management

LIST	to list systems or systemdatas
DELETE	to delete systems or systemdatas
RENAME	to rename system name
PROTECT	to protect or unprotect systems
SYSIN	to input systemdata
TYPE	to type out systemdata
REVISE	to revise systemdata
COPY	to copy systemdata
PICK	to transform coordinate of input and output space
DUAL	to obtain dual system

[2] Systemdata Transformation

COOD	to transform coordinate of state space
CMR	to obtain canonical minimal realization
TF	to obtain transfer function
MRKV	to obtain Markov parameters
DIF	to obtin defferential [difference] equation
IO	to make I/O data realization
DIGIT	to digitize continuous system
CONT	to transform coordinate of dicrete space to continual space

[3] System Analysis

POLE	to calculate system's poles
ZERO	to calculate system's zeros
CCCO	to investigate controllability or observability
RDCT	to execute model reduction
EMR	to calculate epsilon minimum

[4] Controller Synthesis

PLOC	to solve pole assignment problem
OPT	to solve optimal regulator problem
DCPL	to solve decoupling problem
OBS	to obtain observer or Kalman filter
TRACK	to solve tracking problem
MFS	to solve model following servo problem
OSV	to solve optimal servo problem
OSVU	to solve optimal servo problem [augument system]

[5] Simulation

LINK	to link systems
CLPS	to obtain closed system
SMLT	to simulate system's response

[6] Frequency Domain

NYQST	to draw Nyquist or inverse Nyquist diagram
BODE	to draw Bode diagram
LOCI	to draw root locus
BAND	to draw Gershgorin's band or inverse Gershgorin's band

[7] PIPACK (Identification)

GLS	to identify system by generalized least square method
CML	to identify system by conditional maximum likelihood method
SML	to execute CML for symmetric transfer function
LI	to identify system by limited information maximum likelihood method

Table 1 List of DPACS command

store in the database, it should be realized by the computer. To assist the software construction of this real time digital control system, the computer aided control (CAC) named CACOS[10)] has been developed, which shares database with DPACS as shown in Fig.2. The controller realized by CACOS has the form

$$x(k+1) = Ax(k) + Bu(k)$$
$$y(k) = Cx(k) + Du(k)$$

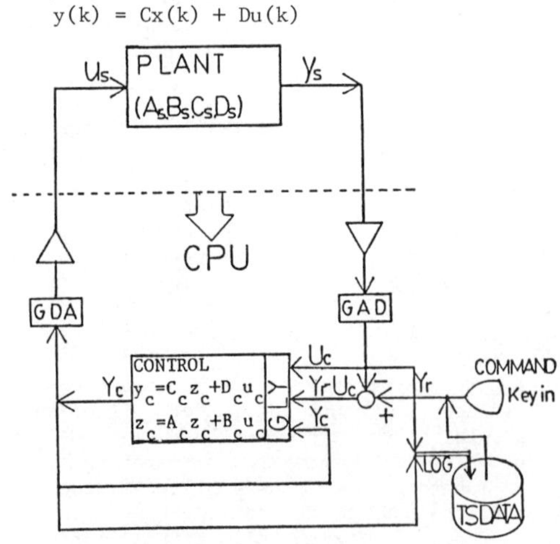

Fig.4 Structure of controller given by CACOS

where all multiplications by zero are eliminated and the program written by p-code is generated by the flow chart shown in Fig.3.

The digital controller designed by CACOS has the structure of Fig.4 and the functions of
1) control of systems
2) logging input-output data and
3) to input the reference signals
in the on-line mode, and of
1) saving and loading the parameters from/to DISK
2) dumping TS data, and
3) typing out parameters
By the combination of DPACS and CACOS, the design and not only thesis projects but also industrial problems have been treated by the system.

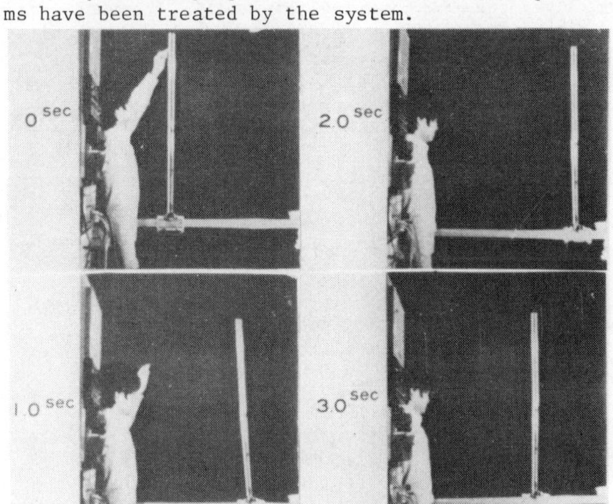

Photo 1. Control of double inverted pendulum to disturbance

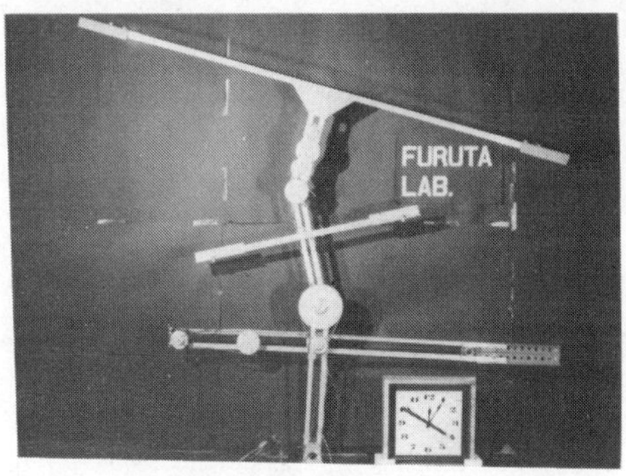

Photo 2. Attitude control of a triple inverted Pendulum

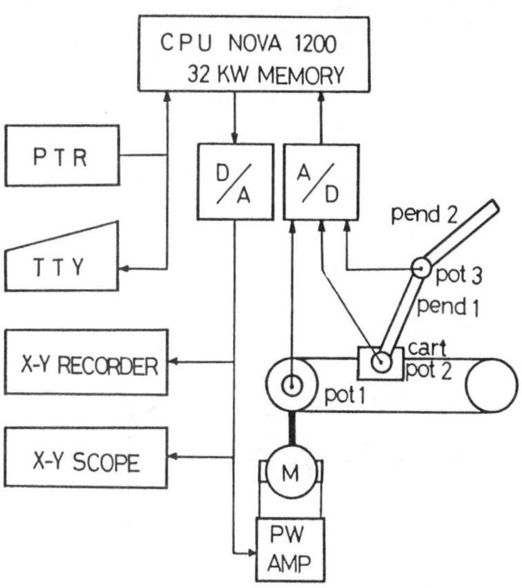

Fig.5 Computer control system for the double inverted pendulum

CONTROL EXAMPLES

At author's laboratory, the control of experimental apparatus has been as thesis projects mainly by the help of CAD. These examples are briefly discussed in this section.

Control of Double Inverted Pendulum[2)]

The control of a double inverted pendulum is one of most excellent examples to convince the power of control to laymen and beginners of control theory. Many students are excited by the stable balancing of a double inverted pendulum. The control system is shown in Fig.5.

Attitude Control of Triple Inverted Pendulum[3)]

Instead using balancing cart, torques are applied to hinges of triple inverted pendulum, where the lowest hinge is free for rotation. Its attitude can be controlled in the balancing condition as shown in Photo 2.

Swing-up of Single Pendulum[4)]

Single pendulum is swing up from the stable condition to the upright unstable inverted position. And at the upright position, it is stabilized by using state feedback using observer.

Control of Magnetic Levitation[5)]

Iron plates connected by plastic bar are levitated by two magnets. The position and angle are controlled as shown in Photo 3.

Control of RC Helicopter[6)]

RC helicopter is known as one of most unstable objects and if it can be controlled well, it can be used for the survey in the dangerous circumstance. The satisfactory dynamic model of an RC helicopter, however, is not established. So the modelling and its evaluation have been done by using an electrically

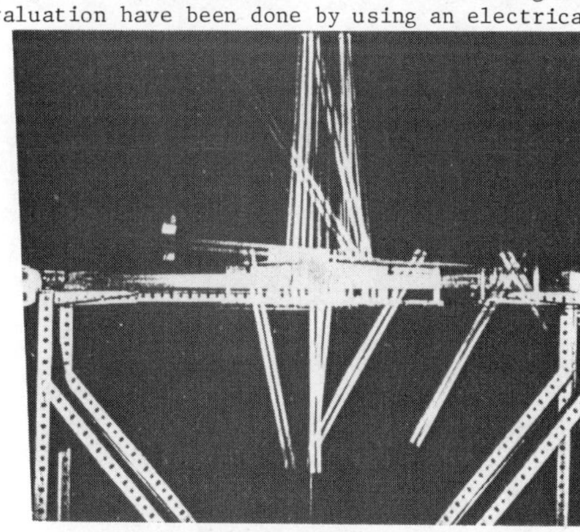

Phto 3. Swing up of a single inverted pendulum

Photo 4. Contro of positions of levitated iron plates

Phto 5. Control of electrically drived helicopter

driven helicopter fixed on a ginbal which is also mounted on a movable arm as shown in Fig.4. Using the system, the evaluation and control of an RC helicopter has been done.

Control of Temperature Distribution of Industrial Furnace[7)]

This is one of examples which are used CAD for the control system design of real industrial processes. The objective is to control the temperature distribution of the industrial furnace to manufacture magnet wire. Excellent results could be obtained.

Path Control of Mechanical System Using Laser Beam[8)]

The stage of the mechanical system shown in Fig.6 is controlled to track the laser beam, which may be used to teach the motion to the mechanical system. Since the control should be done based on the laser beam, the coordinate of the mechanical system is to be transformed into the coordinate system with the laser beam as one base coordinate as shown in Fig.7. In this case the vector x in the ordinary coordinate is transformed into α in the laser beam coordinate system by

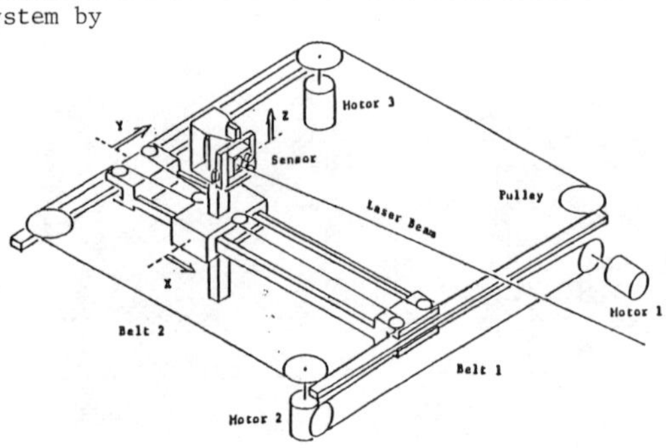

Fig.6 Three-dimensional linear motion mechanical system

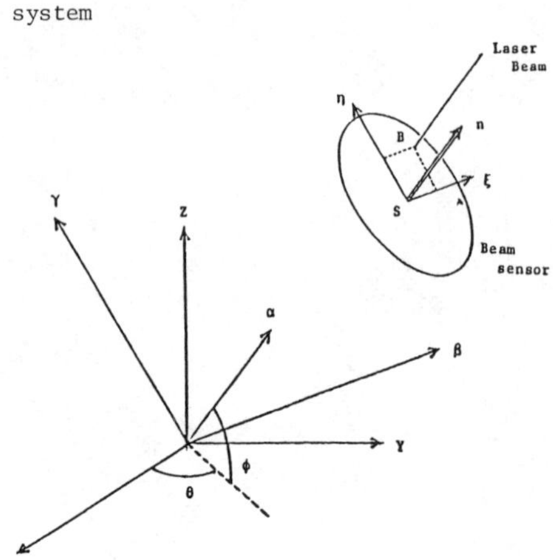

Fig.7 Coordinates

$$x = T\alpha \tag{1}$$

Since the mechanical system is represented by

$$\ddot{x}+A\dot{x} = Bu \tag{2}$$

substituting (1) into (2) makes

$$T\ddot{\alpha}+2\dot{T}\dot{\alpha}+\ddot{T}\alpha+Ax = Bu \tag{2'}$$

Therefore by choosing

$$u = B^{-1}\{2\dot{T}\dot{\alpha}+\ddot{T}\alpha+A\dot{x}-T(K_1\dot{\alpha}+K_2\alpha-Gv)\} \tag{3}$$

(2) yields the linear relation between v and α as

$$\ddot{\alpha}+K_1\dot{\alpha}+K_2\alpha = Gv \tag{4}$$

and the controller for this system is designed easily. The manipulator is highly nonlinear system. However it can be linearized by using the similar technique[9)].

DETERMINATION OF SAMPLING INTERVAL

One problem in the digital control system design is to determine the sampling interval. everybody says that shorter sampling interval gives better results. But how better results can be obtain quantitively by shortening the sampling interval? To get the measure, the quadratic criterion function

$$J = \int_0^\infty (\|x\|_Q^2+\|u\|_R^2)dt \tag{5}$$

for the continuous system

$$\dot{x} = Ax+Bu \tag{6}$$

can be employed. By taking

$$u(t) = u(iT) \text{ for } iT \le t < (i+1)T,$$

the minimum value of J is[11)]

$$\min J = \|x(0)\|_{\hat{P}}^2 \tag{7}$$

where $\hat{P}$ is the positive definite solution of

$$\hat{P} = \Phi^T\hat{P}\Phi+\hat{Q}-(\Phi^T\hat{P}\Psi+\hat{S})(\Psi^T\hat{P}\Psi+\hat{R})^{-1}(\Psi^T\hat{P}\Phi+\hat{S}^T)$$

where

$$\Phi = e^{AT}, \quad \Psi = \int_0^T e^{As}Bds$$

$$\begin{pmatrix}\hat{Q} & \hat{S}\\ \hat{S}^T & \hat{R}\end{pmatrix} = \int_0^T e^{\begin{pmatrix}A^T & 0\\ B^T & 0\end{pmatrix}s}\begin{pmatrix}Q & 0\\ 0 & R\end{pmatrix}e^{\begin{pmatrix}A & B\\ 0 & 0\end{pmatrix}s}ds.$$

Instead of (7), the largest eigenvalue of $\hat{P}$ can be employed as the measure for the sampling interval. Considering the delay due to the computation, similar results are also derived. Concerning the design of digital control, there are still many theoretical problem given in the practical situation.

CONCLUSION

This paper presents the possibility to use CAD for the education of control engineering. The results show that the possibility is very prosperous and should be accelerated by the develop of computers and CAD.

REFERENCES

1) K.Furuta et al.(1979). Computer Aided Design Program for Linear Multivariable Control Systems. IFAC 1st CAD Symp.
2) K.Furuta et al.(1978). Computer Control of a Double Inverted Pendulum. Comp.& Elec.Engr., vol.5.
3) K.Furuta et al.(1984). Attitude Control of a Triple Inverted Pendulum. Int.J.Control, vol.39.
4) S.Mori et al.(1976). Control Unstable Mechanical System. Int.J.Control, vol.23.
5) K.Furuta et al.(1982). Computer Aided Design and Computer Control Using Minicomputer. IFAC 2nd CAD Symp.
6) Y.Ohyama et al.(1984). The training Simulator of RC Helicipter. ISDS Tokyo
7) K.Furuta et al.(1982). Control System Design for Furnace by Using CAD. IFAC Symp., Delhi.
8) K.Furuta et al.(1984). Path Control of Three Dimensional Linear Motion System Using Laser. IECON '84
9) K.Furuta et al.(1984). Robust Control of a Robot Manipulator with Nonlinearity. Robotica, vol.2
10) K.Furuta et al.(1984). Computer Aided Control and Real Time Control Program Organizing System "CACOS". IFAC Congress, Budapest.
11) B.C.Kuo.(1980). Digital Control Systems. Holt, Rinehart and Winston.
12) R.Kondo et al.(to appear). Sampled-data Optimal Control of Continuous System to Quadratic Criterion Function Taking Account of Delayed Control Action.

CAD IN ROBOT CONTROL EDUCATION

P. Kopacek, I. Troch and K. Desoyer

Technical University of Vienna, Vienna, Austria

Abstract. The authors have been teaching since 5 years fundamentals of Industrial Robots and Handling Devices for students of Mechanical Engineering, Electrical Engineering as well as Applied Mathematics and Computer Sciences. These fundamentals include mechanics, model building, simulation and control.

For the practical education mainly in laboratory courses two personal computers, a little hydraulic driven robot and a hybrid computer system for simulations is available. At a lower level the students solve the kinematic and kinetic equations with the personal computer for different structures of robots. The hybrid computer system is applied for simulations of the controlled robot. The hydraulic driven robot can be controlled by means of a control box with pushbuttons for movements (e.g. up/down, rotate and grip and for start/stop). Another programming possibility for robot movements consists by using a special language which is different for each robot. Both possibilities are little efficient for education at a higher level. Therefore the control box of our robot is replaced by a personal computer for advanced studies. For all laboratory experiments the necessary software packages are written in dialoge form like a CAD system.

Keywords. Robots, Education, Computer Control.

INTRODUCTION

The number of applications for industrial robots and handling devices will essentially increase in the nearest future. A mechanical production without these devices will be nearly impossible. Therefore a start for education in this field seems to be necessary in time for workers and engineers of all educational levels.

An industrial robot consists of mechanical and electrical parts as well as digital computers for control. For its operation and service special educated people have to be available on time. This is very easy possible for the mechanical and electrical parts. Well educated electricians need only special knowledge for programming the robots, but for hard- and software problems specialists will be necessary.

In Austria only at Technical Universities knowledge about this field is taught today. Therefore graduates at compulsory vocational schools (Pflichtberufsschulen) industrial technical schools (Höhere Techn.Lehranstalten) or Technical Colleges have only the possibility to pass various postgraduate courses and particular lectures.

At the Technical University of Vienna students of the above mentioned subject have to pass two so called "study emphasis". For example in Mechanical Engineering there are 15 of them. Each consists of

- a special lecture (2 hours/week) not obligatory
- a seminar course (3 hours/week)
- a special laboratory course (2 hours/week)
- a written study work (7 or 3 hours/week).

Usually the student prepares the masters thesis (Diplomarbeit) - necessary for the "Dipl.Ing." - in one of these fields. For further informations of the Austrian education system especially for Automatic Control see e.g. (Kopacek,Genser,Troch,Weinmann,1984).

Since 2 years study emphasis are given by our working group (Desoyer,Kopacek,Troch,1985a) dealing with the structure, the kinematics and kinetics, the control and the utilization of robots and handling devices. The goal of this special education is giving students of different fields (e.g.Mechanical Engineering, Electrical Engineering, Technical Mathematics and Informatics) an overview about the types of robots, the programming possibilities and problems in application. Furthermore for an efficient utilization of robots special knowledge is necessary concerning the kinematics e.g. for the transformation of robot coordinates into world coordinates and vice versa, the kinetics e.g. for calculating "mathematical" models for the dynamic behavior and especially the control. Starting from linearized model equations and simple linear "conventional" control algorithms an overview is given about various models (nonlinear, time-varying, multivariable) for the robot and various "advanced" control algorithms. The main topics of the lecture are collected in Tab.1.

For the practical education in form of laboratory courses mainly for programming and design of control algorithms two robots as well as one hybrid computer system and two personal computers are available. One of the robots is mainly suitable for research activities while the second a hydraulic driven teach robot is only for laboratory courses.

DESCRIPTION OF THE TEACH ROBOT

The robot arm has a spheric working area (kinematic structure RRR) and six degrees of freedom (i.e.five axes plus grip). It is driven hydraulically with a pression of 8 bar. The working area is determined by the "shoulder" rotation through 90 deg. with vertical movement through 90 deg.; the "elbow" rotation through 90 deg. and the wrist rotation through 180 deg. with vertical movement through 60 deg. All arms are equipped with position sensors with an accuracy of 256 steps over the whole working range. The maximum lifting capacity amounts up to 2 kg with a maximum dimension of 50 mm; the maximum reach is 1000 mm.

The processor unit contains an 8-bit processor defining each movement by an 8-bit word.A displayboard gives visual indication of the control mode and audible indication of incoming data. Learning facility by which movements set by the control box are inserted to form a remembered sequence of operations with up to 32 steps. By means of an interface board position detector signals are multiplexed and transferred to an A/D converter. The output signals of the board actuates the solenoid valves of the hydraulic system. An RS 232 interface for connection of external computers is also available. Normally the robot arm is controlled by means of a Control Box. It contains pushbuttons for various movements i.e. up and down, rotate and grip, start, step,slow,delete, insert,play,clear,sequence,pause,wait,reset and edit.The accuracy of positioning is determined by dividing the working range of each axis in 256 steps according to the 8-bit processor.

For advanced work the control box can be replaced by an external computer (e.g. a personal computer). In our case for control a cheap personal computer is used. It is a 16-bit computer with 64 K RAM and a 20 K ROM. For programs written in Assembler 52 K and for programs written in BASIC 38 K are available. Further possible programming languages are different extended versions of BASIC (EX-BASIC,SIMONS BASIC...) as well as PASCAL,FORTH ... For storage of programs serves a Floppy-Disk-Station (5/4"). A monitor as well as a little printer is also available.

TOPICS OF THE LECTURE "INDUSTRIAL ROBOTS AND HANDLING DEVICES"

- design of Industrial Robots and Handling Devices
- examples and applications
- components (kinematic structure, drives, gripper, sensors, control equipments)
- fundamentals of mechanics and mathematics for robots
- models for the dynamic behaviour (model building)
- model simplification (simulation)
- position and position control
- programming and programming languages
- "advanced" control algorithms (optimal and adaptive control)
- criterias for utilization
- developping trends

Tab.1 Lecture on Industrial Robots

LABORATORY EXPERIMENTS

The dynamic behavior of a robot with three degrees of freedom may be written as a matrix equation (Desoyer,Kopacek,Troch,1985b)

$$\underline{k}_{Dr} = \underline{M}(\underline{p})\,\ddot{\underline{p}} + \underline{n}(\underline{p},\dot{\underline{p}}) + \underline{K}_f\,\dot{\underline{p}} + \underline{k}_g$$

with

$\underline{p} = [p_1,p_2,p_3]^T$ ... vector of robot coordinates (the components are either lengths or angles depending on the kinematic structure)

$\underline{k}_{Dr}$ ....[3x1] vector of drives (the components are either forces or moments)

$\underline{M}(\underline{p})$....[3x3]matrix (the elements are masses or masses multiplied with lengths or lengths-squared)

$\underline{n}(\underline{p},\dot{\underline{p}})$..[3x1] vector of Coriolis- and Centrifugal terms

$\underline{K}_f$ .....[3x3] matrix of coefficients of friction forces (simplified proportional to the speed)

$\underline{k}_g$ .....[3x1] vector of gravitational forces or -moments.

At present for robots with one rotatory and two translatory degrees of freedom (kinematic structure RTT) and with three rotatory degrees of freedom (kinematic structure RRR) programs for these equations are available for the personal computer described above. These two types of robots are choosen because approximately 75% of robots applied today have these structures.

This first "laboratory" experiment has the goal imparting the students a feeling for the connections between the construction parameters (dimensions, masses, moments of inertia) and the desired values of coordinates, speeds and accelerations and the drives. The computer program is designed in a dialogue form and has the following structure:

a) choosing of the kinematic structure (RTT or RRR)

b) entry of the construction parameters; parameter sets for commercial available robots are stored on discettes.

c) entry of the values of coordinates, speeds and accelerations

d) initialisation of the program

e) after a short time the result is available in the form e.g. for a RRR type as shown in Fig.1.The print out shows the values of the 3 drive moments with and without the influence of the gyroscopic moments of the driving motors as well as the differences

f) now the computer asks for new values of coordinates or if desired other construction parameters.

With this software package it is possible to study the behavior of the drives as well as the influence of the construction parameters especially of eccentricities in the joints. In this way the student gets an insight in the connections between the variables. Furthermore it is possible to add the software for different drives and gears for each axis.

The second laboratory experiment for the students deals with simulation of robots. A similar program package is available for the hybrid computer of the Technical University of Vienna. By way of contrast to the program for the personal computer this serves mainly for studying varios position control

```
TRAEGHEITSMOMENTE DER ANTRIEBSMOTOREN...

...BERUECKSICHTIGT :

        MM( 1 , 0 : 1 / 1 ) =-175.785428 NM
        MM( 1 , 0 : 2 / 1 ) = 189.881959 NM
        MM( 1 , 0 : 3 / 1 ) = 871.47423 NM

...UNBERUECKSICHTIGT :

        MO( 1 , 0 : 1 / 1 ) =-179.949024 NM
        MO( 1 , 0 : 2 / 1 ) = 179.5006 NM
        MO( 1 , 0 : 3 / 1 ) = 883.440489 NM

DIFFERENZ DER OBIGEN MOMENTE :

      DIFF( 1 , 0 : 1 / 1 ) = 4.16359586 NM
 IN % VON MO( 1 , 0 : 1 / 1 ) =-2.31376407 %
      DIFF( 1 , 0 : 2 / 1 ) = 10.3813593 NM
 IN % VON MO( 1 , 0 : 2 / 1 ) = 5.78346778 %
      DIFF( 1 , 0 : 3 / 1 ) =-11.9662595 NM
 IN % VON MO( 1 , 0 : 3 / 1 ) =-1.35450657 %
```

Fig.1 Example for a print out

algorithms. The software package consisting of an analog part for simulating the robot and a digital part simulating the controllers is developed modular; i.e. it is possible to simulate the kinematic structures RTT and RRR with different drives (electric direct current motor, hydraulic or pneumatic drives) with different gears. Usually the gears are modelled as simple proportional elements or amplifiers. If it is necessarv the masses and moments of inertia may be taken into consideration. This leads to second order lag elements for the dynamic behavior of the gears.

For control different program modules on the digital part of the hybridsystem can be activated. There is the possibility to use conventional control algorithms (P-,PI-,PD-,PID-) in a digital recursive way as well as advanced digital control algorithms including state space controllers. For the determination of optimal controller "parameters" a few of "optimization" algorithms mainly for advanced controllers are implemented also at the digital part.

For the student this "laboratory" experiment is accessible in form of a CAD procedure. The first two steps a), b) are the same as in the last example. The next are

c) selecting kind of drive and entry the parameters for each axis

d) selecting the kind of the model of the gear

e) selecting the control algorithms. Entry other of the controller parameters or of the appropriate optimization algorithm

f) entry the cartesian coordinates of the starting and the final point

g) the result is plotted in form of a curve in the corresponding plane

As example Fig.2 shows a typical test result for a RTT-type robot with constant lifting (z-coordinate) in the x-y-plain.Both axis are controlled by proportional controllers. As indicated in the figure the gain of the φ-direction is 100 while the gain of the controller in the r-direction is varied from 30 to 240. Clearly the form of the patterns depend mainly on the values of the gain in the r-direction.

In contrast to the two described experiments the following are carried out on teach robot. With the test equipment, described before, the student has in a first stage the possibility to program the robot movements by means of the control box in a teach-in mode. With the pushbuttons on the panel of the box the desired points in the working area are approached and the positions of the joints are stored in the memory of the control box. If all necessary points are stored, the complete sequence will be traced one time or continuous. Clearly the movement between the points cannot be influenced. The resulting path is determined by the time points for opening the solenoid valves. Typical tasks in this education stage are various operations with small cylinders or cubes simulating real workpieces.

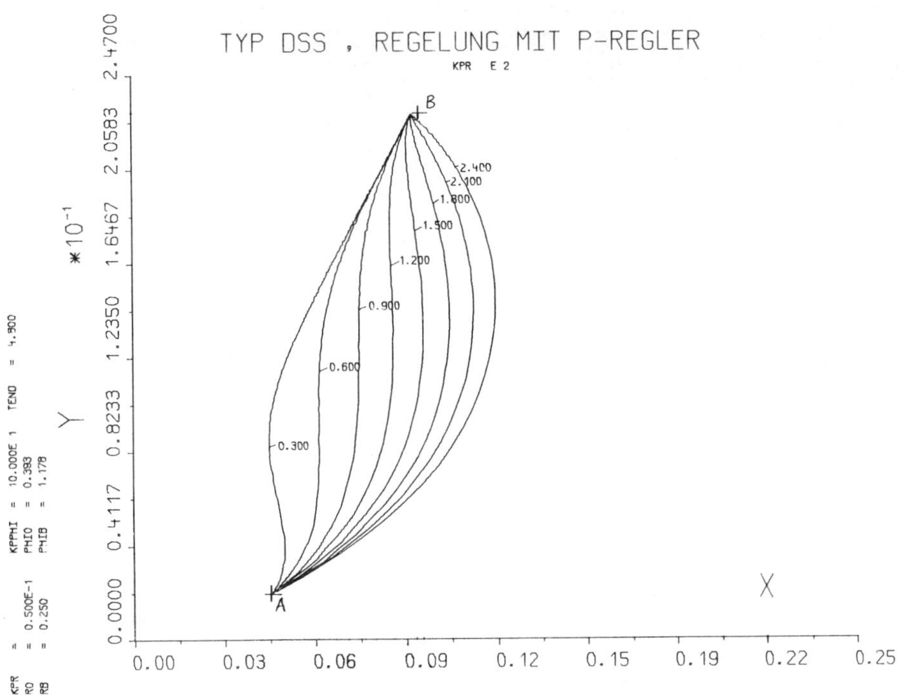

Fig.2 Example for paths with P-controllers.

In an advanced education stage the problems have to be more complicated. Therefore as pointed out above the control box is replaced by a little personal computer. This simple computer has the advantage that many students have the same model at home and so the initial period is reduced to a minimum. The standard software written by our students consists of program modules for actuating the 5 degrees of freedom. Furthermore it contains a program for calculating the cartesian coordinates for each point programmed in the Teach-in mode. This program serves also for feeding the personal computer with the cartesian coordinates of our desired point.

Usually this kind of teach robot is not suitable for control experiments especially with this computer. Nevertheless one of the goals of this test equipment is to collect first experiences in this relatively new field of education. For the position control of the axis different control algorithms are available in form of software modules. These modules are conventional control algorithms (P-, PI-, PD-, PID-) in digital recursive form as well as advanced digital algorithms. The latter are derived from a z-transfer function of the form

$$F_R(z) = \frac{q_o+q_1z^{-1}+\ldots q_\nu z^{-\nu}}{p_o+p_1z^{-1}+\ldots p_\mu z^{-\mu}}$$

The necessary software package is developed in dialoge form like a CAD procedure. It consists of the following steps

a) selection of the points (command values) in cartesian coordinates simultaneously for all axes

b) the computer calculates the robot coordinates belong to this points and prints out them on the monitor

c) selection of control algorithms and the controller parameters

d) during the movement the command as well as the actual position is shown on the screen.

This experiment allows a direct comparison of simulation results and the practice because the teach robot with the drives, gears and control algorithms may be also simulated on the hybrid computer. The student gets a feeling of differences between the behavior of the real system and the simulated system.

SUMMARY

A short overview concerning the educational activities of our working group in the field of robots and handling devices at the Technical University of Vienna is given. Students of different subjects have the possibility to choose courses in this field. In this paper some of the laboratory experiments in connection with CAD are described. In all of them the CAD procedures allow a direct dialog with the computer for designing e.g. control algorithms with appropriate parameter values for the given robot. The first are calculations with a personal computer dealing mainly with the kinematics and kinetics of different types of robots. Students can see the movement of the joints and the actual values of the necessary drive forces and drive-moments for the joints. The computer allows easily to study the influence of distinct construction parameters at the kinematics and kinetics of the robot. This knowledge can be applied simultaneously at the hydraulic driven teach robot with a spheric working area (kinematic structure TTT). By means of a control box it is programmable in a teach in mode. Other experiments at this level are e.g. lifting and stacking of cubes or little cylinders by means of this robot.

For robot control experiments a big hybrid computer system (EAI-Simstar) as well as a personal computer for control of the teach robot is available. The necessary software packages are written in a modular form using different CAD procedures for designing of the "controllers" as well as the representation of the curves of movement of different points. It must be noted that the whole software is written by students during advanced laboratory courses and Diplomatheses (Diplomarbeiten).

Further activities using CAD in robot control education will go in the direction of a 3 dimensional representation of the resulting paths. At present the paths are only represented in two dimensions. Therefore it is necessary to plot two or three curves in different planes. Moreover a movable picture of the robot in form of lines for each link at the monitor screen seems to be necessary for simulation. This offers the possibility to study e.g. collision problems.

REFERENCES

Kopacek, P., R. Genser, I. Troch and A. Weinmann (1984). Automatic Control Education in Austria. Preprints of the 9th IFAC World Congress Budapest. Vol. 4, pp.112-115.

Desoyer, K., P. Kopacek and I. Troch (1985a). Information about the working group on "Industrial Robots" at the Technical University of Vienna, in "Digital Systems" for Industrial Automation.

Desoyer, K., P. Kopacek and I. Troch (1985b). Industrieroboter und Handhabungsgeräte (in German) R.Oldenbourg, München.

COMPUTER AIDED DESIGN OF CONTROL SYSTEMS, AN INTEGRATED APPROACH

M. Mansour, M. Rimvall and W. Schaufelberger

Institute of Automatic Control, Swiss Federal Institute of Technology (ETH), CH-8092 Zürich, Switzerland

Abstract. In this paper on the use of CAD packages in control education we will concentrate on special requirements to be put on CACSD-tools in a student environment. In a first part, we survey the possible use of computers in the design cycle and discuss the incorporation of computers into the educational process. Different approaches towards an integrated CACSD-facility are presented in a second part. Several aspects important in the design of new CACSD-tools are then discussed. In particular, the design of a flexible user interface and the incorporation of graphics are treated. Finally, we provide some implementational considerations.

Keywords. Control education, computer-aided control systems design; computer-aided instruction; man-machine interface; student environments.

INTRODUCTION

The use of computers in control education is already widespread and it is certain to increase in the future. Computers are and will be used to improve understanding of the basics of control engineering in many ways, i.e.

- demonstrations during lectures
- exercises
- laboratory assignments
- student projects.

The use of computers in control education has become vital as computers are changing the whole field of control engineering. The design and implementation of advanced control systems are impossible without the use of computers, and the control education has to prepare the students for this new situation.

As we have only a very restricted amount of space to outline our ideas, this paper will be focused on the use of CAD in control education and the design of CACSD tools to be used by inexperienced users (e.g. students).

COMPUTERS IN THE DESIGN OF CONTROL SYSTEMS

If we focus on the design cycle as shown in Fig. 1, we realize that computers are essential in almost all of the design process steps and that therefore the use of computers must be introduced at an early stage into the educational process.

Some typical applications of computers for the different stages in the design process are made explicit in the following.

Modelling is probably the least automated step of all. Here the equations of the physical system have to be obtained. Network or bond diagrams as well as energy balance and similar principles have to be used. These are often combined with simulations.

Packages: Bond diagrams (TUTSIM), ENPORT, MACSYMA

Identification is the process of obtaining the parameters for a given model structure by the use of optimization techniques. Measurement signals must be used in this process.

Packages: IDPAC, CTRL-C

Representation. Several representations are generally in use, i.e.

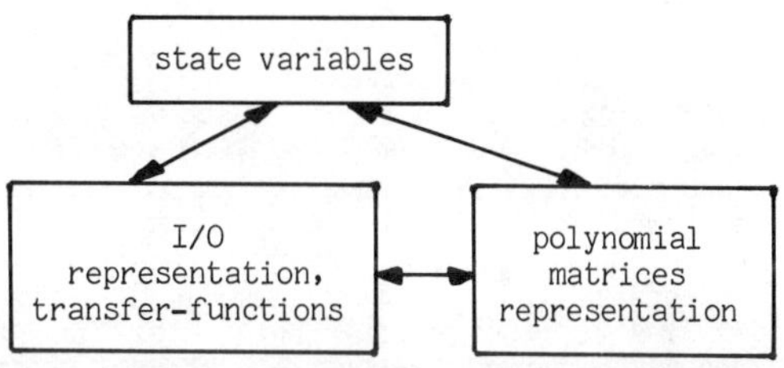

It is often useful to change the representations or to use special forms such as observable or controllable normal forms. The job to change the representation is easily carried out by computers.

Languages and systems: MATLAB, MACSYMA, MODPAC, IMPACT

Design requires tests (observability, controllability) and optimizations (LQG-Problem) or poleplacements. A repetitive sequence of design and analysis is usually used to optimize the system. This involves heavy interactive work for the computer.

Packages: CTRL-C, IMPACT, MATRIX$_X$

Analysis is mostly carried out by interactive simulation, requiring the corresponding software. Tests for stability and sensitivity analysis are also best carried out by computers.

Packages: CSMP, ACSL, IMPACT, MATRIX$_X$

Implementation requires real-time languages and operating systems as well as development support.

Languages: Modula-2, Portal, Ada, Pearl

Development systems: Ada/KAPSE, Jackson Design, Mascot (see Mascot Suppliers Association 1980)

We are convinced that it is important to use adequate tools from the very beginning in the educational process and we therefore advocate the introduction of the corresponding methods into the curriculum.

THE EDUCATIONAL VALUE OF CACSD

Design is by its very nature an iterative process. Several trials have to be made until a reasonable control system results. If done by pencil and paper, this requires much work (drawing frequency responses, calculating dominant pole responses) and the methods used are only approximate ones. This can be considerably improved by the use of computers. Methods that are not important for the understanding of the subject matter (e.g. the manual drawing of Root-Locus curves) can be removed, giving time for the introduction of additional control concepts. Thereby, the essential facets of control engineering can be brought to bear over a shorter time-span.

Moreover, many problems arising in control theory are solvable only by numerical methods. Through the use of CACSD packages, these problems can now for the first time be treated in the exercises of intermediate control courses. For example, in conventional, pencil and paper exercises the students almost exclusively work on linear (linearized) systems. Through the use of CACSD packages supporting linear design techniques as well as nonlinear simulations the students can compare the system behaviour of the regulated linearized system with that of the regulated original, nonlinear process after the controller design. In Fig. 2, we indicate possible uses of an integrated CACSD package in an introductory course.

However, although the use of CACSD indisputably opens up new perspectives in control education, an indiscriminate replacement of conventional exercises with CACSD sessions may prove contraproductive. Several manual schemes, which can be performed much faster using numerically better methods on a computer than by hand, often help the students to get a better understanding of the underlying theories during their use. For example, many CACSD packages support algorithms transforming systems from the time to the frequency domain and vice versa. Nevertheless, we feel that it is pedagogically better to let the students perform this translation by hand a few times using simple examples, as this also gives an illustration of the physical meaning of this translation. Moreover, the student should always have means of roughly controlling the results he gets from the computer.

From our own experience, we conclude that the use of CACSD tools enables an increase of the total scope of a course, but that during carelessly formulated CACSD exercises the students often overlook the actual theoretical/numerical difficulties and loose the overview of the treated subject matter.

THE NEED FOR AN INTEGRATED CACSD-FACILITY

We have seen in a previous chapter that CACSD tools exist for all phases in the design sequence, but that each single package normally is suited for only one or two of these phases. However, particularly in a student environment, the use of a single, multi-purpose package is advantageous. Conceptually, the development of such a multi-purpose CACSD facility can be approached on three levels:

- several independent programs communicating over a common data base
- several control packages connected over a common data base and a common user interface
- a fully integrated CACSD package

The simplest approach is the one using a common data base over which intermediate data can be transported from one part of the facility to another. It is mainly used when already existing packages are to be interconnected (a typical example is the link between the CTRL-C design package and the ACSL simulation language). The main drawbacks of this approach are that the users have to know several different systems, that no consistent data structures exist and that information is often lost in the transfer or has to be added after the transfer (an example is run-time information to be included in the analysis and simulation parts of the design suite, but redundant in and incompatible with the linear design parts).

The approach of combining many individual programs over a common user interface has been successfully used in several larger CACSD projects, for example in the Lund control suite surveyed in Åström (1983), which is built upon a common macro handler (INTRAC, Wieslander 1980). This macro handler gives the user access to a basic set of general commands (edit/execute macros, change/show parameters). The other parts of the package then add specific context-adapted commands, like commands for identification (IDPAC), polynomial operations (POLPAC) and simulation (SIMNON). Using this approach, the user needs to know only one system - the common interface. A further advantage is that the individual parts of the package can be designed quite independent of each other using their own data structures. However, the context-dependency of the available commands and data structures makes an overview of the full package and all its possibilities difficult, especially for beginners.

The last, fully integrated approach differ from the previous ones in that the user is presented with a single, context-independent interface from which all phases of the design sequence can be performed. This means that at each point of time the user is presented with a much more powerful, but not necessarily more complex, tool. In fact, the very integration of the different segments into one interface can make the system easier to use. For example, it is possible to design the package in such a way that the commands for the simulation of linear and nonlinear systems look exactly the same, despite the fact that these simulations have nothing in common from an algorithmic point of view.

In our opinion, this third approach is ideal for a student environment. Not only does the integrated interface shorten the total familiarization time for the students, it also precludes all errors deriving from conceptual differences in the packages (e.g. varying specifications of closed/open loops or different tolerance indications). An example of such an integrated system is the CACSD package IMPACT (presently under development at our institute, see Rimvall 1983 and Rimvall and Cellier 1985). IMPACT will give access to algorithms for all parts of the design sequence (except for the implementation/real-time control) using a common set of data structures. Moreover, in addition to the standardized interface to all algorithms, IMPACT provides the user with a full-fledged algorithmic command language and a wide selection of applicable data structures. This makes the package user extendable and therefore useable not only in education but also in a research environment.

REQUIREMENTS FOR AN INTEGRATED CACSD FACILITY

In this chapter we will discuss some requirements which we feel are imperative for CACSD packages to be used in the educational process. During the last years, much work has been presented in this field, for example in the following areas:

- numerical algorithms
- man-machine interaction
- interactive program design
- design languages
- simulation techniques

For an overview see Cuenod (1979), Herget and Laub (1982), Anonymous (1983a) as well as Herget and Jamshidi (1985). These recent results must be taken into consideration when selecting or designing a CACSD package. Moreover, advanced work on the use of "expert systems" in CAD of intelligent control systems may lead to future control workstations (Taylor and Frederick 1984). Such "expert control systems" would be an excellent tool in control education, as the knowledge base could be used to guide the student through the whole design sequence.

Each of the following sections will concentrate on one aspect we feel to be particularly important in the design of new CACSD-tools.

An algorithmic interface

When designing a new interactive system, one of the first decisions must be to decide, in which form the man-machine interaction is to take place. This decision should not be taken lightheartedly, as the mode of interaction determines the user-friendlyness, and thereby also the user acceptance, of the system; although this interface is not the brain of any CAD-system, it certainly serves as both eyes and mouth.

The mode of interaction also influences the structure of the kernel controlling the package. In particular, the data-structures of the kernel are very closely knitted to the user interface. As any late changes in the central data structures are the worst of all possible nightmares for any software developer, the design of the interactive interface should be done carefully, so that no later modifications need to be done.

Apart from some more exotic ways of communication, like speech input and natural language input, four basically different ways of interactive input exist:

- question-and-answer method
- menu-driven operation
- command-language communication
- graphical input.

Of these four, the first three work with alphanumeric information, which, at least in principle can work on any alphanumerical terminal. The graphical input is particularly interesting during the definition of systems and system structures, However, it requires special hardware in form of a graphical terminal for the graphic echo and some graphical input device. Graphical input will be discussed in some detail in a later chapter.

For CACSD packages primarily designed to solve numerical problems, the three alphanumeric input modes are of interest. A basic difference between these methods exist: the menu and the question-and-answer methods let the computer be in charge of the conversation, whereas the command language method gives the user almost total control. In the last few years, several menu-driven interactive programs using a "mouse" or joystick as input device have emerged. These systems are extremely handy and speedy for dexterous users. However, until a hardware standard has been set, such systems are not very portable. On the other hand, the speed of portable menu-driven systems can be compared with that obtained by question-and-answer methods.

If the only design goals are minimal learning time and maximum accessibility by non-specialists, the question-and-answer method is most certainly the right answer. However, this method gets very tiresome after a while, as the user always can anticipate the next question, but cannot speed up the input. As an example, compare the conversation in INTOPS (Agathokolis and coworkers 1979) (a somewhat old-fashioned interactive program for control system operations) with the equivalent commands in IMPACT. A BODE diagram of the transfer function

$$G(s) = \frac{1}{s^3 + 9s^2 + 5s + 9}$$

is produced in both examples. INTOPS question-and-answer conversation:

```
P>      OP CODE = ENTER
P>      NAME = NUME
P>      COMMENT = NUMERATOR
P>      ORDER = 0
P>      P( 0) = 1
P>      NUME         NUMERATOR
P>      P( 0) =  0.10000E+01

P>      OP CODE = ENTER
P>      NAME = DENO
P>      COMMENT = DENOMINATOR
P>      ORDER = 3
P>      P( 0) = 9
P>      P( 1) = 5
P>      P( 2) = 9
P>      P( 3) = 1
P>      DENO         DENOMINATOR
P>      P( 0) =  0.90000E+01
P>      P( 1) =  0.50000E+01
P>      P( 2) =  0.90000E+01
P>      P( 3) =  0.10000E+01

P>      OP CODE = BODE
P>      GH NUMERATOR = NUME
P>      GH DENOMINATOR = DENO
P>      NUMBER OF FREQUENCY VALUE = 100
P>      OMEMAX = 1000.
P>      OMEMIN = .1
```

(user input is underlined).

Command language input of IMPACT:

```
BODE (1/<9^5^9^1> //DOMAIN=LOGDOM(.1,1000.,100) )
```

or a little more verbose but better readable:

```
S = <^1>;
G = 1 / (S**3 + 9*S*S + 5*S + 9);
FREQ = LOGDOM(.1,1000.,100);
BODE (G //DOMAIN=FREQ )
```

Anyone familiar with both systems will save a factor 10 in time as well as in number of input lines when he uses IMPACT to construct the BODE diagram.

On the other hand, the advocates of menu-driven and question-and-answer interaction justly claim that their methods are specially advantageous for users who are unfamiliar with the system. In our example, the user of INTOPS needed to know only the existence of the two commands ENTER and BODE, whereas (apparently) the user of IMPACT must know how to form a transfer-function, and how to specify a variable as parameter of a function. In a student environment, even this low level of complexity can be cryptic and we have therefore put quite some efforts into solving this problem at our institute.

We believe that the basic interface must be command-driven for speed and flexibility. However, in order to simplify the system for inexperienced users, a query mode should be included. Whenever the user wants to use an unknown function (command), he can switch over to this query mode and thereby give the initiative to the system. He will thereafter be asked for the value of each parameter of the function. Furthermore, each question can be answered with HELP for further information and defaulted values can be skipped. A more detailed description of such an interface can be found in Rimvall and Bomholt (1985) of these proceedings.

Data structures

One of the most serious drawbacks of many control packages is their lack of adequate data structures, many CACSD packages support the complex matrix as the only data structure. On the other hand, control scientists usually work with structures like polynomial matrices, transfer-functions and linear system descriptions. For this reason, and because the absence of proper data structures in a large software package cannot easily be remedied at a later stage (as such a remedy would require extensive changes to the central data structures, and thereby a recoding of large sections of the package), great attention must be given to this problem at the outset of any package design.

A minimal set of data structures in an integrated control package should include:

- complex matrices
- polynomial matrices (*)
- transfer-function matrices (*)
- linear system descriptions (4 matrices)
- domains (one-dimensional structure used for range descriptions)
- trajectories (for time histories, signal and plot descriptions)
- nonlinear system descriptions (ORD's)

The structures marked (*) should be available in factorized as well as nonfactorized form. A full set of functions transferring systems from one representation to another (including linearization) should exist.

Of course this high number of different structures increases the implementational complexity of the package. This will be discussed in a later section on implementational considerations.

Graphics in CACSD

The use of graphical output (step responses, Bode plots and so on) has become an indispensable part of all modern CACSD packages. In a student environment, the frequent use of plots to verify results and give the students an intuitive feeling for systems behaviour is imperative, and in the basic courses where everything still can be "calculated by hand" it is often the only reason for using a computer at all. Unfortunately, most packages only support special graphs adapted to the basic algorithms of the package, giving the user little freedom in constructing his own, problem-adapted plots. This hampers the design sequence where critical information often can be obtained by comparing different curves or mathematically manipulated plot data (plot state x1 versus state x2, plot the difference of output signals y1 and y2 versus the time, plot the error time history using a logarithmic scaling). Here the packages CTRL-C and IMPACT are especially flexible, using their inherent data structures for the description and manipulation of graphs. Moreover, CTRL-C offers the user an interactive, "GKS-like" (Anonymous 1982) plot interface for a very general plot facility.

Whereas graphical output is employed in almost all CACSD packages, the use of graphical input is still in its infancy. Only few packages allow for a graphical input; the MATRIX_X control package from Walkers and coworkers (1982) and the CACSD project of King and Grey (1984) let the user interactively enter nonlinear systems in block diagram form. The TESS simulation environment from Pritsker & Associates (1984) allow for an interactive input of network descriptions of discrete systems. This kind of user-friendly interfaces are particularly well suited in a student environment, as users with little experience with CAD or sparce knowledge of control theory is greatly assisted by the use of graphical symbols rather than complex descriptions using formulas or program structures. When larger systems can be entered using a hierarchical approach (with zooming), the graphical input becomes particularly interesting, also see Elmquist (1982). However, the widespread use of graphical input is hampered by the hardware presently available:

- The connection of graphical input devices like joysticks and mice is seldom foreseen by standard graphical terminals. No standard for these devices exists.

- Graphical terminals with high resolution and fast refresh rates are needed. Ideally, bit-map displays with intelligent graphical processors as found on workstations like the Apollo Domain, Sun or MODULA-computer should be used. On these terminals, the refresh rate is fast enough for zooming and fast deletion/insertion of complex, hierarchical structures.

Furthermore, a very nice combination of a graphical input facility and an alphanumerical algorithmic environment of an integrated CACSD package can be obtained on the modern workstations supporting windowing techniques. Concurrent sessions and multiple windowing can be used to separate these two modes, which are incompatible on most other terminals.

The required graphical facilities, together with the needed computational power, dictate the hardware requirements for CACSD applications. Modern workstations provide excellent graphics and at the same time give their users enough computing power (in the magnitude of a standard "VAX-780" unit) on his own desk - an ideal environment for the dedicated control scientist. However, the price per computer-connenction is rather high for the workstations and comparable to that of a multi-user minicomputer environment (another very common configuration in CACSD applications). In the educational process, where literally hundreds of students have to be given access to computers for their (control-) exercises, more inexpensive solutions have to be found. Economical alternatives are the modern personal computers (e.g. IBM-PC). These provide the user with decent graphics and enough computing power for introductory to intermediate control exercises. Moreover, packages useable in control exercises have already emerged for the PC's (e.g. the new PC version of MATLAB).

Implementational considerations

All programs and packages must be well-tested, error-free, and numerically reliable. The problem area may be restricted to educational use, but no compromising should be done with respect to program quality.

Until now, practically all CACSD packages and CACSD libraries have been implemented in FORTRAN. This insures a fair portability, as FORTRAN compilers are available on virtually all computer systems and ANSI FORTRAN standards exist. However, due to the diminishing cost of hardware and the soaring cost of software, software engineering aspects like reliability, portability, and maintenance costs will play a more important role in future implementations of larger application packages. These aspects certainly do not favour FORTRAN, but indicate that modern languages like Pascal, Modula-2 or Ada (Anonymous 1983b) are better suited for CACSD applications in the long run. Moreover, logical languages like LOGO, LISP and PROLOG are to be considered in projects where control design algorithms are combined with expert system or artificial intelligence environments; good overviews of these languages can be found in Winston and Horn (1981) for LISP and Clocksin and Mellish (1981) for PROLOG.

The availability and use of these "new" languages open up new dimensions for CACSD-developments. However, although a combination of languages within one integrated CACSD package would enable an optimal use of the strong features of each language, such a mixture would also lead to severe portability problems. Therefore, for an integrated and algorithmic approach to CACSD as discussed in this contribution, we advocate the use of only one implementation language whenever possible. Moreover, for several reasons we consider Ada to be the best choice today (possibly together with LISP or PROLOG for knowledge system features):

In Ada (as in other languages, e.g. Pascal) the user can define his own structured data types. This is very useful to CACSD programs, permitting us to implement program structures directly corresponding to the structures used in control theory.

Through the use of Ada's data abstraction, we can hide the details on e.g. data types or algorithm implementations from other parts of the program. This feature is a cornerstone in Ada's aim at robust programming, and should be used to modularize larger software projects.

Ada is per definition portable, there may not exist any sub- and/or super-set of Ada with that name. Furthermore, the language is rich enough to support any algorithmic CACSD package.

A current disadvantage of Ada is that no Ada-libraries of e.g. mathematical algorithms are available. However, such libraries are planned and should emerge on the market in the near future (Anonymous 1984). Moreover, Ada allows for a so-called "PRAGMA INTERFACE" to access programs written in other languages.

CONCLUSIONS

In this paper we have discussed the requirements for a CACSD package in modern undergraduate control courses. We have seen that the integrated approach (using one multi-purpose package for the whole design cycle) is particularly important in a student environment. The students use of such a package is often too sporadic, and their overview of control theory (from an theoretical as well as algorithmical point of view) too limited to allow the introduction of different dedicated packages. Moreover, we have discussed the need for a flexible algorithmic interface, for data-structures useable in control theory and for good graphical interfaces. Finally, we indicated the importance of good software engineering and well-suited implementational tools.

REFERENCES

Anonymous (1982). Graphical Kernel System (GKS) - Functional Description. Draft International Standard ISO/DIS 7942 Version 7.02, August 9, 1982. Available through ANSI, NY.

Anonymous (1983a). IEEE Control Systems Society Symposium on Computer-Aided Control System Design. Abstracts. MIT, Massachusetts, September 28-30.

Anonymous (1983b). ANSI/MIL-STD 1815 A, Reference manual for the Ada programming language (January 1983).

Anonymous (1984). ACM Ada letters, 4(1984)3 p.68.

Åström, K.J. (1983). Analysis and Design of Control Systems - A Perspective. IEEE Control Systems Magazine. 3(1983)2. pp. 4-16.

Clocksin, W.F. and C.S. Mellish (1981). Programming in Prolog. Springer Verlag. 279 pp.

Cuenod, M.A. (Ed.)(1979). Preprints first IFAC Symposium on CAD of Control Systems. Zurich. Pergamon Press.

Elmqvist, H. (1982). A graphical approach to documentation and implementation of control systems. Proc. 3rd IFAC/IFIP Symposium on Software for Computer Control, SOCOCO82. Madrid, Spain.

Herget, C.J. and M. Jamshidi (1985). Advances in Computer-Aided Control Systems Engineering. To be published in North-Holland Press.

Herget, C.J. and A.J. Laub (Eds.) (1982). Special Issue on Computer-Aided Design of Control Systems. IEEE Control Systems Magazine 2(1982)4.

King, R.A. and J.O Grey (1984). A Graphical Man-Machine Interface for CAD and Simulation of Dynamic System. In Proc. EUROCON 84, 6th European Conference on Electrotechnics, Computers in Communication and Control. Brighton. 26-28 September 1984.

Little, J.N., A. Emami-Naeini, and S.N. Bangert (1984). CTRL-C and matrix environments for the computer aided design of control systems. In A. Bensoussan, J.L. Lions (Ed.), Proc. 6'th International Conference on Analysis and Optimization (INRIA). Lecture notes in Control and Information Sciences 63, Springer Verlag. pp. 191-205.

Mascot Suppliers Association (1980). The official handbook of MASCOT. Computer Standards Section, Royal Signals and Radar Establishment, Malvern, Worchestershire WR14 3PS, England.

Moler, C. (1980). MATLAB, User's Guide. Department of Computer Science, University of New Mexico, Albuquerque, USA.

Pritsker & Associates (1984). The TESS User's Manual. P.O. Box 2413, West Lafayette, IN 47906, USA. 515p.

Rimvall, M., (1983). IMPACT, Interactive Mathematical Program for Automatic Control Theory, A Preliminary User's Manual. Institute for Automatic Control, ETH Zurich, Switzerland.

Rimvall, M. and F.E. Cellier (1985). A Structural Approach to CACSD. In M. Jamshidi and C.J. Herget (Ed.), Advances in Computer-Aided Control Systems Engineering. North Holland, Amsterdam.

Rimvall, M. and L. Bomholt (1985). A Flexible Man-Machine Interface for CACSD Applications. In Proc. 3rd IFAC symposium on Computer Aided Design in Control and Engineering Systems. Kopenhagen, July 31 - August 2 1985.

Taylor, J.H. and D.K. Frederick (1984). Expert System Concepts for Computer-Aided Control Engineering. General Electric Technical Information Series, Report 84CRD127. May 1984.

Walker, R., C. Gregory, and S. Shah (1982). MATRIX , A Data Analysis, System Identification, Control Design, and Simulation Package. IEEE Control Systems Magazine, 2, 30-37.

Wieslander, J. (1980). Interactive Programs - General guide. Dept. of Automatic Control, Lund Institute of Technology, Lund, Sweden. Report CODEN: LUTFD2/(TFRT-3156)/1-30/(1980).

Winston, P.H. and B.K.P. Horn (1981). LISP. Addison-Wesley Publishing Company. 430 pp.

Control subjects	Use of CACSD-tools in exercises
LINEAR SYSTEMS	
state variable and transfer function representation	provide translation algebraically and numerically
normal forms	generate and compare normal forms
controllability and observability	generate tests for observability / controllability, for nontrivial examples assist physical interpretation
stability	automatic stability tests
LINEAR CONT. SYSTEMS	
root loci, controller design statical and dynamical behaviour	use graphical features simulations, comparisons with dominant pole responses
LINEAR DISCRETE SYSTEMS	
introduction to discrete/sampled systems	comparisons between behaviour of continuous and sampled systems, transformations between continuous and discrete representations
NON-LINEAR SYSTEMS	modeling, linearizations

Fig. 2. The use of CACSD-tools in control exercises at the ETH

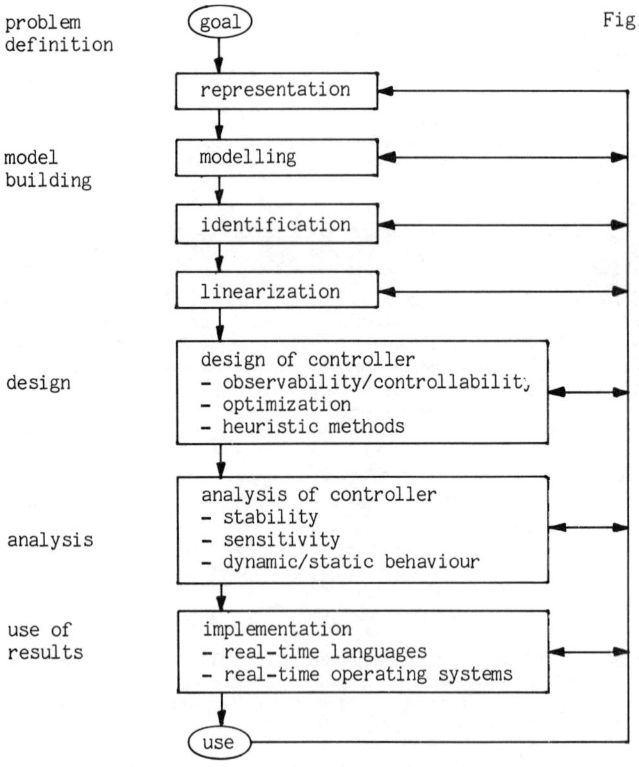

Fig. 1. The design cycle in control theory.

CAD IN CONTROL ENGINEERING EDUCATION

H. Unbehauen

Department of Electrical Engineering, Ruhr-University, 4630 Bochum, FRG

Abstract. The paper describes the role of software packages for computer-aided control system analysis and design (CACSAD) in control engineering education. It is shown that CACSAD-systems are becoming an indispensible tool in education, because even sophisticated methods are more easily understood by the students. Realistic problems can be solved within reasonable time, especially the laboratory experiments. As an example of a CACSAD-package, the KEDDC-system developed by the author and his coworkers, is described. It is shown how the KEDDC-system is introduced into the education of control engineers at the electrical engineering department of Ruhr-University Bochum.

Keywords. CAD in control engineering; software packages; electrical engineering curriculum; real-time application; control system analysis and synthesis.

INTRODUCTION

The university level engineering education in the Federal Republic of Germany offers usually a five-years course leading to a Dipl.-Ing. degree, e.g. in electrical or mechanical engineering. Normally electrical engineering students specialize in one or two topics out of a large variety of electrical engineering subjects during the last two years. One of these topics might be the area of control engineering. Control engineering is offered in the departments of electrical, mechanical, aeronautical and chemical engineering, and in a few cases also in some other sections.

In the past some very powerful CAD-packages had been developed for solving control engineering tasks. Since these packages are used extensively in several universities during the final diploma projects, it is but natural to consider how these modern tools can be integrated into a systematic education of control engineers. This development will be described in this paper using as an example the Department of Electrical Engineering of the Ruhr-University Bochum. The control group (ESR) in this department has developed a very powerful CAD-system, called KEDDC (Unbehauen 1975, Schmid and Unbehauen 1979, Schmid 1984) in the past 11 years. This system is available for students as a tool for solving problems connected with the lecture and laboratory courses as well as to the day-to-day work during their first and final diploma projects. Thus, the students are able to solve very complex problems in control engineering within a reasonable time. Because of the interactive mode, the application can be learnt in a few hours without special knowledge about the details of the software.

SHORT DESCRIPTION OF THE KEDDC-PACKAGE

Originally the KEDDC package had been developed by the author and his coworkers during research projects for analysis and synthesis of control systems as well as for direct digital control. KEDDC represents a combined package for solving CAD-tasks and for real-time direct digital control. This means that a controller can be designed and simulated while in the background a DDC-program works on a real process using a common data base. Thus it is possible to replace an active DDC-program by a newly designed algorithm at any time.

KEDDC is a comprehensive CAD-package designed to cover a wide range of control system engineering tasks. It contains modules for process identification, system analysis, controller design, simulation and controller implementation based on a broad variety of modern as well as classical approaches for SISO and MIMO systems. The user-friendly interface, graphic output and state-of-the-art numerical algorithms give the control engineer more power at his fingertips than ever before. Based on the KEDDC framework, developed at Ruhr-University Bochum, KEDDC is available in different configurations and versions. It can be easily tailored and adopted by the user, as it is written in FORTRAN. Thus it represents an open system. KEDDC can be used in the form of stand-alone programs or in the form of a comprehensive control engineering library. It provides a system which minimizes the engineering and programming resources required for the complete circle of system identification, controller design and design verification. The programming system has the following features:

- simple command-driven dialog combined with questions and answers,
- hierarchy of independent program modules,
- control engineering library,
- reliable numerical algorithms drawn from recent research in numerical analysis,
- device independent graphics output based in the ACM/SIGGRAPH core system standard,
- transformation of system representation forms using alternative sets of strategies or numerical algorithms,
- signal analysis and filtering,
- off-line and on-line identification using deterministic and stochastic approaches (20 methods),
- controller design (15 approaches),
- observer design (pole placement, Kalman filter),
- simulation (general and special purpose simulators),
- adaptive controller design (5 approaches),
- controller implementation (simulation, real time),
- standard tasks centralized in 'Manager' programs.

To provide the necessary flexibility for solving

the wide variety of problems, the systems may be represented in 10 forms in the KEDDC:

- series of deterministic or stochastic input and/ or output signals,
- series of auto- or crosscorrelation values,
- discrete values of the pulse response,
- discrete values of the step response,
- transfer function (matrix) in s-domain,
- transfer function (matrix) in z-domain,
- discrete values of frequency responses,
- matrices for continuous state space,
- matrices for discrete state space,
- polynomial matrices.

A variety of transformations between these forms is possible.

The KEDDC system requires no special programming knowledge. Help facilities and menu selection lists are available. It has, thus, become an indispensible tool not only to the advanced control engineer but also to the students of automatic control.

SOME FUNDAMENTAL DEMANDS FOR CACSAD-SYSTEMS IN CONTROL EDUCATION

The main object of CACSAD-systems should be to support the analytical and the graphical techniques, and thereby relieving the control engineer of laborious calculations. In control education, CACSAD-systems are becoming more an important supplement to traditional courses rather than a self-study unit for computer-aided instruction and learning. The limitation of computer-aided instruction and learning is not acceptable in teaching control engineering, because the student cannot use the packages to solve his own problems.

One of the main advantages of CACSAD-systems for control education is that even sophisticated methods are more easily understood, accepted and applied by the students. For this reason, the program packages must be easy to learn and to use. HELP-functions for understanding and entering the different layers of the program structure at any time as well as menu selection lists for the interactive use should be available. The algorithms must be reliable. Better simulation capabilities should exist within a CACSAD system. Graphics terminals used in a CACSAD-system should be supported by simple ergonomic principles, as e.g. good readability in an easily understandable form.

For teaching control, a CACSAD-system must be used as a tool at different levels. As the application of a CACSAD-system saves time for the student, he can look at much more realistic plant models and examples, which could not otherwise be solved by pencil and paper in reasonable time or for which only a numerical solution is possible. More complicated problems, as e.g. higher order systems, parameter estimation, multivariable linear and nonlinear systems can, therefore, easily be investigated. Special effects of parameter variations such as e.g. variation of controller parameters, sampling time, weighting factors etc. can thus be considered.

CACSAD-systems are becoming an indispensible tool in control education for two purposes:

- The student becomes much closer to real engineering problems with reasonable effort.
- The student must be aquainted with tools, which he will use tomorrow in industry.

When laboratory experiments are combined with CACSAD, mathematical models can be compared with experimental data. On-line parameter estimation can be used to verify the models which in turn can be used for design. Furthermore it is highly desirable to interface CACSAD-techniques to real-time experiments in a control laboratory.

THE ROLE OF CACSAD IN THE DIPLOMA-COURSE

During their education in the electrical engineering department of Ruhr-University, the students are introduced to CACSAD at different levels. In the 6th semester (3rd year) every electrical engineering student starts with a basic course on "Control Systems 1" (4 hours per week). Students, who want to specialize in control engineering continue in the 4th year with further lecture and laboratory courses of the control group. In a period of about two years, the students come into contact with CACSAD at different levels, starting from laboratory courses up to the final diploma thesis. Table 1 gives a short survey about the usage of the CACSAD-package KEDDC.

KEDDC is first used in the Control Laboratory 2. This course consists of 8 experiments, wherein 3 of them KEDDC is used. In the first experiment the students will be acquainted with the design and implementation of digital compensator controllers. The theoretical background is known from lectures. As this experiment is the first touch with KEDDC, an extensive supervisory control by the teaching assistant is necessary. The students have to design and evaluate different types of compensator controllers. The plant to be controlled is simulated on an analog computer and the controller is designed and implemented on a process computer. The students learn computer dialog, controller implementation for DDC operation. They begin to learn about real-time experimentation and evaluation of experiemts using CACSAD-techniques.

In the second experiment the students apply the knowledge gained from the first experiment to the attitude control of a laboratory-scale vertical take-off aircraft. Digital controllers have to be designed and simulated for this unstable plant. The best controller has to be selected and tested in real-time. Only a minor assistance is necessary.

In the third experiment students, starting with a mathematical model of an overhead crane, use CACSAD to design and evaluate real-time experiments with LQ-controllers and observers for angle and speed. KEDDC is extensively used to compare controllers designed using different weightings.

The examples from a laboratory course show, how CACSAD-tools are used in student education. The scope of a CACSAD-system should be at least the same as of all control curricula. KEDDC supports all methods and techniques which are taught by the control group. The students can come to the computer terminal without any programming knowledge and can leave with a complete solution, well documented in terms of listings and diagrams.

With increasing use of CACSAD-packages in control engineering education, a need has arisen to establish a special course in computer-aided analysis and design of control systems. In order to involve the students also in the development of CACSAD-systems it is necessary that they get a deeper knowledge in software engineering, to understand the details of data structures, data handling etc. Furthermore their background in numerics is consolidated.

In the CACSAD-course (item 9 in table 1) the students are introduced to CAD-techniques. Using typical examples they learn the elements of KEDDC. The lecture part of this course is sketched as follows:

TABLE 1. Courses in Control Education and their Relations to CACSAD

Item	Course	Hours/ week	Content	Usage of KEDDC
1	Control Systems	4	theory of linear continuous-time systems, analysis and synthesis, stability, compensation	-
2	Control Lab 1 parallel to item 1	3	introduction to practical applications of classical control methods, first touch with simple digital controllers	-
3	Control Systems 2	3	state-space theory, multivariable systems, digital control, non-linear systems	-
4	Control Lab 2 parallel to item 3	3	experiments for advanced digital control and identification, process computer application on small laboratory-scale processes	tools for design of digital controllers
5	Control Systems 3	3	system identification, parameter estimation, adaptive control, optimal control	for demonstration purposes
6	Simulation techniques	3	analog, digital and hybrid simulation, use of simulation systems and simulation languages	simulation tools for demonstration
7	Simulation Lab parallel to item 6	3	experiments on analog, digital and hybrid computers	simulation tools
8	Seminar	2	varying control topics	depends on topic
9	CACSAD	3	introduction into CACSAD-systems, numerical methods, applications to control system analysis and design	classroom exercises on terminals
10	Study Project	4-6 months 1-2 days/ week	cooperation with research assistants and staff, small projects drawn from recent research	extensive use and new developments
11	Diploma Project (Thesis)	6 months full-time	cooperation with research assistants and staff, original research drawn from actual research projects, development of new CAD-techniques	extensive use and new developments

- Introduction into CACSAD. Goals and scope of CACSAD.
- Numerical representation of dynamical systems.
- Transformations between different representation forms.
- Repetition of numerical methods with respect to CACSAD-problems, such as solving systems of linear equations, rank calculation, singular values, eigenvalues, system transformation, optimization.
- Design of interactive system using graphics.
- Automation of classical approaches for SISO-systems, such as frequency-domain techniques using graphics, root-locus plot, z-transformation.
- Digital simulation of dynamical systems, integration techniques, design of digital simulation systems and simulation languages.
- Techniques for controller synthesis, pole assignment, LQ, LQG design, observer design, solving Riccati or Liapunov equations.
- Introduction into KEDDC as an example of a CACSAD system.

Parallel to the lecture, classroom exercises are offered and the student can use KEDDC for solving these CACSAD-problems on intelligent graphics terminals. He will be introduced to dialogue handling using simple problems. After getting started - this will be after the 2nd session - the student has to solve problems. The background and the solution from the control engineering point of view will be sketched, but he has to get his results independently. We observed that these exercises are very well accepted by students as they are very much interested in this area. The number of students exceeded our expectation by a factor of two.

As KEDDC is used both in education and as well as in research, the students can cooperate with the staff on actual research problems during their studies and diploma projects (item 10 and 11). The basic idea is the following. Students, who are involved in small research projects, e.g. developing new control techniques, are adding only their specific parts to the system, without reprogramming the part which their predecessors had already done. The student can concentrate completely on his problem. Reliable parts will remain in the system and will be available for the entire KEDDC-community. Most of the parts of KEDDC have been added in this way. Thus KEDDC remains as an open system, which can be extended at any time.

EXAMPLE OF A TERMINAL SESSION (Schmid, 1985)

To illustrate the dialog of the package, a hardcopy of a session during classroom exercises has been included in this section. This example shows some sections of the original dialog (Table 2). Student input is underlined, command lines are printed in bold letters and command menus are framed. Graphics output appears on the screen at the phrase '... MAPPED TO VIEWPORT ...'. Here graphics output is not exactly placed in chronological order with the dialog. The hardcopy is printed at the end when the image is complete.

The problem is to design a LQ-controller for a turbo-generator plant (Fig. 1). A plant model was given for this 9-th order system with two inputs and two outputs. In order to explain this session, section numbers are added to the left margin. Section (1) starts with the Frequency Manager. The command menu is shown typing??. It appears in a special window. Principle gain plot is requested but rejected as explained using HELP at (2). The principle gains for a system read from file TGM are plotted at (3). At (4) Matrix Management is interlaced to generate a topological printout and a 3D-mesh surface at (5) for the system given in state-space representation. Simulation is simply

TABLE 2. Original dialogue of a terminal session

```
(1) KEDDC
    KEDDC: FR
    FRMG : ??

    ########################## Frequency Management ##########################Page 1
       SYSTEM:                                 FURTHER CALCULATIONS
    RE  Read system from file                  PO  Popov-frequency response
        G(s),G(z),u(t),y(t),G(jw)              IV  Inverse frequency response
        g(t),h(t),(A,B,C)                      WC  Magnitude/phase for spec. freq.
    LI  List system datas  | WH Write step     RA  Magnitude or phase margins
       FREQUENCY TABLE:   response on file     TT  additional deadtime
    GO  Generate frequency table               FH  Step response from freq. resp.
    EO  Edit frequency table                   CONNECTIONS:
    RO  Read frequency table from file         S2  Read 2nd system from file
    IO  Input frequency table using dialog     ++  Parallel connection
    LO  List frequency table                   **  Series connection
    WO  Write frequency table on file          //  Negative feedback connection
       FREQUENCY RESPONSE:                     GRAPHICS:
    FG  Calculate frequency response            BO  Bode plot complete
    RF  Read frequency response from file       BM  Bode plot magnitude
    IF  Input frequency response using dialog   BP  Bode plot phase
    EF  Edit frequency response                 NY  Nyquist plot
    LF  List frequency response                 NI  Nichols plot
    WF  Write frequency response on file        DH  Step response
    FM  Create frequency response matrix        TR  Toggle curve tracking on/off
        on file                                 NP  Change no. of dots per dia.
    ST  Status protocol                         PG  Principal gain plot
    Standardfilenames see next page

    FRMG : PG
    /FR-004 THERE IS NO FREQUENCY TABLE !

(2) FRMG : HE
    /FR-004 THERE IS NO FREQUENCY TABLE !
    You have initiated an operation which requires a frequency
    table. This table isn't defined.
    Read table from file with RO, or specify table interactively
    with IO, or generate it automatically with GO. The last one is
    the most comfortable way if the frequency range is initially
    unknown.
    FRMG : GO
    RECOMMENDED FREQUENCY RANGE FROM   1.79416    TO   899.2309
    FREQUENCY RANGE FROM, TO   = ? 1,1E3
    NUMBER OF FREQUENCY VALUES = ? 100
    FRMG : PG
    FREQUENCYFACTOR = ?
    FILENAME = ? TGM::DA
    ~~~~~~~~~~~~~~~~~~~~~~~~~~~~~~~~~~~~~~~~~~~~~~~~~~~~~~~~~~~~~~~~
    file=TGM   ::DA, type=UE
    created using MMGR  by SCHMID on  4. 9.1983 at 11.35
    TURBOGEN. MODELL 1 MIMO
    ~~~~~~~~~~~~~~~~~~~~~~~~~~~~~~~~~~~~~~~~~~~~~~~~~~~~~~~~~~~~~~~~
    IT IS A TRANSFER MATRIX
    NUMBER OF INPUTS =  2, NUMBER OF OUTPUTS =  2
(3) MAXIMUM PRINCIPAL GAIN IS MAPPED TO VIEWPORT  4, CURVE  1
    MINIMUM PRINCIPAL GAIN IS MAPPED TO VIEWPORT  4, CURVE  2
(4) FRMG : @@
    KEDDC: MM
    MMGR : RS
    FILENAME = ? TZM::DA
    ~~~~~~~~~~~~~~~~~~~~~~~~~~~~~~~~~~~~~~~~~~~~~~~~~~~~~~~~~~~~~~~~
    FILE=TZM   ::DA, type=MA
    created using MMGR  by SCHMID on  9. 9.1983 at 11.46
    TURBOGEN. STATE SPACE MODEL A2Y3
    ~~~~~~~~~~~~~~~~~~~~~~~~~~~~~~~~~~~~~~~~~~~~~~~~~~~~~~~~~~~~~~~~
    3 BLOCKS,  3 MATRICES IN 1ST BLOCK
    BLOCKNO., MATRIXNO. = ? 2
    MMGR : MS
    DO YOU WANT GRAPHICS ? Y
    MATRIX A
    + 0 0 0 + 0 + 0 -
    0 + 0 + 0 0 0 - 0
    0 0 + - 0 0 0 0 0
    0 0 + + 0 0 0 0 0
    0 0 0 0 + - - 0 0
    0 0 0 0 + + - 0 0
    0 0 0 0 + + - 0 0
    0 + + - 0 0 0 + 0
    + 0 0 0 + + - 0 +
    MATRIX B
    + +
    - -
    0 +
    0 +
    - 0
    + 0
    + 0
    - +
    + +
    MATRIX C
    0 0 0 0 0 0 0 1 0
    0 0 0 0 0 0 0 0 1
(5) MATRICES MESH SURFACE IS MAPPED TO VIEWPORT  2, CURVE  1
```

```
    MMGR : EX
(6) FRMR : @@
    KEDDC: DG
    DIGSI: @@
(7) KEDDC: MM
    MMGR : RS
    FILENAME = ? TZM::DA
    ~~~~~~~~~~~~~~~~~~~~~~~~~~~~~~~~~~~~~~~~~~~~~~~~~~~~~~~~~~~~~~~~
    FILE=TZM   ::DA, type=MA
    created using MMGR  by SCHMID on  9. 9.1983 at 11.46
    TURBOGEN. STATE SPACE MODEL A2Y3
    ~~~~~~~~~~~~~~~~~~~~~~~~~~~~~~~~~~~~~~~~~~~~~~~~~~~~~~~~~~~~~~~~
    3 BLOCKS,  3 MATRICES IN 1ST BLOCK
    BLOCKNO., MATRIXNO. = ? 2
    MMGR : DM
    SAMPLING INTERVAL = ? .2
    MMGR : EI
    DO YOY WANT GRAPHICS ? Y
     ------------ EIGENVALUES ------------
      NO  REALPART        IMAGINARYPART  S/I
       1   .385317          .285704       -
       2   .385317         -.285704       -
       3   .462243          .308177       -
       4   .462243         -.308177       -
       5   .644391          .250971       -
       6   .644391         -.250971       -
       7  2.131015E-02      .437934       -
       8  2.131015E-02     -.437934       -
       9   .715286         0.00000        -
    EIGENVALUES ARE MAPPED TO VIEWPORT  5, CURVE  1
    MMGR : NU
    DO YOU WANT GRAPHICS ? Y
     --------------- ZEROS ---------------
      NO  REALPART        IMAGINARYPART  S/I
       1  -2.03490         0.00000        +
       2  -1.73257         0.00000        +
       3   .770414         0.00000        -
       4   .368101          .463010       -
       5   .368101         -.463010       -
       6   .449659          .360620       -
       7   .449659         -.360620       -
    ZEROS ARE MAPPED TO VIEWPORT  5, CURVE  2
    MMGR : EX
(8) DIGSI: RE
    FILENAME = ? TGM::DA
              .
              .
              .

    DIGSI: SI
    SIMULATION STEP SIZE = ? .1
    SIMULATION TIME      = ? 10
    SIGNAL U ( 1) IS MAPPED TO VIEWPORT  1, CURVE  1
    SIGNAL U ( 2) IS MAPPED TO VIEWPORT  1, CURVE  2
    SIGNAL Y ( 1) IS MAPPED TO VIEWPORT  3, CURVE  1
(9) SIGNAL Y ( 2) IS MAPPED TO VIEWPORT  3, CURVE  2
```

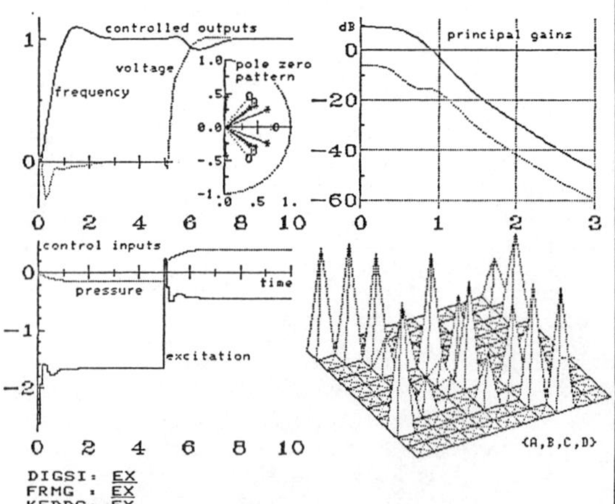


```
    DIGSI: EX
    FRMG : EX
    KEDDC: EX
```

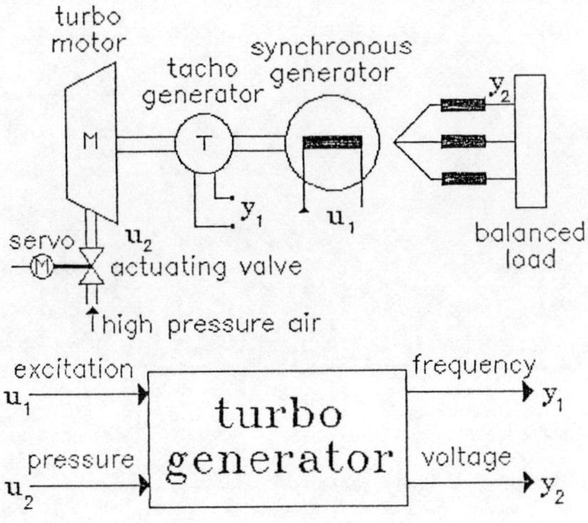

Fig. 1. The turbogenerator plant

interlaced at (6) and then a second Matrix Management at (7) to display the pole zero pattern of the discrete-time system to be controlled. At (8) simulation is continued. After some configuration dialog, which is omitted here, the screen shows the results of all sections at (9). The graphics viewports are numbered from left to right beginning at the bottom of the screen. The viewport number (5) is the small one which is inside of number 3.

CONCLUSIONS

Education of control engineering in CACSAD-software has become almost a necessity. CACSAD-tools will bring the student closer to engineering realities with modest effort. It seems, however, necessary that he is still aquainted with independent analytical techniques, which can be easily applied by simple hand calculation in order to check his results with respect to order of magnitude. Thus, showing a sound scepticism against computer results, CACSAD-techniques represent a very efficient tool to supplement the traditional course work. There is no doubt, that the control engineering student using this tool can better learn by comparison about the efficiency of different methods for selected test examples.

A number of fully developed CACSAD-packages is available today, which meet the educational requirements quite well. Of course, more improvements can still be done, as e.g. more intelligent dialog systems would be advantageous for novices. On the other hand more intelligent dialog interface should also be available for a student if CACSAD-packages are used as knowledge base in forthcoming expert systems.

REFERENCES

Schmid, C. (1985b). KEDDC - A Computer Aided Analysis and Design Package for Control Systems. In M. Jamshidi and C.J. Herget (Ed.), Advances in Computer Aided Control Systems Engineering. North-Holland, Amsterdam.

Schmid, C. (1984). KEDDC - A computer-aided analysis and design package for control systems. Proceed. Workshop "Computer aided control system design", Institute of Measurement and Control, Brighton (UK), pp. 50-55.

Schmid, C., H. Unbehauen (1979). KEDDC, a general purpose CAD software system for application in control engineering. Prep. 2nd IFAC/IFIP Symposium on Software for Computer Control (SOCOCO), Prag, paper C-V.

Unbehauen, H., C. Schmid, F. Böttiger, B. Bauer, B. Göhring (1975). KEDDC, ein kombiniertes Prozeßrechnerprogrammsystem zum Entwurf und Einsatz von DDC-Algorithmen. PDV-Bericht, KFK-PDV 37.

THE INTERACTIVE SYSTEM BLAISE FOR CONTROL ENGINEERING

F. Delebecque and S. Steer

INRIA (Institut National de Recherche en Informatique et an Automatique), Domaine de Voluceau, BP 105 Rocquencourt, 78153 Le Chesnay Cédex, France

Abstract. We give a description of the capabilities of the Blaise system which has been designed as an interactive tool in computer aided engineering. This system implements a full programming language working on a data basis adapted to control problems. It has been designed to achieve extensibility, portability and numerical efficiency.

Keywords. Control engineering, CAD systems, Numerical computations.

INTRODUCTION

An increasing need has appeared in such fields as teaching, industrial design or research to have powerful CAD tools in control engineering [1] [2].

One can observe (at least) three different kinds of needs :

- a reliable and documented software library of mathematical programs for modeling, analysis and design,

- an interactive system allowing usage of the library with as small as possible programming effort,

- facilities to make formal calculus.

To a large extent, these three domains can be developed independently, before being possibly integrated in an expert system.

The Blaise system that we shall briefly describe in the sequel has been designed to meet the constraints of an interactive system in the above framework.

SPECIFICATIONS

The main features which have appeared to us as essential for the design of an interactive system have been the following :

i) A high level programming language. Indeed a menu driven package seems too rigid to answer to the large amount of problems which arise in control engineering.

ii) A data structure adapted to control systems problems. The system should conveniently operate on such objects as matrices, polynomials, polynomial matrices, dynamical systems, etc ...

iii) Expandability by an easy integration of new functions, implementing (or updating) a new algorithm.

iv) Portability, although this constraint is less severe with the advent of low cost hardware.

v) Reliability and numerical efficiency.

These last three points and the fact that most of available packages are written in Fortran have imposed the choice of that language for the implementation of the system.

vi) Finally one can require that the system should be able to deal with non linear problems (Optimization, simulation, ...).

THE BLAISE SYSTEM : MAIN FEATURES

Our starting material has been MATLAB [3], a choice already made by other groups [4,5,6]. Indeed MATLAB is very well suited for the above points iii, iv, v. In addition it is reasonably small (about 4000 lines of fortran) and also it (almost) defines an interactive programming language, thus partly filling the above requirement i). Its syntax, close to the usual mathematical language makes it very attractive to control engineers.

These characteristics have been kept in the Blaise System. To achieve the requirement of a general purpose system we have been led to improve it in the following senses :

- a comprehensive data basis,
- parametrized "macros",
- additionnal powerful general purpose numerical tools.

These choices have been made to efficiently deal with the huge variety of control problems and algorithms. For instance multivariable linear systems can be described by several different representations (state space, transfer matrix, matrix fractions, ...).

General purpose requirement implies that the potential users must have some experience in control : the user has to choose representation of his problem, the algorithm to solve it and the system provides him with all the necessary tools. However for teaching purposes the system can rapidly solve standard problems by macros used as "black-boxes".

The data structure

We have completely modified the data structure of MATLAB in order to define other kinds of variables than complex matrices. In particular we have distinguished between real and complex matrices and have introduced the new types "character string", "polynomial matrix" and "macro".

The major interest to introduce new types of

variables rests upon the possibility to employ the usual operators + - * / ... with a sense depending on variable's type, and the capabilities to conveniently represent complex entities.

In particular one may think useful, in control design, to have at disposal rational function matrices. However the knowledge of the numerical properties of algorithms involving such objects is limited (poles-zeros cancellation problems, ...). In addition these problems are often more relevant to formal calculus (e.g. factorisation of polynomials with rational coefficients ...) for which specific software already exists (MACSYMA, REDUCE, ...). For these reasons this particular type of variable has not been introduced in the system.

As far as numerical computation is required a strong emphasis is put on state space representation. "Compound variables" allow one to easily represent linear systems as triple of matrices or as a pair of polynomial matrices. "Compound variables" allow also to represent much more complicated linear systems such as parameter varying linear systems (A(x), B(x), C(x), D(x)) where A,B,C,D are now matrix valued macros.

Macros

To answer to the large amount of algorithms in Control Engineering we have been led to develop the notion of macros. These macros admit any type of variables as input or output parameters and may refer to any previously defined variable . For the user, there is no difference between a macro and a basic function of the system. In particular these are totally recursive.

In addition, the use of macros to define algorithms allows the user to pause at some crucial stages of the algorithm, to check intermediate results, to choose in view of these the most appropriate next stage or to set decision parameters.

For debugging purposes it is also possible to set breakpoints in macros.

The general philosophy of the system requires a large use of macros. In this respect, we have been led to write a compiler of macros which allows to skip the interpretation time. Usually compiled macros run two or three times faster than interpreted ones. Of course the use of a compiled macro is totally similar to interpreted one. Thus, the efficiency of the system is based on an "optimal" choice between general purpose functions (written in Fortran) and the macros (written in Blaise).

For example it is very well known that ther exist a large variety of Riccati equations (discrete or continuous time systems, explicit or implicit systems, singular or regular systems, ...). They can all be easily solved (i.e. via very short macros) once one has at disposal an algorithm for the "ordered" computation of generalized eigenvectors [7] (basic function).

To be more specific, suppose one wants to solve the equation :

$$P = F'PF - F'PG(R + G'PG)^{-1} G'PF + Q$$

The corresponding macro would be :

```
// <P> = ricd (F,G,R,Q)
   <n,n> = size (F)
   I = eye (n)
   0 = 0*I
   A = <I, (G/R)*G' ; 0, F'>
   B = <F,0 ; -Q, I>
   U = stab (A,B, 'disc')
   P = U(n+1 : 2*n,:) / U(1:n, :)
// end
```

The general purpose functions

We shall not describe here the complete set of functions available in Blaise.

They implement :

- major linear algebra algorithm (Linpack-Eispack),
- control oriented tools such as
  - lyapunov and sylvester equations
  - matrix exponential
  - frequential SISO plots
  - matrix pencils calculations
  - zeros computations ...,
- simulation algorithms (ODE)
- optimization algorithms,
- graphics.

Much care has been taken to implement general purpose functions : the use of optional input or output parameters allows a simple usage of these functions without loosing anything of the available results. In this respect the main difficulty here lies in an adequate choice of the optional parameters. An obvious example is the graphic command.

Furthermore, the set of function is very easily extensible, the amount of work required is equivalent to writing a simple driver routine.

Non linear problems

Generally the computer description of non linear problems is only possible through a program.

Many CAD systems can only deal with predefined non linearities such as saturation ... Since the macros are elements of the data basis, the Blaise system can handle these objects. In particular macros can be input or output parameters of functions or macros, they can be run inside the calling function . In this case it looks like the fortran "external".

A significant application is the ability to define as a macro the non linear functions which arize in optimization [8] or ODE [9] problems without any change to the optimizer or ODE solver.

For these particular problems, we have managed the feasability to directly call fortran routines instead of macros in order to improve the efficiency if macros, even compiled, are too much time consuming.

Example

Consider the following system :

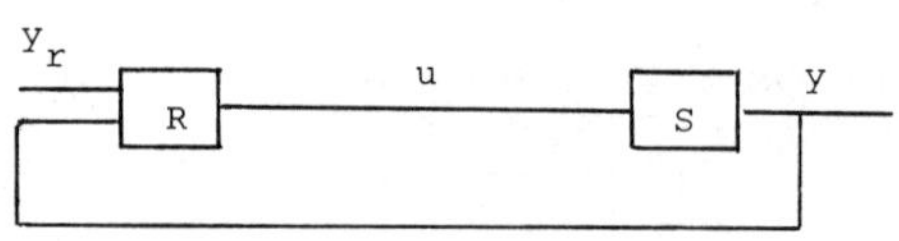

In the above figure :

. S is a non linear continuous time system described by :

$$\dot{x} = f(x,t,u)$$
$$y = g(x,t,u)$$

. R is a non linear discrete time regulator described by :

$$z_+ = f_r(z,t,y,y_r)$$
$$u_+ = g_r(z,t,y,y_r)$$

h the sampling period.

Assume that one wants to compute y at the sampling instants (o,h, ..., n.h).

To simulate such a system one would first have to define the macros (f,g,f_r,g_r) which describe the system e.g.

```
// <xdot> = f(t,x,u)
   :
   :
   xdot = ...
// end
```

Then one could write the following short macro which uses the general purpose ode solver.

```
// <y> = Simul (n,h,x_o,y_o,z_o,y_r)
   y = y_o              // x_o,y_o,z_o initial states
   t_1 = 0
   // loop on sampling instants
   for k = 1 : n, //
               t_o = t_1
               t_1 = t_1 + h
               // one steep of discret
               u = g_r(z_o,t_o,y_r(t_o),y_o)
               z_o = f_r(z_o,t_o,y_r(t_o),y_o)
               // integration of continuous sys-
                  tem over [t_o,t_1]
               x_o = ode(x_o,t_o,t_1,f)
               x_1 = g(x_o,t_o,u)
               // saving output a time t_1
               y = <y,y_o>
   end
//end
```

In this example several interesting features of the system may be underlined :

- The macros f,g,f_r,g_r may implement any type of (static or dynamic) non linearity and may use any type of variable (matrix, macro, ...).

- y_r may be, wihtout any change, either the vector defining the value of regulation input at sampling instants or the macro defining a continuous time signal.

- Some parameters are sent as argument $(y,n,t,x_o,$ $y_o,z_o,y_r)$ and others as global variables $(g_r,f_r,$ $f,g,\ldots)$.

- The general purpose ODE function may receive several optional parameters to specific the integration method, tolerances, jacobian, ...

- Here f is defined by a Blaise macro, but it is also possible to define it by a Fortran routine compiled and linked with the system.

Interface with MACSYMA

As already mentioned and at the stage of development achieved a need for non numeric computations has appeared.

Immediate examples are the computations of Jacobians for ODE solvers, computations of adjoint equations in control problems, analysis of transfert matrices, ...

We have developped some specific programs in MACSYMA and small interfaces which permits to send variables of the Blaise data basis to MACSYMA. We have in mind to automatically generate Blaise macros implementing analytical formulas derived from MACSYMA's symbolic calculations. Of course this interface is totally MACSYMA dependent...

CONCLUSION

The Blaise system is now largely defined and seems to be rather well adapted to the numerical solution of control problems in an user-friendly environment. It is made of a core of about 12000 Fortran lines and a library of about 100 000 lines.

A significant effort must be done for the definition of new functions through the development of standard library of macros and also a documented and reliable library of Fortran routines.

This last point which is probably the most crucial one largely uncompasses the framework of our project and seems to imply a cooperation in a larger framework and the pursuit of important efforts to unify the basic algorithms which make the basis of any CAD system, efforts already initiated by other groups [10],[11].

REFERENCES

[1] K.J. Aström : Computer Aided Modelling, Identification and Control System Design : A perspective IEEE-CSS Magazine (Nov. 1983).

[2] IEEE-CSS Magazine : Special Issue on Computer Aided Design of Control Systems, Dec. 82.

[3] MATLAB by C. Moler. University of New Mexico, Technical Report C581-1, 1982.

[4] Control C and Matrix Environments for the CAD of Control Systems : J.N. Little, E. Emami Naeni, S.N. Baugert, Proceedings INRIA Conf. on Analysis and Optimization of Systems, Nice 1984.

[5] $Matrix_x$: A Data Analysis, Systems Identification, Control Design and Simulation Package IEEE-CSS, Vol. 2,4 (Dec. 1982).

[6] IMPACT, M. Rimvall, F. Cellier, Proceedings INRIA Conf. on Analysis and Optimization of Systems, Nice, 1984.

[7] P. Van Dooren, A generalized eigen value approach for solving Riccati equations, SIAM J. Sci. Comp., Vol.2, (1981).

[8] C. Lemaréchal : La norme Modulopt, Internal Report INRIA (1981).

[9] Hindmarsh A.C. : Lsode and Lsodi, two new initial value ODE solvers, ACM Signum Newsletter, Vol. 1T (1980).

[10] SLICE : M.J. Denham, C.J. Benson, Internal Report 81/3, SEECS, Kingston Polytechnic (1981).

[11] SYCOT : Systems and Control Tools. Internal report WGS, Eindhoven, Dec. 1983.

AN INTEGRATED ENVIRONMENT FOR COMPUTER AIDED CONTROL SYSTEMS ENGINEERING

S. D. Goodfellow and N. Munro

Control Systems Centre, UMIST, Manchester, M60 1QD, UK

Abstract. The engineering of software for Computer Aided Control Systems Engineering (CACSE) using the concept of software tools is discussed. A set of tools are presented which will provide an infrastructure upon which a fully integrated CACSE package could be readily built. Emphasis is placed on the man-machine interface. An input and an output translator are introduced, and the interaction between them is discussed. An action manager, to control the CACSE functions of a package is presented. The control of retained data within a package using a data manager tool is discussed.

Keywords. Computer-aided system design, Computer graphics, Computer software, Distributed processing, Man-machine systems, Software engineering

INTRODUCTION

The next few years will see an important evolution for engineering software in general and Computer Aided Control Systems Engineering (CACSE) software in particular. Engineering software must take advantage of the hardware currently available to enhance the man-machine interface. The software must be engineered so that it is readily adaptable to future technological change in this area. Software engineering also provides a number of techniques which may be applied to CACSE software. These techniques will produce a structure for the software which reduces the effort required to maintain and update software packages.

Software engineering using the concept of software tools (Kernighan, 1976) provides an attractive method of generating CACSE packages. This paper introduces a set of tools which form the infrastructure of a CACSE package. The term infrastructure is used here to describe a combination of a package's skeleton and it's basic functions which support the applications actions. These infrastructure functions include man-machine interface management and data management. The tools described provide an base upon which an integrated CACSE package maybe readily built.

Aims

The use of a CACSE software may be classified into two types, operation and development. CACSE software is used in operation mode by control engineers, research workers, students, etc. as a tool in the engineering process. Such packages are used in development mode to generate new functions or to combine existing functions for a particular engineering task. Any one user may be involved in operation or development. Indeed, many users will do both.

An analysis of the requirements of software provided for control engineering may be made (Jordan, 1985). The tools described in this paper are intended to fulfil those requirements.

The main aim of the infrastructure under discussion is to improve the computational environment for the users of CACSE software, whether they are operating the package or developing the software. The operation of a CACSE package requires an environment in which control engineering may proceed with little or no hindrance from unfriendly and inflexible software. The development of a CACSE package requires an infrastructure and environment into which new functions may be introduced in a rapid and robust manner.

Three secondary aims may be identified.

1) The user must be given complete control over the operation of the software. The chain of command will run down from the user through the man-machine interface to the infrastructure and applications actions. The user's subordinates; i.e. the various parts of the software; may advise on certain courses of action and assist the user in all possible ways, but may not take overall control.

2) CACSE software, particularly it's infrastructure, must be structured in a way which minimises the complexity of both introducing new algorithms to the package and modifying existing algorithms. The administrative details of the infra-structure tools and low level applications actions will be hidden from the developers of the applications software.

3) Data must be structured in a manner which is natural in a control engineering environment. The management of data is a function of the infrastructure which is hidden from the applications software and it's users, whether they are operating or developing the software.

System Requirements

The infrastructure envisaged will need to be implemented as a number of concurrent tasks communicating with each other. This is particularly necessary if the man-machine interface modules of the infrastructure are to provide support for the sophisticated user interfaces, which are required for CACSE software. The computing system upon which the infrastructure is built must also support the issue of system commands from within a program, so that compilation and loading of individual applications actions may be managed from within the package.

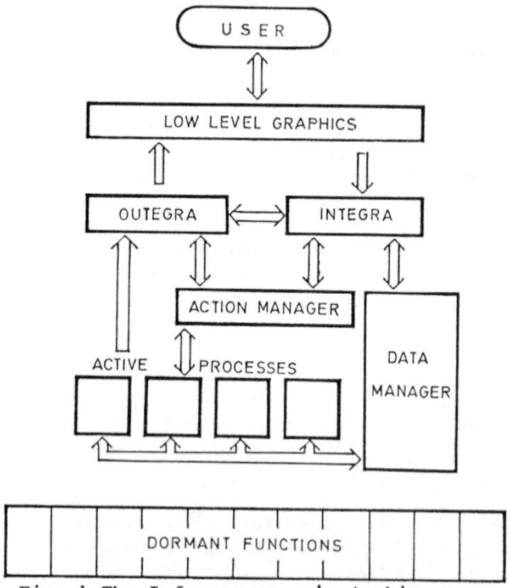

Fig. 1 The Infrastructure's Architecture

THE INFRASTRUCTURE'S ARCHITECTURE

The diagram in Fig.1 illustrates the principal software tools and their relationship to each other. The modules illustrated are separate tasks, they may share the same processor, or they may be implemented in a parallel or distributed processing environment. The principal software tools include an action manager, a data manager, and two tools which form the man-machine interface; an input translator and an output translator. The action manager controls the operation of applications actions. The data manager controls and stores all data other than that held by an action during it's execution. The man-machine interface tools provide mappings from complex user input to data for the applications actions and from internal data to external representations on the output devices available to the user.

Functional Description

The user operating a package built on the infrastructure described here will be aware of only two of the tools, these are the input translator and the output translator. Therefore the user's conceptual view of the software's structure is not necessarily related to the actual structure of the software. For example, the user may wish to view a data object held by the data manager. The user's request is mapped into a command to the data manager by the input translator and the stream of data which forms the reply is mapped to a representation on the output hardware available via the output translator. The representation used would be designed for the user and may not reflect the structure of the stored data.

An individual applications action will be activated as a task, which may be concurrent with other applications tasks. The action may receive input via the action manager, read and write data from the data manager and send output to the output translator. The application action will have no direct contact with the user or the database.

Communications

Communication between the modules may be provided in one of two ways. In general this communication will be implemented using pipes, between the processes, as indicated in Fig. 1. The use of pipes allows the infrastructure to be readily implemented in a distributed environment. However, in certain circumstances one or more of the pipes may become a bottleneck, restricting the performance of the system. The communications links from the data manager to the applications actions will carry particularly large amounts of data. Therefore, when a particular computing system allows, the data manager will communicate with the applications actions via shared memory.

MAN-MACHINE INTERFACE

The environment provided to the user by the infrastructure will depend crucially on the man-machine interface tools. The user's view of the software will be controlled within the man-machine interface. The current state of the applications data and actions is represented to the user through the output translator and to a lesser extent by feedback from the input translator. The user's view of the actions available at a particular state is a function of the input translator.

The tools described here provide the foundation upon which a sophisticated man-machine interface may be readily built. These tools meet two important requirements. Firstly, they do not assume any particular format of man-machine communication. Secondly, they provide a translation facility which may be modified by the user at runtime, the user will be in control of the translation process. For example, the user may alter the representation of a particular set of data from a linear plot to a polar plot or a list of pairs of values.

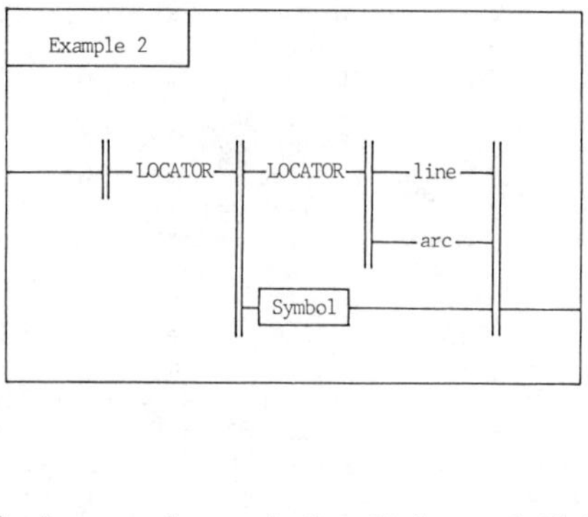

Fig. 2 An example graph for dialogue definition.

Input Translation

The input translator, called INTEGRA (INput of TExt and GRAphics) is briefly described here. A full functional specification has been given by Goodfellow and Munro (1985). INTEGRA uses a dialogue description to control the translation process. The dialogue description is based on state transition networks, represented to the dialogue developer as a set of graphs. An example graph is illustrated in Fig. 2. The graphs provide various levels of abstraction from a single character or graphical action through to complex commands and languages. The graphs are entered and edited graphically both before execution and at runtime.

The translator uses the dialogue definition to check the syntax of input. The translator may follow several paths concurrently, therefore it provides a pattern matching capability as well as the ability to trigger more than one stream of actions with the same input.

A variety of tools are provided for the user including macro languages, synonyms, prompting, menus, graphical icons, help, walkthrough, tracing, and customising. The developer is provided with facilities to report on the use of the application. The translator provides for the use of expert system techniques to provide adaptive man-machine interfaces in the future.

Output Translation

The output translator receives data in it's internal form and converts it to a representation which may be readily understood by the user. The representation used will vary, depending on the output devices available and the user's customising of the system to his own personal tastes. The output translator relieves the application actions of responsibility for output formating at a much higher level than graphics packages such as GINO (CADCentre, 1983).

The mapping of internal data to an external representation forms a natural part of systems such as SMALLTALK-80 (Goldberg, 1983) and has been investigated as part of a strongly typed distributed virtual memory (Hughes, 1984).

The authors are investigating the use of state transition type graphs to control the representation of data. The representation chosen would be dependant on the type of data object and the current state of the application. The graphical language used would be consistent with that used by INTEGRA.

Translator Interaction

The above description of the two translators assumes they work independently, however this is not the case. In a number of cases the input translation is dependent on the current state of the output. For example, the user may wish to pick an object from those displayed on an output device. At the lowest level,the input translator receives a pair of coordinates from an input device. To identify the object at that position on the screen it sends the coordinates to the output translator, which returns the object identifier. It is the identifier that forms the input to the syntax analyser. The translators interact with each other whenever they need to share information or coordinate their activities.

ACTION MANAGER

The action manager will be responsible for initiating and controlling applications and system actions. The action manager will start with a number of dormant tasks. When an action is ordered by the input translator, one of the tasks is given the relevant code to execute. The input translator may send the action manager data to forward to the action during its execution. The user may also request the status of a particular action, or order the action terminated. The action manager responds by sending messages to the action or signalling it to terminate if it refuses to halt.

DATA MANAGER

The data manager will control the retained data. This data will include all the applications data except for data held by an action during it's execution. The data will also include a large amount of system data including the source code for all the applications and system actions and the dialogue definitions.

Security is a key issue when such a large amount of information is retained. Each user will have a security code, giving him, and any actions he initiates, certain rights. Each user will only have access to some of the source code and dialogue definitions, the user is not explicitly informed of the existence of data for which he has no access rights. It is particularly important that the user is limited to information which he is likely to use, so that the useful information is not obscured in a mountain of other data.

Function

The structuring of data into objects has been recognised as an important requirement for CACSE software (Maciejowski 1984). The data manager will hold data as structured objects, an object maybe be made up of several other objects. A dictionary will be kept of the objects retained. Each user may inspect a copy of the dictionary, listing objects to which he has access rights.

The data manager will receive requests for data objects from the active actions and the man-machine interface tools. The data manager responds by making the data available to the requesting task. The data may either be sent to the task using a pipe or if the data manager shares memory with the requesting task, it will be sent the address of the data in the shared memory. These two alternatives are equivalent to passing arguments by value and by address respectively.

The data manager will have to deal with several concurrent tasks at any one time. Concurrent access to the data will be controlled using semaphores and request queues (Holt, 1984). Any number of tasks may read the same data object concurrently. However only one task may have write permission for a data object at a time. Whilst a task has write permission for a data object tasks requesting to read the object will be given a copy of the data as it was when the writing task was given write permission. The reading tasks are also informed that the data is being modified.

User Access

The man-machine interface will provide the user with several ways of accessing the data retained by the data manager. The user may, of course, request the output of a particular object by name, the output translator will reply with a representation of that object. From any given object, the user may browse around the database, following the structure up to an object's super-object or down to an object's sub-objects. The browsing system is obviously useful for inspecting control system descriptions and dialogue definitions. The third method of access will provide a relational ability, objects may be accessed that match a set of attributes specified by the user

CONCLUSIONS

The infrastructure described in this paper will provide the CACSE software developer and the control engineer with a structured set of tools. The emphasis is on hiding any unnecessary information and making every facility readily available at runtime. The environment provided by the infrastructure will have many of the attributes of systems such as SMALLTALK without the disadvantage of a large overhead in execution time.

To summarise, the infrastructure will provide :-

A sophisticated man-machine interface tool set which uses dialogue definitions which are modifiable at runtime, and enables mixed numerical, graphical, and textural information to be readily input and output.

A data manager which can control structured data objects available to concurrent tasks.

An action manager which controls the application actions and some system functions which are executable concurrently.

The infrastructure should allow CACSE packages to be built which integrate the whole control engineering process from system modelling and identification through design to commissioning. The CACSE packages built on this infrastructure will be readily portable to distributed computing systems. In many cases the support of new technological features, particularly in the man-machine interface, will automatically be inherited by packages when the infrastructure is updated.

REFERENCES

CADCentre (1983). GINO-F Reference manual, CADCentre Ltd., Madingley Road, Cambridge, U.K.

Goldberg, A., D. Robson (1983). Smalltalk-80: The language and it's implementation, Addison-Wesley

Goodfellow, S. D., and N. Munro (1985). INTEGRA, an input translation facilty for computer aided control systems engineering. 3rd IFAC/IFIP Symposium on computer aided design in control and engineering systems, Denmark, July 1985.

Holt, R. C. (1983). Concurrent Euclid, the UNIX system, and Tunis, Addison-Wesley.

Hughes, J. W., M. S. Powell (1984). A strongly typed distributed virtual memory.
In D. Juce (Ed.), Distributed computer systems. Peter Peregrinus.

Jordan D. (1985) Specification of a man-machine interface for control system computer aidied engineering. Report 630, Control Systems Centre, UMIST Manchester, M60 1QD, U.K.

Kernighan, B. W., P. J. Plauger (1976). Software tools. Addison Wesley.

Maciejowski, J. M. (1984). Data structures for control systems design. EUROCON'84, Brighton, U.K. September 1984.

AN EXPERT SYSTEM ARCHITECTURE FOR COPING WITH COMPLEXITY IN COMPUTER-AIDED CONTROL ENGINEERING

J. R. James*, J. H. Taylor and D. K. Frederick*****

**Department of Electrical Engineering, United States Military Academy, West Point, NY 10996, USA*
***General Electric Corporate Research and Development, Schenectady, NY 12345,USA*
****Electrical, Computer, and Systems Engineering, Rensselaer Polytechnic Institute, Troy, NY 12181, USA*

Abstract. We discuss an architecture for an expert system to assist a control engineer in coping with the complexity of computer-aided control engineering, with major emphasis placed on implementation. Issues to be treated include: rule base organization, knowledge to be contained in various rule bases, a mechanism to switch between rule bases as the problem solution proceeds, a protocol to coordinate the symbolic computations of the inference engine with the numeric computations of conventional analysis and design software, support of non-monotonic reasoning to permit retracting and revising steps in the design process, and conversion between numeric and symbolic data. We present our approach to these issues, and give examples of rules used to represent operational knowledge and to guide the solution of the problem.

Keywords. Artificial intelligence; expert systems; computer architecture; computer-aided system design; control system synthesis.

1. INTRODUCTION

Control engineering problems and the software tools with which to solve them are becoming ever more complex. The development of more powerful analysis and design procedures together with the rewards for their effective use serve to motivate the creation of these new tools. Unfortunately, as their number, capabilities, and complexity increases, the possibility that any individual will be able to apply the available range of techniques and software effectively is decreasing. We believe that the expert systems approach provides a solution to this difficulty.

The use of expert systems to represent and apply expert knowledge in a particular domain is becoming increasingly widespread (Hayes-Roth, 1983 and Wallich, 1984). Our recent efforts (Taylor, 1983 and 1984 a,b; James, 1985) have focussed on the use of expert systems to aid a control engineer in exploiting available software to achieve an acceptable design. In this research, we use General Electric's DELPHI inference engine in combination with the following conventional analysis and design software tools:

a. the Cambridge Linear Analysis and Design Program (CLADP) (Edmunds, 1979), which performs design, analysis and simulation functions for both single-input, single-output and multiple-input, multiple-output systems in either the time or frequency domain,

b. SIMNON (Elmqvist, 1977), which supports the simulation of nonlinear systems,

c. extensions to SIMNON (Taylor, 1982), which provide equilibrium finding and linearization about the equilibrium point for nonlinear system models, and

d. a routine to categorize eigenvalues, to provide the basis for diagnosing the time-domain characteristics of the system model by the expert system.

In addition, we developed a set of LISP functions which support the various actions described in Sections 2 thru 4 below, and extended CLADP and SIMNON to support the exchange of data between the expert system and the external processes.

Integrating the numerical capabilities of the above software with the knowledge representation and deductive mechanisms of the DELPHI inference engine results in a high-level combination of large-scale numerical processing with large-scale symbolic processing. During the execution of a design sequence, we use the expert system (written in FRANZ LISP and running under the EUNICE operating system) to control the start-up of the various numerical routines (written in FORTRAN and running under the VAX/VMS operating system), direct the transformation and exchange of data required to proceed with the problem, and implement the interface with the user. As many as four external VAX/VMS sub-processes are created, and external programs are started, used, and stopped by the expert system through these sub-processes. To illustrate the scope of this activity, more than a hundred individual commands may be issued by the expert system as control system compensators are designed. Such a software system thus achieves a transfer of much of the burden of the complexity of the control engineering problem from the shoulders of the design engineer to the expert system.

This paper is organized as follows: In Section 2 we discuss a partitioning of an expert's control engineering knowledge into various rule bases and the organization of these rule bases. The implementation of the design process and the role of a switching function in that implementation are discussed in Section 3. The protocol used to coordinate the symbolic manipulations of the inference engine with the numeric computations of the conventional software is treated in Section 4. We conclude in Section 5 with an outline of the key issues and our approach to addressing them, and salient conclusions.

2. KNOWLEDGE PARTITIONING

The rule-base architecture is shown in Fig. 1. The production rules which comprise this structure were developed to realize the concepts presented in the Computer Aided-Control Engineering (CACE) architecture of Taylor (1984b). The present structure differs slightly from what was originally proposed. The principal difference is that the design process has been implemented in one rule base rather than two, to eliminate unnecessary rule-base switching.

As the number of design procedures grows and the complexity of the other portions of the expert system increase, partitioning the knowledge into modules will be an essential requirement to reduce the time required to execute various phases of the analysis and design process. In addition, modularity will make the expert system easier to develop, modify, and maintain.

The six operational rule bases of Fig. 1 are integrated through a set of supervisory rules (meta-rules) which checks the current state of the problem solution and decides what needs to be done next. This implementation of the expert system architecture was also suggested in (Taylor, 1984b). Invoking the next applicable rule base is accomplished by use of a switching function which is further discussed in Section 3.3.

The knowledge contained in the rule bases (Fig. 1) is as follows:

SUPERVISOR: The function of this rule base is to permit the user to select a training or design session, and to direct its execution, Fig. 2. This selection allows any of the sets of rules to be invoked individually (the TRAINING option), or aids the user in carrying out the overall process (the DESIGN option). The training option makes use of archival lists of facts associated with the design of lead-lag precompensators for a third-order, type-one plant, to allow the user to explore the workings of each individual rule base. Specific functions are: starting and stopping the session; permitting the user to select a training or design session; selecting the appropriate rule base to continue the session when one set of rules has completed its work (during a DESIGN session); saving the current list of facts if the user desires to interrupt the session and continue at a later time; and permitting the user to revise the problem formulation as the result of further analysis and design. (This is an example of non-monotonic reasoning. That is, we are able to "revise a belief" that specification development has been completed, and return to update the specifications.)

MODEL: This rule base will either aid the user in building a CLADP description of a linear model, or request the user to enter the file name of a previously-developed CLADP model. Similar ability to define a nonlinear SIMNON model followed by equilibrium finding and determination of a linearized model is also supported.

DIAGNOSE: The function of this rule base is to analyze and characterize the model of the plant. Currently s-domain information is provided on: the stability of the open-loop system; whether the system is type-zero, one or two; whether the plant can be approximated as an overdamped or underdamped second-order system; and whether the plant has a dominant pole or pole pair. Also, the following frequency-domain information is provided: open-loop bandwidth (type-zero system); uncompensated gain and phase margin; low-frequency gain (type-zero system) or velocity constant (type-one system); closed-loop bandwidth with unity feedback; closed-loop bandwidth with 15 dB of gain margin; approximate bandwidth achievable with two leads; and approximate low-frequency gain or velocity constant achievable with two lags.

CONSTRAIN: The function of this rule base is to enable the user to enter constraints. These rules currently support frequency-domain parametric constraints on the compensator dc gain, phase lead added by the leads, or low-frequency gain added by the lags.

DESIGN: This rule base aids the user in applying CLADP procedures to design lead-lag precompensation for a single-input, single-output linear plant. A detailed description of the heuristic used to automatically design the compensator is given in (James, 1985). A synopsis of the heuristic is as follows:

a. Add lead precompensators to adjust the open-loop phase angle at the desired closed-loop bandwidth. This places the high-frequency portion of the Nichols locus in approximately the right region.

b. Add or adjust a constant-gain precompensator to meet the gain margin.

c. Adjust the actual closed-loop bandwidth by making incremental changes in the lead compensator pole/zero ratios.

d. Repeat steps b. and c. until the gain margin and bandwidth specifications are met. When both conditions are satisfied, the high frequency portion of the Nichols locus is correct.

e. Add lag precompensation to adjust the low frequency gain (for a type-zero system) or velocity constant (for a type-one system). The lag(s) are placed so the high frequency portion of the Nichols locus is not disturbed significantly.

SIMULATE: This rule base facilitates the simulation of the closed-loop system. CLADP is used to provide the step response of the system with a linear plant model, and SIMNON is used to perform more general and realistic simulations of the response of the system with linear or nonlinear plant models.

3. IMPLEMENTING A STRUCTURED PROCESS

The user directs the design process of the expert system by providing names of data files, menu selections, and responses to queries. While the user is in charge of the problem solution, the expert system implements a specific train of thought to aid in achieving the solution (e.g., it checks that reasonable specifications are entered before starting the design of a compensator).

The knowledge summarized in Section 2 is needed to implement such a train of thought. This knowledge is symbolically represented in a partitioned sets of rules and a list of facts. A discussion of the syntax of production rules and lists of facts and how they can be used to represent knowledge can be found in (Taylor, 1984b) or (Johnson, 1983).

The expert system carries out the design process by applying the appropriate rule base to the problem. This is achieved by recognizing when the current rule base has completed its tasks, switching to the supervisory rule base, and determining which rule base should be invoked next. Implementing the transition from one rule base to another in DELPHI involves a careful hand-over of the current list of facts from one rule base to the next. A LISP switching function has been written to perform these actions.

To convey a clearer understanding of how we have controlled the logical flow associated with the design process, we will first discuss forward chaining through a rule base, then discuss backward chaining, and then present some lessons learned about switching between rule bases.

3.1 Forward Chaining

Most frequently, the inference engine uses the list of facts to determine the current situation and then take the appropriate action(s) to proceed with the next step in the process. This inference mechanism is called forward chaining. For example, the rule which recognizes that the lead-lag design criteria have been met and writes the facts that will display summary information to the user, direct the expert system to provide a Nyquist plot of the compensated system, and then return to the supervisory rule base, is as follows:

```
(Rule_130 (" design for GM, lfg, clbw is done ")
 (WHEN
  (TRUE (SYSTEM TYPE . ZERO)
  " Have a type-zero system ")
  (TRUE (PHASE-AT-OMEGA-BW VALUE . ADJUSTED)
  " Have adjusted the phase at omega bw ")
  (TRUE (EXCESS-GAIN-MARGIN VALUE . -3<..< +3DB)
  " Excess gain margin is ok ")
  (TRUE (LOW FREQUENCY . ADJUSTED)
  " Low frequency gain is ok ")
  (TRUE (CLOSED-LOOP-BW VALUE . ADJUSTED)
  " Closed-loop bandwidth is ok ")
  (TRUE (GAIN-MARGIN FINAL-VALUE . (? GM))
  " The final gain margin is known ")
  (TRUE (CLBW FINAL-VALUE . (? CLBW))
  " The final closed-loop bandwidth is known ")
  (TRUE (LFG FINAL-VALUE . (? LFG))
  " The final low-frequency gain is known ")
  (TRUE (DESIGN-FACTS VALUES . ALL-ENTERED)
  " We have entered all required facts "))
 (THEN
  (CLEAR (DESIGN-FACTS . ALL-ENTERED)
  " Prevent looping on this rule ")
  (SCREEN (?? ?? . ??)
  " Clear the terminal display for the user ")
  (DISPLAY (?? ?? . ??)
  " "
  " LEAD-LAG design specifications are met: "
  " "
  " gain margin        = ?GM  dB "
  " bandwidth          = ?CLBW  rps "
  " low-frequency gain  = ?LFG  dB ")
  (ACKNOWLEDGE (L-L-DESIGN CONDNS-MET . ACKNLD)
  " Wait for the user to acknowledge info ")
  (WRITE (PROVIDE FINAL . NYQUIST-PLOT)
  " Fire the rule to provide a Nyquist plot ")
  (WRITE (SYSTEM-SESSION REQD . FROM-DESIGNLL)
  " Move to the Supervisory rule base ")))
```

The nine conditions in the premise of the above rule (i.e., those elements between WHEN and THEN) are used to recognize that the lead-lag design has been successfully completed for a type-zero system. Comments associated with each fact are enclosed in double quotes. The six statements in the conclusion (i.e., those elements following THEN) perform those actions needed to continue with the design process or are used to fire other rules which will perform such actions. Some quoted material (e.g., "Prevent looping on this rule") comprise comments, while some phrases are displayed to the user (e.g., "LEAD-LAG design specifications are met:"); this is governed by the verb (CLEAR or DISPLAY, respectively). Symbols are assigned values by the inference engine (e.g., (? GM) may have the value 19.5 dB), and those values are also displayed to the user (e.g., line 5 of the DISPLAY action above results in the message

gain margin = 19.5 dB

being issued). The forward chainer starts work as soon as a rule has its premise satisfied, and continues until no rule has its premise satisfied.

3.2 Backward Chaining

A rule can also recognize that the current situation calls for establishing a goal to be achieved by the inference engine. This inference mechanism is called backward chaining. For example, the rule which repeatedly checks for the completion of the major steps in the design process, and thus is the heart of Fig. 1, does so by invoking the backward chainer to verify a list of facts:

```
(Rule_300 (" Design process being checked ")
 (WHEN
  (TRUE (SYSTEM SESSION . BEGUN)
     " You have selected a DESIGN session "))
 (THEN
  (CLEAR (SYSTEM SESSION . BEGUN)
     " Prevent looping on this rule ")
  (PROVE (MODELING ASSISTANCE . PROVIDED)
     " Has the system model been entered? ")
  (PROVE (DIAGNOSIS ASSISTANCE . PROVIDED)
     " Has the system been diagnosed? ")
  (PROVE (CONSTRAINT ASSISTANCE . PROVIDED)
     " Are system constraints entered? ")
  (PROVE (SPECIFICATION ASSISTANCE . PROVIDED)
     " Are system specifications entered? ")
  (PROVE (DESIGN ASSISTANCE . PROVIDED)
     " Has the design met the specs? ")
  (PROVE (SIMULATION ASSISTANCE . PROVIDED)
     " Is simulation of the CL system done? ")
  (ACKNOWLEDGE (ADVISE DESIGN-VERIF . DONE)
     " "
     " Control system design and verification
    " is finished. You may wish to alter the
    " given model, constraints, and/or specs
    " and redo the design, or repeat another
    " portion of the process, or exit. ")
  (WRITE (DESIGN-SESSION MENU . REQUESTED)
     " Allow user to modify or quit. ")))
```

This rule is fired each time the supervisory rule base is loaded. The backward chainer considers each PROVE statement in turn. If it is already established in the list of facts, the backward chainer moves on to the next statement; the first hypothesis that has not already been established will be picked by the inference engine for testing. The supervisory rules are formulated so that this results in invoking the appropriate operational rule base. The inference engine will proceed in this way until the entire set of hypotheses has been satisfied and therefore the goal can be said to be achieved.

Generally, this rule results in the progression MODEL, DIAGNOSE, CONSTRAIN, SPECIFY, DESIGN, SIMULATE. However, this is not a rigid regimen: if during the design process it is determined that a specification cannot be met and should be changed, the fact (SPECIFICATION ASSISTANCE . PROVIDED) can be cleared by a rule in the design rule base. Other facts can be written to indicate which specification needs attention. When the supervisory rule base tries to verify that (SPECIFICATION ASSISTANCE . PROVIDED) is a fact, it will no longer be in the list of facts. Thus, the inference engine will again invoke the the specification rule base and work to validate this fact.

Comparing the premise of this rule to that of the example used for forward chaining, we see that only a single condition is needed to recognize that the steps in the design sequence need to be verified. The PROVE verb is used to indicate to the backward chainer that the associated fact (e.g., (MODELING ASSISTANCE . PROVIDED)) is a hypothesis to be verified. The backward chainer will then check to see if the fact is already known to be true or false. If the fact is not known, then the backward chainer will proceed to look for rules which will write this fact in their conclusions and work to establish the premise of those rules. This involves switching in the modeling rule base and proceeding until (MODELING ASSISTANCE . PROVIDED) is true.

3.3 Using a Rule-Base Switching Function

Rules in each operational rule base recognize that it is time to return to the supervisory rule base to proceed with the overall problem. The facts are then saved in a file, the supervisory rule base is loaded as the current rule base, and the facts are made available to the supervisor. The next applicable rule base is selected by the supervisory set of rules. The six operational rule bases are contained in separate files which are invoked using the switching function. This process continues until the expert system can do no more or the user elects to end the session. As we have seen above, Rule_300 performs this top-level control of the process.

This approach is useful if the problem being addressed can be partitioned into "phases" with well-defined initial and final conditions, and if the phases are of sufficient complexity that the overhead involved in switching is warranted. The knowledge partitioning discussed in Section 2 satisfies these conditions.

The exit criterion for a rule base must not be satisfied until all required actions have been completed. This condition is met when the pattern which fires the rule in which the switch function is called occurs only for that rule. We normally achieve this by executing a switch as the last action taken by the first rule in a rule base.

The list of facts is passed as a queue from one set of rules to the next, and DELPHI starts to check rules for firing as facts are being loaded. In this case, one must be careful not to start any actions in a new rule base until the entire list of facts has been entered. (Otherwise some facts may be lost if a subsequent switch occurs before all of the old facts are entered, or the new set of rules will not behave properly since all of the facts would not be available.) Therefore, each rule base is designed such that its start-up rule cannot be fired until the last fact written by the prior rule base is loaded.

The seven rule bases now total approximately 300 rules; using a command file to run through a typical design sequence takes about thirty minutes on an unloaded VAX 11/785. About half of this time is required to compile the rules as we switch between operations. This time would be reduced significantly if the DELPHI inference engine supported accessing a compiled version of the rules.

4. PROTOCOL FOR COORDINATING SYMBOLIC AND NUMERIC COMPUTATIONS

The operation of the expert system in concert with conventional analysis and design software requires that:

a. VAX/VMS sub-processes be created to start, run and stop external programs which are sources of data for the expert system,

b. a two-way transformation of data be set up between the symbolic representation of the expert system and the numeric representation of these external programs, and

c. coordination of the operation of the expert system with external programs be established to prevent the expert system and the external programs clashing (e.g., trying to provide information to the user at the same time, or returning to symbolic manipulation before numeric data has been completely supplied).

The DELPHI inference engine has a LISP function available to create a VAX sub-process which uses mailboxes for the input and output ports of the sub-process. Read and write functions are also available to communicate with the sub-process via the mailboxes. We use the mailboxes to start up programs (Fig. 3), and then transmit commands to them and receive data from them using files.

We have implemented a protocol which coordinates the expert system with external programs and provides support for two-way communication via files. Fig. 4 is a timing diagram for running CLADP which depicts the actions taken to implement the protocol. A detailed discussion of this timing diagram follows:

a. The user starts the expert system (point A on Fig. 4) and directs its activity until data is needed from an external program. The expert system then writes the required CLADP commands to the file EXPOUT (points B,C), creates a VAX/VMS sub-process, starts CLADP through the sub-process, and begins to wait (point D) until it detects that CLADP has created the handshake file (point J).

b. Once CLADP is started, it reads the sequence of commands from the file EXPOUT (points E,F), performs the required calculations, writes the results to the file EXPIN (points G,H), creates the handshake file (point I), and begins to wait (point J) until it detects the deletion of the handshake file (point P).

c. When the presence of the handshake file is detected (point J), the expert system reads the symbolic data written to the file EXPIN by CLADP (e.g., (GAIN-MARGIN DB-VALUE . 8.5), points K,L), and continues with the inference process until additional data is needed. At this time, the sequence repeats as described above (points M thru Q).

This sequence of writing commands to EXPOUT and symbolic information to EXPIN while coordinating the operation based on the existence of a handshake file continues until CLADP is no longer needed. At that time the expert system can stop CLADP and the VAX/VMS sub-process is then available to run another program.

Coordinating several processes is identical to this outline; for each process there is an associated file set for handshake and input and output data transfer. Symbol/numeric transformations are required throughout this process: The expert system must know how to write CLADP and SIMNON commands in order to achieve the required results, and CLADP and SIMNON were extended to convert numerical data back into symbolic form (e.g., (GAIN-MARGIN DB-VALUE . 8.5) has to be written into the file EXPIN in the above example).

5. CONCLUSION

The application of the expert systems approach has proven to be useful in coping with the complexity of computer-aided control engineering (CACE) procedures. By reducing the burden on the design engineer to recall the details of using software packages effectively, and by containing the heuristic procedures involved in various analysis and design approaches, an expert system can be an effective aid in the CACE process.

Partitioning the CACE problem into distinct well-defined subprocedures and corresponding rule bases is one key to addressing the complexity issue. This can be achieved by "modeling" the computer-aided control engineering effort along functional lines (Taylor, 1984a,b). In implementing this feature, we have found that switching between rule bases has several benefits, namely:

a. Top-level goals can be reevaluated continually by the supervisor, based on the results achieved by the different operational rule bases. For instance, if work reveals that a bandwidth specification cannot be met under the given conditions, the supervisor may re-invoke the SPECIFY rule base and ask the user to reduce the bandwidth specification.

b. Incremental design is supported, since the rule bases are functionally arranged and the list of facts can be saved to initiate another session. For example, the user can complete a model and diagnosis session, save the facts, and restart the expert system at a later time. Several specification sets and solutions can be studied, starting from the same initial state.

c. Rule bases can be written and debugged separately, and then integrated into the overall structure. The time required to make trial runs is greatly reduced, and errors can be identified more quickly.

d. The supervisor is used to limit the scope of search, so the running time of the complete expert system should be kept to reasonable bounds even as the number of procedures supported by the expert system increases.

Using the switching technique to implement the rule base partitioning requires careful attention to detail, as indicated in Section 3.3.

We have presented architectural and implementation issues involved in developing an expert system to create a higher-level, flexible environment to help the control engineer solve CACE problems. However, it should be emphasized that the development of an expert system is not a panacea. The creation of an expert system has its own set of difficulties, the most burdensome being the creation of the architectural implementation and knowledge base. The process of translating human expertise into a rule base (knowledge acquisition) is not well understood (Hayes-Roth, 1983). However, we believe that "modeling" the computer-aided control engineering effort along functional lines (Taylor, 1984a,b) is a necessary first step in knowledge representation and acquisition that greatly alleviates this problem. The reduction in the difficulty of the problem presented to the user justifies the increased complexity of the software tool which implements the design environment.

ACKNOWLEDGEMENT

The research reported in this paper has been substantially aided by the contributions of others. Dr. Piero Bonissone has assisted from the beginning of the project, providing the initial inference engine and his experience with expert systems development. More recently, Dr. Melvin Simmons, Dr. Dale Gaucus, and Steven Kirk assisted in implementing the expert system using the DELPHI inference engine. Also, David Kassover has helped in making modifications to CLADP and SIMNON. These individuals were employed by General Electric Corporate Research and Development (David Kassover as a consultant). Their support is gratefully acknowledged.

REFERENCES

Edmunds, J. M. (1979). Cambridge linear analysis and design program. *IFAC Symposium on Computer Aided Design of Control Systems*, Zurich.

Elmqvist, H. (1977). SIMNON - An interactive simulation program for nonlinear Systems. *Proceedings of Simulation 77*, Montreux.

Hayes-Roth, F., D. A. Waterman and D. B. Lenat (1983). *Building Expert Systems.* Addison-Wesley, Reading, MA.

James, J. R., D. K. Frederick and J. H. Taylor (1985). On the application of expert systems programming techniques to the design of lead-lag precompensators. *Proceedings of the Control 85 Conference*, Cambridge, UK.

Johnson, H. E. and P. P. Bonissone (1983). Expert system for diesel electric locomotive repair. *Journal of FORTH Applications and Research, 1*, No. 1.

Taylor, J. H. (1982). Environment and methods for computer-aided control systems design for nonlinear plants. *Proceedings of the Second IFAC Symposium: CAD of Multivariable Technological Systems*, West Lafayette, IN. 361-367.

Taylor, J. H., D. K. Frederick and A. G. J. MacFarlane (1983). A second- generation software plan for CACSD. *Abstracts of the IEEE Control System Society Symposium on CACSD*, Cambridge, MA.

Taylor, J. H., D. K. Frederick and J. R. James (1984). An expert system scenerio for computer-aided control engineering, *Proceedings of the American Control Conference*, San Diego, CA. 120-128.

Taylor, J. H. and D. K. Frederick (1984). An expert system architecture for computer-aided control engineering. *IEEE Proceedings, 72*, 1795-1805.

Wallich, P. (1984). Technology '84 software. *IEEE Spectrum, 21*, No. 1, 47-49.

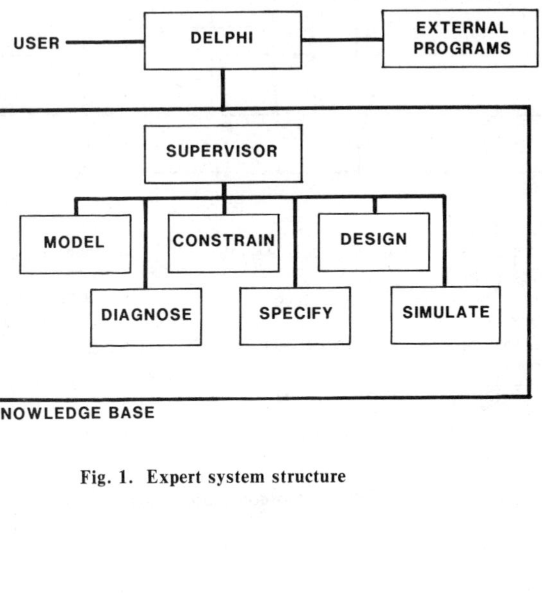

Fig. 1. Expert system structure

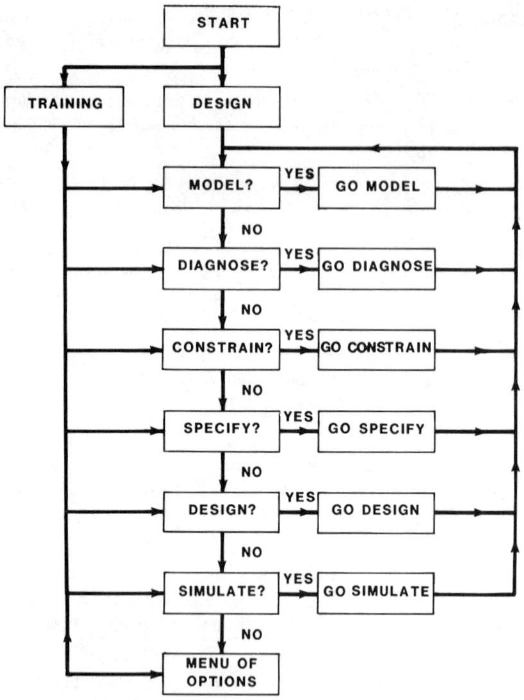

Fig. 2. Flow chart for the supervisor set of rules

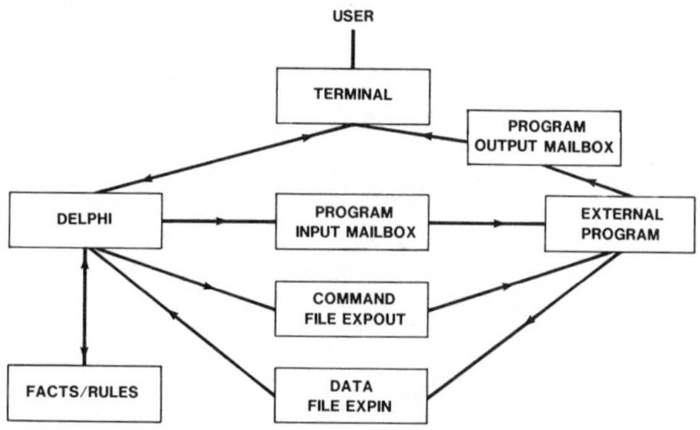

Fig. 3. Running external programs

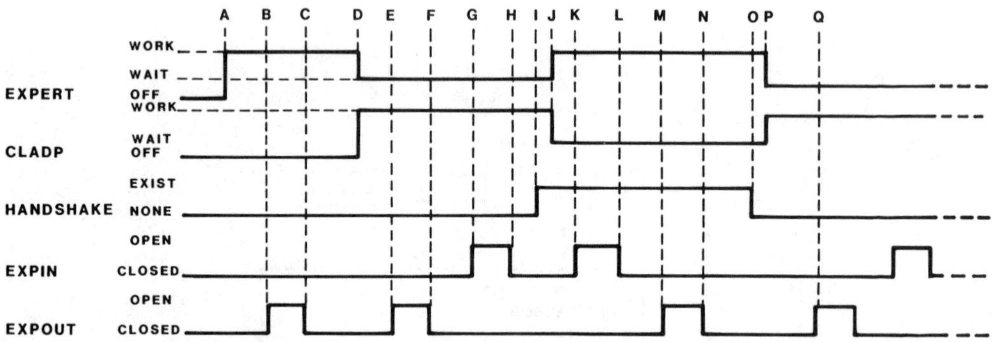

A - USER BEGINS EXPERT
B - EXPERT OPENS EXPOUT AND WRITES COMMANDS
C - EXPERT CLOSES EXPOUT
D - EXPERT BEGINS CLADP VIA MAILBOX AND BEGINS TO WAIT
E - CLADP OPENS EXPOUT AND READS COMMANDS
F - CLADP CLOSES EXPOUT
G - CLADP OPENS EXPIN AND WRITES RESULTS
H - CLADP CLOSES EXPIN
I - CLADP CREATES AND CLOSES HANDSHAKE
J - EXPERT SYSTEMS RETURNS TO ACTIVE STATE, CLADP BEGINS TO WAIT
K - EXPERT OPENS AND READS EXPIN
L - EXPERT CLOSES EXPIN
M - EXPERT OPENS EXPOUT AND WRITES COMMANDS
N - EXPERT CLOSES EXPOUT
O - EXPERT DELETES HANDSHAKE
P - CLADP RESUMES WORK, EXPERT WAITS
Q - CLADP OPENS EXPOUT AND READS COMMANDS

Fig. 4. Protocol timing diagram

COMPUTER-AIDED CONTROL ENGINEERING ENVIRONMENT FOR NONLINEAR SYSTEMS ANALYSIS AND DESIGN

J. H. Taylor

Control Technology Branch, General Electric Corporate Research and Development, Schenectady, NY 12345, USA

Abstract. We report on recent progress in developing a computer-aided nonlinear control system analysis and design environment based on sinusoidal-input describing function (SIDF) methods. In particular, two major additions have been made to our CAD software for nonlinear controls during 1984: a simulation-based program for generating amplitude-dependent SIDF input/output models for nonlinear plants, and a frequency-domain nonlinear compensator design package. Both of these are described in detail. This software can treat very general nonlinear systems, with no restrictions as to system order, number of nonlinearities, configuration, or nonlinearity type. An overview of the application of this software to the design of controllers for a realistic, nonlinear model of an industrial robot is presented in Taylor (1984), which serves to illustrate the use of these tools. Based on the software presented here, the use of SIDF-based nonlinear control system analysis and design methods is substantially easier to carry out.

Keywords. Nonlinear control systems, describing functions, control system synthesis, computer-aided design.

1. INTRODUCTION

The basis of this work has been established in earlier publications. In particular, Spang (1982) outlines the basic CAD software tools that were available before the start of this program, Taylor (1982) describes several extensions to that software suite that are required for performing conventional analysis and design for nonlinear systems (equilibrium finding, standard linearization) and required interfaces to create a functionally integrated environment, and Taylor (1983, 1984) establishes the theoretical basis for the work described here.

There are two major additions to our CAD software for nonlinear controls that we developed during 1984: a simulation-based program for generating amplitude-dependent SIDF input/output models for a nonlinear plant, based on constant-amplitude sinusoidal driving signals; and a frequency-domain nonlinear controller design module based on the theory cited above. The specific approach taken in developing this software was:

a. to extend the nonlinear simulation package SIMNON (Elmqvist, 1977), to allow the direct generation of SIDF models (amplitude-dependent frequency response models) using simulation and Fourier analysis methods, as described in Section 2; and

b. to create a Nonlinear Controller Synthesis Program (NCSP) by adding a module to CLADP (Edmunds, 1979) that accepts externally-generated frequency response models (SIDF models or experimental data), and supports nonlinear control system design as described in Section 3.

The basic functional architecture of our environment is portrayed in Fig. 1. The new elements in that figure are the quasilinearizer (SIDF model generator) and the nonlinear design module.

Taylor (1983) provides an overview of SIDF models for nonlinear systems and the overall approach to controller design, and Taylor (1984) outlines and illustrates the methodology for multi-model nonlinear controller design by SIDF inversion. Although the discussion that follows centers on SIMNON and CLADP, the same basic principles can be applied to implement these features in any nonlinear simulation and frequency-domain design package in a straightforward way.

The software developed under this program specifically implements the functions required to carry out two nonlinear control system design techniques from Taylor (1983, 1984), i.e., the *one-model SIDF-based linear controller design method* and the *multi-model nonlinear controller design approach based on SIDF inversion.* These new capabilities provide the basis for dealing with nonlinear systems by adding control system design approaches based on the behavior of the nonlinear system for signals having amplitudes that correspond to the actual anticipated operation. The amplitude dependence of a nonlinear system is a key characteristic that often must be considered in the analysis and design of nonlinear systems. This issue is distinct from the dependence of nonlinear system behavior on operating point, which can often be characterized by a family of standard linearized models about various operating points.

2. A SIMNON-BASED SIDF MODELING PROGRAM

2.1 Functional Overview

Software for generating amplitude-dependent transfer functions describing the input/output (I/O) behavior of nonlinear systems excited by sinusoidal inputs has been developed by extending SIMNON and installing a new built-in system SIDFGEN. The system SIDFGEN provides a sinusoidal driving signal for the nonlinear system, and contains state variables which are Fourier integrals of a selected plant output variable. The SIMNON command set has been extended to support this activity.

The nonlinear plant model can be a single system, or any other arbitrarily interconnected set of subsystems. The only restrictions are that the nonlinear system must be stable, and that the desired SIDF I/O relation must be single-input single-output. If the system is not stable, then the user must stabilize it with the appropriate feedback. If it is desired to obtain a *matrix* SIDF I/O model (e.g., to describe a plant with two inputs and two outputs), then the same principles outlined below can be used, except that the sinusoidal signals applied to each input must be of different but nearly equal frequencies and the Fourier analylsis must involve integration over an integral number of cycles of both signals. We have successfully determined 2 × 2 SIDF models using input frequencies related by the ratio 4:3; this work will not be described further due to space limitations. Note, however, that one cannot obtain matrix SIDF I/O relations by exciting a nonlinear system one channel at a time, because of the failure of superposition in nonlinear systems; also, exciting both channels with sinusoids of the same frequency is unworkable, because it is impossible to distinguish the effects of multiple inputs on each output.

The input and output of SIDFGEN are designated y_{sig} and $u_{\cos}$, respectively. The signal $u_{\cos}$ drives the plant, and the output of the plant is connected to y_{sig}. Once SIDFGEN and the nonlinear plant model are interconnected appropriately, the following excitation is provided by SIDFGEN:

$$u_{\cos}(t) = u_0 + a\cos(\omega t) \qquad (1)$$

and, by integration of the state variables in SIDFGEN, Fourier integrals for period k = 1, 2, ... are calculated:

$$I_{m,k} = \int_{(k-1)T}^{kT} y_{sig}(t)\exp(-jm\omega t)\,dt; \quad m = 0, 1, 2, 3 \qquad (2)$$

where

$$T = 2\pi / \omega \qquad (3)$$

The real integral for $m = 0$ captures the constant or "d.c." component of the response, and the complex-valued first-harmonic integral $I_{1,k}$ defines the desired I/O transfer function $G_k(j\omega;\, u_0,\, a)$, as follows:

$$y_{0,k}(j\omega;\, u_0,\, a) = I_{0,k} / T \qquad (4)$$

$$G_k(j\omega;\, u_0,\, a) = \omega I_{1,k} / a\pi \qquad (5)$$

The remaining higher harmonic Fourier integrals may be calculated to provide the user with a measure of the importance of nonlinear effects and with some insight as to the validity of neglecting higher harmonics. This information is also presented in pseudo-transfer function form,

$$G_{m,k}(jm\omega;\, u_0,\, a) = \omega I_{m,k} / a\pi; \quad m = 2, 3. \qquad (6)$$

The integrals for $m = 0$, 1 are sampled every period of the sinusoidal input and checked for convergence; when satisfactory convergence is obtained, the simulation is stopped and the Fourier integrals are used to define the I/O relation as in Eqn. (5). Convergence testing is required in this procedure, because transients generally occur in the simulation, and the Fourier integrals are not meaningful until the contributions of these transients have decayed to become small compared to the steady state response. The mechanics of this procedure are discussed in more detail in Section 2.3. Once convergence is achieved, the desired I/O relation is obtained and written to a file to serve as a basis for controller design using NCSP.

2.2 Command Structure

Three new commands have been added to implement SIDF model generation:

a. Frequency selection: The frequencies for which $G(j\omega;\, u_0,\, a)$ will be evaluated are determined by the FREQ command. Four variant forms are:

```
FREQ LIN <ω_min> <ω_max> <nw> [ -ADD ] [ /<fname2> ]

FREQ LOG <ω_min> <ω_max> <nw> [ -ADD ] [ /<fname2> ]

FREQ MAN <ω_1> <ω_2> < ... > [ -ADD ] [ /<fname2> ]

FREQ FILE <fname1> [ -ADD ] [ /<fname2> ]
```

where the notation < . > denotes a numeric value and [.] an optional element. The choices are: LIN ("linear"), yielding *nw* values of ω with equal spacing; LOG, resulting in *nw* values of ω with equal logarithmic spacing, MAN ("manual"), which accepts the user's list of frequencies directly from the command line, and FILE, which uses the frequency list in an existing file <fname1>.F specified by the user. In all cases, the user may supply a filename <fname2> for storing the frequency list; this file can be used in subsequent SIDF model generation. The option -ADD results in *merging* the newly-specified list with the existing frequency list. If either MAN or FILE is selected, or if the -ADD option is used, then the frequencies are organized into ascending order. The combination of these capabilities allow the user to generate a very finely-tuned frequency list.

b. SIDF model generation: Obtaining the SIDF model is controlled by the FRESP command:

```
FRESP [ <T_sim> ] [ <dt> ] [ -PART ]
```

which results in evaluating the I/O characterization or frequency response $G(j\omega;\, u_0,\, a)$. The optional arguments $<T_{sim}>$ and $<dt>$ are a maximum simulation time and suggested integration step, both of which are discussed in Section 2.3 on convergence. The option -PART results in just the first two integrals being evaluated [Eqn. (2), $m = 0$, 1] and the corresponding first harmonic and dc characterization [Eqns. (4, 5)] being obtained; otherwise all integrals [Eqn. (2), $m = 0$, 1, 2, 3] and I/O relations [Eqns. (4-6)] are evaluated. As the frequency response is being generated, the user is informed of progress by messages of the form FRESP PROCESSING $w =$ <value> as each frequency is considered.

c. Error control parameter setting: The user can control convergence (Section 2.3) by changing the error control parameters using the new command FERR. For example, FERR EPSDC <value> sets ϵ_{dc} to the desired value. Other keywords EPSM, EPSPHI, EPSY, NCYC correspond to the symbols in Section 2.3 in the obvious way.

One SIMNON command has been extended to facilitate the use of this software: DISP, or display. New options are DISP FREQY, to show the list of frequencies defined via FREQ; DISP FREQR, to display the frequency response data; and DISP ECPAR, to show the error control parameters.

Many existing SIMNON commands are useful in SIDF model generation. In particular:

a. Parameter setting: The PAR command may be used to set the parameters a and u_0 in Eqn. (1) to the desired values. The default values for these parameters are 1.0 and 0.0, respectively. If the user inadvertently sets a = 0.0, which is meaningless, then a warning message is issued.

b. Integrability: Several commands in SIMNON can be used to ensure that the nonlinear simulation model can be integrated for sinusoidal excitation of the amplitude and frequencies selected: ALGOR and ERROR. The first selects the integration algorithm, and the second sets the integration error control parameter. Further control can be exercised by the specification of $<dt>$ in the command FRESP, which is a suggested integration step size. The issue of plant model integrability is completely in the hands of the user exercising the existing control mechanisms, as it is in the use of any nonlinear system simulation program. We generally recommend using fourth-order Runge Kutta with self-adjusting step size.

The following example illustrates many features outlined above (everything following " is a comment):

```
SYST NA1 SIDFGEN NACON
"  NA1 is a plant model, NACON connects it
"  to SIDFGEN; set a and u0 appropriately:
PAR A: 1.23
PAR U0: 0.0456
"  Define the frequency list:
FREQ LOG .1 10 7
"  Obtain the SIDF model:
FRESP -PART
"  Display the results:
DISP FREQR
```

results in the display

FREQUENCY RESPONSE, A = 1.23000 U0 = 0.04560

ω	Real G	Imag G	y0
0.1000000	1.029311	-0.1029510	0.01929734
0.2154435	1.011241	-0.2178886	0.01970793
0.4641589	0.9382116	-0.4354827	0.02126725
1.000000	0.7308754	-0.7311587	0.02753362
2.154435	0.4335728	-0.9357857	0.04494397
4.641588	0.2131995	-0.9972967	0.09036508
10.00000	0.0950871	-0.9590450	0.01944327

2.3 Convergence

The convergence testing of the Fourier integrals is a major concern in any attempt to automate I/O transfer function model generation. Since we are dealing with nonlinear simulation, where anything can and often does happen, there is no simple or foolproof answer. Therefore, we will only discuss convergence on a simplified level. Further detail may be found in Taylor (1985).

First, we have had to introduce five new error control parameters in order to provide the user with the required control over convergence. These are ϵ_{dc}, ϵ_m, ϵ_ϕ, ϵ_y, and N_{cyc}. The first three of these are error bounds for the dc component, magnitude, and phase, respectively; ϵ_y defines "small" output signal amplitude in the context of the system being studied; and N_{cyc} provides one mechanism to limit the number of cycles of the input sinusoid that will be used. The first two parameters are relative error bounds and thus must be between 0.0 and 1.0; ϵ_ϕ and ϵ_y are absolute parameters and must be in the ranges $0. < \epsilon_\phi < 30.$ degrees, $0. < \epsilon_y$; N_{cyc} must be greater than two.

Somewhat loosely speaking, the Fourier integrals are said to have converged when the d.c. component $y_{0,k}$ and the magnitude and phase of the transfer function G_k, denoted M_k and Φ_k respectively, have converged in the sense that $y_{0,k}$, M_k, and Φ_k evaluated over the previous ((*K*-1)st) period of the simulation and the latest (*K*th) period satisfy the following conditions:

$$err_{dc} = \frac{|y_{0,K} - y_{0,K-1}|}{|y_{0,K}|} < \epsilon_{dc}$$

$$err_m = \frac{|M_K - M_{K-1}|}{|M_K|} < \epsilon_m \tag{7}$$

$$err_\phi = |\Phi_K - \Phi_{K-1}| < \epsilon_\phi$$

where the default values for the error bounds are $\epsilon_{dc} = \epsilon_m = 0.05$ (these are relative error bounds of 5%) and $\epsilon_\phi = 2$ degrees.

In general terms, there are several things that can prevent convergence. Most often, more simulation time is required; one must inspect y_{sig} to determine if that is so, and if so, increase either N_{cyc} or T_{sim}. The first approach changes the number of cycles that will be used before the convergence algorithm gives up (the default is $N_{cyc} = 5$), while the second changes the maximum simulation time (see FRESP above; the default is an arbitrary value of 10 (seconds or whatever unit of time is used)). Since the simulation will stop after N_{cyc} periods or T_{sim} seconds, *whichever is longer*, it is clear that N_{cyc} governs the run at low frequencies, while T_{sim} limits the simulation time at high frequencies. These two simulation control parameters are provided to help ensure that FRESP does not get "stuck" endlessly integrating a difficult system, to minimize the possibility that the user will be compelled to CTRL-Y the program to regain control; therefore, we recommend that these parameters be set carefully.

Another problem is that there may be a very slow mode which keeps the dc component from converging. If the user is not concerned with the dc component (often it is not important), then convergence can be obtained without an increase in simulation time by increasing ϵ_{dc}.

The next most common convergence problem is that the output sinusoidal component of y_{sig} is "small" in one of the following senses and should be checked:

$$|y_{sig}| < \epsilon_y,\ k = K-1,\ K$$

$$\text{or} \tag{8}$$

$$M_k < MG_{max}\epsilon_y,\ k = K-1,\ K$$

where MG_{max} denotes the maximum value of the magnitude of $G(j\omega;\ u_0,\ a)$ for the previous frequencies processed. The first test is used only for the first frequency considered; the second test reflects a better convergence test based on the prior accepted values of $|G|$ rather than on the arbitrary absolute error parameter ϵ_y alone. In both cases, $G(j\omega;\ u_0,\ a)$ is declared to be *undefined* (nothing is written to the frequency response file). In dealing with these situations, the first consideration should be ϵ_y, since this is an arbitrary absolute parameter for the first frequency processed by FRESP. The default value (1.E-6) may not have any meaning; if so, then a change is in order using the FERR command. On the other hand, there may be *no* periodic signal component in y_{sig}. This is especially likely if there are biases and saturations in the model. Then it must determined why this is so and whether or not the frequency response has any meaning under those circumstances.

In conclusion, we reiterate that making this procedure completely foolproof is difficult. In particular, additional care must be taken to handle the dc convergence error correctly when y_0 or G_k are zero or very small in some sense.

3. NONLINEAR CONTROLLER SYNTHESIS PROGRAM (NCSP)

The objective was to create a new module NCSP that interfaces with CLADP (Edmunds, 1979) and supports nonlinear control system design. The discussion that follows does not assume a detailed familiarity with CLADP; the reader is referred to Edmunds (1979) for further information. The basic functions of NCSP are:

a. to accept and display a new system description type (frequency response data);

b. to generate a curve-fitted linear analytic model (in transfer function (ratio-of-polynomial) form, based on minimum mean square error curve fitting (Lin, 1982)) corresponding to frequency response data (Section 3.1);

c. to permit the use of existing CLADP capabilities for designing linear controllers in the frequency domain; and

d. to provide a nonlinear controller synthesis capability based on a pre-existing linear controller design (obtained using the capability c. above or otherwise), a corresponding set of frequency response models generated by SIDFGEN (Section 2) for the controller in cascade with the nonlinear plant for various controller input amplitudes, and automatic nonlinear gain synthesis via SIDF inversion [Section 3.2; Taylor (1984)].

This approach to nonlinear design is best suited for dealing with situations where the amplitude-dependence of the nonlinear plant model is of primary concern, rather than situations where different operating points lead to different linearized models. The NCSD routines we have implemented currently handle only zero operating points; the extension to arbitrary values is obvious but not simple to implement.

Like most frequency-domain design programs, CLADP accepts linear models in either Laplace (ratio-of-polynomials) form or state-space (A, B, C, D) form. We extended the model definitions to include direct frequency response models of the sort generated by FRESP (or taken directly from laboratory measurements). This simply involves reading the frequency response data from a file and storing it for further use.

3.1 System Model Conversion by Curve Fitting

Linear compensators based on the plant frequency response model may be directly designed using the classical frequency domain approaches without further model manipulation, or the user may employ any other linear design approach by using the FIT routine to generate ratio-of-polynomial models that best approximate the FR data in the minimum mean square error sense. Since the original data is in the frequency domain, it is most direct to convert to Laplace form; if the user desires to obtain a state-space model, then the appropriate conversion can be made.

The curve-fitting routine we implemented is based on the algorithm given in Lin (1982). Our tests have shown that this approach works well, as long as there is not a large range of ω in which the slope of the magnitude is not an integral multiple of 20 dB/decade. [This phenomenon can be caused by nonlinear effects, see Taylor (1984); it is particularly troublesome in the context of curve fitting if it occurs at low frequencies.] In such cases, the user can control this problem by the appropriate restriction of the frequency range.

In our implementation, the details of the fitting procedure can be specified in advance by using the following commands: NUM <value> (numerator order), DEN <value> (denominator order), BIAS (toggle bias removal switch on/off), FIXN (toggle switch to fix the numerator s^0 coefficient to be unity) and FREQ (define frequency range for fitting). Ordinarily, the FIT routine provides a model in which the denominator s^0 coefficient is unity; however, that must be changed if the user has a type-one plant model (one having a pole at $s=0$). If the FIT command is issued directly, the user is asked to supply this information.

The user may find it necessary to iterate in obtaining a meaningful fit to the frequency response data. For example, it may be necessary to vary numerator and denominator order (NUM, DEN), to try restricting the range of frequencies used by the fit algorithm (FREQ), etc., until an adequate fit is obtained.

The command LCD has been installed to provide access to the CLADP linear controller design approaches. The use of this capability may result in a linear controller design that provides satisfactory performance, in which case the user has successfully carried out the one-model SIDF-based linear controller design method (Taylor, 1983). If this is not the case, then the systematic approach outlined in that paper moves on to *multi-model nonlinear* controller design by SIDF inversion. We have implemented one of these approaches, as follows:

3.2 Direct Synthesis of Nonlinear Controllers

The basic idea (Taylor, 1984) is to process a set of frequency-domain models for the nonlinear plant in cascade with a linear compensator designed as above and synthesize a *static nonlinearity* so that the forward path of the resulting control system (see Fig. 2) will be as insensitive to input (error signal) amplitude e as possible, where the set of values $\{e_k\ ,\ k =\ 1,\ 2,\ ...\}$ is selected by the user, based on the study of the performance of the linear controller that lead to the decision to seek a nonlinear controller and on anticipated operating conditions.

This procedure involves six steps:

a. select one plant operating regime defined by input amplitude a_0 and generate the $G_0(j\omega;\ a_0)$ model using SIDFGEN, Section 2;

b. design a linear controller of any type denoted $C(j\omega)$ (use CLADP - the primary concern is to achieve good speed of response or bandwidth and transient response); create a SIMNON model of $C(j\omega)$;

c. combine $C(j\omega)$ with the nonlinear plant model and SIDFGEN, drive it with sinusoidal signals of d.c. value e_0 and sinusoidal amplitude e_k, $k\ =\ 1,\ 2,\ ...$ and use SIDFGEN to generate the transfer function models denoted $\{C(j\omega)\ G_k(j\omega;\ e_0,\ e_k)\}$ or, more simply, $\{CG_k\}$;

d. take the set of models $\{CG_k\}$ and for each k determine the precompensating static gain K_k required to force $K_k CG_k)$ to just avoid a specified M-circle, as illustrated in Fig. 3;

e. pass the set of gains $\{K_k(e_k)\}$ to the SIDF inversion routine for nonlinearity synthesis; and

f. validate the nonlinear controller design via simulation.

This approach thus results in designing a *general* compensator [in terms of its dynamic response $C(j\omega)$] with *one* nonlinearity that is automatically synthesized. The nonlinearity to be placed in series with $C(j\omega)$ (in front of $C(j\omega)$ in Fig. 2) is synthesized using a number of existing tools; the new elements are items d. (the M-Circle Algorithm) and e. (the SIDF Inversion Algorithm). The latter routines are described below.

3.2.1 Quasilinear gain evaluation (the M-Circle Algorithm). This technique proceeds as follows: The input for this procedure is frequency response data for CG_k, in k files. The segments that define each curve are processed to determine the gain K_k so that $K_k CG_k$ just touches the M circle, as shown in Fig. 3. The details are omitted for the sake of brevity; note, however, that if the resulting K_k is infinite, the user is so informed and a recommendation that a smaller value of M be tried will be issued if CG_k goes into the half plane $U\ <\ 0$ at all. (If CG_k is always in the right-half-plane (is positive real), then the "gain margin" is infinite, which is surely conservative; generally there is no problem that would necessitate synthesizing a nonlinear controller is such is the case.) As mentioned previously, the output of this procedure is a set of error amplitude-gain pairs $\{e_k,\ K_k\}$ from which a precompensating nonlinearity is synthesized.

3.2.2 Controller Nonlinearity Synthesis by SIDF Inversion. For each value of e_k there exists a linear compensator design that differs only in the gain values K_k. This information serves as the basis for nonlinear controller synthesis, as follows: Given the gain versus amplitude relation of the form $K_k(e_k)$ that is to be achieved by a single nonlinearity, adjust the parameters of a specified piecewise-linear nonlinear function $f_{pwl}\ (e)$ so that the SIDF of f_{pwl} provides the best fit to the gain/amplitude relation $K_k(e_k)$ in the minimum-mean-square-error sense. We defined f_{pwl} in fairly general terms as shown in Fig. 4. The parameters to be adjusted are the slopes, S_1, S_2, the breakpoint, δ, and the step discontinuity, D.

The user is allowed to restrict the nonlinearity by fixing any of the parameters of f_{pwl}; only free parameters will be adjusted for mean square error minimization. For example, fixing

$$S_1\ =\ 0.0,\ \ S_2\ =\ 0.0$$

restricts f_{pwl} to being a relay with deadzone, of arbitrary output level D and deadzone width 2δ. The parameters shown in Fig. 4 allow enough degrees of freedom that most reasonable gain/amplitude relations can be fit with decent accuracy; allowing the designer to restrict this freedom should permit the user to arrive at a practical design.

In summary, denoting the parameters that define $f_{pwl}(e)$ by the vector p,

$$p^T = [S_1, S_2, \delta, D]$$

we developed a routine for evaluating the SIDF of $f_{pwl}(e,p)$, and the parameter adjustment is done using MINPACK (More, 1980). The output of MINPACK is the parameter set p^* for the nonlinearity in the controller, which completes its definition. It should be observed that the gain set $\{K_k(e_k)\}$ must be well-conditioned, in the sense that the values of e_k must cover a reasonable range of e (e.g. $\{e_k\}$ = ({1.2, 3.3, 5.4, 7.5} rather than $\{e_k\}$ = {1.2, 1.35, 1.50, 1.65}); the reason for this is that the SIDF of the nonlinearity f_{pwl} cannot change abruptly with small amplitude changes, so a closely-spaced gain set will usually be meaningless. Also, the user should be aware that quite different nonlinearities may have very similar SIDFs [cf. Gelb (1968), Figs. B.1 and B.2, where the SIDFs for a 3-level quantizer is almost identical to the SIDF of a limiter for all normalized input amplitudes greater than 1.5]. Therefore, the user may have considerable latitude in choosing the nonlinearity to implement.

A second output is provided so that the nonlinearity can be incorporated directly in a simulation of the closed loop system. This is a SIMNON text file <fname>.T, where <fname> is selected by the user, that models the controller nonlinearity; the parameter values are written according to the results obtained by the nonlinearity synthesis procedure. The user is then ready to simulate the plant plus nonlinear controller in either open or closed-loop configuration by supplying the appropriate connecting system.

For further illustration, see Taylor (1984).

4. SUMMARY AND CONCLUSIONS

The primary motivation for the effort described above is to simplify the use of SIDF-based methods for analysis and design of nonlinear systems, and to transition this technology to control engineers in other parts of General Electric. It is the author's hope that this description will allow others to achieve the same objectives.

ACKNOWLEDGEMENT

Kevin Strobel and Ralph Quan, both employed at GE CRD and students at the Rensselaer Polytechnic Institute in 1983 and 1984, contributed a great deal to the success of this effort.

REFERENCES

Edmunds, J. M. (1979). Cambridge linear analysis and design program. *IFAC Symposium on Computer Aided Design of Control Systems*, Zurich.

Elmqvist, H. (1977). SIMNON - An interactive simulation program for non-linear Systems. *Proceedings of Simulation '77*, Montreux.

Lin, P. L., and Y. C. Wu. (1982). Identification of multi-input multi-output linear systems from frequency response data. *Trans. ASME, 104*, March, 1982.

More, J. J., B. S. Garbow, and K. E. Hillstrom. (1980). User Guide for MINPACK-1, Argonne National Laboratory, Report No. ANL 80-74.

Spang, H. A. III. (1982). The federated computer-aided control design system. *Proc. 2nd IFAC Symp. on CAD of Multivariable Technological Systems*, West Lafayette, IN, 121-129.

Taylor, J. H. (1982). Environment and methods for computer-aided control system design for nonlinear plants. *Proc. 2nd IFAC Symp. on CAD of Multivariable Technological Systems*, West Lafayette, IN, 361-367.

Taylor, J. H. (1983). A systematic nonlinear controller design approach based on quasilinear system models. *Proceedings of the American Control Conference*, San Francisco, CA, 141-145.

Taylor, J. H., and K. L. Strobel. (1984). Applications of a nonlinear controller design approach based on quasilinear system models. *Proc. American Control Conference*, San Diego, CA, 817-824.

Taylor, J. H. (1985). Software tools for nonlinear controller design. GE Corporate Research and Development Report No. 85CRDnnn.

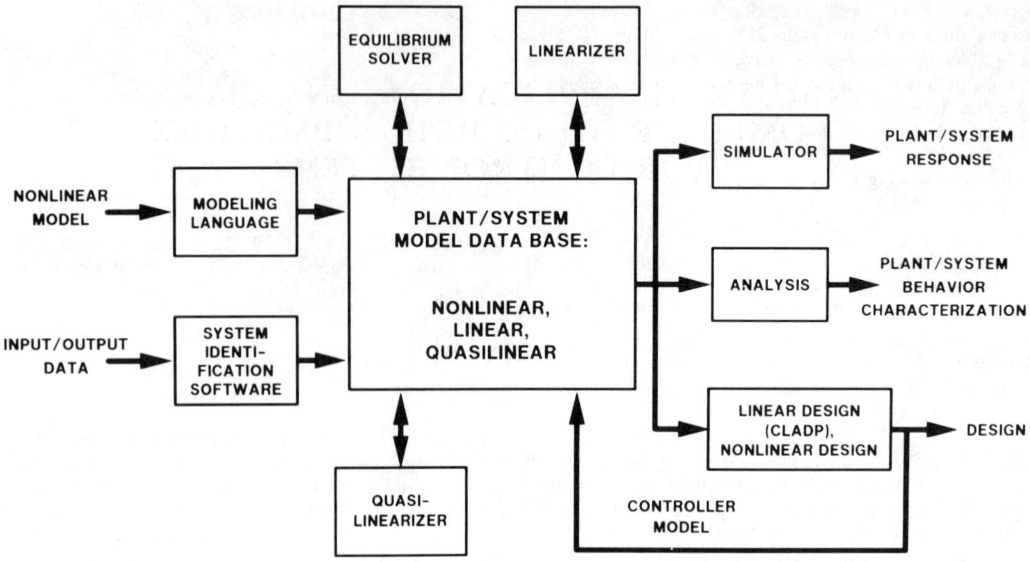

Fig. 1. CAD Environment for Nonlinear System Design.

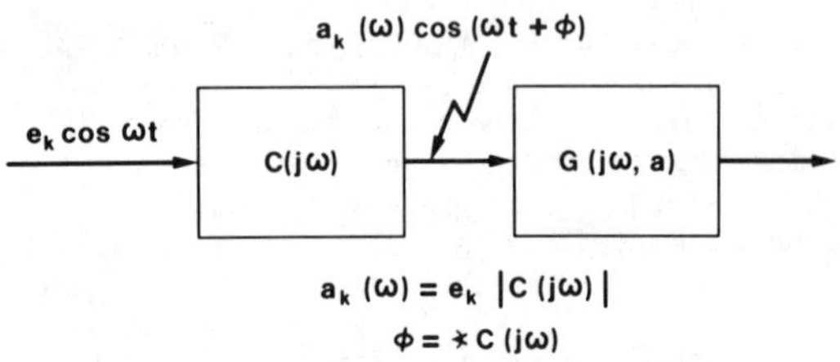

Fig. 2. Compensated Forward Path

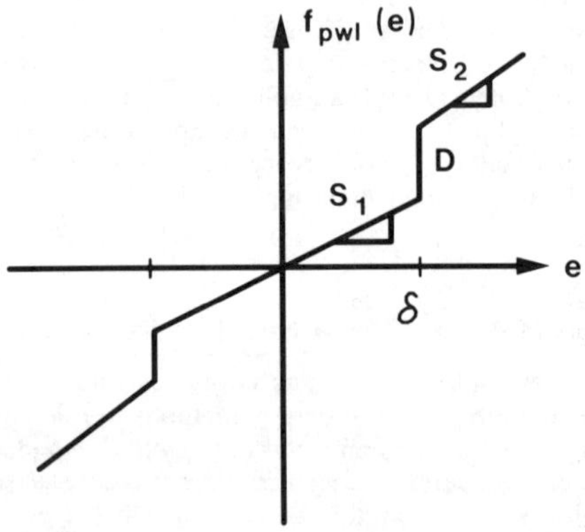

Fig. 4. Piecewise-Linear Compensator Nonlinearity

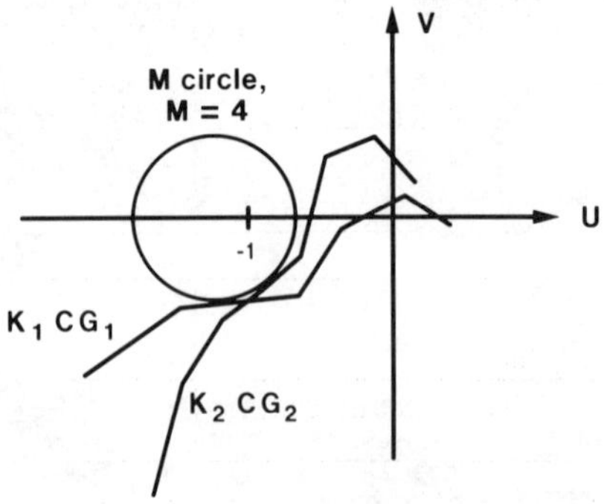

Fig. 3. Illustration of $K_k(e_k)$ Generation

INDUSTRIAL APPLICATIONS IN JAPAN OF COMPUTER AIDED DESIGN PACKAGES FOR CONTROL SYSTEMS

M. Araki

Department of Electrical Engineering 11, Kyoto University Yoshida, Sakyo-ku, Kyoto 606, Japan

Abstract The present status of use of CAD packages for control systems in Japanese industries is surveyed based on two questionnaires done in 1983 and 1984. Lists of CAD packages, statistics about applications, and criticisms and expectations of engineers at industrial practice are reported. Some of recent researches are also introduced.

Keywords CAD, Control System Design, Japanese Industries.

§1. Introduction

The purpose of this report is to make a survey of the present status of use of CAD packages for control systems in Japanese industries. The contents are based on two questionnaires :

QI : Investigation of Use of CAD Packages for Control Systems--done in 1984 by the author

QII : Enquete on Use of Modern Control Theory in Industrial Practice--done in 1983 by the 2nd subcommittee of the Committee for Industrial Problems of SICE (Society of Instrument and Control Engineers)

The outlines of the questionnaires are in Appendix.

The report consists of 6 sections. In §2, CAD packages which appeared in QI will be summarized. In §3, how the packages are used will be investigated. In §4 we proceed to study what are the probelms with the present packages, and what are expected for the future packages. In §5 some of recent researches will be picked up. §6 is for concluding remarks.

§2. CAD packages Used in Japanese Industries

In this section, we summarize the CAD packages which appeared in QI. We include simulators in the lists, too. Packages are classified into 3 categories :

Cat.1 Packages introduced from some company, organization, or researchers outside the users' company. (Table 1)

Cat.2 Packages made in the user's sections. (Table 2)

Table 1. CAD Packages of Cat.1 (QI)

Name	Controlled Object	Main Functions	Frequency of Use	Date
a. Packages for Mini & Large Computers				
ACSL		Simulation	a,b	
CONPAC-Group				
DPACS	Multivariable	Control System Design (State Space)	b×3,c	1977
DPACS-F [1]	Multivariable	Control System Design (Freq. Domain & State Space)	a×2	1977
PIPAC [2]	Multivariable	Identification (AR, ARMA, and other)	b	1978
CONPAC [3,4]	Multivariable	Control System Design & Identification	a,b×5,c,d	1978
CSMP [20]		Simulation	a×5,b,c	
CSMP III [20]		Simulation	a,c	
CSPL		Simulation	a	
DDS		Simulation		1964
DDS-OPT [5.21]		Simulation & Parameter Optimization	a	1970
IDCOM	Multivariable	Identification (Impulse Response) and Control System Design (IDCOM-logic)	b	
Linear Systems Analysis [22]	Single Variable	Control System Analysis (with Root Loci)	c×2	
RANDAP	Single Variable	Random Data Analysis (AR Model)	c	
RANDAP II [23]	Multivariable	Random Data Analysis (AR Model)	b×2	1975
SLCS IV [24]		Simulation	a×3	1979
TIMSAC [6]	Multivariable	Identification (AR Model) & Control System Design (Optimal)	d	1972

b. Packages for Personal Computer				
MCSP [7]		Simulation	b	
* [8]	Single Variable	Control System Design (Freq. Domain & Optimal)	c	1982
*	Single and 2 Variable	Control System Design (Freq. Domain, State Space & Adaptive)	b,c	1979

Table 2. CAD Packages of Cat.2 (QI)

Name	Controlled Object	Main Functions	Frequency of Use	Date
a. Packages for Mini & Large Computers				
BACS II [9]		Simulation	a×2,b×2	1969
CASE-CTL [19]	Multivariable	Identification and Simulation	a	1980
CONSAP		Control System Analysis (with Nonlinear Simulation)	a	1983
CSDP [10]	Multivariable	Control System Design (Freq. Domain & Optimal)	b	1978
ISAC [11]		Control System Analysis (with Nonlinear Simulation Emphasis is on Discrete-Continuous Complex Systems)	a	1983
LACE [12]		Control System Analysis	a,b×2	1984
MARS [13]	Multivariable	Control System Design (includes DPACS), Identification, Special Purpose Simulation & μ-Processor Program Generator		1984
MCSL		Simulation	a	1974
MIMOS [14]	Multivariable	Control System Design (State Space)	a	1984
SAPL [17,25]	Multivariable	Control System Design (Freq. Domain & Optimal) & Identification (AR Model)	a	1983
*-I [18]	Multivariable	Control System Design (Freq. Domain & Transfer Fn.)	a	1984
*-II	Multivariable	Control System Design (State Space)	a	1984
*		Simulation	a×2,b	
*		Simulation of Nonstationary Large-Scale Systems	a×3,d	
*		Simulation of Nonstationary Small-Scale Systems	a×2	
* [15]	Single Variable	Control System Design (Freq. Domain)	c	
*	Multivariable	Control System Design (State Space) & Identification	d×2	
* [15]	Multivariable	Identification	c	
*	Multivariable	Identification	b,c	
b. Packages for Personal Computer				
PRODECOS [16]	3 Variable	Control System Design (Freq. Domain & State Space, Nonlinear Simulation)	a	1984

Cat.3 Packages made in the users' section and used in a single or a few sections.(Table 3)

It should be noted that distinction between Cat.1 and Cat.2 is not directly related to whether the package is on sale or how consolidated it is. Actually, some of packages of Cat.1 are not on the market while some in Cat.2 are ready for sale. Distinction between Cat.2 and Cat.3 was subtle but made based on "How widely used?", "Published or not?", "Ready for realease?", etc. All special purpose (from control viewpoint) packages were put in Cat.3. There are several packages which could not be included in the lists because of lack of information or restrictions by contracts. The packages are arranged in alphabetical order in each table.

Table 1 includes the packages of Cat.1, where the languages and symbols in each column means the following :

Name : "*" is filled in when the package is not named or the name is not released.

Controlled Object : "Single Variable" or "Multivariable".

Main Functions : "Control System Analysis"={Linear simulation and display of frequency responses. Some packages include nonlinear simulation and/or display of root loci}; "Control System Design"={Analysis functions are included, too. As for design, possible methods are indicated in parenthesis};"Freq. Domain"={Design by vector loci, Bode diagrams, Nichols chart and/or root loci, and parameter tunning of PID controller in the single variable case. The INA & DNA methods are included in addition in the multivariable case.};"State Space"={Pole assignment, optimal control with respect to the infinite-time quadratic-integral criterion, and observer design. Sometimes, decoupling by state feedback is included};"Optimal"={Optimal control with respect to the infinite-time quadratic-integral criterion};"Transfer Function"={Model matching and similar methods}
(Note that the above shows the standard menu and there are several exceptional cases.)

Frequency of Use : "a"=very frequently, "b"=once a month on the average, "c"=several times a year, "d"=less
For instance, "a×3,b" means there were 3 sections who use the package very frequently, and 1 section who uses the package once a month on the average.

Date: It was often difficult to know (or difine) when a package was born. But it seems important to give some indication about ages. So, date of birth was *estimated* from various sources. The readers are requested not to give too much credit to these dates.

Table 3 CAD Packages of Cat.3 (QI)

Functions	No. of Packages	Frequency of Use
Simulation Special purpose	15	a×11,b×4
General purpose	4	a×2,b×2
Control System Analysis	2	a,c
Control System Design		
Single Variable, Freq. Domain	5	a×2,b,c×4
Single Variable, Transfer Fn.	1	b
Multivariable, State Space	2	a,b
Multivariable, Optimal	1	a
Multivariable, Freq.	2	b,d
Identification	2	c,d

As for functions of the packages, distinction between "Control System Analysis" and "Control System Design (Freq. Domain)" was sometimes difficult. We followed the emphasis found in the pamphlet or other documents. The relation of the packages of "CONPAK-group" is as follows. DPACS is the first version and includes basic design functions in state space. DPACS-F is the 2nd version including frequency-domain methods and other utility functions. Both of them were developed on a minicomputer. PIPAC is the identification part of CONPAC, which is for large computer and has functions "DPACS-F + Identification". Concerning PIPAC, note that the programs introduced in [2] is the older version. RANDAP uses some routines of TIMSAC and RANDAP II is the multivariable version.

Table 2 includes the packages of Cat.2. The languages and symbols mean the same as in Table 1.

Table 3 is the statistics of the packages of Cat.3 Meanings of the words in the "Functions" column are similar to those of Table 1. But, naturally, most of the packages of this category are somewhat specialized and covers a little narrower range than those in Tables 1 & 2. One of the "Control System Design (Freq. Domain;Single Variable)" packages includes the function of making a linear model based on the results of nonlinear simulation supplied by another simulation packages. Several of special purpose simulators are equipped with parameter optimization and documentation functions. References [27-30] and about packages of Cat.3.

§3. Present Status of Usage

First, let us investigate popularity and actual utility of each function included in CAD packages. In Tables 4 & 5, statistics were taken from QI about sections who own packages with a certain function and about applications in which a certain function is actually used. These statistics, together with the respondents' comments, could be summarized as follows :

(i) 2 thirds - 3 fourth of sections have "Freq. Domain" and "State Space Design" packages, and approximately half own "Indentification". These numbers seem to reflect the "need + expectation" of engineers at industrial practice.

(ii) "Simulation" has been the major tool of design from early periods and this tendency will continue. Main reasons of heacy dependence on simulation are : (a) complicated (nonlinear, hybrid, nonstationary, etc.) nature of the actual objects, (b) ease of communication with engineers in other fields, and (c) that, once engineers have got experienced with the object, simulation becomes much straight forward design method.

(iii) "Freq. Domain Design" packages started as private programs in sections. They were improved or renewed according with replacement of computers, and are now kept and frequently used at majority of sections who specialize in design of control systems.

(iv) Rapid introduction of "Modern Control Theoretical Design" (i.e. "Transfer Fn. Design", "State Space" and "INA & DNA") packages were promoted by the development of CONPAC (actually, half of those packages reported in QI are of CONPAC-Group). Now, they are used by wide spectrum of engineers while complaints about manuals and user friendliness were mentioned.

(v) As for "Identification", TIMSAC seems to have played an important roll in popularization of the statistical method. It is now well-known to and actually practiced by engineers, especially those who are dealing with thermal, chemical, or cement processes.

(vi) In several industries, packages which cover as wide range of analysis/design functions as CONPAC and, at the same time, fit better to their own requirements are being (or have been) developed. A typical way is to combine existing design packages (e.q. CONPAC) with other programs such as special purpose simulators, documentation routines and program generators.

Next, let us study in what situation and in what form the "Advanced" methods are used. In Tables 6-11, data from Part 1 of QII are shown. Here, "Advanced" methods include any methods other than PID and phase-lead-phase-lag compensators. About the statistics of Tables 6-10, we can point out the following :

(vii) Although "Pure Delay" is the top reason for the difficulty (Table 6), such modern methods for time-delay systems as the process-model control [31], finite pole assignment [32] and algebraic

Table 4 (QI)

Which Functions Are Popular? -- Percentage of Sections Who Own Packages with a Certain Function

Total = 43 sections

Function	%
Simulation Special P.	30%
General P.	79
Control System Analysis/Design	
Freq. Domain & PID Tuning (Single V.)	74
Transfer Fn. Design	40
State Space	72
INA & DNA	40
Smith Controller	2
Identification	56

Table 5 (QI)

Which Functions Are Actually Used? -- Number of Applications in Which a Certain Function Was Used

Total = 105 cases

Function	Cases
Only Simulation	48 cases
Analysis/Design Methods	
Freq. Domain & PID Tuning (Single V.)	36
Transfer Fn. Design	1
State Space	16
INA & DNA	4
Model Reduction	1
Identification	13

Table 6 (QII)

What Was the Main Problem of Control

Pure Delay	24%
Disturbance	21
Interaction	17
Response	16
Stability	11
Others	9

Table 7 (QII)

What Was the Purpose of Introducing Advanced Methods

Quality	34%
Energy	24
Safety	15
Automatization	11
Efficiency	10
Others	5

Table 8 (QII)

Which Size of Computer Was Used for Analysis & Design

Large	51%
Mini	22
Personal	20
Others	7

Table 9 (QII)

Which Type of Controller Was Used

DDC	42%
Process Computer	39
General Purpose µ-Processor	10
Others	9

Table 10 (QII)

Which Method Was Used for Modeling

AR or ARMA	45%
Theoretical Empirical	35
Parameter Estimation	17
GMDH	3
Transfer Fn.	52%
State Space	48

design method [33] did not appear in the list of the tools for solution(Table 11). This may be partially because the sampled-data control is believed to solve pure-delay problems, and partially because the above methods are still "too modern" to be acknowledged as usable methods at industrial practice

(viii) Tables 8 and 9 say that general purpose µ-computers are not used very often either for analysis/design nor for implementation on the average. (But the situation seems to be changing rapidly. Refer to the comment at the end of the section and also to §5).

(ix) Table 10 indicates popularity of AR and ARMA models.

Table 11 is a statistics similar to Table 5. The difference exists in that Table 5 was taken from QI which asked about "application of CAD packages" whereas Table 11 is from QII which asked about the applied method directly. From the similarity and difference of the two tables we can find the following :

(x) "Optimal" and "Kalman Filter" are the most useful among the methods included in the standard menu of CAD packages, and "Improved PID" and "Observer" follow as the second group.†

(xi) But there are several other methods which are not included in the standard men but are used more, or at least comparatively, often as "Optimal". Namely, "Adaptive, Learning & Optimalizing Control" is applied most frequently among all, and "Feedforward" and "Docoupling"†† are considerably.

The above suggests that development of convenient tolls for the design of "Adaptive", "Feedforward" and "Decoupling" controllers is neccessary.

Lastly, let us pick up the respondents' comments which are concerned with the present status of usage of CAD packages and have some generality. Concerning simulation :

By any means the control systems are tested by simulation at the final stage.(QI)

Simulators which can accept block diagram description of systems are preferred.(QI)

To use simulation for the determination of size and frequency of the inputs for identification was the main reason of success.(QII)

Concerning methods of modeling :

Modeling and model reduction are both conducted based on physical consideration.(QI)

Parameter estimation was carrid out based on the data from physical/chemical experiments.(QII)

#Refering to the results of theoretical analysis, plan of experiments using actual plants was made, and then a practical model was constructed based on experimental data.(QII)

Concerning methods of control or design :

Advanced methods which are actually applied are "improved PI", "feedforward", and "model analysis".(QII).

Most of the processes can be controlled satisfactory by improved PID.(QII)

Concerning the main reasons of success :

Enough data for modeling/indentification were available.(many in QII)

#Use of theoretical models.(QII)

#Repetition of simulation combined with other methods.(many in QII)

Concerning other items :

Personal computers seem to be frequently used in fields for analysis and re-adjustment. Handy and powerful software is expected.(QI)

#Education is not enough to make application of advanced methods possible.(QII)

§4. Advantages, Problems and Expectations

In this section let us investigate advantages and problems of present packages, and needs and expectation for future packages.

Table 11 (QII)

Which Method of Control or Design Was Applied

Adaptive, Learning & Optimalizing	13 cases
Optimal	12
Kalman Filter	10
Feedforward	9
Decoupling	8
Improved PID	7
Observer	4
Selector Control	3
INA	2
Pole Assignment	1
Smith	1
Fuzzy Control	1

† In several cases of answering "Kalman Filter", the respondent seems to have meant some parameter estimation method in the modeling stage. So, it was not combined with "Observer".

†† Most of the 8 cases shown in Table 11 are the feedforward-type decoupling associated with INA/DNA method though there are some cases which may imply the decoupling by state feedback. The design method of the former does not usually appear in the menu.

As for advantages, most of the respondents who actually applied their CAD packages answered that packages were very useful, while some answered that they were useful but not enough (QI). Several mentioned that general application of modern control theory was first enabled by the introduction of CONPAC (QI).

As for problems, there were many complaints about user friendliness, as usually the case with any package software. Most of the opinions were concerned with the way of data input, and preference for inputting in the form of block diagrams (graphical) was mentioned by several respondents. Concerning the organization of packages two opposite types of preferences were shown. Namely, some prefer that "input of block diagram" - "conversion to state-space description" - "model reduction" - "control system design" - "simulation" - "indication for implementation" are organized as a series of tasks and a user can automatically reach the destination without considering too much about what is the next step. But some other like that each function (such as solution of Ricatti eq.) is prepared as a separate module and a user can use them in any combination. These requirements are opposite but not contradictory, and it would not be difficult to meet the both by one package. Some mentioned that they want "to compute transfer zeros", "to treat models including algebraic relations", etc. These requirements could be solved by adding small modules to the present packages.

Next, let us proceed to such problems which lead to the needs and expectations for future packages. There were many comments about the problems of nonstationarity, nonlinearity, pure delays and distributed parameter systems (QI & QII). These factors are said to be causing serious difficulties in all stages of identification, simulation, and design, and more powerful algorithms and usable identification, control and design methods are expected. Next notable suggestions are those about difficulty of choosing a control method and those about on-line computation. Concerning the first point, one of the problems with modern control theory is that too many possibilities exist and comparison can be done, at present, only after carrying out all of them, which requires much time and knowledge. Importance of a guide about the choice and limitation of methods was emphasized and application of knowledge engineering was suggested (QI). Concerning the second point, expectation for on-line identification, on-line computation of the optimal gain and on-line analysis in the case of sudden (unexpected) disturbance were mentioned (QI & QII). Other than the above, the following were expected.

#Design packages which can deal with larger (his present package: state variables ≤ 40) systems.

Faster personal computer.

Table 12. Expectation for Future Packages (QI)

A: More advanced functions are needed.
B: More user friendly packages are expected.
C: Present packages are enough.
D: Even present packages are not used often.
E: Other (unnecessary, not considered, etc.)

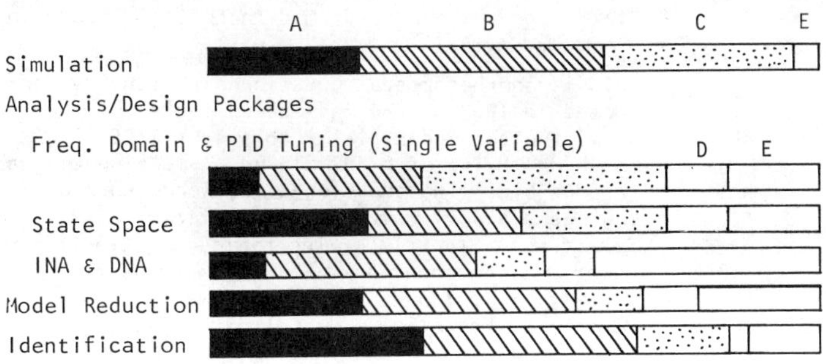

In Table 12, statistics about needs and expectation for improved versions of CAD packages is shown (QI). Table 13 shows the top ten items which was nominated as "most accessible (i.e. easiest to apply)" by engineers at industrial practice, and Table 14 lists the same of "most prospective" (QII). To be exact, Table 13 was made based on answers to the question : "whether you can apply the following items 'immediately', 'after studying a little', or 'after studying a lot'." Table 14 were made from the answers to the question : "whether the following items are thought 'effective and practical', 'may be effective but not practical', or 'not either'."

Table 13 (QII)

Most Accessible Items
1. Feedforward
2. Response Simulation
3. Transfer Fn. Model
4. Improved PID
5. Analysis of Model & Simulation
6. Smith Controller
7. Stability Criterion
8. Selector Control
9. Decoupling
10. Optimalizing Control

Table 14 (QII)

Most Prospective Items
1. Analysis of Model & Simulation
2. Feedforward
3. Response Simulation
4. Transfer Fn. Model
5. Improved PID
6. Decoupling
7. Adaptive, Auto-tuning & Learning
8. Theoretical/Empirical Model
9. Optimalizing Control
10. Statistical Model

§5. Researches

Let us briefly look at some of the recent researches about CAD packages in Japan.

We already introduced that, in industries, research and development to combine a CAD package with other programs such as a μ-processor program generator and to make an "all-through" package is in progress. The "computer aided control" system CACOS, which was developed in Tokyo Institute of Technology, has gone a step further [34]. This package was produced based on the understanding that realization of a control system includes 4 essential steps: modelling, controller design, implementation for experiments, and re-adjustment of parameters (and, sometimes, of the control scheme itself). CACOS is an "all-through, on-line" package which enables a mini-computer to carry out all the above 4 steps on line. It includes DPACS (together with the function of PIPAC) as the core, but carries out the discrete-time design only, and has ability to cope with nonlinearities by modification.

Other class of researches are about CAD packages on personal computers. In addition to publication/release of PRODECOS [16] and DTCS/P [39], which cover state space and other design functions, researches are reported on improvement of the input method [36], on a package for an improved DNA and pseudo-diagonalization method [38], on packages for root-loci [37] and root-loci bands [40], and on the educational use [39].

As researches which are potential to offer powerful tools for future packages, we should cite the application of the symbolic and algebraic manipulation. At present, researches are limitted to the use of REDUCE, which maniputates algebraic equations. The present author and colleague applied it to the automatic linearization of nonlinear systems [41], Senda et al. did the similar for the analysis and design of robots [42], and Takahashi et al. used it for the analysis of linear systems [43].

§6. Conclusion

We can summarize and conclude as follows :

(a) As for "Single Variable, Freq. Domain Analysis/Design & PID Tuning", 74% of sections of industries already own some packages, and it could be said that problems are almost done except improvement of user-friendliness.

(b) As for "State Space Analysis/Design", 72% of sections own some package, but still more advanced functions are desired.

(c) As for "INA & DNA", evaluation was splitted.

(d) As immediate needs, development of packages for "Adaptive, Auto-tuning & Learning", "Feedforward", "Decoupling" and "Optimalizing" controls are expected. Expecially for "Adaptive, Auto-tuning & Learning" and "Optimalizing", computer-aided-control systems (i.e. on-line design & control systems like CACOS) rather than mere CAD systems will be strongly required.

(e) Practical indication about utility and limitation of various modern control theoretic methods and a guide for the choice of them are also required.

(f) Researches on algorithm to simulate nonlinear systems with pure-delays, and on design/control methods for nonstationary, nonlinear, and/or time-delay systems are strongly expected.

(g) At present, development of all-through on-line packages, use of personal computers for design, and application of symbolic and algebraic manipulation are going on.

References

[1] K.Furuta & H.Kajiwara (1979), J. SICE (Society of Instrument and Control Engineers), Vol.18, pp.777-786
[2] K.Furuta et al. (1981), Automatica, Vol.17 pp.755-762
[3] K.Furuta et al. (1980), IFAC Computer Aided Design of Control Systems (edit. M.A.Cuenod), pp.267-272, Pergamon
[4] K.Furuta et al. (1982), IFAC Theory and Application of Digital Control (edit. A.K.Mahalanbis), pp.527-532, Pergamon
[5] T.Mitsumaki et al. (1972), J. SICE, Vol.11, pp.455-467
[6] H.Akaike & T.Nakagawa (1972), Statistical Analysis and Control of Dynamic Systems, Tokyo: Science Ltd.
[7] T.Matsuzaki (1982), Interface, Vol.8, pp.183-197 ; (1983), Information Science, Vol.2, pp.49-54
[8] T.Hara & F.Kobayashi (1982), J. JAACE (Japan Association of Automatic Control), Vol.26, No.6 Separate Volume, pp.44-47
[9] T.Matsuura & M.Toyama (1969), Trans. Inform. Process. Society of Japan, Vol.10, pp.216-226
[10] T.Yamauchi et al. (1980), Prep. 24th JAACE Conference, pp.101-102
[11] M.Tanuma et al. (1984), Prep. IFAC 9th World Congress, Vol.VIII, pp.208-213
[12] T.Matsuura & M.Kataoka (1984), Trans. Inform. Process. Society of Japan, Vol.25, pp.85-91
[13] PET-B of Nippon Steel Corp. (1984), Rep. No.88-9-1, Committee for Measurement, Iron & Steel Inst. Japan,
[14] M.Kataoka & N.Terai (1983), 27th National Cof. of Inform. Society of Japan, pp.1509-1510
[15] H.Kurihara et al. (1974), J. Japan Petrol. Inst., Vol.17, pp.292-297
[16] H.Kimura et al. (1985), CAD Programing for Control Systems on PC9801, Nikkan-Kogyo
[17] Y.Nishikawa et al. (1984), Automatica, Vol.20, pp.321-332
[18] M.Funabashi (1981), Prep. Tokyo Branch Conf. of IEEJ
[19] T.Shigemasa et al. (1985), Toshiba Review, to appear
[20] CSMP-III Manuals (SH19-7001-2, GH18-6013-1, GH18-6014), IBM (1974)
[21] DDS-OPT Manual, Central Research Lab. Hitachi Ltd. (1970)
[22] Linear Systems Analysis-Manual, Hewlett-Packard Desktop Computer Division (1980)
[23] RANDAP-II Manual (Research Rep. RC-SC30 Data Computation Subcommittee), Japan Society of Mechanical Engineers (1976)
[24] SLCS IV Manuals (70AR-0131, 70AR-0132), Fujitsu Ltd. (1979)
[25] SAPL Manual, Fuji ElectricCo. (1983)
[26] M.Shima et al. (1975), Tech. Rep. Mitsubishi Heavy Industries Ltd., Vol.12, No.6
[27] T.Fujiwara et al. (1981), Tech. Rep. Mitsubishi Heavy Industries Ltd., Vol.18, No.2
[28] R.Tagawa (1981), Prep. IFAC 8th World Congress Vol.III, pp.125-130
[29] Y.Kurosaki et al. (1984), Tech. Rep. Kawasaki Heavy Industries, Vol.84, pp.19-26
[30] N.Nakatsuchi et al. (1984), Prep. 28th JAACE Conf., pp.151-152
[31] K.Watanabe & M.Ito (1984), J. JAACE, Vol.28 (Special Issue on Control of Time-Delay Systems) pp.269-277
[32] K.Watanabe (1984), ibid., pp.309-317
[33] Y.Yamamoto (1984), ibid., pp.299-308
[34] K.Furuta & O.Yamano (1984), Prep. IFAC 9th World Congress, Vol.VIII, pp.173-178
[35] M.Kaneda & H.Kuraoka (1984), Proc. IECON'84, pp.237-242
[36] T.Kawabata et al. (1983), Prep. 27th JAACE Conf., pp.251-252
[37] K.Takeuchi et al. (1983), Prep. 26th Joint Automatic Control Conf. of Japan, pp.71-72
[38] B.Kondo et al. (1984), Prep. 4th Simulation Technology Conf., pp/13-16
[39] Y.Matsuo (1984), Prep. 22th SICE Conf., pp.387-388
[40] Y.Okuyama (1984), Prep. 27th Joint Automatic Control Conf. of Japan, pp.5-6
[41] M.Araki & H.Yukawa (1984), Prep. 28th JAACE Conf., pp.229-230
[42] Y.Senda et al. (1984), Prep. 27th Joint Automatic Control Conf. of Japan, pp.391-392
[43] T.Takahashi (1984), ibid., pp.253-254
[44] I.Hashimoto & T.Takamatsu (1981), Proc. CPC-II, pp.148-185

Appendix Outline of Questionnaire Inquiries

QI : "Investigation of Use of CAD Packages for Control Systems"

This is a postal questionnaire inquiry conducted in 1984 by the author. We sent 3 forms (Form 1: General status in the company, Form 2: Status in each section, Form 3: Applied Examples) to one person of one company and asked him to distributed them to related sections, branches, etc. 18 companies out of 24 answered, the fields of which are: Electric & Elctronic = 6, Engineering = 2, Heavy Industry = 3, Iron and Steel = 5, and Oil Refinery = 2.

QII : "Enquete on Use of Modern Control Theory in Industrial Practice"

This is another postal questionnaire inquiry conducted in 1983 by the 2nd Sub-committee of the Committee for Industrial Problems of SICE, for which Mr. M.Ito of Mitui Mining & Smelting Co. was responsible. The questionnaires were sent to 92 engineers and 47 answered, whose fields are Cement = 1, Electric & Electronic = 6, Electric Power = 2, Engineering = 5, Food = 1, Heavy Industry = 3, Iron & Steel = 16, Medicine = 1 Paper = 2, Petrolium = 2, Petro-chemistry = 6 Photo Film = 1, Water Supply = 1

INDUSTRIAL APPLICATIONS IN THE NETHERLANDS OF COMPUTER-AIDED DESIGN PACKAGES FOR CONTROL SYSTEMS: TRENDS AND PRACTICE

P. P. J. van den Bosch* and A. J. W. van den Boom**

**Delft University of Technology, Department of Electrical Engineering, PO Box 5031, 2600 GA Delft, The Netherlands*
***Eindhoven University of Technology, Department of Electrical Engineering, PO Box 513, 5600 MB Eindhoven, The Netherlands*

Abstract: In this paper we will describe and illustrate some applications of CACSD packages in Dutch industry. Some typical usages of CACSD in industry will be illustrated. Further we will analyse the results of an inquiry with respect to the use and need for CACSD in industry and universities which was held in 1982/1983 in the Benelux and we will resume the activities of a working group in the Benelux which aims to promote CACSD.

Keywords: Computer-aided design, Computer-aided control system design, Computer control, Control engineering computer applications, Identification, System Analysis.

1. INTRODUCTION

For some time mathematical theories for the analysis and design of linear systems have been in existence. For linear multivariable systems in particular, a large number of reliable algorithms has been developed and derived from theory. In spite of the availability of these reliable algorithms, existing computer programs often do not satisfy requirements such as a standardized computer language (Fortran), numerical robustness, portability and documentation.
Based on this undesirable situation the Working Group on Software (WGS) was appointed in 1981 in The Netherlands to investigate present applications and future requirements of Computer-Aided Design of Control Systems (CACSD).
In 1982 WGS made an inquiry among a large number of Dutch private enterprises, universities and governmental research institutes with respect to their present hardware and software and their need or desire for software (Working Group on Software, 1982).
A similar inquiry was held in Belgium in 1983 which resulted in the same conclusions (Working Group on Software, 1983a). Since 1983 the WGS has been a Benelux initiative.

In this paper we will discuss the results of this inquiry. Moreover, we will illustrate the use of CACSD programs in industry with a number of typical examples.

2. RESULTS OF THE INQUIRY

The majority of the respondents have their own digital computer installations. The relative number of analog computers or their modern successors, the parallel digital computer, has been reduced considerably and are only used for very time-consuming simulation studies or relatively simple laboratory set-ups for simulation and/or real-time control. The type of digital computer varies considerably, ranging from 8-bit microcomputers to time-shared mainframes. Superminis and mainframes in general support the mathematically and/or control-theory-oriented subroutine libraries such as NAG, IMSL, EISPACK, LINPACK and ORACLS.
The use of microcomputers, especially the 16-bit ones, in CACSD applications is, in 1982, not yet impressive. In spite of the availability of many new programming languages, such as Pascal, Ada and Modula-2, Fortran is still widely used and accepted for technical and scientific work. The majority of subroutine libraries for control system design are Fortran oriented. Consequently, Fortran is still the major programming language for CACSD.

The most important conclusion of this inquiry was that good, reliable and portable basic software is needed more than large packages for solving general control problems, since the demands of the individual users differ too much.

There is a small number of generally-accepted interactive design packages in use in industry in The Netherlands, such as TUTSIM, PSI, TRIP, SATER and PRIMAL. These packages are discussed in the Appendix.

Many companies have developed their own programs for their own applications, where the program is usually strongly application oriented and therefore not suitable for more general use. One reason they develop their own programs is that the appropriate software is not available. Another reason is that a dedicated program yields a better performance and/or better satisfies the stated requirements than generally-available software. Still, there was a great need for a set of standardized, reliable and accurate subroutines for solving mathematical and/or control-oriented problems.

This more or less describes the situation as it was in 1982/83. Recently, other programs are being developed for use in industry, for example CONTROL-C and MATRIXx.

3. CLASSIFICATION OF INDUSTRIAL USE OF CACSD

In order to be able to classify the use of CACSD programs in industry it is valuable to distinguish the use of subroutine libraries, analysis and design programs for linear systems, simulation programs and real-time programs for the following three situations:

1-Research centers
Large enterprises, for example Shell and Philips, have at their central research centers their own groups

dealing with control theory. These groups are interested in and use advanced topics of modern control theory. These groups have the knowledge and potential to understand modern developments in both theory and numerical analysis. They have facilities to develop their own CACSD programs or adjust available programs to their present needs. At a lower level in the organization smaller groups deal with practical control problems. If there are frequent and intensive contacts with the central reseach centers, modern control theories can be realized in applications, e.g. optimal control and identification. If these contacts are few or even non-existent such a local control group can be considered as a group of one of the two following categories.

2-Control groups
Smaller enterprises cannot afford too much theoretical research. Their control groups are involved more with the production process. These groups have to deal with many practical problems and are less open to advanced theories. Their analysis and design is application oriented. They will preferably use interactive programs for linear systems and for simulation. Owing to the many essential non-linearities in a process, simulation is their main analysis and design tool.

3-Small enterprises
Small enterprises have only a very small group for control-oriented problems, often only one person, mostly for instrumentation. In such a group there is not sufficient knowledge of control theory to be able to use existing analysis or design programs. Design is mainly based on intuition, rules of thumb and simulation.

It turns out that the first category (research center) is mainly interested in reliable, well-tested and documented subroutine libraries and interactive programs that satisfy high standards with respect to facilities, flexibility and numerical reliability. The second category has confidence in interactive, analysis and design programs for linear systems and simulation programs. The third category only considers interactive simulation programs.

In all three categories some applications require an interactive, real-time program for on-line measurements. These measurements could be processed later by appropriate off-line analysis programs.

We can combine these findings in a table to illustrate which CACSD facilities were being used in which groups in 1982/83.

Table 1. CACSD facilities according to type of enterprise in 1982/83.

	research center	control group	small enterpr.
Subroutine library	+	-/0	-
Linear analysis and design programs	+	+	0/-
Simulation programs	+	+	+
Real-time programs	0	0	0/-

\+ required and used
0 used sometimes
\- generally not used

As indicated in table 1, simulation is the most generallly-accepted tool for computer-aided control system analysis and design.

The above sketched situation of different kinds of industrial control groups would not be complete without the role and position of university control groups, and especially their relation to industrial control groups being taken into consideration. Traditionally the universities and consequently their control groups (which are mainly located in the Universities of Technology of Delft, Eindhoven and Twente and in the University of Groningen) have been financed directly by the government. Financing through research contracts with industry has been limited. Contacts with industry existed, of course, but has not played a dominant role.

Because of the cutbacks in direct state financing of universities and the consequent stimulation of contract research, many more contacts with industry have been established; especially those projects with expected short- term results and short-term applicability in industry were favorites. It even became possible to obtain subsidies for theses types of projects.

Because of these developments, which roughly cover the past decade, the university groups are now more involved in industrial research projects. The contacts are mainly with research centers of large enterprises and control groups of medium-sized and smaller enterprises and incidently with very small enterprises. There is a slowly growing awareness in smaller and very small enterprises that their lack of facilities for research can be compensated by contacts with university groups.
It is through these contacts that modern control theories and their derivatives (the control software) can be applied in smaller and very small enterprises.

It was already noted in the previous section that the use of various aspects of modern control and system theory have not yet become common practice in smaller and very small enterprises.
Although a very extensive use of these theories in these types of enterprises is not feasible even in long term, it is safe to predict that a broader use than that of today - which is almost negligible - is in the near future possible and desirable. But for a successful introduction of modern control and system theories into small enterprises, the useful results of these theories should be presented in a much more tractable form to the users.
These results should be embedded in expert systems which are understandable to a large class of users and which have a high degree of automation with respect to fault detection and capability of recognizing the applicability of modern theories to the practical problems at hand.

Of course this sounds somewhat futuristic, with many serious problems remaining to be ironed out. Still, the possibility of non-experts using computers in an useful way was not esteemed very probable some twenty years ago.

As a consequence of the inquiry and the considerations outlined above, the Working Group on Software has settled on an objective, namely to realize, document and distribute a subroutine library for control system analysis and design. This subroutine library is called SYCOT (software library of SYstem and COntrol Tools), (Working Group on Software, 1983b). For that purpose many international contacts have been and are being made with groups with similar goals in The United Kingdom, West-Germany, France, Belgium and Sweden. As a long range goal, the Working Group will attempt to realize interactive design packages based on this subroutine library, which should evolve in time to generally-applicable expert systems.

4. APPLICATIONS

In this section we will illustrate the application of CACSD in industry with a few typical examples. These examples do not cover all applications but have been selected by the authors owing to their involvement with industrial applications.

Aero-space industry:
In the aero-space industry many advanced CAD/CAM programs are being used. If we restrict our attention to control problems, we note that dedicated programs are often developed for one type of satellite or airplane. These efforts are justified because studies on such applications take many months to years. Then, a dedicated program can yield a higher performance than a general-purpose program. For smaller problems, for example the analysis and design of a navigation system for an airplane, interactive design programs, such as TRIP and PSI, can be and are being used (Van den Bosch and Geels, 1984).
Sometimes the algorithms of a program are used in another, problem-oriented environment. For example, in studies concerning the analysis of turbulence and vibrations around an airplane wing, the frequency responses are measured by dedicated hardware. Consequently, these frequency responses could be analyzed with TRIP. However, in order to get a better communication between measurements and identification program, the complex-curve fitting program of TRIP has been isolated and incorporated into another program. So, generally available programs are incorporated in a dedicated application program.

Chemical processing industry:
Some chemical enterprises frequently use PSI for modeling and identification. A minicomputer measures time responses of a chemical plant. These responses are measured after disturbing the process with a deterministic type of input signal, such as a step or block. Consequently, a limited number of measuring points are sufficient for the modeling process. These measuring points are stored in a function generator of PSI. Consequently, a model is estimated in PSI by adjusting the parameters of a model, such that the response of the model to the same input signal which was applied to the process equals, as well as possible, the measured response of the process, as illustrated in figure 1.

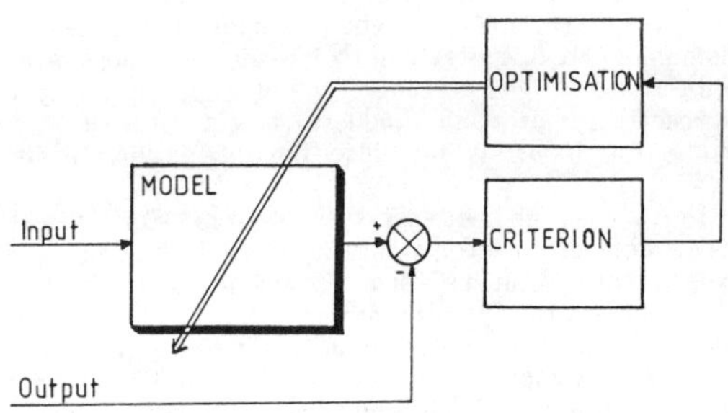

Figure 1. Identification with simulation and optimization.

Non-linear optimization techniques are used for adjusting the parameters of the model. Such an approach requires much more calculation time compared with, for example, the least-squares method. However, particularly in modeling a process, it is very valuable and almost necessary to use as much a priori information as possible. Any information, for example that dealing with non-linearities, the structure of the model, already known values of some parameters, etc., can be incorporated in the model and simulated properly, whereas mathematically-oriented methods cannot often take advantage of this a priori information.

These chemical enterprises and many others judge for their problems the simulation and optimization approach to be more valuable than an analytically-oriented approach based on least-squares methods. The increase in calculation time is not a disadvantage during the modeling and identification phase. For adaptive control this approach is too slow.
It should be stressed that this kind of modeling with a limited number of measuring points (100 - 1000) is only possible if sufficient information for modeling the process is indeed contained in the data. This approach can however not be applied if the signal-to-noise ratio is too small. This situation can arise if we are not allowed to excite a process with a relatively large external input signal. Then time-consuming measurements of normal operating signals become necessary. A posteriori analysis, for example with the aid of correlation techniques, may yield the required model.

Ship control design:
Several enterprises dealing with autopilots for ships use simulation and optimization as their primary analysis and design tool to realize appropriate controllers. Linear design techniques were not able to deal with the inherent non-linear behavior of the steering machine. Neglecting these non-linearities in the design phase resulted in a diminished performance (Van Amerongen and Van den Bosch, 1985).
Many times it turned out that a design based on linear system theory could yield the appropriate start values for the controllers or for the model to be identified. Completing the study with some form of simulation improves the design considerably.

Glass production industry:
Indentification programs, as available in SATER, have been used for constructing models of the glass fabrication process, including the model of the distrubances. These models have been developed for controller design, resulting in an increased product yield of the process. As in a previous example the interactive program SATER was not used directly but the parts of the package were incorporated in an adapted version in a dedicated application.

Biomedical applications:
Identification methods have frequently been used in biomedical applications. We will only refer to the estimation of the closing impedance of the aorta, which is a necessary information for design of aortic valves. In vivo signals of aortic pressure and flow of dogs were used for estimating the order and the parameters of the Windkesselmodel of the impedance (Rooyakkers, 1982).

5. CONCLUSION

Several aspects related to industrial applications of CACSD in The Netherlands have been dealt with;
- a short description of developments in The Netherlands with respect to CACSD software and their applications in industry.
- an inquiry held in 1982 in The Netherlands and in 1983 in Belgium to clarify the need for CACSD software.
- a Benelux initiative (WGS) started in 1981 to stimulate and coordinate CACSD developments in the Benelux.
- the WGS efforts to introduce a basic subroutine library SYCOT as a solid basis for future CACSD developments.
- a number of examples of application of CACSD in industry.

6. REFERENCES

Amerongen, J. van and P.P.J. van den **Bosch** (1985). Computer-Aided Design of Ship-Control Systems. Proceedings IFAC Symposium on Computer-Aided Design of Control and Engineering Systems, Copenhagen, July 1985. Pergamon Press London.

Boom, A.J.W van den and R. **Bollen** (1984). The Identification Package SATER. Preprints 9th IFAC World Congress, Budapest, July 1984, Vol VIII, (190-195).

Bosch, P.P.J. van den (1975). TRIP - TRansformation and Identification Package for Interactive Computer Aided Control System Design. Journal A, Vol 16, no 3 (107-117).

Bosch, P.P.J. van den (1983). Interactive System Analysis and System Design Using Simulation and Optimization. Proceedings IFAC Symposium on Computer Aided Design of Multivariable Technological Systems, Lafayette, 1982. Pergamon Press London (225-232).

Bosch, P.P.J. van den and D. **Geels** (1984). An Accurate Navigation System for a Laboratory Aircraft. Journal A, Vol 25, no 4 (229-235).

Meerman, J (1980). TUTSIM, Software for the Simulation of Continuous Dynamic Systems for Small and very Small Computer Systems. Proceedings Simulation '80 Conference, Interlaken 1980.

Rooyakkers, J.C. (1982). Estimation of the Aortic Closing Impedance using in vivo signals. M.Sc Thesis, EE Dept., Eindhoven University of Technology.

Working Group on Software (1982). Programs for System and Control Theory in The Netherlands. Eindhoven University of Technology.

Working Group on Software (1983a). A Subroutine Library in Control Engineering (Belgian Chapter). Eindhoven University of Technology.

Working Group on Software (1983b). Implementation and Documentation Standards for the basic subroutine library SYCOT, Eindhoven University of Technology.

APPENDIX

TUTSIM (Meerman, 1980) is an interactive block-oriented simulation program for 8-bit and 16-bit microcomputers developed by the Twente University of Technology. It has been implemented in assembly language in low-cost microcomputers such as Commodore-64, Apple II and IBM-PC.

PSI (Van den Bosch, 1983) is an interactive, block-oriented simulation program for 16-bit microcomputers, mini- and superminicomputers with many advanced features like optimization, generally only found with statement-oriented simulation programs. PSI has been developed by the Delft University of Technology. It is written in Fortran 77 and implemented on 16-bit microcomputers (IBM-PC, TULIP I, etc), 16-bit minicomputers (PDP 11, HP 1000, etc.) and superminis (VAX, PRIME).
PSI is in use in the chemical and biomedical processing industries, process engineering companies, the blast-furnace and steel-making industries, research centers of power utilities, navy, food-processing industries, electrotechnical, electronic and glass production industries, research institutes, etc., in both The Netherlands and abroad. Besides it is used in many educational and research institutes.

TRIP (Van den Bosch, 1975) is an interactive analysis and design program for linear single-input, single-output systems developed at the Delft University of Technology. It distinguishes seven different system representations, namely the continuous and discrete transfer functions (H(s) and H(z), respectively), continuous and discrete state-space descriptions, continuous and discrete frequency responses and time responses. Owing to the linearity of all representations, almost any system representation can be calculated from any other one. For example, identification (Chidambra estimator, least-squares method, complex-curve fitting) and designing with the aid of frequency responses, root loci or via optimal state feedback are supported. Moreover, eigenvalues and the solution of a Lyapunov equation can be solved. Systems up to tenth-order can be used and a maximum of 128 measuring points are allowed.
TRIP is in use in the chemical process industries, process engineering companies, blast-furnace and steel-making industries, research centers of power utilities, navy, electrotechnical, electronic and glass production industries, research institutes, etc. Moreover, TRIP is in use at many educational and research institutes in both The Netherlands and abroad.

SATER (Van den Boom and Bollen, 1984) is an interactive modeling and identification program for linear, single-input, single-output systems. It contains several recursive identification methods such as least squares, (suboptimal) instrumental variables, maximum likelihood, extended matrix and Tally estimators. It constructs ARMA models of process and disturbance noise dynamics. Several order-testing methods have been incorporated. Furthermore additional features which are related to identification have been added such as signal analysis, conversion of time discrete models into continuous models and analysis methods like Bode, Nyquist and root-loci plots.
As a package it is primarily meant for inexperienced users, e.g. in the courses, the interaction mechanism is of the question-and-answer type with default answers. SATER is written in Fortran and implemented on minis (PDP 11) and Superminis (VAX), and was developed at the Eindhoven University of Technology. University control research groups are the main users of the package, mainly due to its type of interaction, although industry, e.g. in the glass-production industry, has also have been involved.

PRIMAL is an interactive system for real-time experimental determination of models. The experimentor can perform experiments in real time and analyze the results simultaneously, having at his disposal various on-line and off-line methods for modeling, signal analysis, identification and control design. The system is currently being developed by the Eindhoven University of Technology. The software is written in Fortran for minis (PDP 11) and superminis (VAX).

INDUSTRIAL APPLICATIONS OF CAD PACKAGES FOR CONTROL SYSTEMS IN THE PEOPLE'S REPUBLIC OF CHINA

Chen Zhen-Yu

Automation Research Institute of Metallurgical Ministry, People's Republic of China

Abstract. This paper introduces the development of CAD techniques for control systems in P.R. China. A general description has been made about the industrial applications of CAD packages for control systems. Some examples have been illustrated about the MIMO discrete-time systems, LQG systems, SISO control systems, self-tuning control systems and adaptive prediction systems. Some prospect ideas have been described in conclusion.

Keywords. CAD package; industrial automation; control systems.

1. INTRODUCTION

The techniques of CAD for control systems had been developed gradually at the seventies in P.R. China. At the initial stage, a lot of calculating programs of single and multi variable control systems as well as simulation softwares had been constructed to aid for design and analysis of control projects (Chen, 1979; Han, 1979; Liu et al. 1979; Wang, 1972; Wang et al., 1979; Wu, 1978; Xu et al., 1976). Those softwares were special kinds of programs which must were used by experts and were always written by themselves. Anyone who wanted to use those programs, he must read over the whole programs which always contructed as subroutines, and must wrote a main program as well as the another subroutines which can described the appointed jobs. These circumstance got some limitations of the possibility to spread the applications of control technology into the engineering areas, especialy the modern control theory contained a lot of complex algorithms, and the engineers got difficulty to be familiar with all the details of algorithms. When entering the eighties, through the domestic and foreign technical exchanges for which some activations were orgenaized by Chinese Association of Automation, the CAD techniques of control systems have developed more quickly. A lot of CAD packages have been constructed in the universities and institutes, which have the features including management systems to dispatch all the functions within packages (Bao et al., 1982; Chen et al., 1985; Gan, 1982; Gao et al., 1984; Han, 1980; Han, 1982; Han, 1983; He, 1982; Lu et al., 1984; Ruan, 1982; Tu et al ., 1982; Wang et al., 1984; Xiong et al., 1982). There have interactive dialog system to simplify the operations of the users, who can get annotations from packages during every operating step. And also have command system to support the efficient running which looks like batch process, even the users want to compare the results of different choices of means.

2. APPLICATIONS OF CADCS

The techniques of CADCS are going to progress and also to carry forward the applications in P.R. China. Following, I will introduce the general situations about the applications of CADCS.

MIMO Discrete-time Systems

The linear multivariable stochastic system can be described as (Ruan, 1982)

$$y(k)=A_1y(k-1)+\cdots+A_py(k-p)+B_0u(k-1-d)+\cdots+B_pu(k-p-d)+e(k) \tag{1}$$

where noise

$$e(k)=V(k)+A_1V(k-1)-\cdots-A_pV(k-p)+B_0W(k-d)+B_1W(k-1-d)+\cdots+B_pW(k-p-d) \tag{2}$$

The equivalent state space equation can be described as

$$x(k+1)=Ax(k)+Bu(k-d)+BW(k-d) \tag{3}$$

and the measuring equation

$$y(k)=Hx(k)+B_0u(k-d)+B_0W(k-d)+V(k) \tag{4}$$

The optimal filter of state x(k) which may satisfies the equation (3) and (4) can be described as

$$\hat{x}(k)=\hat{x}(k/k-1)+K_k\Delta y(k) \tag{5}$$

where K_k is the gain matrix of Kalman filter and

$$\Delta y(k)=y(k)-H\hat{x}(k/k-1)-B_0u(k-d)$$
$$\hat{x}(k/k-1)=A\hat{x}(k-1)+Bu(k-1-d)$$

The quadratic performance criteria

$$J(u)=E\sum_{k=d}^{s}\left[x^T(k)Q_0x(k)+u^T(k-d)u(k-d)\right] \tag{6}$$

and the steady-state optimal feedback control

$$u(k)=L_d\hat{x}(k)+\sum_{i=0}^{d-1}M_iu(k+i-d) \tag{7}$$

where L_d and M_i are the functions of steady state optimal feedback coefficient matrix L.

A MIMO discrete-time CAD package had been constructed, and the main features of the algorithm functions are described as above. This CAD package had been used for project design of optimal control for cover-type annealing furnaces (Ruan et al., 1981). These furnaces were used for annealing of alloy steel coils. The scheme of furnace is illustrated as Fig. 1.

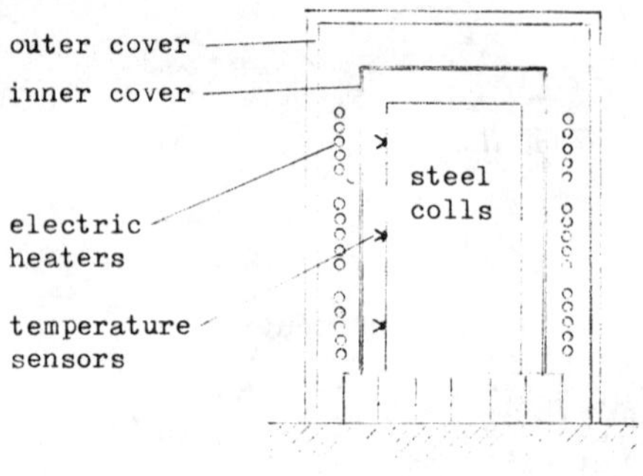

Fig. 1. Scheme of annealing furnace

The furnace was heated by electric power, and can described as a 3 input 3 output system. The practical measuring had been made for collecting the step response data series of temperature outputs y1, y2, y3. Then used the CAD package to carry out the identification and modelling, the calculating procedures illustrates as Fig. 2.

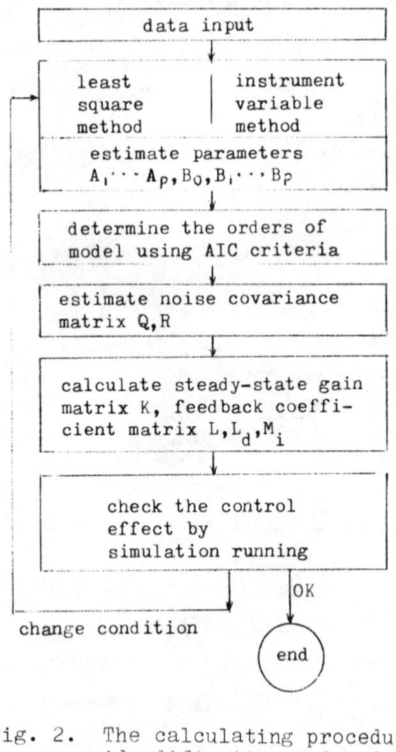

Fig. 2. The calculating procedures of idendification and modelling

When the process model had been constructed, the CAD package can carried on simulation study as well as real time control running, which is illustrated as Fig. 3. This furnaces control systems had been implemented in Shanghai Iron and Steel Factory, and have got more fine performance to satisfy the production objects then common instrumental systems.

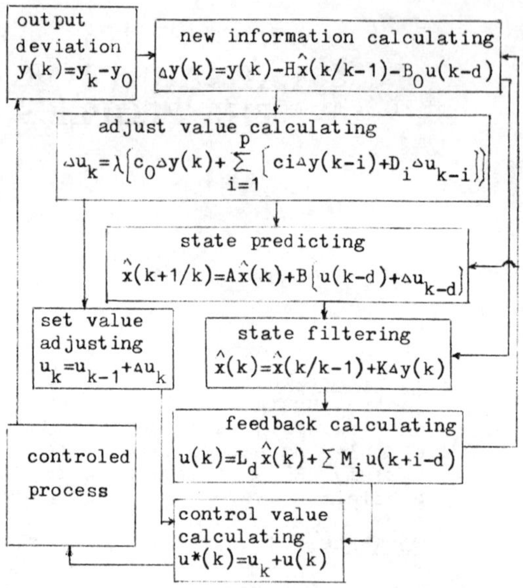

Fig. 3. The block diagram of real time control running

This CAD package also had been used for project design of temperature control systems of creep-testing furneces (Ruan et al. 1981). The control priciple and system construction were similar as the above example and the practical control performance were also satisfactory.

The another example of using this CAD package was the project design of electric power control system of electroslag remelting process (Ruan et al., 1984). Even that was a single variable system, the design as well as the practical control performance had got fine results.

LQG Systems

The Linear-Quadratic-Gaussian problems generally can be described as state space epuation and measuring equation (Gan et al. 1982)

$$\dot{x}(t)=Ax(t)+Bu(t)+w(t) \tag{8}$$

$$y(t)=Cx(t)+v(t) \tag{9}$$

The quadratic performance criteria

$$J=E\left[x^T(t_f)P_f x(t_f)+\int_{t_0}^{t_f}(x^T(t)Qx(t)+u^T(t)Ru(t))dt\right] \tag{10}$$

For LQ problems the optimal feedback control law is given by

$$u(t)=-K(t)x(t) \tag{11}$$

$$K(t)=R^{-1}B^TP(t) \tag{12}$$

where $P(t)$ can be obtained by solving Riccati differential matrix equation

$$-\dot{P}(t)=P(t)A+A^TP(t)-P(t)BR^{-1}B^TP(t)+Q \tag{13}$$

For steady-state the equation (13) becomes algebraic equation form

$$PA+A^TP-PBR^{-1}B^TP+Q=0 \tag{14}$$

And the corresponding control law is given by

$$u(t)=-Kx(t) \quad (15)$$

$$K=R^{-1}B^{T}P \quad (16)$$

For Kalman-Bucy filter, the calculating equation is given by

$$\dot{\hat{x}}(t)=A\hat{x}(t)+Bu(t)+G[y(t)-x(t)] \quad (17)$$

The block diagram of LQG CAD package is illustrated as Fig. 4.

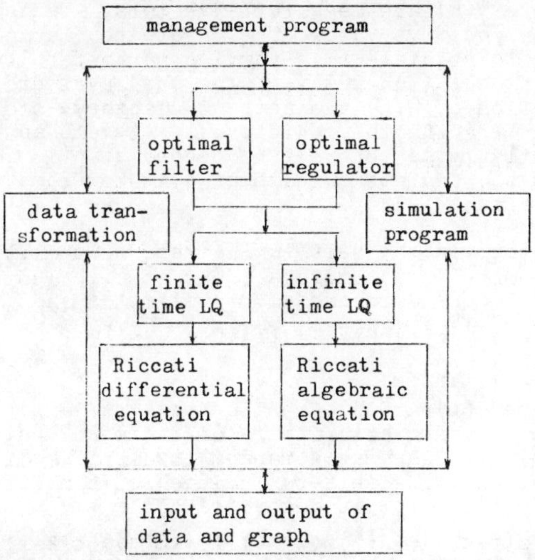

Fig. 4. The block diagram of LQG CAD package

This CAD package had been used for project design of wide range speed regulating servo system (Su et al., 1983) as Fig. 5.

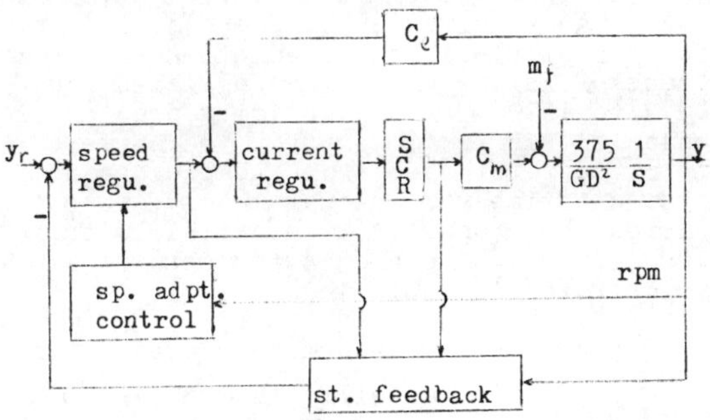

Fig. 5. Scheme of wide range speed regulating servo system

The speed regulating range was 0.1 rpm to 1500 rpm, and the speed up time was ≤170 ms from 0 to 1500 rpm, the brake down time was ≤170 ms from 1500 to 0 rpm, the reversing time was ≤250 ms from +1500 to -1500 rpm, the regulating overshoot < 2%.

The another example of using the LQG CAD package was the project design of effective anti-disturbance speed regulating system (Gao et al., 1983). The block diagram of speed regulating system is illustrated as Fig. 6.

There had added a state feedback transient compensator into the system to improve the anti-disturbance performance. When the step load of 100% rate had been applied into this system, the danymic speed drop $\Delta n \leq 0.6\%$ the resuming time $t \leq 0.8$ s.

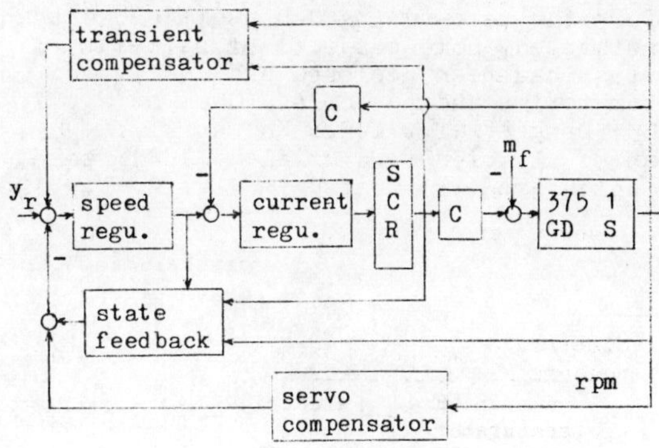

Fig. 6. Scheme of anti-disturbance speed regulating system

This CAD package also had been used for design of some carriage control systems of vehicles.

SISO Control Systems

There have wide spread demands to design the SISO control system in the real engineering problems, therefore the SISO CAD package had been constructed which included a lot of useful algorithms of time domain and frequency domain methods (Tu et al., 1982). The application examples have been in many cases. One of the user's department introducing some examples are illustrated as Fig. 7, Fig. 8 (Li et al., 1983).

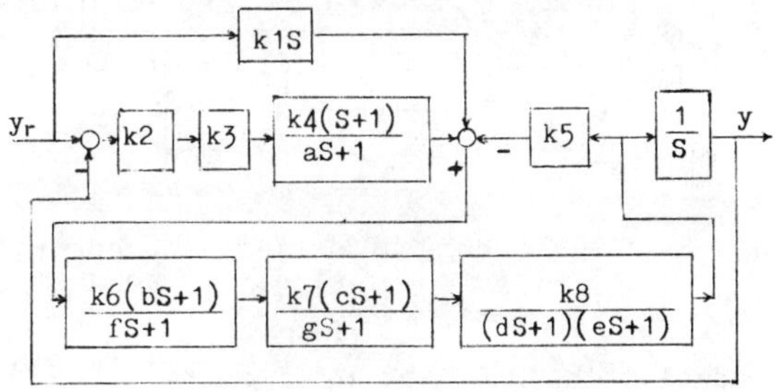

Fig. 7. Block diagram of high precision servo system

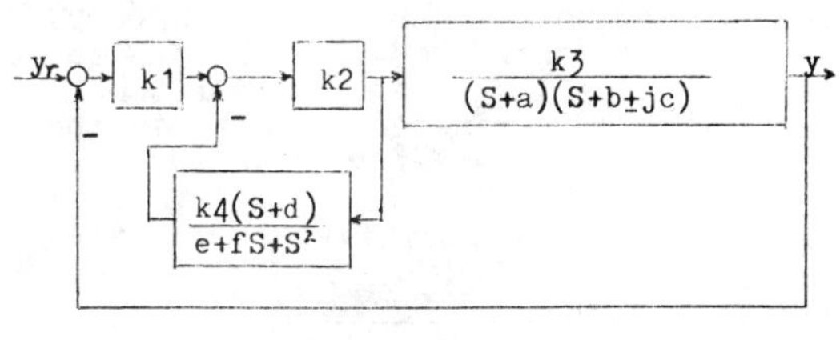

Fig. 8. Block diagram of servo system of on line robot

All of the users had appreciated that they had saved many times for using CAD package then their own calculating, and had advantage to compare their different plans of design from various viewpoints at a short period of computer running.

Self-tuning Control Systems

Before the self-tuning CAD package had been constuucted, some projects of self-tuning systems had been designed and applicated in laboratories and factories (Chen et al., 1983; Deng et al., 1982; Shu et al., 1982; Yuan et al., 1980; Wu et al., 1984), some of which are illustrated in Fig. 9, Fig.10.

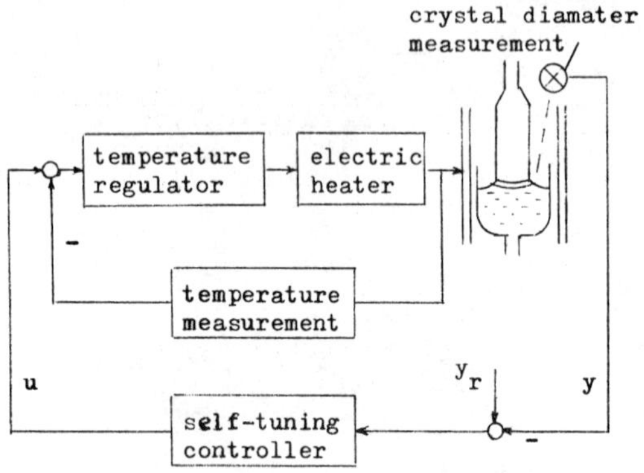

Fig. 9. Scheme of self-tuning control system of silicon crystal drawing process

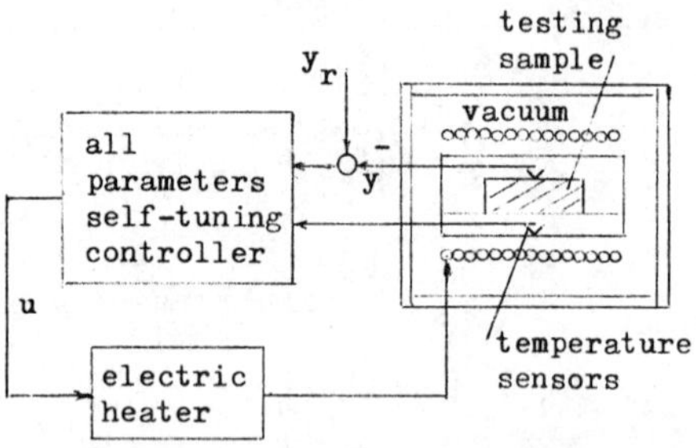

Fig. 10. Scheme of self-tuning control system of electric vacuum testing equipment

Gradually the CAD package of self-tuning control systems have been constructed in universities and institutes (Cao et al., 1984; Chen et al., 1985; Yin et al., 1984). These CAD packages contain the algorithms of SISO and MIMO self-tuning control systems, which include self-tuning regulator; self-tuning controller; pole assignment self-tuning controller; all parameters self-tuning controller, and the feedforward terms are applied into the algorithms. The algorithms of parameters estimation include least square; extended least square; maximum likelihood etc, and can use variable or nonvariable forgetting factors methods.

Adaptive Prediction Systems

The adaptive prediction CAD package contains two kinds of algorithms. One is the multi-level progressive order method (Han, 1982), which considers the dynamic system model as

$$y(k)=f\left[Y_{k-1},U_k,\theta,k\right]+v(k) \tag{18}$$

where y(k) are outputs of n dimensions, u(k) are inputs of p dimensions, θ are time-varying parameters, v(k) are random noises.

Equation (18) also can be rewritten as CARMA form

$$y(k)=\varphi(k)^T\theta(k)+v(k) \tag{19}$$

For single variable time-varying system, the estimated values

$$\hat{\theta}(k)=\hat{\theta}(k-1) + \frac{\delta}{\|\varphi(k)\|^2}\varphi(k)\left[y(k)-\varphi(k)^T\hat{\theta}(k-1)\right] \tag{20}$$

where $\varphi(k)$ are measurable vectors, δ is certain positive value. The first step is going to predicte time-varying parameters by using equation (20), if the one level prediction of $\theta(k)$ are not satisfactory, then can be going to predicte multi-levels as similarly as before. The second step is going to predicte outputs of systems from equation (19).

Another is the self-tuning recursive multi-step prediction method (Deng et al., 1983), which considers the MIMO systems as

$$A(q^{-1})z(t)=B(q^{-1})u(t)+D(q^{-1})y(t) + \xi(t) \tag{21}$$

where z(t) are p dimention vectors, u(t) are deterministic inputs of r dimentions, y(t) are measurable random inputs of m dimentions. The self-tuning recursive multi-step predictor is described as

$$\hat{z}(t+k/t)=\Sigma\hat{A}(t)\hat{z}(t+k-i/t)+\Sigma\hat{B}(t)u(t+k-i) +\Sigma\hat{D}(t)\hat{y}(t+k-i/t) \tag{22}$$

$$\hat{y}(t+k/t)=\Sigma\hat{\Phi}(t)\hat{y}(t+k-i/t) \tag{23}$$

where $\Phi(t)$ are estimated parameters.

This CAD package had been applicated for predictions of the productivity of oilfield, the production, worth and marketing of light industry commercial products etc (Deng et al., 1983; Guo et al., 1984; Han, 1983; Han et al., 1983). One example was for the oilfield, the prediction was on the bases of data , which was collected from 1972 to 1979. The productivity prediction equation was in time-varying parameters form, so first used this CAD package to predicte time-varying parameters, then predicted the productivity of oilfield from 1980 to 1981. The prediction results are illustrated in table 1.

Table 1. The productivity prediction of an oilfield

time	real productivity k tons/day	predicted productivity k tons/day	relat. error %
1980.3	41.97	42.22	-0.59
1980.6	41.21	41.63	-1.02
1980.9	40.51	41.07	-1.38
1980.12	39.87	40.55	-1.7
1981.3	39.28	40.05	-1.96
1981.6	38.72	39.57	-2.19
1981.9	38.22	39.13	-2.38
1981.12	37.75	38.71	-2.54

The average error was 1.72%.

Another example was the prediction of plastic production of a province. The prediction results are illustrated in table 2.

Table 2. The prediction of plastic production

time	production tons	investment k yuan
1972	9.40	1230
1973	12.43	1530
1974	14.32	2170
1975	14.32	2790
1976	18.81	4960
1977	26.55	2030
1978	31.29	4890
1979	31.67	620
1980	33.60	3700
1981	37.50	3350

When this CAD package had been used for predicting the time-varying parameters and the plastic production, the average prediction error of 1980 and 1981 was 4.2%, which was obtained by using the real data from 1972 to 1979.

The above descriptions show a part of industrial applications of CADCS in PRC, hoping that will give out a general impression about this fields.

3. PROSPECT

Even the techniques of CADCS have been developed and applicated in some main engineering areas, but there are still not familiarized by many engineers, so it needs making efforts to popularize this techniques for engineering applications, through organizing seminars, technique exchanges and another possible ways. The engineering experiences have been shown that the CADCS techniques become the bridges between the theory and pratice, and it also can be forecasted that these kinds of bridges will give more play to act on the engineering fields. We also hope to hold much more technique exchanges in the international range, and use the advanced equipments as far as possible. Besides to develope and enrich the contains of CADCS packages, the combination of control techniques and special applicated techniques will provides another wide working area. For example the biomedical identification CAD package have been constructed for doctors, who want to use the identification techniques but can not carry on any particular calculations, this CAD package provides as confortable instruments and have been appreciated by doctors. The extend of discipline of CADCS is also a broad problem, which is not limited into the engineering area, especially the control science have permeated into the sociology, economics nowadays. Such as system planning, system programming as well as large-scale system, there have a lot of problems providing to be developed concerning with CADCS and its applications. The techniques of data base and expert system are also useful for CADCS.

4. REFERENCES

Bao Yuan-Lu,et al. (1982). CAD of the multivariable feedback system based on modern frequency domain methods. Information and Control, Vol.11. No.6. 24-33.

Cao Shu-Guang, et al. (1984). CAD package of self-tuning control systems. Treatise of Shan Xi Mechinery College.

Chen Zhen-Yu. (1979). Simulation study of automation systems of cold tandem mills. Metallurgy Automation, No.1. 9-19.

Chen Zhen-Yu, et al. (1983). Self-tuning regulator for roll eccentricity of cold tandem mills. 4th IFAC Symp. on Automation in Mining, Mineral and Metal Processing, Aug. 22-25, 1983. Helsinki, Finland.

Chen Zhen-Yu, et al. (1985). Self-tuning CAD package and its application. 3rd IFAC Symp. on Computer Aided Design in Control and Engineering Systems, July 3 -Aug. 2, 1985. Lyngby, Denmark.

Deng Zheng-Duan, et al. (1982). Self-tuning control of silicon crystal drawing process. Report of Shan Xi Machinery College.

Deng Zi-Li, et al. (1983). Multivariable self-tuning recursive multistep predictors and their applications. Acta Automatica Sinica, Vol.9. No.4. 241-247.

Gan Hua-Lei. (1982). LQGSP-A CAD software package for the LQG systems. Information and Control, Vol.11. No.6. 11-16.

Gao Long, et al. (1983). The LQ design of effective anti-disturbance speed regulating system. Report of Automation Department of Tsing Hua University.

Guo Yi-Xin, et al. (1984) CAD package for adaptive prediction. Report of Heilongjiang Institute of Applied Mathematics.

Han Jing-Qing. (1979). Canonical form of linear systems and calculation of feedback systems. Annual Meeting of Theory Com. of Chinese Association of Automation, May.10-17, 1979. Xia Men, China.

Han Jing-Qing. (1980). The theory of multivariable control systems -- polynomial matrix operations and frequency domain methods. Treatise of Institute of Systems Science, Academia Sinica.

Han Zhi-Gang. (1982). Multi-level progressive order method. Treatise of Heilongjiang Institute of Applied Mathematics.

Han Zhi-Gang. (1983). A new method of dynamic system prediction. Acta Automatica Sinica, Vol.9. No.3. 161-168.

Han Zhi-Gang, et al. (1983). An adaptive forecast about the production, worth and marketing of light industry commercial products. Information and Control, Vol.12. No.1. 14-18.

He Huan-Xi. (1982). The computer aided design software package of the multivariable control system. Information and Control, Vol.11. No.6. 50-58.

Li Jie-Xin, et al. (1983). Applications of SCIT CAD package. Report of Shen-Yang Institute of Automation, Academia Sinica.

Liu Zhi-Jun, et al. (1979). The method of digital continuous system simulation. Acta Automatica Sinica, Vol.5. No.4. 316-324.

Lu Gui-Zhang, et al. (1984). CAD software package for identification, simulation and control. Treatise of Nan Kai University.

Ruan Rong-Yao, et al. (1981). Optimal feedback control for cover-type annealing furnaces. The Symp. Proceeding of the Sino-US Bilateral Meeting on Control Systems, Science Press of China.

Ruan Rong-Yao, et al. (1981). The mathematical model of temperature control for creep-testing furnaces. ibid.

Ruan Rong-Yao. (1982). Software for

identification and optimization of multi-input multi-output discrete-time systems. Information and Control, Vol. 11. No.6. 17-23.

Ruan Rong-Yao, et al. (1984). A mathematical model for computer control of the electroslag remelting process. Acta Automatica Sinica, Vol.10. No.3. 247-252.

Shu Di-Qian, et al. (19820. Multivariable self-tuning control of electric furnaces. Report of Beijing Iron and Steel College.

Su Zhong-Fei, et al. (1983). CAD of wide range speed regulating system. Report of Automation Department of North-East Technology University.

Tu Qi-Lie, et al. (1982). An introduction to SCTI, a CAD software package for control system design. Information and Control, Vol.11. No.6. 43-49.

Wang En-Ping. (1972). Recursive methods of optimal linear filter. Mathematical Practice and Cognition, No.6.

Wang Kia-Sang, et al. (1979). A mathemetical model of computer controlled creep-testing furnaces. Acta Automatica Sinica, Vol.5. No.1. 6-14.

Wang Zhi-Bao, et al. (1984). The design of a software package for CAD of control systems. Control Theory and Applications, Vol.1. No.4. 28-34.

Wu Jie. (1978). An algorithm of inertial navigation control system. Journal of South-China Institute of Technology, Vol.6 No.2. 75-79.

Wu Hong-Xin, et al. (1984). All parameters self-tuning control method. Annual Meeting of APCOM of Chinese Association of Automation. Xi-An, P.R.China.

Xiong Guang-Leng, et al. (1982). IDENT- A computer aided program for the identification of linear systems. Information and Control, Vol.11. No.6. 34-42.

Xu Jian-Pian, et al. (1976). Adaptive filtering. Journal of Fu Dan University, No,2. 22-38.

Yin Yi-Xin, et al. (1984). CAD package of multivariable self-tuning control systems. Treatise of Beijing Iron and Steel College.

Yuan Zhu-Zhi, et al. (1980). Self-tuning adaptive control of two installations in chemical industry. 6th IFAC/IFIP Digital Computer Application to Process Control, Preprints, 161-166.

COMPUTER AIDED DESIGN OF CONTROL SYSTEMS IN THE UNITED KINGDOM

O. L. R. Jacobs

Department of Engineering Science, University of Oxford, Oxford OX1 3PJ, UK

ABSTRACT Current work in U.K. is surveyed, on uses of digital computers to aid design of feedback controllers. Most of the work is concerned with methodology but it is all motivated toward practical applications. There are grounds for believing that favourable conditions exist for the methodology now to evolve into engineering practice.

KEYWORDS Computer-aided design, Computer control, Control engineering, Computer applications, Identification, Multivariable control systems, Nonlinear control systems.

1 INTRODUCTION

This paper summarises results of an attempt to survey current work on Computer Aided Design of control systems for industrial processes in the United Kingdom, with emphasis on practical applications. Computer Aided Design (CAD) is interpreted in a fairly wide sense as referring to almost any use of a digital computer to aid design of a feedback controller, excluding adaptive control which already has its own large literature (Åström 1983). Primary sources of information consisted of private communication with known workers, and a timely workshop on Computer Aided Control System Design (Inst. M. & C., 1984). The resulting summary may well not be comprehensive in its coverage of all relevant projects, but is nevertheless thought to be representative of the range of current work in U.K.

The survey did not, in the event, discover many published case-histories of applications. Most reported activity concerned methodology. This is surveyed here against a background of methodological categories which are outlined in Section 2. Section 3 contains the actual survey. The overall impression, stated in a concluding Section 4, is of work in progress over a broad front with good prospects for reversing the conclusion of an earlier survey that "the real breakthrough among industrial users has not taken place" (Tysso, 1981). What seems now to be happening is more like infiltration by many routes than breakthrough in any single direction.

2 TOPICS IN CAD METHODOLOGY

CAD methodology can conveniently be presented in five categories:-

(i) Type of mathematical model used as a basis for design

(ii) Type of computing job constituting the design

(iii) On-line or off-line CAD

(iv) Uses of CAD packages

(v) Technology of CAD

These are discussed in Sections 2.1 – 2.5 respectively.

2.1 Mathematical Models

At the heart of most design procedures is some model whose function is to predict the consequences of design decisions. In CAD the model is usually mathematical, and for control systems the object to be modelled is the controlled process. Two main types of mathematical model of controlled processes, "structural" and "behavioural", provide starting points for CAD procedures.

Structural models reflect prior knowledge of structure of the modelled object. They consist of equations from physics or chemistry or physiology or whatever science is relevant to the controlled process. They are not necessarily linear or deterministic, and can be at whatever level of complexity or simplicity is thought necessary to capture significant features of controlled process behaviour.

Behavioural models reflect behaviour rather than underlying structure. They are usually linear, whether or not the modelled object is linear, and can be regarded as the result of fitting linear equations to observed behaviour. If the object really is linear, the linear behavioural model can correspond completely to the structural model. However most real controlled processes are more or less nonlinear and behave in ways which cannot be completely predicted by a linear model. The usefulness of linear behavioural models is thus problem-specific.

Computers can aid derivation of either type of mathematical model,usually by processing plant data so as to obtain estimates of some chosen parameters. For structural models the estimated parameters would be physical or other coefficients determining the structural equations. For behavioural models, coefficients in linear equations can be fitted to observed behaviour by procedures which have come to be known as System Identification (Isermann, 1981).

2.2 Computing Jobs

Once a mathematical model is given, there are various jobs which a computer can do to aid design. Most comprehensive, would be a fully automated design synthesis where the user had only to specify a mathematical model and a performance criterion, and the computer would then determine a suitable controller. Adaptive control algorithms usually include some such automated design, operating on a behavioural model obtained by System Identification. Adaptive control is not discussed in detail here, beyond noting the general conclusion that it can give good results when the controlled process is sufficiently linear that a linear behavioural model reliably predicts performance.

Optimal control theory, using performance criteria in the form of scalar cost functions, offers a form of fully automatic design algorithm. For linear mathematical models with quadratic criteria LQ optimal theory can be used to determine linear controllers. However, accurate nonlinear models of practical control engineering problems seldom lead directly to soluble problems in optimal control.

Less ambitious, but more popular, is mechanisation of design calculations determining quantities such as stability margin or closed-loop pole position in linear systems. Computer aid of this type provides a natural extension to traditional trial-and-error design methods. The user remains free to direct the design, with the computer doing the heavy calculations. Much recent U.K. work has been in this area of CAD.

Another useful type of job is simulation, especially when the controlled process is so nonlinear that classical linear design methods cease to be directly relevant. Simulation, where the computer predicts controlled behaviour by solving non-linear equations of a structural model, is probably the type of CAD most widely used in industry. In the words of one industrial correspondent "Dynamic simulation has proved to be an invaluable tool in the design of control systems for many process operations, particularly where interaction of process controls and nonlinear behaviour of equipment bring about the problem of instability". Simulation packages range from the general-purpose, like CSMP, SIMNON (Elmquvist, 1977) and ACSL (Havranek, 1984) mostly not developed in U.K., through suites devoted to some class of interconnected unit process, to those which concentrate on some specific nonlinear problem.

2.3 Mode of computing

The computer used for CAD of a control system may or may not be on-line to the controlled process, and may or may not be required to perform design calculations in real process time. Of the jobs in Section 2.2, adaptive control has the most demanding on-line real-time requirements, whereas classical designs and simulations are usually done off-line and to some non-operational time-scale.

Routes from off-line CAD to on-line computer control are discussed in Section 3.3.

2.4 Uses of CAD

Two principal uses for CAD in industrial processes are (i) to improve operations of existing plant, and (ii) to contribute to specifying new plant. For existing plant the CAD procedure could include interaction with the plant to help specify a mathematical model or to test specific controllers. For new plant there may need to be more emphasis on simulation of structural models.

A third use is as a teaching aid. In higher education, most control engineering courses include some CAD, both as an object of study for engineers and as an aid to understanding classical design calculations. Simple CAD packages can also be used to train other technical staff (King,1984), not necessarily graduates in Control Engineering.

2.5 Technology of CAD

CAD of control systems has a technology which is developing in several areas. Theoretical work produces new design criteria, such as multi-variable generalisations of classical stability margins, or H^{∞} norms, which can be calculated by a computer. Algebraic and numerical work makes it possible for a computer to perform safe and robust manipulations of linear mathematical models, including order-reduction and linearisation of nonlinear models. Computing science contributes to the structure of interaction between CAD-user and computer, particularly to development of facilities for graphical input from the user, for example to specify block diagrams on a VDU. It should also contribute to specification of database structures for mathematical models and of data interfaces for communication between CAD-computer and controlled process. Current computer technology offers the possibility of implementing CAD in a desk-top or portable microcomputer.

At almost all levels, there is scope for standardisation. For example in data protocols, in database structure, in user-interaction, in programming language, in specifying CAD functions. However, advantages of standardisations have always to be weighed against the long term evolutionary benefits of variety.

3 SURVEY OF CURRENT U.K. WORK

Current U.K. work on CAD of control systems can conveniently be surveyed in three sections.

(i) Off-line computer packages where a linear controller is designed to match a given linear mathematical model of the controlled process. In the rest of this paper, as elsewhere, the description "CAD" is abused as referring specifically to such procedures.

(ii) Off-line packages where controller design uses simulation of a given nonlinear, probably structural, mathematical model.

(iii) Integrated packages, combining linear CAD and simulation and possibly having communication from and to the controlled process, with facilities which may range from system identification to on-line computer control.

These are discussed in Sections 3.1-3.3 respectively. Examples quoted are biased toward process control because the original focus of the survey was directed to CAD applications in that area.

3.1 Off-line linear CAD

Probably the best-known British CAD packages are off-line linear packages such as CLADP - the Cambridge Linear Analysis and Design Programs (Maciejowski and MacFarlane, 1982) and the UMIST package (Munro, 1979). Written in FORTRAN, they mechanise a comprehensive set of classical and generalised design calculations for linear controllers of linear (or linearised) controlled processes. Mathematical models in these packages can be specified either in time-domain in the form of state-space matrices, or in frequency domain as transfer function matrices or frequency-response data. Powerful routines for manipulating matrices and polynomials provide robust transformations between the two domains. Models of specific problems, together with their controllers, can be stored in disc files for future reference. Controller designs are assessed by classical frequency-response and root-locus techniques as well as by their multivariable generalisations due to MacFarlane (1980), Rosenbrock (1974) and others. The assessment makes much use of graphic displays. Other facilities include: some optimal control and Kalman filter programs; equal support for both continuous-time and discrete-time designs; approximation routines to give a low-order version of a higher-order mathematical model; command macros to allow the user to define a sequence of commands. Work is in progress to include additional new design criteria as these are developed.

Both packages have been the subject of some tens of man-years of development, and are commercially available. They have been acquired by major industrial customers in Britain and abroad. Case histories exploiting the full power of CLADP have not yet been published. Initial applications, for example to single loops in steel mills or to a pair of interacting loops in a chemical plant, appear mostly to be of a tutorial nature. Users are familiarising themselves with the powerful facilities for classical single-loop design work as well as beginning to explore the multivariable generalisations. The UMIST package has several reported applications (referenced in Munro, 1982) such as its use to improve performance of an ore-crushing system having as many as three inputs and three outputs, at an overseas goldmine (Koudstaal and coworkers, 1981). How far and fast these packages will infiltrate general control engineering useage, remains to be seen. One determining factor may be price of the software and its supporting hardware.

That there exists optimism about the usefulness of off-line linear design packages is indicated by the significant number of packages which have been developed to perform similar functions. These have been produced by other university groups, and by individuals in industry such as Dutton (1984) at British Steel. No packages have been quite so professionally developed and marketed as those mentioned above, but there is much other work on technology of CAD which will contribute to the next generation of off-line packages. This includes:-

(i) Provision of graphical interactions at the input interface between user and computer, so that mathematical models can be specified in block diagram format. Work is in progress on this at Salford University (Gray, 1984) and also at UMIST (Goodfellow and Munro, 1984) where an input translator is under development to enable a computer to accept mixed graphical, numerical and textural input.

(ii) Research at Cambridge (Maciejowski, 1984) on data structures which would easily support CAD functions.

(iii) Introduction of new and computable theoretical criteria on stability and robustness of linear closed-loop performance. Semi-infinite optimization methods, where performance criteria are in the form of nondifferentiable, infinite-dimensional constraints, have been developed at Imperial College (Polak and Mayne, 1984, Polak and coworkers, 1985). Research into use of H^{∞} norms (Francis, Helton and Zames, 1984) is in progress, for example at Cambridge (Glover, 1985) and at Oxford (Foo and Postlethwaite, 1984)

(iv) Microcomputer implementation of linear CAD. Single-loop routines are readily implemented, for example in BASIC on an Apple (Thompson, S. 1984, Lawrence, 1984). A micro version of the UMIST inverse Nyquist array package is implemented on a Sirius (Munro, Bowland and Hopkins, 1984). To what extent a more comprehensive package such as CLADP could usefully be implemented on a microcomputer, remains to be seen. It seems likely that as microcomputers become increasingly powerful and inexpensive, CAD will become more and more widely available at an acceptably low starting price.

3.2 Off-line simulation

Of the computing jobs mentioned in Section 2.2, simulation is probably the most widely used in industry as an aid to designing feedback controllers. In one major chemical engineering company, for example, several man-years each year are devoted to use of a general-purpose simulation package for modelling and investigating control of specific processes. The commercially available simulation software is not discussed here because mostly it was not developed in U.K.

Microcomputer implementations are, however, under development and offer various levels of comprehensiveness. Most ambitious, is a package from Sheffield (Linkens, 1984), which uses extensive overlaying to transfer to a 380Z all the facilities of the local minicomputer package. A similar, but perhaps less comprehensive, transfer at Salford (Hay, 1984) offers a package to run under CP/M or MS-DOS, which therefore can run on an IBM-PC. A simpler package (Clements, 1984) runs on the BBC computer.

Several other U.K. simulation packages are dedicated to aiding design of controllers for processes within some specific engineering field. One approach has been to provide a library of unit processes and standard controllers which can be called up and interconnected in a user-language closely related to the chosen class of engineering problems. This has been done in FORTRAN for, for example, chemical engineering processes (Sebastian and Murray, 1984) and hydraulic actuators (Dorey, 1984).

Another approach has been to concentrate on a single notoriously difficult, because nonlinear, process and provide a package where performance can be simulated under a wide variety of control laws. This has been done at Oxford for regulation of pH in a continuous stirred tank reactor (Jacobs and Badran, 1983). A selection of conventional and novel algorithms for classical control, adaptive control and on-line estimations was included so that their performances and cost-effectiveness could be compared. The comparative results are problem-specific, depending on parameter values, disturbance magnitudes, and nonlinear characteristic of the simulated process. The package PHDESIGN has, in collaboration with major chemical engineering companies, had both the uses of Section 2.4: it helped to improve control of existing plant as well as to guide the specification of new plant.

3.3 Integrated packages

An obvious further development is creation of integrated packages in which routines for linear CAD can communicate directly with simulation software. Nonlinear controlled processes can then be simulated for use as test-beds to evaluate linear controllers designed by the CAD routines. Furthermore, linearisation routines can be provided to extract from the nonlinear simulations a linear mathematical model which will be a starting point for the linear controller design.

Communication with the controlled process, in either or both of two directions, can also be provided as part of an integrated package. Data from the process would be used to establish the mathematical model, as discussed in Section 2.1. Communication in the other direction, from CAD computer to plant, becomes important when the plant includes an on-line computer which will implement the designed controller. Good engineering practice then requires a machine route, such as could exist in an integrated package, to download the designed controller to the on-line computer.

Existing U.K. integrated packages have mainly been created by industrial users who have brought together routines from different sources. Dowty Electronics, who design electronic controllers for a wide range of hydromechancial systems, have integrated routines for linear CAD, simulation, and identification using recorded plant data (Foss, 1984). The Central Electricity Generating Board (CEGB), who make extensive use of computing for the analysis and design of power plant and its associated control systems, have a similar package (Thompson M.A., 1984) which includes a route to their own high-level language specifying power plant control systems. These packages both include CLADP.

Experience in creating current integrated packages draws attention to a need for standardisation in the environment of CAD, simulation, data-handling and control routines. The standardisation should be effective at all levels of communication with packages; for example in data-base structures, data-transfer and real-time signal protocols, and user interactions. Without some such standardisation it will become increasingly difficult to create generally accessible integrated packages exploiting future developments by individual workers.

A more homogeneous integrated CAD system has recently been developed by the Manchester University Company VUMAN (Sandoz, 1984). It consists of two packages, designed to communicate with each other according to standard protocols accessible also to other systems. One package performs off-line system identification and linear CAD to establish control system structures which will be implemented by the other, on-line, package. This performs many standard functions of an on-line process control computer, such as signal processing, data-logging, alarm monitoring, control-loop configuration, sequence control, mimic-diagram displays. In addition it can implement advanced feedback and feedforward control schemes, and has an on-line tuning facility for adjusting controller gains. The on-line package is implemented in read-only-memory (ROM) for a PDP11 computer, using a BBC computer as an intelligent colour graphics terminal. It does not at present include general-purpose simulation software, so its use is primarily of type (i) from Section 2.4, to improve operations of existing plant. The system is currently undergoing full-scale industrial evaluation at two major chemical engineering companies, at CEGB, and elsewhere.

A different type of integrated system has been developed at Oxford (Tuffs, 1985) as an aid to implementing and commissioning self-tuning controllers. This package provides a modular FORTRAN environment within which any one of a repertoire of adaptive control algorithms can be interfaced on-line to control a full-scale process. Facilities in the package include, continuous calculation of PID control action as an alternative to adaptive control, signal conversion and conditioning, data-logging and real time display, off-line simulation using a linear mathematical model of the controlled process, simple English-based run-time commands for operating the controllers. The main objective of this package, as also of the Oxford simulation package in Section 3.2, is to make available a variety of advanced controllers. The modularity of the package has allowed many different workers to install and test their self-tuning controllers with a minimum of programming effort. The package can be implemented in a PDP LSI11 and has been used succesfully for on-line control of two full-scale industrial processes as well as of a number of pilot plants.

4 CONCLUSIONS

Current U.K. work on CAD of control systems is still largely concerned with methodology. The absence of many reports of succesful case-histories could be partly attributable to commercial secrecy but is more likely a reflection of the time needed for new methodology to evolve into engineering practice.

A major factor affecting evolution rate is the starting price as perceived by cost-conscious industry. Until recently this price has appeared to be of order tens of thousands of pounds to purchase or develop a package, plus one or more highly-qualified man-years to learn to use it. Simulation packages, commercially available from outside U.K., have been found worth this price as an aid to designing controllers for nonlinear processes. Linear CAD packages, mostly a U.K. development, are still under assessment.

One reason why time is needed to determine suitable applications for the new linear CAD packages was summarised in the conclusion of an earlier review (Hashimoto and Takamatsu, 1981): "most chemical process systems have been designed ...safe and self-regulatory processes which do not pose any difficult control problems. Control systems, therefore, can be designed very easily by applying classical methods...CAD systems which are necessary for the design of advanced control systems are scarcely utilized at present". Existing processes were designed to match existing classical methods of controller design. For new processes, linear CAD packages offer advanced control of multivariable interacting processes, but the poor financial climate of recent years has inhibited construction which might exploit this offer. Another factor affecting uses of linear CAD packages is competition from adaptive control, which provides alternative procedures for controlling linearisable processes.

Microcomputer implementations may bring about a significant reduction of starting price. Such reduction could make the linear CAD packages competitive for classical single-loop design. This might provide the route by which linear CAD, with its facilities for more advanced designs, infiltrates industrial practice. In any case, microcomputer implementation is a promising line of development, actively pursued in U.K.

Amongst existing packages, ACSL is popular for general-purpose simulation and FORTRAN persists as a widely used language. CLADP and the UMIST package have emerged as the dominant linear CAD packages although during recent years, whilst linear CAD evolved in U.K., there was inevitably much duplication of effort. Now, there seems no reason why what has been achieved should not be consolidated within a single standardised environment. This is particularly necessary in connection with the promising integrated packages of Section 3.3.

To establish an environment which will be sufficiently professional in its standardisation and support as to command general acceptance, is a major undertaking beyond the resources of most organisations. So the Science and Engineering Research Council (SERC), which is the main body allocating U.K government money to U.K. research, has recently initiated a program to meet the need. The principal objective of this program is to achieve a standardised environment for CAD of control systems, providing user-friendliness, easy communication between packages, fertile conditions for new developments, and easy transfer of useful work to industry. A starting point is likely to be the UNIX-based DELIGHT environment (Polak and coworkers, 1982).

Work in U.K. on methods for using computers to aid design of feedback controllers has evolved over a broad and varied front. It includes development of general multivariable linear design criteria and linear CAD packages, microcomputer implementations of off-line packages, creation of integrated packages, development of packages which make available a variety of advanced controllers, and several full-scale industrial trials. The work is all motivated towards practical applications and this next stage of evolution may benefit from a temporary retreat away from variety in favour of the standardisation of the SERC initiative. There is ground for optimism that this initiative, coupled with continuing reductions in the cost of computing power, will soon help the new methodology to find its rightful place in engineering practice.

5 ACKNOWLEDGEMENT

Thanks are due to many colleagues in U.K. and authors of referenced publications who provided helpful responses to the author's queries.

6 REFERENCES

Åström, K. J. (1983). Theory and applications of adaptive control. Automatica, 19, 471-486

Clements, R. (1984). BCSSP. Inst. M & C (1984), q.v., p243

Dorey, R. E. (1984). Computer simulation of control systems. Inst. M & C (1984), q.v., 173-186

Dutton, K. (1984). CLASIC-Developing a system analysis design and simulation package. Inst. M & C (1984), q.v., 99-103

Elmquvist, H. (1977). SIMNON. Proc. Simulation '77, Montreux 1977

Foo, Y. K. and I. Postlethwaite (1984). An H^{∞}-minimax approach to the design of robust control systems. Syst. & Cont. Letters, 5, 81-88

Foss, A. M. (1984). DICAST: Integrated use of CACSD. Inst M & C (1984), q.v., 129-135

Francis, B. A., J. W. Helton and G. Zames (1984). H optimal feedback controllers for linear multivariable systems. IEEE Trans. AC, 29, 888-900

Glover, K. (1985). Robust stabilization of linear multivariable systems and relations to approximation, CUEL Report, in preparation.

Goodfellow, S. D. and N. Munro (1984). An input translation facility for control systems CAE. Inst. Mec. (1984), q.v., 21-32

Gray, J. O. (1984). Nonlinear system design, an integrated approach. Inst. M & C (1984), q.v., 39-47

Hashimoto, I. and T. Takamatsu (1981). New results and the status of computer-aided process control systems design in Japan. Chem. Proc. Cont., 2, ?-185

Havranek, W. A. (1984). Inst. M & C (1984), q.v., 165-168

Hay, J. L. (1984). Inst. M & C (1984), q.v., 159-164

Inst. M & C. (1984). Computer aided control system design. Preprints of Workshop, Brighton, Inst. Meas. and Cont.

Isermann, R. (Ed.) (1981). System identification. Pergamon

Jacobs, O. L. R. and W. A. Badran (1983). A computer package for designing pH regulation systems. Oxford Univ. Eng. Lab. Rept., 1506/83

King, M. J. (1984). Interactive process control training aids. Inst. M & C (1984), q.v., 105-108

Koudstaal, J., D. B. Hulbert, M. Braae and G. I. Gossman (1981). The application of a multivariable controller to an industrial grinding circuit. Proc. VIIIth IFAC congress, Vol. 22, 210-215

Lawrence, P. J. (1984). Control system design toolkit for personal computers. Inst. M & C (1984), q.v., 115-120

Linkens, D. A. (1984). muPSI - A microcomputer simulation language for continuous and discrete dynamic systems. Inst. M & C (1984), q.v., 153-158

MacFarlane, A. G. J. (ed.) (1980). Complex variable methods for linear multivariable feedback systems, Taylor and Francis

Maciejowski, J. M. (1984). Data structures for control system design, EUROCON 84, Brighton

Maciejowski, J. M. and A. G. F. MacFarlane (1982). CLADP: the Cambridge linear analysis and design programs. Cont. Syst. Mag., Dec., 3-8

Munro, N. (1979). The UMIST control systems design and synthesis suites. IFAC Symp. CAD, Zurich

Munro, N. (1982). The inverse Nyquist array design method. Chap. 5 of Design of modern control systems, Eds. Bell, Cook and Munro. Peter Peregrinus

Munro, N., B. J. Bowland and C. Hopkins (1984). MINAD - Micro inverse Nyquist array design. Inst. M & C (1984), q.v., p 221

Polak, E., P. Siegel, T. Wuu, W. T. Nye and D. Q. Mayne (1982). DELIGHT MIMO: an interactive optimization-based multivariable control system design package. Cont. Syst. Mag., Dec, 9-14

Polak, E., D. Q. Mayne (1984). Theoretical and software aspects of optimization-based control system design. Proc. VI the Int. Conf. Anal. & Opt. of Syst., Nice, 175-190. Springer-Verlag.

Polak, E., D. Q. Mayne and D. M. Stimler (1985). Control system design via semi-infinite optimization. IEEE Proc., to appear.

Rosenbrock, H. H. (1974). Computer aided control system design. Academic Press

Sandoz, D. J. (1984). CAD for the design and evaluation of industrial control systems. Proc. IEE, 131D, 125-139

Sebastian, D. J. G. and S. J. Murray (1984). Inst. M & C (1984), q.v., 193-197

Thompson, M. A. (1984). Computer aided control system design software in the Central Electricity Generating Board. Inst. M & C (1984), q.v., 143-146

Thompson, S. (1984). Microcomputer-based CACSD software for continuous single-input single-output systems. Inst. M & C (1984), q.v., 79-83

Tuffs, P. S. (1985). FAUST. Oxford Univ. Eng. Lab. Rept., 1568

Tysso, A. (1981). New results and the status of computer-aided process control system design in Europe. Chem. Proc. Cont., 2, ?-211

INDUSTRIAL APPLICATIONS OF CAD-PACKAGES FOR CONTROL SYSTEMS IN THE USSR

B. C. Tamm*, M. Oit, R. Raud** and J. Vain****

**Tallinn Technical University, Tallinn, 200026 USSR*
***Institute of Cybernetics, Estonian Academy of Sciences, Tallinn, 200026 USSR*

Abstract. Different CAD-packages for control systems, designed and used in the Soviet Union are treated. Control algorithm design and control system design, being the essential parts of CAD-package development are described in more detail. References are made to the contributions which discuss a number of particular packages as well as their application areas.

Keywords. Computer-aided design; control algorithm design; control system design; software development.

INTRODUCTION

This paper deals with CAD-packages for control systems designed and used in the USSR. Since the development goes through control algorithm design and control system design, those two stages are described in more detail. References are made to the contributions where the problem under discussion is treated in greater depth.

The paper concludes with an overview of the application areas of the packages discussed.

The survey does not claim to give a complete coverage of the results obtained and the packages designed and used in this country.

CONTROL ALGORITHM DESIGN

Control algorithm design is the part of control system design which starts from the mathematical description of the dynamical behaviour of the control object and results in the desired regulator parameters. The latter may serve as input data for the design of the control system itself. Mathematical and computational problems are the prime concern in the design and the reasons for computer use lie here. Functionally we can distinguish several stages in the design (Fig. 1).

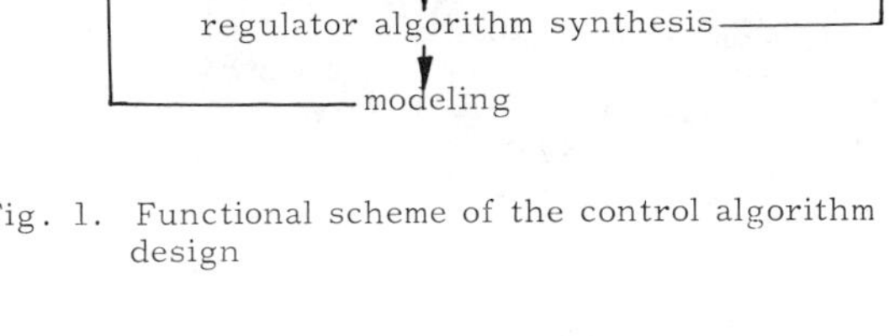

Fig. 1. Functional scheme of the control algorithm design

Object specification. In this stage we have to specify object variables (control inputs. disturbances, controllable and uncontrollable outputs), determine their allowed values, possible application areas and other qualities. This part is certainly indifferent to the place of investigation, and the approach in the USSR is quite the same as elsewhere.

Mathematical description. The description of the object dynamical behaviour may be obtained theoretically, using the corresponding physical, chemical or mechanical relations between object variables. Such models are commonly too complicated and need some simplification (e.g. linearization).

In addition, some packages have special subsystems for approximation, e.g. GAMMA-IM (Aleksandrov and others, 1981), AVANS (Andrievskii and others, 1983). Mathematical model can also be calculated from the measured values of the input and output variables. The identification process is one of the most complicated problems of control theory. Quite different identification methods are used in CAD-packages (Abezgauz and others, 1981, Smirnov and Sazonova, 1981), but there are also some packages without identification possibilities at all, e.g. package GAMMA-IM. Depending on the method used we get system description in the form of state-space equations, ARMA models, Markov parameters or in other forms. None of the description types is given more preference, each one is equally used. State-space description is used in packages SINTEZ (Arutjunov and Brdzozovski, 1981), COMICS (Jaaksoo, 1984), frequency methods in a package elaborated at Yerevan Polytechnic Institute (Preprints, 1984, p.286), ARMA models packages developed at Moscow Institute of Control Problems (Preprints, 1984, p. 284-285), networks in packages SPADE (Kukk, 1982) and DIPSTRUP (Melnikov, 1981). In some packages several description methods are allowed: in system DISPAS (Sadykov and Soloveitshik, 1981) - state-space description and transfer functions, in package SPRINT (Smirnov and Sazonova, 1981) - frequency characteristics and ARMA models.

Control problem statement. It means formulating your control problem via object description, and determining the desired dynamics and control criteria. The quadratic control criterion seems to be most popular in the CAD system. There are also some multicriterial systems (Solodovnikov and Zverev,

1984, Grigorenko, 1983). The exact problem statement is the basis for the subsequent design stages.

Analysis. Usually we analyze object stability, controllability, observability, and also interactions. Several packages consist only of analysis and modeling stages of design, e.g. SPADE, AINS (Ptitchkin and Batura, 1981). Some new interaction analysis techniques are included in package COMICS.

System structure design. In accordance with the results of the quality and interaction analysis, we design system structure and specify all the regulators for which we have to calculate parameters.

Control algorithm synthesis. This is the most computative part of the whole process. If the desired control law cannot be obtained by the fixed object parameters, then the choice of the regulator type or the control system structure is inappropriate. Almost all the well-known regulator algorithm types are presented in CAD systems, more frequently P, PI, PD, PID algorithms, also modal, dead-beat and adaptive regulators. Special methods for adaptive system design are used (Derevitski, 1982, Andrievski and others, 1983). In system DISPAS the regulator is not determined, and the algorithm is synthesized by the prescribed dynamics instead.

Modeling. Modeling of the designed control system dynamics is the final check for control algorithms. Depending on the control process type, modeling might occur quite complicated to accomplish. There are special theoretical approaches to modeling in CAD systems (Lobusov, 1982) and special packages for modeling, e.g. DISNEM (Osipov and others, 1981). If modeling results are not satisfactory, then there is something wrong with the control problem statement. If the requirements to object behaviour are satisfied, we have got the necessary initial data (structure and regulator algorithms) for the control system design.

CAD systems for algorithm design include on each occasion subsystems called IDENTIFICATION, ANALYSIS, SYNTHESIS and MODELING and some others, depending on the purpose and possibilities of the system. Differences between these CAD systems are initiated by:

- control object classes (continuous, discrete, stochastic, etc.)
- object mathematical description methods (state space equations, frequency methods, etc.)
- regulator algorithm types (optimal, modal, adaptive, etc.)
- design process type (fixed, interactive).

CONTROL SYSTEM DESIGN

The development of control systems can be divided into the following stages.

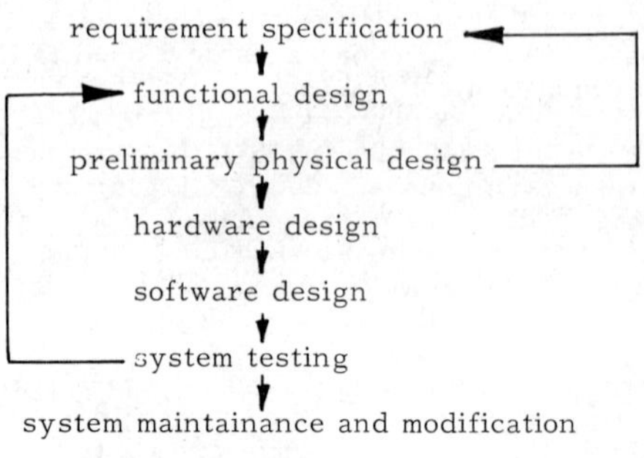

Fig. 2. Functional scheme of the control system design

Requirements Specification. Control system CAD-packages developed in the USSR are mainly oriented to the formal analysis of the requirements specification and specification languages of these packages are therefore close to the used formalism rather than oriented towards the user.

For example, the package (Motus and Lomp, 1984) is based on the conceptual model of communicating sequential processes and on the formalism described in (Quirk and Gilbert, 1977; Motus and Kaaramees, 1983).

Another package (Pranevicius, 1982) developed at Kaunas Polytechnic Institute uses the notation of aggregate theory and enables to specify operational parameters as in (Motus and Lomp, 1984). In addition a package (Pranevicius, 1982) specifies algorithms of aggregates.

Problem oriented constructs are developed in ASM specification language of a package (Burovoi and Uriadov, 1983). For the specification in ASM, the package generates a simulation model of the control system in the GPSS language. This package has been applied for a number of industrial systems (e.g. for the control of concentration technology of copper nickel converter matte).

Functional design. To verify the functional design, static and dynamical analysis have to be distinguished. These two kinds of checks were first suggested in (Biewald and others, 1979).

Time invariant properties such as deadlocks, stability, termination and others are examined by the state analysis, whereas dynamical aspects of a given specification are examined by simulation (Pranevicius, 1982) or less frequently by analytical modeling (Motus and Lomp, 1984; Motus and Vain, 1982).

The packages developed for requirement specification (Motus and Lomp, 1984; Pranevicius, 1982; Burovoi and Uriadov, 1983) are also available for functional design due to the ability to model system behaviour as a time function. Another package DSIM (Kostin and others, 1984) should be noted here. This package uses E-net formalism for simulation of computer networks. One of the most advanced packages with static and performance analysis tools is described in (Pranevicius, 1982). Unfortunately for the aggregate model no ready methods exist to prove timing correctness analytically.

Timing properties are readily provable only in package (Motus and Lomp, 1984). It has been successfully applied for oil cracking catalyst production process description (Motus and others, 1984) and is being used for specification and design of an automatic test system for semiconductor devices. At the end of the functional design stage we get an abstract network of interacting functional tasks described in terms of some formalism.

Preliminary physical design. In this stage the main design decisions are coupled with the transformation of abstract process network into a set of physical (software/hardware) modules and selection of algorithms for them.

For the first rough estimation of preliminary design decisions, analytical modeling methods based on queuing theory and particularly queuing network models are mainly used. A good example is a package supporting synthesis and analysis of large scale distributed structures elaborated in Leningrad (Zakharov and Lohmotko, 1981). At the end of the preliminary physical design we get a list of tasks for a detailed design of hardware modules and different software parts.

For automated design of software there is a package ASTRA (i.e. An Automated Structural Technology for Algorithm Design),Novosibirsk (Zavialov and others, 1983). ASTRA has a specification language for multi-level structural design of programs - SMP close to essential Russian. By means of SMP one can specify the initial algorithm which then is to be translated into one of the output programming language codes: Assembler of ELEKTRONIKA-60 micro, FORTRAN, Cobol, Assembler of EC computer. The package has also a macrogenerator, debugging, testing and documentation tools.

Control system hardware detailed design. It has long tradition in the USSR (Glushkov, 1962; 1965; 1981). Some of these theoretical results are supported with CAD/CAE systems (Glushkov and Gusev, 1975). Functionally these methods and systems are somehow covering the whole area starting from conceptual design (Landau and Pogosjanz, 1983; 1984; Glushkov and Gusev, 1975), logic design and modeling (Goldina and others, 1981; Glushkov and Gusev, 1975; Rjabov and Lakshin, 1980), physical design (Rjabov and Lakshin, 1980; Petrenko and Tetelbaum, 1979), placement and routing for printed circuit boards and VLSI design (Petrenko and Tetelbaum, 1979).

Not all of these systems have advanced designer interface and technological service. One of the reasons why they are not broadly used is the fact that they operating on mainframe computers under large operating systems. Therefore they are accessible to hardware designer only by specially trained persons.

The recommended hardware description languages for modeling on different levels of detailization (Goldina and others, 1981, Rjabov and Lakshin,1980) only have textual interface and are not syntactically and semantically related to each other.

An ARM2-01 system where CAD-packages for physical design are intergrated with specialized graphic hardware relates to the physical design area. These packages offer flexible graphic interface for a hardware designer (Zabara and Verpinski, 1983). There is another CAD-package for physical design worth mentioning - MAGISTER (Oihman and others, 1983), running on SM-4 minicomputer with graphic peripherals.

Software development. In software development the picture is completely different. The main results here are not embedded in theoretical conclusions, but different software packages are produced instead. These packages, being cross-systems, are dedicated for control systems software implementation and are mainly running on mainframe computer under general purpose OS (Raud and Tamm, 1982, Raud and Smeljanski, 1984). With a few exceptions they are offering support for coding (mainly on the level of cross-assemblers with macrocapabilities) and functional debugging.

There is one system which claims to full coverage of control system software life cycle support - RUZA (Shtrik, 1984). Unfortunalately RUZA does not offer a complete set of tools for real-time debugging and testing. Though it is a complicated system, it is good for supporting large control systems.

For medium and small projects we have good experience with another software development environment - SERP (Raud and Smeljanski, 1984), offering some support for real-time software coding, interactive debugging, test data generation and program testing. SERP is a cross-system too and therefore does not deal with software/hardware integration and system testing.

There are two development systems in production. One of them is SO-4 (Malinovski and others, 1982) which has the same functional capabilities as INTEL INTELLEC earlier versions. The other one ARM2-05 (Milner, 1984) is a network, including SM-4 minicomputer as a central unit with disk storage and up to four K580 based microcomputers SM-1800 as terminal units. Terminals are equipped with in-circuit emulators and PROM programmers for software development. ARM2-05 supports also microprogram development having cross-microassemblers and specialized in-circuit emulator units for K589 K1804 based architectures.ARM2-05 offers a full set of software development support tools. ISIS-II and CP/M compilers are running on ARM2-05 also.

Since it takes time to produce sufficient volumes of special hardware to support software development, cross-systems will still play a significant role for some time, despite some shortcomings, such as: hardware/software development are isolated from each other and integration cannot be supported on the software side, lack of autonomy for a programmer.

Software maintainance and modification. Software maintainance and modification have some kind of remote support only in system RUZA.

INDUSTRIAL APPLICATIONS OF CAD-PACKAGES

The most advanced application of CAD-packages is computer-aided design of flying objects control systems. For such problems package DISPAS is oriented, which is today in use for airplanes and other flying objects control system design in many design bureaus. Another package DIASIM (Preprints, 1984, p. 321-322) elaborated at Moscow Aviation Institute, like the previous one, is used for airplane numerical control system design.

Quite wide-spread is the application of CAD-packages for chemical process control system design. Packages DSS (Preprints, 1984, p.175-177), COMICS, UNISAM (Kondrashin and Krutchin, 1981), SYNSAU (Preprints, 1984, p. 303-304), ESKORT (Podtchukaev and Iljin, 1983) deal with such problems, e.g. using SYNSAU, a system for acrylonitrile synthesis control is designed. There are also applications in metallurgy, e.g. using a package (Diligenski and Tumanov, 1981) aluminium alloy hot rolling control system is designed. Some applications are in the oil industry. CAD-packages are used for oil boring stations control system design (Preprints, 1984 pp. 305-306; 318-320), also for oil refining process control design (Preprints, 1984, p. 287-288).

Applications are mentioned in the field of discrete control systems computer-aided design, e.g. package DISTRUP is used for control system design in radioelectronics. Some more application areas are as follows: packages IISPUO (Preprints, 1984, p. 95-96), SAMP (Preprints, 1984, p. 22-24) are used for management systems design; package AVRORA (Smirnov and Malov, 1981) - for municipal information management systems design; packages ANSIN and MADINESE - for electrical network design; package MODUS-(Preprints, 1984, p. 61-63) - for nuclear power station control system design, package COMICS - for grain dryer system design.

The development and use of CAD-packages for control systems are expanding in the USSR in line with the advances of automatic control methods and computer control in new technology and management.

REFERENCES

Abezgauz, M.M., R.I. Maizus, and A.S. Grinberg (1981). Identification of industrial systems in in the computer-aided design process. In V.A.Trapeznikov (Ed.) Computer-aided control system design. Finansy i Statistica, Moscow. pp. 182-202. (in Russian).

Aleksandrov, A.G., A.A. Markov, and T.A. Petrov (1981). Dialogue computer-aided design package "GAMMA-IM". In Preprints of the 3rd All-Union conference on computer-aided design of automatic and automatized technological process control systems. Ivanovo Electrotechnical Institute, Ivanovo. pp. 149-150. (in Russian).

Andrievski, B.R., V.V.Kasatkin, A.A. Spiridonov, V.N. Utkin (1983). Architecture of adaptive systems CAD-package AVANS. In Preprints of the All-Union conference on adaptive systems theory and application. Leningrad. p.324 (in Russian).

Arutjunov, S.K., and A.E. Brdzozovski (1981). Computer-aided control system design system "SINTEZ". In Preprints of the 3rd All-Union conference on computer-aided design of automatic and automatized technological process control systems. Ivanovo Electrotechnical Institute, Ivanovo. pp. 139-140. (in Russian).

Biewald, J., E. Joho, S. Jovalekic, and H. Shelling (1980). Application of the specification and design technique EPOS to process control problems. In Proceedings of the 6th IFAC/IFIP Conference on digital computer applications to process control. Dusseldorf. pp.517-522.

Burovoi, I.A., and V.M.Uriadov (1983. Automated synthesis of distributed real-time computer system simulation models. In Modular informatical-computational systems; Preprints. Irkutsk. pp. 50-52. (in Russian).

Derevitski, D.P. (1982). Prototype method for adaptive system computer-aided design. In Cybernetical problems. Actual problems of adaptive control. Council for Cybernetics, Acad. of Sciences of the USSR, Moscow. pp.68-90 (in Russian).

Diligenski, N.V., P.V. Sevastjanov, and N.V. Tumanov (1981). Automation of optimal rolling regime design. In Preprints of the 3rd All-Union conference on computer-aided design of automatic and automatized technological process control systems. Ivanovo Electrotechnical Institute, Ivanovo. pp. 127-128. (in Russian).

Epshtein, V.L. (1983). ARIUS. Computer-aided design of data bases and software packages; Preprint. Institute of Control Problems, Moscow. 27 pp. (in Russian).

Glushkov, V.M. (1962). Synthesis of numerical automata. Fizmatgiz, Moscow. 390 pp. (in Russian).

Glushkov, V.M. (1965). Automata theory and computer structure design. Kibernetika, 1, 3-11. (in Russian).

Glushkov, V.M. (1981). On formal transformations of algorithms. Lecture Notes on Computer Science, 122, 430-440.

Glushkov, V.M., and V.V.Gusev (1975). Program tools for analog-digital system modeling. Naukova Dumka, Kiev. 237 pp. (in Russian).

Goldina, L.L., I.L. Landau, and G.M. Pogozjanz (1981). Automated development system of digital devices. Computers in Socialist Countries, 119, 33-42. (in Russian).

Grigorenko, V.P. (1983). Modeling and optimisation in computer-aided design systems. In Modeling and optimisation of design; Preprints. Tallinn. pp.12-16. (in Russian).

Jaaksoo, U.J. (1984). Computer aided package for multivariable control system design, using state variable methods. Mashinostroenie, Moscow, 40 pp. (in Russian).

Kondrashin, A.V., and V.V. Krutchinin (1981). Universal system for computer-aided modeling of complex technological processes (UNISAM). In Preprints of the 3rd All-Union conference on computer-aided design of automatic and automatized process control systems. Ivanovo Electrotechnical Institute, Ivanovo. pp.102-103. (in Russian).

Kostin, A.E., V.M.Iljuchenkin, and V.F. Schaugin (1984). The presentation and modeling of protocols by means of E-nets. In Preprints of the 9th workshop on computer networks, vol. 1.1, Pushchino. pp.14-19. (in Russian).

Kukk, V. (1982). System SPADE: Principles of Operation. Transactions of Tallinn Technical University, 535, 3-17. (in Russian).

Landau, I.J., and G.M. Pogosjanz (1983). Hierarchical modeling method of modeling digital hardware and microprocessor systems. In CAD and modeling of microprocessor systems. Znanie, Moscow. pp.55-60. (in Russian).

Landau, I.J., and G.M. Pogosjanz (1984). Abstract data types for digital hardware modelling language construction. Proc. of INEUM, 102, pp. 17-22. (in Russian).

Lobusov, E.S. (1982). Using of imitational modeling, hybrid computers and stands for CAD of control systems. Mashinostroenie, Moscow. 56 pp. (in Russian).

Malinovski, B.N., A.I. Slobodjanuk, and S.D. Pogorely (1982). A K580 IK80 based microcomputer and debugging system software. USIM, 3, 30-34. (in Russian).

Melnikov, A.V. (1981). Dialogue system for structural design of complex control and process control systems. In Preprints of the 3rd All-Union conference on computer-aided design of automatic and automatized technological process control systems. Ivanovo Electrotechnical Institute, Ivanovo. pp. 153-155. (in Russian)

Milner, A.D. (1984). Linguistic Support of Development System ARM2-05; Prerequisites, concepts, survey. In Microprocessor programming. Tallinn. pp. 30-41. (in Russian)

Motus, L., and K. Kaaramees (1983). A model based design of distributed control systems software. In Proceedings of the 4th IFAC Workshop on DCCS. Pergamon Press, London. pp. 93-101.

Motus, L., and A. Lomp (1984). A tool for specifying distributed computer control system's software dynamics. In Preprints of the 9th IFAC World Congress, Budapest. vol. 2. pp. 144-148.

Motus, L., and J. Vain (1982). A set of tools for designing and evaluating communication protocols in industrial computer networks. In Preprints of the 3rd IFAC/IFIP Symposium on Software for Computer Control. Madrid. pp. 101-109.

Oihman, E.G., J.V. Novodzenov, and J.V.Zuzin (1983). Principles of personal workstations for printed circuit boards. Pribory i sistemy upravlenia, 11, 4-6. (in Russian).

Osipov, B.L., A.Yu. Variash, and A.A.Kurtasov (1981). Special software for continuous-discrete control system modeling. In Preprints of the 3rd All-Union conference on computer-aided design of automatic and automatized technological process control systems. Ivanovo Electrotechnical Insittute, Ivanovo. pp. 93-94. (in Russian).

Petrenko, A.I., and A.J. Tetelbaum (1979). Formal construction of computer hardware. Sovetskoe Radio, Moscow. 256 pp. (in Russian).

Podtchukaev, V.A., and S.A. Iljin (1983). Dialogue system "Escort" for computer-aided synthesis of linear control systems. In Elaboration and inculcation of automatic and automatized systems for continuous and continuous-discrete technological process control. All-Union

Scientific-Technical Society, Moscow. pp.69-70. (in Russian).

Pranevicius, H. (1982). Models and methods for computer system investigation. Mokslas, Vilnius. 228 pp. (in Russian).

Preprints. 9th All-Union Conference on control problems. (1983). VINITI, Moscow. 535 pp. (in Russian).

Preprints. All-Union Conference on computer-aided design of control systems (1984). VINITI, Moscow, 341 pp. (in Russian).

Ptitchkin, V.A., and M.P. Batura (1981). Moment equation methods in computer-aided analysis of nonlinear control systems. In Preprints of the 3rd All-Union conference on computer-aided design of automatic and automatized technological process control systems. Ivanovo Electrotechnical Institute, Ivanovo. pp.99. (in Russian).

Quirk, W.J. and R. Gilbert (1977). The formal specification of the requirements of complex real-time systems. Harwell, AERE, R-8602. 57 pp.

Raud, R.K., and B.G.Tamm (1982). Microcomputer programming: Survey. Programmirovanie, 5, 31-43. (in Russian).

Raud, R.K., and R.L. Smeljanski (1984). Cross-systems: present states, main problems. In Preprints of All-Union Conference on industrial technology of software development, application and management. Sverdlovsk. pp. 250-251. (in Russian).

Rjabov, G.G., and G.L.Lakshin (1980). Logic level modeling in KASPI-EVM. In Computer modeling. Znanie, Moscow. pp. 62-66. (in Russian).

Sadykov, F.R., and G.J. Soloveitchik (1981). Dialogue system for design of automatic control systems (DISPAS), verse 2.1. In Methodical guide for using programs from MAI computer centre fund. Moscow Aviation Institute, Moscow pp. 17-39. (in Russian).

Shtrik, A.A. (1984). RUZA - a cross-system for automated software development for micro- and control computers. Microprocessor Devices and Systems, 2, 46-49. (in Russian).

Sivtshov, V.I., and N.A. Tchulin (1982). Computer aided synthesis of control systems using control theory (frequency methods). Mashinostroenie, Moscow. 56 pp. (in Russian).

Smirnov, A.V., and A.D. Malov (1981). Interactive system for computer-aided solving of operative tasks AVRORA. In Problems of elaboration of support tools for computer-aided design systems. Moscow. pp. 59-69. (in Russian).

Smirnov, M.V., and T.G. Sazonova (1981). Computer-aided package SPRINT. Subsystem for identification. In Preprints of the 3rd All-Union conference on computer-aided design of automatic and automatized technological process control systems. Ivanovo Electrotechnical Institute, Ivanovo. pp. 144-146. (in Russian).

Solodovnikov, V.V., and V.J. Zverev (1984). Using of control theory methods and multicriterial control methods for computer-aided design of technological process control systems. Mashinostroenie, Moscow. 48 pp. (in Russian).

Tshvetkov, V.D., A.I. Petrovski, and A.A. Tolkatchev (1984). Problem-oriented languages for computer-aided technological design. Nauka i technika, Minsk. 192 pp. (in Russian).

Zabara, S.S., and G.J. Verpinski (1983). Graphic development station. In Computers in design and industry. Mashinostroenie, Moscow. pp. 191-202. (in Russian).

Zakharov, G.P., and V.V.Lohmotko (1981). Structural optimisation of data transmission systems with packet communication. Council for Cybernetics, Acad. of Sciences of the USSR, Moscow. 64 pp. (in Russian).

Zavjalov, Yu. S., V.A. Skorospelov, V.M. Gamidov, N.N. Pavlov, and P.A. Turuk (1983). Adaptive system ASTRA for geometric information processing. In T.M.R. Ellis and O.J. Semenkov (Ed.) Advances in CAD/CAM. North-Holland Publishing Company, IFIP. pp. 399-407.

INTEGRA, AN INPUT TRANSLATION FACILITY FOR COMPUTER AIDED CONTROL SYSTEMS ENGINEERING

S. D. Goodfellow and N. Munro

Control Systems Centre, UMIST, Manchester M60 1QD, UK

Abstract. A software tool for input translation called INTEGRA (INput of TExt and GRAphics) is presented. INTEGRA is able to translate a stream of mixed graphical and textural input. The acceptable input is defined by the applications developer using extended transition networks represented as graphs, which are entered and manipulated graphically. INTEGRA provides a means of controlling the fine detail and a method of abstracting away from it using nested graphs. A variety of interaction tools are supplied to the user including macro languages, synonyms, prompting, menus, help, walkthrough, tracing, and customising. INTEGRA provides for the future development of expert system based adaptive man machine interfaces.

Keywords. Computer graphics, Computer software, Man-machine systems, Software engineering.

INTRODUCTION

Computer Aided Control Systems Engineering (CACSE) software provides the engineer with a set of tools which may be applied to the analysis, design and implementation of control systems. The basic techniques of CACSE have been well tried and tested through two generations of CACSE software. A new generation of software is now being developed which will significantly improve the engineer's design environment. The engineer will be able to concentrate on the engineering task rather than the detailed application of a particular piece of software.

The working environment provided by CACSE software is primarily a function of it's man machine interface. To provide the appropriate design environment a sophisticated man machine interface is required, supporting graphical and textural input in a consistent manner. The interface will also have to be able to adapt to the engineer's experience and skill. This paper concentrates on the input aspects of the man machine interface, in particular an input translator tool is described called INTEGRA.

Interaction Styles

A wide variety of interaction styles are used in conversationally based software. However, there are two fundamental modes of operation; selection mode and command mode. A software package which uses the selection mode of interaction provides a restricted selection of input from which the user must choose a response. The selection of input may either be presented to the user in the form of a menu or a question may be asked to which the user responds, in the event of an erroneous response the user is prompted with the selection available for a correct response. A well designed interaction of this type will suit the unfamiliar or infrequent user of CACSE software, particularly when the functions involved require complicated dialogues.

The skilled user often finds the selection mode of operation somewhat pedantic. For him, the command mode of interaction offers distinct advantages. A package using the command mode of operation allows the user to enter any reasonable sequence of commands and data at each point during the conversation. In this mode the user has to take responsibility for the path through the applications program. This facilitates the use of more powerful and sophisticated CACSE functions. The command mode is not easy for unskilled users to operate. Often some combination of the selection and command modes of interaction will suit a large majority of users from unskilled to expert.

Input Translation

Input translation code may be either embedded in the applications code or implemented as a separate unit alongside the applications code, with a specified interface to the application. The former is the case in most existing conversationally based software. Input translation code appears in the applications code immediately before the code which requires the user response. The UMIST control systems computer aided design suite (Munro, 1979) implements a selection mode of interaction using embedded translation code.

A number of packages exist which provide an input translator as a separate unit to be inserted alongside the application. These include YACC (Johnson, 1975), INTRAC (Wieslander, 1978), GILT (CADCentre, 1981), and CANAL (Cartwright, 1984). YACC generates a parser from a set of dialogue rules specified by the applications designer. The resulting parser may be used in conjunction with a lexical analyser such as one produced using LEX (Lesk, 1979) to provide an input translator. A similar parser generator was used to create the parser for Delight (Nye, 1983). GILT and CANAL use a dialogue definition based on extended transition networks, a textural description of the networks or graphs is specified by the applications designer. A fixed lexical analysis is used which can recognise a limited number of textural atoms; CANAL's lexical analyser can also recognise some rudimentary graphical input. Although neither CANAL nor GILT assume a particular mode or format of interaction, both are limited by their lexical analysis and basic parsers.

REQUIREMENTS OF AN INPUT TRANSLATOR

An input translator is a software tool which at runtime translates user input into applications action lists. A dialogue definition describes the legal syntax. The translator may support a number of interaction tools for the user, such tools are hidden from the application. In particular an input translator is required which may readily be applied by the developers of CACSE software, and provide a significantly higher functionality for the man machine interface.

A suitable input translator will :-

1. translate mixed graphical, numerical, and textural input.

2. use a dialogue definition schema which is readily understood by typical developers and users of CACSE software.

3. have a minimum of restrictions on the acceptable sequences of input. Any restrictions must form part of the syntax of dialogue definition.

4. provide a rich variety of interaction tools including high level languages, menus, prompting, synonyms, help, etc.

5. readily support existing dialogue techniques and interaction styles.

6. have a well defined interface with the applications code, which allows the dialogue to be modified without necessitating alteration of the applications code.

7. provide tools to assist the dialogue designer to debug the dialogue and record the use of the dialogue, and hence the underlying applications actions.

Due allowance must be made for issues of portability between varieties of machine from single user workstations to large multi-user systems, distributed systems must also be allowed for.

None of the input translation packages listed above, YACC, INTRAC, GILT and CANAL, provide all the facilities listed above. CANAL comes the closest to meeting our requirements because it supports rudimentary graphical input. However it does not meet requirements 2 and 3 or provide the flexibility we require.

INTEGRA

The authors are currently developing a two pass input translator called INTEGRA - INput of TExt and GRAphics. Existing input translators are essentially three pass translators consisting of lexical analysis, syntatic analysis, and semantic analysis. The lexical analysis partitions the input sequence into atoms; an atom being the minimal unit of input recognised by the syntatic analyser. The syntatic analyser parses the input, and the semantic analyser carries out the meaning of the input, that is the indicated application program actions.

INTEGRA is a novel two pass parser; the lexical analysis is omitted. INTEGRA's syntatic analysis operates with the natural input atoms. The dialogue definition schema is simplified because a separate lexical analysis definition is not necessary. INTEGRA will be described by first specifying the dialogue definition schema, followed by the translator whilst introducing further aspects of the dialogue definition schema, and finally INTEGRA's interaction tools are specified.

DIALOGUE DEFINITION SCHEMA

The dialogue is defined using extended parallel transition networks, referred to here as graphs. The graphs may be entered, displayed and edited graphically both at compile time and runtime. The presentation of the dialogue definition in a graphical form provides a representation which is readily understood. The graphs define acceptable sequences of input atoms.

Table 1 A list of INTEGRA atoms.

Generic	
SINCHAR	A single specified character e.g. A, -, /, 4, a.
SINLET	A single specified letter case ignored e.g. Aa read A or a.
RANCHAR	A range of characters from the ASCII sequence e.g. A-Z, g-j, 0-9.
INTERRUPT	A single specified interrupt, probably a control character, a range of interrupts will be available.
LITERAL	String of characters specified e.g. All modes, 'Aa' the string is surrounded in quotes if confusion with another atom is possible.
LITLET	String of characters specified, no discrimination on case.
Specific	
ANYCHAR	Any single character.
NOMATCH	A transition containing the atom is followed if there is no matching transition from the state.
NULL	The empty string, a transition containing this is always followed.
LOCATOR	A single pair of co-ordinates returned by a pointing device.
PICK	A GKS segment name.
CHOICE	A choice from a selection.
VALUATOR	A value from an input device other than the keyboard.

Atoms

An atom is a minimal chunk of input such as a character or a single graphical action. A list of typical atoms is given in Table 1. Generic atoms are specified when they are placed in a graph, these include SINCHAR - a single specified character. The generic atoms LITERAL and LITLET are not minimal chunks of typed input, they are available to ease the task of specifying literal strings of characters. The specific atoms listed in Table 1 include all the graphical atoms such as LOCATOR -a single pair of coordinates resulting from the operation of a pointing device. The graphical atoms are closely related to the GKS input devices (Hopgood, 1983). Similar atoms could be specified for other graphical input environments.

The atoms used by INTEGRA are too fine for most input specification, particularly textural input. To aid the specification of input, molecules are specified which provide a level of input chunks similar to that of output from lexical analysers. Molecules are defined by graphs as sequences of atoms. These graphs are available to the user so that the molecule definition may be changed and extra molecules may be added.

Graphs

A graph is used to define the acceptable sequences

of atoms. A particular sequence is legal if there is a matching path through the relevant graph. A graph consists of a number of states interconnected by transitions. The states are represented by a pair of vertical lines and the transitions by a connected sequence of horizontal and vertical lines. The states represent points in the input stream immediately before an input atom.

A transition represents the input atom or atoms required to move from the initiating state to the receiving state, the two states connected to the transition. The transition contains a condition which may either be an atom or a subgraph. A path through a graph is a sequence of transitions, the initiating state of a transition on the path must be the same as the receiving state of the previous transition.

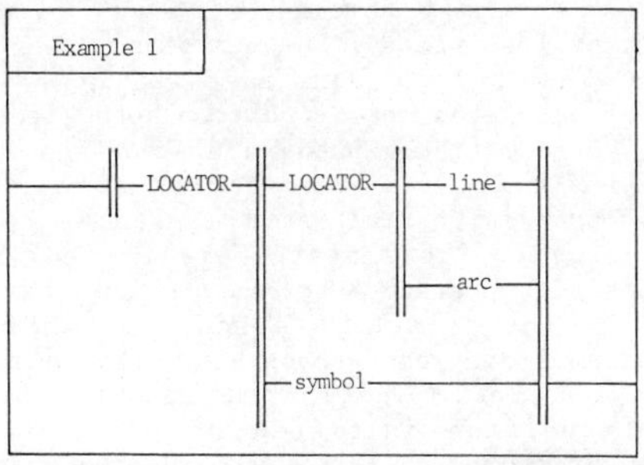

Matching input - LOCATOR LOCATOR line
LOCATOR LOCATOR arc
LOCATOR symbol

Fig. 1 An INTEGRA graph.

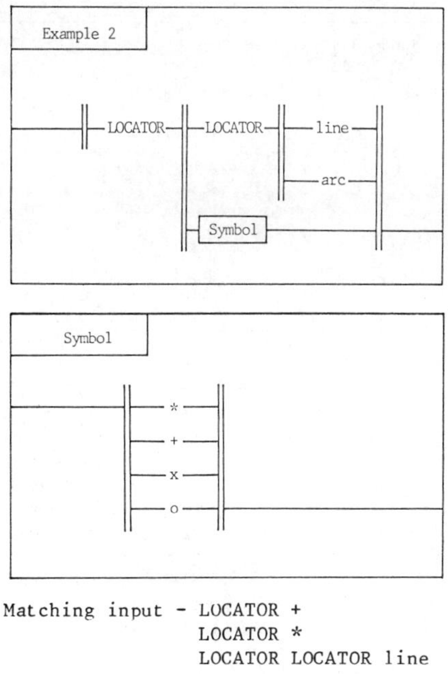

Matching input - LOCATOR +
LOCATOR *
LOCATOR LOCATOR line

Fig. 2 A graph containing a subgraph.

A matching path through a graph is a path for which the conditions on each transition are met in correct sequence by the input. A condition may be either an atom or a graph. An atom condition is met when the atom of input following the state matches the condition. A graph condition is met when the sequence of atoms of input following the state form a matching path through the graph. A graph is entered along a transition connected to the left hand edge and a transition connected to the right hand edge defines the exit. A condition referencing graphs is differentiated from one referencing an atom by placing the condition in a rectangular box.

The graph given in Fig. 1 illustrates the variety of acceptable input that even a simple graph may allow. The three possible sequences of matching input are listed. They represent the input required to place a line, an arc or a symbol on the screen. The operator may use a pointing device once, followed by typing the text "symbol" to position a symbol on the screen. Alternatively, the pointer may be operated twice, followed by typing either "line" or "arc" to position a line or arc on the screen.

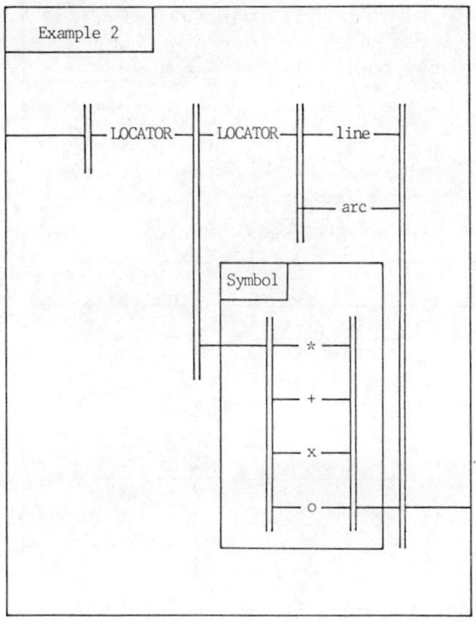

Fig. 3 The effective insertion of subgraphs

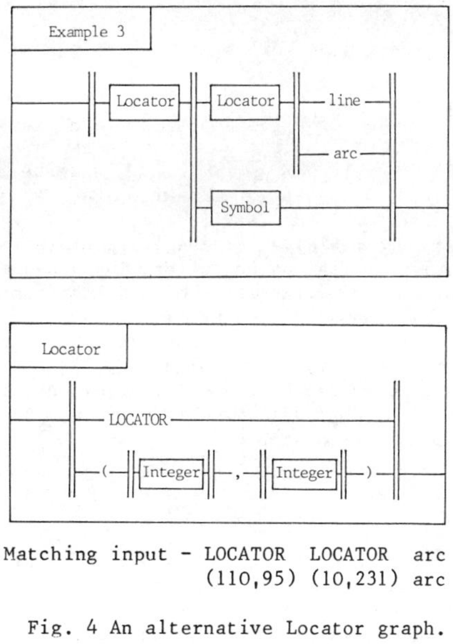

Matching input - LOCATOR LOCATOR arc
(110,95) (10,231) arc

Fig. 4 An alternative Locator graph.

Figure 2 illustrates the graphs Example 2 and Symbol. Example 2 is similar to the graph Example 1 illustrated in Fig. 1, instead of typing "symbol" the user may specify which symbol is required from a selection of four. The graph uses a subgraph call to Symbol to represent the choice of transitions at that point. The alternative representation of the two graphs given in Fig. 3 illustrates the action of a subgraph call. The subgraph is effectively inserted at the point of call in the calling graph. Subgraphs are particularly useful for reducing the complexity of graphs and avoiding repetition.

Recursion is allowed and is particularly useful for expression translation.

Figure 4 illustrates the use of graphs to provide alternative input sequences to produce the same results. In this case the location of a point on the screen may be specified using a pointing device or by typing its location. The graph Integer called from Locator is a standard molecule.

Table 2 A list of INTEGRA molecules

WORD	Any sequence of letters.
TEXT	Any sequence of characters, other than specified delimiters
FILENAME	A syntactically correct filename.
REAL	A Real number within language range.
DOUBLE	A double precision number within language range.
INTEGER	An integer number within language range.
COMPLEX	A complex number within language range.

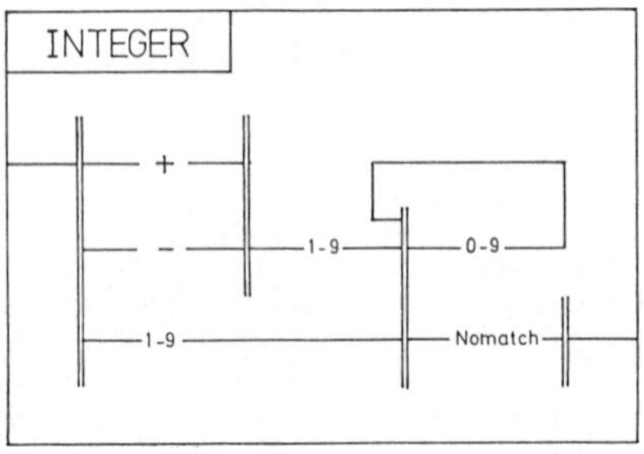

Fig. 5 An integer molecule.

Molecules

A molecule is a graph which specifies a standard form of input. A list of INTEGRA molecules is given in Table 2. The developer and user have access to the molecule graphs which may be edited and deleted or appended to the list. A typical integer molecule graph is illustrated by Fig. 5. The integer is defined as an optional sign followed by a nonzero digit which may be followed by a sequence of digits. The receiving state of a transition may be any state in the graph, so transitions may be used to form loops. The matching path may pass through a state or transition more than once. Loops are normally implemented with an exit condition. In this case, the loop is followed until an atom which is not a digit is encountered.

TRANSLATION

INTEGRA receives the input in a single stream of atoms, arranged in the order of initiation. The translator passes through the single stream of mixed input twice, the translation phase and the action phase. The translation phase is concurrent with the user's input actions allowing INTEGRA to react instantaneously in a conversational manner with assistance to the user, in the form of help, prompts, menus, etc. Action signals may be specified in the dialogue definition at states. When an action signal is encountered during translation, the translation halts and any remaining input is queued for the next translation phase. If the translation is not in error the action phase is initiated. During the action phase the matching path discovered by the translation is followed from beginning to end, and the actions specified at transitions or states are carried out.

Translation Phase

The INTEGRA translator uses the dialogue graphs to generate a tree of matching paths. At each state, the next atom of input is compared to the conditions on the transitions initiated by the state. All matching transitions are followed by adding the transitions to the translation tree. If a state is encountered with no matching transition initiated from it, the branch ending in the state is lopped off at its root.

The example of a translation tree given in Fig. 6 illustrates the lopping process. After the translation of the first atom, there is only one matching path from state 1 to state 2. The translation of the second atom finds two matching paths from state 1 through state 2 to state 6 (1,2,6) and from state 1 through state 2 to state 3 (1,2,3). The third atom must be compared to conditions on states initiated from both state 6 and state 3, three matching paths are found (1,2,6,8), (1,2,6,7), (1,2,3,4). The fourth atom does not match any transitions initiated at states 7 and 4, so the paths to these states are deleted, lopping off branches from state 2 to state 4 and state 6 to state 7 at their roots. A no match at a state causes an implicit back-tracking to the most recent fork in the tree. The fourth atom does match a single transition initiated at state 8. The sequence of atoms "ABC+" has a unique matching path (1,2,6,8,5) through the graph Example 4. Although it is not necessary for the translator to obtain a unique matching path, it is recommended that dialogue designers use the translator in this mode.

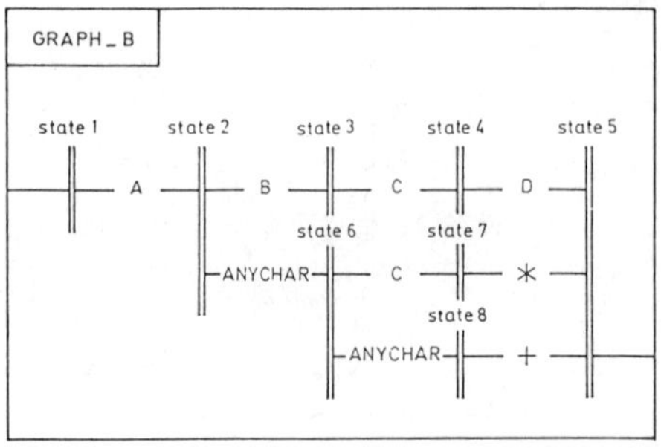

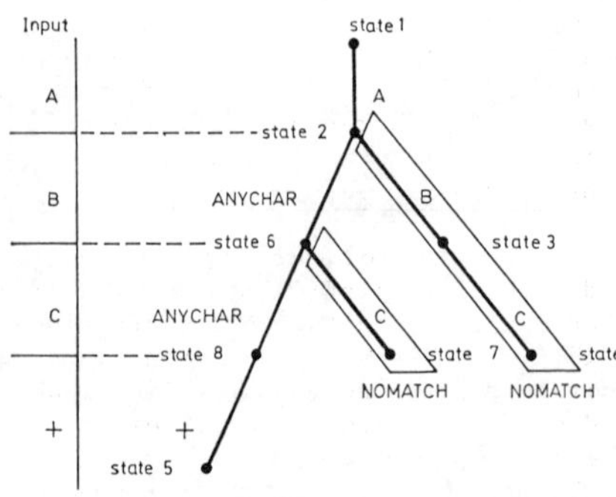

Fig. 6 A translation tree.

Parallel graphs are allowed. When they are used the input is compared to transitions initiated from the current state(s) in each of the parallel graphs. A graph may be spawned from an active graph, the spawned graph is activated at its first state.

Similarly, an active graph may be deactivated by collecting it in another active graph.

Each state and transition may have a page of information associated with it. The page contains information used in both the translation and action phases. Its schema is illustrated in Fig. 7.

During the translation, applications actions may be carried out. These serve two purposes, to collect input data and to check for erroneous input. The application translation actions must not alter or delete applications data because the translation phase may be aborted in the event of an error and repeated. No controlled method of undoing translation actions is provided.

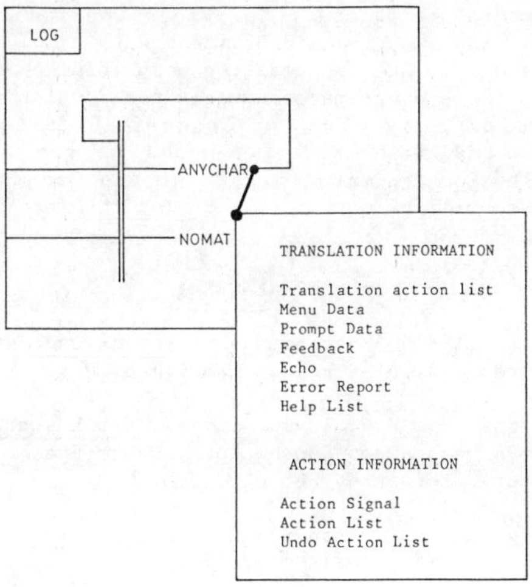

Fig. 7 Transition or state information.

The state or transition information page may contain information for menus, prompts, feedback, echoing and help. The translation phase uses this information to assist the user during his input actions. The use of parallel graphs confuses the provision of such assistance. The assistance is given for the context of the states receiving the transition matching the last input atom, unless it is specifically indicated that another context should be used.

Action Phase

The action phase is initiated by the translation phase encountering a state with an action signal on its information page. Conventional interaction styles use a return key to signal actions for a line of text input. The inclusion of a transition containing a return character condition received by a state containing an action signal will model the conventional style.

The action phase follows the matching path discovered by the translation phase inspecting the information page for each state and transition. The actions listed on the information page for the action phase are carried out. Applications actions are listed on the page in one of three ways. A block of source code may be entered as an applications action. INTEGRA is then responsible for managing and calling the block. Alternatively, a call to a subprogram may be specified. INTEGRA is then responsible for providing a block of code to carry out the call. The third alternative is to provide a message which INTEGRA sends to the applications subprogram ACTION. This third method is particularly useful because it can readily be implemented in a distributed processing environment as well as a conventional one.

The applications actions may send messages to the translator which treats them in a similar manner to user input. This facility is useful for error responses and for implementing expert systems. In the former case it is often useful to provide an error recovery dialogue. Following an error, the message from the application will move the translator to a defined error recovery dialogue graph.

The implementation of expert systems that do not sit between the user and the applications program is an interesting possibility. The specification of INTEGRA realises an expert system as a module of the package that communicates with INTEGRA using messages. An expert system could be used for dialogue management, reacting to a user who is floundering by providing more assistance and steering the dialogue towards a more explanatory style. It would also react to an impatient skilled user by steering the dialogue towards a more terse style. Applications expert systems that embody heuristical information are an important field of research (Taylor, 1984). INTEGRA provides hooks for such systems and ensures that the user is always in control. The expert systems along with the rest of the applications subprograms are subservient and dialogue driven.

DIALOGUE DEVELOPMENT TOOLS

Dialogue development and maintenance is supported by two major tools: tracing and reporting. The tracing tool will give the developer a graphical trace of the translation process; it is particularly useful during debugging and as a learning aid. The reporting facility will provide the developer with feedback on the use of the dialogue and application.

The reporting facility will collect statistics on the use of the dialogue, including detailed information on the use of sections specified by the developer. The occurrence of input errors will be recorded, especially errors repeated consecutively. To protect the users, the information will be anonymous unless the user wishes to reveal himself. The user will be able to send messages to the reporting facility and switch off the reporting of his use of the package if he so wishes. This facility will be extremely important to the development of dialogue styles and applications functions, because it provides information that is not easily obtained from users directly.

INTERACTION TOOLS

Many of the interaction tools, such as help, menus, etc., form an integral part of the translation phase. The design philosophy of INTEGRA calls for tools to be represented as graphs. High level languages and synonyms are implemented in this way. Walk-through and customising are also introduced.

A standard high level language will be provided by parallel graphs containing the language syntax; these graphs will be available at all sensible points in the dialogue as well as during the creation of macros. A set of primitive actions will be present which provide flow control, labelling, expression evaluation, assignment and macro management. The developer may create an alternative high level language using graphs containing these actions.

Synonyms are created by the developer and user in the form of a graph that recognises a synonymous string and pushes back on to the input the string that is recognised by the rest of the dialogue. A primitive push back action is provided.

The walk-through facility will allow the user to try out a dialogue without applications actions. The use of this facility is limited because, applications actions are allowed to steer the dialogue to a limited extent.

Customising gives the user the opportunity to tailor the dialogue to suit his preferences. INTEGRA stores a file of information for each user, usually on the user's own file area. This file contains a history of the user's operation of the system and information about the way the user prefers the system to be initialised. The file will be available to the user and any package executed by the user. The user may give permission to others, particularly the dialogue developers, to access the file. The execution of an application using INTEGRA will consult the relevant file and customise the dialogue to the user's specification. The history contained in the file will be particularly useful to a dialogue expert system, giving information on the areas of dialogue in which the user is skilled or unskilled.

THE APPLICATIONS ENVIRONMENT

The provision of INTEGRA as part of the infrastructure of a CACSE package has major implications for the infrastructure as a whole. The organisation of the applications code, data management and output production are all positively affected.

INTEGRA forces a modular structure on to the applications code. Each applications action has a block of code which must be relatively independent of the remainder of the code. A library of common subprograms may also be provided.

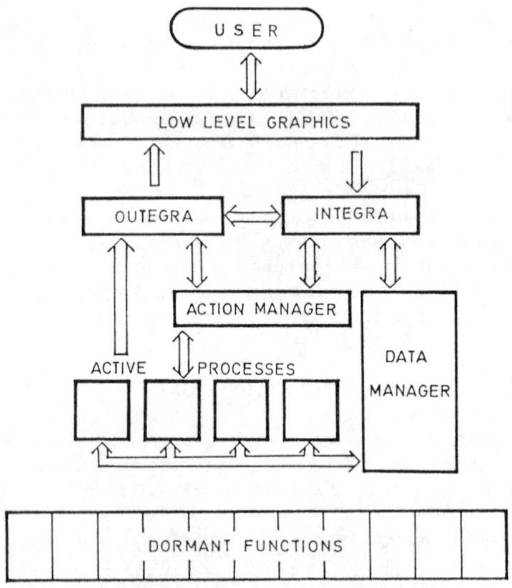

Fig. 8 An infrastructure.

The database and data management code must be a separate module of the program. To obtain a realistically user friendly system, the data must be directly accessible to the user via INTEGRA by object name.

The facilities provided for input by INTEGRA may also be provided for output production. Output involves the mapping of machine orientated data into a user friendly form. A facility for managing the mapping of output using graphs in the manner of INTEGRA is under development by the Control Systems Centre at UMIST.

The infrastructure described by Goodfellow and Munro (1985) provides the sort of environment needed by INTEGRA. Figure 8 illustrates the infrastructure required.

CONCLUSIONS

The present and future needs of the CACSE community are not fully served by any existing input translator. INTEGRA provides a dialogue definition schema that is readily understandable by engineers in the community. INTEGRA can be used to implement a mixture of common dialogue formats using graphical and textural media and it provides the facilities needed to develop and experiment with new ideas in the dialogue. The interface between INTEGRA and the applications programs encourages the development of modular programs that may be configured to suit distributed systems. The needs of the CACSE community for the input aspect of the man-machine interface will be met by INTEGRA for the foreseeable future.

REFERENCES

CADCentre (1981). The command processor user guide, CADCentre, Madingley Road, Cambridge, U.K.

Cartwright, P., C.W. Taylor (1984). CANAL starter guide, PA Management Consultants, Cambridge Laboratory, Melbourn, SG8 6DP, U.K.

Goldberg, A., D. Robson (1983). Smalltalk-80: The language and it's implementation, Addison-Wesley.

Goodfellow, S.D., N. Munro (1985). An integrated environment for computer aided control systems engineering, 3rd IFAC/IFIP Symposium on computer aided design in control and engineering systems, Denmark, July 1985.

Hopgood, F.R.A., D.A. Duce, J.R. Gallop, D.C. Sutcliffe (1983). Introduction to the Graphical Kernel System (GKS), Academic Press.

Johnson, S.C. (1975). Yacc - Yet Another Compiler -Compiler. Computer Science Technical Report No. 32, Bell Laboratories, Murray Hill, New Jersey, USA.

Lesk, M.E., E. Schmidt (1979). Lex - A lexical analyzer generator. In UNIX Programmers Manual, vol. 2B, 7th Edition. Bell Laboratories, Murray Hill, New Jersey USA.

Munro, N. (1979). The UMIST control systems design and synthesis suites. IFAC Symposium on Computer Aided Design of Control Systems, Zurich, August 1979.

Nye, W.T. (1983). Delight: An interactive system for optimisation-based engineering design. Memorandum No. UCB/ERL/M83/33, Electronics Research Laboratory, University of California, Berkeley.

Taylor, J.H., D.K. Frederick. (1984). Expert system concepts for computer aided control engineering. Report No. 84CRD127, General Electric Company, Corporate Research and Development, Schenectady, New York, USA.

Wieslander, J., H. Elmqvist (1978). INTRAC: A communication module for interactive programs: Language Manual. CODEN LUTFO2/(TFRT-3149)/1-060/(1978), Dept. of Automatic Control, Lund Institute of Technology, Lund, Sweden.

COMPUTER AIDED DESIGN AND VALIDATION OF INTERACTIVE IMAGES FOR A PROCESS CONTROL SYSTEM: AN INTEGRATED, DATA-BASE FOUNDED APPROACH

B. Heilbronn and D. Pichot

Electricité de France, Direction des Etudes et Recherches, 78400 Chatou, France

Abstract. The conception of the graphic, interactive images used in the computerized control rooms cannot be separated from the whole set of data describing the process itself and its control system. We have then chosen to integrate, inside a general purpose, not specialized data base, an interactive image design system. The software, including the graphic code generation, is realized in the DBMS data manipulation language itself.
Such an approach improves the data consistency checking at all the stages of conception ; moreover, it leads to a more flexible system, as the specifications (representations and interactivity of the different types of objects) can be stored as data in the data-base and easily modified by the designer. The same data-base, used for the CAD system, can also be used for certain kinds of simulations, and later to generate configuration data for the real-time system. The use of general-purpose DBMS in CAD systems leads to very flexible tools and is not restricted to interactive image design applications.

Keywords. Computer-aided design ; computer graphics ; data-base ; man-machine systems ; nuclear plants ; process control.

INTRODUCTION

In the conventional control rooms, the various informations and means of actions given to the operators are, schematically speaking, dedicated to separate control-room equipments. They are then available at any moment, but scattered on a large area and difficult to relate to each other.
The use of computers allows to regroup on graphic displays the informations and means of action that the operators need at a given moment. In the French power plant control room "S3C" project (Guesnier, 1983), to the specifications of which the CAD system described here is related, various kinds of graphic interactive images are used : plant circuits representations for actuators operation, regulation diagrams, plant functions or situations characteristic values following up, various aids to the operators,...
The interactive images become the main medium given to the operators to know the state of the process and perform actions on it. In a nuclear power plant control system, hundreds of such images are used ; moreover, new images or existing images modifications are likely to be needed in the life time of the power plant : a CAD system is really necessary.
These images have no signification by themselves : they are a part of a whole, the control system, extending from the process instrumentation (sensors, actuators) to the operator himself. This general unity must be taken into consideration since the conception phases. For that reason, we have chosen not to use a specialized CAD system with a graphic oriented data-base, but to realize such a system completely integrated in a general-purpose data-base management system. There is then no separation between the interactive images description and other data of the project.

BASIC PRINCIPLES

Images and Objects

Besides the fixed graphical elements (such as pipes or other links, fixed texts, etc...), an interactive image is composed of graphical representations of different types of objects : these objects may be directly related to the process itself (sensors, actuators such as pumps, valves,...) or more elaborated (validated measurements, graphs, functional representations of plant subsystems etc...). In any case, an internal representation of each of them will have to be processed in the control system computers.

Four Types of Data

For image conception, we will consider that the graphical representation of the objects in the images are in fact the result of four types of data, each of them being elaborated at a different phase of the project :

- the objects fixed descriptions : type, name and other associated texts, links with other objects, links with the process low-level control system,... These data are used, not only for the graphical images, but for the whole control system design.
- the objects state or analogic variables : their values have a sense only in the real-time system, and default values are given in the design phases ; they may be changed, however, for simulation and verification purposes.
- a limited number of normalized parameters specifying the representation of a given object in a given image : size, orientation, colour, position.

- specification data, describing the possible graphical representations of each type of object, the way they depend on the variables, the authorized sizes and orientations, etc...

The image CAD system manages these four types of data in the same data-base. During image conception, the objects description data and the graphical representations specification data are known by the system ; only a few parameters are asked to the designer for each object representation, their consistency with the specifications is immediately checked, and a standardized graphical representation is generated.

BASIC SOFWARE IMPLEMENTS

The Data-Base Management System

We wanted it to be a general-purpose, well tried and portable DBMS, able to take into account the rather complex structure of the different types of data. We have chosen to use SOCRATE-CLIO (SYSECA, 1984), a network-type DBMS providing large capacities of data structuration in terms of entities and relations ; two main types of relations may be used between entities : the first one is based on entity imbrication (hierarchy) with a finite 1-to-n cardinality, and the other one, using chaining mechanisms, achieves infinite 1-to-n relation with no limitation at all (Fig. 1).

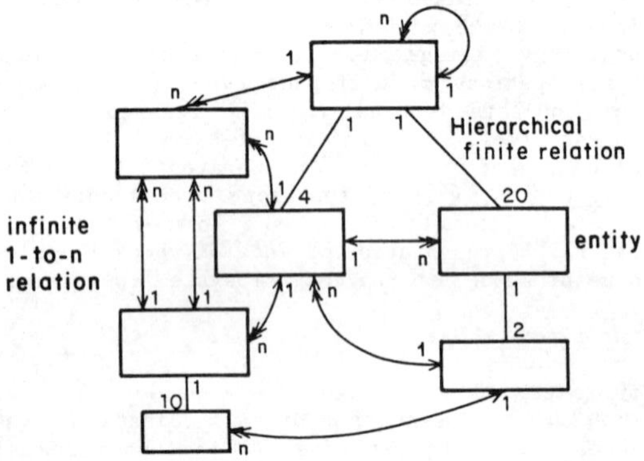

Fig. 1 : example of relations in a Clio data-base

A structure definition language is used to describe the entities characteristics and relations. The Data Manipulation Language is also a part of the DBMS ; the same language is used to read and update the data ; it is a high-level language, with advanced control structures and rather natural formulations using directly the characteristics and relations identifiers declared in the structure description.

This language can be used directly to consult or update the data-base, or as a programming language : a software realization is then a set of macros or compiled programs, written in that language and stored by the DBMS in the data-base itself.

The Graphic basic Software

The graphic code generated by the data-base is an intermediate-level, terminal-independant code called INTGR, which has been defined by EDF inside the LGIVD (graphic industrial software for display and dialog) system (Beaucourt and others, 1980). In the real-time environment, this code is generated by application modules but, in the CAD system, it is directly generated by the data-base (Fig. 2).

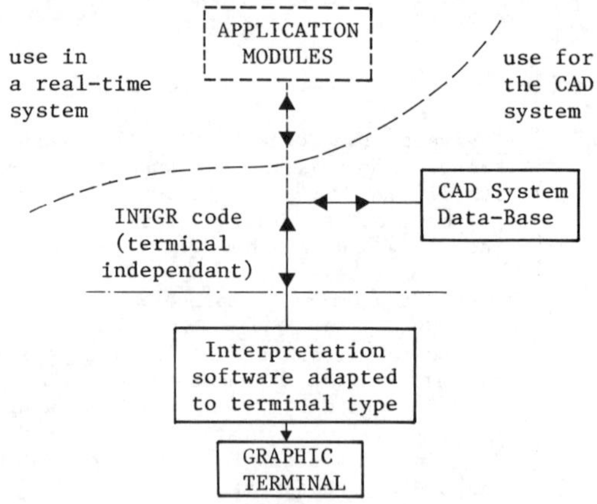

Fig. 2 : Intermediate Graphic Code generation and interpretation

The images are organized into two levels of decomposition : segments and subsegments. An INTGR image generation or animation list consists of a succession of orders of segment creation or suppression, subsegment replacement in a segment, position or colour modification, etc...
A basic interpretation software, adapted to each type of graphic terminal, stores the segment descriptions, interpretes the INTGR orders and generates the terminal low level graphic code. It also interpretes the operator's actions and sends the corresponding messages to the upper levels.
An image generated by the CAD system will have the same structure as a real-time interactive image, with all the designation and modification possibilities. Simulation functions can then easily be implemented in the CAD system.

THE DATA-BASE STRUCTURE

The Graphic Specification data Structure

These data include the definition of graphic representations, size and orientation parameters, and graphic class decomposition.

Graphic representations. If we take the example of the motor-drive valve representation specified in the project (Fig. 3), we notice that it leads to a decomposition into three parts : a fixed part (the motor-drive representation), another one (the valve "body" representation) whose shape depends on the value of a state variable, and a third part (the valve name) whose aspect (here the colour) depends on the value of another state variable.

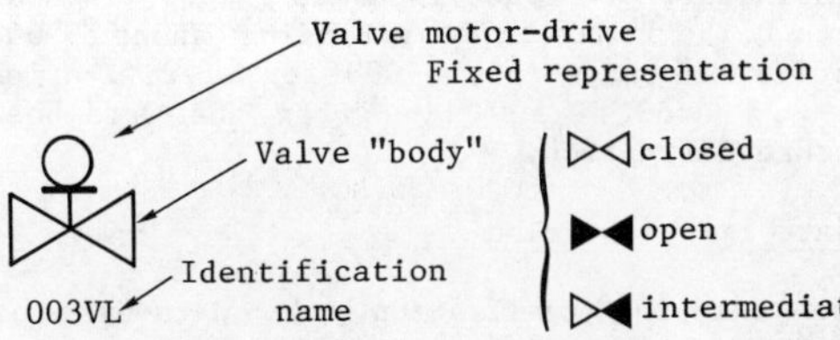

Fig. 3 : a Motor-drive Valve Representation

In the INTGR graphic code, each of these parts will be represented as a subsegment (it must be changed if the corresponding state variable value changes), and the whole representation will be a segment.
This decomposition and state variable value dependence finds its expression in the data-base structure (Fig. 4).

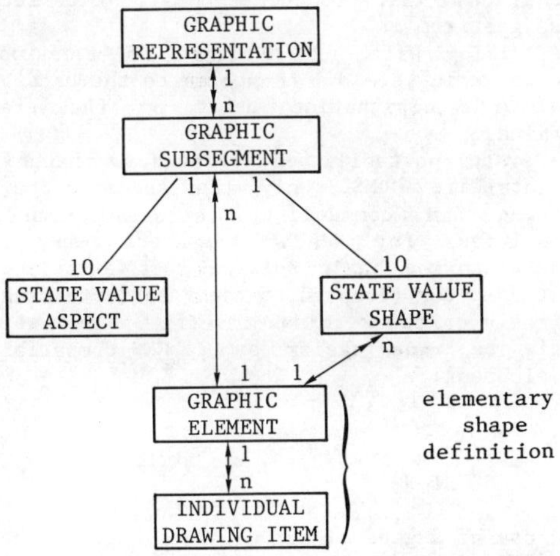

Fig. 4 : Graphic Representation Structure

Elementary shapes are defined in the data-base by entities called "graphic element", each of them being related to an unlimited number of "individual drawing item" entities. If a graphic subsegment has a fixed shape, the corresponding entity realization is directly related to a "graphic element". In the other case, the state dependence is expressed by an imbricated entity "state value shape", with one realization for each possible value of the considered state variable, each one having a relation with a "graphic element". These relations are n-to-1, for the same elementary shape definition may be used in different graphic representations (for instance, the same valve "body" elementary shapes may be used in a motor-drive or a pneumatic valve representation).

Size and orientation parameters. Specific entities are used to represent the allowed sizes and orientations for a given representation, and the associated parameters : number, x-scale and y-scale for size, mnemonic, rotation and symetry for orientation, and other parameters defining the object name possible places on the representation, the mask used for designation, etc... The same specification may be shared by several representations : this is expressed by a 1-to-n relation.

Graphic class decomposition is represented by a hierarchichal part of the data-base structure (Fig. 5). For instance, the class "valve" divides into "two-way valves" and "three-way valves" subclasses, each of them having several types (manual, motor-drive, pneumatic...). For each type, a limited number of representations are allowed, each of them pointing to a graphic representation definition by a n-to-1 relation.

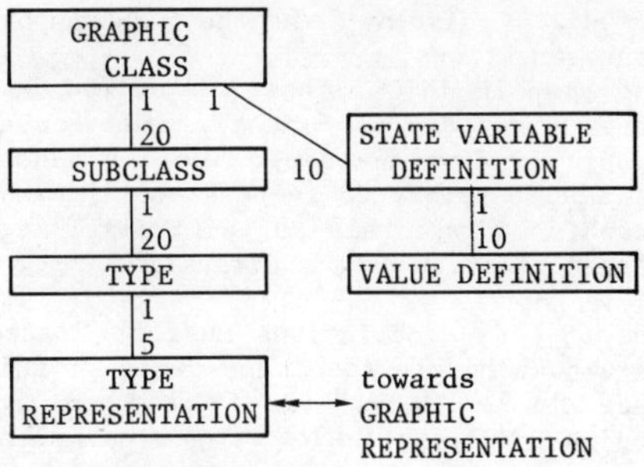

Fig. 5 : Graphic Class Decomposition

A graphic class definition also includes the state variables definitions (name, possible values, default value...). The same state variables are used by all the graphic representations associated to a given graphic class.

The Image Description Structure

The objects fixed descriptions (including the identification of their graphic class, subclass and type) are stored in an entity "object" of the data-base. As an object may be represented in several images, there is a m-to-n relation between "object" and "image", expressed by two 1-to-n relations and an intermadiate entity "object-in-image" (Fig. 6). There is one realization of that entity for each object representation in a given image, or for each image where a given object is represented. The corresponding parameters (size, orientation, position, colour) are stored in it, as well as the corresponding graphic segment identification name and a relation to the "graphic representation" used for it.

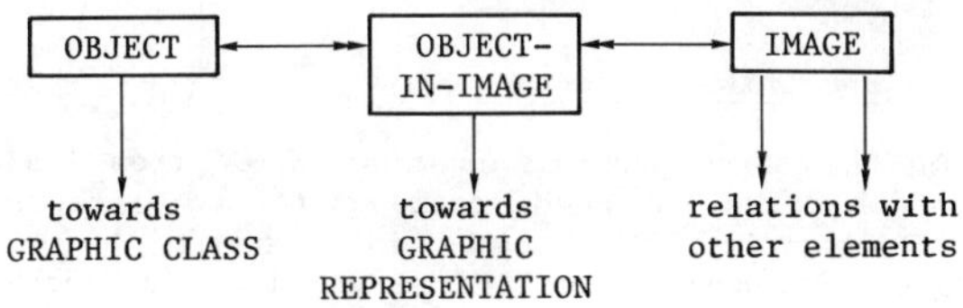

Fig. 6 : Image Description Structure

The complete description of an image also includes other elements, such as links (pipes,...) or fixed texts.
All that part of the structure corresponds to application data, updated during object description or image construction/modification phases. There is always a strict matching between an image graphical aspect and its logical description in the data base.

THE CAD SYSTEM FUNCTIONS AND IMPLEMENTATION

Main Functions

Image design. A set of conversational programs are used for image design. The designer may add, modify or suppress any element in an image (objects representations, link drawings, fixed texts, fixed graphic elements). The design process is interactive, with objects designation or coordinates plotting on the graphic terminal, and immediate display of the added or modified representations.
The graphic INTGR lists are stored in the data-base ; they may be directly used for quick image display ; from the image logical description, it is also possible to re-generate the corresponding graphic lists : this is particularly useful after a specification data modification.

The objects descriptions must be loaded into the data-base before the image design : this is often made in a "batch" way, from external files. Logical descriptions of images may also be loaded in that way ; in that case, the "regeneration" program is used to create and store the corresponding graphic lists and the images can be interactively modified.

Specification data editing. The graphic classes, graphic representations and size/orientation parameters sets may be created or modified by conversational programs. The graphic elements can be created interactively (on the graphic terminal) and modified in a conversational way.
A hierarchichal structure of menus is used to guide the designer in specification data editing as well as in image design. The logical description of each part of the specification data can be listed individually or globally.

Data verification. A first error detection test is made at the data loading phase (with error reports in batch mode and wrong parameters re-asking in conversational mode). The system structure itself ensures a good consistency level, for instance it is impossible to have a mismatch between an object graphic representation and the object type. Global verification programs are however necessary to test the whole data consistency and completeness, for instance to check that every object is at least represented in one image. Owing to the data-base structure, such verification programs are very easy to realize.

Sofware Implementation

The data-base programs must be able, from image descriptions and specification data, to write the corresponding INTGR lists in a memory buffer. Buffer reading and writing mechanisms exist in the Socrate-Clio DBMS, but only in a fixed format mode. A limited number of specific non-standard programs have been realized to allow sequential writing in buffers. These programs are used by a first level of Socrate macros realizing graphic primitives (vector, move, circle, fill,...) and segment or subsegment headings. Upper levels are graphic items interpretation, graphic subsegment and graphic representation generation programs. In image design mode, this latter program is called by a conversational program which asks the necessary parameters to the designer, accesses the proper specification data to know the representation structure and some parameters used in graphic generation (scale numbers, rotation matrix,...), and updates the images logical description in the data-base.

The whole sofware extent is about 15 000 lines of Socrate-Clio source code including about 9 000 lines for image design, 5 000 for specification data, and about 1 500 lines for the data-base structure definition.

Hardware Implementation

In the existing realization, the data-base is supported by a 16 bit Bull Sems Solar minicomputer with a multifunction (time sharing, real-time and batch) executive system. The graphic terminal is a 1024 x 1024 TV scanning "Aydin Controls" terminal, with a multiprocessor (2 Intel 8086) architecture. The basic graphic software used to interprete the INTGR lists has been implemented on that terminal for real-time projects, and was then available for the CAD system.
The data-base runs on the time-sharing function of the computer (the CAD system is itself multi-user), and activates real-time communication tasks (developped for the real-time projects) to send the graphic lists to the terminal, or to receive messages from it.
The dialog with the designer is made on an alphanumeric terminal connected to the data-base, and with designation means on the graphic terminal.
The system portability is related to that of the Socrate-Clio DBMS. Only the specific buffer-writing and communication programs must be re-written for each computer. An INTGR interpretation basic sofware must exist on the host computer (adapted to the graphic terminal) or directly on the terminal itself. Its portability could be made easier by a GKS-compatibility development.

CONCLUSION

The use of a general-purpose data-base as the main part of a graphic interactive image CAD system was at first motivated by the necessity of an open system, allowing cross-verifications on a great variety of data. The integration of the main part of the realization inside the data-base itself, including complex specification data, shows that existing network-type DBMS are well-adapted to rather complex industrial problems. It would be interesting, however, to estimate what the solution would have been with a relational DBMS, or with the functions of future "generalized" DBMS.
The interactive structure of images generated by the CAD system described in this article makes them ready-to-use for simulation functions. Such a function has effectively been realized, on the same data-base, to simulate the access mechanisms to images and operations provided for the control room. This shows that the general philosophy, on which the present CAD system is grounded, is not limited to interactive image design and could be applied to many other fields.

REFERENCES

Beaucourt, D., E. Delalleau, B. Henry, Y. Lelous, F. Venard (1980). LGIVD : Logiciel Graphique Industriel de Visualisation et de Dialogue. Congrès AFCET, Nancy, 24-27/11/1980 Hommes et Techniques, 1980, 655-664.

Guesnier, G. (1983). Etudes de Salle de Commande et de Controle-Commande du projet N4 (S3C) : présentation générale. EDF internal documentation.

SYSECA (1984). CLIO Overview. Groupe SYSECA. Saint-Cloud - FRANCE

A FLEXIBLE DATA INTERPRETER FOR COMPUTER AIDED DESIGN AND SIMULATION OF DYNAMICS SYSTEMS

R. A. King and J. O. Gray

Department of Electronic and Electrical Engineering, University of Salford, Salford M5 4WT, UK

Abstract. A general purpose graphics based man-machine computer interface is described with particular reference to its use in the design of control systems and the simulation of dynamical systems. The software requirements for such an interface are outlined and the structure of a real time graphics operating system (ARGOS) which meets these requirements is discussed. The operating system allows complex interconnected linear and nonlinear systems to be described in simple diagrammatic form for direct interpretation by existing algorithms for analysis in both the time and frequency domain. The graphics input procedures are applicable to a number of existing computer aided design packages.

Keywords. Computer aided design; man-machine systems; graphical data interpreter.

INTRODUCTION

This paper is concerned with an outline of some recent work on the development of a flexible graphics based input interpreter for computer aided control systems design and system simulation packages. A typical design suite has the modular structure shown in Fig. 1, where data concerning system configuration and system parameters are entered via a keyboard, according to well defined syntactical rules which require a rigid adherence for correct interpretation by the algorithms resident in the various analysis and design modules. Increasingly, however, emphasis is placed on the creation of more anthromorphic user interfaces and, in an engineering context, the use of graphical or scratch pad methods for data input is a natural development which is now aided by the availability of powerful, modern terminal hardware. As an example a feedback configuration such as that shown in Fig. 2 can be assembled on screen from a menu of block primitives and, using cursor control in conjunction with a suitable graphics data base, changes in individual elements or system topography can be easily made.

The interesting problem is, of course, that of interpretation so that suitably formatted data can be supplied to the existing algorithms for the generation of responses in the time or frequency domain, and the dimension of this problem depends on the degree of flexibility required in the definition of the set of graphics symbols used. It may be desirable, for example in Fig. 2, to represent the separable nonlinear elements in a variety of forms such as a polynomial or a piecewise linear approximant while the linear elements may be transfer functions (derived directly or inferred from ω domain plots) or alternatively sets of transition matrices. Some systems may not be amenable to a simple decomposition into separable linear and nonlinear elements as shown and a more natural representation might be a set of nonlinear differential equations, each equation or subset of equations being associated with a particular component of the overall configuration. Indeed some existing simulation routines require just such an interpretation. An additional requirement will almost certainly be the representation of systems such as that shown in Fig. 3, which contain both continuous and discrete signals.

In general what is needed is a software structure which implicitly links user interface operation with algorithmic functions and which allows a wide degree of freedom in the choice of system description. Such a structure must incorporate both syntactical and semantic checks to ensure a correct interpretation for each system description and yet present a modus operandi which is both simple and intuitively obvious in use. The overall requirements for a graphics based input interpreter are considered here and a real time graphics operating system (ARGOS) is proposed which can create a software environment to meet such requirements. The paper represents an extension of earlier developments [1][2] which are associated with a comprehensive design suite for nonlinear feedback systems [3][4].

REQUIREMENTS FOR A GRAPHICAL INPUT INTERPRETER

A natural way of describing complex interconnected dynamical configurations is to use diagrammatical representation which allows a large amount of information to be compressed into a structure which is indicative of the functional form of the original system. Branching and other topological relationships such as parallelism can be more easily represented than by using formal written syntax and historically this was (inter alia) one of the attractions of analogue simulation procedures. In hardware terms computer realisation of diagrammatic format can be realised by using a digitiser or tracker ball device in conjunction with a screen and keyboard. For useful development an associated software structure must be designed which is capable of creating, displaying, manipulating and interpreting this format. Each diagram can be constructed from a menu of symbols which, in turn, can be derived from a set of graphics primitives. Such primitives and their constructs need a comprehensive graphics data base both for creation and storage and this base must be common to all the facilities at a number of different levels. Thus it must be possible to address a group of graphics files without regard to their content or call up a symbol irrespective of its size or complexity.

Each graphical symbol can imply a dynamical function which is more or less complex. To enable a symbol to represent a function, translaters will have to be devised to translate it into a computable operation. Analysis will have to be performed in terms of syntactical and semantic rules which relate specifically to a graphics environment to produce a mathematical description of the symbol which is relevant to the programmed algorithms. Separate interpreters will have to be developed to meet a range of input requirements. Different dialects will have to be identified and appropriate rules formulated. Such interpreters will however share the same fundamental data types and many of the same syntactical rules. Interpreters should be able to exchange data in a similar way in which Fortran and Pascal programs can be combined in many operating systems. Software structures will have to be devised which are parallel to those used in language compilers with expansions to cope with an essentially graphical environment.

One of the most important aspects of a graphics based environment will be the use of the position of a symbol within a diagram to imply function and connectivity relationships. The latter must be clearly and unambiguously identified by a specific nodal interpreter, or nodal system with clearly defined rules for acceptable relationships and which has a direct analogy to the use of blocks and labels in conventional computer languages. Such a nodal system will constitute a framework in which the syntax of a graphical structure can be developed. In the absence of the linear structure of a text file on which a syntax can be superimposed, an equivalent graphics form must have inherent ordering so that rules can be applied.

Such software structures will require an efficient file management system to handle the large amounts of data being processed and it is important that adequate help, information and error recovery routines are developed to complement the special requirements of a graphical environment. Such routines must aid the user and protect against system failure. As the overall software system is considered to be a development environment both help and error facilities must be capable of evolving with design requirements.

A general purpose controller for creating pictures at a terminal from the data base description must be developed. Ideally this should be both processor and terminal independent and capable of taking advantage of the full range of facilities available on more sophisticated terminals. Within a graphics environment it is possible to imbed many levels of complexity and interpretation within the same diagram and the use of a range of colours is an example of how this may be achieved. Various levels may operate in parallel and thus any operating system should be capable of supporting parallelism both within the data base and in system interpretation. Such parallel processes can be identified by the principle of levels and each process must be capable of intercommunication and synchronising - in a system sense - coordinate actions. Such a concept should be independent of the machine's ability to operate in parallel and the software system outlined here has been designed to run on a single processor.

ARGOS - A GRAPHICAL OPERATING SYSTEM

The basic requirements outlined above have been incorporated into a real time graphical operating system now being developed at Salford University. A general structural diagram of the operating system is given in Fig. 4. The sequence of operations performed is analogous to that associated with the compilation and construction of a conventional text file except that the file is created using the graphics editor and represents the system in esentially diagrammatic form. The file is then entered into the interpreter process which consists of various stages of nodal interpretation, syntax and semantic tests, interpretation for a particular algorithm, numerical operation and finally presentation of the results in a specified format. Some of the routines are described in more detail below.

Nodal interpreter

This stage is required to convert the graphics file into a sequence of interlinked and analysable stages. The graphical language uses position to establish the operations, interconnections and relationships. This removes any requirement for element labelling. The nodal interpreter is required to analyse the picture with relation to these positions and create a graph type map of the inter-relationships. This is required to replace the inherent relationships that would be exhibited by a text file's linear organisation. The nodes are key positions in the graphical language which are sought out by the interpreter. The inter-relationships and connections of the nodes are identified and stored in a separate data-structure, the node data base.

Node data base

Associated with the Nodal Interpreter is the data-base of node positions and inter-connections. This structure is the equivalent of the linear organisation of syntax elements in a conventional language syntax. This data structure is used as the basis for the syntactical, semantic and interpreter analyses of a symbol. The position information will be required to identify appropriate help and error messages with respect to the parts of the symbol. All the stages of the interpretation can call on the node data as required to develop the analysis.

Syntax tests

Graphical syntax rules will be required in a form suitable for both the graphical and the ordinary programming environment. The syntactical structure of a symbol is the equivalent of a syntax of an ordinary language. It is used to identify the phrasing of the description into the graphical equivalents of sentences made up of the graphical equivalents of nouns, verbs etc and their ordering. Using the node data-base the syntax tester will identify that the symbol is made of appropriate syntax elements. These should also be arranged in a manner that obeys the laws of the syntax structuring.

Semantic tests

Allied to the syntax rules will be the semantic rules for actually interpreting the graphical programs. After the syntax of a symbol has been tested and proved correct, it must be tested for meaning. The semantic tests identify that a syntactically correct statement has a comprehensible form. These tests will be apropriate for the task being analysed and the limitations of the numerical package being used.

Interpreters

The conversion of syntax and semantic forms to applications language format is achieved by the interpreter. This is equivalent to the machine code of the graphical language. All the dialects of syntax and semantics will ultimately be translated into instructions at this level. The translations will be into a form that can be transferred to the numerical problem module for processing.

It is possible that different numerical processes require varying data formats for correct operation and thus a structure with multiple interpreters as shown in Fig. 5 may have to be devised. It may ultimately be more convenient to devise a general purpose intermediate language to enable the framework shown in Fig. 5 to be optimally designed so that a range of numerical process or algorithmic modules can be serviced. An analogous concept is the use of the P code pseudo machine language used by many PASCAL compilers in conventional computation. A translator between the intermediate language and each target numerical process will have to be developed but this will be easier to accomplish than the development of a complete interpreter for each numerical process.

It is perhaps important to note that an intermediate language of this type would also greatly reduce the effort involved in attaching the graphics input interpreter to a range of software packages now available from university and commercial sources and widen the applicability of the procedures described.

Numerical processes and graphical output

These modules form part of the conventional design suite architecture and except for the generation of error and help signals to the appropriate data busses do not react with the input environment once data has been accepted from the interpreter module.

HARDWARE REQUIREMENTS

Software development has been based on a virtual machine and terminal concept and written in PASCAL. This ensures a high degree of hardware independence and ease of transportability across a wide range of machines and terminal types. The approach has required the development of a terminal independent graphics kernel which can utilise a range of devices from simple terminals to intelligent workstations with local memory and segmentation facilities. There are similarities with the recently defined GKS concept [5] but with enhancements to segmentation to include data base support by the host computer.

The software has been developed using a Tektronix/PRIME 550 combination and is currently being transferred to the Control Group's DEC VAX 11/750 machine with a view to ultimately hosting the software on a portable MICROVAX for demonstration purposes.

EVALUATION AND REQUIREMENTS

This work evolved through discussions with industrial colleagues and the realisation that a wide application of the analysis and design techniques developed within the university required data input and data manipulation procedures which were more readily usable in an industrial design environment. Clearly engineers will be attracted to design aids which are readily available in a standard format, simple to use, robust in operation, widely applicable and extendable to meet new design requirements. For many, (but by no means all) systems a graphics format provides an intuitively powerful medium for handling data associated with complex interconnected configurations and is a development which parallels the availability of relatively inexpensive graphics peripherals. Using colour variations and scrolling techniques a high degree of system complexity can be accommodated while still maintaining clarity of presentation.

As indicated above the graphical input interpreter was developed for use with control systems analysis and design software available at Salford University. However the procedures could be applied to any other existing or proposed design suite for dynamical systems and the intermediate language now being developed will simplify the task of linking with alternative algorithms.

A graphical operating system of this type is not necessarily restricted in use to the interpretation of dynamical system performance in either the time or frequency domain and could have applications to any engineering, economic or planning strategy in which graphical representation is a useful, efficient and obvious medium for data input.

CONCLUSIONS

The requirements for a graphical input interpreter suitable for use with interconnected dynamical systems representation have been met by the development of a graphical operating system. Quite complex configurations can be represented with the present limit being determined by the available algorithms for analysis and design. Although the requirements were relatively easy to define, the realisation, even to the present stage of development has proved to be a non trivial exercise and has involved a range of expertise associated with graphical and algorithmic software development, control systems design and simulation studies.

REFERENCES

1. Gray, J.O. (1984). Nonlinear Feedback System Design. An Integrated Approach. Proc. Inst. M.C. Workshop on Computer Aided Design of Control Systems. University of Sussex. Sept., 39-47.
2. King, R.A. and Gray, J.O. (1984). A Graphical Man/Machine Interface for CAD and the simulation of Dynamical Systems. Proc. 6th IEE Conf. on Electrotechnics, Brighton. Sept., 42-46.
3. Gray, J.O. and Brown, L.S. (1984). A Computer Aided Design Suite for the analysis and design of nonlinear control systems. Trans. Inst. M.C.. Vol.6, No.1, 13-25.
4. Gray, J.O. and Nakhla, N.B. (1981). The Prediction of Limit Cycles in Multivariable Nonlinear Control Systems. Proc. IEE, Vol.128, Pt.D No.5, September, 233-241.
5. Hopgood, F.R. et al (1983). Introduction to the Graphics Kernel System. Academic Press.

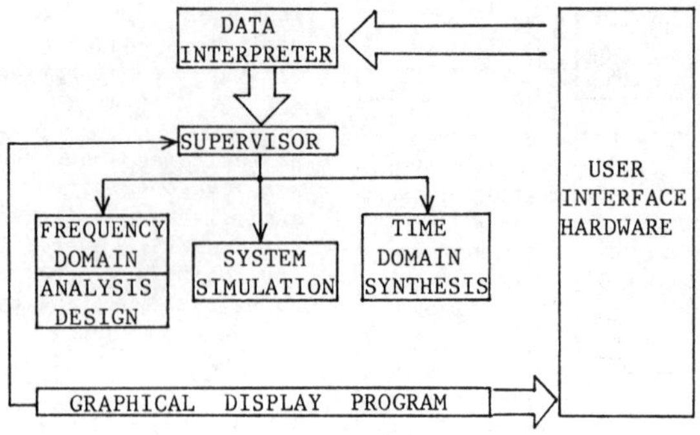

Fig. 1 Integrated Design Suite General Structure

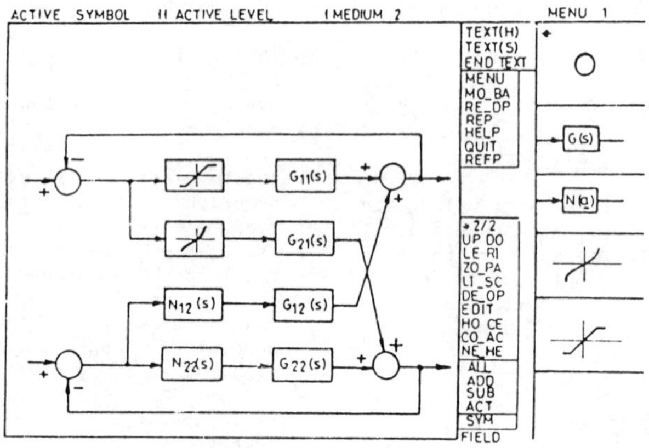

Fig. 2. Typical System Block Diagram Representation

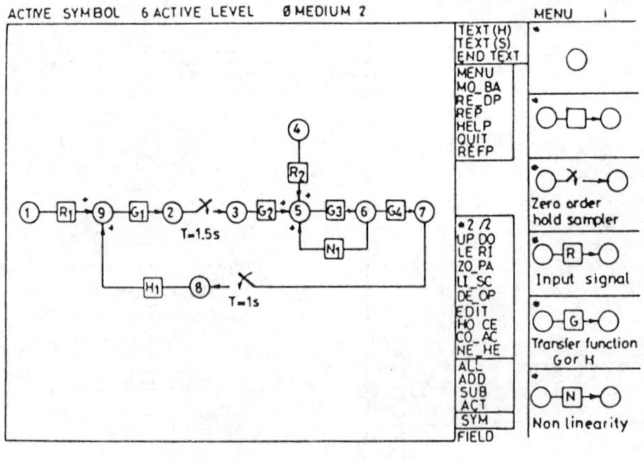

Fig. 3. System Signal Flow Graph

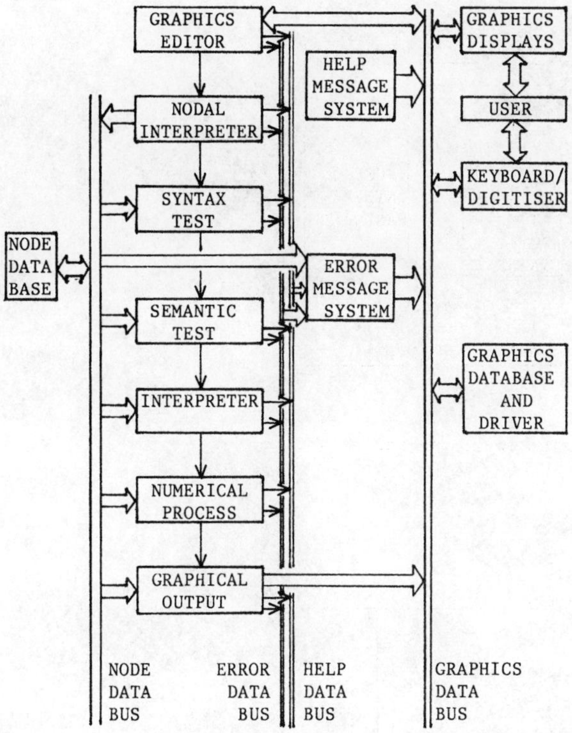

Fig. 4 Graphical Symbol Interpreter and Processor

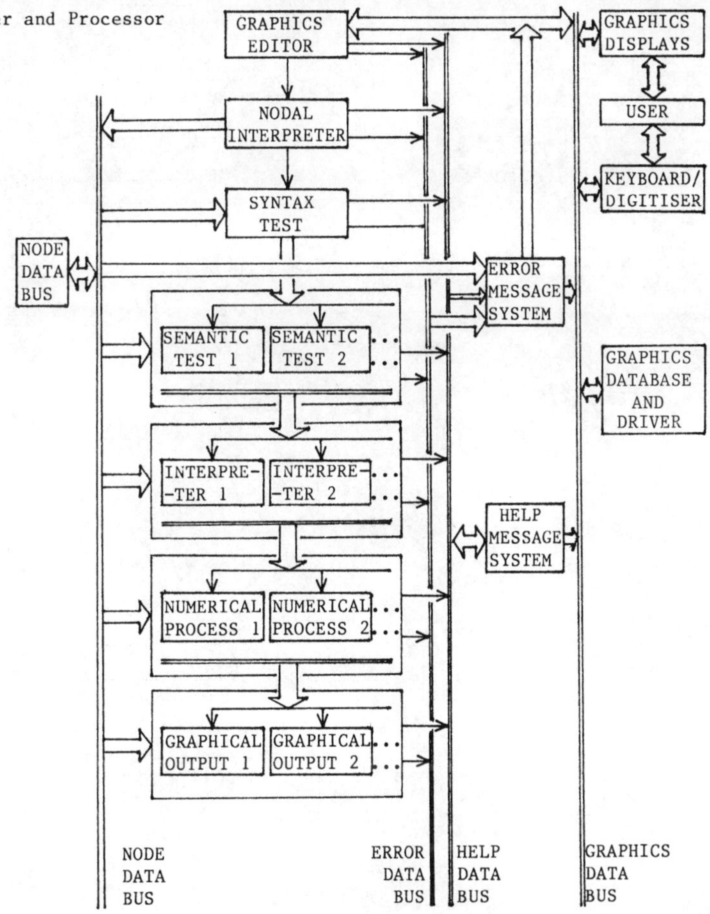

Fig. 5 Graphical Symbol Multiple Interpreter with Cross Compilation

AN INTERACTIVE COMPUTER GRAPHICS ENVIRONMENT FOR CAD OF CONTROL SYSTEMS IN PARAMETER SPACE

P. Putz and M. J. Wozny

Center for Interactive Computer Graphics, Rensselaer Polytechnic Institute, Troy, NY 12180-3560, USA

Abstract. A newly developed interactive graphics CAD package for linear multivariable control systems in the parameter space framework is presented. By direct geometric interpretation of stability, performance, robustness, integrity, actuator constraints, and other practically important objectives, the 3D capabilities of high-performance interactive computer graphics systems can be employed to design directly in the space of free system parameters. Intelligent strategies help to access high-dimensional parameter spaces by decompositions or heuristic search for display subspaces. The structure and computer realization of the package is discussed with special emphasis on the unique user-computer interaction features. Design examples show application of the main techniques for up to 15-dimensional parameter spaces and illustrate the potential as an engineering analysis and design environment.

Keywords. Linear systems; stability; parameter space design; computer-aided control system design; computer graphics; pole region assignment.

1. INTRODUCTION

The parameter space design approach for linear multivariable control systems as formulated by Ackermann (1980) is based on a direct geometric interpretation of design objectives like stability, performance, robustness, integrity, soft control, etc., in the space of free design parameters. We present a newly developed interactive graphical computer-aided control system design package incorporating this approach and recent work at the RPI Center for Interactive computer Graphics (Putz, 1984, 1985) on modeling and display of 3D parameter space regions and strategies for high-dimensional parameter spaces arising from realistic multivariable systems.

High-performance computer graphics devices are used to display admissible regions of up to 3 free design parameters at a time. By interactive manipulation of these regions and selection of design combinations for which immediate closed-loop analysis is provided, direct insight into design tradeoffs is extracted from the geometric properties of parameter space. Systematic strategies are offered when high dimensionality makes unguided choice of display subspaces impractical.

2. SUMMARY OF PARAMETER SPACE THEORY

We will now summarize the theoretical methods used for parameter space region modeling. Some results were already given in Putz (1984) and a comprehensive development is contained in Putz (1985).

2.1 Terminology

We consider finite-dimensional linear plants with constant state feedback:

$$\dot{\underline{x}} = A.\underline{x} + B.\underline{u} , \qquad \underline{u} = -K.\underline{x} \tag{2.1}$$

with $\underline{x} \varepsilon R^n$, $\underline{u} \varepsilon R^m$, $A \varepsilon R^{nxn}$, $B \varepsilon R^{nxm}$, $K \varepsilon R^{mxn}$. The initial stability and performance specifications are formulated as a "nice stability region" Γ (subregion of the complex plane in which the closed-loop eigenvalues shall be placed). We wish to describe all feedback gains K which nicely stabilize the system (2.1). For SISO systems with m=1, $B=\underline{b}$, and $K=\underline{k}_n^T$ we denote this subregion of the n-dimensional design parameter space spanned by the elements of $\underline{k}_n$ as $K_\Gamma(n)$:

$$K_\Gamma(n) = \{\underline{k}_n \ \varepsilon \ R^n \mid \sigma(A-\underline{b}.\underline{k}_n^T) \subseteq \Gamma\} \tag{2.2}$$

where $\sigma(A)$ is the spectrum of eigenvalues of A.

2.2 SISO Parameter Space Region Modeling via Algebraic Inequalities

As a first modeling approach, we can model $K_\Gamma(n)$ as an algebraic inside/outside function $h(\underline{k}_n)$ (a set of inequalities in the design parameters) such that

$$(\underline{k}_n \varepsilon \ K_\Gamma^{(n)}) \iff (\underline{h}(\underline{k}_n) \geq \emptyset) \tag{2.3}$$

The functions $\underline{h}(.)$ can be derived in an automated procedure which essentially involves conformal mapping of Γ onto the left half-plane and application of algebraic Routh-Hurwitz stability criteria on the transformed characteristic polynomial. This problem has been discussed by Sondergeld (1983), but the automated procedure developed in Putz (1985) to generate $\underline{h}(.)$ recursively for most practically meaningful Γ is new. Note that (2.3) also models lower-dimensional slices of $K_\Gamma(n)$ when certain elements of $\underline{k}_n$ are fixed.

2.3 SISO Parameter Space Region Modeling in Recursive Parametric Form

Putz (1984) describes an alternative, parametric approach to model the boundaries of $K_\Gamma(n)$. In particular, algorithms are given to display 2D and 3D slices when (n-2) or (n-3) elements of $\underline{k}_n$ are fixed. This is used to represent admissible regions in parameter space such that the designer can interactively select specific points based on secondary criteria such as robustness or actuator constraints.

Since the choice of a particular 2D or 3D display subspace is fundamental for the design flexibility and highly nontrivial for practical applications, intelligent strategies had to be developed for systematic guidance. They are discussed in the sequel.

2.4 Heuristic Search for Optimal Display Subspaces

We select p=2 or 3 free elements $\underline{k}_p \varepsilon R^p$ in which we want to see the admissible solution region. To invoke the modeling procedure from Section 2.3, we need a suitable slice $\underline{k}_{n-p}$ = const. Based on the heuristic that the "best" display subspace encloses the maximum admissible volume, we can formulate an optimization problem to automate this selection:

$$\max_{\underline{k}_{n-p}} V_p\,(\underline{k}_{n-p}) \qquad (2.4)$$

That is, we are looking for the $\underline{k}_{n-p} = \underline{k}_{n-p}^*$ which cuts the biggest p-dimensional volume V_p out of $K_\Gamma(n)$ (or of the robust intersection of several such regions). Putz (1985) shows how the hypervolume in any subspace can be approximated by a Monte-Carlo-type method when the inequality model (2.3) is used, and how heuristic search strategies like hill-climbing or branch-and-bound (see Winston, 1977) are employed to get close to the maximum with minimum effort. Note that for realistic specifications, Γ and therefore $K_\Gamma(n)$ are always bounded, hence the optimization problem is well defined.

2.5 Decomposition into Subspaces of Invariant Eigenvalues

Another strategy to guide SISO parameter space design was originally proposed by Ackermann (1982). In an extension, we can select a p-dimensional subspace such as to make (n-p) selected eigenvalues invariant (uncontrollable). Due to system linearity, this process can be repeated and the poles can be shifted 2 or 3 at a time.

This is a systematic decomposition with high intuitive appeal and allows permanent insight into the design tradeoffs. Moreover, in the formulation of Putz (1985) the modeling algorithms at each stage are extremely fast, numerically robust (orthogonal transformations) and do not distort the geometric interpretation of the displayed regions with respect to control magnitude, quantization effects, or sensor failures.

2.6 Sequential Decomposition of MIMO Parameter Space Design

For more than one input m>1, the dimension of the parameter space is m.n and linearity of the parameter space mapping is lost. Now it is even more important to give systematic procedures to deal with the many degrees of freedom and to use the modeling and display algorithms developed for the single-input case.

As outlined in Putz (1984), the SISO algorithms immediately apply when K has rank 1 (dyadic feedback) or the free elements are chosen from the same row or column of K. Practically, both may be too restrictive, especially when no guidance is available for selection of the fixed parameters.

Putz (1985) has derived a method to decompose the MIMO parameter space design problem into sequential single-input problems of reduced dimensionality. It is based on a pole placement technique due to Munro (1979) and synthesizes a full-rank feedback K from m dyadic matrices such that at each stage a certain number of eigenvalues can be re-designed in parameter space while the others are kept invariant. This gives natural and easy access to high-dimensional problems and still provides the design flexibility due to the geometric interpretations of parameter space.

3. STRUCTURE OF THE CACSD PACKAGE

Fig. 1 shows the modular structure of the interactive CACSD package based on the parameter space principle and reflects the typical stages of a design process.

System Data I/O. Handles I/O of all relevant system matrices (A,B,K) to and from terminal or disk files. The current input data file is displayed as part of the status information.

Pole Region Editor. This translates the primary design specifications (stability, damping, overshoot, bandwidth for continuous and discrete time systems) into the data structure of the nice stability region in the complex plane. The regions are composed interactively from primitives, can be displayed, deleted, rescaled, stored, and retrieved from disk files. The current pole region data file is displayed.

Projection Strategies. For SISO and MIMO systems, (sequences of) display subspaces are suggested by the methods discussed in the previous sections. Search for the subspace with maximum admissible volume, decomposition into subspaces of invariant eigenvalues, rank-1 feedback, and decomposition into dyads are provided and can be accessed in an interactive fashion.

Parameter Space Region Modeling. Given the data of the system, pole region, and display subspace, this module processes the algorithm of Section 2.3 and generates a geometric model of the desired 2D or 3D admissible parameter space region slice. The geometric data structure is very generic (polygons or polyhedra described as linked edge lists) to facilitate rendering on any vector or raster display device. The resolution of polygon approximation can be varied, and a topological consistency check is implemented which tests if the polygons are closed to detect missing solutions.

Details of this implementation are described in Lau (1985).

Parameter Space Region Display. The parameter space region data structure can be displayed by a very simple driver program on a variety of devices (interpreted as either vector list or polygon data). Interactive rotation, scaling, deletion, and overlay of 2D and 3D regions allow excellent understanding of even the most complicated shapes.

Feedback Design. From the displayed admissible region, the designer now selects specific points based on additional objectives. 2D and 3D points can be either specified numerically or graphically (lightpen or tablet). The data structure is rich enough to support region intersection for automatic display of robust solutions, norm minimization for soft control, or finite neighborhood placement for quantization effects.

Analysis. Open- and closed-loop eigenvalues, eigenvectors, and time response simulation are available interactively and provide the essential indicators for judging and tuning the design. This immediate link between parameter space points and closed-loop system behavior as well as ease of iteration between the modules (see the possible paths in Fig. 1) are the crucial factors to obtain insight into design tradeoffs.

4 COMPUTER REALIZATION OF THE DESIGN ENVIRONMENT

The basic package is implemented in FORTRAN 77 at the RPI Center for Interactive Computer Graphics on a PRIME 750 superminicomputer system under PRIMOS with IMLAC 6220 vector refresh terminals running DYNAGRAPHICS graphics language. In the current stage the package consists of 250 subroutines with a total of 17000 lines of code (extensively commented). Some special features which are of generic interest in CACSD are outlined in more detail below.

4.1 Software Engineering Principles

The task of producing and maintaining a reliable, manageable, and widely useful software package of that size and complexity is nontrivial. Modularity (decomposition into largely independent parts which communicate via well-defined interfaces) is essential for development by a software team, to control overall complexity, and to hide implementational details which are liable to change (e.g., when better numerical analysis routines become available). The main modules of our package and their interaction are shown in Fig. 1.

Closely coupled with modularity is extensibility. It is extremely easy in our structure to add more pole region primitives, new projection strategies, analysis techniques, or design criteria. This should prolong the useful lifetime of the package and enable tailoring to users' particular needs.

Software portability is still limited whenever graphics is present. We have attempted to maximize device independence by isolating all graphics commands in small tool primitives (file I/O is independent of the operating system in FORTRAN 77). These tools perform commonly used tasks (such as displaying a new menu, selecting a menu item, displaying status or error messages) and are easily emulated on a different graphics system (typically less then one page of code). This use of functional abstraction is a generally recommendable practice and makes all main subroutines easily readable and fully portable.

4.2 Numerical Analysis Issue

The modeling algorithms of Section 2.3 draw heavily on matrix manipulation and linear algebra. For this and other very sensitive operations (eigensystem computation, orthogonal decompositions) we have used an extension of the EISPACK subroutine library (Smith, 1976). The time response simulation is taken from ORACLS (Armstrong, 1980). Other commonly used routines such as rank or adjoint computation were written and collected in a mathematical library. All computations are done in double precision.

One essential part of the algorithm for which no directly usable software seemed to exist concerned the solution of a set of nonlinear equations. The problem is aggravated because the functions are not continuous (asymptotic singularities occur), no analytic gradient information is available, and all roots in a given interval have to be computed. A theoretically rigorous solution was found based on interval analysis and is described in Moore (1977). For its implementation, an interval arithmetic library was written. See Lau (1985) for details.

4.3 User-Computer Interaction

This is one of the strongest points of the package. The dialog is via self-explanatory hierarchical menus with lightpen selection. The screen is permanently divided into functional areas for status, prompt, and error messages, menus, data input, and three graphical display areas (see Fig. 5). This consistent layout makes user orientation very easy since each piece of information appears always at the same place. While the menu items change after each selection, the options HARDCOPY, HELP (on-line explanation of the current situation), and EXIT are available at all times. As much as possible, wrong user input is prevented which creates a very robust, comfortable environment in which even novices can solve problems rapidly without significant training delays.

The most outstanding feature in our user interface is certainly graphics interaction. Since all the design concepts and solution criteria are in geometric form, this interface is the crucial factor for the success of the parameter space technique. In the structure of Fig. 1, Region Display and Feedback Design require the highest graphical interactivity. The designer must understand the implications of the admissible regions in parameter space and must have utmost flexibility in manipulating and analyzing them.

For 2D regions, the capabilities of any graphics terminal are certainly sufficient. On the IMLACS we have implemented auto-scale (when the region is first drawn, it is scaled to just fit in the display area) with manual zoom on any part of the region. Together with a lightpen- or arrowkey-driven crosshair for positioning design points, this is a completely satisfactory environment.

3D regions, however, are not adequately understood in the above graphics environment, although auto-scale, zooming and rotation are available. For example, the unfamiliar shapes cannot be visualized in wireframe representation, rotation is too slow, and no mechanism for 3D positioning is available. This is why we have used other, higher performance graphics devices. Shaded color rendering with hidden surface removal on raster screens gives the best appreciation of the object's shape. The CATSIG software (Falk, 1984) for the ADAGE RDS-3000 "Ikonas" display allows real-time rotation and interactive coloring and shading (positioning of the light sources) in 512 x 512 pixel resolution and is an ideal environment for initial study of the admissible regions. The generic geometry data structure generated by the CACSD modeling algorithm made it very easy to use CATSIG (only a 200-line driver program was required).

Interactive 3D positioning for design is provided by an Evans & Sutherland PS 300 high performance vector graphics terminal. The wireframe model is simultaneously displayed in three different views to facilitate orientation. Rotation and scaling is done in real time from dials. A cutting plane can be interactively moved through the object (by virtue of hardware depth-clipping) and a crosshair can be positioned in the resulting 2D cross-section while the corresponding 3D coordinates (the design values) are displayed. Besides positioning, this unique feature also allows 3D information to be viewed as dynamically changing 2D slices which is easier to comprehend for certain applications (e.g., location of common intersection for robustness). Again, the required driver software was written in less than one man-week including familiarization with the E & S graphics language.

A full appreciation of the achievable interactivity and flexibility for control system design is only possible in a live demonstration. We can only convey some of it by the following examples.

5. DESIGN EXAMPLES

5.1 Automatic Search for Display Subspace

We will first demonstrate the heuristic search method from Section 2.4. Consider a 5th order discrete-time SISO system where the requirement is simply stability (Γ is a unit circle). We want to display a 3D admissible slice in (k_1, k_2, k_3)-space and look for the values of k_4, k_5= const. which maximize the enclosed volume.

A hill-climbing algorithm produced the solution $k_4^* = 1.75$, $k_5^* = -1$. To verify that this is a global maximum, the plot of Fig. 2 was compiled which shows the volume $V_{123}(k_4,k_5)$. Note that this is essentially a graphical representation of 5D information where 3 dimensions are compressed into a scalar indicator (the admissible volume). The practical benefit is the range of k_4 and k_5 for which meaningful 3D parameter space design is possible.

Based on this suggestion, several "parallel" 3D slices were modeled and are displayed in Fig. 3. (for $k_5 = 1$ and $k_4 = 1, 1.75, 2.5$). They give a slight 4D impression, but the wireframe models are difficult to comprehend. Fig. 4 shows how shaded raster rendering greatly enhances understanding of such unusual shapes.

5.2 SISO Decomposition of 10D Parameter Space

Let us now illustrate the decomposition into subspaces of invariant eigenvalues for a 10th order SISO sampled-data system. The design specification requires a damping ratio of at least 0.3 which corresponds to the spiral nice stability region in the bottom left display of Fig. 5. Also shown there are the open-loop eigenvalues. The system is unstable, but 2 eigenvalues are already inside Γ. We therefore have to move 8 poles and will decompose this task into 3 steps.

First, we select the 3 poles at -2.53, -0.96 $\pm$ j0.61. The program finds the 3D display subspace in which the other 7 eigenvalues are invariant and displays the admissible region in the main area. A design point from this region is (3.3, 3.3, 9) and was found using the E & S PS300 terminal. The system after step 1 has closed-loop poles as in the bottom right of Fig. 5. Indeed, we have moved the 3 selected eigenvalues inside and left the others untouched. We still have 5 poles outside.

For the second step, we relocate the eigenvalues 1.05, 0.82 $\pm$ j0.84 and move them inside Γ by a choice of (0, 1.7, -4.3). The 3D region for this step looks similar to the one from step 1 and is not shown. The resulting pole configuration is the bottom-left plot of Fig. 6.

For step 3 we have the two poles -0.37 $\pm$ j1.06 left to move. This can be done in the 2D slice shown in the main area of Fig. 6. To demonstrate the validity of the small gain interpretation, we select the marked point which is closest to the origin. Indeed, the two poles move the shortest distance necessary to enter Γ (smallest control effort sufficient to Γ-stabilize the system). The final poles are shown in the bottom right. Note that we have designed a 10D system with continual graphical insight and utmost economy, but never had to use parameter spaces of dimension greater than 3, and never needed guesswork or arbitrary decisions.

5.3 MIMO Decomposition of 15D Parameter Space

As an example for the method described in Section 2.6 we consider the continuous-time system used by Munro (1979) as Example 4. It has 5 states and 3 inputs, the complete parameter space has therefore dimension 15.

We want a full-rank feedback gain to place the eigenvalues in the region in the bottom of Fig. 7 (damping > 0.7, < 5% overshoot, rise time > 2.5, bandwidth < 3). The open-loop poles are all Ø.

Since there are 3 inputs, our procedure has 3 steps. In step 1, the algorithm only considers the first input with which one eigenvalue is controllable. This corresponds to a 1D-parameter space region, the interval [0.71,2.12] in the Figure. A design choice of 2 gives the pole configuration of the bottom left in Fig. 7.

Step 2 takes the first two inputs. A dyadic feedback is constructed which makes the already assigned pole uncontrollable. There is freedom to move a second eigenvalue, and the corresponding 1D admissible region is [0.82,2.45]. A design value of 1 produces the picture in the bottom right of Fig. 7. One more pole is in Γ.

In step 3 (with all 3 inputs) the whole system is controllable. Again, a dyadic feedback is found to keep the two good poles fixed. There remains a 3D parameter space region as in Fig. 8. The bottom of the figure shows the effect of two different choices. In any case the first two poles are invariant and the other 3 are inside Γ. For both designs the total feedback matrix was found to have full rank 3.

6. CONCLUSION

We have presented a novel computer-aided control system design package based on a fascinating combination of parameter space design theory and the capabilities of state-of-the-art computer graphics devices. For linear multivariable systems with low to medium dimensionality of the state space it represents a unique tool to integrate many contradictory objectives and synthesize simple control laws in an interactive environment.

The effective use of graphics for geometric interpretation of systems concepts and for visualization of complex high-dimensional relationships is believed to have substantial impact beyond the specific application discussed here. The full extent of possible user-computer communication via this medium is far from being fully exploited, but the very rewarding experience so far strongly encourages further research and experimentation. In that sense, the ideas and principles outlined in this paper are hoped to be of generic interest for the CAD community.

ACKNOWLEDGEMENT

Support by NSF grant ISP70-20240 and the Center for Interactive Computer Graphics Industrial Associates Program is gratefully acknowledged. Software development of the package owes much to the contributions of Edmund Lau, Harry Bims, Jim Andre, and Mary Beausoleil.

REFERENCES

Ackermann, J. (1980). Parameter Space Design of Robust Control Systems. IEEE Trans. Autom. Contr., AC-25, 1058-1072.

Ackermann, J., and S. Tuerk (1982). A Common Controller for a Family of Plant Models. IEEE Conf. Dec. Contr., 240-244.

Armstrong, E.S. (1980). ORACLS - A Design System for Linear Multivariable Control. Dekker, New York.

Falk, E. (1984). CATIA Shaded Image Generator (CATSIG) - Project Report, Tech. Rept., TR-84017, Center for Int. Comp. Graph. (CICG), RPI, Troy, NY.

Lau, E.Y. (1985). Interactive CAD Software for Parameter Space Region Modeling and Display. MS Thesis and Techn. Rept., CICG, RPI (to appear).

Moore, R.E. and S.T. Jones (1977). Safe Starting Regions for Iterative Methods. SIAM J. Num. Anal., 14, 1051-1065.

Munro, N. and S. Novin-Hirbod (1979). Pole Assignment Using Full-Rank Output Feedback Compensators. Int. J. Syst. Sc., 10, 285-306.

Putz, P. and M.J. Wozny (1984). Parameter Space Design of Control Systems Using Interactive Computer Graphics. Proc. Amer. Contr. Conf., 105-114.

Putz, P. (1985). Interactive CAD of Control Systems in Parameter Space. PhD Thesis and Techn. Rept., CICG, RPI (to appear).

Sondergeld, K.P. (1983). A Generalization of the Routh-Hurwitz Stability Criterion and an Application to a Problem in Robust Conroller Design. IEEE Trans. Autom. Contr., AC-28, 965-970.

Smith, B.T., et.al., (1976). Matrix Eigensystem Routines-EISPACK Guide. Springer, New York.

Winston, P.H. (1977). Artificial Intelligence. Addison-Wesley, Reading.

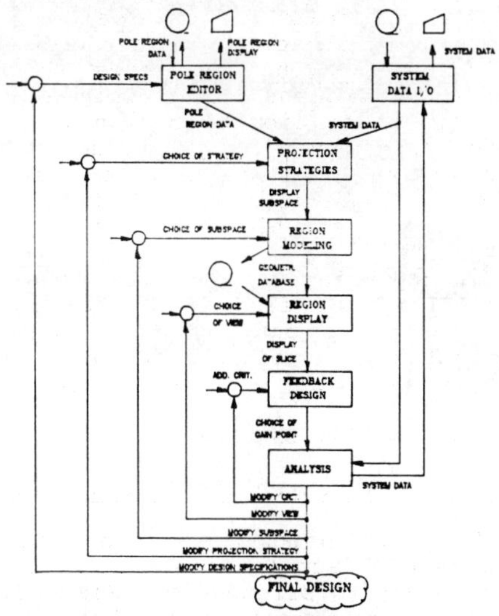

Fig. 1. Structure of the CACSD Package

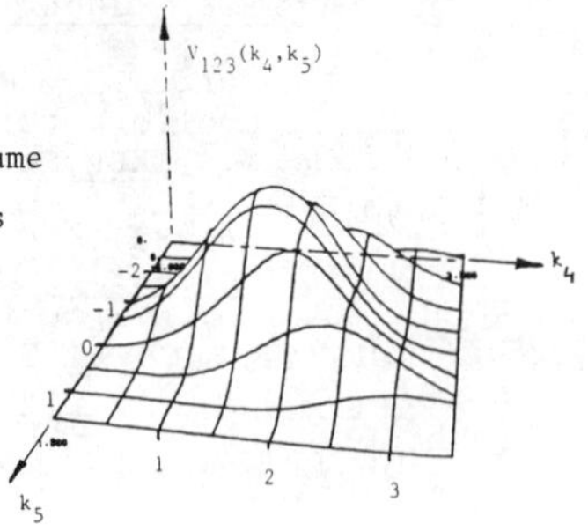

Fig. 2.
Admissible Volume
in 3D Subspaces

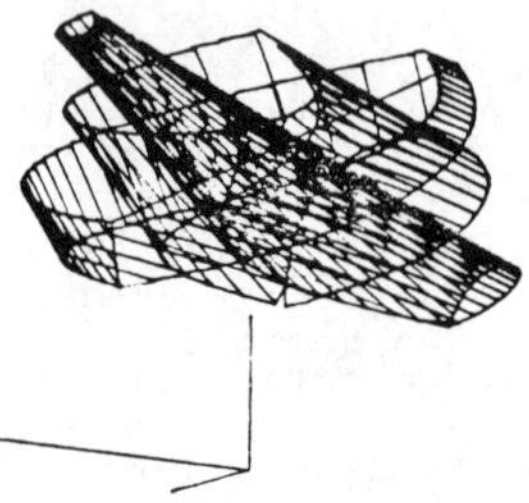

Fig. 3.
3 Parallel
3D Slices
from 5D
Parameter Space

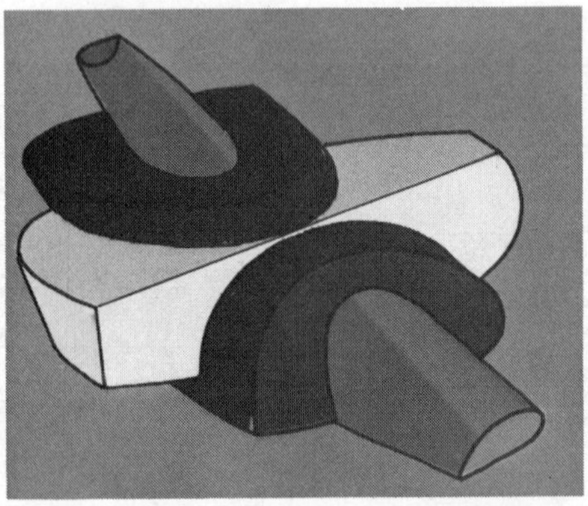

Fig. 4. Raster Rendering of Fig. 3

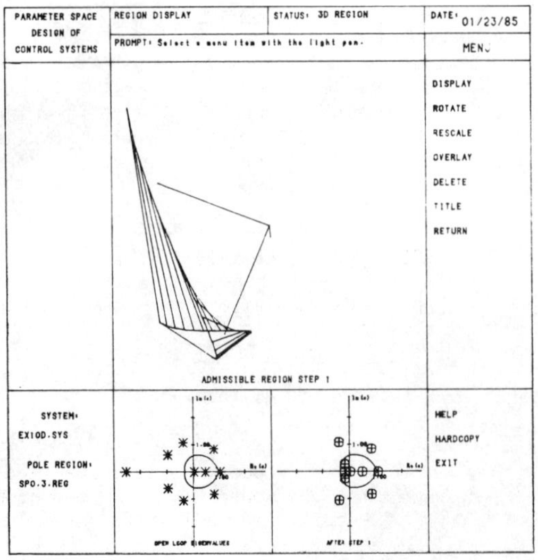

Fig. 5. Step 1 of Design Example 2

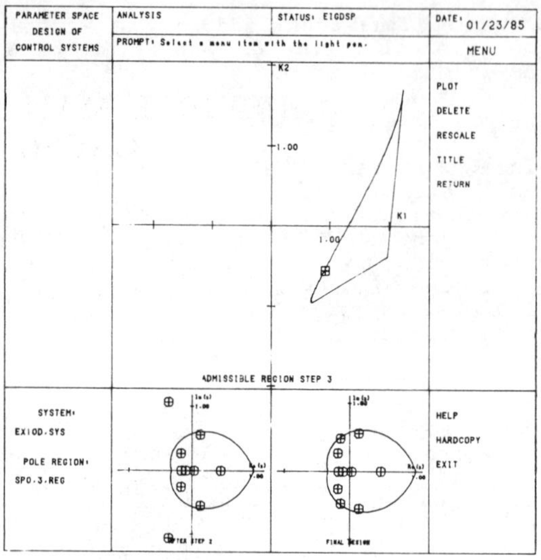

Fig. 6. Steps 2 and 3 of Design Example 2

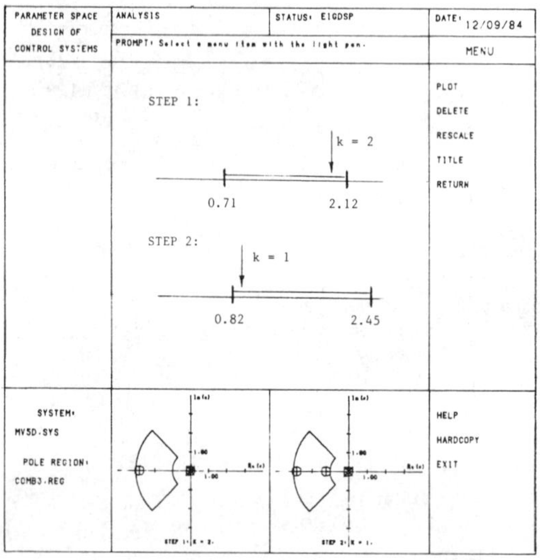

Fig. 7. Steps 1 and 2 of Design Example 3

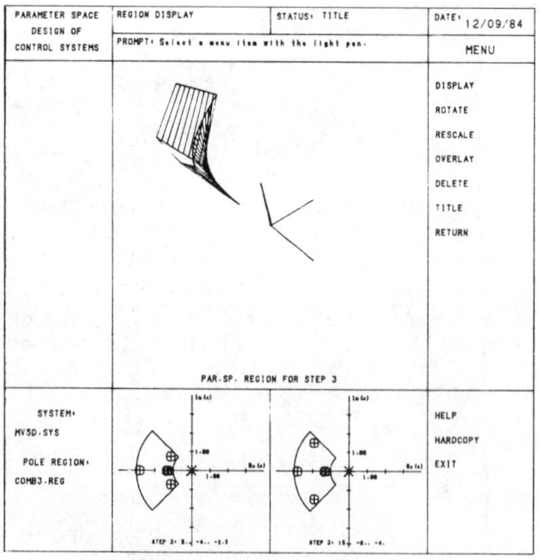

Fig. 8. Step 3 of Design Example 3

A FLEXIBLE MAN-MACHINE INTERFACE FOR CACSD APPLICATIONS

M. Rimvall and L. Bomholt

Institute for Automatic Control, Swiss Federal Institute of Technology (ETH), CH-8092 Zurich, Switzerland

Abstract. In this paper a man-machine interface well suited for Computer Aided Control Systems Design (CACSD) is presented. We will start with presenting the general command-driven interface of IMPACT, an interactive CACSD-package presently implemented in the new programming language Ada. In particular, we will elaborate on its versatile macro concept (functions/procedures). The interface of IMPACT uses a syntax similar to that of Ada and is flexible enough for the algorithmic type of problems found in CACSD. A facility of concurrent sessions not unlike those found on many modern workstations is presented. Thereafter, the macro and session concepts are combined to form the "query" option, a feature intended to be used to guide the inexperienced user in a more controlled manner.

Keywords. Computer-aided control systems design; computer-aided instruction; man-machine interface; command-driven interface; Ada.

INTRODUCTION AND OVERVIEW

In this paper the flexible man-machine interface of IMPACT (Interactive Mathematical Package for Automatic Control Theory), a new interactive CACSD-package is presented.

IMPACT is presently being implemented at the Swiss Federal Institute of Technology (ETH), Zurich, Switzerland (Rimvall, 1983; Rimvall and Cellier, 1984a). At this time, the kernel (controlling the interactive user dialogue) and data administrator (handling the dynamically used data structures) exist. Presently, the necessary control algorithms are developed/collected and included into IMPACT. The package is already internally used at the ETH and will soon be generally available.

IMPACT is implemented in the new programming language Ada (ANSI, 1983). This is a rupture with the tradition of writing all portable scientific software in FORTRAN, yet brings several implementational advantages. See Rimvall (1984b) for a discussion on Ada, in this contribution we will see an example of how the parallel tasking of Ada can be used to increase the structurability of the IMPACT man-machine interface.

The development of IMPACT is made with the objective of serving a very heterogeneous group of users. These range from the undergraduate student, who must solve trivial control problems without any previous contact with interactive CAD-packages, to the skilled control researcher with programming experience, who wants to use IMPACT for the analysis and synthesis of complex control systems. At a first glance, it seems impractical for this wide range of users to use the same man-machine interface. Whereas the inexperienced student will get an easy access to the system over a question-and-answer or menu-driven interface (which is slow and cumbersome but requires a minimal knowledge of the used system), the experienced researcher needs a much more versatile tool; as it is impossible to include every conceivable control application algorithm in IMPACT, it must be possible for the user to make his own extensions by interactively defining his own algorithms. However, this requires a structured, more intelligent, command driven interface. On the other hand, two alternative interfaces to the package would presume that each user belonged to one of the above extremes (which is certainly not the case), and does not give the beginner a smooth transition to more advanced usage of the package.

Therefore, to accommodate a full range of users, the following concepts have been developed for IMPACT: An ample set of data structures usable in control theory together with a full-fledged, structured command language containing most elements found in higher computer languages gives the advanced user the flexibility he needs. In particular, a versatile macro facility allows the user to interactively implement his new control algorithms (in form of functions or procedures). A session concept enables the user to maintain several concurrent working-areas, imperative for retaining the overview while solving larger control problems.

To help the user with only basic knowledge of IMPACT and/or control theory, a "query" option has been introduced. By each command where the user selects this option, the system will take the initiative and guide the user through a question and answer dialogue especially adapted for that particular command. The query option is available by all system-defined commands/functions and can also be included by the user for his own new, interactively defined, macros.

The similarity between the IMPACT command language and Ada enables the user to fairly easy "translate" interactive IMPACT macros to Ada coded procedures/packages to be incorporated into IMPACT.

BASIC COMMAND AND DATA STRUCTURES

In this chapter, we will give some examples showing the simplicity of the basic level of IMPACT. At the same time, a short presentation of the data structures available in IMPACT will be given (also see Rimvall and Cellier 1984a).

In the interactive command language of IMPACT, as well as in any other procedural language used to describe algorithmic actions, the assignment statement is the most commonly occurring language construction. The full expression parser of IMPACT allows the user to construct expressions of any complexity using the normal infix mathematical notation and standard function calls. Therefore, when a first-time user is introduced to IMPACT, he can immediately start calculating simple results without further instructions:

```
>> --This is a comment.">>" is the IMPACT
>> --prompt.";" suppresses resulting output
>> MY_VAR = 1;
>> T = -5.6;
>> YOUR_VAR = 2*MY_VAR*(SIN(75)+ABS(T));
```

However, as IMPACT supports quite complex data structures, the IMPACT assignment statement is much more versatile than that of e.g. Pascal, yet simple and transparent in its structure. As in MATLAB (Moler, 1980), we normally work on complex structures, for example complex matrices, and therefore allow for the following types of constructions:

```
>> I = SQRT(-1);

>> A = <1 , 2+3*I;
->      0 , MY_VAR>
```

resulting in

```
A     =
    1.0           2.0 + 3.0*i
    0.0           1.0
```

Whereas many CACSD-packages perform all their operations on complex matrices (Little and others, 1984; Walker and co-workers 1982), IMPACT supplies the user with a range of data structures common in control theory, e.g. polynomial and transfer-function matrices, linear system descriptions (four dimensionally tuned matrices), domains and trajectories (for range specifications, time histories and plots) as well as nonlinear system macros (for the simulation and linearization of nonlinear systems). The operations on these structures are kept simple through a very natural notation. For example, the input line

```
>> Q = < 1^2^1 , 2^1 >
```

will result in the polynomial row vector

```
Q(p)       =
   1. + 2.*p + 1.*p**2              2. + 1.*p
```

Alternatively, a longer but better readable way of entering the polynomial matrix Q would be to first define the variable P as

```
>> P = <^1>;
```

Thereafter Q can be entered as

```
>> Q  =  <1 + 2*P + P**2, 2 + 1*P >;
```

The polynomial matrix Q was entered in non-factorized form. Of course it could also have been entered in factorized form:

```
>> QF = <-1|-1, |-2>
```

resulting in

```
QF(p)   =
     (p + 1.)*(p + 1.)             (p + 2.)
```

Transfer-function matrices can be obtained by dividing two polynomial matrices element by element. For example:

```
>> NUMER = <P,1>;
>> TRAFUN = NUMER ./ Q
```

will result in the structure

```
TRAFUN(p) =
           p                              1
  -------------------              ---------
  1. + 2.*p + 1.*p**2              2. + 1.*p
```

which of course also can be obtained by

```
>> TRAFUN = < ^1/(1^2^1), 1/(2^1)>
```

THE IMPACT COMMAND LANGUAGE

As no general CACSD-package can include every conceivable control algorithm, especially not if it is to be used as a tool in scientific research, the user must be supplied with an interface flexible enough to let him extend the package according to his own needs. In particular, it must be possible for the user to assemble existing base algorithms to form more powerful or more general algorithms. To provide the user with an interface flexible enough, many modern interactive CACSD-packages are command driven in the sense that the user controls the flow of action of the package using an (often quite complex) command language. Although such a command language could be developed from scratch (giving the developer full freedom of design), we have decided to build the command language of IMPACT around a syntax similar to that of Ada. Ada was an excellent template for a language to be rich enough to describe almost any control algorithm. Moreover, as Ada becomes a wide-spread language in the near future, many users will find the interactive dialog with IMPACT very familiar and will require only a short time to get acquainted with the package interface.

As an example of the IMPACT command language, let us consider the problem of solving the Riccati Equation

$$\dot{x} = A*x + B*u \qquad \int_0^\infty (x'*Q*x + u'*R*u)dt \overset{!}{=} MIN$$
$$y = C*x$$

After defining the matrices A, B, Q and R, this problem can be solved by a simple algorithm described in Potter (1966):

```
a = <...>; b = <...>; q = <...>; r = <...>;

<v,d> = EIG(<a, -b*(r\b'); -q, -a'>);
k=0; n = DIM(a);
FOR j IN 1 .. 2*n
  LOOP
   IF d(j,j) < 0
    THEN k=k+1; v(:,k) = v(:,j);
   END IF;
  END LOOP;
p  = REAL(v(n+1..2*n,1..k)/v(1..n;1..k));
fc = -r\b'*p
```

In this algorithm, we first calculate the eigenvectors and eigenvalues of the Hamiltonian. We store the eigenvalues diagonally in d and the eigenvectors as columns in v. Thereafter we extract the columns corresponding to negative eigenvalues and finally we can calculate the feedback coefficients fc. We note the similarity with Ada, the main differences derive from a notation also used by MATLAB and include the use of '=' for assignment statements , '<' and '>' to describe the matrix structures and ':' to form substructures (e.g. to form column vectors out of matrices). We note the presence of structural commands like FOR- and WHILE-loops, IF-statements and BEGIN-END blocks.

MACRO FEATURES IN IMPACT

In the last chapter we have seen an example on how sequences of basic statements can be put together to form control algorithms. However, the system would not be particularly user-friendly if you were forced to enter all these statements every time you wanted to perform the algorithm. Therefore, IMPACT supports macro facilities, permitting you to only once describe entities/sequences of action to be used repetitively. In this chapter we will present the function and procedure macros, however, IMPACT also supply the user with string macros for more general "low-level programming" and system macros to describe nonlinear systems.

The above algorithm for solving the RICCATI equation could be included in a function using the following syntax:

```
Function RICC(a,b,q,r);
  v,d,p : MATRIX;
  k,n   : INTEGER;
BEGIN -- RICC
  <v,d> = EIG(<a, -b*(r\b'); -q, -a'>);
  ..
  ..
  p = REAL(v(n+1..2*n,1..k)/v(1..n;1..k));
  RETURN -r\b'*p;
END RICC;
```

The declaration of local variables is optional in IMPACT. Such declarations may be used to enhance the programming security (if one variable is declared, all variables on that level must be declared) and to introduce type checking. However, during interactive sessions, variable declarations are normally omitted, as they interfere with the flexible form of action normally taken during interactive work.

After the Riccati function has been defined, we can access it as a normal function

```
>> ffcc = RICC(aa,bb,qq,rr)
```

Characteristic for many control algorithms are that they seldom appear in only one version. Often, this multitude of versions is caused by structural variations in the treated systems. Other times, these structural variations do not influence the algorithm itself, but lead to particular parameter selections (affecting error tolerances, type of input parameters and so on). In IMPACT, the combination of defaulted formal parameters and named rather than positional actual parameters can be used to take care of both algorithmic and parametrical variations. For example, in our Riccati equations the Q and R matrices are often chosen to be unity matrices. In IMPACT, we could shorten many calls to RICC if we made the unity matrices default for Q and R:

```
Function RICC(a,b,q=eye(dim(a,1)),
                r=eye(dim(b,1)));
  v,d,p : MATRIX;
  ..
```

We have now defined Q and R to be square unity matrices of correct dimensions. These default values will be used whenever no other values are given for the parameter in question. In the default declaration we make use the fact that the actual assignment of default values is performed at the time of call, when the dimensions of the A and B matrices are known.

After thsi redeclaration of function RICC, it can be called with a shortened parameterlist whenever the unity matrix is to be used for Q or/and R. For example, if Q is a unity matrix, we would call Riccati using

```
ffcc = RICC(aa,bb,,rr)
```

or, a little more readable

```
ffcc = RICC(aa,bb,R=rr)
```

It is always psible to specify named rather than positional parameters (useful for extra clarity by documentation):

```
ffcc = RICC(A=aa,B=bb,R=rr)
```

Similar to defaulted structures, algorithms requiring error-tolerances are often coded with default tolerances, which can be overridden whenever needed. The function for pseudoinverse therefore uses the machine tolerance as an annulable default:

```
>> pia = PINV(a);
```

but we can at any time replace this default by an alternative value:

```
>> pia = PINV(a,TOL=1.e-5);
```

Another example of defaulted parameters is the procedure DELETE: This procedure can take any number of parameters, each of which is the name of a variable to be deleted. A non-specified parameter is ignored, and certain string parameters change the meaning of the procedure completely:

```
>> DELETE(pia)      -- deletes the variable pia
>> DELETE(a,b,r,q)  -- deletes a,b,r,q
>> DELETE('all')    -- deletes all variables
>> DELETE()         -- legal call, no action
```

THE USE OF PARALLEL SESSIONS

Everybody remembering the times when research still was made using pencil and paper, can certify the advantage of using two or more sheets of paper concurrently. On one sheet the actual results (e.g. theorems and proofs) were painstakingly noted down. On the other sheet(s) intermediate calculations were performed, alternative paths were examined, numerical examples were tested and so on. When you were ready, the scratch-sheet(s) were discarded, leaving only the clean results. Then the computer became widespread and you started to write large FORTRAN programs, possibly using subroutines from standard algorithmic libraries. As award for and proof of your strenous programming work, you received the results (together with program listings, memory dumps and intermediate results) on oversized sheets of papers which did not fit into any loose-leaf binder, turning your book-shelves into paper-shelves. Then, suddenly, you got your first interactive CACSD package (most likely MATLAB (Moler, 1980), which really isn't a CACSD package, but you were so happy to have this first numerically stable, interactive, algorithmic package that you really didn't care). You could now work on your CRT-terminal and all intermediate results appeared only on the terminal screen. At the end of your session, you saved only the main results on an external file to be printed, made an exit from the program and all intermediate results disappeared.

Indisputably, these new interactive CACSD-packages have become invaluable tools in modern control research/development. However, they have not yet brought back the comfort of your scratch-sheet. You normally work interactively with a single set of variables, all residing in "one big bucket". After an hour or so of work, you have most certainly lost the overview of your variables, especially since you are forced to give each of them a meaningless, four-letter name. The only remedy is to delete every intermediate variable (giving you the function of an eraser instead of a scratch-paper).

The man-machine interface of IMPACT avoids this accumulation of variables and gives the user his scratch paper back by introducing a concept of sessions. Each session is a logical work-area, conceptually comparable to the directory structures of many operating systems (Digital, 1982) or to the windowing techniques of modern workstations like Apollo Domain and Sun. It is possible to open several session during one call to IMPACT, these sessions are then concurrent; the user can switch between the sessions at any time and each session is completely independent of the others. In each of these sessions, the user can define local variables, guaranteeing that the number of current variables is kept to a minimum and that related data structures are kept together. Moreover, the session feature does not add any unreasonable complexity to the basic IMPACT system described in the previous chapter. In fact, the novice user can ignore the session facility to start with and start structuring his work-environment only as he gets more familiar with the system.

To get an intuitive feeling how sessions can be used, let us consider a user working on a control problem in both the frequency and time domain. Upon entering IMPACT, the user is met by the prompt of the MAIN session:

```
>>
```

He now decides to form three additional sessions: FREQ, TIME and SCRATCH. The first of these new sessions is created with the command

```
>> SPAWN(FREQ)

%IMPACT-MESSAGE, Global session MAIN is left.
%IMPACT-MESSAGE, Global session FREQ is created.
%IMPACT-MESSAGE, Global session FREQ is entered.
%IMPACT-MESSAGE, No variables are imported to FREQ.

FREQ>>
```

The new prompt FREQ>> will remind the user that he has left the main session. He now proceeds creating the sessions TIME and SCRATCH. Thereafter he can use the command SWITCH to jump between between the different sessions:

```
SCRATCH>> SWITCH(FREQ)

%IMPACT-MESSAGE, Global session SCRATCH is left.
%IMPACT-MESSAGE, Global session FREQ is entered.

FREQ>>
```

For typographical reasons, we have in the above and following examples assumed that you work on a normal, dumb terminal with IMPACT running under any "standard" operating system. However, on a modern, intelligent workstation supporting windowing techniques (e.g. Apollo Domain), you would have each session on a separate window of your terminal, see fig. 1. The SWITCH command is then redundant, as the same effect can be obtained simply by moving your cursor from one window to another. Moreover, the implementation of concurrent sessions is done using the tasking facility of Ada, guaranteeing a fairly portable implementation despite the heavy hardware/firmware host-dependency of any windowing techniques. The modular structuring of Ada programs is furthermore used to separate and minimize the program sections describing the host-system dependencies.

We have discussed how a problem can be decomposed into several subproblems, and how each of these subproblems can be treated in separate sessions to avoid large accumulations of variables. However, sometimes the user needs a variable from another session. This can then be directly accessed using a dot-notation. For example, if the user wants to work in the frequency domain (session FREQ) on the transfer-function equivalent of a linear system description SYS1 defined in the time domain (session TIME), he can access and transform this system directly (TRANS transforms a linear system description to a transfer-function) by:

```
FREQ>> sys2 = TRANS(time.sys1)
```

or, alternatively, from session TIME:

```
TIME>> freq.sys2 = TRANS(sys1)
```

In the case where a new scratch session is started for some intermediate calculations, it is sometimes meaningful to import some or all variables from another session:

```
SCRATCH>> IMPORT(freq,'all')
```

Finally, the command DELETE can be used to close active sessions and delete any local variables. If the current session is deleted, the user will find himself in the MAIN session:

```
SCRATCH>> DELETE(scratch)

%IMPACT-MESSAGE, Global session SCRATCH is closed.
%IMPACT-MESSAGE, All local variables are deleted.
%IMPACT-MESSAGE, Global session MAIN is entered.

>>
```

SESSIONS VERSUS MACROS

In this short section, we will give a graphical representation of the session and macro concepts to be used in the following chapters. Furthermore we will show that these two concepts from an implementational point of view are practically identical.

We have already seen how the macro concept (functions/procedures) was used to define logical areas where new, local variables could be created. For example, when we called the Riccati function, a new logical level was temporarily created for the local variables. Graphically, we represent the call to Riccati as a new vertical level under the main session level (we denote the present point of calculation with a star):

```
 MAIN
   |
RICC (*)
```

Note that the local level RICCATI is eliminated as soon as the function is left. However, when a new session is opened, a permanent logical area is created. This area we represent by a new horizontally parallel level:

```
MAIN-----FREQ (*)
```

Naturally, these two representations can be combined:

```
MAIN-----FREQ-----TIME-----SCRATCH
                    |
                 RICC (*)
```

Note that all vertical lines are temporary, all horizontal are permanent.

Although conceptually different, from an implementational point of view macros and sessions differ only slightly. The main differences are:

- Normally calls to macros include a list of variables and, in the case of functions, terminate by returning a value.
- After a macro is called, IMPACT finds the instructions to be executed in the macro description (stored internally in IMPACT or externally on a file). After a new session is selected, the new instructions are still read from the terminal.

THE QUERY FEATURE

With the potent data structures and command language of IMPACT, the advanced control scientist is given a very powerful algorithmic environment, which he can further adapt to his own needs. On the other hand, if first time users are directly presented with the full IMPACT package, they will most certainly be stunned by the complexity of the package. Many CACSD-packages try to resolve this problem by including an interactive HELP facility. However, whereas such help is excellent once you have a general overview of the package and only need information on a particular subject, during an introduction to the package it is as pedagogic as a 200 page reference manual. Of course, IMPACT does support an interactive HELP, but to prevent the initial shock it also gives the user a gradual introduction. A tutorial presents only the simplest language elements, e.g. how to create variables and how to call standard functions. Moreover, as even this might be too complicated for a beginner unfamiliar with the standard concepts of control theory, a query-mode has been introduced. In this mode, the initiative is transferred from the user to the system. Through a guided conversation, the system will determine the correct action to take.

Assuming that an inexperienced user wants to use the for him new function RICC. He would then call the function using the //QUERY qualifier, forcing IMPACT to enter the query mode. Thereafter the user will be asked for values of each parameter. Optional (defaulted) parameters need to be specified only when another default value is to be changed (user input has been underlined):

```
TIME>>FF = RICC(//QUERY)
RICC>>The function RICC solves a Riccati equation.
RICC>>A Riccati equation is defined by the A and B
RICC>>matrices of a linear systems and an
RICC>>integral involving two weighting matrices
RICC>>Q and R (often unity matrices) where
RICC>>
RICC>>.             /inf                      !
RICC>>x=A*x+B*u and |   (x'*Q*x + u'*R*u)dt = MIN
RICC>>              /0
RICC>>
RICC>>System matrix A (NO DEFAULT): <0,1;0,-1>
RICC>>Input matrix  B (NO DEFAULT): <0;1>
RICC>>Weight matrix Q (DEF=Unity) : <1,0;0,0>
RICC>>Weight matrix R (DEF=Unity) :

FF =
   1    0.7321
```

Especially for functions with many parameters, this facility is very useful. Moreover, if the user is uncertain on the meaning of a particular parameter, he can enter a HELP for further information.

```
RICC>>System matrix A (NO DEFAULT): HELP
RICC>>Please enter a matrix describing the state
RICC>>connections of a linear system. This matrix
RICC>>is the A matrix in the system description
RICC>> .
RICC>> x=A*x+B*u
RICC>> y=C*x+D*u
RICC>>
RICC>>System matrix A (NO DEFAULT):
```

If at this point the users has forgotten the name of an already created system matrix, or if he has to perform some intermediate calculations, one of the options available to him would be to open another interactive session through the command SPAWN, and thereby get access to the general HELP facility, directory command or any other IMPACT statements. He can thereafter return the generated state matrix directly to the RICC query function:

```
RICC>>SYSTEM: SPAWN

%IMPACT-MESSAGE, Local session RICC_1 is started.
%IMPACT-MESSAGE, All variables in TIME are imported
%IMPACT-MESSAGE,    to RICC_1 as local copies.

RICC_1>> HELP
 ..
RICC_1>> A = A_MAT(sys1); -- extract A from sys1
RICC_1>> RETURN A

%IMPACT-MESSAGE, Local session RICC_1 is closed.
%IMPACT-MESSAGE, All local variables are deleted.
%IMPACT-MESSAGE, 1 parameter is passed back to
%IMPACT-MESSAGE, queried function RICC.

RICC>>Input matrix  B (NO DEFAULT):
```

The user has three choices to close session RICC_1: With RETURN 'value', he deletes the intermediate session RICC_1 and at the same time gives back a value to the querying function. The commands RETURN and DELETE(RICC_1) both close RICC_1 and force IMPACT to return to the original query for A. The equivalent meaning in this case of RETURN and DELETE(RICC_1) is a good example of the previously explained close connection between functions (left with RETURN) and sessions (closed with DELETE). The following diagram also illustrates these resemblances:

```
MAIN_SESSION-----TIME-------FREQ-----SCRATCH
                  !
              RICC(//QUERY)
                  !
               "query"---------RICC_1
                                 !
                               A_MAT (*)
```

In session RICC_1 it is of course also possible to import foreign variables and to switch to other sessions. However, the RICC_1 session only exists as long as session TIME, and we can only return to the queried function RICC by deleting session RICC_1.

QUERY MODE AND USER MACROS

In the last chapter we saw an example on how the query option was used by the system-defined function RICC. We now have to ask ourselves the following obvious question: How does IMPACT know what information to give the user when RICC has been interactively defined by the user just a few minutes ago? Well, the answer is as obvious. Either IMPACT does not know, in which case the query-mode only prompts for the parameters using the formal parameter names (which is not much help for the person not familiar with the function), or we have told IMPACT what to ask. The definition of RICC is therefore extended once again:

```
Function RICC(a,b,q=eye(dim(a,1)),
                  r=eye(dim(b,1)));
INFO --RICC
  General('The function RICC solves ...');
  General('A Riccati equation is     ...');
  ..
  Param_name('A');
  Param_short('System matrix A (NO DEFAULT)');
  Param_long('Please enter a matrix     ...');
  Param_long('connections of a linear  ...');
  ..
END INFO; -- RICC
  v,d,p : MATRIX;
  k,n   : INTEGER;
BEGIN -- RICC
  ..
  RETURN -r\b'*p;
END RICC;
```

The same method is used for the system functions/procedures defined in Ada. In fact, a close resemblance exists between the macros coded in IMPACT and the functions/procedures implemented in Ada. This is very advantageous both for the IMPACT developer and for the advanced IMPACT user: As the user input, and any functions described in the command language, have to be interpreted rather than compiled, the execution of complex algorithms described by command language is rather slow. During the development of new algorithms, this is offset by the time not spent on compilations. However, to obtain shorter execution times, any complex algorithm should be compiled and incorporated into the package itself as soon as it is developed and tested. As both types are similar, such a transition can be made with a minimum of recoding.

SUMMARY

We have in this contribution discussed several aspects of man-machine interfaces in CACSD. We have stated the need, in particular in a research environment, for a highly structured, command driven interface with which it is possible to develop new control algorithms based upon standard primitives. We have seen examples of the IMPACT command language, which together with a modular session concept used to structure the user working environment gives us an excellent control research environment. Finally, we have seen how the same complex interactive environment can be made accessible to novices and sporadic users through a query-option.

REFERENCES

ANSI/MIL-STD 1815 A (1983). Reference manual for the Ada programming language.

Digital Equipment Corporation (1982). VAX/VMS command language user's guide, version 3.0. Chapter 2. DEC, Maynard, Massachusetts, USA.

Little, J.N., A. Emami-Naeini, and S.N. Bangert (1984). CTRL-C and matrix environments for the computer aided design of control systems. In A. Bensoussan, J.L. Lions (Ed.), Proc. 6'th International Conference on Analysis and Optimization (INRIA). Lecture notes in Control and Information Sciences 63, Springer Verlag. pp. 191-205.

Moler, C. (1980). MATLAB, User's Guide. Department of Computer Science, University of New Mexico, Albuquerque, USA.

Potter, J.E. (1966). Matrix quadratic solutions. SIAM Journal of Applied Mathematics, 14, 496-501.

Rimvall, M., (1983). IMPACT, Interactive Mathematical Program for Automatic Control Theory, A Preliminary User's Manual. Institute for Automatic Control, ETH Zurich, Switzerland.

Rimvall, C.M., F.E. Cellier (1984a). IMPACT - Interactive Mathematical Program for Automatic Control Theory. In A. Bensoussan, J.L. Lions (Ed.), Proc. 6'th International Conference on Analysis and Optimization (INRIA), Lecture notes in Control and Information Sciences 63, Springer Verlag. pp. 578-597.

Rimvall, M. (1984b). On the use of Ada in CACSD. In Proc. Workshop on computer aided control systems design, Brighton September 19-21 1984. Institute of Measurement and Control, London.

Walker, R., C. Gregory, and S. Shah (1982). MATRIX$_x$, A Data Analysis, System Identification, Control Design, and Simulation Package. IEEE Control Systems Magazine, 2, 30-37.

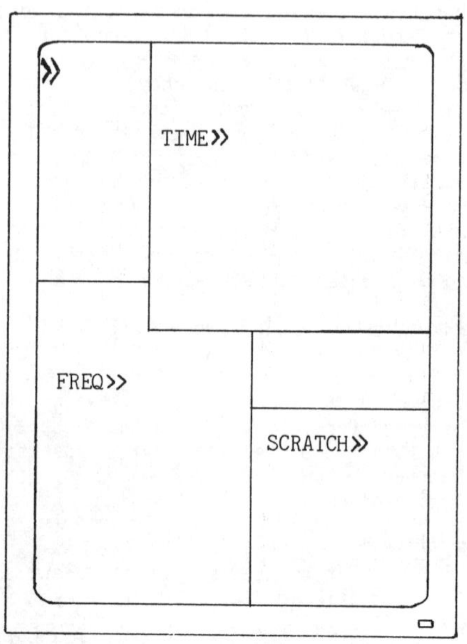

Fig. 1. IMPACT sessions on a workstation supporting virtual windows.

EVALUATION OF COMPUTER SYSTEMS FOR CONTROL SYSTEM DESIGN

P. P. J. van den Bosch* and Guo Shuling**

**Delft University of Technology, Department of Electrical Engineering, PO Box 5031, 2600 GA Delft, The Netherlands*
***Beijing Institute of Control Engineering, People's Republic of China*

Abstract. We will discuss and evaluate a number of different computer systems for computer-aided control system design, ranging from superminis (VAX) to microcomputers (IBM PC). This evaluation will deal with price, software support, calculation speed, memory size and graphical output facilities.

Keywords: Computer-aided design, Computer-aided system design, Control engineering computer applications, Computer evaluation, Computer graphics, Computer hardware, Computer language.

1. INTRODUCTION

The widespread availability of the digital computer to a large group of control theorists and technicians has had a significant impact on Computer-Aided Control System Design (CACSD). Up to now, there have been a number of sophisticated CACSD programs which can improve the efficiency of analyzing or designing control systems. A great deal of software has been developed for CACSD; in the past this was mainly intended for large, centralized computers, but during the seventies the emphasis swung to minicomputers and in this decade both to superminicomputers and to microcomputers. Each of these installations has its own performance-price ratio. In this paper we will review a number of digital computer installations with respect to their ability to be used for CACSD.
Therefore, we will first derive a number of requirements to be posed on an CACSD environment, namely the application software, the hardware and the compiler of the computer installation.

2. REQUIREMENTS

In this section we will discuss a number of subjects that are important in designing and/or using CACSD software, namely

1 user interface
2 documentation
3 compiler
4 calculation speed
5 memory requirements
6 linear design programs.

1. User interface:
If a program has attractive and reliable mathematical algorithms, it may still be inferior with respect to control system design when the interaction between the program and its user is not accepted by the user. This interaction is determined by a number of factors, but especially by the communication between the user and the program and the presentation of graphic information (Van den Bosch and Bruijn, 1979). Only an interactive program can support a designer-oriented environment.
Either a command language or a question/answer approach can take care of the communication between the program and the user. In general, a command language is selected for CAD programs because it yields the necessary flexibility for most users.

However, inexperienced users lack the guidance of a question/answer approach. By means of extensive support capabilities, clear error messages and a macro-facility, some guidance can be incorporated into a command language. A macro-facility combines separate commands to obtain a higher-level command language which is easier to use, but is, at the same time, less flexible (Wieslander, 1979).

In designing control systems, graphic representations of the system behavior are of paramount importance. Although numbers are much more exact, design considerations mainly deal with graphic representations of a system. For example, linear optimal state feedback, output feedback or pole placement are well-established, mathematically-oriented methods for control system design. However, whether or not such a design meets the ultimate design requirements cannot be judged by looking at only the value of the criterion or at the feedback matrix. In general, only time (or freqency) responses offer enough information to make an evaluation of the ultimate system behavior possible.
So, a graphical display or a plotter is almost unavoidable when analyzing or designing systems.

2. Documentation:
Programs produce large amounts of data. In order to keep track of all this data, it is necessary to have good documentation facilities on all printed output. So, all plots and output printed for later use need some headings that contain information concerning date, time, user and file names.

3. Compiler:
In spite of the availability of many new programming languages, such as Pascal, Ada and Modula-2, Fortran is still widely used and accepted for technical and scientific work. The majority of subroutine libraries for control system design are Fortran oriented. Consequently, a Fortran compiler has to be available. This compiler has to satisfy at least the Fortran 66 or Fortran 77 standards or a generally-accepted subset of this language. Although double-precision complex numbers are not incorporated in the Fortran 77

standard, this data type is very useful and almost necessary in many algorithms.

4. **Calculating speed:**
Simulation and non-linear optimization pose severe requirements on the calculating power of the computer. Consequently, the calculating speed of computers used in an interactive environment is very important. Other design programs, for example those based on pole placement, require less calculating speed.

5. **Memory requirements:**
Design programs can be quite large. Especially when high-order systems are allowed, much memory has to be available, for example 200k to 1M bytes of internal memory. Overlay or virtual memory techniques reduce the requirements for physical memory. Although the overlay technique increases the size of a program that can be run on a minicomputer, it restricts the largest module to a predefined value. For example, with a PDP 11 this maximum size is 64k bytes. Even if more memory is available, the largest module of programs that have to be in memory at the same time is still limited by this amount of 64k bytes. Virtual memory techniques really increase the size of programs. This may be needed when dealing with MIMO systems or simulation programs with a large database for the simulation model.
Owing to their limited memory capacity of 64k bytes and their lack of overlay or virtual memory techniques, the 8-bit microcomputers are not suitable for large CAD programs.

6. **Linear design programs:**
For many years the theory of control system analysis and design has been developed with little or no consideration of the reliability of the underlying numerical calculations. In spite of this lack of interest, these algorithms still work satisfactorily for low-order systems, especially if double-precision accuracy has been implemented. However, the growing interest in the application of multivariable system theory to large and complex systems has exposed the computational deficiencies. This emphasizes the need for efficient, reliable and numerically stable algorithms. Recently, much research has been devoted to linear algebra to derive and describe reliable and numerically stable algorithms. Denham et al. (1983) indicate that, at least at present, concentration on the use of system models which involve the manipulation of real matrices is more likely to yield numerically robust algorithms than is the case for models involving polynomial manipulations. This favors the use of state-space models instead of matrices of tranfer functions. At present there is a class of generally-accepted subroutine libraries, such as LINPACK (linear algebra) and EISPACK (eigenvalues) which can be used as a reliable basis for computer-aided design programs. The subroutine package ORACLS (Armstrong, 1980) supplies the user with Fortran subroutines for designing with the Linear Quadratic Gaussian (LQG) methodology.
Research is going on in several places in the world to extend these subroutine libraries for general use in interactive analysis and design packages, for example SLICE (Denham et al., 1983) and SYCOT (Van den Bosch and Van den Boom, 1985).

In table 1 we have illustrated the influence of using either single or double precision numbers for calculating the eigenvalues of a matrix. The eigenvalues are calculated with the aid of the programs BALANC, ELMHES and HQR taken from EISPACK. The eigenvalues have to be -1, -2, -3, .. , -10.

This simple example illustrates the importance of using double precision numbers even if we use reliable and robust algorithms. In particular, we have to use double precision numbers if we utilize the coefficients

Table 1. Influence data type.

Single precision RE	IM	Double precision RE	IM
- 0.9999,	0.0000	-1.0000,	0.0000
- 2.0019,	0.0000	-2.0000,	0.0000
- 3.1459,	0.3290	-3.0000,	0.0000
- 3.1459,	-0.3290	-4.0000,	0.0000
- 4.6426,	1.6349	-5.0000,	0.0000
- 4.6426,	-1.6349	-6.0000,	0.0000
- 7.2879,	2.8287	-7.0000,	0.0000
- 7.2879,	-2.8287	-8.0000,	0.0000
-10.9227,	1.7578	-9.0000,	0.0000
-10.9227,	-1.7578	-10.000,	0.0000

or roots of polynomials.

In conclusion, the main requirement for linear design programs is the availability of efficient, reliable and numerically robust algorithms for the analysis and design of linear systems and a software environment that supports double precision real numbers.

3. COMPUTER CONFIGURATIONS

At the present time many computer configurations can be selected to realize a computer-aided design environment. We can distinguish among superminicomputers, minicomputers and 16- or 32-bit microcomputers. Each installation has its own capabilities, support and price. In this section we will compare these installations for computer-aided design applications. We will discuss a VAX 750 with VMS, the PDP 11/23 with RSX-11M, the TULIP 16-bit microcomputer with MS.DOS and the IBM PC 8/16-bit microcomputer with PC.DOS. The VAX and PDP are manufactured by Digital Equipment Corporation and are equipped with an appropriate floating-point processor. The TULIP has Intel's 8MHz 8086 processor and the Intel's 8 MHz 8087 floating-point processor and is manufactured by the Dutch company Compudata in Den Bosch. The IBM PC has Intel's 4.77 MHz 8088 and 8087 processors. Both use the Microsoft Fortran V3.2 compiler.

In this section we will discuss a number of software and hardware aspects of a computer installation with respect to CACSD purposes.

Fortran compiler

Although many Fortran compilers claim to satisfy the Fortran 77 standards, a number of modifications are still required to convert a program from one computer to another one. These differences mainly concern the difference in dealing with input/output to and from the user terminal and to and from the disk.

If we compare the calculation speed of an application program on several installations we compare, in fact, the combination of hardware and software. The software part is mainly influenced by the efficiency of the code produced by the Fortran compiler. The efficiency of these compilers may differ from operating system to operating system. But even one compiler can be utilized in many different ways. In order to illustrate the effect of utilizing one compiler in its many different methods of operation, we have selected the Microsoft Fortran 77 compiler, version V3.2. This compiler can be used for almost all MS.DOS computers and the IBM PC. This type of computer now dominates the 16-bit microcomputer market. The results of this comparison are shown in the appendix.

Operating system support

There is a difference in quality and support of the compilers available for the operating systems. The Fortran compilers of VMS and MS.DOS support both the data types of double precision complex numbers. The Fortran compiler of RSX lacks this data type. The MS.DOS operating system still does not have the refinements and facilities of, for example, VMS or RSX-11M. Two facilities could improve its value for CACSD usage, namely a better support of run-time errors and the facility to interrupt a program. Both will now be explained.

If an exception (under- or overflow) occurs during a mathematical operation, for example during multiplication, addition, sine calculation, etc., the run-time system can handle this exception in different ways.
-An error can destroy the program, so that the only possibility for the user is to restart the computer,
-it can stop the program and return control to the operating system, or
-it can detect the kind of error an supply the user with a useful error message, and then continue operation with a predefined action.

The last possibility is the most suitable, and moreover standard procedure with VMS and RSX-11M. Sometimes, running a program with MS.DOS may result in one of the first two situations. The likelihood of influencing the error recovery procedure with MS.DOS is much less than with VMS and RSX-11M.

If a program runs with MS.DOS, it cannot be stopped by a user unless the programmer of that program has taken special actions. For example, starting a simulation run with a very small integration interval can mean that a large time will elapse before the simulation stops. Such a run cannot be stopped without resetting the computer.
Of course, the programmer can take special actions to avoid this undesirable situation, but this requires programming with assembler language.

Program development

Up to now we have considered the computer system from the point of view of running CACSD programs, in which case only calculation time, memory size and graphical capabilities are important. However, there are also CACSD programs, e.g. simulation programs, that require one or sometimes more editing, compilation and link steps. Then, other requirements become important, especially disk-access time and disk space.
In decending order with respect to their access times we can distinguish RAM disks, winchester disks and floppy disks. A RAM disk is a part of the internal memory that is declared and used as a disk. The VAX and PDP computers have a winchester disk. Microcomputers in general utilize floppy disks or a winchester disk. When program development becomes important, a winchester is almost necessary. It is interesting to note that linking of a large program can be executed with VMS and MS.DOS much faster than with RSX. With RSX we need overlays, which are avoided with MS.DOS. For example, linking of PSI requires about 1 to 2 minutes with a VAX, about 3 minutes with a TULIP with a winchester disk, about 10 minutes with a TULIP I with 800 k floppies and utilizing a RAM disk and about 10 to 30 minutes with a multi-user RSX system equipped with a winchester disk.

The size of the programs to be stored on one floppy disk may be a limiting factor to be able to develop programs. In table 2 we illustrate the size of the Fortran V3.2 compiler and linker of Microsoft.

This indicates that developing Fortran programs on an IBM PC with 360k floppy disk drives may be a tedious job. Different floppies have to be used just for the operating system, the screen editor, the compiler and the linker, which requires a frequent exchange of the floppies in the drives. Then, the 800k

Table 2. Size of compiler and linker of Microsoft

Compiler:	FOR1.EXE	120k
	PAS2.EXE	97k
Linker:	LINK.EXE	40k
Libraries:	FORTRAN.LIB	86k
	MATH.LIB	35k
Total		378k

floppy disk drive of the TULIP I and the 1.2M floppy disk drive of the IBM PC/AT are more attractive.

The capacity of a floppy disk drive can be a limiting factor for the storage of large CACSD programs.

Graphics Capabilities:

In our comparison we have equipped the VAX and the PDP with the VT241 color terminal. The IBM PC uses IBM's color graphics adapter, although other graphic adapter boards have now become available (e.g. Hercules, Tecmar). A comparison of the graphics capabilities of the two microcomputers is shown in table 3, namely the number of different colors, the resolution (in dots) and the possibility to use two separate display memories: one alphanumeric memory for text (the communication between program and user) and one graphics memory for the graphics to be displayed on a screen.

Table 3. Comparison of graphics capabilities.

Name	colors	resolution	sep.mem.
VT241	4	800*240	no
TULIP I	8	768*288	yes
IBM PC	4	320*200	no

This table is illustrated in figure 1. In figure 2 we show the difference in the number of pixels that can be draw on the screen. This number not only depends on the resolution but also on the number of colors.

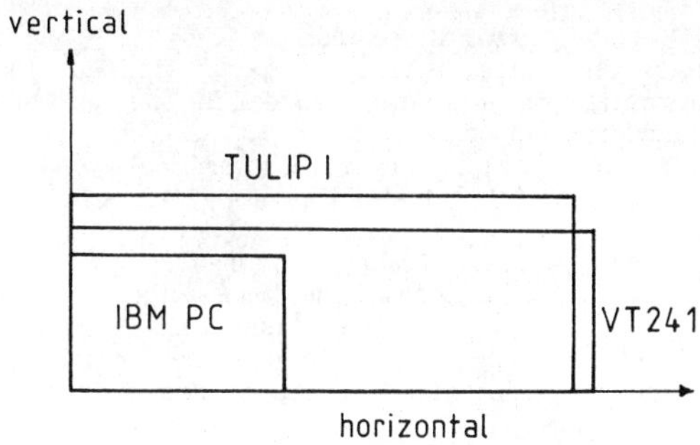

Fig. 1. Graphics resolution for 1 color.

Calculation speed

Table 4 compares the calculation speed of a number of computer installations running the block-oriented simulation program PSI (Van den Bosch, 1983). Such a program has a relatively large number of integer operations in order to calculate appropriate pointers in the program. Still, these numbers give a good comparison of the computing power of the different processors. All computers are used in a single-user mode, which is not common practice for the multi-user systems. Moreover, we have indicated whether or not the computer is equipped with a separate, dedicated hardware floating-point processor.

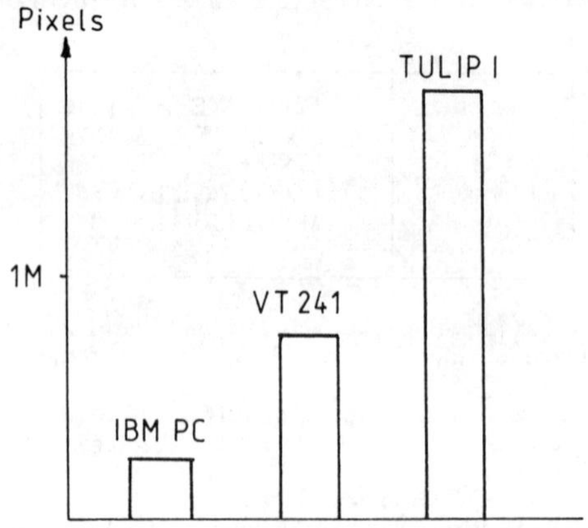

Fig. 2. Number of pixels.

Table 4. Comparison of calculation times.

Computer	hardware FPP	relative calc. time
VAX 750		1.0
microPDP11/73		1.8
microVAX I		2.1
PDP 11/23	yes	3.2
TULIP I	yes	3.6
microPDP11/23	no	4.1
IBM PC	yes	8.6
TULIP I	no	9.8
IBM PC	no	23.5

This table is illustrated in figure 3. With a solid line we assume that each computer installation has only 1 user. With the dotted line we assume that the VAX has 4 users and the microVAX and PDP has 2 users.

Overall comparison

Table 5 illustrates the price per terminal, the maximum size of a program loaded in memory and the calculation speed. The price per terminal is the price of the computer system with terminals, divided by the number of terminals. This price is normalized for the IBM PC. For example, 4 CAD stations can be realized with 4 TULIP I computers or one VAX with 4 terminals. In such a case, the VAX solution is six times more expensive. The calculation speed is normalized for a VAX 750 with one user, running a simulation program. With four users this speed can drop to 0.25.

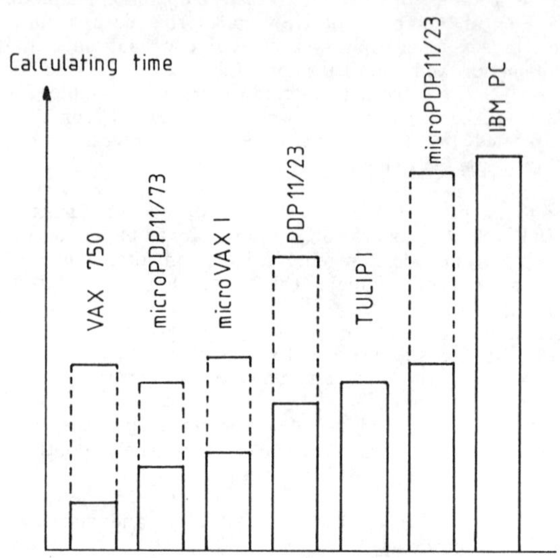

Fig. 3. Relative calculation time.

These tables indicate that a decentralized solution with separate 16-bit microcomputers yields an attractive price-performance ratio. Depending on the number of users of the multi-user VAX, a TULIP can be up to three times slower or a little faster, it has better graphics capabilities and it costs much less than a solution based on a VAX. Although the calculating speed of a microcomputer may be less, response times to commands of the user are shorter than in a multi-user environment. The IBM PC offers a less attractive performance. In 1985 the IBM PC/AT has become available. This computer uses the 6 MHZ 80286 processor of Intel. From the point of view of calculation speed this new computer can be compared with the TULIP with its 8MHz 8086.

As soon as a decentralized solution has been selected, the administration becomes more complicated. It takes more time and effort to update all decentralized computers with the newest versions of the software. Moreover, the exchange of data, models and files becomes more complicated, namely via floppy disks instead of via the common large disk. On the other hand, with a decentralized solution each user is responsible for his or her own files and models. If he is willing to work with older versions of the software, there will be fewer conversion problems.
An important improvement can be obtained by using a Local Area Network (LAN) to connect all decentralized microcomputers with each other. Then, the exchange of files and data among several users, all connected to the same LAN, can easily be realized.

Table 5. Comparison among computers.

Name	backgr. memory	number of terminals	price per terminal	maximum program size	calculation speed
VAX 750	60M	4	6	virtual memory	1-0.25
microVAX I	31M	2	3	virtual memory	.5-.25
PDP 11/23	10M	2	2	64k+overlay	0.3-0.15
TULIP I	2*800k	1	1	896k	0.3
IBM PC	2*360k	1	1	640k	0.14

At our laboratory we have decided to decentralize the CAD facilities from a multi-user environment to a number of stand-alone computers or work stations. We have selected the TULIP I computer for this purpose, but fast computers of other manufacturers exist as well, for example the M24 of Olivetti, the Apricot and Wang computers.

Future developments

Even more powerful microprocessors are already announced for 1985 (Zorpette, 1985). In particular, the 32-bit microprocessors such as the MC68020 of Motorola and the NS32332 of National Semiconductor Corporation are very promessing with respect to calculating power. The 16.7MHz MC68020 has a calculating power approaching 4 million instructions per second (MIPS) while the NS32332 will reach 2.5 to 3 MIPS.
These numbers are quite impressive compared with other announcements in 1985 of superminicomputer systems, such as the VAX 8600 of Digital Equipment Corporation (4 MIPS), the MV/10000SX of Data General (3.6 MIPS) and the Prime 9955 with about 3.5 MIPS.
Depending on their price and the ability to maintain their calculating power, incorporation of these powerful 32-bit microprocessors in 32-bit microcomputers can yield an even more attractive work station for CACSD than the present 16-bit microcomputers.

CONCLUSION

Many requirements for CACSD have been discussed. These requirements pose severe restrictions in selecting an appropriate digital computer for CACSD applications. We have compared superminicomputers, minicomputers and 16-bit microcomputers.
At this moment (1984/1985), the 16-bit microcomputer yields a very attractive performance-price ratio, which makes it a new and interesting work station for the control engineer. New defacto standards on hardware (IBM PC) and software (MS.DOS, UNIX) offer new possibilities for the exchange of data and programs among the control engineering community. We are of the opinion that a decentralized solution with a microcomputer for each user will become even more attractive in the near future.

REFERENCES

Armstrong, E.S. (1980). ORACLS, A Design System for Linear Multivariable Control, Marcel Dekker Inc., New York.

Bosch, P.P.J. van den and P.M. **Bruijn** (1979). Requirements and Use of CAD Programs for Control System Design. Proceedings IFAC Symposium on Computer Aided Design of Control Systems, Zurich 1979. Pergamon Press London (459-464).

Bosch, P.P.J. van den (1983). Interactive System Analysis and System Design Using Simulation and Optimization. Proceedings IFAC Symposium on Computer Aided Design of Multivariable Technological Systems, Lafayette 1982. Pergamon Press London (225-232).

Bosch, P.P.J. van den and A.J.W. van den **Boom** (1985). Industrial Applications in The Netherlands of Computer-Aided Design Packages for Control Systems. Preprints IFAC Symposium on Computer-Aided Design of Control and Engineering Systems, Copenhagen 1985. Pergamon Press London.

Denham, M.J., C.J. **Benson** and T.W.C. **Williams** (1983). A Robust Computational Approach to Control System Analysis and Design. Proceedings IFAC Symposium on Computer Aided Design of Multivariable and Technological Systems, Lafayette 1982. Pergamon Press London.

Wieslander, J. (1979). Design Principles for Computer Aided Design Software. Proceedings IFAC Symposium on Computer Aided Design of Control Systems, Zurich 1979. Pergamon Press London (493-496).

Zorpette, G. (1985). Microprocessors. IEEE Spectrum, Vol 22, no 1 (53-55).

APPENDIX

Influence of compiler parameters

We will discuss the performance of the Microsoft Fortran 77 compiler, version 3.2. This compiler can be used in many different ways, partly defined by compiler metacommands and partly by linking it with a different library. We show calculation times for the following eight mathematical operations:

- Tdo an empty DO-loop
- Tint multiplication with integer arithmetic
- Treal multiplication with real arithmetic (4 bytes)
- Tdp multiplication with double precision numbers (8 bytes)
- Tcn multiplication with complex numbers (2*4 bytes)
- Tdpcn multiplication with double precision complex numbers (2*8 bytes)
- Tsin calculation of sine function with real numbers of 4 bytes
- Tdsin calculation of sine function with double precision numbers of 8 bytes.

These variables Tdo, Tint, Treal, etc. are expressed in seconds for 1000 times the indicated operations with a TULIP I computer. Their relative size can be used for the calculating times on other computers, such as the IBM PC. The variable "Size" indicates the size of the executable program that realizes the above mentioned operations, expressed in bytes.

Metacommands

The compiler metacommands influence the way the compiler produces the object code of a Fortran program. We will only discuss the influence of the following metacommands.

- STORAGE:4 realizes the standard of Fortran 77. All variables, both real, integer and logical, are stored in integer and logical, are stored in 4 bytes.
- STORAGE:2 realizes that a real variable occupies 4 bytes, and integer and logical variables 2 bytes each.
- FLOATCALLS directs compiler to generate calls to subroutines in the emulator library for floating-point operations.
- NOFLOATCALLS causes the compiler to generate in-line interrupt instructions for floating-point operations.

Libraries

Besides the Fortran library (FORTRAN.LIB) one of the following libraries has to be selected. These libraries are used by the linker to produce an executable program.

- MATH.LIB This library uses the standard IEEE floating-point accuracy of 80 bits. The resulting program can be run on computers both with and without an 8087 processor. If an 8087 is present it will be used.

Table 6. Comparison of compiler parameters.

Metacm.	Lib.	8087	Tdo	Tint	Treal	Tdp	Tcn	Tdpcn	Tsin	Tsdp	Size
FLT. 4	MATH.	+	.016	.072	.080	.084	.422	.457	.46	.46	47172
FLT. 2	MATH.	-	.016	.072	.574	.608	2.641	2.800	14.22	14.24	47172
FLT. 2	MATH.	+	.008	.030	.074	.076	.422	.441	.44	.46	46836
FLT. 2	MATH.	-	.008	.030	.570	.600	2.641	2.801	14.20	14.22	46836
NOF. 2	MATH.	+	.008	.030	.042	.044	.418	.441	.44	.46	46756
NOF. 2	MATH.	-	.008	.030	.752	.782	2.641	2.801	14.22	14.24	46756
FLT. 2	ALTM.	+	.008	.030	.204	.406	1.223	2.203	2.16	7.84	44944
FLT. 2	ALTM.	-	.008	.030	.204	.406	1.219	2.199	2.16	7.84	44944
FLT. 2	8087.	+	.008	.028	.060	.062	.359	.379	.44	.46	37958
NOF. 2	8087.	+	.008	.030	.042	.044	.359	.379	.44	.46	37958
Ratio big/small			2.0	2.6	17.9	17.8	7.36	7.39	32.3	30.9	1.24

ALTMATH.LIB This library uses 64 bit accuracy for intermediate results. Even if an 8087 floating-point processor is available, it is not used.

8087.LIB This library generates only code for the 8087. If this device is not present, the program will not run. All mathematical operation will utilize 80 bit accuracy.

Calculation time and program size

Table 6 yields a comparison of the different ways of utilizing the Microsoft Fortran 77, V3.2 compiler. Moreover, we have indicated whether or not an 8087 hardware floating-point processor is present.

As can be seen in table 6 the introduction of the double precision data type for reals does hardly increase the required calculation time for mathemathical operations for the 8086/8087 processors and the Fortran V3.2 compiler of Microsoft. With MATH.LIB and 8087.LIB this increase is very small. Only programs linked with ALTMATH.LIB show some additional calculation time.

In applications like a block-oriented simulation program with its many integer calculations for the calculation of pointers, the introduction of the 8087 floating-point processor reduces calculation-time requirements by a factor of 4 (using MATH.LIB). If we compare the fastest codes of both, i.e. with the 8087 we use NOFLOATCALLS and 8087.LIB and without the 8087 we use FLOATCALL and ALTMATH.LIB, then the 8087 yields an increase in calculation speed by a factor of 2.5. Moreover, it reduces memory requirements by about 10%.
With special calculations that use heavily goniometric and logarithmic functions the increase will be more dramatic, as illustrated in table 6.

Two examples of real programs are used in table 7 for comparing the influence of the selection of the library on the size of the executable module. PSI is a block-oriented simulation program with relatively much data and TRIP is an analysis and design program for linear systems with relatively much code (Van den Bosch and Van den Boom, 1985).

Table 7. Comparison of program sizes (bytes).

Program	NOFLOATCALLS 8087.LIB	FLOATCALLS ALTMATH.LIB
TRIP	326560	368560 (+13%)
PSI	181834	198268 (+ 9%)

INTEGRATED COMPUTER SUPPORT FOR THE DEVELOPMENT AND PROJECT MANAGEMENT OF SOFTWARE/HARDWARE SYSTEMS

P. Göhner

Gesellschaft für Prozeßrechnerprogrammierung mbH, Kolpingring 18a, 8024 Oberhaching b. München, FRG

Abstract:

The fact that industrial computer control systems consist more and more of complex software and hardware components and it isn't possible to seperate software and hardware at the beginning of the design enforces integrated development tools, supporting software development as well as hardware development.

Although the development of software components and the development of hardware components are different at the realization level they are interrelated to a high degree in the main parts of the development, project management and maintenance.

This paper illustrates the main disadvantages of using separate tools. It shows the requirements for an integrated tool and discusses the concept of the EPOS system as an example of how these requirements can be satisfied. Results of applying EPOS to industrial projects are reported, which show that quality and reliability have been improved considerably.

Keywords
Computer-aided design; computer-aided system design; computer software; computer programming; parallel processing; process control; programming languages; project management; specification.

INTRODUCTION

The development of software/hardware systems, e.g. as in the automation of technical systems, is essentially determined by three factors:

- the creativity of the developer
- the development methods
- the development tool

Although in early years of data processing the developer's creativity was almost solely the decisive factor in the success of a project, in the last few years, it has been clearly demonstrated, that without the application of suitable tools and development methods complex software/hardware systems are not realizable.

For this reason, in recent years tools have been created, which make it possible, to carry out work intensive activities in the development process with computer support to reduce the manual error sources.

The first generation of these development tools consists of independant **tools** of different deliverers, each supporting only one part of the development process. The main difficulties using these randomly connected tools in a project were the overlapping application areas and the non-existence of a predefined, well structured application ordering.

To avoid the overlapping application the second generation consists of **tool boxes** containing different tools in some kind of order. Each tool uses its own information system and its own user interface. The transformation between the different information systems has to be done manually. Therefore it was very difficult to guarantee the consistency of the information over the whole life cycle of the project. Another handicap using tool boxes was the variety of user interfaces.

The third generation of development systems combines adapted tools closer in form of **tool systems,** using a solid data base and one user interface. In the industrial application of these tool systems, it has been shown, that to reduce the development costs, to improve the quality, to cope with the complexity and to realize an efficient project control it isn't enough to support the development of software using a tool system, to support the development of hardware using another tool system and to support the project management using a third tool system.

To produce computer control systems, that means integrated software/hardware systems, it is necessary to give integrated support in the technical development of software /hardware, the project management as well as the product management. Although, only a few years ago, such a **production environment** for integrated software/hardware systems appeared unthinkable, some computer support systems have now been conceived, which attempt to realize these advanced aims.

In the following section, the requirements for an integrated support system to be used in the development of software/hardware systems will be stated. As an example of an integrated support system several features of the Engineering and Project-management Oriented Support system EPOS are explained. Results of applying EPOS to a great number of industrial projects are reported.

REQUIREMENTS FOR AN INTEGRATED SUPPORT SYSTEM

1. Integrated support throughout all development and maintenance phases

The development process includes three main activities:

- requirements engineering
- system design
- implementation

Requirements engineering includes, first, determining and defining the problem statement for the system to be developed and, secondly, establishing a problem solving concept breaking down the user requirements into technical oriented requirements.

During the system design the technical problem solving concept is transformed into a software/hardware system. The process of system design begins with system structuring, which partitions the whole system into subsystems, which declares the interfaces to the outside and the internal interfaces, and which defines the relations between the subsystems. After the system has been structured first a rough design and then a detailled design of software and hardware follow.

Implementation includes on the one hand transforming the detailed software design into programs coded in a specific programming language, and on the other hand transforming the detailed hardware design into circuits. The implementation process also includes testing the software and hardware subsystems, integrating software and hardware and validation of the total system.

Although the development phases are very different in nature they are to a high degree interrelated. Between successive phases there are transitions in both directions. The main transition is the top-down direction because the results of the former phases are the starting point of the following phases. In all projects there are iterations and it is necessary to go back and to redesign the former levels.

It is therefore indispensable to have computer aided transitions between the different phases and computer checkable relations between them.

The main disadvantages of using different tools to support requirements engineering, system design and implementation become obvious:

- missing computer assistance in checking consistency and completeness of the information belonging to different phases during the whole life cycle

- missing computer assistance in transforming the specifications of the former phases to the following phases.

To realize an integrated support system throughout all development phases it is necessary to have computer support in

- specifying the information belonging to requirements engineering, system design and implementation

- checking requirements specifications, system design specifications and program/hardware specifications for consistency and completeness

- transforming the requirements specifications into system design specifications, the system design specifications into program specifications

2. Integrated support of software and hardware development

The increasing number of embedded systems, which was one of the main causes for the development of the Ada programming language enforces more and more a combined design of software and hardware components. To realize an integrated software/hardware development it is essential to have integrated development tools, which allow the description of the realization independent structuring of the system as well as the specification of the software and of the hardware structure.

Therefore it is necessary to have means to describe, to analyze and to document on the one side software design objects, i.e. modules, procedures, tasks, data, interfaces, on the other side hardware design objects, i.e. hardware components, signals, bus connections, computers, and most important the relations between the software and the hardware design objects that means the relations between the logical structure and the physical structure of the system. The main difficulties using separate design tools for the development of software and hardware components result from the enforced early and final decomposition in software and hardware parts and the lack of analyzing relations between the software and the hardware structure.

3. Integrated support of different design methods

Software/hardware systems like process automation systems are complex structures which allow different viewpoints because of their composition of different types of elements.

For the development and realization as well as understanding of such systems, it is necessary to know:

- the sequence of functions

- the module structure

- parallel activities and the corresponding triggering events

- the dataflow

- the datastructure

- the structure of the hardware system

Various types of development methods can be derived from these different viewpoints. Depending on which of the above are dominant the following methods can be disinguished:

- function-oriented development methods

- module-oriented development methods

- event/process-oriented development methods

- dataflow-oriented development methods

- datastructure-oriented development methods

- hardware-oriented development methods

Which of these design methods is the best? The results of a lot of published investigations of this question show that there is no "best" method. Each method has its specific advantages and disadvantages, which depend upon the respective problem as well as on the respective development phase. Therefore it is good practice to select the design method according to the respective problem, and if necessary, to change the method during the design, or to use different design methods in different subsystems.

To realize an efficient and general purpose development support system it is essential to integrate the different design methods in a method base and to give the developer the opportunity to select the suitable method according to his task.

4. Integrated support of development and project management

Although the project management and the development tasks are different in nature they are interrelated to a high degree. The development reports about the current state of development including problems and estimations to complete and proposes change requests. The project management plans, monitors and controls the project using the development information to release work packages, to fix deadlines and to give cost limitations.

Just as time-out and delays in a feedback system can lead to poor control behavior or even to instability, the management of a project also depends on whether the information between project management and developers flows accurately and without delay.

The disadvantages of the use of separate tools for technical development and for management support result from the insufficient exchange of information between these tools, the missing computer-assistance in checking consistency and completeness between the planning of tasks and their implementation and the frequently subjective and inaccurate project reports with the latest information available only at the end of a report period.

5. Integrated product management

Product management includes the management of all temporary and final results during the whole life cycle of the product and consists of version control, variant control, configuration control and quality control .

Version control is storing and handling of changes and of different versions of the product during the life cycle i.e. changes of the requirements specifications, the design specifications and the project management specifications.

Variant control is storing and handling of different variants of the system that means the system exists in different types at the same time.

Configuration control includes the procedural aspect of product management that means the fixing of base lines and the handling of change requests during the whole life cycle.

Quality control includes all activities to ensure the correctness of the results of the technical development i.e. the consistency of the requirements specifications with design specifictions, the correct realization of design specifications in form of programs and hardware.

In addition to the strong connections between project management and technical development there are also strong connections between project management and product management and between technical development and product management.

For this reason it is necessary to have an integrated product management tool, which works directly on the data base of the technical development and of the project management. The main disadvantage using a separate product management tool is the repeated and redundant storing of information, whereby it is very difficult to ensure that the developers use the right version corresponding to a baseline fixed by the management.

6. Integrated test and evaluation system

Besides the development of the operational software-hardware system another important part of the technical development is the development of the corresponding test and evaluation system, that means the specification of test requirements, test strategies, test procedures, test programs and test results. An integrated test and evaluation system gives the opportunity, to check the technical specifications against the test and evaluation specifications already during the design (rapid prototyping). Additionally there are computer assisted relations between the operational system and the test system during the whole life cycle, i.e. if you change some part of the operational system the computer states the corresponding tests to be performed.

FUNDAMENTALS OF THE ENGINEERING AND PROJECT-MANAGEMENT ORIENTED SUPPORT SYSTEM EPOS

EPOS was developed in order to provide all the participants in a project, planners, programmers, electrical engineers, project leaders - with a "slave" EPOS workstation computer. EPOS provides the project participants with integrated computer support for:

- producing a consistent, complete, and unambiguous customer needs document and/or system requirements document
- developing software, especially realtime software, which is well-structured, error-free and well-documented
- developing hardware systems and - in the case of an automation project - the sensor and actor systems.
- project and configuration management
- product assurance and quality control.

The EPOS system can be applied from the beginning of a system's life-cycle (i.e. starting with the definition of the problem statement in the customer needs document and/or system requirements document). It offers easy-to-learn verbal, graphic, and formal means for describing all phases of problem-solving and system development on a hierarchy of levels.

Furthermore, EPOS also provides users with the necessary means to describe factors which influence project planning and execution, and to describe intermediate results. The formal representation needed for computer support is supplemented by natural language and graphic descriptions, which simplify communication among project personnel.

All the descriptive information which the various users enter using EPOS' description languages ("specification languages") is stored in a <u>single</u> central data base. EPOS evaluates the contents of <u>this</u> data base with special evaluation and analysis programs: the EPOS checks whether the requirements defined in the system requirements document were considered; it analyzes the various design levels and determines at an early stage whether contradictions, incompletions, or other specification or design errors have occurred. It relieves the planner of most of the time-consuming tasks of producing thorough and complete documentation; and it immediately updates the entire documentation whenever changes in the problem-statement or in the problem-solving concept make this necessary. Moreover, EPOS supports system managers and project leaders in planning a project, tracing its development, and controlling its progress.

All together, the EPOS system includes seven components: three description languages ("specification languages"), three evaluation program systems, and a communication system. The seven EPOS components are:

- The specification language **EPOS-R** which uses decision tables and an adaptable degree of formality. It is used to define the requirements in the customer needs document and system requirements document, for constructing models, and for developing a technical conceptual solution to the problem at hand.

- The specification language **EPOS-S**, used to describe system design (including software and hardware design). It offers the possiblity of automatic code generation.

- The specification language **EPOS-P**, used to describe information related to project management and configuration management (e.g. project structure plans, work packages, new configurations).

- The program system **EPOS-M**, which offers computer support for management activities related to project management and configuration management.

- The program system **EPOS-A** for the early detection of specification and design errors and for producing test and result reports.

- The program system **EPOS-D** for producing the entire documentation (texts, lists, graphical representation) according to the information entered by means of EPOS-R, EPOS-S and EPOS-P. The following kinds of diagrams can be automatically generated: hierarchy diagrams, data structure diagrams, dataflow/controlflow diagrams, Nassi-Shneiderman diagrams, flow charts, Petri nets, project structure diagrams, hardware block diagrams, module connections diagrams.

- The communication system **EPOS-C** for simple, user-oriented communication between the user and the EPOS work-station computer.

EPOS has been under development at the University of Stuttgart, Germany, since 1973. After experimental use by students during 1978 and 1979, it has been used since 1980 in a lot of industrial applications, where it was installed and is maintained by the GPP software house of Munich. At the beginning of 1985 more than 75 EPOS installations had been delivered by GPP.

REALIZATION OF THE REQUIREMENTS FOR AN INTEGRATED SUPPORT SYSTEM IN EPOS

1. Integrated support throughout all development and maintenance phases

EPOS supports the development of software/hardware systems from requirements specifications down to coding and circuit layout.

After sufficient decomposition the design specification can be transformed automatically to program code for arbitrary programming languages as well as to inputs for hardware computer aided design systems. During the whole life cycle of a computer control system EPOS assures the consistency between the design specifications, the documentation and the program code because it is possible to retransform changed, optimized program code and to correct the design specifications and the documentation automatically. Therefore maintainability is guaranteed.

2. Integrated support of software and hardware development

EPOS offers description means to specify the logical structure of the system as well as the realization structure. On the higher design levels the developer describes logical relations between actions, data, interfaces and events and decides during the design, in which way the different parts of the systems are realized by software, hardware and human interaction. It is possible to decompose the software structure as well as the hardware structure and to state relations between these different viewpoints.

The realization oriented view is documented in form of hardware block diagrams, the logical view in form of dataflow charts, Petri nets, Nassi Shneiderman diagrams and function block diagrams. A graphical representation of the connections between the realization structure and the logical structure is in work.

3. Integrated support of different design methods

Appropriate to the goal as an engineer's aid and tool for common realtime systems, EPOS is not centered on one specific design method. It provides different methods in a method base from which the method suitable to the engineer and his problem can be selected. EPOS enables the use of the design methods described above, i.e. EPOS has facilities to design function-oriented, module-oriented, event/process-oriented, dataflow-oriented, datastructure-oriented, or device-oriented and forces the method rules of the respective design method selection with computer assistance.

A series of consistency tests can be made depending on the selected design method. For example, in a dataflow-oriented method open ends can be found, in a module-oriented method violation of visibility rules are analyzed, in an event/process-oriented method synchronization errors are detected.

For each design method there are graphical techniques which emphasize the characteristic of a method and its dominant objects. A function-oriented design is shown clearly in diagrams which present the hierarchy structure as well as the flow structure (hierarchy-flowchart). For the module-oriented method, hierarchy diagrams and module connection plans are mainly employed. An event/process-oriented design is illustrated by Petri nets, a dataflow-oriented design by dataflow diagrams. The relationships of a data structure-oriented method are shown best by data structure diagrams, whereas hardware block diagrams showing the structure of components and their connections present a device-oriented method especially well. Each of these diagrams can be generated by EPOS with computer assistance.

For each design method there are "good" and "bad" designs according to how well certain norms are kept for the respective methods. For example, in a module-oriented design many or few modules can be defined, and they can have more or fewer connections. That means that a norm can be defined for comparison with an actual design. The developer receives computer provided information on the quality of his design and can make specific changes in the structure. These metric analyses are presently in development.

Since not each design method is equally efficient in the various development phases, the EPOS system enables a change from one design method to another with a computer-aided transformation of already existing specifications. During this transformation the relations are changed to complementary relations, i.e. the current subset is changed and the designer receives information which specifications have to be added to go on using a new design method.

4. Integrated support of development and project management

EPOS supports development as well as project management. On the development part EPOS allows the specification, analyzing and documentation of user requirements, solution strategies, system structuring, software/hardware design down to coding and hardware realization. On the project management part EPOS allows the specification, analyzing and documentation of work breakdown structure in form of subprojects and work packages, project organization structure, time and cost limitations, progress information, resource planning, change management and error management.

But the most important feature of the EPOS system in this area is the integrated support of development and project management because it is possible to evaluate relations between development specifications and project management specifications in a computer aided way.

For example the access rights of different teammembers during the development can be fixed and controlled. It is possible to identify which requirements and which part of the technical development belong to a distinct work package. In this way the completion of a work package can be controlled because the EPOS system can reject the completion announcement if there are any requirements not fulfilled or some open ends in design, i.e. some code parts are missing. Another important report produced automatically is the objective representation of technical progress and completion status of the project.

5. Integrated product management

All objects described with the help of the specification languages, i.e. management objects, requirements specifications, design objects, are assigned a version identification part and a variant identification part. The version identification part includes a multiple-digit version number in addition to the date and author of the change/new entry. The variant identification part includes one or more variant identifiers and allows the definition of different variants of the system.

EPOS realizes a single source concept also for version and variant control. It is possible to put in a specific version and/or variant identifier and to work with the corresponding specifications, for example to analyze the corresponding project management specifications, to produce the corresponding code file or to change a corresponding requirement. Other features of the EPOS product management system are the showing of differences between versions, the checking of changes, the fixing of base lines and the management of change requests (configuration control). The product management system is in development and will be delivered to EPOS users at the end of 1985.

6. Integrated test and evaluation system

Beyond the available analysis system for the detection of formal specification errors in the requirements and design specifications EPOS is extended to realize a design simulation system for rapid prototying. A first version will be delivered at the end of 1985, which allows the static simulation of the design specifications.

All activities belonging to the project dependent test system are summarized in the "auxiliary design" of the EPOS system. Today it is possible to specify test requirements, to specify, to analyze, to record the development of test programs and to describe relations between the functional system and the test system.

The evaluation of the relations is in development and will be delivered in 1985, for example the identification of functional parts, corresponding to test programs, the formulation of test results and the realization of test management, i.e. if there are changes in the functional system, which tests have to be performed again.

USE OF EPOS IN INDUSTRIAL PROJECTS

Application fields

EPOS has been used in very many different projects. The range of applications of EPOS stretches from the development of small assembler programs for products with large numbers of items to large systems, which are realized by different firms with different development computers in different programming languages, from the development of complex, security-critical real time systems to the development of commercial programs, from the development of pure software systems to the development of integrated software/hardware systems.

EPOS is used in the development of new systems, in the extending of existing systems which were originally developed without the aid of EPOS, as also in the follow-up documentation of unstructured program systems.

The main fields of application of EPOS lie in the development of

- control systems for aircraft and areo-engines
- navigation systems for aircraft and land vehicles
- automatic control systems for rotating printing machines
- systems for data processing in ships and for the organisation of documentation of ship movements
- directing systems in local traffic
- management information systems with distribution, exchange and presentation of information
- systems for the dealing with commercial problems
- automatic control systems for test benches
- automatic control systems in local traffic areas
- nuclear power plant monitoring systems
- systems with security-critical components in local traffic areas
- injection and antiblocking systems in the motorcar industry
- user software for the radiotelephony and radiotelegraphy sectors
- off-shore-systems

Experiences in the application of EPOS

The basic content of the experience report is:

In practical application EPOS has proved successful. It essentially contributes a systematic, structured procedure in the development of software and hardware and supports both the developer and the project manager.
In detail, the comments of EPOS users can be summarized as follows:

- On the whole EPOS is applied without difficulty and is easy to learn. The use of design-level models, the transition from device technical to function oriented design, and the project accompanying documentation assume a certain discipline, which can easily be acquired with training and practice.

- It has been shown, that an approximately ideal "top-down"-procedure can be achieved in every instance of pure software problems. Projects, in which a great deal of hardware is involved, also require, in many cases, "bottom-up" procedure, too.

- EPOS aids the quality protection of software and assists generally in the reduction of technical hazards, clarifying the project, improving communication between fellow-workers and detecting design errors. This makes EPOS a suitable instrument for security proof by examining authorities.

- The design descriptions are in general easy to read, consistent and logically constructed, such that new personnel can be trained relatively fast. EPOS is also a good help in project management, as the field of work and the formulation of the problem can be more easily defined in relation to each other. As a result, the project and its state of development are seen with much greater clarity.

- Using EPOS, complex, firm engulfing, large projects are kept under control. As EPOS is available for different computers with functionally identical EPOS systems, it is possible to transfer EPOS specifications from one computer to another without problem.

- EPOS permits computer aid throughout a project and makes development both visible and comprehensible. There is a growing tendency for many clients to dispense with informal progress reports and demand instead current EPOS specifications as proof of achievement.

- The acceptance of EPOS has been considerably promoted by the possibility of active cooperation of EPOS users in its further development. In this way requirements, specific to a firm, can be realized in an integrated development environment.

FOR MORE INFORMATION

Baur, P. (1983). On the Use of a Computer-aided Specification Tool to Support the Development and Licensing of Safety-Related Systems. Proceedings of IFAC/IFIP/IFORS-Conf. on Control in Transportation Systems. Baden-Baden, Germany, April 1983 Pergamon Press, Oxford, New York 1983.

Baur, P. (1983). Fault Avoidance and Fault Removal Features of the Computer-Aided Development Support System EPOS. Proceedings Third IFAC/IFIP workshop, Safety of Computer Control Systems SAFECOMP '83, Cambridge U.K. Sept. 1983 (Ed.: J.A. Baylis). Pergamon Press, Oxford, New York 1983, pp. 103-110.

Biewald, J., Göhner, P., Lauber, R., Schelling, H. (1979). EPOS-A Specification and Design Technique for Computer Controlled Real-Time Automation Systems. Proc. 4th Int. Conf. Software Engineering, München 1979 IEEE Comp. Soc., Los Alamitos, Cal., 1979, pp 245-250.

Biewald, J., Göhner, P., Schelling, H. (1980). Real-Time Features of EPOS: Formulation, Evaluation and Documentation Proc. IFAC/IFIP Workshop on Real-Time Programming, Leibnitz, Austria, April 1980 (Ed. V.H. Haase). Pergamon Press, Oxford, New York 1981, pp 95-101.

Biewald, J., Joho, E., Jovalekic, S., Schelling, H. (1980). Application of the Specification and Design Technique EPOS to a Process Control Problem. Proc. IFAC Symposium on Digital Computer Applications to Process Control, Düsseldorf, Oct. 1980 Pergamon Press, Oxford, New York, pp 517-522.

Ghassemi, A., Reinshagen, K.-P. (1983). Erfahrungen beim industriellen Einsatz des Spezifikations-und Entwurfssystems EPOS zur Automatisierung von Tiefdruck-Rotationsmaschinen (Experiences with the industrial application of the specification and design system EPOS for the automation of rotating printing machines). Regelungstechnische Praxis rtp vol. 25 (1983) pp. 110-114 and 156-159.

Göhner, P. (1984). Methoden zur Entwicklung von Realzeitsystemen und ihre praktische Anwendung in EPOS, Prozeßrechnertagung 1984, Springer-Verlag Berlin, Heidelberg, pp. 325-335.

Göhner, P. (1984). EPOS - gestern, heute, morgen eine rechnergestützte Software/Hardware-Entwicklungsumgebung 4 Jahre im industriellen Einsatz. Compas '84, Berlin 1984, pp. 1213 - 1224.

Göhner, P. (1984). Ingenieurgerechte Rechnerunterstützung für die Entwicklung und Pflege von Hardware/Softwaresystemen. 11. Internationaler Kongreß Mikroelektronik München 13. - 15. November 1984, pp. 491 - 500.

Joho, E. (1982). An Interactive Simulation System - A New Component of the Development Support System EPOS IFAC/IFIP Symp. Software for Computer Control, Madrid, Spain, Oct. 1982, (ed. G. Gerrate, E.A. Puente) pp 163-168.

Lauber, R. (1979). Strategies for the Design and Validation of Safety-Related Computer-Controlled Systems. First European Symp. on Real-Time Data Handling and Process Control, Berlin 1979. North-Holland Publ. Co. Amsterdam 1980, pp 305-310.

Lauber, R. (1980). Design Methods for Computer Controlled Real-Time Automation Systems. Proc. IFAC Symposium on Digital Computer Applications to Process Control, Düsseldorf, Oct. 1980 (Ed. R. Isermann, H. Kaltenecker), Pergamon Press, Oxford, New York 1980, pp 25-34.

Lauber, R., (1981). Computer-Aided Development of Real-Time Computer Systems. Anais do 1° SICOP Simposio em Controle de Processos por Computador, Rio de Janeiro, Brasil, May, 1981, pp 105-116.

Lauber, R., (1982). Development Support Systems IEEE Computer 15 (1982) May, pp 36-46.

Lauber, R., (1982). Specification Languages and Computer Aided Development Support Techniques to Achieve Reliable and Safe Systems: Present Status and Future Directions SAFE COMP Workshop 1982, Purdue University Lafayette, USA, Oct. 1982.

Lauber, R., (1982). Impact of a Computer Aided Development Support System on Software Quality and Reliability Proc. 6th Int. Computer Software & Applications Conf. COMPSAC 82, Chicago, Nov. 1982. IEEE Comp. Soc. Press, Los Angeles 1982 pp 248-256.

Lauber, R. and Lempp, P. (1983). Integrated Development and Project Management Support System. Proceedings 7th Int. Comp. Software & Applications Conf. COMPSAC '83, Chicago, Nov. 1983. IEEE Computer Society Press, Los Angeles, 1983.

MODELLING SYSTEM SPECIFICATIONS ON A FINITE STATE MACHINE

H. Hvelplund

Department of Advanced Information Processing, Computer Resources International A/S, Vesterbrogade 1 A, DK-1620 Copenhagen V, Denmark

Abstract. CODE (COntrol DEfinition language) is a formal language designed and used for implementation of control strategies in sequence control applications, such as found in industrial control. CODE treats processes as finite state machines (FSM), by defining transition conditions that determine when a state change will occur, and the consequent transition action. A control definition consists of a hierarchical model, each layer of which consists of a number of FSMs. The CODE concept is proposed for use in a formal specification language, that would form part of a specification work bench, including syntax-checking editors, consistency and completeness checks, automatic simulation and documentation systems. This work bench will provide a powerful tool for specification and development of process control and other systems. It will help increase the effectiveness of software designers and engineers; it will also contribute to increased formality in system specification, helping to eliminate a major source of error.

The CODE concept is thus part of the current trend towards increased automation of software development. It can be used to develop formal system specification languages that are application-specific, and that can be part of a semi-automatic system generation process, in which the system specification is transformed directly to the implementation. Maintenance is then performed on the specification, which is reimplemented with computer assistance.

Keywords. Application orientated languages; control system analysis; control system synthesis; finite automata; modelling; process control; prototyping.

INTRODUCTION

Developments in computing systems and process control applications are constantly reducing the ratio of hardware cost to function, leading to a much broader range of applications. Higher scales of integration of electronic components will perpetuate this trend. In the past, computers were more expensive than people; now, people are more expensive and they are scarce - there is a shortage of adequately trained staff. Issues such as the speed with which control software can be produced, the increased complexity of control algorithms and their reliability have become of major importance.

Control software development paradigms need to be found which augment the effectiveness of the costly and limited supply of people who produce and maintain software.

Investigations of software projects show that:

- Only one third of all faults detected during the life of a project are caused by programming (coding) errors. Most faults are caused by specification and design errors, traceable to poor understanding or to misinterpretation of requirements. (EPOS PRIMER, 1983).
- Most programming errors are found by tests before release; but the vast majority of specification and design errors are not found until after release.
- The cost of correcting specification and design errors is, typically, about 100 times more than for programming errors (Boehm, 1981), and this cost increases with the length of time they remain undetected.

More effort needs to be spent in the specification phase, which must be considered as a separate stage in software development.

Specification

A specification describes the external behaviour of the system at an early stage of its development life cycle. The specification should describe what the product should do, not how it is to do it.

The specification must be used to obtain the approval of the product's prospective users, to ensure the product's designers understand what they are expected to design and implement, and to provide a basis for system validation.

The specification must be carefully checked to eliminate omissions, inconsistencies, ambiguities, and unsuitability.

Major specification problems arise because the customer who orders a software system often does not know precisely what he wants. The customer and supplier discuss ambiguous, incomplete and sometimes contradictory specifications, and both feel they understand one another. They describe system specifications informally, in natural language.

This may aid users in understanding the system, but the inherent ambiguities of natural languages make them inadequate for system specification, and lead to misinterpretations, unforseeable costs, delays and often unsatisfactory results. And it is difficult to check a natural language specification for completeness, consistency and verifiability.

These problems can be solved to some extent by use of a formal system specification language. There has been a lot of effort in this area and the feasibility of software support for formal languages has been demonstrated. Examples are PDL (Caine, 1975), SDLA (Demetrovics, 1982), EPOS (Lauber, 1982), ESPRESSO (Ludewig, 1982, 1983), PCSL (Ludewig, 1980), RSL (Davis, 1982), RTRL (Davis, 1982; Taylor, 1980), RLP (Davis, 1982).

This article outlines a practical formal specification language based on CRI's COntrol DEfinition language CODE.

CODE CONCEPT

CODE was designed by CRI to assist engineers to define, implement, and document control strategies. CODE is based on a set of assumptions concerning the way processes are controlled, how a control structure is built, and how control definitions can be implemented. The details of control differ for different processes, but certain universal features can be derived, and these can be used to establish a general language for the definition of control.

CODE is based on the principles that:

- Control definitions are specified as interacting processes.
- A process is defined by its states, by the conditions for transitions between states and by the actions taking place during transitions. When so defined, a process can be considered as a finite state machine (FSM).
- Control definitions can be structured in a hierarchy of finite state machines, based on a concept similar to the ISO-OSI model, describing different logical layers in a data communication network. The bottom level of the hierarchy is responsible for the direct operation of the plant; upper levels are responsible for the strategy of coordination of lower levels to realize the purpose of the plant.

A sequential process, for instance, is composed of concurrent sets of sequential activities, each corresponding to a transformation on one or more of the attributes of the objects it acts upon.

Partitioning a system into a set of interrelated FSMs is a well known concept which has been successfully applied to many applications, e.g., human decision modelling, and engineering fields. It is easier to construct physical phenomena from an informal understanding of a process when they are represented as sequences. It is only necessary to consider successive states and the events that indicate the state transitions of the process. These transitions are subject to only a few well-defined disturbances, and their consequences can be foreseen. In the worst case, a disturbance might modify the expected behaviour by taking the system into an exceptional but represented state.

Examples of sequence control applications are:

- Design of digital electronic modules based on asynchronous and synchronous sequence networks.
- Protocols in data communication networks, where sequences control synchronisation between communication nodes in the network. An example is the format and protocol language (FAPL) used for description of the SNA protocol (IBM, 1980).
- Important areas of software engineering are based on sequence control; in this context, called "finite automata". Finite automata theory plays an important role in areas such as construction of compilers, use of Petri nets to facilitate specification of parallel activities, augmented transition networks (ATNs), and so on.

The concept of sequence control is well known within process control applications and the manufacture of computer hardware and software.

Basic Structure of CODE

In order to control a sequence of events the FSM must be able to remember where it is in that sequence. This is the basis for the STATE concept in CODE. The transition from one STATE to another occurs when a transition condition (TC) is met. The transition is accompanied by a transition action (TA).

Figure 1 shows the basic structural element for a sequence. The formal notation used for this is:

```
S1   TO   S2   WHEN TC
               WITH TA
```

which should be read as:

> The sequence controller will change from state S1 to state S2 when the transition condition TC is met, with the transition action TA taking place.

A transition condition is a predicate that evaluates to either true or false. If the predicate evaluates to true, the transition condition is met and the transition will take place. A predicate can include the value of process parameters, parameters local to the FSM, timer values, and so on. Compound transition conditions can be defined using logical operators such as AND, OR, EQ and NOT and relational operators such as >, <, =.

A transition action is composed of one or more simple actions. A simple action can be either a statement or a call to a built-in function. Statements are of the form:

```
X : = Y + Z * W;
A : = B OR C AND (NOT D);
KV01 : = ON;
```

They can be used to perform actions on the plant being controlled, e.g., opening and closing valves. Calls to built-in functions are of the form:

```
RUN ( SEQUENCE 1 );
```

They can be used for control of other sequences and for communication with the operator.

The concepts described so far cover the most important part of the CODE notation, which also includes the following features:

- A state action (SA) can be defined for a state. An SA is bound to the state in such a way that no matter how the state is entered the actions of the SA will be executed. The SA can, for instance, be used for the common part of a number of transitions to the same state (e.g. an error state).
- Transition conditions (TCs) for the pseudo state ANY are checked before the checks of the TCs of the current state are executed. If one of the TCs of the ANY state is met, the cor-

responding path is taken. The ANY state can be used to check general error conditions.

- The pseudo state RESET is a predefined state which is entered when a sequence is reset. A state action can be defined for the RESET state.

- A generic FSM can be defined, which can be shared by a number of instances. This facility can be used to build a library of standard FSMs, thus encourageing reusability.

Implementation of Control

The CODE concept includes an interpreter program which uses the control definition language directly to implement actual control. This interpreter is based on the following assumptions:

- FSMs are executed successively, following a synchronous activation scheme (that is, with fixed time intervals).

- For each activation of an FSM, the program checks whether the transition condition for the current state is met, if so, the transition will take place with the transition action.

- An FSM is activated only once for each activation cycle. This means that for a given FSM, only one state transition can take place in an activation cycle.

- The sequence of the FSMs within an activation cycle is the same for each activation. The activation sequence can be specified during definition of the FSMs.

These assumptions simplify the control definition, since they make it unnecessary for the user to deal with hazard problems which would occur in an asynchronous implementation of the control strategy.

FSM HIERARCHY

Ideally the control definition should be decomposed into a multilayer model, in which each layer consists of a number of FSMs, and the layers are separated by logical interfaces. From layer _n_ the FSMs at layer _n_-1 are seen as an ideal image of the process being controlled, and it is possible for the FSMs at layer _n_ to operate on the FSMs at layer _n_-1 in the same way as FSMs at the lowest level operate on the process being controlled. Each FSM performs a single, clear, well-defined function.

Three different operational modes are defined for an FSM: ACTIVE, RESET and PASSIVE. In ACTIVE mode the FSM is testing conditions and performing actions. In RESET mode the FSM is inactive in the initial state (start state) and performs neither checks nor actions. In PASSIVE mode the FSM is inactive in the state in which it received the command to become passive. An FSM can be set in ACTIVE, PASSIVE or RESET mode from another FSM via a function call of a built-in function, or by operator intervention. Within an FSM it is possible to determine whether another FSM is in RESET, PASSIVE or ACTIVE mode. Figure 2 shows a simple hierarchy of FSMs.

EXAMPLE

A requirement specification is input to the control specification. The requirement specification normally includes a piping and instrumentation (PI) diagram and a process description in natural language. Figure 3 shows a PI diagram for a coffee machine used in this example. (The example is chosen to illustrate facilities in the CODE programming language, and is not representative of industrial process control.) Figure 4 shows the structure of the corresponding control definition. The superior FSM (CMACHINE) controls a set of subordinate controllers that are responsible for the different phases of the process of brewing coffee:

FILLING:	Fill water into the boiler
HEATING:	Heat water in boiler
FEEDCOF:	Feed coffee beans into the filter
EXTRACT:	Extract coffee from coffee beans
EMPTY:	Empty coffee buffer tank into vacuum jug

The function of the superior FSM (CMACHINE) is established by the outlined division into subordinate FSMs. The CMACHINE FSM starts the subordinate sequences in succession. A subordinate sequence starts when the preceding sequence has finished (is passive). The superior sequence depends on the fact that each subordinate sequence sets itself into passive mode when it has finished its function.

Filling water in the boiler is controlled by the FSM FILLING, shown in Figs. 5 and 6.

RELATED EXPERIENCE

The CODE concept has been integrated as an implementation language in a general purpose process control system, PROCOS. This implementation language has been applied successfully to the definition of a number of large process control applications, such as the fuelling system for the European heavy launcher, ARIANE. This practical experience has shown that CODE is a very powerful concept, in which it is easy to structure and express process control definitions. It has also been found that users who are not software specialists can easily grasp the idea of CODE and have a spontaneous understanding of a control definition in CODE. New users are able to learn to use the system in a short time simply by using it. The CODE concept also relieves the user from the difficult job of handling details of interprocess communication, synchronisation, mutual exclusion of software tasks, and so on.

CODE AS A SPECIFICATION LANGUAGE

In view of the problems that occur within the specification phase, it seems natural to exploit the concepts of CODE in a system for specification of control strategies, since:

- The CODE concept is well known and understood by users in both the process control and software development areas.

- Experience has shown that CODE is a powerful concept, which has been applied successfully to the implementation of control strategies for many large process control applications.

- The CODE concept includes an implementation model which can be used to build an executable specification. This creates the possibility of verifying and validating a product during the specification phase of its life cycle (a definition of the terms verification and validation is given in Boehm (1984)). The specification can then become a prototype of the specified system. This approach leads to the software development paradigm shown in Fig. 7.

- o Definition of control strategies in CODE can be checked for completeness and consistency. A control definition is complete when all of its parts are present and each part is fully developed and consistent, that is, its provisions do not conflict with other parts of the specification (Boehm, 1984).
- o Use of an FSM concept similar to the one defined by CODE for modelling a specification has been suggested in a number of papers (Davis, 1982; Salter, 1976).
- o CODE allows a modular structure of the control definition in a hierarchy of logically-separated layers.
- o CODE relieves the user of the need to handle details of software implementation (interprocess communication, synchronisation, etc.).

In order to cover the most important aspects of the specification, the specification work bench must include interactive syntax checking editors, a consistency and completeness check module, a simulator, and a documentation system.

Interactive Syntax Checking Editors

Editors can be produced to provide a family of formal application specific languages. Each language is formal in that it possesses a unique, well defined and documented syntax and semantics. It also possesses "knowledge" of a particular application area, and thus is rich in the terminology of that area. These languages combine the formality of operational specification languages with the high readability of technical natural languages.

The editors should drive the collection of information in an efficient way, using the form filling concept. This guides the user through the necessary steps of a definition. Context and domain analysis techniques are used to define interrelationships between different information fields in a form and to define a dynamic check of the consistency of information in a form. Forms, vocabularies, and the knowledge needed to represent interrelationships can be tailored to a particular application area.

Simulator

The formality of a specification that is based on an operational model enables it to be executed as a program, albeit slowly. Except for speed, the behaviour of this program will be identical to that of a valid implementation. The specification itself can serve as a system prototype, enabling the user to experiment with the prototype until the requirements are matched in an appropriate manner. Experience shows that such prototypes normally lead users to improved perceptions of their needs. The resulting systems are thus more responsive to those needs.

Ideally a simulator should be formed as an interactive interpreter working in a dynamic environment, where the effects of modifications can be shown to the user immediately, using the appropriate level of abstraction. The system thus provides an exploratory environment.

Documentation

The documentation system should provide facilities for structured, modularised documentation of the specification. Examples of such documentation include the complete specification documentation, subsections of the complete specification, lists of identifiable components that are ordered by various criteria, and so on.

FUTURE

A software paradigm has been characterised as having a formal specification that is created and validated by its users. This specification then becomes a prototype of the specified system. Ideally, maintenance of the product should be based on the specification. The specification is reimplemented after each revision, and so the implementation of a specification must be fast, reliable and inexpensive.

The next step in the development of an automation-based software development paradigm for the process control area will be the automation of the implementation process (Balzer, Cheatham and Green, 1983). This will allow maintenance to be performed on the specification rather than on the implementation. The revised specification can then be reimplemented with computer assistance. Maintenance becomes a simple continuation of the process by which the system was initially developed. Maintenance can thus be performed by modifying the least complex and most localized form of the specification that is also closest to the user's conceptual model. This is important, since maintenance occupies 80 to 90 percent of total development effort (Balzer, Cheatham and Green, 1983). The tremendous backlog of user demand for changes to existing systems remains unanswered because it is time-consuming and expensive to make these changes without the use of automation. An automated paradigm would include all phases of the product's life cycle. Figure 8 shows the automated specification development paradigm.

CONCLUSIONS

A specification work bench based on the ideas outlined in this paper would be a powerful tool for use in a wide range of applications, including large process control systems, design of customised chips for testing chip functions before implementation, specification and testing of systems for which uncertainty necessitates the use of an explorative environment, in which a number of iterations can be carried out before the product is implemented, and so on.

REFERENCES

Balzer, R., T. E. Cheatham Jr., and C. Green (1983). Software technology in the 1990s: using a new paradigm. IEEE Computer, Nov. 1983.

Boehm, B. W. (1981). Software engineering economics, Prentice-Hall.

Boehm, B. W. (1984). Verifying and validating software requirements and design specification. IEEE Software, 1, 1, 75-88.

Caine, S., E. K. Gordon (1976). PDL - a tool for software design. AFIPS Conf. Proc., 44.

Dansk Automationsselskabs Konference (1982). Specifikation og programmering af sekvensstyringer. 31.3.1982.

Davis, A. M. (1982). The design of a family of application oriented requirements languages. IEEE Computer, May 1982.

Demetrovics, J., E. Knuth, and P. Rado (1982). Specification meta systems. IEEE Computer, May 1982.

EPOS PRIMER (1983). Institute for Control Engineering and Process Automation of the University of Stuttgart and GPP Gesellschaft fur Prozessrechner-Programmirung mbH.

IBM (1980). Systems network architecture format and protocol reference manual: architectural logic, 3rd ed. Report no. SC30-3112-2, Appendix N.

Lauber, R. J. (1982). Development support systems. IEEE Computer, May 1982.

Ludewig, J. (1980). Process control software specification in PCSL. IFAC/IFIP Workshop on Real-Time Programming, V. Haase, Ed. Graz, April 1980, Pergamon.

Ludewig, J. (1982). Computer-aided specification of process control systems. IEEE Computer, May 1982.

Ludewig, J. (1983). ESPRESSO - a system for process control software specification. IEEE Transactions on Software Engineering, SE-9, 4, July 1983.

Salter, K. G., (1976). A methodology for decomposing system requirements into data processing requirements. Proceedings of Second International Conference on Software Engineering, October 1976.

Taylor, B.J. (1980). A method for expressing the functional requirements of real-time systems. IFAC/IFIP Workshop on Real-Time Programming, Leibnitz, Austria, April 1980.

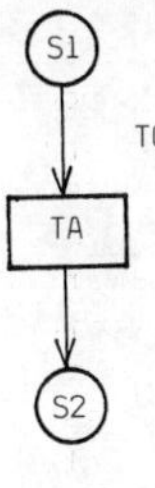

Fig. 1. Basic structural element

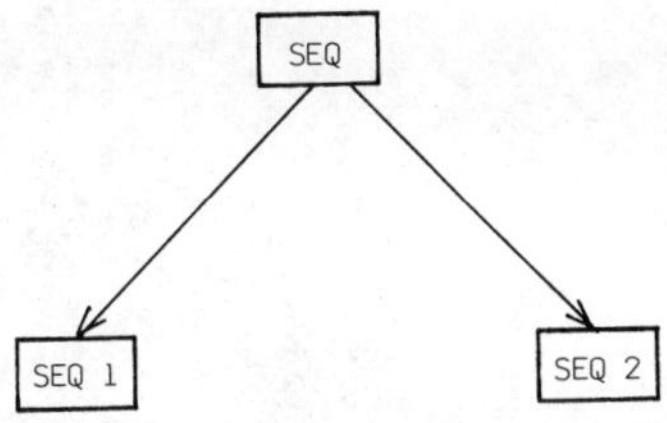

Code for sequence:

```
S1  TO  S2  WITH  RUN(SEQ1)

S2  TO  S3  WHEN  PAS(SEQ1)
            WITH  RUN(SEQ2)

S3  TO  S1  WHEN  PAS(SEQ2)
            WITH  HLD(THIS)
```

Fig. 2. Hierarchy of FSMs

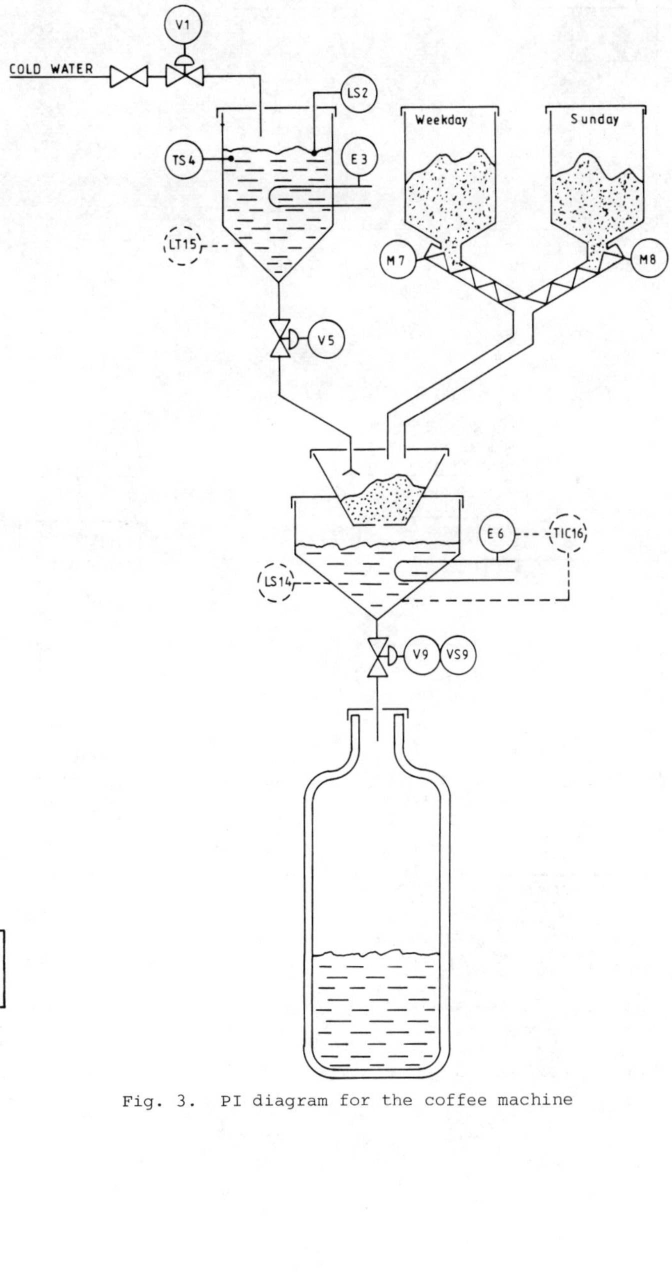

Fig. 3. PI diagram for the coffee machine

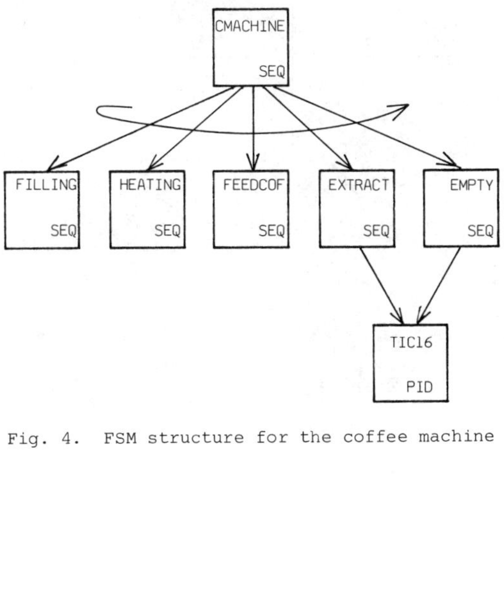

Fig. 4. FSM structure for the coffee machine

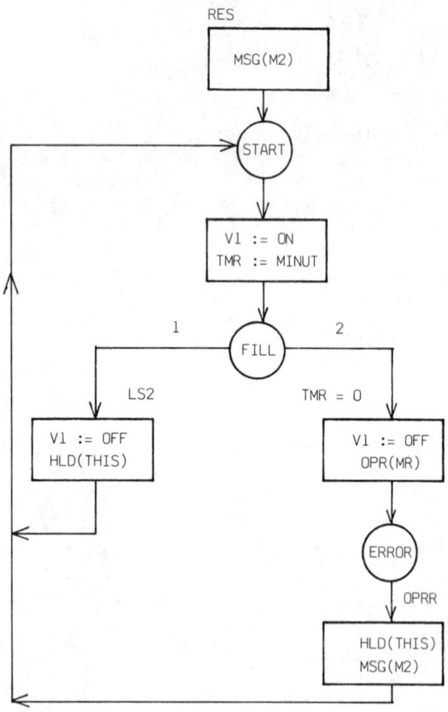

Fig. 5. State diagram for FSM FILLING

```
RES                             MSG(M2)

START   TO   FILL     WHEN      V1 := ON;
                                TMR := MINUT

FILL    TO   START    WHEN      LS2
                      WITH      V1 := OFF;
                                HLD(THIS);

        TO   ERROR    WHEN      TMR = 0
                      WITH      OPR(M1);
                                V1 := OFF

ERROR   TO   START    WHEN      OPRR
                      WITH      HLD(THIS);
                                MSG(M2)

M1 = "CHECK VALVE V1 OR V2"
M2 = ""
```

Fig. 6. Code for FSM FILLING

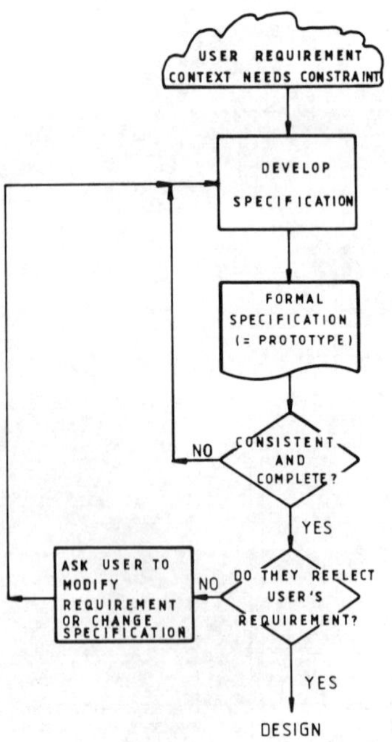

Fig. 7. Software development paradigm

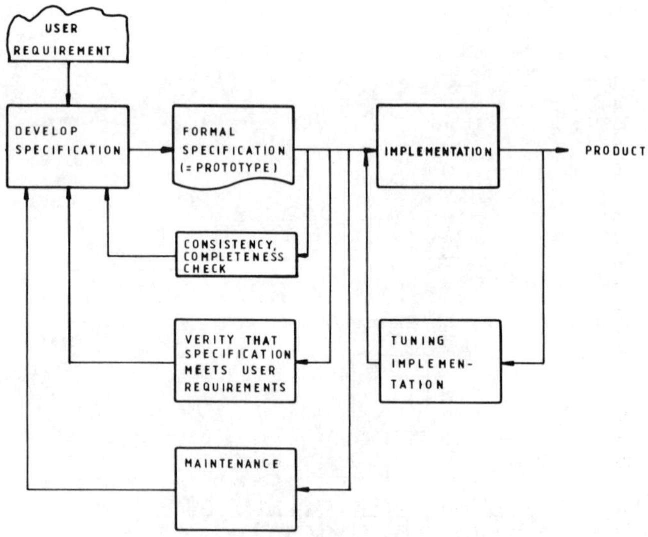

Fig. 8. Automation based specification development paradigm

SUNS: THE SUSSEX UNIVERSITY NONLINEAR CONTROL SYSTEMS SOFTWARE

D. P. Atherton, O. P. McNamara, M. D. Wadey and A. Goucem

School of Engineering and Applied Sciences, University of Sussex, Brighton BN1 9QT, UK

Abstract. The paper outlines the capabilities of the Sussex University Nonlinear Software. The programs based on two theoretical techniques, the describing function and Tsypkin's method, evaluate limit cycle solutions in both continuous and discrete feedback systems. The software allows the display of solution waveforms at the input to each nonlinearity and an exact orbital stability criterion, for continuous systems, is used for the limit cycles predicted by the Tsypkin method. Most of the software is for fixed structure canonical feedback systems although one of the DF based programs allows analysis of systems of a free structure. The software may also be used for frequency domain design where gain and phase margins of compensation schemes may be calculated by finding the minimum amount of additional gain or time delay required for a system to limit cycle.

Keywords. Limit cycles, nonlinear control systems, describing functions, relay control, computer-aided design, sliding states,sampled data systems, digital systems.

INTRODUCTION

This paper describes software developed at the University of Sussex for the study of nonlinear feedback control systems. The algorithms, which are based on extensions of describing function (DF) methods and Tsypkin's method, can be used to determine periodic modes in both continuous and sampled data control systems. Since most instabilities of interest in nonlinear control systems occur as limit cycles the software can be used to determine the stability of a system. This can be done accurately using the Tsypkin methods and also using the extensions of the DF method, which allow iterations on the DF solutions using FFT techniques. The programs can also be used to design compensators by classical techniques since the amount of phase shift or gain which may be added to a loop before it becomes unstable can easily be determined. This approach, of course, assumes that the compensator structure is known, and this can invariably be done using a linear design technique on a linearised form of the nonlinear system.

The programs, whose various features will be described in the following sections, have all been written in FORTRAN 77 and are normally used on a PRIME 550 running PRIMOS or VAX 11/780 running VMS. Graphics facilities are essential for satisfactory use of the programs and they use the GINO graphics library. The programs also make use of algorithms from the NAG library and control library SLICE. Several examples are given to illustrate a variety of applications of the software.

USER INTERFACE

All the programs in the suite run interactively with the useful feature of a user friendly interface. Prompts are displayed at each stage of user input where help messages may be easily obtained if required. Each run consists of two distinct phases, the data entry and calculations.

The software is mainly for the single or multivariable system of figure 1. However, one of the programs allows for a free structure where the system may be composed of any configuration of linear and nonlinear elements. The system description data is obtained via a question and answer procedure. Nonlinearities are specified by two fields, its type defined by an integer and its characteristic, typically defined by three or four parameters. Apart from the programs specifically for relay systems the software supports many nonlinearities. The linear elements are specified as transfer functions which are composed of four data fields. These are the gain and time delay, which are real numbers, and the numerator and denominator polynomials. The polynomials are entered as character strings in terms of 'S'.

There is also other data specific to each program, for example the Tsypkin based programs require an initial guess for the limit cycle parameters. The data entry phase also allows for amendment of a previously entered system description and display of the currently entered data.

Once a satisfactory system description has been entered the calculation phase is started. This is where the analysis of the system is undertaken. It requires far less user input and involves numerical algorithms and graphic displays which are an essential feature of the software. For example, the DF method is implemented using cursor graphics to obtain an initial evaluation of possible limit cycle solutions. After each stage of the calculations a number of options are available which are listed in a menu and chosen by a simple character code.

DF BASED ROUTINES

SISO systems of the type shown in figure 1, with any of a wide variety of nonlinearities can be analysed using a describing function based program (McNamara, 1984). A number of limit cycle prediction methods are provided including the single sinusoid DF, and a multiple input DF which is evaluated using fast Fourier transform techniques.

The limit cycle equations are solved using Newton Raphson techniques, with the single sinusoid DF also able to be solved graphically using a cursor. A well known limitation of any DF method is the uncertain accuracy of any limit cycle prediction. To overcome this, iterative methods are provided which usually can accurately verify any predicted limit cycle, and obtain a more accurate solution. Plots of actual limit cycle waveforms and spectral information are also readily available.

The above iterative methods have also been implemented to study free structured MIMO nonlinear systems. The program MV allows the system to be defined by up to ten five input-output blocks. Each element of each block can be either linear, nonlinear, unity, zero or a sample and zero order hold device. This free structure allows the analysis of a wide variety of complex systems such as SISO systems with multiple feedback paths, MIMO systems, and sampled data systems, where limit cycles which are odd or even subharmonics of the sampling frequency can be investigated. Systems with samplers operating at different, but harmonically related, sampling frequencies can also be studied. Indications of phase and gain margins can be obtained using the programs. These methods allow accurate predictions of limit cycles in many cases where the single sinusoid DF has failed to make a prediction, and in cases where the DF has incorrectly predicted a limit cycle.

TSYPKIN BASED PROGRAMS FOR CONTINUOUS SYSTEMS

The software includes a suite of programs based on Tsypkin's method which is for exact determination of periodic modes in relay control systems. Extensions of the method, which in some cases give approximate results, have also been incorporated to handle more general systems and situations. When the results are approximate the accuracy, which in most cases is good, can be evaluated and improved.

For relay systems (Atherton, 1981, 1982) the starting point in the analysis is to assume a specific periodic pulse type waveform at the relay output. The response of a linear plant to this waveform can then be derived using Fourier analysis. Although this yields an infinite series for the expressions corresponding to the relay input, closed form expressions are available for a set of simple transfer functions (Atherton, 1981). For more complex transfer functions closed form expressions for the relay input can be obtained by making a partial fraction expansion. Applying the switching conditions to produce the assumed relay output yields a set of nonlinear algebraic equations in terms of the limit cycle parameters. An efficient iterative routine is used to solve these which requires an initial solution guess as a starting point. Once found the parameters completely define the periodic relay input waveform. An important part of the solution procedure is to construct and plot this waveform to check that it does produce the assumed relay output as it is possible to obtain invalid solutions with additional unassumed crossings of the switching boundary (Atherton and Wadey, 1981).

As only stable limit cycles exist in practice it is important to be able to assess the stability of any solutions obtained from the Tsypkin method. A necessary and sufficient condition (Balasubramanian 1981) is used where the stability is assessed from a transition matrix associated with the equation of first variation, about the limit cycle solution, over the first half period. This can be set up by virtue of the fact that the exact limit cycle solution is known and the variational equation is piecewise constant. For a stable limit cycle n-1 of the eigenvalues of the transition matrix must have magnitudes less than unity where n is the order of a state space representation of the linear part of the system. One of the eigenvalues necessarily has the value -1 which is due to the periodic solution.

The program SYMRLY is available for the basic single loop autonomous relay system of figure 1 where the relay output is assumed to be a single pulse/half period. This means there are one or two switching conditions to produce the assumed relay output, depending on whether the relay has a dead zone. For one or two unknowns the nonlinear equation solver converges to a solution, if one exists, for most initial guesses. Once a solution has been found the corresponding relay input waveform may be plotted and its stability assessed using the exact criterion described above.

Under certain conditions it is possible to obtain limit cycles in relay systems with a sliding region (Utkin, 1977). In this case the assumed single pulse/half period relay output gives an invalid solution waveform, the form of which indicates the presence of sliding. Specifically the relay input reverses its direction at a switching which is due to the large discontinuity of its derivative. To obtain accurate solutions the sliding is idealised by motion along the switching boundary and the resulting equivalent control, at the relay output, to produce this input is derived. This control is modelled by a periodic linear segmented waveform for which closed form expressions for the relay input can be found for a given linear plant. As before the limit cycle parameters are found by imposing the appropriate switching conditions (Wadey, 1983). The solution is approximate but the accuracy can be improved by increasing the number of linear segments modelling the relay output over the sliding region. This, however, adds to the number of nonlinear equations which have to be solved making computation time longer and increasing the difficulty in finding an initial guess which yields convergence to a solution. In practice two or three linear segments often give accurate results and the accuracy is readily observed by comparing the actual equivalent control, over the sliding regime, with its linear segmented approximation.

The load module SLIDE to SYMRLY uses the above method for finding partially sliding limit cycle solutions. Initially the relay output over the sliding regime is modelled by one linear segment. This yields the minimum number of nonlinear equations to solve and the initial solution guess can be based on the invalid single pulse/half period solution obtained from SYMRLY. If this yields a solution the accuracy is improved using an automated method, where the number of linear segments is increased one at a time and the new solution guess is based on the previous solution. This ensures fast convergence at each stage of iteration and the process can be terminated when the desired accuracy is achieved.

As instabilities in nonlinear systems often exhibit themselves as limit cycles the method can be used as a frequency domain design technique. That is compensation schemes for elimination of any existing limit cycles in a system can be assessed. The program COMPRLY uses the simple phase advance network $G_c(s) = (1+sT)/(1+s\alpha T)$ for stabilisation of a single loop autonomous system where the relay has a dead zone. The parameters α and ω_c are requested input from the user, where $\omega_c = 1/\sqrt{\alpha}T$ is the frequency at which the maximum phase lead occurs. The smallest additional gain and phase for the system to limit cycle are then determined, the latter by addition of a time delay. These values correspond to the gain and phase margins of the compensated system.

The introduction of a sinusoidal forcing signal makes a variety of modes of oscillation possible. The program FOROSC handles this situation and allows evaluation of subharmonic and multipulse forced oscillation solutions. As additional input it requires specification of the forcing signal and type of oscillation being searched for, that is the number of pulses/half period and subharmonic order. Again once a solution has been found the relay input waveform can be displayed and the stability is assessed by applying the exact criterion to a two loop model of the forced system. For multipulse oscillations there are more nonlinear equations to solve and a good initial guess is more important. For this reason it is often useful to employ a search procedure for solutions by varying the forcing signal's amplitude and frequency. This often reveals the point when a single pulse mode switches to a multipulse mode.

A more general method is to approximate nonlinearities by a quantised characteristic. The program QTZR uses this where the nonlinearity must be single valued and odd symmetric. The quantiser can then be constructed from several relays with dead zone in parallel. The output of each relay is assumed to be a single pulse/half period and exact limit cycle solutions are obtained for the quantised systems. The number of nonlinear equations to solve for the limit cycle parameters increases with the number of quantised steps. It is therefore better to start with fewer steps to establish an initial estimate of the solution. This is particularly true for nonsaturating type characteristics where an input amplitude must be specified in order to construct a quantised fit. The program allows iteration on to a more accurate solution by respecifying the assumed input amplitude and increasing the number of quantised steps. This method gives very accurate solutions for saturating nonlinearities and is also useful for characteristics defined by a set of points. The program provides plots of the limit cycle waveform and the nonlinearity with its quantised fit. The latter is useful for assessing the accuracy of the solution. The exact limit cycle stability criterion for relay systems is also used.

The program for multivariable systems is MVREL (Wadey, 1984) where the system structure is of the form shown in figure 1. The nonlinearity may contain either relay or saturation elements, however, the solutions are only exact when N is composed of relays. For saturation elements the assumed periodic output is a linear segmented waveform. The limit cycle solutions are again obtained from satisfaction of the set of nonlinear algebraic equations corresponding to the conditions to produce the output of each nonlinearity. Once a solution has been found plots are available for the input waveform to each nonlinearity. These are important for checking the solution validity and also provide an accuracy check when the system contains saturation elements. The necessary and sufficient stability criterion can still be applied although this requires an exact solution. However, the linear segmented saturation output often gives sufficient accuracy and the validity of the results are readily observed from the closeness of an eigenvalue to -1.

DISCRETE SYSTEMS

A number of programs have been developed for the prediction of limit cycles in sampled data systems of the type shown in figure 1, where a sample and zero order hold precedes the nonlinearity. For such a system limit cycles usually occur at subharmonics of the sampling frequency. Single loop systems can be analysed by a DF based program SAMDF and by programs ZSAM, ASAM (Goucem 1984, McNamara 1983), which use an extension of Tsypkin's method for continuous relay systems, to give exact limit cycle solutions. The extension is possible because the output of the zero order hold, and hence of the nonlinearity, is the summation of a number of single pulse waveforms of unknown height which occur at different known times. The unknown limit cycle parameters with a fixed sampling frequency are then the pulse heights for a given subharmonic limit cycle. The program ZSAM also allows the introduction of a quantiser and discrete controller for digital compensation. Once a subharmonic limit cycle has been found there are routines available to calculate the amount of gain or phase needed to eliminate a given limit cycle mode. The exact method has been implemented for the analysis of MIMO nonlinear sampled data systems as well as for MIMO systems with multirate sampling. The only restriction being that the sampling frequencies in any loop must be integral multiples of each other. All of the exact methods can predict both even and odd subharmonics and plots of exact limit cycle waveforms can be obtained.

APPLICATIONS

Several examples are given in the following which have been chosen to illustrate the typical problems of nonlinear feedback systems, which can be examined, and also show the diversity of the applicability of the software.

Example 1

Here we consider the application of the Tsypkin method to a single loop relay feedback system. In the classical approach formulated by Tsypkin any limit cycle solution is found either graphically or from the solution of one or two nonlinear algebraic equations depending whether the relay has not or has a dead zone. To check that the solution obtained is correct it is necessary to obtain, as available in the software, the relay input solution waveform to determine that it does not cross the switching levels at other than the assumed times. Incorrect solutions are occasionally obtained, typically when the plant G(s) is highly resonant or when the degree of the denominator of G(s) is only one greater than the numerator. In the latter case a limit cycle with sliding may be possible. This situation is illustrated here where we consider a system with an ideal relay and a plant transfer function

$$G(s) = (s^4-0.7s^3+2.1s^2-0.7s+1)/(s^5+0.7s^4+10.1s^3+4.7s^2+9s).$$

Figure 2 shows the solution obtained by direct application of the Tsypkin method. Because the relay input reverses immediately the relay switches the relay input does not produce the assumed relay output, which is the dotted line in the figure. This behaviour usually indicates that if a limit cycle exists it may possess sliding, that is the relay switches on and off very rapidly giving a non-zero average output which is generated by an input which is essentially zero throughout the sliding period. The program gives a message to indicate the solution is invalid and suggests a search for a solution with sliding. This involves the solution of extra nonlinear algebraic equations since the unknown relay output during sliding is approximated by one or more linear segments. The approximate solution using one linear segment is shown in figure 3 and the nearness of the linear segment to the actual relay output is shown in figure 4. More accurate solutions can be obtained if desired. Figure 5 shows the relay input, which can now be seen to be essentially zero over the sliding period, implying that the solution is very accurate. In this case four linear segments are used to approximate the relay output during sliding.

Example 2

An interesting example, which has been considered in the literature before (Freeman and Barney 1963) consists of a single loop with unity feedback and a feedforward path which contains a saturation, a transfer function $G_1(s)$, a further saturation and a transfer function $G_2(s)$. The saturations are assumed to be ideal with unit linear gain and saturation levels of ±1.

$G_1(s) = 10^4(s+1)/(s+100)^2$ and $G_2(s) = 2/s^2(s+2)$.

The system may be regarded as a position control with both error and torque saturation. The phase advance compensator $G_1(s)$ has been designed to provide gain and phase margins of 28.0db. and 53.8° for linear operation. The problem of concern is how does the system behave for large inputs when both nonlinearities are saturated. Using individual DFs for the two saturations obviously indicates stable behaviour but it is also readily apparent that $G_1(s)$ is not a low pass filter and that high frequencies produced by the first nonlinearity will be passed to the second.

Limit cycle analysis of this example can be easily carried out by the harmonic balance section of the free structured program.MV. The harmonic balancing proceeds by opening the loop before the first saturation and injecting a signal, which corresponds to the error in this example, composed of up to ten harmonics. The resulting fedback error signal is then calculated using the F.F.T. and the injected and fedback harmonics compared. A Newton Raphson procedure is employed to automatically generate guesses for the harmonics in the error signal until a balance is obtained. If only one sinusoid is injected, then the ratio of the fundamental amplitude in the fedback error to the injected amplitude is the effective DF of the complete loop. For this example the error limit cycle waveform turns out to be sinusoidal, even though the input to the second saturation is quite distorted. A single sinusoid harmonic balance yields a limit cycle solution with amplitude a = 2.893, and frequency ω = 0.6468 rads/sec. The input signal to the second saturation shown as the solid line in figure 6, with the dashed waveform being the error signal. This example shows the advantage of using the F.F.T to obtain harmonic balance solutions for free structured systems, where in general it is extremely difficult to write down an analytical expression for even a single sinusoid DF of a combination of nonlinear and linear elements, and almost impossible for several sinusoids.

The multivariable Tsypkin program was also used to predict this limit cycle. Initially the output of each saturation, in the linear regime, was modelled by one linear segment. This gave parameters which could be used as an initial guess for a solution employing more linear segments, thus giving greater accuracy. With the "one segment approximation" initially used, several solutions fairly close to each other were found. This was probably due to the coarseness of the approximation which was revealed using more linear segments when only one solution was found. In addition the exact limit cycle stability criterion required a high degree of accuracy to be valid and therefore yield one eigenvalue of -1. When an accurate solution was used one of the eigenvalues was greater than unity in magnitude implying that the limit cycle was unstable. This was confirmed by digital simulation.

Example 3

Here we consider the stability of a 2 x 2 multivariable system with a plant G(s) having

$G_{11}(s) = 5(1-0.4s)/s(s^2+s+1)$, $G_{12}(s) = 1.3/(s+1)$,

$G_{21}(s) = 1.5\alpha/(s+2)$ and $G_{22}(s) = 8/s(s+1)^2$, and

N a diagonal matrix of two relays with dead zone of ±1 and height ±1. A controller KI is assumed prior to N and the maximum value of K for stable operation is required. With α = 1 and K = 1 two limit cycles, one stable and the other unstable, were found. The inputs to the relays were approximate sinusoids and as K was decreased the limit cycles converged, disappearing for K = 0.27. Because the relays have dead zones it is possible for an oscillation to exist in one loop only (i.e. it is not large enough to switch the second relay). For the uncoupled loops the limit cycle in loop one disappears for K = 0.22 and in loop two for K = 0.376. Thus for the coupled system an oscillation exists in loop one only for K = 0.22 and this is the maximum value for stable operation. On the other hand with α = 10 a limit cycle existed in both loops until K was reduced to 0.19.

Example 4

As another use of MVREL for a multivariable system we consider a 2x2 plant, G(s), with elements

$G_{11}(s) = 3.5/s(s^2 + s+1)$, $G_{12}(s) = -5.9/s(s+1)$,

$G_{21}(s) = 6.4(s+1)/s(s+2)(s+3)$, $G_{22}(s) = 6.6/(s^2+0.5s+1)$ and N a diagonal matrix of two ideal relays with unit height. Searching for a periodic mode near to the oscillatory frequency of the first loop gave two solutions, one of which was acceptable. The other solution, however, was invalid as the relay input waveform in loop two possessed more zero crossings than assumed. Indications were that a limit cycle with three pulses in loop two to one pulse in loop one might be possible. This type of solution was then sought, which yielded a valid solution with the resulting relay input waveforms shown in figure 7. Another one to three pulse mode limit cycle was found and an invalid solution with additional unassumed crossings in loop two was also obtained. The invalid solution indicated the possibility of a multipulse mode where the output of relay one and two were one and five pulses/half period respectively. Two solutions of this form were found and also an invalid solution indicating that a seven pulse mode in loop two might also be possible. Again two solutions, where the output of relay two was assumed to be seven pulses/half period, were found when this mode was investigated.

The exact, eigenvalue criterion was used to predict the stability of all the limit cycle solutions discussed above. They were all unstable apart from the one to three pulse mode, shown in figure 7, and the single pulse mode. The latter, however, was close to instability, indicated by the eigenvalue magnitudes. These results were confirmed by digital simulations as the stable one to three pulse mode was easily found although initial conditions extremely close to the single pulse mode were required to obtain this limit cycle. None of the unstable modes could be simulated which underlines the importance of being able to assess the stability of limit cycle solutions.

Example 5

As a simple example of the application of the exact routines based on Tsypkin's method for a discrete system we consider the stability of a sampled data system with sampler operating at 1 Hz, zero order hold, ideal saturation nonlinearity with limit levels ±1, and plant $G(s) = K/(s+1)(s+2)$. The system was determined to be stable for $K<12.3$ since with K above this value a second subharmonic limit cycle was found. For K = 15 the limit cycle could be eliminated by a phase lead of around 86° or an increase of approximately 10% in the sampling frequency. With a unit gain quantiser with quantisation levels of 0.1 included after the sampler, the system required a further reduction in gain to K = 11.69 for stability.

Example 6

As a final example the limit cycles in a multivariable sampled data system are studied. The system is a 2 by 2 MIMO system consisting of a sample and zero order hold in each error loop, followed by a nonlinear and linear block. The nonlinear block has elements N_{11}, N_{22} which are ideal saturations with unity slope and limit levels of ±1 and elements N_{12}, N_{21} which are ideal relays with unit height. The elements of the linear block are $G_{11}(s) = 10/s+3$, $G_{12}(s) = 10/(s+1)^3$, $G_{21}(s) = 10/s^2+.3s+1$, $G_{22}(s) = 10/s(s+1)$. There are a wide range of sampling frequencies for which there are a variety of subharmonic limit cycles possible. For example if both samplers are synchronised and have sampling periods of $T_1 = T_2 = 0.5$ seconds, then both second and fourth subharmonic limit cycles are possible. The second subharmonic has positive heights of h = 4.19 in loop 1, and h = 0.084 in loop 2. The fourth subharmonic is shown in figure 8, with the crosses marking the sampling instants. If a multirate system is considered with the sampler in loop one having a period of $T_1 = 0.5/3$, and in loop 2 $T_2 = 0.5$ seconds then a limit cycle is obtained which is a sixth subharmonic in loop 1, and a second subharmonic in loop 2. That is the limit cycle which is shown in figure 9, has a frequency of 2 π rads/sec. in both loops. The results can be found using either MV or ZSAMV.

CONCLUSIONS

The paper has outlined the software suite, SUNS, developed at the University of Sussex for the analysis and design of nonlinear feedback systems. The algorithms, which are based on the describing function or Tsypkin's method, can be used to evaluate periodic modes in continuous and discrete feedback systems. The relative stability of a system can be assessed by evaluating the amount of gain or phase shift which when added to the system will cause it to limit cycle. Several examples of applications of the software have been given to illustrate the scope of the various procedures available. In particular the program MV, which is a describing function based routine, has been shown to be applicable to a variety of system types and configurations.

REFERENCES

Atherton, D.P., 1981, "Stability of Nonlinear Systems", Ch.6, Research Studies Press, John Wiley, Chichester.

Atherton, D.P., 1982, "Oscillations in Relay Systems", Trans. Inst. Meas. and Control, Vol.3, No.4, pp. 171-184.

Atherton, D.P. and Wadey, M.D., 1981, "Computer Aided Analysis and Design of Relay Systems", IFAC Symposium in CAD of M.V. Tech. Systems, Purdue, USA, pp. 355-360.

Balasubramanian, R., 1981, "Stability of Limit Cycles in Feedback Systems Containing a Relay", IEE Proc. D, Control Theory and Applications (1) 128, pp. 24-29.

Freeman, E.A. and Barney, G.C., 1963, "Divergent oscillations and their excitation in control systems with two saturation type nonlinear elements", IEE Proc. Vol. 110, No.6, pp 1096-1106.

Goucem, A., 1984, "Limit Cycles in Discrete Systems", M.Phil Thesis, University of Sussex, England.

McNamara, O.P., 1983, "Programs for limit cycle analysis in SISO nonlinear sampled data systems", University of Sussex Control Engineering Reports Nos. CE/S/13, CE/S/17.

McNamara, O.P., 1984, "A Describing Function based Program for the Analysis and Verification of Limit Cycles in SISO Nonlinear Systems", University of Sussex Control Engineering Report No. CE/S/22.

Utkin, V.I., 1977, "Variable Structure Systems with Sliding Modes", IEEE Trans. on Automatic Control, Vol. AC-22, No.2, pp. 212-222.

Wadey, M.D., 1983, "An Approximate Means of Predicting Sliding Limit Cycles and Evaluating their Accuracy", University of Sussex Control Engineering Report, No. CE/T/7.

Wadey, M.D., 1984, "MVREL, a program for Limit Cycle Analysis in a class of Nonlinear Multivariable Control Systems", University of Sussex Control Engineering Report, No. CE/S/20.

ACKNOWLEDGEMENTS

The authors wish to acknowledge support of this research by the U.K. Science and Engineering Research Council.

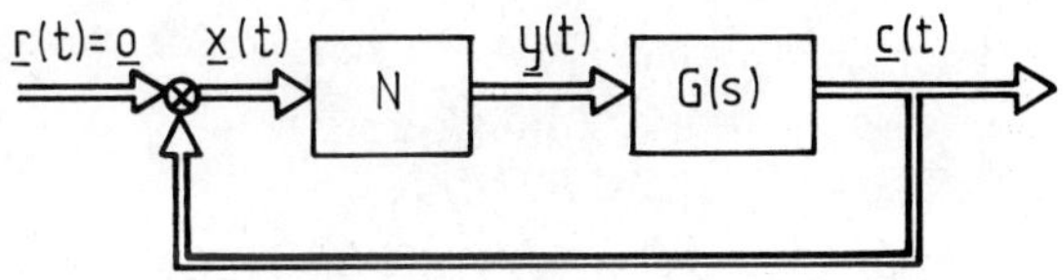

Fig. 1. Block diagram of canonical system.

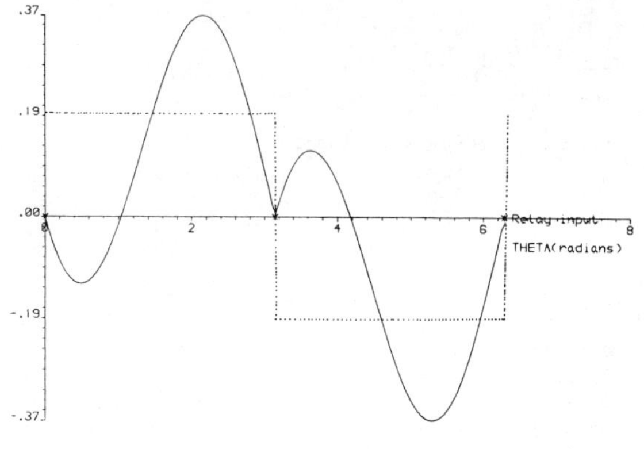

Fig. 2. Incorrect solution to Example 1.

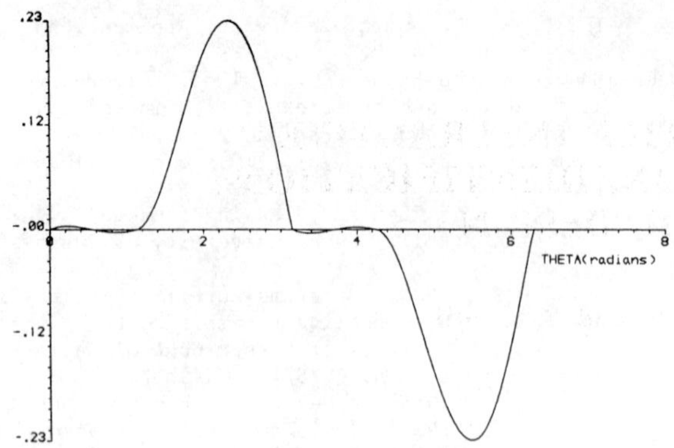

Fig. 3. First sliding approximation solution.

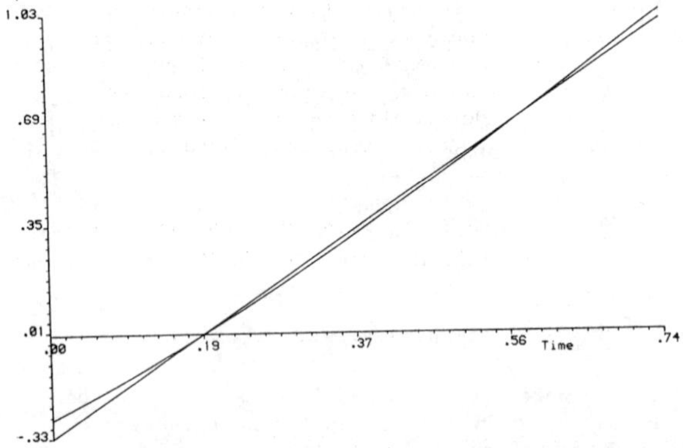

Fig. 4. Approximate and exact output during sliding.

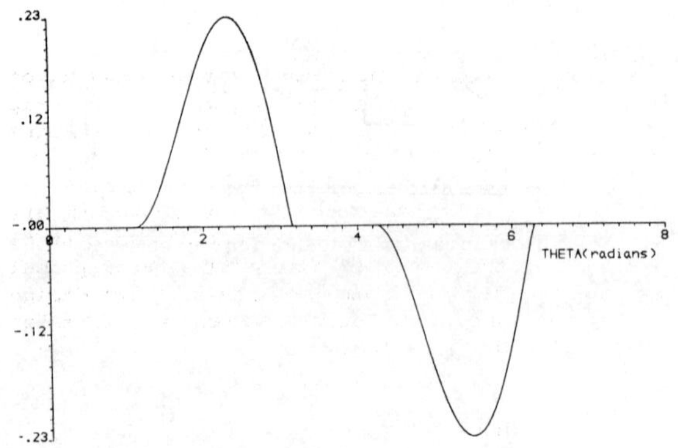

Fig. 5. Sliding solution for Example 1.

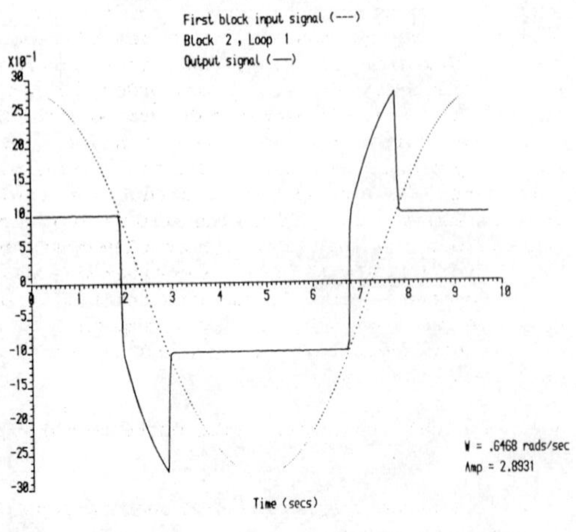

Fig. 6. Limit cycle in Example 2.

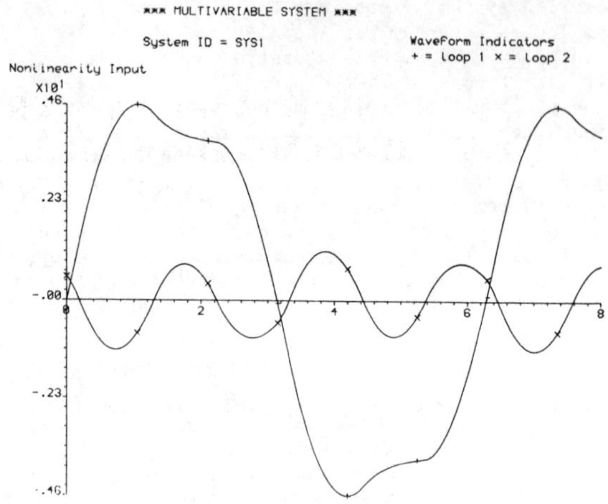

Fig. 7. Limit cycles in Example 4.

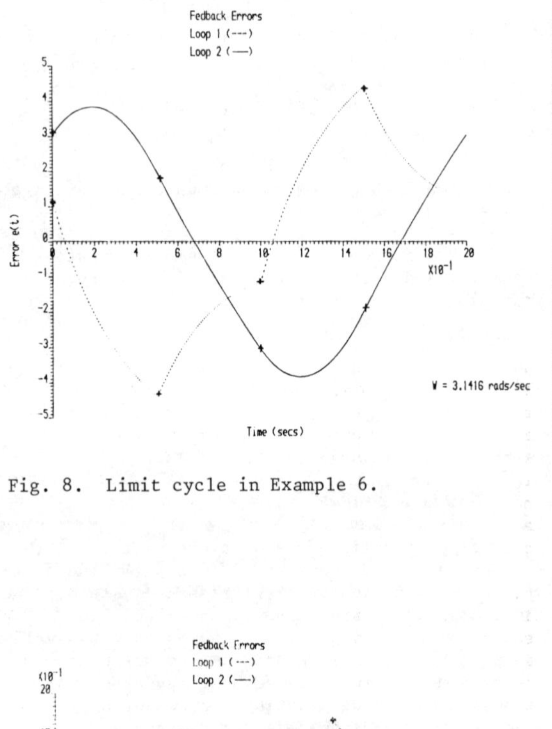

Fig. 8. Limit cycle in Example 6.

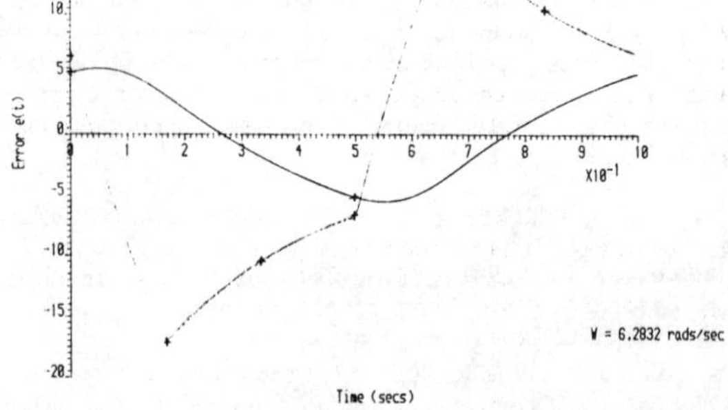

Fig. 9. Limit cycle with multirate sampling.

SIRENA +: A VERSATILE INTERACTIVE SYSTEM FOR SIMULATION, IDENTIFICATION AND CONTROL DESIGN

A. Barraud, P. Laporte and S. Gentil

Laboratoire d'Automatique de Grenoble, Ecole Nationale Supérieure de Grenoble, BP 46, 38402 Saint Martin d'Hères, France

Abstract. This paper deals with an interactive system for simulation, identification and control of linear systems described by state-space or transfer-functions, in a continuous, discrete or sampled representation. Based upon a university prototype, SIRENA + has now reached an industrial quality level. Written in FORTRAN-77, this system has been implemented on a large variety of computers. Robustness has been ensured by a high quality mathematical software developed from and in a similar way as some well known libraries (Harwell, Nats projects, Matlab, ...). Its modular structure (700 routines / 60 000 lines) greatly enhances its maintenability and evolution.

The system is user friendly, with an on-line help and error checking, files saving options, and on-line graphical features. Commands are prompted adaptively according to the working context and to the user skillness.

SIRENA + performs all the steps necessary for control design : simulation, identification and control.

Keywords. Computer aided design, Control system analysis, Identification, Modelling.

INTRODUCTION

The Automatic Control Laboratory of Grenoble has developed research work and applications in all of the process control fields for twenty years, in collaboration with industrial firms. This results in experience and practice in all control design areas - simulation, modelization, identification, control analysis and synthesis- and in the elaboration of high level software also used as Computer Aided Education tools for Automatic Control students.

This experience is now resulting in the implementation of an industrial Computer Aided Control System Design Package, in collaboration with a company. The first industrial version, SIRENA, only performed the simulation systems defined by a block description, but the structure of the software was designed in such a way as to permit further extentions.

Today, SIRENA + is really what is known as a CACSD System, since it performs interactively all the steps of the control design for systems described by state space or transfer functions, in a continuous, discrete or sampled representation, involving some "classical" non linearities.

- the simulation commands allow to define the systems, with open or closed loops, in an arbitrary way, without any a priori limitation in complexity.

- the user can choose between several powerful identification methods based either on an equation error or an output error, processing time-response or frequency-response data.

- control design is based on an enlarged LQG theory, enabling the user to define easily the closed loop performance, including constrained control.

Furthermore, SIRENA + presents all of the features of a CAD software : it is user friendly with on-line help and error checking, like saving options and on-line graphical tools.

In this paper, the three main parts of the package are described : simulation, identification and control design. Then the numerical, interactive and graphical aspects are discussed and then the software design guidelines are evoked. A concrete example illustrates some of the SIRENA + features.

SIMULATION

SIRENA + is a versatile computer aided design package. It is fully interactive ; the user operates in a very friendly way to describe one or more dynamical systems, (continuous, discrete or sampled). Each system can be decomposed into a set of elementary blocks (transfer function, non linear expression, state space representation of any order) connected in an arbitrary way. All elements described under SIRENA + are recognized by their name (inputs, outputs, auxiliary signals, elementary blocks, formal parameters, etc...). Appropriate instructions allow the user to modify, at any time, the structure of a system (add a loop, an input, change an elementary block, ...) or to link it with previously created systems. In order to have a nice man machine communication, an adaptive on line help has been implemented which efficiently supports beginners and advanced users, during working sessions.

Any numerical simulation requires to go through the following steps :

- structural and formal descriptions of the systems to be simulated.

- description of the inputs. These inputs are chosen from built-in signals (impulse, step, ramp, sine wave, PRBS, PRTS, uniform and gaussian noise) or come from user data files (in an ad-hoc format).

- time or frequency ranges according to simulation domain.

At this time, SIRENA + checks that all the parameters are numerically known. If it is not the case, the user is prompted to give appropriate numerical values (or range of values) for each formal parameters (for example : position of pole, zero order, PRBS register numbers, sampling rate value, ...).

Technically, SIRENA + operates under several modes according to the type of the systems described.

- Fully linear dynamical systems, if described by transfer functions, are formally and numerically reduced to one open loop transfer function (Mason's rules) ; otherwise the simulation is performed step by step according to the global path.

- Sampled and non linear systems are computed step by step, or partly reduced for performances.

- The continuous non-linear blocks are simulated with variable steps, and the linear ones with constant steps. These steps are automatically computed. Numerical aspects are described in more details below.

Computed signals can be displayed as numerical values (terminal, line-printer) or graphically. In the later case, all computed signals are automatically displayed in a standard way, including Bode, Black and Nyquist plots for the frequency domain. The corresponding default options can then be modified by the means of an on line graphical editor (new labelling, zooming, screan configuration, ...).

If the results obtained are considered meaningfull for other computations or comparisons (identification, control design, ...) they can be saved as new built-in SIRENA data files. These files can also be accessed (read only) by other user programs.

Before leaving SIRENA +, the user can make several actions. All the work done under SIRENA + is recorded in a data base (systems described with their structure, parameters values, saved data files obtained from simulation, ...). This data base can be completely or partly cleaned, or completely saved. The user has the possibility to resume his work or to begin a new one every time SIRENA is logged in.

IDENTIFICATION

Sirena + contains a relatively large number of identification algorithms which mainly belong to two classes of methods : output error or equation error minimization, and apply to frequency responses or time responses of MISO systems (Multiple Inputs - Single Output) (Astrom and Eykoff, 1970). Multiple Outputs systems are considered at this stage as a set of Single Output ones ; a global approach has not been implemented owing to the difficulties of handling multivariable identification methods without some skillness.

With y as an output, u_1 as an input, s as the Laplace operator, z^{-1} as the delay operator and k as a discrete sampling-time, the following three models can be used to define a system

$$A(z^{-1})\, y(k) = \sum_{i=1}^{r} B_i(z^{-1})\, u_i(k) \qquad (1)$$

$$y(k) = \sum_{i=1}^{r} \frac{B_i(z^{-1})}{A_i(z^{-1})}\, u_i(k) \qquad (2)$$

$$\frac{y(s)}{u(s)} = \frac{B(s)}{A(s)} = H(s) \qquad (3)$$

A(.) and B(.) are polynomials of order n and m respectively, the constant term of A(.) being normalized to 1. In the following, Θ is the unknown vector containing the A(.) and B(.) parameters.

The minimization of an output error, M-O-E, which is a very general method can be applied to the three types of models. For models (1) and (2), it processes sampled time response of the system ; model (1) corresponds in fact to model (2) with a common denominator, so that :

$$A(z^{-1}) = \prod A_i(z^{-1}).$$

If $y_m(k)$ represents the measurements and N the number of sampling times, the following criterion is minimized :

$$\Theta_{OE} = \arg\min \left(\Sigma\, (y_m(k) - y(k))^2\right.$$

For model (3), the method processes (at present) the frequency response of the system ; the minimization criterion is :

$$\Theta_{OE} = \arg\min \left(\Sigma\, |H(jw_k) - (v_k + jx_k)|^2\right.$$

where v_k and x_k represent the real and imaginary parts of the experimental NYQUIST plot at pulsation w_k.

It does not present any difficulty to implement the output error method for time response of model (3), this is part of the ungoing extensions of the package.

The minimization of an equation error is applied to model (1), owing to the three methods : Simple-Least-Squares, S.L.S., Extended-Least-Squares, E.L.S. (Don carli, de Larminat, 1978), and Maximum-Likelyhood, M.L. In this last case, besides polynomials $A(z^{-1})$ and $B(z^{-1})$, a filter $C(z^{-1})$ is used for noise modelling.

To get into the identification modulus, the user has to type MODELISATION ; the nature of the system (CONTINUOUS/DISCRETE) has yet been defined. The measurement set is supposed to be pre-processed in the SIGNAL-PROCESSING modulus (filtering, centering, averaging, correlating : input-output correlation gives the impulse response of the system if the input is a white noise or a PRBS and by these means the user gets qualitative informations on the system : delays, oscillating poles, ...).

Then the user has the choice between model (1) and (2) if the system is discrete. He has to fix the delay with respect to each input and the orders n and m of the polynomials A(.) and $B_i(.)$. He then selects among the methods applying to the chosen structure. Afterwards, identification is running : matrix handling or non-linear programming, depending on the method. In this latter case, initial parameters values have to be provided to start the search in the parameter space ; the user can supply these values if he has a previous knowledge of the model or the initialization can be automatic. In that instance, the S.L.S. algorithm is activated and its results are used to initialize the

M.O.E. one, ant the E.L.S. results are used for M.L. initialization.

In addition to the values of the parameters, the package provides extra-information to help the user to decide if the results are good or if another run is necessary : standard-deviations of parameters, values of poles and zeros, gains, optimal value of the output error ; AKAIKE's criterion, graphical plot of model simulation and measurements (Laporte et al, 1985).

Then the user can repeat the identification, starting again from the model definition or the choice of the method, Fig. 1, or leave the MODELIZATION modulus.

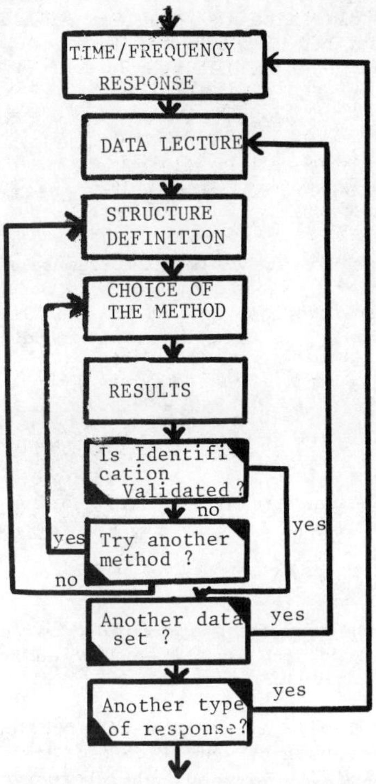

Fig. 1 : The modelization modulus

CONTROL DESIGN TOOLS

Some of the facilities implemented in Sirena +, in order to perform control synthesis, mainly via state space representation will be discussed now. Pole zero evaluation, pole placement, and classical linear quadratic control will be successively considered.

Pole/Zero Evaluation.

The problem is trivial for transfer function representation, since poles and zeros are computed as roots of polynomials. There exists some excellent routines to solve this problem. Now, let us consider a state space model $\Sigma(A, B, C, D)$. Then the poles of the system are computed as the eigenvalues of A. The computation of the finite zeros, which is much more difficult, is done according to Van Dooren method, 1982. Let $S(\lambda)$ be the system matrix associated with Σ :

$$S(\lambda) \quad \begin{bmatrix} \lambda I - A & B \\ C & D \end{bmatrix} \qquad (4)$$

Now $S(\lambda)$ is transformed by unitary transformations P and Q to the generalized upper Schur form :

$$P.S(\lambda).Q = \begin{bmatrix} A_r-\lambda B_r & X & X & X \\ & A_f-\lambda B_f & X & X \\ & & A_i-\lambda B_i & X \\ & & & A_l-\lambda B_l \end{bmatrix} \qquad (5)$$

where $A_r - \lambda B_r$ and $A_l - \lambda B_l$ are non square pencils with no zeros (right and left minimal indices of S), $A_i - \lambda B_i$ is a regular pencil with no finite zero and $A_f - \lambda B_f$ is a regular pencil whose generalized eigen values are the finite zeros of $S(\lambda)$.

Pole Placement

This technique is only available for SISO systems. Sirena implements the Targa's method, 1981, which essentially transforms the (A, B) pair in Schur form via orthogonal transformation and lastly computes the feedback gain in a stable way and comes back to the initial basis.

Linear Quadratic Control Algorithms

Let Q and R be the weihting matrices respectively associated with the state **x** and the input **u**. The following hypotheses hold :

$$R = R^T > 0 \quad ; \quad Q = L^T \geq 0$$

(A, B) is stabilisable ; (L, A) is detectable.

Then the Riccati equation that must be solved has a unique solution P such that $P^T > 0$.

$$L^T . L \geq 0$$

$$P^T = P > 0$$

In order to compute the matrix state feedback G, hence P, it is now necessary to distinguish between discrete and continuous systems.

- Continuous case :

It is well known that

$$G = VP \; ; \; V = B\,R^{-1}\,B^T$$

$$P : PA + A^TP - PVP + Q = 0$$

Let H be the hamiltonian matrix :

$$H = \begin{bmatrix} A & -V^T \\ -Q & -A \end{bmatrix} \qquad (6)$$

From this step, Sirena + runs two algorithms

a) Laub's method (1979) $P = F_{21}\,F_1^{-1}$

$$\text{with} \quad F = \begin{bmatrix} F_1 & F_{12} \\ F_{21} & F_2 \end{bmatrix}$$

$$\text{such that} \quad F^THF = \begin{bmatrix} H_1 & H_{12} \\ 0 & H_2 \end{bmatrix} \qquad (7)$$

i.e. F is an orthogonal transformation which transforms H into the real **ordered** Schur form (i.e. the spectrum of H_1 is asymptotically stable while the spectrum of H_2 is unstable).

b) Barraud's method (1979) $P = F_2 F_{12}^{-1}$

$$F = 1/2 [I - \text{sign}(H)] \quad (8)$$

where sign(H) is the matrix sign function computed via an accelerated Newton algorithm.

- Discrete case :

Now :

$$G = (B^T PB + R)^{-1} B^T PA \quad (9)$$

$$P : A^T PA - P - A^T PB(B^T PB + R)^{-1} B^T PA + Q = 0 \quad (10)$$

$$P : A^T PA$$

The matrix equivalent to H is now a symplectic matrix S which can be written :

$$S = U^{-1} L = \begin{bmatrix} 1 & V \\ 0 & A^T \end{bmatrix}^{-1} \begin{bmatrix} A & 0 \\ -Q & 1 \end{bmatrix} \quad (11)$$

If A is regular (in practice well conditionned) S is explicitely known :

$$S = \begin{bmatrix} A + VA^{-T}Q & -VA^{-T} \\ -A^{-T} & A^{-T} \end{bmatrix} \quad (12)$$

a) Laub's method, 1980, $P = F_{21} F_1^{-1}$

In order to define F, let us consider the generalized eigenvalue problem $L = \lambda U Z$. This problem can be changed into generalized real ordered Schur form, i.e. L is the real ordered Schur form and U upper triangular. The orthogonal transformation is F.

b) Barraud's method $P = F_{21} F_1^{-1}$

$$F = \text{sign} \begin{bmatrix} 1+A & V \\ -Q & 1+A^T \end{bmatrix}^{-1} \begin{bmatrix} A-1 & -V \\ Q & 1-A^T \end{bmatrix} + \begin{bmatrix} -1 \\ 0 \end{bmatrix} \quad (13)$$

It must be emphasized here that this result can be directely derived from the continuous case via a bilinear transformation, and holds even if A is singular.

NUMERICAL, STRUCTURAL AND COMMUNICATION ASPECTS

Since Sirena + is an industrial product, robustness, efficiency, transportability, maintenability and evolution are some of the required qualities for it. They result in constraints in all the steps of the software design, but the time spent in this way is widely recovered when new developments are going on.

As Sirena + is user-oriented, the interactive and graphical features have been fully studied, in connection with the computer aided teaching experience of the L.A.G.

Numerical Aspects

High quality numerical algorithms have been used to ensure robustness and accuracy ; most of them are parts of well known librairies : BLAS (Lawson et al, 1972), LINPACK (Dongarra et al, 1979), EISPACK (Garbow et al, 1977), HARWELL (Hopper, 1980).

The simulation method depends on the type of the system : for discrete systems, it only consists in simple inner products ; the linear continuous systems case is handled like the previous one, after the system, described by a transfer function or a state space representation, has been transformed into a discrete form :

Let :

$$\dot{\underline{x}} = A\underline{x} + B\underline{u} \qquad \underline{y} = C\underline{x} \quad (14)$$

be the continuous state space representation of the system, either provided by the user or obtained from the transfer function. Matrix exponentiation (Pade approximation) gives the equivalent discrete system :

$$\underline{x}_{k+1} = \Phi \underline{x}_k + \Gamma \underline{u}_k \qquad \underline{y}_k = C\underline{x}_k$$

with

$$\left[\begin{array}{c|c} \Phi & \Gamma \\ \hline 0 & 1 \end{array}\right] = \exp \left[\begin{array}{c|c} A & B \\ \hline 0 & 0 \end{array}\right] h \quad (15)$$

where h is the sampling step provided by the user or calculated according to a stability and precision criterion (Barraud, 1981).

In case of non linearities or when the previous method fails because of an ill conditionning, the equations are integrated with Runge Kutta Feldberg (Forsythe et al, 1977) or the Gear (1971) algorithms. Finally, the sampled systems simulation combines both techniques.

The system identification mainly involves a Least Squares problem, and two cases have to be considered. In the linear unconstrained case, the criterion is a quadratic function of the parameters set and the explicit solution is calculated through an algorithm based on sequential orthogonal transformations (Givens rotations), which is numerically stable and has small memory requirements (Lawson an Hanson, 1974 ; Don Carli and De Larminat, 1978).

In the non-linear unconstrained case, the Least Squares problem is solved by a quasi-Newton algorithm, which requires analytic gradients of the criterion (Powell, 1975).

Most of the matrix computation involved in the control design algorithms come from LINPACK, EISPACK and collected algorithms from ACM. The commandability - stabilisability (observability - detectability) tests which can be performed automatically, or at the user requests, are performed according to Varga's idea, 1981.

The reliability of Sirena + is enhanced by a general error checking at each step of the calculus : whatever user's problem no breakdown can occur.

Implementation Aspects

Sirena + is entirely written in Fortran 77 to ensure its transportability -only a small kernel of 200 lines out of 60 000 lines is device dependant- and in order to easily include external libraries. It has now been implemented on a large variety of computers, ranging from Micromega 32 to CDC Cyber.

The structure is modular (700 modules) and the program is fully annotated to enhance maintenability and evolution. New methods or algorithms can be implemented very easily and Sirena + is currently under progress.

Interactive and Graphical Features

Beside the numerical qualities, user interface is a very important feature in a CAD system : the way in which the dialogue is provided and the system outputs appears -results, error messages, graphical plots- must be fully studied.

Sirena + provides a question-and-answer command mode which does not require any previous knowledge : it is well adapted to beginners or casual users. In addition, an on-line help is always available with the reserved character "?" and all the inputs of the user are checked. For advanced users, answers can be abreviated or even anticipated when the next questions are known ; it is also possible to go backwards in the scheduled dialogue with the reserved character "£".

Every calculus is illustrated either with numerical tables or with graphical plots. In the latter case, the standard presentation, adapted to the working context, can be immediatly modified by the user with an on-line graphical editor which provides a large set of interactive tools.

Sirena + also provides a file saving option to keep the results for further processing.

Clearly, Sirena + has been designed to be user friendly and easily accessible : the users concerned by it are students or engineers who mostly want to get a lot of reliable results rapidly. So, unlike other university CASCD systems, Sirena + does not provide all the existing methods but only an efficient subset, neither all the possible interactive facilities, such as a command-driven mode, which would require to learn another language.

Example

As an example, the pilot plant of the L.A.G. has been used. It consists in a distillation column controlled by a network of microcomputers. This column is composed of 9 plates of diameter 150 mm ; the feed flow is lead to one of the intermediate plates. The boiler contains 18 resistors (42 KW) to provide variable heating. Once condensed, a part of the vapor goes out of the column, forming the product, while the other part is internally recycled : this is known as internal reflux. This system is MIMO (Fig. 2).

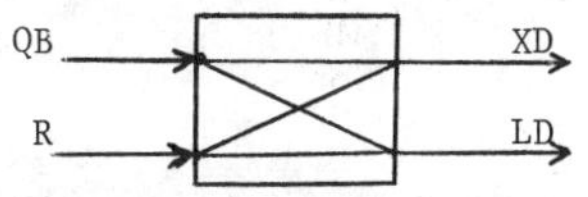

Fig. 2 : Pilot plant structure

XD represents the concentration of the final product, LD its flow rate. QB is the heating power and R the internal reflux. Every output is coupled to every input.

Identification

Fig. 3 shows an identification run applied to a response of XD to a PRBS applied on QB. The graphical output comparing the process response with the model response appears on Fig. 4.

Every SISO transfer function is identified in the same way. Careful attention is payed to the selection of delay K, and of n and m, degrees of A(.) and B(.).

Simulation

The simulation of the transfer function between XD and the two inputs is shown on Fig. 5 where computer prompts are underlined. Simulation are used at this step to validate the model behaviour, using data that have not been used for the identification, for instance.

Control

An LQ control is then computed. Fig. 6 shows the use of this part of Sirena +. Obviously a lot of simulation can then be run in order to choose the best feedback gain and possibly check the effect of some non linearities neglected by the model.

CONCLUSION

Is Sirena + a new product, or is it another C.A.D. package among many others ? In fact, there are many hundred of packages in the world, devoted to this field. Several papers have already been devoted to exhaustive surveys. Sirena + is not a general purpose simulation tool as for example Dynamo, CSL, ACSL, CSMP and many others. It is not a language nor a software tool as Delight, Nye 1982. It is not a set of "main programs" more or less linked together Cypros, Tysso, 1979, Simnon, Modpac, Synpac, Idpac, Wieslander 1980. Sirena + has not been developed from MATLAB, Moler 1980, see Matrix_x, CTRL-C,... Its facilities look like KEDDC, Schmid 1983. Sirena + is a new product in the CAD area. It manages a complete data-base implicitely created by the user.

Any number of systems can be recognized by Sirena +, linked together, modified, cleared... Up to now 70 % of Sirena + is devoted to this data-base management, and alphanumeric / graphical dialogue. This part has approximatively reached its optimal efficiency.

Most of the present developments are concerned with new facilities : formal and numerical reduction of state space models path, petri-net simulation and management of a knowledge data-base firstly in the identification field : Sirena + must become an expert system in the C.A.D. field.

REFERENCES

Astrom, K.J., Eykoff, P., (1970). System Identification : A Survey. **IFAC Symposium on Identification and System Parameters Estimation**, Prague.

Barraud, A. Y., (1979). Investigations autour de la fonction signe d'une matrice-application à l'équation de Riccati. **RAIRO Automatique**, vol. 13, nº 4, pp. 335-368.

Barraud, A., Y., (1981). More on the conversion problem of discrete-continuous models. **Appl. Math. Modelling**, vol. 5, pp. 414-416.

Don Carli, C., De Larminat, Ph., (1978). Analyse de la stabilité globale d'un algorithme d'identification récursive des systèmes linéaires stochastiques discrets. **RAIRO Automatique**, vol. 12, n° 3, pp. 269-276.

Dongarra, J.J., Moler, C.B., Bunch, J.R., Stewart, G.W., (1979). Linpack user's guide. **SIAM Philadelphia**.

Emami-Naeisi, A., Van Dooren, P., (1982). Computation of Zeros of Linear Multivariable Systems. **Automatica**, vol. 18, n° 4, pp. 415-430.

Forsythe, G.E., Mallon, M.A., Moler, G.B., (1977). Computer Methods for Mathematical Computations. **Ed. Prentice Hall**.

Garbow, B.S., et al (1977). Matrix Eigensystems Routines - EISPACK Gird Extensions. Lecture notes in **Computer Science**, vol. 51, Springer Verlag, Berlin.

Gear, G.W., (1971). Numerical Initial Value Problem in O.D.E. **Ed. Prentice Hall**.

Hopper, M.J. (1980). Harwell Subroutine Linbrary. A catalogue of subroutines. **A.E.R.E. R 9185, UKAEA Harwell**, England.

Laporte, P. Barraud, A., Gentil, S. (1985). An Interactive Computer Aided Identification Package. **IFAC Symposium on Identification and System Parameters Estimation**, York, England.

Laub, A.J., (1980). A Schur method for solving algebraic Ricatti equations. **I.E.E.E. Trans. on Aut. Cont.** vol. AC-25, n° 4, pp. 631-641.

Lawson, C.L., Hanson, R.J., Kincaid, D.R., and Krogh, F.T. (1977). Basic Linear Algebra Subprograms for FORTRAN Usage. **SAND 77**, 0898, Sandia Laboratories, U.S.A.

Moler, C., (1980). Matlab user's guide. Department of Computer Sciences, University of New-Mexico, U.S.A.

Nye, B., Tits, A., (1982). **Delight for Beginners**, Report, Electronic Research Laboratory, Berkeley, U.S.A.

Pappas, J., Laub, A.J., (1980). On the Numerical Solution of the discrete Time Algebric Riccati equation. **I.E.E.E. Trans. Aut. Cont.**, vol. AC-25, n° 4, pp. 631-641.

Petkov, P., Christov, N., Konstantinov, M. (1984). Numerically Stable Algorithm and Program for Pole Assignment of Linear Single Input Systems. **6th International Conference on Analysis and Optimisation of Systems**. Nice, France.

Powell, J.D., (1975). Some Global Convergence Properties of a Variable Metric Algorithm for Minimization without Exact Line Searches. **C.S.S. 15 Report AERE Harwell**.

Schmid, C.H.R., (1983). **CAD of Adaptive Systems Using KEDDC**, Report, Ruhr University, Bochum, R.F.A.

Tysso, A., (1979). CYPROS : Cybernetic Program Packages, **IFAC Symposium on CAD of Control Systems**, Zurich.

Varga, A. (1981). Numerically Stable Algorithm for Standard Controllability Form. **Electr. Letters**, vol. 77-2, pp. 74-75.

```
!!!! FONCTION(S) DE TRANSFERT EN W=Z**-1
!!!! POLYNOMES ORDONNES PAR PUISSANCE CROISSANTE DE W

**** DENOMINATEUR ****
**** DEGRE   N =   1

        COEFFICIENTS                    ECART TYPE

         .100000E+01                    .000000E+00
        -.836238E+00                    .243532E-01

        POLES

    PARTIE REELLE    PARTIE IMAGINAIRE      MODULE

     .836238E+00       .000000E+00        .836238E+00

**** NUMERATEUR ****
**** DEGRE   M =  0
**** PUISSANCE DE W EN FACTEUR K =  3

        COEFFICIENTS                    ECART TYPE

        -.118232E+00                    .140658E-01

GAIN STATIQUE     :    -.721973E+00

**** NUMERATEUR DU FILTRE SUR LE BRUIT ****
**** DEGRE   L =  1

        COEFFICIENTS                    ECART TYPE

         .100000E+01                    .000000E+00
         .329852E-01                    .119775E+00

       * ZEROS

    PARTIE REELLE    PARTIE IMAGINAIRE      MODULE

    -.329852E-01       .000000E+00        .329852E-01

>>>> ERREUR D'EQUATION EN % :    32.0552

>>>> ERREUR DE SORTIE EN %  :    43.4257

>>>> CRITERE D'AKAIKE       :   -17.5106
```

Fig. 3 Identification

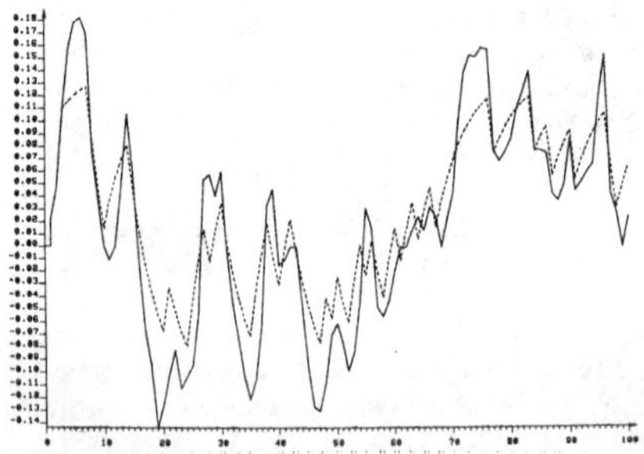

Fig. 4 Graphical output

```
TYPE : DISCRET
NOM  : S1

PARAMETRE-SYSTEME : QB
NATURE : POINT-PAR-POINT
NOM DU FICHIER : SG-C:SYM
FORMAT DE LECTURE : 3(E12.4)
NOMBRE DE POINTS : 104
NOMBRE D"ELEMENTS PAR ENREGISTREMENT : 3
POSITION DU POINT : 1

PARAMETRE-SYSTEME : R
NATURE : POINT-PAR-POINT
NOM DU FICHIER : SG-C:SYM
FORMAT DE LECTURE : (3E12.4)
NOMBRE DE POINTS : 104
NOMBRE D"ELEMENTS PAR ENREGISTREMENT : 3
POSITION DU POINT : 2

PARAMETRE-SYSTEME : H
NATURE : TRANSMITTANCE
COEF-GAIN : .10000000E+01
FORME-NUMERATEUR   : POLYNOME
COEFFICIENTS : .10000000E+01
FORME-DENOMINATEUR : RACINE
VALEUR DES RACINES : (0)

PARAMETRE-SYSTEME : H1
NATURE : TRANSMITTANCE
COEF-GAIN : -.10399999E+00
FORME-NUMERATEUR   : POLYNOME
COEFFICIENTS : .10000000E+01
FORME-DENOMINATEUR : RACINE
VALEUR DES RACINES : (.83599997E+00)

PARAMETRE-SYSTEME : H3
NATURE : TRANSMITTANCE
COEF-GAIN : .96999973E-01
FORME-NUMERATEUR   : POLYNOME
COEFFICIENTS : .10000000E+01
FORME-DENOMINATEUR : RACINE
VALEUR DES RACINES : (.93499994E+00)
```

Fig. 5 Simulation

RECENT ADVANCES IN THE L-A-S SOFTWARE USED IN CAD OF CONTROL SYSTEMS

S. P. Bingulac*, P. J. West and W. R. Perkins****

**Electrical Engineering Department, Virginia Polytechnic Institute and State University, Blacksburg, VA 24061, USA*
***Coordinated Science Laboratory and Department of Electrical and Computer Engineering, University of Illinois, Urbana, IL 61801, USA*

ABSTRACT
This paper reviews the basic features of a high-level, interactive, conversational language called L—A—S (Linear Algebra and Systems). Particular emphasis is placed upon recent developments of this Computer-Aided Design (CAD) language. Unlike the conventional menu-driven approach of many CAD packages, L—A—S provides the user with considerable flexibility in the development and maintenance of linear algebra and control algorithms. L—A—S is both a language and an interpreter. Users of L—A—S write programs while interacting with the interpreter. Programs may be edited, stored, recalled, and executed with the same or a different data set. L—A—S data structures include matrices, polynomials (continuous or discrete), and matrices of polynomials. In this paper, we briefly review the basics of the L—A—S language. Then, we concentrate on those features of L—A—S which have been added within the last year. These include the Multiple Operator Statement (MOS), L—A—S user-defined subroutines, and the most-recent "echo" feature for interaction with the interpreter. Examples accompany the text to illustrate each new feature.

KEYWORDS Computer-aided Design, Programming Languages, Control Engineering Computer Applications, Control System Analysis/Synthesis

INTRODUCTION

This paper reviews the basic features of a high-level, interactive, conversational language called L—A—S (Linear Algebra and Systems). Particular emphasis is placed upon recent developments of this Computer-Aided Design (CAD) language. Unlike the conventional menu-driven approach of many CAD packages, L-A-S provides the user with considerable flexibility in the development and maintenance of linear algebra and control algorithms. L-A-S is both a language and an interpreter. Users of L-A-S write programs while interacting with the interpreter. Programs may be edited, stored, recalled, and executed with the same or a different data set. L-A-S data structures include matrices, polynomials (continuous or discrete), and matrices of polynomials.

L-A-S is composed of a large collection of FORTRAN subroutines (about 20,000 lines of code). Furthermore, L-A-S employs software such as the EISPACK and LINPACK packages to insure that the algorithms used for solving problems are of the highest degree of numerical integrity and precision. In addition, L-A-S is available in both single and double precision versions. The software structure of L-A-S is very modular. Hence L-A-S can and has been implemented on computer installations ranging from mini's to high-speed virtual memory machines. Moreover, modularity implies that extensions and enhancements of the L-A-S software are readily incorporated. The syntax and semantics of L-A-S are straightforward so that familiarity with the language may be easily and rapidly attained.

In the normal interactive mode, the user types statements directly to the L-A-S language interpreter. Each statement is either an interpreter command (e.g., put L-A-S into program debug mode) or an operator statement. The former instructs L-A-S to display or modify various status information concerning the current L-A-S work session. The latter invokes the L—A—S language parser which subsequently calls upon the FORTRAN subroutines needed to process some kind of computation. Essentially, input data are combined with one or more L-A-S operators to produce output data. The concept of an operator, as well as the utility of L-A-S operators as algorithm building blocks has been widely documented [1,2,7]. An extensive on-line help facility containing detailed information about L-A-S usage is at the disposal of the user.

The interpreter commands are unique to L-A-S. Once the operator statements are mastered, the interpreter commands permit efficient management and control of the current L-A-S work session. A thorough description of all the interpreter commands will not be pursued here. Rather, we summarize them in Table 3 at the end of the paper.

One of the primary functions of the L-A-S language is in the development, testing, and evaluation of control design and analysis problems. Its usefulness has also been verified in undergraduate and graduate courses treating linear algebra, numerical analysis, and control theory. In non-academic settings, L-A-S has served as a "matrix desk-top calculator" for quick solution of any problem that may be decomposed into a sequence of existing L-A-S operators. The computer resources required by L-A-S are rather modest and may be easily adjusted to a particular computer installation. Presently, L-A-S has been implemented on several different computers at over a dozen locations around the world.

In this paper, we briefly review the basics of the L-A-S language. Then, we concentrate on those features of L-A-S which have been added within the last year. These include the Multiple Operator Statement (MOS), L-A-S user-defined subroutines, and the most-recent "echo" feature for interaction with the interpreter. Examples accompany the text to illustrate each new feature. The references should be consulted for additional information concerning the L-A-S CAD language.

L-A-S OPERATOR STATEMENTS

The fundamental concept behind an L-A-S program is the notion of an L-A-S *operator*. Essentially, operators should be viewed as generalizations of choices on a menu. That is, consider an arbitrary linear algebra or control system menu of allowable computations. Then, an L-A-S operator simply extracts the function performed by any particular choice and attaches a mnemonic name to that function. In other words, L-A-S operators perform some desired calculation on input data, indicate errors if any, and generate output data. Operators are issued to the L-A-S language interpreter along with the names of the data used and generated. Operators are divided into five groups : Input/Output, Data Handling, Linear Algebra, Control Systems, and L-A-S Program Control. Presently, there are more than 100 L-A-S operators, (see Table 1). Furthermore, the L-A-S user can currently define a maximum of 100 matrices with the total number of matrix elements not to exceed 50,000. The maximum order of any particular matrix is not explicitly limited.

L-A-S programs are written by combining one or more operator statements. L-A-S is a *free-format* language. This means that unless a comment is entered, blanks are removed prior to scanning by the interpreter. Thus, the format in which a program is typed is unrestricted or free. Upon receiving a line of text, the interpreter parses and subsequently analyzes the L-A-S statement. If the syntax and semantics of the statement are correct, then the L-A-S driver library calls upon the necessary FORTRAN subroutines to perform the computations.

1. Single L-A-S Operator Statement Syntax

The general structure of a single operator statement is as follows:

label : *input-field* (*operator-field*) = *output-field*. (1)

The symbols :, (,) and = act as field delimiters. The *label*, which is optional, is usually used in conjunction with the L-A-S Program Control group of operators (Table 1). The *input-field* consists of zero or more variables and/or constants. The *operator-field* can be decomposed into 2 subfields, the *operator mnemonic* and the *option* subfields. The decomposition has the following form:

operator − field ::= *op −mnemonic* [, *option*] (2)

where

option ::= E | L | P | T . (3)

In the programming language literature [3,4,5], (2) and (3) represent what are commonly called *language production rules*. The so-called metasymbol "::=" can be interpreted as "is produced by". Thus, (3) may be read as "the symbol *option* is produced by a choice of one of the following letters : E, L, P, or T". In particular,

E means print numbers in an Exponential (E) format,
L means append to the Lineprinter data file,
P means Plot the results on a graphics terminal, and
T means Type the result(s) of this calculation.

Finally, the *output-field* consists of zero or more variables that are defined upon successful completion of the calculation.

It is useful to illustrate the preceeding ideas with a simple example. We emphasize the fact that single L–A–S operator statements resemble, to a high degree, standard linear algebra notation and that the "overhead" notation required by the L–A–S language interpreter is minimal.

Consider the following three linear algebra operations : matrix addition, matrix multiplication, and matrix inversion. Suppose that we are already given two square matrices, A and B. Suppose further that we want to perform the following sequence of calculations:

$$C = A + B \qquad (4)$$

$$D = C * A \qquad (5)$$

$$E = B^{-1} \qquad (6)$$

Notice that we implicitly assume that the assignment statement computes the quantity on the right hand side and subsequently stores it in the array named on the left hand side of the equation, that is, the equal sign acts "right to left". Conversely, L-A-S operators act "left to right". Hence, the L-A-S statements that implement the computations (4), (5), and (6) above are, respectively,

```
A,B (+)  = C     (7)
C,A (*)  = D     (8)
B  (-1)  = E     (9)
```

Because the operator follows the input variables, we say that single L-A-S operator statements are written in a *postfix* notation [5]. Also, the same variable name may be used in both input and output fields.

2. Multiple L-A-S Operator Statements

It is possible to perform more than one operation

per L-A-S statement. Let us continue with our simple example of linear matrix algebra. Suppose that the matrix C in (7) and (8) is computed only as an intermediate quantity. That is, the result of adding A and B in (7) is needed implicitly in the ensuing calculation (8), but not explicitly. Thus, we view C as a temporary variable.

Given this situation, it would be desirable to eliminate the explicit computation of the array C. Not only would this save user memory space, but in addition it would lessen the overall quantity of L-A-S computer code. This savings can be accomplished by entering the following *Multiple Operator Statement* (MOS) to the L-A-S interpreter.

```
A,B (+) ,A (*)  =  D                    (10)
```

To understand the effect of (10), one must understand the L-A-S parsing process. Upon receipt of any statement, the interpreter begins scanning left-to-right. When the first operator is encountered, in this case (+), L-A-S immediately performs the indicated operation. The result is stored in a temporary array, call it T1. Thus after one pass through the MOS, (10) becomes :

```
T1 ,A (*)  =  D                         (11)
```

and (7) has been performed. The scanning process is repeated until all operators are exhausted. Clearly, (10) is equivalent to (7) followed by (8), where the array C has been replaced by the temporary array T1.

USER-DEFINED L-A-S SUBROUTINES

A very useful feature of a programming language is a subroutine capability. The ability to write user-defined L-A-S subroutines permits the construction of structured, compact, and efficient L-A-S programs.

To illustrate this, consider again our simple example. Suppose that the calculations defined by (4), (5), and (6) are to be performed repeatedly throughout some more complex algorithm. Then the definition of a subroutine that performs this task is a rational solution. Assume that, in this subroutine, we wish to use the MOS version of (7) and (8) given by (10). For lack of a more meaningful name, we will call our routine XYZ. Then, the L-A-S subroutine definition for XYZ takes the following form :

```
   A,B (XYZ,SUB)  =  D,E                (12)
 A,B (+), A (*)   =  D                  (13)
         B (-1)   =  E                  (14)
          (RET)   =                     (15)
```

The subroutine XYZ is declared by (12). There are two array names in the input field, (A & B), and two names in the output field, (D & E). The body of the routine consists of (13) and (14). Finally, a return to the calling routine in the L-A-S language is performed by executing the RETurn operator displayed in (15). Typical calls to this subroutine might look like :

```
   A,B (XYZ,SUB)  =  D,E                (16)
```

or

```
   R,S (XYZ,SUB)  =  A1,A2              (17)
```

or even

```
   Y,X (XYZ,SUB)  =  Z,A .              (18)
```

L-A-S subroutines are "call-by-address". Thus, the *address* of the named array is passed to the subroutine and that array is subsequently operated on directly during the execution of the routine.

Table 1 Summary of L-A-S Operators by Groups

INPUT / OUTPUT OPERATORS

(CDI),	(DDM),	(DIM),	(DIS),	(DPM),
(DSC),	(DSM),	(DVC),	(DZM),	(INP),
(NIK),	(OUT),	(PLL),	(PLT),	(RDF),
(RDI),	(RGE),	(TXT),	(WDF),	(WGE)

DATA HANDLING OPERATORS

(CTC),	(CTI),	(CTR),	(EMD),	(MCP),
(RTI),	(SHD),	(SHL),	(SHR),	(SHU),
(TVC)				

LINEAR ALGEBRA OPERATORS

(+),	(-),	(*),	(-1),	and
(ATG),	(CHD),	(CHE),	(COS),	(CUR),
(EGC),	(EGV),	(EXP),	(F*),	(F/),
(GIN),	(INV),	(JFR),	(JFS),	(LOG),
(MEA),	(NRC),	(NRR),	(NSP),	(P+),
(P*),	(PMA),	(PMM),	(PNR),	(RKC),
(RKR),	(RSP),	(S*),	(SIN),	(SLE),
(SQR),	(SVC),	(SVD),	(T),	(TR)

CONTROL SYSTEM OPERATORS

(BOD),	(CLS),	(COT),	(CRE),	(CRL),
(DRE),	(DRL),	(DRO),	(EAT),	(GBN),
(GS),	(KFL),	(LAP),	(LYP),	(MIN),
(MTF),	(NYQ),	(OBS),	(OFP),	(PPL),
(RCS),	(RCT),	(RDS),	(RDT),	(RIC),
(RIK),	(RLC),	(STR),	(TZS)	

L-A-S PROGRAM CONTROL OPERATORS

(ELM),	(IFJ),	(JMP),	(LIS),	(NLI),
(NOP),	(NTE),	(NTR),	(NTY),	(RET),
(STO),	(TES),	(TRA),	(TYP)	

EXAMPLE

In this section, we present a detailed example of how the L-A-S language can be used to implement a discrete-time matrix Lyapunov equation algorithm. Here, we will illustate the form or structure of the L–A–S program. The text that is printed during an execution of an L-A-S program is dealt with in the next section.

The fixed-point, X, of a discrete-time Lyapunov equation satisfies

$$X \;=\; A^T \cdot X \cdot A \;+\; Q\,. \tag{19}$$

where X, and Q are positive semi-definite symmetric matrices. The matrix X can be calculated by the recursive formula

$$X_{i+1} = A^T \cdot X_i \cdot A + Q \tag{20}$$

provided that the iterates, X_i, converge to a limit, X. This occurs if and only if $\lambda_{max}(A) < 1$. Further, we must define a tolerance for convergence of this algorithm. Call the tolerance variable EPS. For the initial guess, we take $X_0 = A$.

The corresponding L-A-S program written entirely with single operator statements is :

```
;
;   Discrete-time Lyapunov Equation
;              Solver
;   Single Operator Statement
;             Version
;
                (INP)      =   A,Q
                (DSC)      =   EPS
              A (MCP)      =   X
              A (T)        =   AT
;
Q :      AT,X (*)          =   ATX
        ATX,A (*)          =   T
          T,Q (+)          =   TT
         TT,X (-)          =   EM
           EM (MEA,T)      =   EMS
           TT (MCP)        =   X
      EMS,EPS (IFJ)        =   R,R,Q
;
R :     A,Q,X (OUT,E)      =
```

The mnemonics deserve some additional explanation. Table 2 contains a one-line description of all operators used in the program above. A complete guide to all existing operators may be found in [8] or more recently [9].

Table 2 Short Description of L-A-S Operators Used in the Lyapunov Algorithm

+	Matrix addition
-	Matrix subtraction
*	Matrix multiplication
DSC	Define SCalars (input from keyboard)
IFJ	IF Jump (conditional jump)
INP	Array INPut from terminal keyboard
MCP	Matrix CoPy
MEA	Maximum Element in Absolute value
OUT	OUTput the results on the terminal screen
T	Matrix Transposition

An MOS version of this same algorithm would look like :

```
;
;   Discrete-time Lyapunov Equation
;              Solver
;   Multiple Operator Statement
;             Version
;
                          (INP)     = A,Q
                          (DSC)     = EPS
                        A (MCP)     = X
                        A (T)       = AT
;
Q : AT,X (*) ,A (*) ,Q (+)          = TT
                     TT,X (-)       = EM
                     EM (MEA,T)     = EMS
                     TT (MCP)       = X
                EMS,EPS (IFJ)       = R,R,Q
;
R :           A,Q,X (OUT,E)         =
```

Now, assume that the discrete-time Lyapunov equation solver algorithm must be performed on several different matrices. Then it would be desirable to make the iterative portion of the computation into an L-A-S user-defined subroutine which could be called as often as needed. Such a routine might take the following form:

```
;
;   Discrete-time Lyapunov Equation
;              Solver
;         Subroutine Version
;
              A,Q,EPS (LYP,SUB)    = X
                    A (MCP)        = X
                    A (T)          = AT
;
Q : AT,X (*) ,A (*) ,Q (+)         = TT
                  TT,X (-)         = EM
                  EM (MEA,T)       = EMS
                  TT (MCP)         = X
             EMS,EPS (IFJ)         = R,R,Q
;
R :                  (RET)         =
```

Then the MOS program written earlier would be equivalent to:

```
;
;   Discrete-time Lyapunov Equation
;              Solver
;   Multiple Operator Statement
;             Version
;   Calls L-A-S subroutine : LYP
;
                 (INP)    =   A,Q
                 (DSC)    =   EPS
;
  A,Q,EPS (LYP,SUB)       =   X
    A,Q,X (OUT,E)         =
```

L-A-S ECHO FEATURE

The previous section has emphasized the structure and form of L-A-S programs. The problem of solving for the fixed-point of a discrete-time Lyapunov equation was formulated and reformulated using various features of the L-A-S language. In this section, we focus on the execution of L-A-S programs.

Suppose we run the MOS version of the Lyapunov algorithm. For brevity, we consider a second order example with

$$A = \begin{vmatrix} 0.5 & 0.1 \\ -0.05 & -0.3 \end{vmatrix}, \quad Q = \begin{vmatrix} 1 & 0 \\ 0 & 1 \end{vmatrix}.$$

Then, if EPS is chosen to be 0.01, we obtain the following run :

```
* ;
* ; Discrete-time Lyapunov Equation
* ;          Solver
* ; Multiple Operator Statement
* ;          Version
```

```
* ;
* (INP)=A,Q
Enter dimensions for  < A > : 2,2
Enter dimensions for  < Q > : 2,2
Matrix < A > ;
 Enter E,R,C,D,M,I,Z,P,N or H for HELP : H
E means Element, R means Row, C means Column,
D means Diagonal, M means Matrix, I means Identity,
Z means Zero, P means Pseudorandom matrix,
 and N means No matrix.
Matrix < A > ;
 Enter E,R,C,D,M,I,Z,P,N or H for HELP : M
      A
 0.5 , 0.1
-0.05,-0.3
Matrix < Q > ;
 Enter E,R,C,D,M,I,Z,P,N or H for HELP : I
* (DSC)=EPS
      EPS
0.01
* A(MCP)=X
* A(T)=AT
* ;
* Q:AT,X(*),A(*),Q(+)=TT
* TT,X(-)=EM
* EM(MEA,T)=EMS
Maximum Element of EM  =  1.2765e+00

* TT(MCP)=X
* EMS,EPS(IFJ)=R,R,Q
* Q:AT,X(*),A(*),Q(+)=TT
* TT,X(-)=EM
* EM(MEA,T)=EMS
Maximum Element of EM  =  1.5936e-01

* TT(MCP)=X
* EMS,EPS(IFJ)=R,R,Q
* Q:AT,X(*),A(*),Q(+)=TT
* TT,X(-)=EM
* EM(MEA,T)=EMS
Maximum Element of EM  =  3.7564e-02

* TT(MCP)=X
* EMS,EPS(IFJ)=R,R,Q
* Q:AT,X(*),A(*),Q(+)=TT
* TT,X(-)=EM
* EM(MEA,T)=EMS
Maximum Element of EM  =  9.3249e-03

* TT(MCP)=X
* EMS,EPS(IFJ)=R,R,Q
* R:A,Q,X(OUT,E)=
      A
 5.00000e-01  1.00000e-01
-5.00000e-02 -3.00000e-01

      Q
 1.00000e+00  0.00000e+00
 0.00000e+00  1.00000e+00

      X
 1.32925e+00  7.17904e-02
 7.17241e-02  1.10868e+00

* END
L-A-S STOP
```

Notice the interactive nature of the L-A-S interpreter. It prompts the user by displaying a "*". After determining that an INP operator was issued, L-A-S asks the user for the dimensions of the matrices A and Q. Then the matrices are entered. Next, the value of EPS is specified. Since EPS is a scalar, its dimensions are known. Finally, L-A-S executes the Lyapunov algorithm, displaying each operator statement as it is performed.

Often, it is desirable to suppress the printing of each line of L-A-S program execution. This may be accomplished by issuing the NLI (No LIsting) interpreter command (see Table 3). Then only comments, and the data of the OUT operator or statements with a "T" or "E" option get printed.

Let us run the same L-A-S program with EPS = 0.0001 while using the NLI command. The result is :

```
* NLI
* ;
* ; Discrete-time Lyapunov Equation
* ;           Solver
* ; Multiple Operator Statement
* ;           Version
* ;
Enter dimensions for  < A > : 2,2
Enter dimensions for  < Q > : 2,2
Matrix < A > ;
 Enter E,R,C,D,M,I,Z,P,N or H for HELP : M
      A
 0.5 , 0.1
-0.05,-0.3
Matrix < Q > ;
 Enter E,R,C,D,M,I,Z,P,N or H for HELP : I
      EPS
0.0001

* ;
Maximum Element of EM  =  1.2765e+00
Maximum Element of EM  =  1.5936e-01
Maximum Element of EM  =  3.7564e-02
Maximum Element of EM  =  9.3249e-03
Maximum Element of EM  =  2.2469e-03
Maximum Element of EM  =  5.5131e-04
Maximum Element of EM  =  1.3384e-04
Maximum Element of EM  =  3.2699e-05
      A
 5.00000e-01  1.00000e-01
-5.00000e-02 -3.00000e-01

      Q
 1.00000e+00  0.00000e+00
 0.00000e+00  1.00000e+00

      X
 1.33221e+00  7.20706e-02
 7.20706e-02  1.10879e+00

* END
L-A-S STOP
```

The savings in terms of the number of lines printed is evident. Also, a more accurate estimate of the fixed-point to this problem is obtained.

Now, it is clear that before program execution proceeds automatically, only a few lines are typed to the L-A-S interpreter at the beginning of the Lyapunov algorithm. It is reasonable to assume that an L–A–S user may repeatedly study a single problem with one or more sets of data. On such an occasion, it would be useful to extract the input that the user enters and store it in a file for retrieval at a later time. This would spare the user from the tedious task of re-entering several variables that never change throughout the study. This concept is captured and realized by the L-A-S "echo" feature.

The "echo" file for the last run is :

```
NLI
2,2
2,2
M
 0.5 , 0.1
-0.05,-0.3
```

I
0.0001

Once a user has an L-A-S program ready for execution, he/she may instruct the interpreter to take input from the "echo" file by simply typing the ECHO command (Table 3). Then, execution begins immediately. If an end-of-file is detected then this condition is indicated to the user and control is transferred back to the normal interactive mode. In fact, the previous two runs were made using the ECHO command, however this was not explicitly shown in the listings.

Table 3 Short Description of L-A-S Interpreter Commands

BEGIN	Begin a new L-A-S work session by clearing internal storage
CHANGE	Change an L-A-S program statement
CONTINUE	Continue the current L-A-S program
DUMP	Dump all matrices to the user
ECHO	Invoke the L-A-S echo feature
ELIMINATE	Eliminate one or more L-A-S program statements
ELMATRIX	Eliminate some matrices from internal storage
END	End this session with L-A-S
HELP	Make an on-line help inquiry
INCLUDE	Include an L-A-S program statement
JUMP	Jump to a particular L-A-S program statement
LIST	List the execution of an L-A-S program
MEMORIZE	Memorize subsequent program statements entered
NAMES	Name all matrices defined thusfar
NLIST	Turn off LIST mode
NMEMORIZE	Turn off MEMORIZE mode
NTEST	Turn off TEST mode
NTRACE	Turn off TRACE mode
NTYPE	Turn off TYPE mode
PRLISTING	Print the current L-A-S program to results file
PROGRAM	Display some lines of the current L-A-S program
RPF	Read an L-A-S Program File
STATUS	Status information about the current work session
TCHANGE	Total (global) CHANGE command
TEST	Enter L-A-S self-test mode. (For porting L-A-S)
TRACE	Enter L-A-S program debug mode
TYPE	Enter output typing mode
VERSION	Inquire about the current version of L-A-S being used
WPF	Write an L-A-S Program File

CONCLUSION

We have described the highlights of the L-A-S programming language. The fundamental concept of an L-A-S operator is introduced. The utility of operators as algorithm building blocks is established. The syntax of single and multiple L-A-S operator statements are discussed. L-A-S user-defined subroutines are mentioned. An entire section of this paper has been dedicated to a detailed example of L–A–S language use. This serves to unify the concepts of previous sections and simultaneously motivate the new L-A-S "echo" feature. Finally, several references [1,2,6-9] have been included should the reader desire additional information about L-A-S.

REFERENCES

[1] S. P. Bingulac and M. A. Farias, "L-A-S (Linear Algebra and System) Language and Its Use in Control Education and Research," *Computing*, Springer-Verlag, vol. 23, 1979, pp. 1-23.

[2] S. P. Bingulac and N. Gluhajic, "Computer Aided Design of Control Systems on Mini Computers using the L-A-S Language," Proc. of 1982 IFAC Symposium on Computer Aided Design of Multivariable Technological Systems, Purdue University, Indiana, September 15-17, 1982.

[3] A. Aho and J. Ullman, *The Theory of Parsing, Translation and Compiling*, Prentice-Hall, Englewood Cliffs, N.J., 1972.

[4] J. Backus, "The syntax and semantics of the Proposed International Algebraic Language of Zurich ACM-GAMM Conference," *Information Processing*, UNESCO, Paris, 1960, pp. 125-132.

[5] T. W. Pratt, *Programming Languages: Design and Implementation*, Prentice-Hall, Englewood Cliffs, N.J., 1975.

[6] S. P. Bingulac, "Recent Modifications in the L-A-S (Linear Algebra and Systems) Language and Its Use in CAD of Control Systems," Proc. of 1983 Allerton Conference on Communication, Control, and Computing, Monticello, Illinois, October 1983, pp. 393-402.

[7] S. P. Bingulac and P. J. West, "Computer-Aided Design of Control Systems : An Example of Working Software," Proc. of 1983 Allerton Conference on Communication, Control, and Computing, Monticello, Illinois, October 1983, pp.423-424.

[8] S. P. Bingulac, J. H. Chow, S. H. Javid and H. R. Dowse, "User's Manual for Linear Algebra and Systems (L-A-S) Language," Report No. 83-EUE-205, (Internal Report), Electric Utility Systems Engineering Department, General Electric, Schenectady, New York, 12345, November 1983.

[9] P. J. West, S. P. Bingulac, and W. R. Perkins, " L-A-S : A Computer-Aided Control System Design Language," *Advances in Computer-Aided Control Systems Engineering*, North-Holland, New York, 1985.

Acknowledgement - This work was supported in part by the Joint Services Electronics Program under Contract N00014-84-C-0149, and in part by the University of Belgrade, Department of Organizational Sciences.

OUTLINE OF A NEW APPROACH TO COMPUTER AIDED DESIGN OF CONTROL SYSTEMS

H. K. Eldeib

Department of Electrical and Computer Engineering, George Mason University, 4400 University Drive, Fairfax, VA 22030, USA

Abstract. This paper presents an outline of an ongoing effort to develop a new approach to computer aided design of control systems. The new approach integrates three computer aided techniques for problem solving: artificial intelligence (expert systems), decision analysis, and standard routines for control systems analysis and design. The new approach will offer guidance to inexperienced users on how to approach a design problem. A special subproblem; the design of a position control system for for a printwheel mechanism is used as an initial design problem that the prototype expert decision aid should solve.

Keywords. Artificial Intelligence; computer aided design; control system analysis; decision analysis.

INTRODUCTION

Computer aided analysis and design is becoming an essential part of control systems engineering. Both universities and research institutes (Astrom, 1983; Edmunds, Jeans, and Maciejowski, 1983; Haerget and Weiss, 1980; Maddux, 1982; Weiss, 1980) as well as industries (Boeing Computer Services; Integrated Systems Inc.; Systems and Control Inc.; and others) have developed computer programs to aid control engineers in analysis and design.

Some of these programs are designed for interactive use and are typically menu driven, while others are designed to use a list of commands instructing the program to perform a sequence of tasks. The first approach is sometimes frustrating since the computer controls the questionning of the user, while the second approach is not very user friendly.

Most of these programs consist of a set of tools for analysis and design such as frequency response computations routines, graphics routines, state-feedback design routines, etc. These tools offer excellent help to experienced designers who can choose an approach to design, i.e. who can choose the appropriate set of tools and the sequence in which to use them.

Less experienced designers, however, may find the choice of a design approach quite challenging. Students of control systems represent a large group of inexperienced designers that may need more help from the computer than the set of tools. There is a need for computer programs that can offer guidance to designers, if needed, on the appropriate approach to solve a given design problem.

Artificial intelligence has been used successfully to develop expert systems that offer needed guidance to physicians in medical diagnosis, to computer engineers in system configuration, as well as to industrial and military planners and decision makers (Buchanan and Shortliffe, 1984; Winston and Prendergast, 1984; and their references). It is our contention that expert systems can offer the needed guidance to control engineers in analysis and design of systems to meet desired specification.

REVIEW OF EXPERT SYSTEMS

Artificial intelligence is the science of developing machines capable of performing tasks that are usually achieved by intelligent humans. Considerable success has been achieved in the last decade in the area of artificial intelligence known as "expert systems". This area is concerned with methods and techniques for constructing man-machine systems with specialized problem solving expertise (Hayes-Roth, Waterman, and Lenat, 1983).

Several applications of expert systems are reviewed in (Buchanan and Shortliffe, 1984; Winston and Prendergast; and their references). The application areas are classified into: diagnosis, design, planning, interpretation, prediction, monitoring, debugging, repair, and control.

An ideal expert system consists of a knowledge base containing a set of facts and rules, an inference engine, and a user interface. In a typical application, the user provides a description of a problem to the system and receives advice and explanation (Buchanan and Shortliffe, 1984). The evolution of an expert system is a gradual process consisting of five stages: Identification, Conceptualization, Formalization, Implementation, and Testing (Hayes-Roth, Waterman, and Lenat, 1983).

EXPERT SYSTEMS IN CONTROL SYSTEMS DESIGN

There is a great potential for use of expert systems in control systems engineering. Possible applications include aiding the control engineer in problem formulation, in modeling of systems, in design of feedback control and/or compensation to meet specified requirements, as well as in implementation of "expert controllers". There is a growing number of reports on new research projects aimed at realizing the existing potential. For example, the use of MACSYMA, an expert system for mathematical manipulations, in the manipulation of Kalman filter matrix equations has been discussed in (AGORA Systems, 1984), and the use of expert controllers in reconfigurable control systems has been oulined in (Stengel, 1984).

In this section we outline a new approach for developing an expert system for analysis and design of control systems. The outline will follow the evolution stages suggested in (Hayes-Roth, Waterman, and Lenat, 1983). Only the first three stages of development will be discussed in this paper, i.e. identification, conceptualization, and formulation.

The Identification Stage:

During this stage the problem area is identified, and the scope of the problem is defined. A small but interesting subproblem may be identified and used to focus the knowledge acquisition process. In our case, the problem at hand is to develop an expert system for aiding in the design of a control system to meet certain requirements. This problem is very broad, and can be considered a long-term objective. To start our project the following simple but representative subproblem has been chosen:
" It is desired to design a position control system for a printwheel mechanism (Kuo, 1982) to satisfy specific performance criteria. The input to the system is a desirable angular position and the output is the actual angular position of the printwheel.

The system consists of a difference amplifier, a power amplifier, an armature controlled dc motor, and a printwheel as a load to the motor. The system is described by a mathematical model whose parameters are to be provided by the user. Acceptable types of mathematical models are rational transfer functions and state variable models.

Performance specifications are given in terms of steady state error and transient response to a step input. Transient response specifications may include percentage overshoot, rise time, and settling time. Acceptable design procedures include choice of the power amplifier gain K, design of cascade compensators such as phase lead, phase lag, or PID, and state feedback compensation.

The system should offer guidance to the user if needed, and should contain all the analysis and design routines needed to implement the chosen design procedures."

The Conceptualization Stage:

During this stage, key concepts, relations, and information flow characteristics needed to describe the problem solving process are explicated, and subtasks, strategies, and constraints are specified. In this paper we will give only a brief discussion of some of the key concepts in the development of our expert system.

The proposed expert system will contain the following components:
1) User Interface: This part of the program will be responsible for conducting an English-like dialog with users. The basic tasks of the user interface are:

- To ask the user for a description of the system and record that description with appropriate labels, e.g. type of model, order of model, etc.

- To ask the user for desired performance specifications, record them with appropriate labels, and pass them to the decision unit (inference engine).

- To inform the user of suggested actions for analysis and design, and ask for his/her approval. If approval is granted, appropriate information should be sent to both user and decision unit.

2) Decision Unit (Inference Engine): This part of the program will be for analyzing of information contained in English-like lists, and suggesting an action involving use of some of the analysis and design procedures available in the knowledge base. An example of an inference engine under development for computer aided design of microprocessor controllers at the Naval Postgraduate school is given in (Carson, 1984). In that system, the desired requirements were supplied using a special computer language, and a translator from this language to lists that are compatible with the knowledge base was developed. Our objective is to avoid the need for translation by a good choice of an expert system building tool.

3) Knowledge Base: This part of the expert system contains a set of procedures and rules that combine standard routines for analysis (Frequency response computations, steady state error calculations, zero and pole finding, etc.) and a set of instructions for design based on experience and rules of thumb. Most introductory textbooks on control systems analysis and design contain some rules of thumbs and heuristics for the design of compensators (Dorf, 1983; Kuo, 1982).

The knowledge base will also contain descriptions of procedures that will enable the decision unit to decide whether a procedure can be used to solve a particular analysis or design problem or not. These descriptions will be based on textbook information and on experience of designers who used these procedures. For example, the following description of a phase-lead compensation network (Dorf, 1983) can be used as a basis for a similar description in the knowledge base:

Name: Phase Lead Compensation.
Function: Adds phase angle near crossover frequency or to yield desired dominant roots in the s-plane.
Results: Increases system bandwidth, and Increases gain at high frequencies.
Advantages: Yields desired transient response,and Can achieve fast transient response.
Useful: When fast rise time is desired.
Not Useful: When phase decreases rapidly near the crossover frequency.

Descriptions of the procedures will be used by the decision unit to choose among procedures, and hence the descriptions and the decision making rules will be designed such that they "talk the same language".

Formalization Stage:

This stage involves mapping the key concepts and relations into a formal representation suggested by some expert system building tool or language. We are currently studying the feasibility of formalization of the proposed expert system using an expert system development language called ROSIE that was developed at the RAND Corporation (Fain and others, 1982). ROSIE operates in an INTERLISP environment on a VAX 11/780. Although it seems that ROSIE is a good choice for building the proposed expert system, the final decision on the language of implementation has not been made yet. A good choice of a language can save a lot of effort and improve the performance of the new system.

ROSIE has been described as an excellent general purpose tool for building prototype expert systems. Its main strength is its English-like syntax, and its powerful pattern matching capability which enables it to access local operating systems and thus other languages (Hayes-Roth, Waterman, and Lenat, 1983). This capability can be utilized to access FORTRAN programs, for example, but this is currently possible only within a UNIX operating system. This capability makes ROSIE particularly attractive since a lot of control systems routines are available in FORTRAN code.

DECESION ANALYSIS IN CONTROL SYSTEMS DESIGN

The process of control systems design can be viewed as a sequential decision making process. The design process involves a number of stages and the designer has to choose some action concerning the design procedure at each stage. There is a cost or reward incurred at each stage that is a function of the decision chosen, the outcome, and the objectives of the designer. The designer's objective is to satisfy several, sometimes conflicting performance and design criteria. This suggests that the design problem can be modeled as a problem in multiple criteria decision making (Starr and Zeleny, 1977).

In an optimal design process, the designer chooses a sequence of design procedures to achieve the design objectives to the best possible extent, as well as to optimize some performance measures for the design process. For example, computer time is costly, and hence one of the design criteria may be minimizing the computer time utilized. Another criterion may be minimizing the cost of realizing and implementing the controller designed.

The integration of decision analysis in control system design requires two prerequisites: the design process should be formulated in a highly structured manner as a decision problem, and some information concerning the costs associated with different actions should be known. This will require a thorough analysis of the design space, the action space, as well as the value system of the designer. There is a great deal of research reported in the literature and in textbooks on decision making procedures, (for example: Bertsekas, 1976; Keeney and Raiffa, 1976; Sage, 1977; and their references). We are not aware, however, of any application of decision aiding to control systems design. The reason seems to be that current approaches to design of control systems may not possess the structure needed for decision analysis.

AN EXPERT DECISION AID FOR CONTROL SYSTEMS DESIGN

Integration of the three problem solving approaches discussed above, i.e. control systems program packages, expert systems, and decision analysis, requires a great deal of basic research in the following issues:

- Formulation of the control systems design problem as a decision making problem.
- Incorporating information about the costs and rewards associated with different actions into the knowledge base.
- Incorporating some decision making mechanism in the decision unit of the expert system in order to optimize the design process.

These topics will be the target of our future research efforts.

CONCLUSION

We have presented an outline of an ongoing research project aimed at developing an expert decision aid for the analysis and design of control systems. The project is still in its early stages, and hence a lot of future research is needed. In this paper we described an approach to developing the expert system by discussing three stages of the evolution stages of the system: identification, conceptualization, and formalization. We suggested some of the topics that should be investigated in order to complete the development of the system.

REFERENCES

AGORA Systems, Inc., "Artificial Intelligence/ Expert Systems for Inertial Navigation Guidance and Control", Technical Report, AGORA Systems,Inc. Gaithersburg, MD, 1984.

Astrom, K. J., "Computer Aided Modeling, Analysis, and Design of Control Systems- A Perspective", <u>IEEE Control Systems Magazine</u>, May 1983, 4-16.

Bertsekas, D. P., <u>Dynamic Programming and Stochastic Control</u>, Academic Press, New York, 1976.

Buchanan, B. G., and E. H. Shortliffe, ed., <u>Rule Based Expert Systems</u>, Addison Wesley, Reading, MA, 1984.

Carson, T. H., "An Input Translator for a Computer Aided Design System", M.Sc. Thesis, Naval Postgraduate School, Monterey, CA, 1984.

Dorf, R. C., <u>Modern Control Systems</u>, Addison Wesley, Reading, MA, 1983.

Edmunds, J. M., S. E. Jeanes, and J.M. Maciejowski "CLADP (Cambridge Linear Analysis and Design Program) User Reference Manual", Technical Report, Cambridge University, England, 1983.

Fain, J., and others, "The ROSIE Language Reference Manual", Technical Rept. N-1647-ARPA, Rand Corp., Santa Monica, CA, 1981.

Fain, J. , and others, "Programming in ROSIE: An Introduction by Means of Examples", Technical Rept. N-1646-ARPA, Rand Corp., Santa Monica, CA, 1982.

Hayes-Roth, F., D. A. Waterman, and D. B. Lenat, <u>Building Expert Systems</u>, Addison Wesley, Reading, MA, 1983.

Herget, C. J., and T. P. Weis, "Linear Systems Analysis Program User's Manual", Technical Rept. UCID-30184, Lawrence Livermore Lab., Livermore, CA, 1980.

Keeney, R. L., and H. Raiffa, Decisions With Multiple Objectives: Preferences and Value Trade-offs, John Wiley, New York, 1976.

Kuo, B. C., Automatic Control Systems, Prentice-Hall, Englewood Cliffs, NJ, 1982.

Maddux Jr., A. S., "Powerful Tool for Design and Analysis of Control Systems", Technical Report UCID-19373, Lawrence Livermore Lab., Livermore, CA, 1982.

Nillson, N. J., Problem Solving Methods in Artificial Intelligence, McGraw-Hill, New York, 1971.

Sage, A. P., Methodology for Large Scale Systems, McGraw-Hill, New York, 1977.

Starr, M. K., and Zeleny, M., Multiple Criteria Decision Making, North-Holland, Amsterdam, 1977.

Stengel, R. F., Artificial Intelligence Theory and Reconfigurable Control Systems", Technical Report 1664-MAE, Dept. of Mechanical Engineering, Princeton University, Princeton, N.J., 1984.

Weis, T. P., "Linear Systems Analysis Program, Programmer's Manual", Technical Report UCID-30183, Lawrence Livermore Laboratory, Livermore, CA, 1981.

Winston, P. H., and K. A. Prendergast, ed., The AI Business, MIT Press, Cambridge, Mass., 1984.

CADOCS — COMPUTER AIDED DESIGN ON MULTIVARIABLE CONTROL SYSTEMS — METHODS AND THE SOFTWARE PACKAGE

H. Hensel

Institute für Regelungstechnik, Technische Hochschule Darmstadt, Schlossgraben 1, D-6100 Darmstadt, FRG

Abstract. A CAD-package serving for the design of discrete-time multivariable control systems is presented. Various design methods for different feedback controller types (DB, PID, SC) are described based on discrete-time parametric process models which are obtained by On-line identification methods. The concept of the software package contains an interpreter and a database offering high flexibility in the design and coordinating the design modules. A design example starting with the identification and ending with a DDC-implementation will be shown.

Keywords. Multivariable discrete-time control systems; computer aided system design; direct digital control; interpreter based software system; parameter estimation, PID control; state space methods.

INTRODUCTION

Modern mathematical design methods for multivariable control systems numerously proposed in the literature often require the application of a computer which may as well be used for the management of the whole design procedure thus considerably reducing the amount of necessary design time. Almost all mathematical design methods require the a priori knowledge of parametric or nonparametric process-models. This modeling is just as difficult and time consuming as the controller design and may as well be supported by computer methods. Especially in the case of parametric modeling efficient parameter estimation procedures have been developed for On-line identification on a process-computer (Isermann, 1981). Moreover using this type of computer permits the implementation of direct digital control.

A software package serving for the topics mentioned above should be very flexible on one side but handy on the other and should support the user on a very high level. The design facilities should be modularized and concatenated by a superior centralized management system. In order to consider varying design requirements the package must be easy modifiable without changing the central management package.

This paper will first describe some algorithms for the On-line identification and the controller design and then continue with the description of a central interpreter based CAD-management package and a management package for DDC-purposes.

The identification and controller design methods described in this paper are based on the discrete-time multivariable process model

$$y_i(k) = \sum_{l=1}^{m_i} [-a_{iil}\, y_i(k-l) + \sum_{j=1}^{r} b_{ijl}\, u(k-d_{ij}-l)], \quad i = 1,\dots,q \tag{1}$$

a simplified p-canonical model with r inputs and q outputs, the order m_i at the i-th output and the discrete deadtime d_{ij} from the j-th input to the i-th output. The model can as well be described in the z-domain by

$$\underline{A}(z^{-1})\, \underline{y}(z) = \underline{B}(z^{-1})\, \underline{u}(z) \tag{2}$$

with the diagonal polynomial matrix $\underline{A}(z^{-1})$ and the general polynomial matrix $\underline{B}(z^{-1})$ which contains the deadtimes additionally.

Forming a state model from Eq.(1) often leads to a nonminimal representation because of possible multiple considerations of eigenvalues or common deadtimes. Therefore a modified algorithm for obtaining a minimal realization given in Nour Eldin (1977) is applied resulting in a state model of minimal order which contains the deadtimes d_{ij}

$$\underline{x}(k+1) = \underline{A}\,\underline{x}(k) + \underline{B}\,\underline{u}(k) \tag{3}$$

$$\underline{y}(k) = \underline{C}\,\underline{x}(k) \tag{4}$$

ON-LINE IDENTIFICATION

In order to obtain parametric models of the process two identification methods (least squares (LS) and correlation analysis and least squares (COR-LS) are used in the package. Both methods are (finally) based on driving the square of the equation error $e_i^2(k)$ to a minimum. For this purpose a more suitable description of the model (Eq.(1)) is given by

$$y_i(k) = \underline{\psi}_i^T(k)\hat{\underline{\theta}}_i + e_i(k), \quad i = 1,\dots,q \tag{5}$$

and the equation error

$$e_i(k) = y_i(k) - \hat{y}_i(k) = y_i(k) - \underline{\psi}_i^T(k)\hat{\underline{\theta}}_i, \quad i = 1,\dots,q \tag{6}$$

with $\hat{y}_i$ and $\hat{\underline{\theta}}_i$ denoting the predicted i-th output and the estimated parameters respectively and

$$\underline{\psi}_i^T(k) = [-y_i(k-1),\dots,-y_i(k-m_i),\, u_1(k-d_{i1}-1),\dots u_1(k-d_{i1}-m_i),\, u_2(k-d_{i2}-1),\dots,u_2(k-d_{i2}-m_i),\dots,u_r(k-d_{ir}-m_i)] \tag{7}$$

$$\hat{\underline{\theta}}_i = [\hat{a}_{ii1},\dots,\hat{a}_{iim_i},\hat{b}_{i11},\dots,\hat{b}_{i1m_i},\hat{b}_{i21},\dots,\hat{b}_{i2m_i},\dots,\hat{b}_{irm_i}]^T \tag{8}$$

Minimizing the loss function

$$V = \sum_{k=m+d}^{m+d+N} e_i^2(k) \rightarrow \min \tag{9}$$

yields the LS-equations

$$\hat{\underline{\theta}}_i = [\underline{\Psi}^T\underline{\Psi}]^{-1} \underline{\Psi}^T\underline{y} \quad , \ i = 1,\ldots,q \tag{10}$$

In the presence of arbitrary stochastic disturbances with zero mean biased parameters can be avoided by using the two stage method COR-LS. This method first computes the autocorrelation function $\phi_{uu}(\tau)$ and the crosscorrelation function $\phi_{uy}(\tau)$. Due to linear system theory the input signal u(k) can be substituted by $\phi_{uu}(\tau)$ and the output signal y(k) by $\phi_{uy}(\tau)$. In the multivariable case the sum

$$u_s(k) = \sum_{j=1}^{r} u_j(k) \tag{11}$$

is used in order to build the functions $\phi_{u_s u_j}(\tau)$ and $\phi_{u_s y_i}(\tau)$ corresponding to Eq.(1). In the second step, Eq. (5-10) with the substituted signals $\phi_{u_s u_j}(\tau)$ and $\phi_{u_s y_i}(\tau)$ are used to estimate the parameters $\underline{\theta}_i$.

Some improvements are: An additional parameter C_i appended to the parameter vector $\underline{\theta}_i$ allows to consider steady state disturbances. Restricting the range of possible model orders and deadtimes and storing the data (in a compressed form) for the model of highest possible order and deadtimes allows to apply an order- and deadtime-search method before estimating the parameters (Maencher and Hensel, 1985). The identification of the whole multivariable process can be carried out simultaneously if special multistage pseudo random binary signals (MPRBS) which are orthogonal are applied to the process inputs (Blessing, 1979).

FEEDBACK CONTROLLERS

Three different types of multivariable controllers can be designed by the package

- generalized deadbeat controllers
- parameter optimized controllers
- state controllers including integral action and observers

Deadbeat Controllers

Consider a square multivariable process (r=q). The desired deadbeat-behaviour of all inputs and all outputs can be achieved by

$$\underline{y}(z) = \underline{P}(z^{-1})\ \underline{w}(z) \quad , \quad \underline{P}(1) = \underline{I} \tag{12}$$

$$\underline{u}(z) = \underline{Q}(z^{-1})\ \underline{w}(z) \tag{13}$$

with the reference signals $\underline{w}(z)$, if $\underline{P}(z^{-1})$ and $\underline{Q}(z^{-1})$ are rational polynomial matrices of finite order. Decomposing

$$\underline{Q}(z^{-1}) = \underline{C}(z^{-1})\ \underline{A}(z^{-1}) \tag{14}$$

and inserting Eq.(13) into Eq.(2) results in

$$\underline{P}(z^{-1}) = \underline{A}^{-1}(z^{-1})\ \underline{B}(z^{-1})\ \underline{C}(z^{-1})\ \underline{A}(z^{-1}) \tag{15}$$

The assumption of the finite order of $P(z^{-1})$ can be meet by choosing $\underline{C}(z^{-1})$ such that the matrix product $\underline{B}(z^{-1})\ \underline{C}(z^{-1})$ becomes diagonal. Hence Eq.(15) becomes

$$\underline{P}(z^{-1}) = \underline{B}(z^{-1})\ \underline{C}(z^{-1}) = \mathrm{diag}\ [p_{ii}(z^{-1})] \tag{16}$$

$C(z^{-1})$ can be calculated by

$$\underline{C}(z^{-1}) = K_c \cdot \mathrm{adj}[z^{\delta}\ \underline{B}(z^{-1})] \tag{17}$$

with $\delta = 1+\min(d_{ij},\ 1\le i,j\le r)$ generating the lowest possible order of the adjoint matrix and therefore of $\underline{P}(z^{-1})$. The gain factor K_c is introduced to satisfy the constraint $\underline{P}(1) = \underline{I}$

$$K_c = \frac{1}{\det[\underline{B}(1)]} \tag{18}$$

If $\det(\underline{B}(1)) = 0$ no deadbeat controller exists. The resulting control law turns out to

$$\underline{u}(z) = \frac{z^{\delta}}{z^{\delta}\ \det[\underline{B}(1)] - \det[z^{\delta}\underline{B}(z^{-1})]} \cdot \mathrm{adj}[z^{\delta}\underline{B}(z^{-1})]\ \underline{A}(z^{-1})\underline{e}(z) \tag{19}$$

Extension of Eq.(17) by a diagonal matrix

$$\underline{C}(z^{-1}) = K_c \cdot \mathrm{adj}[z^{\delta}\ \underline{B}(z^{-1})] \cdot \mathrm{diag}[f_{ii}(z^{-1})] \tag{20}$$

offers additional degrees of freedom, which may be used to minimize an energy cost functional or a control performance index like

$$J = \sum_{k=0}^{n} \sum_{i=1}^{r} (\delta e_i^2(k) + \sum_{j=1}^{r} [\eta u_{ij}^2(k) + \sigma \Delta u_{ij}^2(k+1)]) \tag{21}$$

with the controller order n, the partial manipulating variables $u_{ij}(k)$ from $u_i(k) = \sum_{j=1}^{r} u_{ij}(k)$, the partial manipulating variable deviations $\Delta u_{ij}(k+1) = u_{ij}(k+1) - u_{ij}(k)$, the control errors $e_i(k)$ and the weighting factors δ,η and σ. Hensel (1981) and Bayer (1982) have pointed out that the minimization of Eq.(21) with respect to the free parameters contained in diag $[f_{ii}(z^{-1})]$ leads to a direct analytical solution for the free parameters.

Parameter-Optimized Controllers

The straightforward numerical optimization of the complete multivariable controller is due to the large number of parameters often very time consuming or perhaps may result in unstable control loops. Therefore three different approaches for the design of discrete-time multivariable parameter-optimized PID controllers will be presented. The first approach follows a proposal by Föllinger (1972) reducing the optimization problem to the SISO-case. The second and third approach compute the PID-controller from the design of multivariable optimal state feedback controllers.

Approach 1. Consider the multivariable process given by Eq.(2). A controller structure with the coupling controllers behind of the main controllers is selected (cf.Fig.1). In the first step the coupling controllers are designed as feedforward controllers producing complete decoupling of the process, e.g. with r = q = 3 and $\underline{P}(z^{-1})$ and $\underline{C}(z^{-1})$ denoting the elementary transfer functions of the process and the controller respectively (the argument z^{-1} is omitted for simplification).

$$\begin{bmatrix} P_{11} & P_{12} & P_{13} \\ P_{21} & P_{22} & P_{23} \\ P_{31} & P_{32} & P_{33} \end{bmatrix} \begin{bmatrix} 1 & C_{12} & C_{13} \\ C_{21} & 1 & C_{23} \\ C_{31} & C_{32} & 1 \end{bmatrix} = \begin{bmatrix} P_{11}^* & 0 & 0 \\ 0 & P_{22}^* & 0 \\ 0 & 0 & P_{33}^* \end{bmatrix} \tag{22}$$

Depending on the resulting structure and the parameters of the controllers obtained from Eq.(22) low order coupling controllers of P, PD or PDT_1-structure are selected. The parameters of the latter controllers are tuned by LS-approximation of the dynamic behaviour of the exact decoupling controllers.

Neglecting the remaining couplings the main controllers of P-, PI- or PID-structure are designed in the second step as SISO-controllers for the modified main paths $P^*_{ii}(z^{-1})$, Eq.(22) taking the low order decoupling controllers into account. The parameters are optimized by means of a modified Hooke Jeeves pattern search together with an analytical calculation of the control performance index by means of Parseval's theorem (Aström, 1970)

$$I_i = \sum_{K=0}^{\infty} [e_i^2(k) + rK_p^2 \Delta u_i^2(k)] =$$

$$= \frac{1}{2\pi i} \int [e_i(z)e_i(z^{-1}) + rK_p^2 \Delta u_i(z)\Delta u_i(z^{-1})]\frac{1}{z}dz \quad (23)$$

with the control error $e_i(k)$, the manipulating variable deviations $\Delta u_i(k) = u_i(k) - u_i(\infty)$, the process gain K_p and a weighting factor r. The resulting control performance of the multivariable control loop mainly depends on the degree of decoupling achieved in step 1. Even instability may occur, if the partial decoupled transfer function matrix is not diagonal dominant (Rosenbrock, 1974).

To improve the control performance a third step introducing an iterative scheme known as successive design (Isermann and Hensel, 1985) is applied. Experimental examinations have pointed out that this method converges rapidly in a few successive cycles.

Approach 2. The state model of a square multivariable process (Eq.(3,4) with r=q) is extended by "integral states" which are appended to the outputs $\underline{y}$ of the process, cf. state controller structure 2. For this augmented system an optimal state controller is designed by solving the discrete matrix Riccati equation. The closed state control loop is further augmented by an arbitrary tuned reference signal filter (Fig.2) and stimulated at the reference inputs by a MPRBS proposed above. The parameters of the related PID-controllers with the main controllers parallel to the coupling controllers are tuned by approximating the control error signals $\underline{e}(k) = \underline{w}(k) - \underline{y}(k)$ and the manipulating variables $\underline{u}(k)$ of the state control loop by means of minimizing the cost functional

$$I = \sum_{l=k}^{k+N-1} \varepsilon^T(l)\varepsilon(l), \quad l \geq 2 \quad (24)$$

with

$$\varepsilon(l) = [\varepsilon_1(l), \varepsilon_2(l), \ldots, \varepsilon_r(l)]^T \quad (25)$$

$$\varepsilon_i(l) = u_i(l) - \hat{u}_i(l) \quad (26)$$

$$\hat{u}_i(l) = u_i(l-1) + \sum_{j=1}^{r} \sum_{s=0}^{2} q_{ij} e_j(l-s). \quad (27)$$

The minimization due to the PID-controller parameters is realized by using a modified LS-approach (cf.Fig.2). A detailed description is given in Hensel and Peter (1984).

Approach 3. The fixed controller structure of the PID-type with

$$\underline{u}(k) = \underline{u}(k-1) + \underline{Q}_0 \underline{e}(k) + \underline{Q}_1 \underline{e}(k-1) + \underline{Q}_2 \underline{e}(k-2) \quad (28)$$

is transformed to the state description

with $\underline{e}(k) = \underline{w}(k) - \underline{y}(k)$

$$\underline{r}(k+1) = \underline{A}_c \underline{r}(k) + \underline{B}_c(\underline{w}(k) - \underline{y}(k)) \quad (29)$$

$$\underline{u}(k) = \underline{C}_c \underline{r}(k) + \underline{D}_c(\underline{w}(k) - \underline{y}(k)) \quad (30)$$

Using the observer canonical form the system matrix $\underline{A}_c$ is fixed

$$\underline{A}_c = \begin{bmatrix} \underline{A}_1 & 0 \\ 0 & \underline{A}_r \end{bmatrix} \quad \underline{A}_i = \begin{bmatrix} 0 & 0 \\ 1 & 1 \end{bmatrix} \quad i = 1, \ldots, r \quad (31)$$

and the tunable parameters q_{ijl} are contained in the matrices $\underline{B}_c$ and $\underline{D}_c$. Because of the fixed matrices $\underline{A}_c$ and $\underline{C}_c$ the state model of the process can be augmented by the state model of the controller defining the system inputs $[\underline{u}^T(k), \underline{u}_c^T(k)]^T$ in order to group all tunable parameters in the feedback. This yields

$$\begin{bmatrix} \underline{x}(k+1) \\ \underline{r}(k+1) \end{bmatrix} = \begin{bmatrix} \underline{A} & \underline{B}\,\underline{C}_c \\ \underline{0} & \underline{A}_c \end{bmatrix} \begin{bmatrix} \underline{x}(k) \\ \underline{r}(k) \end{bmatrix} + \begin{bmatrix} \underline{B} & \underline{0} \\ \underline{0} & \underline{I} \end{bmatrix} \begin{bmatrix} \underline{u}(k) \\ \underline{u}_c(k) \end{bmatrix} \quad (32)$$

$$\underline{y}(k) = [\underline{C} \quad \underline{0}] \begin{bmatrix} \underline{x}(k) \\ \underline{r}(k) \end{bmatrix} \quad (33)$$

$$\begin{bmatrix} \underline{u}(k) \\ \underline{u}_c(k) \end{bmatrix} = - \begin{bmatrix} \underline{D}_c \\ \underline{B}_c \end{bmatrix} \underline{y}(k) + \begin{bmatrix} \underline{D}_c \\ \underline{B}_c \end{bmatrix} \underline{w}(k) \quad (34)$$

This is now a standard static output-feedback problem which can be solved by methods with good convergence properties recently given in Wenk and Knapp (1980) and Mäkilä (1984).

STATE CONTROLLERS

All described state controller structures with observers introduce integral action into the state control loop in order to track non zero set points or to eliminate steady state disturbances. The extended controllers and observers are designed by solving the matrix Riccati equation. Therefore the control-loop-structures are described only.

Structure 1. Following a proposol in Isermann (1981) the integral paths are added to the observer, cf. Fig.3a. A reference variable model is introduced describing the reference signals $\underline{w}(k)$.

$$\underline{\nu}(k+1) = \underline{A}\,\underline{\nu}(k) + \underline{B}\,\underline{\gamma}(k) \quad (35)$$

$$\underline{\gamma}(k+1) = \underline{\gamma}(k) \quad (36)$$

$$\underline{w}(k) = \underline{C}\,\underline{\nu}(k) \quad (37)$$

With the control error $\underline{e}(k) = \underline{w}(k) - \underline{y}(k)$ an augmented description is obtained which the observer is designed on.

$$\begin{bmatrix} \underline{\varepsilon}(k+1) \\ \underline{\gamma}(k+1) \end{bmatrix} = \begin{bmatrix} \underline{A} & \underline{B} \\ \underline{0} & \underline{I} \end{bmatrix} \begin{bmatrix} \underline{\varepsilon}(k) \\ \underline{\gamma}(k) \end{bmatrix} + \begin{bmatrix} -\underline{B} \\ \underline{0} \end{bmatrix} \underline{u}(k) \quad (38)$$

$$\underline{e}(k) = [\underline{C} \quad \underline{0}] \begin{bmatrix} \underline{\varepsilon}(k) \\ \underline{\gamma}(k) \end{bmatrix} \quad (39)$$

The state controller is designed for the not augmented system yielding the feedback matrix $\underline{K}_1$ and is extended to

$$\underline{K}_c = -(\underline{K}_1 \quad \underline{I}) \quad (40)$$

Structure 2. The integral paths are appended to the outputs $\underline{y}$ of the process, cf. Fig.3b.

$$\underline{q}(k+1) = \underline{q}(k) - \underline{C}\,\underline{x}(k) + \underline{w}(k). \qquad (41)$$

Hence the augmented system description is

$$\begin{bmatrix} \underline{x}(k+1) \\ \underline{q}(k+1) \end{bmatrix} = \begin{bmatrix} \underline{A} & \underline{O} \\ -\underline{C} & \underline{I} \end{bmatrix} \begin{bmatrix} \underline{x}(k) \\ \underline{q}(k) \end{bmatrix} + \begin{bmatrix} \underline{B} \\ \underline{O} \end{bmatrix} \underline{u}(k) + \begin{bmatrix} \underline{O} \\ \underline{I} \end{bmatrix} \underline{w}(k) \qquad (42)$$

$$\underline{y}(k) = [\underline{C} \;\; \underline{O}] \begin{bmatrix} \underline{x}(k) \\ \underline{q}(k) \end{bmatrix} \qquad (43)$$

The reference signals $\underline{w}(k)$ in Eq.(42) act as disturbances and are therefore not considered in the design. The controller is designed for the augmented system whereas the observer is not extended and therefore designed for the original system, Eq.(3,4).

Structure 3. The integral paths precede the process inputs, cf. Fig.3c. Differing from the first two structures the transfer function matrix of the integral paths is proper but not strictly proper and is described by

$$\underline{z}(k+1) = \underline{z}(k) + \underline{u}^*(k) \qquad (44)$$

$$\underline{u}(k) = \underline{z}(k) + \underline{u}^*(k) \qquad (45)$$

Hence the augmented system is

$$\begin{bmatrix} \underline{x}(k+1) \\ \underline{z}(k+1) \end{bmatrix} = \begin{bmatrix} \underline{A} & \underline{B} \\ \underline{O} & \underline{I} \end{bmatrix} \begin{bmatrix} \underline{x}(k) \\ \underline{z}(k) \end{bmatrix} + \begin{bmatrix} \underline{B} \\ \underline{I} \end{bmatrix} \underline{u}^*(k) \qquad (46)$$

$$y(k) = [\underline{C} \;\; \underline{O}] \begin{bmatrix} \underline{x}(k) \\ \underline{z}(k) \end{bmatrix} \qquad (47)$$

with the new system inputs $\underline{u}^*(k)$. Controller and observer are extended and designed for the augmented system, Eq.(46) as if the integral paths would be part of the process.

The identity observers used in the three structures may as well be substituted by reduced order observers. A detailed description of all structures is given in Hensel (1983).

FEEDFORWARD CONTROLLERS

In the presence of measurable disturbances feedforward controllers may be applied to improve the control performance. Some well known approaches are given in Isermann (1981) for the SISO-case and can be easily extended to the MIMO-case. Approach 1 is the complete dynamic compensation, which cannot always be applied due to the cancellation of the process behaviour. One way to overcome this problem is a reduction of the dynamic compensation to the static case cancelling the gain factors only which often may not result in a satisfactory behaviour. A better performance can be achieved by using a dynamic compensator of prefixed structure and to tune the parameters by means of a parameter-optimization method.

THE CADOCS PACKAGE

The whole package consists of a central data-based CAD-interpreter-management system with an alterable command set,a block diagram generation package, the design modules for identification, controller design and simulation, a (Calcomp-compatible) plot package for high resolution graphic-devices and a flexible real-time package for DDC-purposes (Fig.4).

The CAD-Management Concept

The interpreter-management-system is divided into three parts, cf.Fig.5. The generator system CASSY allows among other features the definition or modification of an arbitrary selectable CAD-phase flow structure or the extension or reduction of the command set and therefore of the range of integrated design modules related to these commands. The results obtained by CASSY are stored in the database (DS) and can then be used by the management system CADCO which is realized as an interpreter with a simple command syntax. Depending on the generated CAD-phase flow structure and the actually activated phase the user can choose a command initiating his next intended action out of a subset of all available commands. This feature helps to avoid senseless design sequences (e.g. controller design before estimation of the process parameters). On the other hand it offers a high flexibility because the user can choose within these subsets and is not guided by a fixed program flow. In addition a special command allows to jump directly to each implemented CAD-phase. Each command-line consists of a 3-character command and optional separators and parameters defining modifications of the command-action or running parameters for the module to be activated.

CADCO not only interprets the command lines and coordinates the activation of the design modules, but as well serves for a user-management thus allowing different users to work with the package. A 4-character password assigned to the user allows CADCO to allocate different priority levels in order to permit or prohibit the usage of special commands and to assign different knowledge levels in order to enable or disable automatic output of help functions during the user´s session. Help functions are design phase titles, descriptions of the commands and the activated command subset and error and system messages. All these helps are stored in the data base with a special 2-character grammalogue defining the text-language; up to 16 different languages can be hold simultanously. For more details see Hensel and Schnell (1984).

The data base consists of three central management files containing system data, user management data and pointers to all other files of the data base (SYSTB), the CAD-phase flow structure (INFIL) and command line parameters (PARAM). User support is realized by five language dependend textfiles (<id> grammalogue) concerning general information (<id> CAD), help-information about the flow structure and the commands (<id> HLP), error and system messages (<id> ERR) and dialog-texts for question and answer parts (<id> DIA) as well as a user dependend (<us> grammalogue) default-value file defining alterable prefixed answers to the questions of the dialog-file (<us> DFT). The user dependend files concerning control engineering topics are the block diagram file containing up to 31 block-diagrams of a maximum complexity of about 8000 elements each (<us> BLD) and four files containing the parameters which describe the dynamics of identified processes and designed controllers (<us> SPR, <us> MPR, <us> SCO, <us> MCO). All user activities and all results obtained from the design modules are stored in a protocol file with alterable name.

All design activities concerning control engineering tasks are oriented on and routed through a central arbitrary definable block diagram which can consist of up to 32 different element types like multivariable processes, special nonlinearities, summation or branch points, etc. displayed on a graphical device (colour or black and white). Each block diagram description record of <us>BLD contains the whole structure definition and graphical representation of the element, pointers describing the linkage to other elements and pointers to the parameter files SPR, MPR, SCO and MCO. The element types are predefined in relative coordinates and stored in special records of the block diagram file which simplifies the definition of new elements.

The results from the identification and the controller design are written to four parameter files (SPR, MPR, SCO and MCO) thus allowing flexible simulations of complex or selected control systems or parts of them. The simulation results are displayed on graphical devices. Comparisons of the control performances of different controller-types or -parameters for the same controller block are supported by displaying dynamically selectable bar diagrams.

The Real-Time Concept

Within the real-time concept every implementable function (e.g. estimation, feedback- or feedforward controllers, filters) is designated as Multi-OPerational Segment further called MOPS which is restricted to a maximum of four inputs and four outputs. This definition allows the real-time management package to handle all functions in a similar manner, only the number of parameters, inputs and outputs differ. The package can handle quasi-simultaneously up to 30 functions with different sampling times which have to be multiples of the basic 200 msec cycle of the package. All MOPS-parameters relevant for management handling are stored in a special "system global area" and can be accessed from CADCO by a special modification routine. All signal values or computational results which have to be saved for further local use are stored in a "program global area" which cannot be accessed from CADCO. As it is possible to activate or deactivate functions thus requiring or releasing parts of the global areas, these are dynamically managed. In the case of deactivation the remaining parameter sets are packed to avoid split up of the areas. Parameter sets of new functions are added subsequent to the occupied space. Some further features are the prevention from integral wind up, open loop mode of each MOPS, Manual-mode of each D/A-channel and a special routine for bumpless closing of control loops. A detailed description of the real-time package is given in Hensel (1985).

A DESIGN EXAMPLE

The design of a multivariable control loop starting with the identification of the process and terminating with the implementation of the designed feedback controller will be shown for a two input, two output testcase simulated on an analog computer. The outputs are disturbed by a coloured noise with a rms-amplitude of 0.3V and a small mean value of about 0.002V.

The whole design (including implementation) is divided into 5 phases. A representative example of a design may be as follows:
In phase 1 a block diagram element representing the process is defined. The On-line identification is then started in phase 2. Due to the disturbances acting on the system the COR-LS method is applied (Fig.6). After a measuring period of 266 sec (sampling time T_0=2sec) orders,deadtimes and parameters of the model are estimated. The fitting of the model is visually checked by comparing the measured and the model cross correlation function (Fig.7). Phase 3 deals with the controller design. The selected controller structure shall be of PID-type and approach 2 is selected. After computing the controller phase 4 enables the simulation of the closed control loop (Fig.8). Phase 5 finally activates the On-line implementation. The performance of the On-line control loop is shown in Fig.9 (the disturbance has been disconnected for comparison purposes).

CONCLUSIONS

Different methods for the estimation of orders, deadtimes and parameters of multivariable processes, the design of feedback and feedforward controllers (DB, PID, SC) have been proposed. A flexible CAD-package serving for On-line identification, controller design, simulation and On-line implementation containing the proposed methods was presented. It is obvious that the high support and the flexibility given by the system may help to improve the design results and to save design time considerably.

REFERENCES

Aström, K.(1970). Introduction to Stochastic Control Theory, Academic Press, New York

Bayer, Th.(1982). Entwurf eines verallgemeinerten Mehrgrößen-Deadbeat-Algorithmus. Diploma thesis, Institut of Control Eng., Techn.Univ. of Darmstadt.

Blessing, P.(1979). Identification of the input-output- and noise-dynamics of linear multivariable systems. 5th IFAC-Symposium on Identification and System Parameter Estimation, Darmstadt.

Hensel, H.(1981). Entwurf eines verallgemeinerten Eingrößen-Deadbeat-Algorithmus mit Güteindex. Internal report, Inst.of Contr.Eng., Techn. Univ. of Darmstadt.

Hensel, H.(1983). Zur Synthese und Simulation von Zustandsreglern mit integralem Anteil. Internal Report, Institut of Contr. Eng., Techn. Univ. of Darmstadt.

Hensel, H. and Peter, K.H.(1984). Multivariable PID controllers designed by linear quadratic state space approach. American Control Conference, San Diego.

Hensel, H. and Schnell, R.(1984). CADCO- Ein lernfähiger Interpreter für regelungstechnische CAD-Systeme. Prozeßrechnertagung, Karlsruhe.

Hensel, H.(1985). A flexible real-time concept for various control requirements. (to be presented at) 7th Conference on Digital Computer Applications to Process Control, Vienna.

Isermann, R.(1981). Digital Control Systems. Springer Verlag, Berlin

Isermann, R. and Hensel, H.(1985). Sequentiel design of decentralized controllers with identification and selftuning methods. 3rd IFAC Symposium on Computer Aided Design in Control and Engineering Systems, Copenhagen.

Mäkilä, P.M.(1984). On the Anderson-Moore method for solving the optimal output feedback problem. IEEE Tr AC 29.

Mäncher, H. and Hensel, H.(1985). Determination of order and deadtime for multivariable discrete-time parameter estimation methods. 7th IFAC-Symposium on Identification and System Parameter Estimation, York.

Nour Eldin, H.A.(1977). Minimalrealisierung der Matrix-Übertragungsfunktion. Regelungstechnik Vol. 25.

Rosenbrock, H.H.(1974). Computer Aided Control System Design. Academic Press, London.

Wenk, C.J. and Knapp, C.H.(1980). Parameter optimization in linear systems with arbitrarily constrained controller structure. IEEE Tr AC 25.

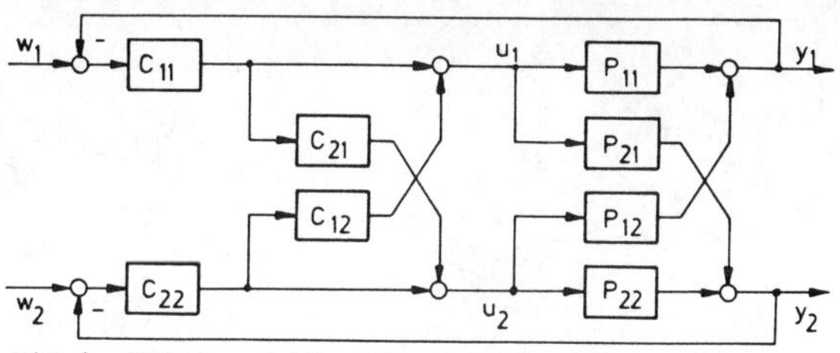

Fig.1. Multivariable PID-control with coupling controllers behind of main controllers

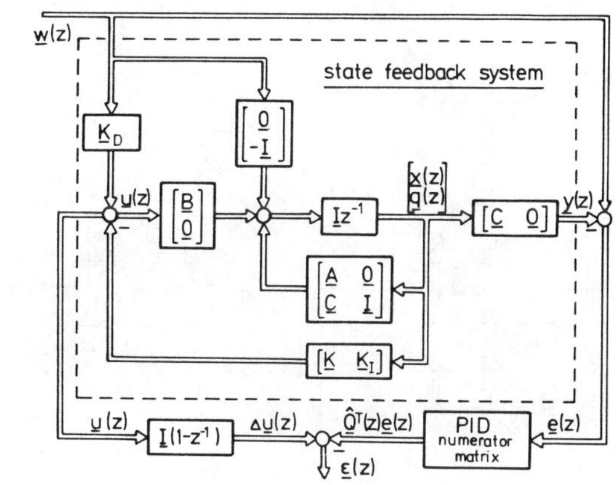

Fig.2. Multivariable PID-controller design by approximating state control loop signals

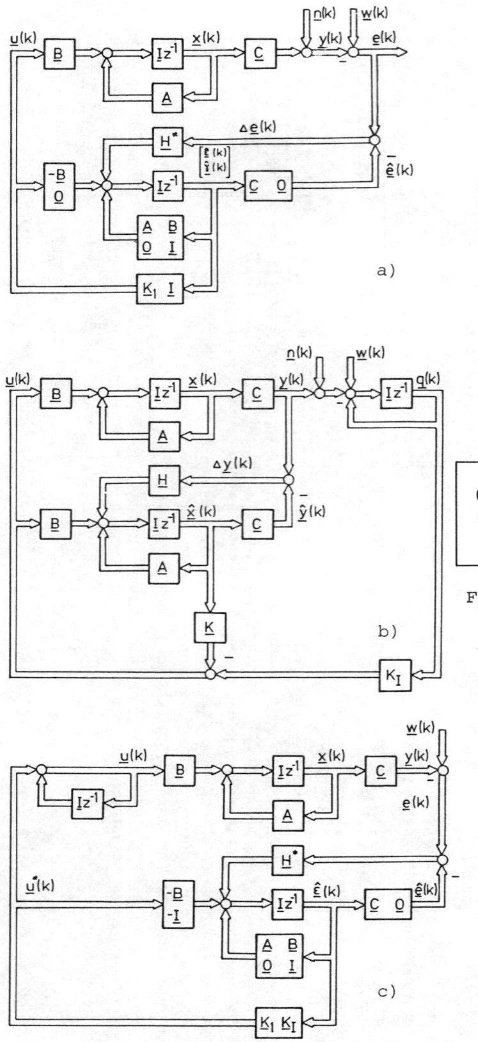

Fig.3. State control with integral action,
a) in the observer
b) appended to the process outputs
c) preceding the process inputs

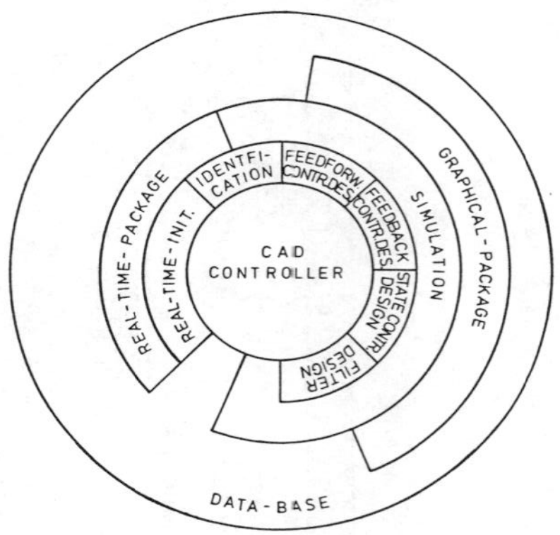

Fig.4. The CADOCS-package

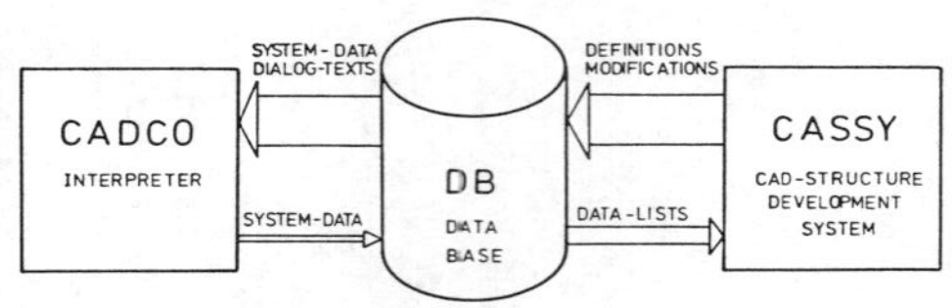

Fig.5. The management system CADCO/CASSY

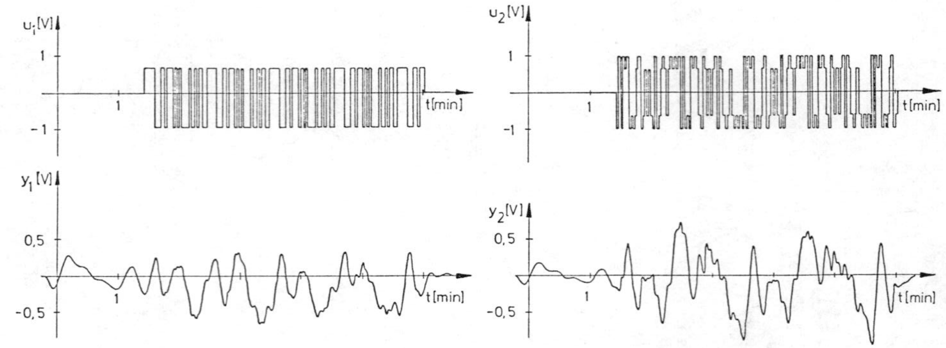

Fig.6. On-line process identification

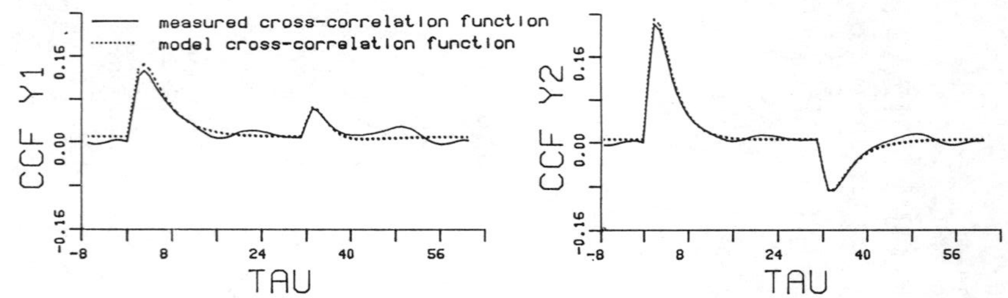

Fig.7. Model validation

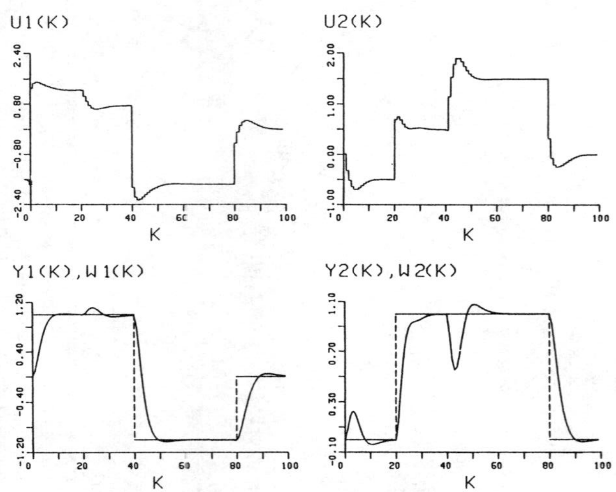

Fig.8. Closed loop simulation

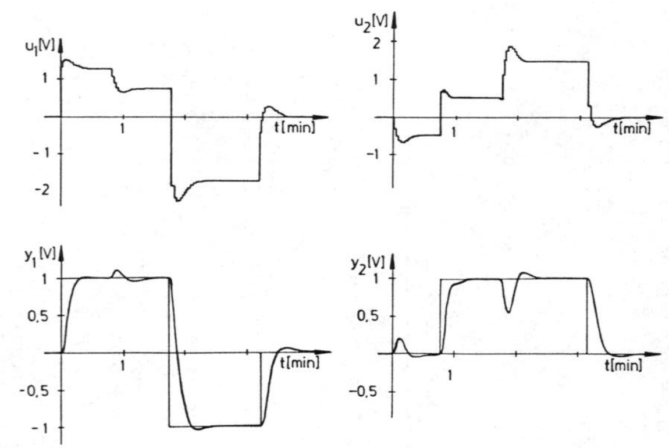

Fig.9. On-line control with the designed controller

INTERACTIVE OPTIMIZATION OF CONTROLLER AND PLANT PARAMETERS IN THE CASE OF MULTIPLE DESIGN OBJECTIVES

J. Lückel, R. Kasper and K. Jäker

Fachgruppe für Automatisierungstechnik der Universität-GH Paderborn, FRG

Abstract. This paper presents a design method for simultaneous optimization of a large number of controller and plant parameters of high-order linear time-invariant systems. User-friendly supervision and control of several design objectives is achieved through an interactive graphics-supported design process. The application of the design package for optimal matching of construction and controller parameters is demonstrated in great detail taking as an example a vehicle model with actively controlled tandem wheel suspension.

Keywords. Computer-aided system design; control system synthesis; optimal control; sensitivity analysis; optimisation; vibration control

INTRODUCTION

Over the last few years the demands placed on dynamic - for example, mechanical - systems have constantly increased. This is reflected among other things in more complex modeling, that is to say, in more precise system description over a large frequency range and a wide range of operating points. Additionally, design requirements have also become more complex, in the shape of several interdependent design objectives that all have to be monitored or optimized simultaneously (Kreisselmeier, 1979).

All this automatically results in greater demands being placed on available design aids. These are required to support the flexible use of complex controller systems of higher dimension such as those found in state space theory or in robust control. Critical applications have also shown up the necessity of studying controller and plant as a unit in order to determine optimal control inputs and measuring outputs or optimal plant construction or to adapt controller and system parameters to a particular task (Lückel, 1982). Where several design objectives are to be optimized simultaneously, scalar figures of merit are replaced by vectorial, and interactive visual control and monitoring of the design process becomes even more essential.

In this paper a design procedure for linear time-invariant plants and controllers is presented which allows of precise alteration of several design objectives on the basis of a sensitivity analysis in the relevant parameter space. In addition to pure parameter optimization, sensitivity analysis also affords an insight into the structure of the design problem. For both deterministic and stochastic experiments, formulating the design problem in terms of system theory allows for flexible modelling of a realistic operational environment at the input to the control system which is to be optimized in dynamic excitation models. At the output, dynamic weighting models support the processing of such plant output quantities as are relevant to a particular problem.

The application of the design package is demonstrated in great detail taking as an example a vehicle model with tandem wheel suspension. A combined optimization of controller and construction parameters will provide information on the extent to which it is possible to transfer the spring action of conventional pneumatic tyres to an active suspension system and to use the tyres solely to transmit drive and guidance forces.

FORMULATING THE DESIGN PROBLEM

The main aim of computer-aided controller design is to replace conventional laboratory hardware investigation of a particular physical problem by computer software simulation, thus either bypassing it altogether or at least restricting it to the final phase of development. To achieve this, the design problem has to be suitably formulated for computer application, allowing the user to put design requirements arrived at by means of actual physical observation into a mathematically exact and numerically manipulable form. However, a compromise has to be found between the accuracy of the simulation on the one hand and the time and effort involved in its realisation on the other. In this paper, we will restrict ourselves to modelling by means of linear time-invariant systems.

System theory aspects

Fig. 1 shows the block diagram of a standard configuration used for systematic formulation of the design problem (Kasper, 1985a). The dynamic behaviour of the system to be designed is reproduced in the plant model. The approximation has to have sufficient bandwidth, which automatically results in models of high order, for example when dynamic effects in actuators and measuring elements are taken into consideration or when elastic structures are modelled in the case of mechanical systems. The practical effects of the limitation to linear time-invariant system models can be reduced if several operating points are

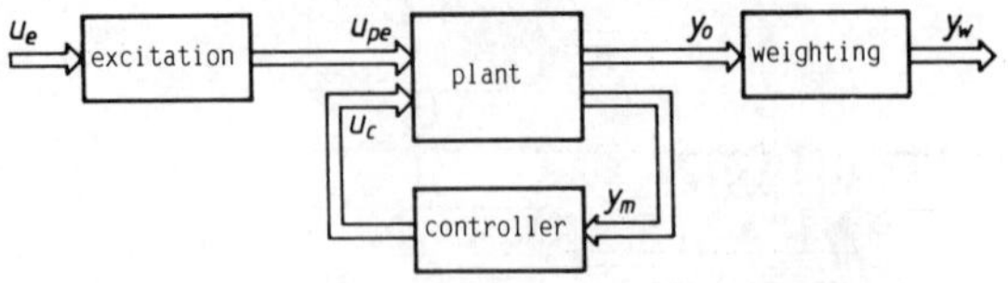

Fig. 1 Block diagram of a standard configuration

included, for example in quasi-stationary examination of time-variant systems or linearization of non-linear ones. Reproduction of the external excitation functions effective at the input of the plant is equally important. It enables the expert to carry out analysis and synthesis in an operational environment which is as realistic as possible. In the case of nonlinear systems the adjustment of defined operating points as a basis for linearization is of particular importance, since in this way it is possible to check whether the limited range of validity has been adhered to. The necessary excitation functions are reproduced in the excitation model, with two different classes of function:

- deterministic excitation functions such as step, ramp and sinus functions, which are generated in the instable part of the excitation model from Dirac impulses and
- stochastic excitation functions, whose spectral distribution throughout the stable part of the excitation model can be fixed almost at will.

The excitation model also serves to reproduce disturbance and reference inputs. What functions the outputs of the excitation model finally take depends on how they are fed into the plant via the input $\mathbf{u}_{pe}$:

The output quantities $\mathbf{y}_o$ which are of relevance to a particular application can be processed dynamically in the subsequent weighting model. For example, certain discrete frequency intervals or individual frequencies may be suppressed or amplified. Characteristic quantities of the weighted output quantities $\mathbf{y}_w$ are used for analysis.

dynamic amplitude factors

- integral mean values of the transient process amplitudes during impulse excitation
- RMS due to noise excitation

static amplitude factors

- polynomial parts of the steady state amplitudes during impulse excitation.

In order to fully exploit all the advantages of controller synthesis which are theoretically possible and practically necessary, the selection of controller structure ought to be as unrestricted as possible, i.e.

- support of the development of multivariable controllers of high order such as those found in state space theory (Riccati controllers, Kalman filters, compensations, disturbance and reference feedforwarding), and
- consideration of requirements which are shown up during practical realisation, such as those arising during digital control with signal processors or from nonlinear effects in the plant.

The controller is constructed as a linear dynamic system of arbitrary structure, consisting of several substructures, the measuring variables $\mathbf{y}_m$ functioning as input and the control variables $\mathbf{u}_c$ as output.

Plant and controller, joined by dynamic excitation and weighting models, form the augmented controlled system

$$\dot{\mathbf{x}} = \mathbf{A}(\mathbf{p})\mathbf{x} + \mathbf{B}(\mathbf{p})\mathbf{u}_e \qquad (1)$$
$$\mathbf{y}_w = \mathbf{C}(\mathbf{p})\mathbf{x} \; ,$$

which is the basis of the design. The parameter vector $\mathbf{p}$ can contain both controller and plant parameters. It is divided into a fixed, operating point-dependent vector $\mathbf{p}_f$ and an optimizable vector $\mathbf{p}_o$,

$$\mathbf{p} = \begin{bmatrix} \mathbf{p}_f \\ \mathbf{p}_o \end{bmatrix} \begin{matrix} \longrightarrow \text{fixed, operating point} \\ \longrightarrow \text{optimizable} \end{matrix}$$

thus allowing various operating points to be included in a linearized treatment of non-linear problems.

Formulating the optimization problem

The scheme of an interactive design concept shown in Fig. 2 simulates practical experimental work in the laboratory. However, both expert and process levels are either supported or simulated off-line with the aid of a computer. In the planning phase, the expert decides on various experiments which appear relevant to the particular application for individual operating points of the augmented system. For each experiment, he determines input and output configuration, specifies the type of excitation (deterministic/-stochastic) and defines objectives important to the application. These combine formally to make the objectives vector

$$\mathbf{z} = \begin{bmatrix} z_1 \\ \cdot \\ z_j \\ \cdot \\ z_{nB} \end{bmatrix} \begin{matrix} 1. & \text{operating point} \\ \cdot & \\ j\text{-th} & \text{operating point} \\ \cdot & \\ n_B\text{-th} & \text{operating point} \end{matrix}$$

where (2)

$$z_j = \begin{bmatrix} z_{j1} \\ \cdot \\ z_{jk} \\ \cdot \\ z_{jnE} \end{bmatrix} \begin{matrix} 1. & \text{experiment} \\ \cdot & \\ k\text{-th} & \text{experiment} \\ \cdot & \\ n_E\text{-th} & \text{experiment .} \end{matrix}$$

Such a procedure has the advantage of allowing individual criteria (components of the objectives vector) to be monitored individually and directly influenced by the strategy. In contrast to this, the conventional concentration of different partial functionals into an overall scalar functional prevents a clear view of individual aspects and only allows indirect influence via alterations in the weighting factors.

At the process level, simulation of individual experiments for each operating point (an explicit numerical simulation can usually be dispensed with in the case of linear systems) takes place in the execution phase along with an analysis calculating the objectives relevant to each experiment. The results of the analysis are used to adjust the free parameters in accordance with a pre-set strategy.

At the expert level, the execution phase involves monitoring the analysis results and where necessary interactively altering the strategy or even defining new experiments and operating points.

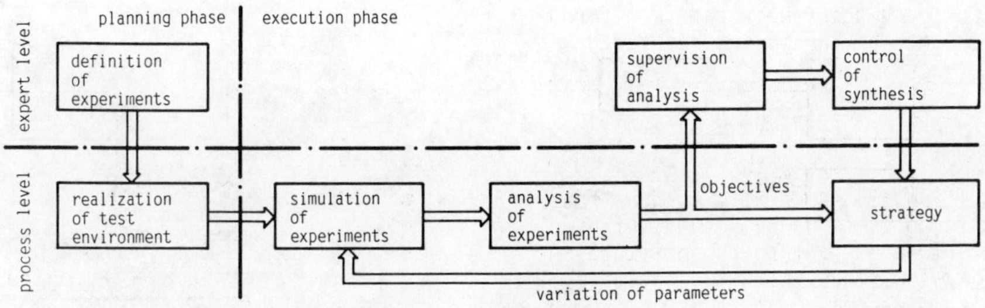

Fig. 2 Basic concept of the design package

SOLUTION OF THE DESIGN PROBLEM

The design task consists in influencing certain elements of the objectives vector in accordance with the application in question. That may mean minimizing or maximizing some of the elements as far as possible, keeping others as they are or moving them into certain pre-determined ranges. The expert´s task consists in monitoring the phases of the analysis and communicating desired changes to the strategy module. Such an interactive procedure allows the course of the optimization to be observed. The relatively low computing time requirement of the strategy enables the design process to be interactively monitored and controlled on the graphics monitor.

Analysis and solution of the linear problem

The general nonlinear design problem, that of moving the components of the objectives vector $\mathbf{z}_i(\mathbf{p}_o)$ for a set of particular operating points $\mathbf{p}_{fj} \hat{=} \text{const}$ ($j=1, \ldots, n_B$) into certain predefinable ranges

$$z_i(p_o) \in \mathbb{Z}_i \tag{3}$$

is reduced to the iterative solution of a sequence of linear subproblems(Kasper, 1985b). In each step of the design the aim is an alteration $\Delta\mathbf{z}$ of the objectives vector $\mathbf{z}(\mathbf{p}_o)$ which enables requirement (3) to be fulfilled. If only small alterations $\Delta\mathbf{z}$ are carried out, the sensitivity matrix

$$\mathbf{S} = \nabla_{\mathbf{p}_o}\mathbf{z} \tag{4}$$

provides a linear approximation of the relationship between the change in objectives vector $\Delta\mathbf{z}$ and parameter vector $\Delta\mathbf{p}_o$:

$$\Delta\mathbf{z} \approx \mathbf{S}' \, \Delta\mathbf{p}_o \; . \tag{5}$$

In each step, the linear sub-task of the design procedure consists in adjusting the possible infinitesimal alterations $\Delta\hat{\mathbf{z}}$ in the objectives vector $\mathbf{z}$ as far as possible to the desired alterations $\Delta\mathbf{z}$. Fig. 3 shows an example of this procedure for two objectives and a two-dimensional parameter space. If this adjustment is described by means of a minimization of a weighted Euclidean vector norm of the error

$$\|\Delta\hat{\mathbf{z}} - \Delta\mathbf{z}\|_{\mathbf{L}} = \text{Min} \quad \text{where} \quad \mathbf{L} = \operatorname*{diag}_{i=1}^{q}(l_i) \; , \tag{6}$$

the unknown alterations of the parameter vector $\mathbf{p}_o$ can be determined by the solution to the Least Square Problem defined in equations (5 and 6). The desired directions $\Delta\mathbf{z}$ of the objectives vector $\mathbf{z}$ and of weightings $\mathbf{L}$ are available as an interactive means of controlling the design process. The desired range was intentionally not pre-set, as in many cases this alters during the design process along with the physical problem. On the contrary, the solubility of the linear subproblem (rank of the sensitivity matrix) defined by equations (5 and 6) shows up any conflict in objectives and thus provides an important aid to deciding on new directions.

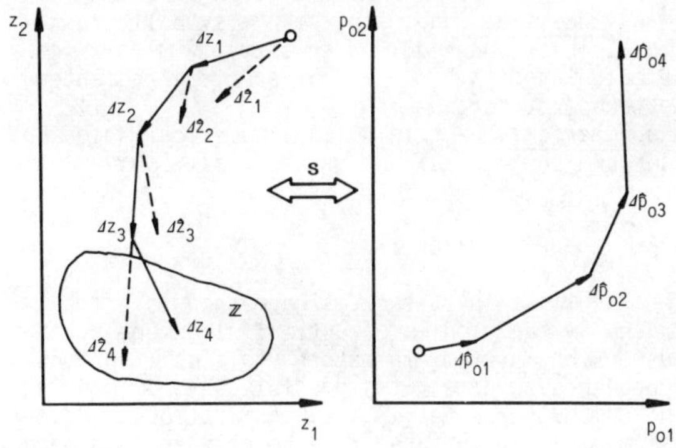

Fig. 3 Schematic representation of the iterative solution to the nonlinear design problem by mean of a sequence of linear subproblems for 2 free parameters and 2 objectives

Nonlinear problem and step size adjustment

For the execution of the design method a step size control which provides the interface from the linear subproblem to the real nonlinear problem is necessary. This has the function of adjusting the real alteration of the parametervektor

$$\Delta\hat{\mathbf{p}}_o = s \, \Delta\mathbf{p}_o \quad \text{where} \quad 0 < s \leq 1 \tag{7}$$

to the requirement given by the linear calculation in such a way as to keep the linearization error

$$\Delta\mathbf{z}_L = \Delta\mathbf{z}_N(\mathbf{p}_o) - s \, \Delta\mathbf{z}$$

$$\text{where} \tag{8}$$

$$\Delta\mathbf{z}_N = \mathbf{z}(\mathbf{p}_o + \Delta\hat{\mathbf{p}}_o) - \mathbf{z}(\mathbf{p}_o)$$

as small as possible for a maximum value of s at each step. The two requirements contradict each other, so that specification of step size has a decisive influence on the effectiveness of execution. To achieve simplicity of implementation, step size control is based on sign equality of both pre-set and actually calculated alterations in the weighted components of the objectives vector

$$\operatorname{sgn}(\Delta z_{Ni}) = \operatorname{sgn}(\Delta z_i) \quad \text{for} \quad l_i \neq 0 \tag{9}$$

Additionally, for each step the relative alteration of the components of the objectives vector

$$\max_{i=1}^{q} \frac{\Delta z_{Ni}}{z_i(\mathbf{p}_o)} \qquad \text{for} \qquad z_i(\mathbf{p}_o) \neq 0 \qquad (10)$$

is limited. Condition (9) especially ensures that during the design process none of the weighted components "deteriorates" and that optimization proceeds over a wide range with a relatively large step size (as many applications have shown).

Calculation of the sensitivity matrix

As the description of the design procedure's basic elements shows, the calculation of the sensitivity matrix **S** is of central importance as a link between the expert's demands (alteration in objectives vector) and execution (alteration in parameters). There are two possible methods of accomplishing this task:

- an analytical one, in which a symbolic representation of the required sensitivity matrix can be calculated with the aid of special Gradienten-kettenoperators and
- a numerical one, in which a numerical solution is found on the basis of QR-decomposition.

APPLICATION

The application of the design method is demonstrated by the problem of optimal alignment of the suspension system in motor vehicles. In most approaches to date, two possible ways of meeting the demand for safety, i.e. as little variation as possible in surface pressure between tyre and road, and for ride comfort, i.e. as little discomfort as possible for the passengers during acceleration, have been investigated. These are:

- improvement and optimal adjustment of the construction of the passive springing systems in common use and of wheel suspension
- use of actively controlled hydraulic suspension in place of the passive system and optimal adjustment of the necessary controller.

For more sophisticated requirements, this separate procedure is no longer adequate. If a satisfactory result is to be achieved, wheel suspension construction and active springing control have to be adjusted to one another.

Model description of the design objectives

In present suspension construction, ride comfort and safety depend mainly on the pneumatic tyres (spring rate C_R) used, which fulfil both guidance and springing functions. If the two combined functions of the conventional pneumatic tyre (generation of friction coefficients to ensure grip and formation of an air cushion to insulate against road irregularities) are separated, it can be replaced by a considerably harder solid rubber tyre. If this is done, neither a conventional spring-damper system (spring rate C_a, damper rate D_a) nor an active damper system alone (controlled hydraulic cylinder with conventional wheel suspension) can bring ride comfort ($\ddot{x}_a$) and safety (F_{dyn}) down to the level of a pneumatic tyre system. For this reason, we suggest replacing conventional wheel suspension by the tandem wheel suspension shown in Fig. 4, which has many of the essential springing properties of the pneumatic tyre. The two wheels with hard rubber tyres (mass M_{R1}, M_{R2}, spring rate $C_{R1}=C_{R2}=4\ C_R$) are coupled to the

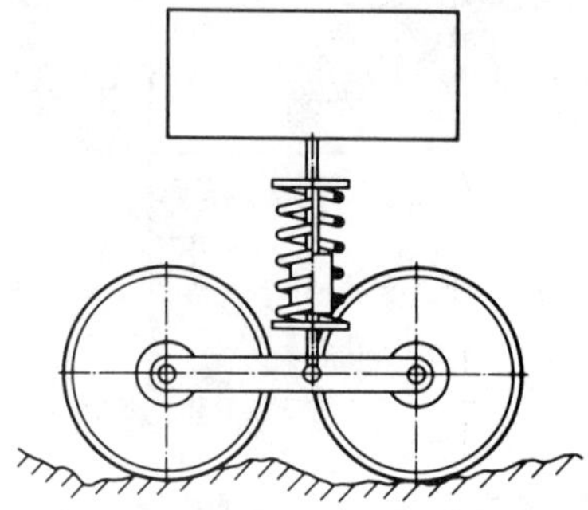

Fig. 4. Scheme of the tandem wheel suspension

damper system of the car body in place of a single wheel at a distance of L_{W1}, L_{W2} by means of a rocker arm (mass M_W, moment of inertia J_W, centroidal distance L_W) and can be rotated (damping rate D_W). The elasticities of the rocker arm construction and of the two wheel-bearings are approximated by linear springs and dampers (C_1, C_2, D_1, D_2).

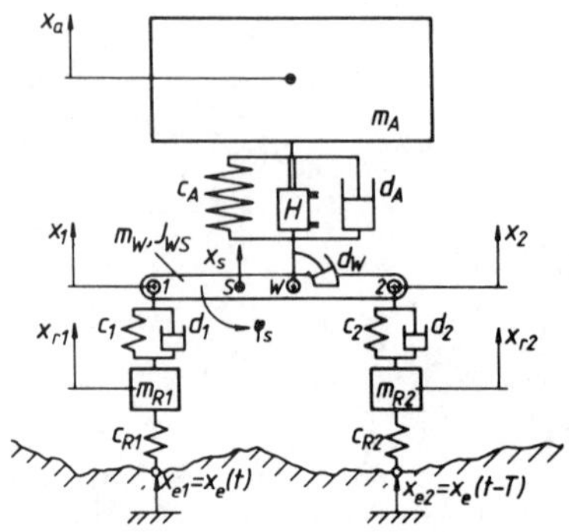

Fig. 5. Physical model of the tandem wheel suspension

To optimize the parameters of the rocker arm construction, the rigid body model represented in Fig. 5 can be described with sufficient accuracy by a linear time-invariant differential equation system of tenth order. For the active vibration-damping, the passive damper is replaced by a hydraulic cylinder (surface area A, volume V). The pressure difference Δp in the hydraulic cylinder (leakage coefficient C_L, bulk coefficient β, piston speed $\dot{x}_k$)

$$(V/\beta)\ \Delta\dot{p} = -\ 2\ C_L\ \Delta p\ -\ 2\ A\ \dot{x}_k\ +\ 2\ Q_v$$

provides damping-force $F_{hy} = A\ \Delta p$.

The necessary oil-flow Q_v is controlled in an electro-hydraulic servo-valve.

$$T_v^2\ \ddot{Q}_v + 2\ D_v\ T_v\ \dot{Q}_v + Q_v = K_v\ U_v$$

The spectral distribution of road surface irregularities for the disturbance input $x_{e1} = x_e(t)$ is formed from white noise in a form filter with the transfer function

$$G_F(s) = \frac{K_s}{(1+T_1\ s)(1+T_2\ s)}$$

Parameters of T_1=2sec, T_2=0.02sec, K_s=0.01 result for an average asphalt surface and a velocity of

$v = 20$ m/s (Mitschke, 1972). When travelling straight ahead, road surface irregularities $x_{e2}(t) = x_e(t - T)$ affect the second wheel with a time lag of

$$T = v/(L_{W1} + L_{W2})$$

The dead time system is approximated by an all-pass filter with the transfer function

$$G_T(s) = \frac{a_0 - a_1 Ts + a_2 T^2 s^2 - a_3 T^3 s^3 + a_4 T^4 s^4}{a_0 + a_1 Ts + a_2 T^2 s^2 + a_3 T^3 s^3 + a_4 T^4 s^4} .$$

The parameters $a_0 = 1072$, $a_1 = 536$, $a_2 = 120$, $a_3 = 13.55$, $a_4 = 1$, results from a Padé approximation of 4th order (Johnsen, 1963).

In order to take passengers' physiological sensitivity to vertical vibration into account, car body acceleration $\ddot{x}_a$ is weighted in the lower frequency range as a measure of ride comfort according to DIN (1984) by means of a formfilter

$$G_W(s) = \frac{K_W (1 + T_{W1} s)}{1 + 2 D_W T_{W2} s + T_{W2}^2 s^2}$$

with the parameters $K_W=0.42$, $T_{W1}=0.107$, $T_{W2}=0.03$, $D_W=0.733$ in the lower frequency range.

The relative wheel load variation RL (ratio of dynamic (F_{dyn}) to static (F_{stat}) wheel load) is used as a measure of driving safety. For tandem wheel suspension, the wheel load variation of both front and rear wheel

$$RL1 = \frac{F_{dyn1}}{F_{stat1}} 100\ \% \quad , \quad RL2 = \frac{F_{dyn2}}{F_{stat2}} 100\ \%$$

and their sum

$$RL = \frac{F_{dyn2} + F_{dyn2}}{F_{stat1} + F_{stat2}} 100\ \%$$

are of importance. The static load carried by each wheel (F_{stat1}, F_{stat2}) can be calculated from the rocker arm geometry. Fig. 6 shows how the subsystems connect up. The total coupled system of 21st order forms the basis for optimization.

Optimization of the actively controlled tandem wheel suspension

The design of the actively controlled tandem suspension takes place in three stages:
- optimization of rocker arm and tandem wheel construction for a given passive springing system. Taking this construction as a starting-point,
- design of a Riccati controller for an actively controlled system and finally
- mutual adjustment of controller and construction.

In all three stages, the primary objective of optimization is to reduce the weighted car body acceleration and relative wheel load variation. In order to monitor and keep to the given physical restrictions of the mechanical (position of car body x_a and wheels x_{r1} or x_{r2}, position and angle of the rocker arm x_s or φ_s), the hydraulic (pressure difference Δp, oil flow Q_v) and the electrical variables (control voltage U_v), these are additionally included.

$$\mathbf{y}_W^T = (\ddot{x}_{aw}, RL, RL1, RL2, x_a, x_{r1}, x_{r2}, x_s, \varphi_s, \Delta p, Q_v)$$

The RMS values of the vector $\mathbf{y}_W$ form the vector σ_W. In order to guarantee a desired range of stability in the optimized system, all eigenvalues are grouped together in the eigenvalue vector λ and included in the objectives vector .

$$z^T = (\sigma_W^T, \lambda^T)$$

1st design stage

To find out what improvement potential there is in the area of construction alone, the first design stage is to investigate optimal adjustment of rocker arm and tandem wheel for a conventional passive springing system with constant car body spring Ca and damper Da. If the possible values of the weighted car body acceleration and the relative wheel load variation for the tandem wheel suspension are compared with the reference values entered in Fig. 7 for a conventional wheel suspension, it becomes clear that the relative wheel load variations, which have risen sharply as a result of the hard rubber tyres, have to be reduced. The weighted car body acceleration can be kept constant during this process. The change in the RMS of these quantities occurring during optimization is given in Fig. 7. Optimization had to be ended after a conjugate complex eigenvalue pair reached the minimum permissible damping ratio, further improvement in the given parameter space thus being impossible.

2nd design stage

After all scope for improvement in construction has been exhausted, in the 2nd design phase the passive damper is replaced by an active spring element. As the car body spring then serves only to carry static loads, its spring rate is reduced to 25%. In order to gain an overview of the improvement potential of the active suspension, a complete state controller

$$\mathbf{u}_c = \mathbf{k}^T \mathbf{x}$$

is designed. As the vector of the controller coefficients can be determined in one step through the solution of the matrix Riccati equation, values for ride comfort and safety improve dramatically.

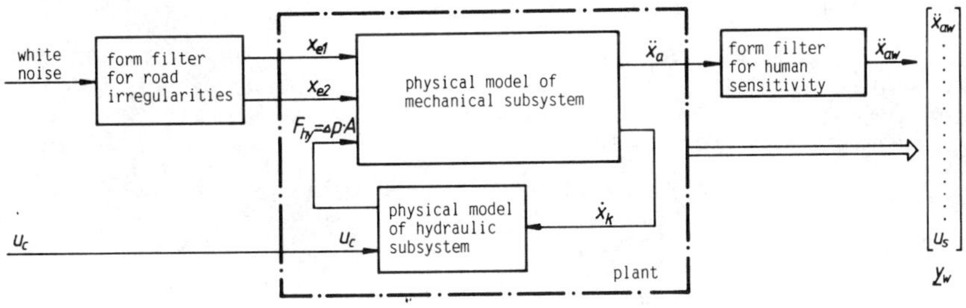

Fig. 6 Blockdiagram of the coupled system used for optimization

3rd design stage

After everything possible has been done to improve construction and controller separately, further improvements in the actively controlled tandem wheel suspension can only be achieved through mutual adjustment of system and controller parameters. There are altogether 30 parameters which have to be optimized in this stage

$$p^T = (p_s^T, k^T) .$$

As the values for the relative wheel load variations have already been reduced to a satisfactory level by the Riccati controller, the most important goal here is to reduce the weighted car body acceleration. The change in RMS values of these quantities occurring in the third optimization stage are depicted in Fig. 7. In this stage, too, the design process was ended when a conjugate complex eigenvalue pair went outside the permissible damping range.

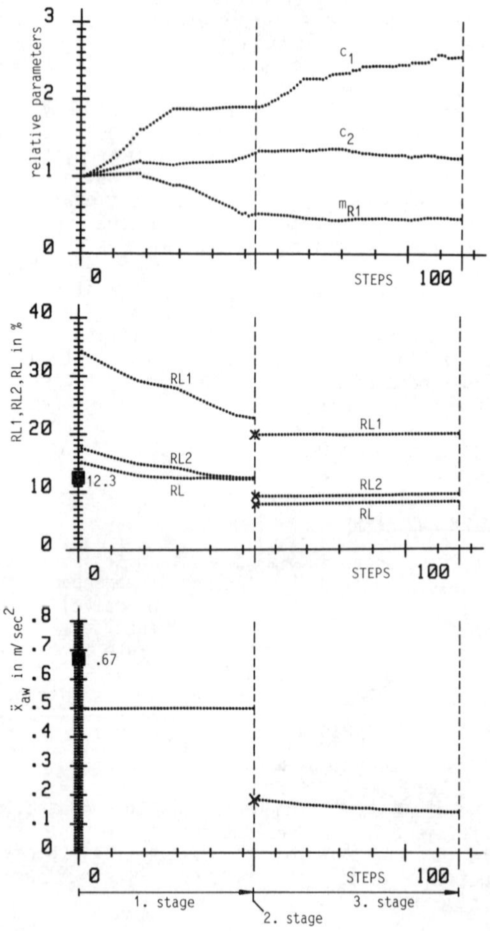

Fig. 7 Alterations of some system parameters and objectives during optimization

■ -Reference values for conventional wheel suspension with pneumatic tyre
× -Riccati controller for actively controlled tandem wheel suspension

SUMMARY

The example gives a clear demonstration of the efficiency of the design package presented here. The ability to influence various objectives individually in the separate stages of design allows the expert to achieve his design requirements directly. The relatively small number of optimization steps (124 steps with approximately the same number of objectives vector calculations) needed despite the large parameter space allows of interactive design control even in the multi-user environment of a mini-computer (Prime 750). As Fig. 7 shows, the alterations of some constructive design parameters during optimization all take place within a range which makes actively controlled tandem wheel suspension appear promising for the future.

REFERENCES

DIN-Norm 45671 (1984). Measurement of occupational vibration immissions; vibration meter; requirements and verification

Johnson, C.L. (1963) Analog computer techniques, Mc Graw-Hill book company

Kreisselmeier G., und R. Steinhauser (1979) Systematische Auslegung von Reglern durch Optimierung eines vektoriellen Gütekriteriums. Regelungstechnik 30, p. 76-79.

Kasper R. (1985a). Enwicklung und Erprobung eines instrumentellen Verfahrens zum Enwurf von Mehrgrößenregelungen. To appear in Fortschrittsberichte der VDI-Zeitschriften, Reihe 8.

Kasper R. (1985b) Grundlagen eines instrumentellen Verfahrens zum Entwurf von Mehrgrößenregelungen. To appear in Regelungstechnik.

Lückel J. und R. Kasper (1983). Rechnerunterstützter Regelkreisentwurf im Zustandsraum mit Softwaremodulen derlinearen Systemtheorie. Tagungsband zum Aussprachetag: "Rechnergestützter Regelkreisentwurf" der VDI/VDE -Gesellschaft am 26. - 27. 9. 1983 in Langen. p. 1-15.

Lückel J., F. Frühauf und R. Kasper (1982) Entwurf von linearen Mehrgrößenreglern mit Hilfe einer 2-Schritt- Optimierung. Festschrift zum 70. Geburtstag von Prof. Dr. Rer. Nat. K. Magnus, München, p. 363-376.

Mitschke M. (1972). Dynamik der Kraftfahrzeuge. Springer Verlag Berlin

SYSLAB: AN INTERACTIVE SYSTEM FOR ANALYSIS AND DESIGN OF LINEAR MULTIVARIABLE SYSTEMS

P. H. Petkov*, N. D. Christov* and M. M. Konstantinov**

**Department of Automatics, Higher Institute of Mechanical and Electrical Engineering, Bl. 2, 1156 Sofia, Bulgaria*
***Institute of Engineering Cybernetics and Robotics, Akad. G. Bonchev Str., Bl. 2, 1113 Sofia, Bulgaria*

Abstract. An interactive system called SYSLAB for analysis and design of linear multivariable control systems in the state space is described. It is developed on the basis of MATLAB - a powerful interactive language for matrix computations which implements high quality software for solving matrix problems. SYSLAB makes possible to solve easily the basic problems of the analysis and design of linear multivariable control systems: stability analysis, investigation of the controllability/observability and reduction into canonical forms, pole assignment and synthesis of optimal linear systems. Only algorithms that are proved as numerically stable are implemented. SYSLAB is written in Fortran and is fully portable.

Keywords. Computer-aided system design; multivariable control systems; canonical forms; pole assignment; optimal systems.

INTRODUCTION

In recent years a great interest has been paid to the computer-aided analysis and design of linear multivariable control systems. Various computational algorithms for continuous and discrete-time systems are developed and implemented in a number of program packages.

The most efficient way for analysis and design of control systems is to solve the corresponding problems in interactive mode using a simple command language. This reduces greatly the user's efforts and makes possible to obtain many solutions for different data in a short time.

In the analysis and design of linear systems in the state space the possibility to perform various matrix computations is very important. The solution of linear equations, eigenvalue/eigenvector computations, QR- and singular value decompositions are encountered in almost every control problem. That is why each interactive system intended for analysis and design of linear systems should possess the possibility to perform such computations.

Fortunately, there is MATLAB - a powerful interactive system for computations involving matrices developed by Professor C.B. Moler (1982). This system provides easy access to high quality software contained in LINPACK (Dongarra and others, 1979) and EISPACK (Smith and others, 1976; Garbow and others, 1977) packages. MATLAB is written in Fortran and is designed to be readily portable. The functions included in this system involve:

- solution of linear equations, finding the inverse, the determinant and the condition number of a matrix;
- computation of eigenvalues/eigenvectors, reduction into Hessenberg, tridiagonal and Schur triangular forms;
- singular value decomposition and computation of the pseudoinverse;
- QR decomposition and solution of linear least squares problems;
- computation of a matrix exponential, and many others.

An important feature of MATLAB is the possibility to count the necessary computational operations and to execute each function with different accuracy in order to measure the effect of the rounding errors. In addition, MATLAB is extensible new functions being added very easily. These properties make MATLAB an unique matrix environment for the analysis and design of linear systems. Consequently, it is used as a basis for development of interactive systems for solving linear control problems (Gregory, Shah and Walker, 1982; Little, Emami-Naeini and Bangert, 1984).

In this paper an interactive system, called SYSLAB, for analysis and design of linear multivariable control systems, developed on the basis of MATLAB, is described. The main features of this system are:

- only computational algorithms that are proved numerically stable are implemented. This makes possible to solve high order and ill conditioned problems;
- all algorithms work with complex matrices which ensures full compatibility with the programs used in MATLAB;
- all programs are developed so that to ensure the possibility of counting the computational operations and to execute the commands with different accuracy, as the functions from MATLAB.

SYSLAB makes possible to solve easily the basic problems of the analysis and design of linear multivariable systems:

- stability analysis and solving the Sylvester and Lyapunov equations (both continuous and discrete);
- reduction into controllable/observable orthogonal canonical form and determination of the dimension of the controllable/observable subspace;
- pole assignment synthesis and design of state observers;
- synthesis of optimal linear systems and optimal filters, solving the corresponding matrix Riccati equations (continuous and discrete).

In the next two sections we describe the SYSLAB functions and the numerical algorithms implemented.

SYSLAB FUNCTIONS

Solving Sylvester and Lyapunov Equations

The following four functions are intended to solve Sylvester and Lyapunov matrix equations in the continuous and discrete-time case.

X = SYLC(A,B,C)

Solves the Sylvester equation

AX + XB = C

where A and B are square matrices.

X = SYLD(A,B,C)

Solves the "discrete" Sylvester equation

AXB - X = C.

X = LYPC(A,C)

Solves the Lyapunov equation

A*X + XA = C

(A* is the conjugate transpose of A).

X = LYPD(A,C)

Solves the discrete Lyapunov equation

A*XA - X = C.

Controllability/Observability Analysis and Reduction into Orthogonal Canonical Form

For a linear continuous

$$\dot{x} = Ax + Bu$$
$$y = Cx$$

and discrete

$$x_{k+1} = Ax_k + Bu_k$$
$$y_k = Cx_k$$

system, the functions

NC = NCON(A,B)

NO = NOBS(A,C)

determine the dimension of the controllable, resp. observable, subspace of the system.

The reduction into orthogonal controllable canonical form (Konstantinov, Petkov and Christov, 1982)

$$A_c = U^*AU = \begin{bmatrix} & A_1 & & \\ A_{21} & & A_2 & \\ 0 & \ddots & & \\ & & A_{p,p-1} & A_p \end{bmatrix}$$

$$B_c = U^*B = \begin{bmatrix} B_1 \\ \\ 0 \end{bmatrix},$$

where U is an unitary matrix, is performed by the function

<AC,BC> = FCON(A,B)

the matrices A_c and B_c being stored in the variables AC and BC.

The function

<AO,CO> = FOBS(A,C)

is used to obtain the orthogonal observable canonical form of the system.

If the transformation matrix which reduces the system into canonical form is required, these functions are used in the form

<U,AC,BC> = FCON(A,B)

<U,AO,CO> = FOBS(A,C).

In case the system is not completely controllable/observable its stabilizability/detectability may be checked by determining the eigenvalues of the uncontrollable/unobservable part of the system matrix using the MATLAB function EIG.

Example 1. The following part of a dialog with SYSLAB illustrates the reduction of a 4th order system into orthogonal controllable canonical form.

```
A =
 3. -2.  4.  7.
 5.  1. -4. -2.
-1.  3. -6.  7.
-2. -3. -8.  1.

B =
 2. -1.
-6.  1.
-2.  4.
 3.  8.

<U,AC,BC> = FCON(A,B)

BC =
 0.8835  9.0554
 7.2263  0.0000
 0.0000  0.0000
 0.0000  0.0000

AC =
-1.9390  4.6841  9.0572  4.2995
 3.3393  0.9439 -0.1298 -5.6317
-8.8919 -1.9597 -0.6777 -3.4959
 0.0000  3.5065 -4.3211  0.6728
```

```
U =
-0.1104  0.2903 -0.6990  0.6442
 0.1104 -0.8438  0.1061  0.5143
 0.4417 -0.3308 -0.6676 -0.4997
 0.8835  0.3071  0.2332  0.2661
```

Pole Assignment Synthesis

The gain matrix K of the control law

u = -Kx,

corresponding to the desired poles stored in the vector W, is obtained by the function

K = POLS(A,B,W).

After the synthesis the actual eigenvalues of the closed-loop system matrix may be found by

EIG(A-B*K).

Example 2. This example demonstrates pole assignment synthesis for the system considered in Example 1.

```
W =
-1.0000 + 1.0000i
-1.0000 - 1.0000i
-2.0000 + 0.0000i
-3.0000 + 0.0000i

K = POLS(A,B,W)

K =
-0.2159 -0.0647 -0.9413  0.7208
-0.2005 -0.0600 -0.8741  0.6693

EIG(A-B*K)

ANS =
-1.0000 + 1.0000i
-1.0000 - 1.0000i
-3.0000 + 0.0000i
-2.0000 + 0.0000i
```

The function POLS may be used also for synthesis of full order observers, working with matrices A*, C* instead of A, B.

Synthesis of Optimal Systems and Solution of the Matrix Riccati Equations

In the continuous-time case the control law

u = -Kx

is determined so that to minimize the quadratic cost

$$J = \int_0^\infty (x^*Qx + u^*Ru)\,dt.$$

The optimal gain matrix

$$K = R^{-1}B^*P$$

where P is the positive (semi) definite solution of the Riccati equation

$$A^*P + PA + Q - PBR^{-1}B^*P = 0$$

is obtained by the function

K = RICC(A,B,Q,R).

The closed-loop system matrix eigenvalues may be located in the variable W by

<K,W> = RICC(A,B,Q,R).

The full form of this function

<K,W,P> = RICC(A,B,Q,R)

allows to get also the solution of the Riccati equation.

In the discrete-time case the quadratic cost

$$J = \sum_{k=0}^{\infty} (x_k^*Qx_k + u_k^*Ru_k)$$

is minimized taking the gain matrix as

$$K = (R + B^*PB)^{-1}B^*PA$$

where P is the positive (semi) definite solution of the discrete Riccati equation

$$A^*PA - P + Q - A^*PB(R + B^*PB)^{-1}B^*PA = 0.$$

In SYSLAB this is done simply by

K = RICD(A,B,Q,R).

The extended forms of this function,

<K,W> = RICD(A,B,Q,R),

<K,W,P> = RICD(A,B,Q,R)

allow to obtain the closed-loop eigenvalues and the solution of the discrete Riccati equation.

Example 3. Consider the synthesis of optimal continuous system using the function RICC. The system matrices A, B are as in the Examples 1 and 2.

```
Q =
 1.  0.  0.  0.
 0.  1.  0.  0.
 0.  0.  1.  0.
 0.  0.  0.  1.

R =
 1.  0.
 0.  1.

<K,W,P> = RICC(A,B,Q,R>

P =
 0.7514  0.1587 -0.0181  0.3170
 0.1587  0.1914 -0.0079  0.0159
-0.0181 -0.0079  0.1150 -0.0375
 0.3170  0.0159 -0.0375  0.2827

W =
-7.4618 - 8.5474i
-7.4618 + 8.5474i
-6.1121 - 4.3563i
-6.1121 + 4.3563i

K =
 1.5375 -0.7672 -0.3311  1.4618
 1.8708  0.1279  0.1702  1.8106
```

The functions RICC and RICD may be used also for synthesis of optimal filters.

Discretization and State Vector Determination

The matrices of a discrete model of a continuous system for a discretization interval DT are obtained by

<AD,BD> = DSCR(A,B,DT).

For given initial condition X0 the solution for the state vector is found in the continuous-time case by

VECC(A,X0,DT,L)

and in the discrete-time case by

VECD(A,X0,L)

where L is the number of the points for which the solution should be obtained.

Additional functions will be included in SYSLAB for synthesis of reduced order observers and dynamic compensators.

THE NUMERICAL ALGORITHMS

As mentioned above, only numerically stable computational algorithms are implemented in SYSLAB. The using of complex versions of the algorithms to ensure compatibility with the programs implemented in MATLAB allows also to obtain programs that are considerably shorter than in real arithmetic.

The continuous Sylvester and Lyapunov equations are solved by the algorithm of Bartels and Stewart (1972). The algorithm of Golub, Nash and Van Loan (1979) is in some cases more efficient but requires a longer program. The discrete equations are solved by the algorithm of Barraud (1977). A modification of these algorithms provides the solution of all four equations by three subroutines only.

The reduction into orthogonal canonical form and the controllability/observability analysis is performed by the algorithm described in (Konstantinov, Petkov and Christov, 1982). For this purpose QR decomposition with column pivoting (Dongarra and others, 1979) is used. If only the dimension of the controllable/observable subspace is required the unitary transformations are not accumulated.

The pole assignment synthesis is done by the algorithm proposed in (Petkov, Christov and Konstantinov, 1984b, 1984c). It includes a preliminary reduction of the system into orthogonal canonical form. Since complex arithmetic is used the real and complex poles are treated in the same manner.

The solution of the continuous Riccati equation is obtained by the algorithm of Laub (1979). The ordering of the Schur form of the Hamiltonian matrix is done easily in complex arithmetic using the programs developed by Stewart (1976).

The discrete Riccati equation is solved by the corresponding generalized eigenvalue problem formulation as proposed in (Pappas, Laub and Sandell, 1980), (Walker, Emami-Naeini and Van Dooren, 1982). The complex version of the QZ algorithm for solving the generalized eigenvalue problem is realized by the programs described in (Garbow, 1978). The ordering of the upper triangular form of the generalized eigenvalue problem is done as suggested in (Van Dooren, 1982), (Petkov, Christov and Konstantinov, 1984a). The solution of the discrete Riccati equation by the inverse generalized eigenvalue problem as proposed in (Walker, Emami-Naeini and Van Dooren, 1982) is not to be recommended since it may be very ill conditioned and time consuming.

The discretization of a continuous system and the determination of the state transition matrix is done by the algorithm of Ward which is one of the best among available algorithms. The algorithm implemented in MATLAB is not used since it exploits the eigenvector matrix which may be very ill conditioned in some cases.

All programs are written in Fortran using double precision. The system is installed on VAX-11/750 (for problems of order 100 or more) and, for lower order systems, on PDP-11/40.

CONCLUSIONS

An interactive system called SYSLAB intended to the computer-aided analysis and design of linear control systems is briefly described. SYSLAB gives an easy access to the recent powerful computational algorithms for linear multivariable systems as well as to the up-to-date high quality software for matrix computations. The system is portable and may be used for solving high order and ill conditioned problems.

REFERENCES

Barraud, A.Y. (1977). A numerical algorithm to solve $A^TXA - X = Q$. IEEE Trans.Automat.Contr.,AC-22,883-885.

Bartels, R.H., and G.W. Stewart (1972). Solution of the matrix equation AX + XB = C. Comm.ACM, 15,820-826.

Dongarra, J.J., J.R. Bunch, C.B. Moler, and G.W. Stewart (1979).LINPACK User's Guide. SIAM, Philadelphia.

Garbow, B.S.(1978). The QZ algorithm to solve the generalized eigenvalue problem for complex matrices. ACM Trans. Math.Soft., 4,404-410.

Garbow, B.S., J.M. Boyle, J.J. Dongarra, and C.B. Moler (1977). Matrix Eigensystem Routines-EISPACK Guide Extension. Springer-Verlag, Berlin.

Golub, G.H., S. Nash, and C. Van Loan (1979). A Hessenberg-Schur method for the problem AX + XB = C. IEEE Trans. Automat.Contr., AC-24,909-921.

Gregory, Ch.,Jr., S. Shah, and R. Walker (1982). $MATRIX_X$: Application to large space structure control design problems. Proc. 21st IEEE Conf.Dec.&Contr., Flo.,Dec.1982,pp.912-917.

Konstantinov, M.M., P.Hr. Petkov, and N.D. Christov (1982). Orthogonal invariants and canonical forms for linear multivariable systems. Proc.8th IFAC Congr. (Kyoto,Aug.1981),vol.3,Pergamon, Oxford,pp.49-54.

Laub, A.J. (1979). A Schur method for solving algebraic Riccati equations. IEEE Trans.Automat.Contr., AC-24, 913-921.

Little, J.N., A. Emami-Naeini, and S.N. Bangert (1984). CTRL-C and matrix environments for the computer-aided design of control systems. Proc.6th Internat.Conf.Anal.&Optimization of Syst., Nice, June 1984, part II, Springer-Verlag, Berlin,pp.191-205.

Moler, C.B. (1982). MATLAB User's Guide. Dept.of Comp.Sci., University of New Mexico.

Pappas, T., A.J. Laub, and N.R. Sandell (1980). On the numerical solution of the discrete-time algebraic Riccati equation. IEEE Trans.Automat.Contr., AC-25,631-641.

Petkov, P.Hr., N.D. Christov, and M.M. Konstantinov (1984a). Remark on Algorithm 590. ACM Trans.Math.Soft.,10, 207.

Petkov, P.Hr., N.D. Christov, and M.M. Konstantinov (1984b). A computational algorithm for pole assignment of linear single-input systems. IEEE Trans. Automat.Contr.,AC-29,1045-1048.

Petkov, P.Hr., N.D. Christov, and M.M. Konstantinov (1984c). A computational algorithm for pole assignment of linear multi-input systems. Proc. 23rd IEEE Conf.Dec.&Contr.,Las Vegas,Dec.1984.

Smith, B.T., J.M. Boyle, J.J. Dongarra, B.S. Garbow, Y. Ikebe, V.C. Klema,C.B.Moler (1976). Matrix Eigensystem Routines-EISPACK Guide. Springer-Verlag,Berlin.

Stewart, G.W. (1976). HQR3 and EXCHNQ: Fortran subroutines for calculating and ordering the eigenvalues of a real upper Hessenberg matrix. ACM Trans.Math.Soft., 2,275-280.

Van Dooren, P. (1982). DSUBSP and EXCHQZ: FORTRAN subroutines for computing deflating subspaces with specified spectrum. ACM Trans.Math.Soft., 8,377-382.

Walker, R.A., A. Emami-Naeini,P.Van Dooren (1982). A general algorithm for solving the algebraic Riccati equation. Proc. 21st IEEE Conf.Dec.&Contr.,Flo.,Dec. 1982,pp.68-72.

Ward, R.C. (1977). Numerical computation of the matrix exponential with accuracy estimate. SIAM J.Numer.Anal.,14,600-610.

INTERACTIVE PROGRAM PACKAGE ANA FOR SYSTEM ANALYSIS AND CONTROL DESIGN

M. Šega*, S. Strmčnik*, R. Karba and D. Matko****

**University Edvard Kardelj, Institute "Jožef Stefan", Jamova 39, 61000 Ljubljana, Yugoslavia*
***University Edvard Kardelj, Faculty of Electrical Engineering, Tržaška 25, 61000 Ljubljana, Yugoslavia*

Abstract. The work contains a brief description of an interactive program package ANA which can be used as a tool for analysis and design of system control. It is designed for linear, time invariant, univariable or multivariable, continuous or discrete systems. It can be of great help in all steps of control design. The usefulness of the package is shown especially through the great interactiveness of programming which enables fast and simple work for different users which are familiar with system theory. For communication the simplest dialogues are used but they are not tedious and time consuming for experienced users because indirect running (BATCH) of the package is possible to repeat the whole design procedure with changed parameters or to repeat a part of the procedure and then proceed in direct way. Program package is built modularly so that its development is easy. All programs are written in program language FORTRAN-IV what enables simple transfer of the package to different computer configurations. In the future some additional extensions are foreseen like introducing digital simulation language in the package and some other approaches to the controller design. This will ensure greater capabilities of the package also in the sense of nonlinear systems treatment.

Keywords. Control theory; computer aided design; computer software; interactiveness; system presentations.

INTRODUCTION

Over the last decade a large amount of work was devoted to the development of computer-aided-design tools (Frederick, 1982) because it is obvious that the variety of control methods and algorithms is so wide that an engineer could not master both areas i. e. the control theory and the necessary programming. Computer-aided-design packages are based on the idea of combining engineer´s intuition and digital computing facilities (Astrom, 1983) as it can be seen from the available literature (Bingulac, Farias, 1977; Emami Naeini, Franklin, 1981; Eugene, 1982; Furuta, Kajiwara, Tsuruoka, 1979; Polak and others, 1982; Walker, Gregory, Shah, 1982).

Nowadays so called interactive program packages for computer aided design (CAD) represent the highest level of digital computer usage. Program sequences for the concrete area are namely integrated. So system engineers can simply use them because the work is interpretatively oriented. Users correspondingly choose appropriate operations and the package executes their numerical parts. Of course the user must be familiar with the implemented methods and with the way of the work with the package. The latter needs relatively short time in comparison with the amount of work required without such a tool. Usage of CAD packages however stresses the engineers´ understanding and knowledge of control theory as well as their intuition and experiences.

The essential contributions of CAD packages can be summarized as follows:

- There is no need to waste a time with programming.
- The time for solving the problem is significantly shortened.
- The data treatment is simplified especially for large scale systems.
- Quick comparison of different design principles is enabled.
- The packages can be of use to wide spectrum of system experts (scientists, engineers and students) and they extend their knowledge (Agathoklis and others, 1979).
- The properties of subroutine libraries remain unchanged and so the programming of new, more complex methods is enabled.

The aim of the work is to introduce program package ANA for analysis and design of system control (Šega, 1984a, 1984b). It is designed for linear, time invariant, univariable or multivariable, continuous or discrete systems. The fundamental principles for computer aided design software development were taken into account, namely the package modularity, flexibility, interactiveness and portability (Wieslander, 1979). It includes all necessary operations from control theory (transformations, analysis, synthesis) and appropriate I/O facilities. The usefulness of the package is shown especially through the great interactiveness. In our opinion the simplest man-machine dialogues in connection with BATCH processing are appropriate for differently skilled users as well as for educational purposes.

CAPABILITIES OF ANA

Figure 1 shows the information scheme of the program package which includes all necessary units (blocks) from the control theory:

- transformations between different system presentations as well as conversions of some presentations in other forms (e.g. canonical forms),
- analysis and verification of systems,
- synthesis of control systems and simulation.

Besides the operations from control theory the package consists also input and output operations. The input block enables generation of the internal data base on the basis of apriori knowledge about the system. On the one hand the output function is the presentation of simultaneous results inside

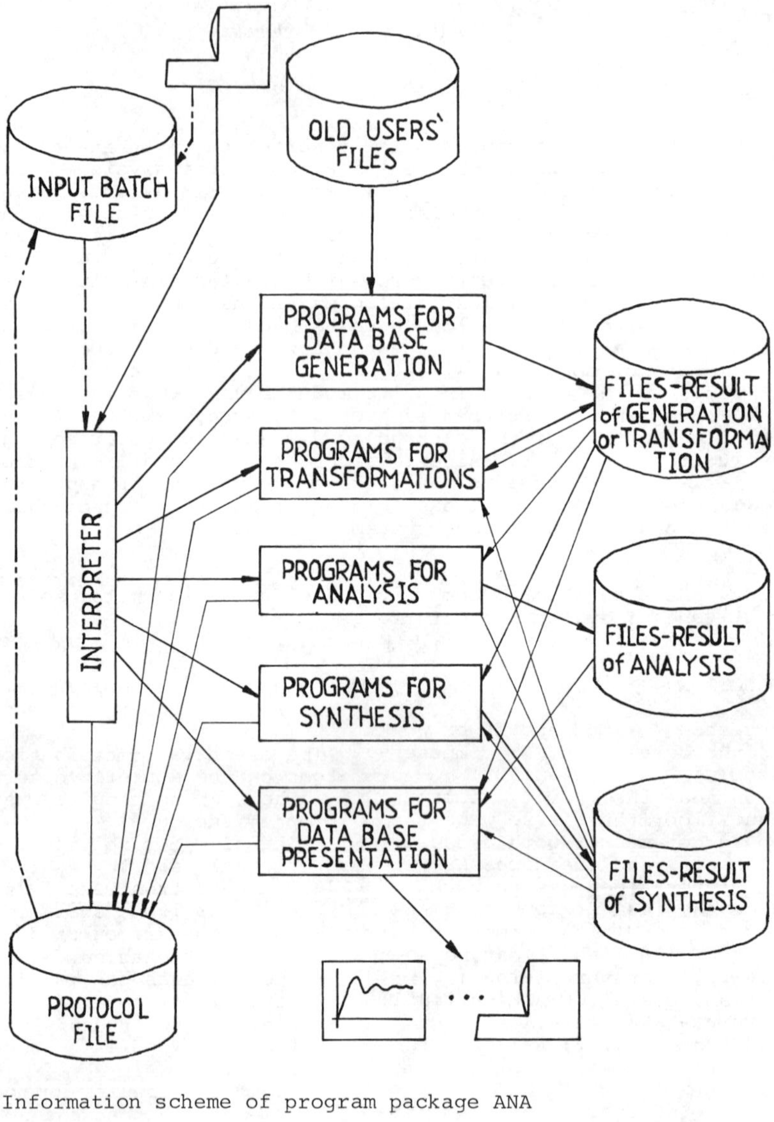

Fig. 1. Information scheme of program package ANA

the programs and the presentation of the contents of resultant files through the output block. On the other hand the output function is so called protocol file which saves all correct commands and data in the same order they have been entered during the work with the package.

Representations of Systems

All operations in the package are founded on following presentations of multivariable systems and they are appropriate coded for the interactive work:

a) in time domain

- sampled data series of inputs and outputs (TTI),
- series of self and cross correlation functions (TKF)[1],
- discrete system response on step inputs (TST)[1],
- discrete system response on pulse inputs (TIM)[1],

b) in frequency domain

- frequency spectrum of inputs and outputs (TFT)[1],
- transfer function matrix for continuous systems (TGS),
- tranfer function matrix for discrete systems (TGZ),

From the users point of view the most important is the interactive way of general control structure definition on the (semi) graphical terminal. In the package it is realized rather simple on the basis of fixed general control structure and simple substructure and their specific connections. Through the interactive definition the data base for so called universal computer simulation program (UNICUS) is built. On it basis UNICUS creates adequate program for simulation of the treated system. As an illustration we give a simple example of interactively defined structure in Appendix.

In the up-to-date version of the package univariable (PID and generalized deadbeat) and multivariable optimal state space controller in connection with Kalman estimator can be designed.

INTERACTIVE CAPABILITIES

For the man-machine communication the simplest dialogues are used. They are question and answer dialogue, dialogue using mnemonics and menu selection dialogue. Because they are well known, we describe only the specific way of work with the internal data base and possibility of BATCH processing with the package.

Files - Data Base

Files that contain system data are results of the operations. In regard of system representation they differ by their contents and because of that their extensions of the name are equal to the described codes of system presentations.

The file which is the result of concrete operation has the same whole name (name and extension) in all executions of this operation. Because there is a danger of deleting some usefull files by repetitive executions of the same operations the interactive change of the file name (only the name, the extension reamins the same) is realised after each operation that generates new file.

For the specific operation predefined form of system data,a file with defined extension is needed of course. So the package searches for files with that extension and user decides which one among them is appropriate.

So organized work is senseful only if the package considers only the user´s files from current running. For that reason the deleting of files with described extensions is realized at the beginning of the work.

Batch Processing

Batch processing is based on protocol file which contains all commands and data that have been used in previous treatment by the package. This file should be only appropriate corrected and then used as so called input batch file.

The elementary possibilities of batch processing are:
- repetition of the unchanged procedure,
- repetition of the whole procedure for changed parameters,
- repetition of a part of procedure by batch processing and than continuation in the direct way,
- documentation of the communication of the procedure.

PROGRAM ORGANIZATION

As it is shown in Fig. 1 the package is divided in several independent programs that are linked together with the interpreter program which enables also the choice of concrete operation. This characteristic and the fact that we store the internal data base on disks ensures a great modularity of the package.

All programs are written in standardized program language FORTRAN-IV and this besides the described fundamental organization enables relatively simple transfer of the package to computers with small dimensions of the memory.

In this moment the package is implemented on minicomputer PDP 11/34 with RT11-SJ operating system. The computer has 64k byte of memory, two disks RK05J and appropriate I/O devices (graphical terminal VT40 or semigraphical terminal KOPA-1000, printer FACIT 4542 and analogue plotter HP7045B).

Software of the up-to-date version is divided according to the function in:
- (sub)programs of the interpreter (cca 35),
- (sub)programs for specific complex operations (cca 100),
- subroutines for linear algebra (cca 15),
- subroutines of common meaning for all (sub)programs (for the input of data; in connection with internal data base and protocol file; etc) (cca 10),
- libraries of subroutines for graphics on the terminal and plotter (cca 20).

A relative small size of memory dictates following limitations for systems:
- The number of inputs and outputs is not bigger than 10.
- The order of polynomials in transfer function matrix is less than 11.
- The order of the system in state space representation is not bigger than 20.
- The number of samples in I/O time series is not bigger than 400.

CONCLUSION

As it can be seen from the paper the up-to-date version represents only the part of our main project. The experiences with the package show that its concept and program organization ensure simple extensions and modifications. Batch running enables comfortable work also for experienced users in spite of the simple dialogues which are on the other hand suitable for unexperienced users. Relatively poor computer equipment will probably involve some problems concerning necessary extensions of the package so we already think about its transfer to more capable computer.

In the near future some extensions are foreseen like introducing digital simulation language what will result in ANA´s greater capabilities in the sense of nonlinear systems treatment. We shall also include some other approaches to the controller design which have been already developed and programmed in the past. So they require only some modi-

- discrete frequency response (TGF)[1],
c) state space description
- for continous systems (TMS),
- for discrete systems (TMZ),
d) special representations based on simulation
- physical system model (real model or analogue computer)[1],
- model realized by digital simulation [1],
- so called user´s block (subroutine).

Transformations

Among described different system representations there are distinguished direct and indirect transformations as well as tranformations between different forms. It is also obvious that some transformations between foreseen system descriptions are not realizable by digital computer.

Table 1 qualifies the transformations between different system presentations that are installed in the package in this moment. In table 1 some direct transformations are presented with special names that are well known in control theory, and the others with simple notation "direct". Transformations which give modifications of the same presentation are not included in the table because of their specificity.

TABLE 1 Overwiev of tranformations between different system representations

from \ to	TTI	TGS	TGZ	TMS	TMZ	TBL
TTI		through TGZ	identification	through TMZ	identification	through TGZ
TGS	through TGZ*		discretization	direct	through TGZ or TMZ	—
TGZ	simulation	direct		through TGS or TMZ	direct	—
TMS	through TMZ*	direct	through TMZ or TGS		discretization	—
TMZ	simulation	through TGZ or TMS	direct	direct		—
TBL	simulation	through TTI	through TTI	through TTI	through TTI	

* in this version; exactly by digital simulation

Analysis

Block for analysis enables stability, controllability, observability and cyclicity tests of systems as well as rank, norms, eigenvalues and eigenvectors computation for systems presented in state space.

For continuous systems presented with transfer function matrix Bode, Nyquist and Rootlocus plots are included in the package.

For verification of system model or controlled system the comparison of sampled data series and/or computed time responses can be realized giving also some statistic performances.

Synthesis

In the block for synthesis there are two possibilities:

a) Figure 2 shows the maximal structure for continuous or discrete linear time invariant control system treatment. In the concrete case user defines interactively the substructure, the only condition is that single parts of complex system (blocks) are presented in state space. The result of the operation is the system described in state space that is equivalent to the control system according to input output behaviour.
b) The simulation of general system structure including precompensators, cascade, feedback and feedforward controllers as well as corresponding observers or estimators. For this treatment it is supposed that the process is presented with one of following descriptions:

[1]Not yet installed

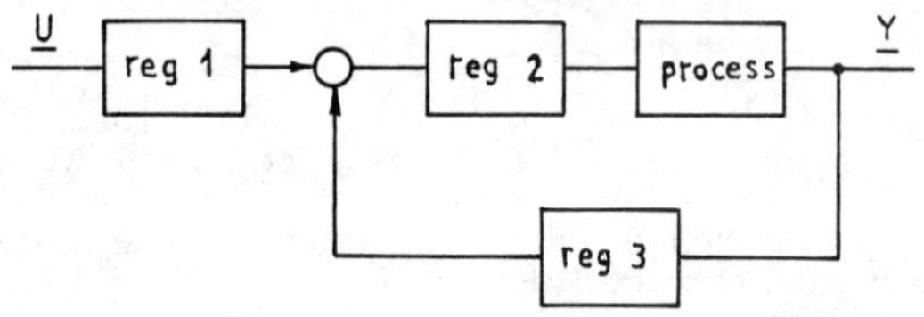

Fig. 2. Maximal structure for linear time invariant control system treatment

- in state space (TMZ or TMS[1]),
- with transfer function matrix (TGZ or TGS[1]),
- by physical model (TMO[1]),
- model realized with digital simulation (TSI[1]),
- with user´s block (TBL).

Because of the frequency of optimal problems the possibility of parameter estimation is included in the treatment of general system structure. It is based on some standardized criterion functions and some installed optimization algorithms. Both criterion function and optimisation algorithm can be also written by users.

fications. These approaches are multivariable controllers based on the pole placement methods, decoupling and inverse Nyquist array technique in combination with corresponding observers and some adaptive controllers in connection with appropriate identification methods.

REFERENCES

Agathoklis,P., F.E. Cellier, M.Djordjević, P.O. Grepper and F.J. Kraus (1979). Educational aspects of using computer aided design in automatic control. Proc. IFAC Zürich Symp., 441-446.

Astrom, K.J. (1983). Computer aided, analysis and design of control system-a perspective. Control Systems Magazine, Vol.3, No. 2, 4-16.

Bingulac, S., and M.A. Farias (1977). L-A-S language and its use in automatic control education and research. Automatika, Vol. 5-6, 213-224.

Emani-Naeini, A., and G.F. Franklin (1981). Intercative computer-aided design of control systems. Control Systems Magazine, Vol.1, No. 4, 31-36.

Eugene, A.L. (1982). LCAP2-linear controls analysis program. Control Systems Magazine, Vol. 2, No. 4, 15-18.

Frederick, D.K. (1982). Software summaries on computer aided control system design software packages. Control Systems Magazine, Vol. 2, No. 4, 37-44.

Furuta, H., H. Kajiwara and K. Tsuruoka (1979). Computer aided design program for linear multivariable control systems. Proc. IFAC Zürich Symp., 267-272.

Polak, E., P. Siegel, T. Wuu and W.T. Nye (1982). DELIGHT. MIMO: An interactive, optimization-based multivariable control system design package. Control Systems Magazine, Vol. 2, No.4, 19-24.

Šega,M.,S. Strmčnik and J. Čretnik (1984a). Analysis and synthesis of systems with program package ANA-1. Proc. ETAN Split Yugoslavia, Vol. 4, 35-42 (in slovene).

Šega,M. (1984b). Computer aided analysis and control design of dynamical systems. Faculty of Electrical Engineering Ljubljana, Yugoslavia (work for M.S. degree in slovene).

Walker, R., C.Gregory and S. Shah (1982). $MATRIX_x$: A data analysis, system identification, control design and simulation package. Control Systems Magazine, Vol. 2, No. 4, 30-37.

Wieslander, J. (1979). Design principles for computer aided design software. Proc IFAC Zürich Symp., 493-496.

APPENDIX

Example 1

A simple system which is given by:

$$G(s) = \frac{1-4s}{1+14s+40s^2}$$

was treated to illustrate the work with ANA. First the analysis in frequency domain was undertaken and as illustration only root locus diagram is shown in Fig. 3. Then the system was transformed to the state space

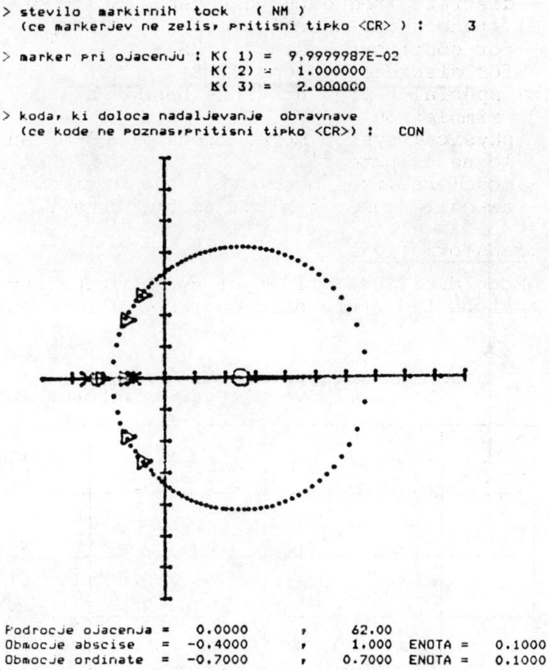

Fig.3. Root locus diagram

and controllability and observability was tested. Discretized system was then simulated (Fig.4).

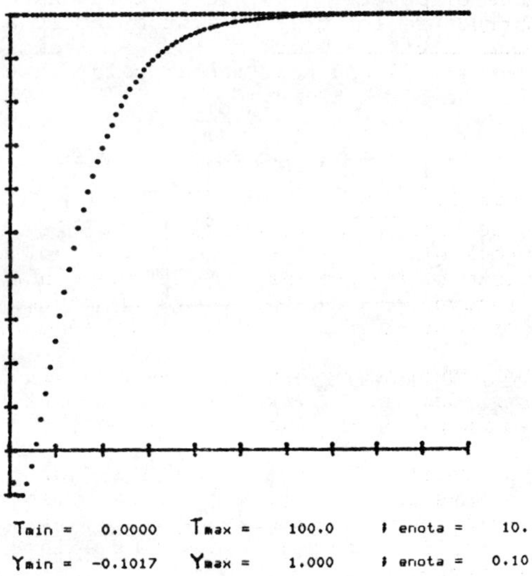

Fig.4. Open loop system

The simulation of closed loop system with appropriate PID controller

$$G(z) = \frac{5.958-10.337z^{-1}+4.492z^{-2}}{1-z^{-1}}$$

is shown in Fig. 5. Though in slovene the listing of the part of communication (the definition of closed loop simulation) is given in Fig. 6 as illustration.

Example 2

Figure 7 shows a regulation scheme for univariable process and Fig. 8 which is s photo of the semigraphical terminal shows the final step of the interactive definition of the structure inside the package.

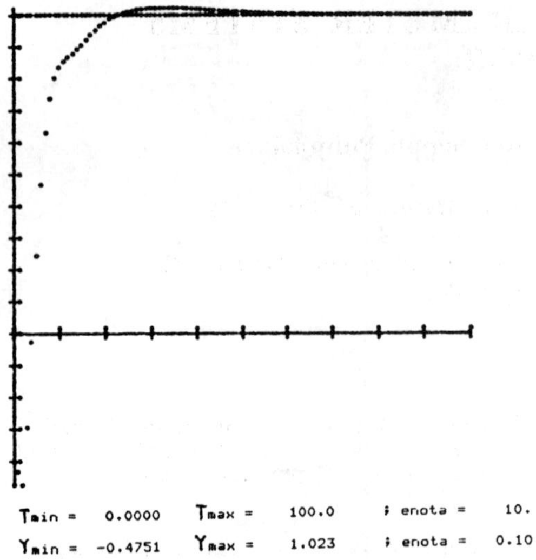

Fig. 5. Closed loop system simulation

```
SINTE: *TMZ

> koda oblike,v katero zelis transformirati
  zapis v prostoru stanj - diskretno -  TMZ
  (ce kode ne poznas, pritisni tipko <CR> ) :   TTI

> datoteka, ki jo zelis uporabiti
  (samo IME,ker je EXT. default-TMZ)
  (ce ne zelis nobene, pritisni tipko <CR>)

          SISTEM
          PROCES
          REGUL     : SISTEM

> vsaj en vhodni signal je podan v datoteki *.TTI
  (ni eden izmed standardnih signalov) (Da ali Ne) :   N

> komentar za datoteko DK2:TRMZTI.TTI (do 62 znakov) :

    SIMULACIJA REGULACIJSKE ZANKE ( VHOD-STOPNICA )

> ime  1. reguliranega  vhoda  :   VHOD

> ime  1. izhoda :   IZHOD

> cas trajanja simulacijskega teka ( Tf ) :     100.0000

  V DATOTEKO SE VPISE VSAK 'I'-TI IZRACUN
  ( cas vzorcenja v datoteki: I*  1.0    )

> podaj parameter za vpis v datoteko ( I )
  ( pritisni tipko <CR>, ce zelis I = 1 )  :    1

> zacetne vrednosti spremenljivk stanja ( Xi(0),i=1, 4 ) :

     0.0000        0.0000        0.0000        0.0000

> koda signala za vhod    VHOD
  (ce kode ne poznas,pritisni tipko <CR>)  :   ST

> amplituda stopnice ( AM )   :    1.000000

> zacetni cas ( TZ ) [ sek ]  :    0.0000000
```

Fig.6. The part of communication (in slovene)

As it can be seen the main control structure on Fig. 9 is rather simplified for this case and two additional substructures are included in Fig. 8 to realize the discussed regulation. It is necessary to stress that the scheme in Fig. 8 is only one of several possible definitions of structure for treated system and it assume that the block M1 changes a sign of the signal and that the transfer function of the block M2 is $(G_{p1}(s)-1)$.

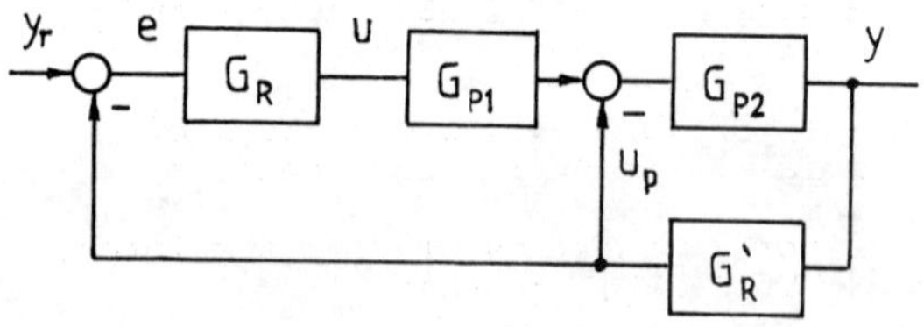

Fig.7. Regulation of univariable process

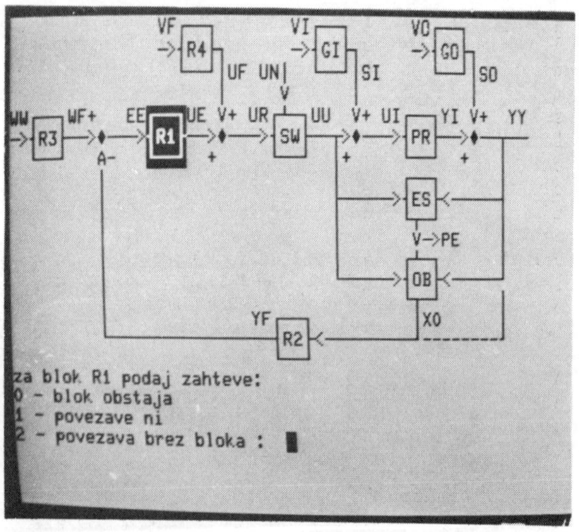

Fig. 9. Fixed general control structure used at the beginning

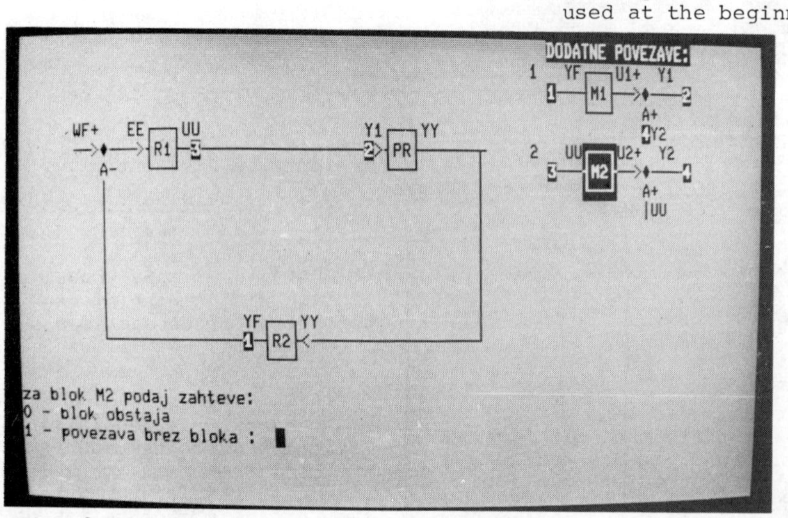

Fig. 8. Interactively defined structure for the system

TDP — A DISTRIBUTED PARAMETER SYSTEMS SIMULATOR

Matti Aaltonen, Juha Tanttu* and Seppo Pohjolainen

Tampere University of Technology, Department of Mathematics, PO Box 527, SF-33101 Tampere 10, Finland
**Tampere University of Technology, Department of Automatic Control, PO Box 527, SF-33101 Tampere 10, Finland*

Abstract. A simple and easy - to - use computer simulator for distributed parameter systems is presented. The simulator is able to treat parabolic and elliptic systems in two - dimensional domains. Two examples are given to demonstrate the use of the simulator.

Keywords. Distributed parameter systems; computer sofware; PID control; simulation.

INTRODUCTION

Many important industrial processes can quite accurately be modelled with partial differential equations (PDE). These models - usually called distributed parameter systems (DPS) - are often based on specific physical, chemical, biological etc. laws, and therefore they are accurate and the parameters, such as temperature, pressure, concentration, etc. are spatially distributed and have a clear physical meaning. From the control point of view an important advantage also is that location of measurements and controllers becomes a natural part of a control problem.

Although distributed parameter systems have certain benefits, they have not been so popular among control engineers. This is certainly due to the fact that existing control theory is finite-dimensional by nature and prefers systems with finite-dimensional state spaces or rational transfer functions.

However, during the 1970's and 1980's theoretical means to control DPS have rapidly increased and they include e.g. observer theory (Bhat, 1976), finite-dimensional compensators, (Curtain, 1984), (Schumacher, 1983), robust controllers (Pohjolainen, 1982) and transfer function algebras (Callier, Desoer, 1978). The need to simulate DPS arises naturally from both theoretical and practical needs.

TDP - A DISTRIBUTED PARAMETER SYSTEMS SIMULATOR

The purpose of this work was to produce a reliable, flexible, easy - to - use DPS - simulator that could be run at a minimum programming cost with various PDE models. Thus it was quite natural to start by looking for a suitable PDE - software to actually solve the process model equations. Our selection was to use a commercially available PDE - solver, called TWODEPEP, produced by IMSL (TWODEPEP, 1982). This program has certain advantages e.g.

- to minimize programming effort it contains a preprocessor which transforms the original problem description "Input Data Set" file into FORTRAN.
- it uses finite-element method with automatic user-controlled mesh refinement
- the description of boundaries is easy
- graphical output (3 - D surfaces and vector fields)
- it is possible to solve systems of partial differential equations.

The major limitations of TDP - simulator, due to TWODEPEP, are that

- only parabolic and elliptic systems in two dimensional domains are allowed
- in general separate partial differential equations must have the same domain.

For a system of two equations the most general form is

$$C_1(x,y,u,v,t)\frac{\partial u}{\partial t} = \frac{\partial}{\partial x}\,OXX\,(x,y,u_x,u_y,v_x,v_y,u,v,t) + \frac{\partial}{\partial y}\,OXY\,(x,y,u_x,u_y,v_x,v_y,u,v,t) + F_1\,(x,y,u_x,u_y,v_x,v_y,u,v,t)$$

$$C_2(x,y,u,v,t)\frac{\partial v}{\partial t} = \frac{\partial}{\partial x}\,OYX\,(x,y,u_x,u_y,v_x,v_y,u,v,t) + \frac{\partial}{\partial y}\,OYY\,(x,y,u_x,u_y,v_x,v_y,u,v,t) + F_2\,(x,y,u_x,u_y,v_x,v_y,u,v,t)$$

for (x , y) in the region R , with the general boundary conditions

$$u = FB\,1\,(s,t)$$

$$v = FB\,2\,(s,t)\ ,$$

for s on a part of the boundary, $s \subset \partial R_1$, and

$$OXX \cdot n_x + OXY \cdot n_y = GB\,1\,(s,u,v,t)$$

$$OYX \cdot n_x + OYY \cdot n_y = GB\,2\,(s,u,v,t)$$

on ∂R_2, where (n_x , n_y) is the unit normal to the boundary.

The initial conditions are

$$u = u0\,(x,y)$$

$$v = v0\,(x,y)\ ,\ t = T0\ .$$

However, TWODEPEP can also solve a single eguation or several simultaneous equations of the above form.

Because TWODEPEP, as such, is capable only of open loop simulation, the following modifications were made to realize feedback actions (Fig. 1).

In general, one has to measure the solution at specific points, defined by the user, and use these measurements with the controller to compute control signals. To elaborate, TDP consists of the following subprograms (Fig. 2).

The subroutine MIMAAR contains definitions of measurement points, dimensions of control and measurement vectors. TWODEPEP data input file contains CONTROL block as a function subroutine. A call to this subroutine will result in a call to the module MIMAAR to read the time and the solution from TWODEPEP at specified points. Then MIKIRJ calls CONTROLLER, which computes the actual control values. The control values are transferred to CONTROL module trough the common block. MIKIRJ also writes the time instant, control and measurement values for plotting.

Plotting is realized by using GPGS - package (GPGS-F User's Guide, 1980), which is more versatile than the original TWODEPEP plotting program, which is available, too.

To run a problem with TDP, the user has to create two files. First, PROCESS.DAT, which contains a description of the process and control as a TWODEPEP input data set, and second, CONTROLLER. OHJ, to define the control algorithm in FORTRAN. When these two files are ready, the user may execute the simulation by using the following commands

TDPINIT
- asks the name of the CONTROLLER FILE, measurement points and the dimension of control vector.

TDPSOLVE PROCESS
- creates from PROCESS.DAT a FORTRAN program PROCESS.FOR, compiles, links and runs the necessary programs.

TDPOINT
- plots the controls and measurements as functions of time.

TWOD3DIM
- plots 3D - surfaces

TDP IN AN ATTACHED PROCESSOR

So far TDP was used in a VAX 11/780 - computer at Tampere University of Technology. Because execution times for systems became rather large, up to ten hours, it was quite natural to try to speed up computation by using a FPS-164 attached processor that was available, too.

Perhaps the easiest way, to speed up a FORTRAN program in FPS-164, is to replace possible parts of the program by APMATH 64 library routines (APMATH Manual, 1982). This library is written in APAL 64 (Attached Processor Assembly Language), and it contains over four hundred relatively fast mathematical routines. Some modules of TWODEPEP could directly be replaced by APMATH 64 routines. In addition, some parts were modified to be better prepared for computation in an attached processor (APFTN 64 User's Guide, 1983). After these improvements TDP became about twice as fast as the original VAX - version. This ratio will even be improved, if the size of the programs becomes large, especially, if the number of partial differential equations to be solved, increases.

The size of the main memory at our installation is only 256K words. Thus the heart of TWODEPEP - a large linear/nonlinear equation solver - typically cannot be stored in the main memory. The resulting I/0 between the main memory and a disk file will slow the speed considerably down, even though FPS-164 may perform asynchronous I/0. With a larger main memory a faster TDP version could easily be obtained by solving the linear/nonlinear equations with APMATH 64 routines.

EXAMPLES

Heating a moving sheet

A sheet is moving through a heating zone (Fig. 3) with velocity v . The problem is to control, with two PI-controllers, heat flux from the heaters, so that temperature at the measurement points (0.2 , 0.8) (0.2 , 0.2) will be requlated to 35 degrees.

The values of the parameters are: v = 5 , h = 0 , α = 1000 . The initial temperature u(x,y,0) = 20 . The TWODEPEP input data file with control block (PROCESS.DAT) may be seen in Fig. 4. Figure 5 shows that CONTROLLER.OHJ file contains two PI-controllers. The results of the run may be seen in Figures 6 and 7.

Control of a nuclear reactor

This problem originates from (Neboyan, Koivo, 1983). Neutron flux in a nuclear reactor may be described as

$$\frac{\partial \Phi}{\partial t} = \frac{k + k_t}{l} + \frac{k^o - 1 - B}{l}\,\phi + \frac{M^2}{l}\,\nabla^2 \phi + \lambda G$$

$$\frac{\partial C}{\partial t} = \frac{\beta \phi}{l} - \lambda C$$

$$T_t \cdot \frac{\partial k_t}{\partial t} + k_t = k_{0t}\,\phi\,,$$

where

$\phi(x,y,t)$ is a normalized neutron flux

$c(x,y,t)$ speed of delayed neutrons

$k_t(x,y,t)$ the effect of temperature feedback

The values of the constants were

$M^2 = 550$ $\quad l = 0.001$

$\beta_0 = 0.0064$ $\quad \lambda = 0.08,\ T_t = 5,$

$k = 1.0075$ $\quad k_{0t} = 0.0128$

The domain of the system is a circle (reactor form above). The four control rods are modelled by gaussian functions of the type

$$\frac{1}{2\pi\sigma^2}\, e^{-\frac{x^2+y^2}{2\sigma^2}} \cdot u(t)\ .$$

The purpose of a P - controller was to stabilize and regulate the system. The solutions are seen in Figures 8 and 9.

CONCLUSIONS

An easy-to-use, flexible, multipurpose simulator for distributed parameters systems has been presented. The possibility to speed up computation with an attached processor was discussed. Two examples were given to show the performance of the TDP-simulator.

In future, the purpose is to develope the control module, so that it could include e.g. multivariable PID - controllers, robust controllers, parametric optimization, and transfer function measurement routines.

ACKNOWLEDGMENT

The authors thank Mrs. Päivi Kuusisto for her careful typing.

REFERENCES

APFTN 64 User´s Guide (1983). Floating Point Systems, Inc., Publication.

APMATH Manual (1982). Floating Point Systems, Inc.

Bhat, M.K.P. (1976). Regulator theory for evolution systems. Ph.d. thesis, department of Electrical Engineering, University of Toronto.

Callier F.M., Desoer C.A. (1978). An algebra of transfer functions for distributed linear time-invariant systems. IEEE Tr. Circuits and Systems, CAS - 25, 651-662.

Curtain R.F. (1984). Finite dimensional compensators for distributed parameter systems with unbounded control and observation. SIAM J. Control Opt. 22, 255-276.

GPGS - F User´s guide 4th. edition (1980). NORSIGD, TAPIR.

Neboyan V., Koivo H.N. (1983). Design of a neutron flux control system by the inverse Nyqvist array method. Int. J. Control 37, 535-552.

Pohjolainen S.A. (1982). Robust multivariable PI - controller for infinite dimensional systems, IEEE Tr. Aut. Control, AC - 27, 17-30.

Schumacher J.M. (1983). A direct approach to compensator design for distributed parameter systems. SIAM J. Control Opt. 21, 823-836.

TWODEPEP User´s manual, (1982). IMSL.

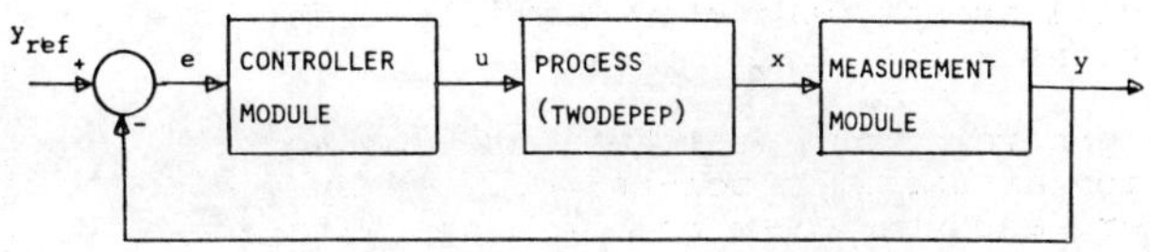

Fig. 1. A feedback loop.

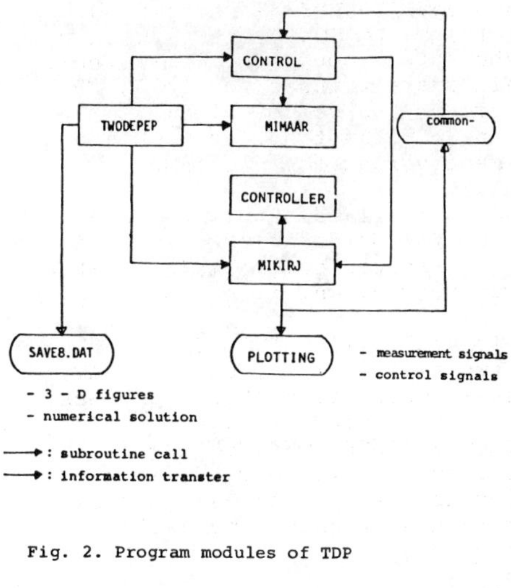

Fig. 2. Program modules of TDP

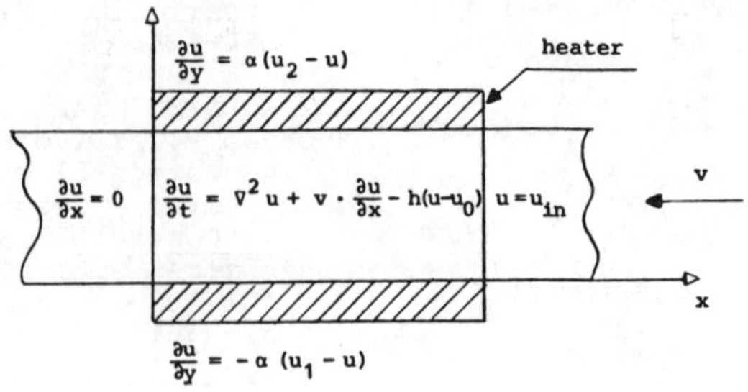

Fig. 3. A sheet moving trough a heating zone with process model equations.

```
            1   150     25596
****
****        Equations for a metal plate which moves
****        through region where the plate is heated
****        from two sides:
****
OXX         UX
OXY         UY
****
F1          -5.*UX - 0.001*U
C1          1.
U0          20.
****
****        Boundary conditions:
****
ARC=-1
FB1         20.
ARC=1
GB1         con1(u,t)
ARC=2
GB1         0.
ARC=3
GB1         con2(u,t)
****
NX          40
NY          20
MWR          8
T0           0.
TF           1.
DT           0.01
NOUT         50
NUPDT        0
****
VXY         2.0,1.0,  0.0,1.0,  1.0,0.5,  0.0,0.0, 2.0,0.0
IABC        1,2,3,    2,4,3,    4,5,3,    5,1,3
I           1,        2,        3,        -1
ADD.
****
            FUNCTION CON1( Y )

            REAL U(2)

            CALL OHJAUS(U)
            CON1 = 1000.*( U(1) - Y )

            RETURN
            END
****
            FUNCTION CON2( Y )


            REAL U(2)

            CALL OHJAUS(U)
            CON2 =-1000.*( U(2) - Y )

            RETURN
            END
****
END.
```

Fig. 4. Input data file for the first example.

```
SUBROUTINE SAATAJA( t, y, n, u, m )

INTEGER    n, m
REAL       t, y(n), u(m)

LOGICAL   first_call
REAL      diff_0(2), diff_1(2), integ(2), y_ref(2)
REAL      g0, g1, old_t

DATA    first_call                / .TRUE. /
DATA    diff_0, diff_1, integ     / 6*0. /

IF ( first_call ) THEN
     g0 = 2.
     g1 = 20.
     y_ref(1) = 35.
     y_ref(2) = 35.
     diff_0(1) = y_ref(1) - y(1)
     diff_0(2) = y_ref(2) - y(2)

     first_call = .FALSE.
ELSE
     diff_0(1) = y_ref(1) - y(1)
     diff_0(2) = y_ref(2) - y(2)
     integ(1) = integ(1) + (diff_0(1)+diff_1(1))/2.*(t-old_t)
     integ(2) = integ(2) + (diff_0(2)+diff_1(2))/2.*(t-old_t)
END IF

u(1) = g0*diff_0(1) + g1*integ(1)
u(2) = g0*diff_0(2) + g1*integ(2)

diff_1(1) = diff_0(1)
diff_1(2) = diff_0(2)
old_t = t

RETURN
END
```

Fig. 5. Controller-file for the first example. The file contains two PI-controllers.

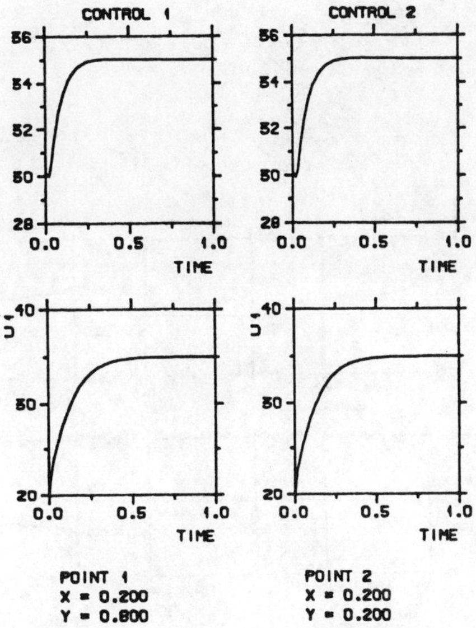

Fig. 6. The controls and measurements as functions of time.

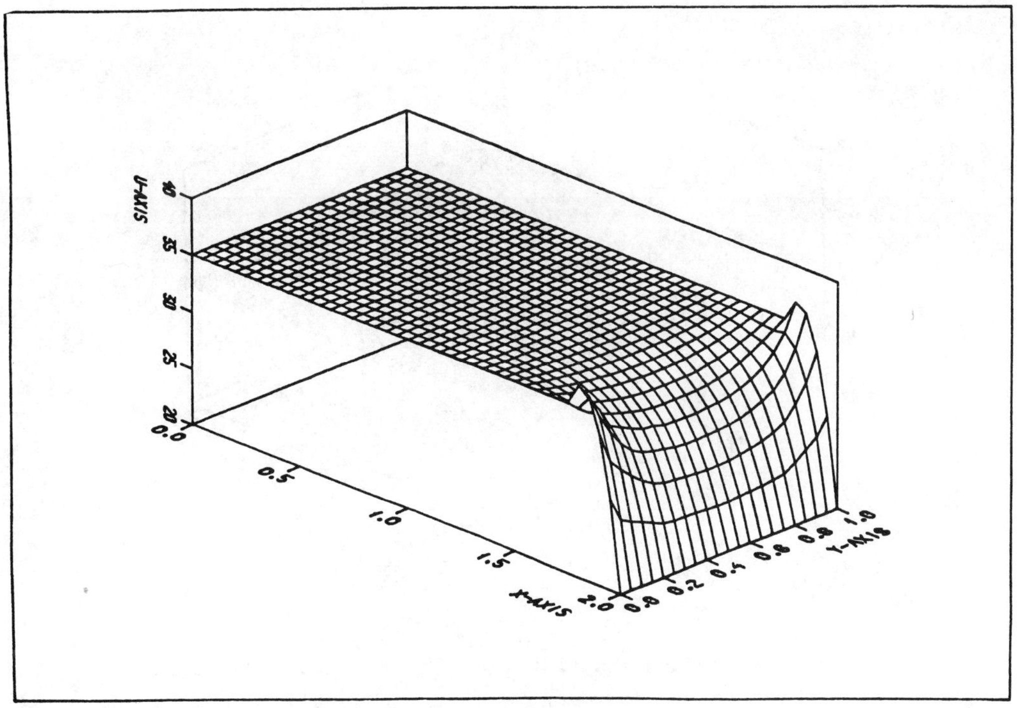

Fig. 7. Temperature profile of the sheet at t = 10.

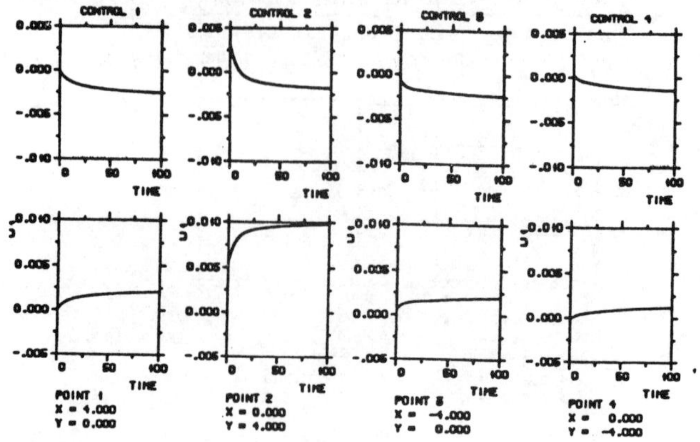

Fig. 8. The controls and measurements as functions of time.

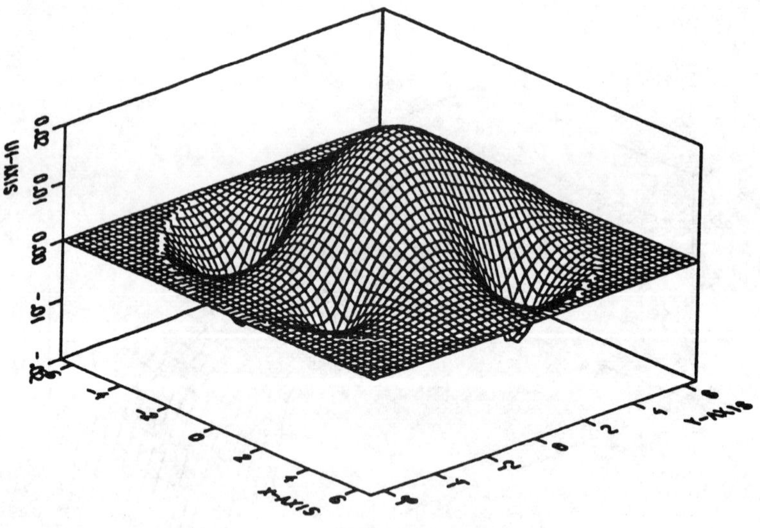

Fig. 9. The neutron flux profile at t=150. The effect of the four control rods to stabilize the system is clearly seen.

SELF-TUNING CAD PACKAGE AND ITS APPLICATION

Chen Zhen-Yu and Fang Li-Zhang

Automation Research Institute of Metallurgical Ministry, People's Republic of China

Abstract. This paper introduces a self-tuning control system CAD package CADST which includes 4 kinds of self-tuning algorithm. Engineers can use the CADST by simple man-machine interactive dialogs to implement the project design. A project of self-tuning thickness control system for reversing cold mill has been implemented by the using of CADST. The thickness output deviations reduce distinctly comparing with convention regulating system.

Keywords. CAD package; self-tuning control; thickness control system.

1. INTRODUCTION

The designs and applications of self-tuning control systems have been developed more then ten years. In order to prpvide a powerful tool for the engineers to design such kind control systems, we have developed a computer aided design software package of self-tuning CADST. In this paper there have an example which is a project of thickness regulating system of reversing cold rolling mill to perform the using of various method to design the regulating system and to analyse the calculating results. Engineers can use the CADST by simple man-machine interactive dialogs to implement the project design, and also can made a simulation stady in detials when a meticulous simulation plant model has been linked into this package.

CADST package includes 4 kinds of self-tuning algorithm: Self-tuning Regulator STR (Astrom et al., 1973; 1977); Self-tuning Controller STC (Clarke et al., 1975; 1977; Allidina et al., 1981); Pole Assignment Self-tuning Controller PAST (Wellstead et al., 1981); All Parameters Self-tuning Controller APST (Wu Hong-Xin et al., 1984) which is a new algorithm and the principal base is that the summing value of all parameters equals unit.

2. CADST PACKAGE

Description of algorithm

The main principal of APST can briefly illustrate as following. The continuous system can be expressed as

$$y^{(n)}(t)+a_{n-1}y^{(n-1)}(t)+\cdots+a_0y(t) = b_{n-1}u^{(n-1)}(t-d)+\cdots+b_0u(t-d) \quad (2.1)$$

where d is time delay. When take the sample of this system by cycle time Δt, the discrete system can be expressed as

$$y(t) = \alpha_1 y(t-1)+\cdots+\alpha_n y(t-n)+\beta_0 u(t-k-1) + \cdots + \beta_{n-1}u(t-k-n) \quad (2.2)$$

where $\quad k = \mathrm{INT}(d/\Delta t) \quad (2.3)$

is the time delay of discrete system.

Theorem 2.1. The summing value of all parameters of discrete system (2.2) equals unit when Δt approaches zero.

$$\lim_{\Delta t\to 0}\left[\sum_{i=1}^{n}\alpha_i + \sum_{j=0}^{n-1}\beta_j\right] = 1 \quad (2.4)$$

The prof see (Wu Hong-Xin et al., 1984). The main procedure is to use the differences to replace the derivatives.

$$y^{(m)}(t) = \left[y(t+m)-\binom{m}{1}y(t+m+1) + \binom{m}{2}y(t+m-2)-\cdots+\binom{m}{m}(-1)^m y(t)\right]/(\Delta t)^m$$

$$1 \leq m \leq n \quad (2.5)$$

where $\binom{m}{i}$ are the coefficients of binomial expansion. Using the equation (2.5) to combine the equation (2.1) can get result equation (2.4). Using the iterative method, equation (2.2) can be expressed as following prediction equation

$$y(t+k+1) = \alpha_1' y(t)+\cdots+\alpha_n' y(t-n+1) +\beta_0' u(t)+\cdots+\beta_{n-1}' u(t-n+1) \quad (2.6)$$

Lemma 2.1. The summing value of all parameters of system (2.6) also equals unit when Δt approaches zero.

$$\lim_{\Delta t\to 0}\left[\sum_{i=1}^{n}\alpha_i' + \sum_{j=0}^{n-1}\beta_j'\right] = 1 \quad (2.7)$$

Consider the condition of practical applications, the functions of STR,STC and APST of CADST package are including the feedforward terms for the detection of measureable disturbances.

The algorithm of APST can be described by model

$$Y(t+1) = \alpha(q^{-1})Y(t)+\beta_0 U(t-k_1) +\beta(q^{-1})U(t-k_1-1)+\gamma(q^{-1})W(t-k_2) +\psi(q^{-1})\varepsilon(t)+\varepsilon(t+1) \quad (2.8)$$

where

$$\alpha(q^{-1}) = \alpha_1+\alpha_2 q^{-1}+\cdots+\alpha_{n1}q^{-n1+1} \quad (2.9)$$

$$\beta(q^{-1}) = \beta_1+\beta_2 q^{-1}+\cdots+\beta_{n2}q^{-n2+1}$$

$$\gamma(q^{-1}) = \gamma_1+\gamma_2 q^{-1}+\cdots+\gamma_{n3}q^{-n3+1}$$

$$\psi(q^{-1}) = \psi_1+\psi_2 q^{-1}+\cdots+\psi_{n4}q^{-n4+1}$$

and $W(t)$ are measureable disturbances, k_2 is the time delay from $W(t)$ to $Y(t)$, $\varepsilon(t)$ is residual, assume $k_2 \geq k_1 \geq 0$.

Control criterion

$$u(t) = -\big[\alpha_1\hat{y}(t+n)+\cdots+\alpha_{k1}\hat{y}(t+1) \quad (2.10)$$
$$+\alpha_{k1+1}y(t)+\cdots+\alpha_n y(t-n_1+k_1+1)$$
$$+\beta(q^{-1})U(t-1)+\gamma(q^{-1})W(t+k_1-k_2)$$
$$+\psi_{k1+1}\varepsilon(t)+\cdots+\psi_{n4}\varepsilon(t-n4+k_1+1)\big]/\beta_0$$

ELS method is used to estimate the parameters, the estimate value of β_0 can be solved from (2.4). In pratical the sampling time Δt only can approaches to a sufficient small value, therefore it must exists some error part in the calculating equation, so equation (2.4) can rewrites as

$$\sum_{i=1}^{n}\alpha_i(t)+\sum_{j=1}^{n-1}\beta_j(t)+\beta_0(t)=1+e(t) \quad (2.11)$$

where $e(t)$ is error term.

Setting $\sum_{i=1}^{t}e^2(i)$ to minimal by the selection of $\beta_0(t)$ can get the iterative estimation equation as

$$\beta_0(t)=\beta_0(t-1)+\frac{1}{t}\big[\Sigma(t-1)-\beta_0(t-1)\big] \quad (2.12)$$

$$\Sigma(t-1)=1-\Big[\sum_{i=1}^{n}\alpha_i(t-1)+\sum_{j=1}^{n-1}\beta_j(t-1)\Big] \quad (2.13)$$

The algorithm of STR can be described by prediction model

$$Y(t+k_1+1) = \alpha(q^{-1})Y(t)+\beta_0 U(t) \quad (2.14)$$
$$+\beta(q^{-1})U(t-1)+\gamma(q^{-1})W(t+k_1-k_2)$$
$$+\varepsilon(t+k_1+1)$$

Control criterion

$$u(t) = -\big[\alpha(q^{-1})Y(t)+\beta(q^{-1})U(t-1) \quad (2.15)$$
$$+W(t+k_1-k_2)\big]/\beta_0$$

where β_0 is preselected.

The algorithm of STC can be described by extended output

$$\phi(t+k_1+1) = P(q^{-1})Y(t+k_1+1)-R(q^{-1})Y_r(t)$$
$$+Q(q^{-1})U(t)+S(q^{-1})W(t+k_1-k_2)$$
$$+\varepsilon(t+k_1+1) \quad (2.16)$$

where $Y_r(t)$ is the reference output.

Prediction model

$$\phi_y(t+k_1+1) + \alpha(q^{-1})Y(t)+\beta(q^{-1})U(t-1)$$
$$+\beta_0 U(t)+\gamma(q^{-1})W(t+k_1-k_2)+d$$
$$+\varepsilon(t+k_1+1) \quad (2.17)$$

where d is the constant component

The sampling equation of extended output can be described by

$$\phi_y(t+k_1+1) = P(q^{-1})y(t+k_1+1) \quad (2.18)$$

Control criterion

$$u(t) = -\big[\alpha(q^{-1})Y(t)-R(q^{-1})Y_r(t)$$
$$+\beta(q^{-1})U(t-1)+\big[\gamma(q^{-1})+S(q^{-1})\big]W(t=k_1-k_2)$$
$$+d\big]/(\beta_0+Q(q^{-1})) \quad (2.19)$$

The algorithm of PAST can be described by model

$$\hat{A}(q^{-1})Y(t) = q^{-k1}\hat{B}(q^{-1})U(t-k_1) +\hat{C}(q^{-1})\varepsilon(t) \quad (2.20)$$

where

$$\hat{A}(q^{-1})=1+\hat{a}_1 q^{-1}+\cdots+\hat{a}_{na}q^{-na} \quad (2.21)$$

$$\hat{B}(q^{-1})=\hat{b}_1 q^{-1}+\cdots+\hat{b}_{nb}q^{-nb}$$

$$\hat{C}(q^{-1})=1+\hat{c}_1 q^{-1}+\cdots+\hat{c}_{nc}q^{-nc}$$

Control criterion

$$u(t) = -\frac{G(q^{-1})}{F(q^{-1})}y(t) \quad (2.22)$$

where

$$F(q^{-1})=1+f_1 q^{-1}+\cdots+f_{nf}q^{-nf}$$

$$G(q^{-1})=g_0+g_1 q^{-1}+\cdots+g_{ng}q^{-ng}$$

is the solution of following equation :

$$\hat{A}(q^{-1})F(q^{-1})+q^{-k1}\hat{B}(q^{-1})G(q^{-1})$$
$$= T(q^{-1})\hat{C}(q^{-1}) \quad (2.23)$$

where $T(q^{-1})=1+t_1 q^{-1}+\cdots+t_{nt}q^{-nt}$ is the closed-loop prespecified polynomial.

All of the above parameter estimation algorithms of self-tuning are using the variable forgetting factors method (Fortescue et al. , 1981).

Construction of CADST package

This software package is constructed by a lot of modules and with layers (Fig. 2.1) which include all the funtions as described above. There have been built a management program to implement the dispatching for all functional software modules. The selection of functions can easily appointed from the menu, and any parameters of the self-tuning control system also can be easily exchanged during every running through simple man-machine interactive dialogs, which are wholly charged by management program. The whole programs are written by Fortran language, and it is easy to make any extension.

When the user applies this package to design the engineering control systems, if the control object is a simplified process model as linear difference equation, then this simplified model can be linked as the block "Process", the whole procedure of the self-tuning control system design can be going on. If the user wants to investigate the control effects in detials, then the package can links a meticulous simulation plant model with the block "Plant" (such as Fig. 4.2) to implement the simulation running.

The same project job can uses any kinds of self-tuning algorithms which have been constructed in this package. It gives much convenience for the user who can applies any kind of algorithm and to compare the running results.

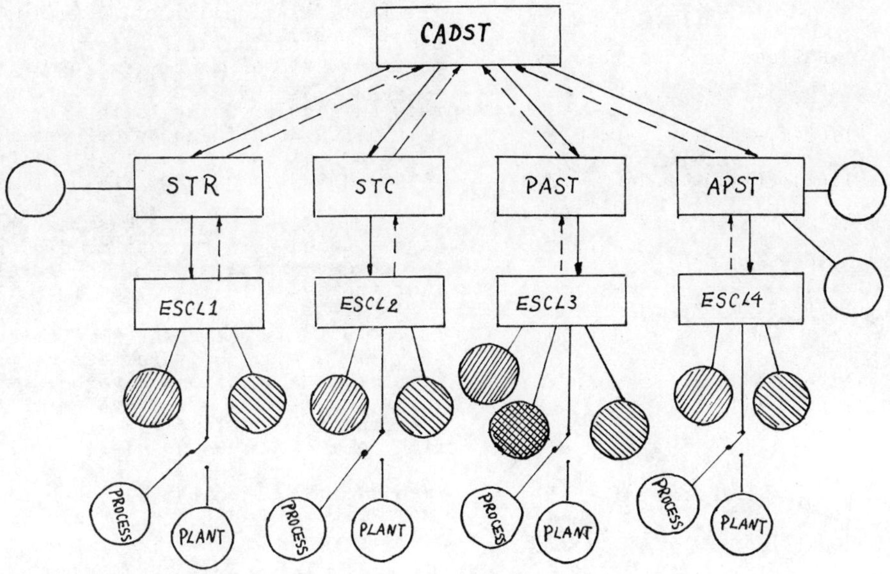

Fig. 2.1. Construction of CADST package

This package also can be used as teaching aid for the university.

3. NUMERICAL CALCULATING EXAMPLE

Example 1. A nonminimum phase system

$$y(t)-1.5y(t-1)+0.49y(t-2) = 0.4u(t-1)-0.401u(t-2)+\xi(t) \qquad (3.1)$$

where $\xi(t)$ is the zero mean, variance $\sigma^2=1$ uncorrelated Gaussian random sequence. The APST algorithm has been applied to this system. The system outputs is illustrated as Fig. 3.1(a) and the system inputs and estimated parameters are illustrated as Fig. 3.1(b) and Fig. 3.1(c).

Example 2. A correlated noise system

$$y(t)-1.6y(t-1)+0.7y(t-2) = 0.7u(t-2) -0.5u(t-3)+\xi(t)+0.4\xi(t-1) \qquad (3.2)$$

All of the STR, STC, PAST and APST algorithms have been applied for this system. The system outs of four algorithms are illustrated as Fig. 3.2(a) to 3.2(d). The mean variances of system outputs at t=1000 steps are illustrated in table 3.1.

Table 3.1. Comparison of mean variances

Algorithms	STR	STC	PAST	APST
Mean variance	5.13	8.66	5.62	2.88

It can be obtianed from example 1 and 2, that outputs of STR system are strongly fluctuated at the beginning several tens steps, by this influences the mean variances of outputs can not decrease rapidly. The outputs of STC which has been weighted on control signal are obviously less fluctuated at the beginning steps, but mean variances of the outputs are not optimal. The outputs of PAST are quite stable if the poles assignment are defined stablely. The outputs of APST have the least fluctuations, either the mean variances or the convergence of parameters are both in best performances,

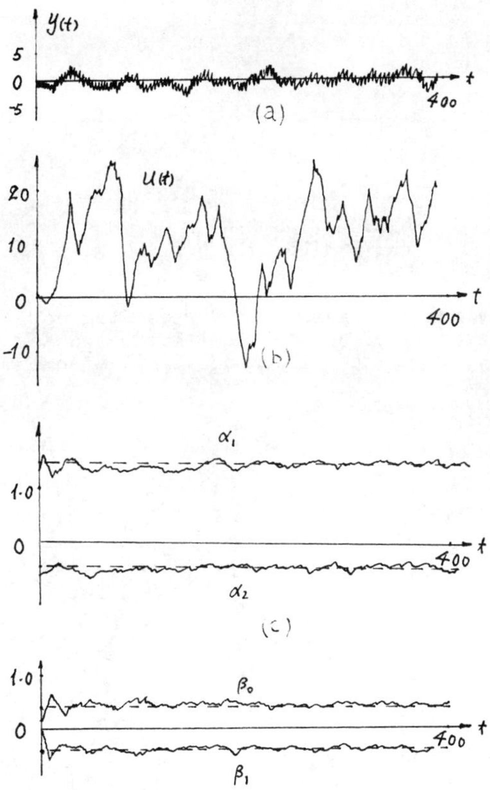

Fig. 3.1. Calculation results of example 1.

it has been illustrated that APST algorithm can get very good control property when the sampling time Δt is sufficient small.

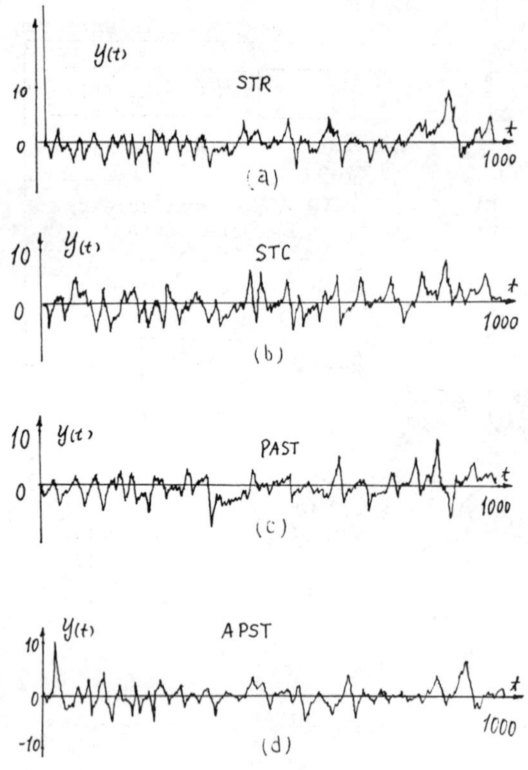

Fig. 3.2. Calculation results of example 2.

4. THE DESIGN OF THICKNESS REGULATING SYSTEM OF REVERSING COLD ROLLING MILL

The system configuration of reversing cold rolling mill of a plant is illustrated as Fig. 4.1. And the meticulous simulation block diagram of rolling mill is illustrated as Fig. 4.2 (links with "Plant" block during simulation runing, ref. Fig. 2.1). The former conventional thickness regulating system is worked by the principal of PI with correcting factor. Owing to the large time constant of hydraulic mechnical screwdown system, the thickness output deviations are rather unsatisfied (see Fig. 4.3(d)).

After the CADST package has been built, we have adopted this package to design the thickness control system. The results illustrate as following (see Fig. 4.3).

Fig. 4.3 illustrates the simulation results of various kinds of thickness control systems, the calculations have been applied under the rolling speed 3.33 m/s, the 5th path, and the sampling time is Δt=0.268 s., time delay d_1=0.558 s., d_2=1.15 s.

When the STR algorithm has been applied, the input data of CADST are : (ref. Eq. (2.9), (2.14))

$$n1=2,\ n2=3,\ n3=3,\ k1=2,\ k2=4,\ \beta_0=20$$

The variable forgetting factors method have been used, and the lower limit λ_{min}=0.95. (Following are the same case)
Fig. 4.3(a) illustrates the thickness output deviations.

When the STC algorithm has been applied, the input data of CADST are : (ref. Eq. (2.9), (2.16))

$$n1=2,\ n2=4,\ n3=3,\ k1=2,\ k2=4,\ P(q^{-1})=1,$$
$$R(q^{-1})=0,\ S(q^{-1})=1.5-1.5q^{-1}$$

Fig. 4.3(b) illustrates the thickness output deviations.

When the PAST algorithm has been applied, the closed-loop prespecified polynomial $T(q^{-1})=1-0.5q^{-1}$, and the input data of CADST are ; (ref. Eq. (2.21),(2.23))

$$na=3,\ nb=2,\ nc=0,\ k1=2,\ nf=4,\ ng=3$$

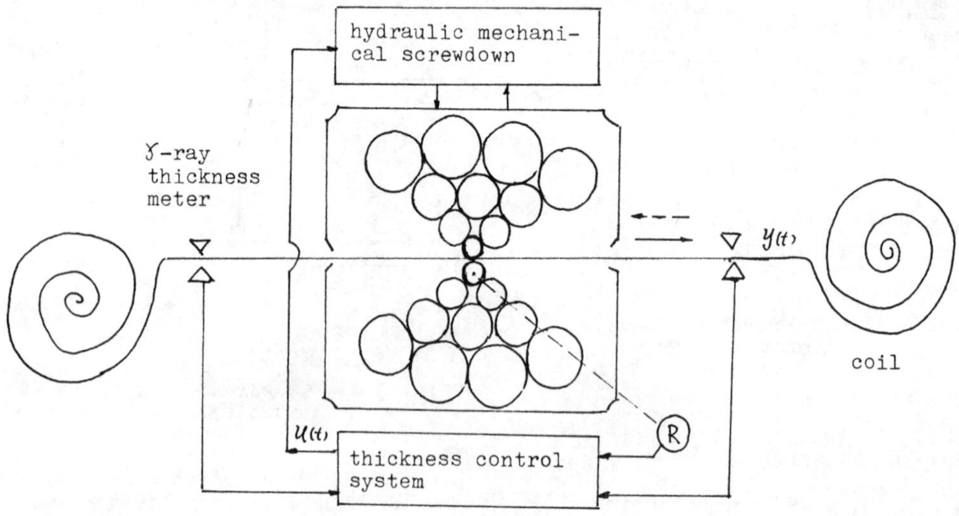

Fig. 4.1. System configuration of reversing cold rolling mill

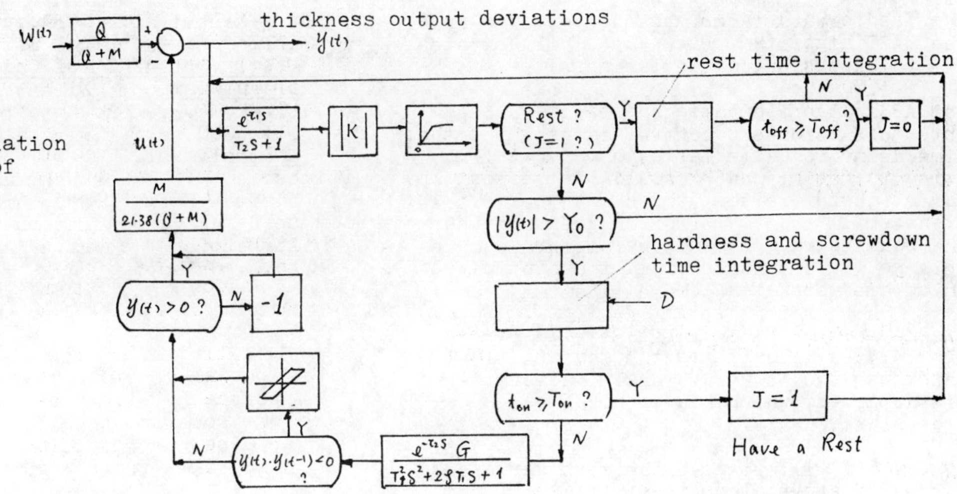

Fig. 4.2. Simulation block diagram of rolling mill

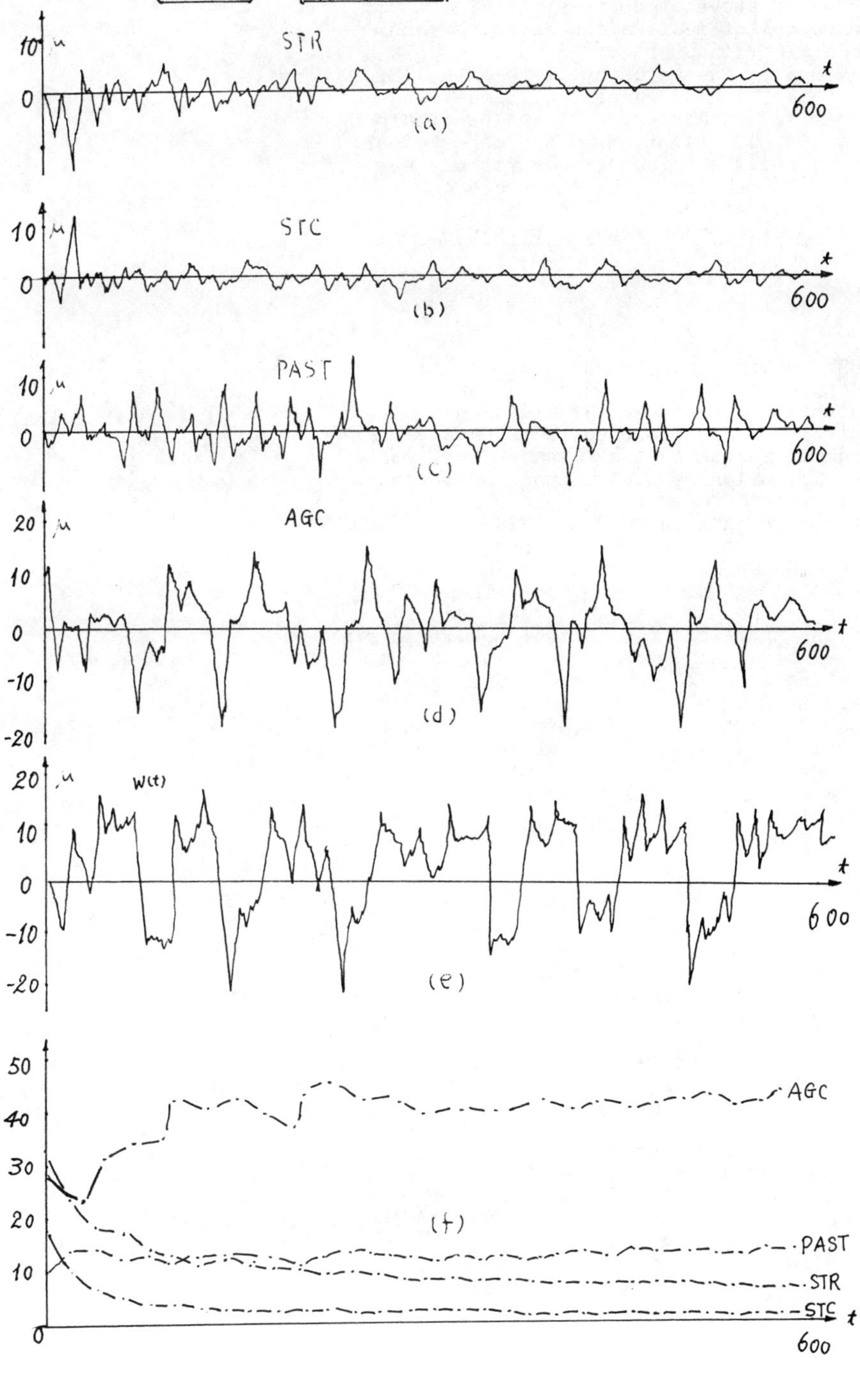

Fig. 4.3. The simulation results of various kinds of thickness control systems

The feedforward terms can not be added into the PAST algorithm. Fig. 4.3(c) illustrates the thickness output deviations.

The APST algorithm also has been applied into this thickness control system, but the time delay is quite large in this system, owing to the k1 and k2 would not be too large, therefore the Δt can not be sufficient small and the error of relation where the summing value of all parameters equals unit would be enlarged, and the control results are worse than conventional AGC system.

Fig. 4.3(e) illustrates the measureable thickness feedforward signals, as same as disturbances, which had been collected from plant.

Fig. 4.3(f) illustrates the mean variances curves of above algorithms, At 600 steps, the values of mean variances of various kinds algorithms are : The value of STR equals 6.7; the value of STC equals 1.8; the value of PAST equals 12.7; the value of former AGC equals 42.2. It can be obtained that the STC algorithm is the more better method for the thickness control of reversing cold rolling mill. It is obviously that the feedforward terms are very active to reduce the thickness disturbances.

5. CONCLUSION

CADST package includes several main self-tuning algorithms, it is useful and convenience for engineers as a design tool of control systems. This software package constructs reasonably and is easy to extend. The simulation results of thickness regulating system of reversing cold rolling mill illustrate that the STR and STC algorithms which have measureable thickness feedforward signals have good performance to reduce the thickness deviations. The PAST algorithm has not included the feedforward action, therefore the control results are not as good as STR or STC. Owning to the sampling time Δt can not be sufficient small in APST algorithm for this thickness control system, so the control results are unsatisfaction. But the numerical calculating examples have been illustrated, if the system delay is not large and the sampling time Δt is sufficient small, the APST algorithm can get very good control performance.

6. REFERENCES

Astrom, K.J.,and Wittenmark, B. (1973). On self-tuning regulators. Automatica, Vol. 9. No.2, 185-199.

Astrom, K.J., Borrison, U. et al. (1977). Theory and application of self-tuning regulators. Automatica, Vol. 13. No.5, 457-476.

Clarke, D.W., and Gawthrop, P.J. (1975). Self-tuning controller. Proc. IEE, Vol. 122. Pt.D, No.9, 929-934.

Clarke, D.W., and Gawthrop, P.J. (1979). Self-tuning controllre. Proc. IEE, Vol. 126. Pt.D, No.6, 633-640.

Allidina, A.Y., and Hughes, F.M. (1981). Self-tuning tuning control for systems employing feedforward. Proc.IEE, Vol. 128. Pt.D, No.6, 283-291.

Wu Hong-Xin et al. (1984). All parameters self-tuning control method. Annual Meeting of APCOM of Chinese Association of Automation, Xi An P.R. China

Wellstead, P.E., and Sanoff, S.P. (1981). Extended self-tuning algorithm. Int.J. Control, Vol. 34. NO. 3, 433-455.

Fortescue, T.R., Kershenbaum, L.S., and Ydstie, B.E. (1981). Implementation of self-tuning regulators with variable forbetting factors. Automatica, Vol.17. No.6, 831-835.

INTERACTIVE GRAPHIC DESIGN OF SIGNAL FLOW DIAGRAMS FOR SIMULATION

H. Domeisen, B. Dorn, T. Sulzer and S. Studer

Institut für Mess- und Regeltechnik, ETH-Zentrum, CH-8092 Zürich, Switzerland

Abstract. This project deals with the man-machine interface and its applications to control problems. The objective was a program which allows the user to define any physical system like heat exchangers or turbines in an efficient and simple way for simulation. Systems defined by a block diagram can simply and efficiently be entered by use of a special CAD program. This efficiency results in using graphic symbols instead of differential equations. With an inital menu, predefined elements can be chosen and are placed by a cursor to compose the system. Another menu containing a selection of functions is used to give each element its functional meaning. There is no need for typing in differential equations and a very clear documentation is produced by the computer. The two-dimensional block diagram is then transformed into a one-dimensional instruction sequence controlling the simulation. Different systems can be stored as permanent files for reuse or modification. Various stored systems can be freely combined to form a new one. An extended version of the program package is able to do real time simulation and process control.

Keywords. Signal Flow Diagram, Simulation, CAD, interactive, Menu technic, real time.

1. INTRODUCTION

There exist many simulation packages for different applications. Usually the systems are defined by differential equations or by function blocks. Most packages are used in batch mode. This results in some disadvantages:

- the user must be familiar with the simulation language before using it
- usually, there is a special effort to prepare the documentation
- it is often difficult to find errors in the simulated system.

The program package DOSU was built to use only the signal flow graph for the definition of the system and takes it as the basis for simulation. Like this, most of the mentioned disadvantages can be eliminated.

The graphic presentation of the system permits a better survey and is used for documentation with no further effort. Interactive input and definition of the system with dynamic menu technic show all the actual possibilities at any time and permit checking the input data.

After input of all the needed informations the signal flow diagram is transformed into a suitable form for simulation.

DOSU was designed for linear and nonlinear systems with lumped parameters. It has a real time part for simulations on process computers and direct coupling with external systems, e.g. direct digital control of a continuous process.

In this study, primarily the interactive design of the signal flow diagram together with the definition of functions and parameters has been faced.

2. SIGNAL FLOW DIAGRAM

The signal flow diagram is a special kind of drawing, which helps to show the mathematical and physical relationship between different signals. It is used as a substitute for equations.

Using signal flow diagrams, it is very easy to follow a signal path and to see how it will be changed and linked with other signals by different functions.

The aim was to develop an interactive program which allows to design and interprete signal flow diagrams, e.g. for simulations and process control without the need of a common computer language, like Pascal. For that reason the inner coherence of those designs had to be examined.

2.1 Structure of Signal Flow Diagrams

Signal flow diagrams consist of just a few structure elements which define the signal flow and contain the mathematical functions. In a signal flow diagram consisting only of structure elements without any functions, it is already possible to see how the signals are linked.

An example:

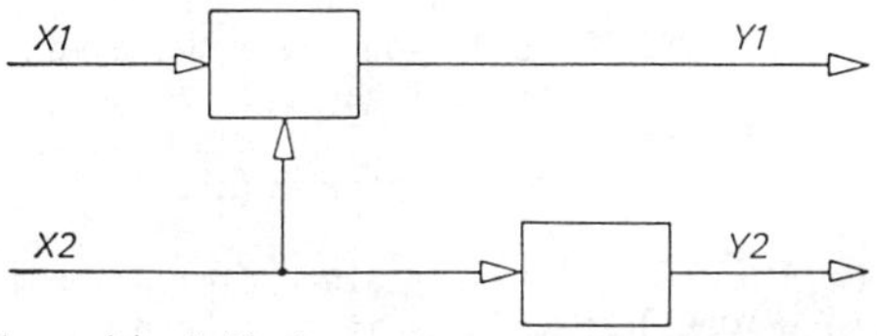

It is evident that
y1 = f(x1,x2)
y2 = f(x2)

CADCES-G*

If the functions are added to the structure of the signal flow diagram, the mathematical relationship between the signals is known too:

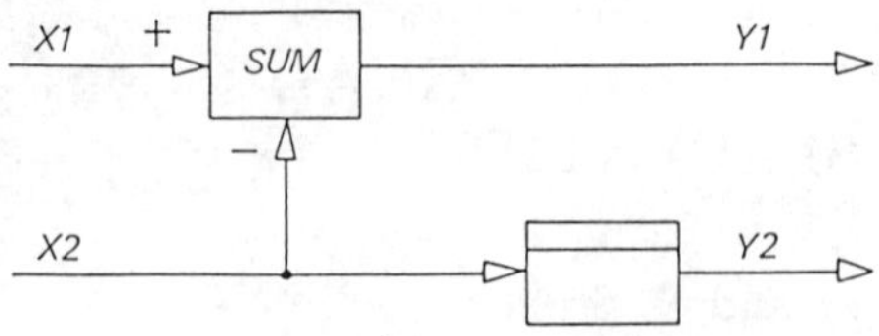

y1 = x1 - x2
y2 = C·x2

As structure elements we define:

- input/output

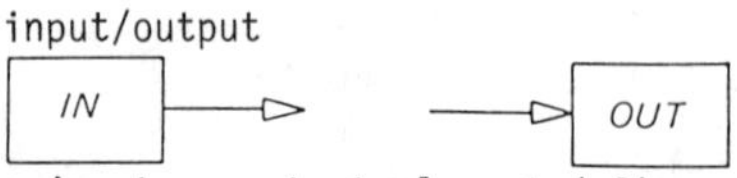

The input or output element defines the connections of the signal flow diagram to its environment (environment means A/D and D/A converters, switches, recording instruments...).

- connection (paths, including corners)

- crossing

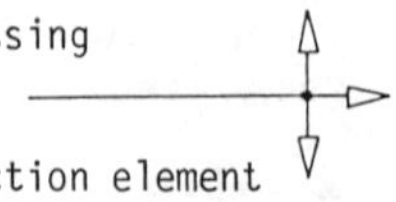

- function element

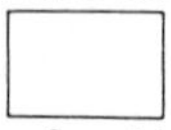

The function element has one or more input signals and one output signal. It is important that the number of connections corresponds to the function within the function element.

It is useful to distinguish function elements with different numbers of input signals to get the possibility to exchange functions with the same number of variables without changing the structure of the signal flow diagram. Functions with more than two variables can always be splitted into a combination of functions with two variables. This is the reason why only function elements with one or two input signals are used. They are called function element 1 and function element 2.

- function element 1

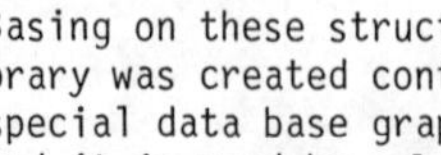

  this structure element may contain functions like logarithm, integration...

- function element 2

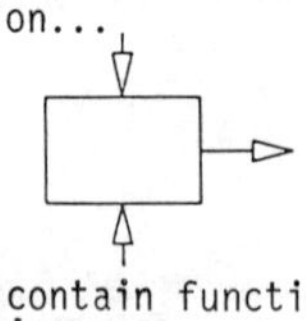

this structure element may contain functions like summation, multiplication...

Basing on these structure elements a structure library was created containing the elements. In a special data base graphical information is stored and it is used by algorithms, like compilation (see chapter 2.2). The solution was a two-dimensional array. Every component has a defined relation to a point of the design plane and contains graphical, and mathematical information.

If the user now chooses a structure element from the library (menu) and places it somewhere on the design plane, the element will be drawn in that position and simultaneously stored in the array. To connect elements it is only necessary to select the connection symbol from the structure library and mark the initial and the final point.

This gives the possibility to generate the whole structure of a signal flow diagram.

2.2 Compilation of Signal Flow Diagrams

To transform the two-dimensional array with the structure information into a computable list of instructions, a compilation program is needed.

The compilation program produces a list of procedures with the addresses of all variables.

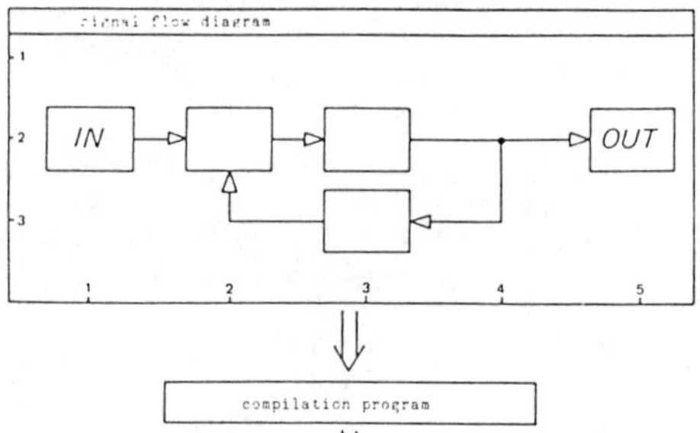

instruction list (list of procedures)								
x- coordinate	y- coordinate	structure element	address of input variable 1	address of input variable 2	address of output variable	mathematical function	address of parameters	step number
1	2	IN			1			1
2	2	fnc 2	1	4	2			2
3	2	fnc 1	2		3			3
3	3	fnc 1	3		4			4
5	2	OUT	3					5

Fig. 1 Compilation of Signal Flow Diagram

As the example shows, the compilation program does not care about the mathematical functions because they are not stored in the two-dimensional array, they are put directly into the instruction list. It is very important, that the compilation program gives the correct step number to each structure element.

For the compiler it was better to develop a heuristic algorithm rather then an analytic one. An imaginary observer who moves along the signal flow, first comes to an input element. This element will yield the first line in the instruction list. Now the observer follows the signal and leaves the input element. If the next element is a crossing, he chooses one direction and sets a mark to know later on, that he already has taken this way. If he now finds a function element "1", he just creates a new line in the instruction list. The observer numbers the signals he finds on his way. These numbers correspond to the variable addresses in the instruction list and to the components of an array which summarizes all signals of the signal flow diagram.

If the observer comes to a function element "2", he checks whether it is the first time he gets to this element or not. If yes, he must find out whether the second input signal is a feed back or not. If it is a feed back, he passes and creates a new line in the instruction list. If not, he searches for a crossing he already passed, but which still has one ore more directions without a mark. If he does not find such a crossing, he looks for a new input element. If the observer comes to a function element "2", he can pass because he already knows the address numbers of both input signals.

In this manner the observer walks through the whole signal flow diagram until he has reached all output elements. Now the instruction list is complete, exept that the mathematical functions and their parameters are still missing.

2.3 Definition of Functions and Parameters and Specification of Input and Output Elements

To complete the instruction list with functions it is useful to have a menu containing all functions the user needs. Then it is possible to choose the function from the menu and to mark the requested structure element in the design plane. In this manner the instruction list will be completed, until all structure elements contain functions.

Every function has an address in the instruction list which defines the relation to a parameter array. The parameters themselves are added as soon as the structure elements contain their functions. It is obvious that a change of any function of the same structure type and any parameter is possible, without the need of a new compilation.

To execute the instruction list, some information is still missing. First of all, the user must choose if he wants to run a simulation only with test input signals, or to connect the computer to a process and run a real time computation. Both versions need a further specification of input and output elements. If the user wants to connect the signal flow diagram to an external process, he can interactively connect all input elements to the A/D converters and the output elements to the available D/A converters. It is also possible to connect output elements to a printer or to a recording instrument and to define some of the input elements as test signals which were produced by the computer.

2.4 Execution of the Instruction List

The last thing the user has to do, is to define the computation time interval. For the real time application it is very useful to determine the minimal possible interval by the computer, because it can vary in a wide range (20...200 μsec), depending on the size of the signal flow diagram.

All information for the execution is complete now. The calculation program can start to execute all functions in the instruction list in the sequence of the step numbers. For every line the program calls a procedure which corresponds to the actual function. All required variables for this procedure are available by calling those components of the variable array which correspond to the addresses in the instruction list. Reaching the end of the instruction list, the program restarts with the first step number. This procedure is repeated up to the end time of the simulation.

3. THE PROGRAM PACKAGE D O S U

DOSU is a utility-program for system designers who work with signal flow diagrams. These diagrams are configured on the screen and stored for later use.

They are used for documentation and computation. DOSU transforms the signal flow diagram into a machinereadable form, a "programming" is not necessary.

Special features of DOSU are: The dimension of the signals can be choosen between one and five. The menus depend on information already stored in a signal flow diagram. The menu is restricted to the necessary components which prevents wrong manipulations.

3.1 Implementation

The program is implemented on a Minicomputer with floating point arithmetic and 64kbyte memory. The graphic display is a Tektronix 4010 storage tube. As curser input we use a location-proportional joy-stick (not the better known speed-proportional, Similiar to a "mouse"). Documentation is done on a plotter as well as on printer. Connection with a process is realised via A/D and D/A converters.

3.2 Program Structure

System Operating This part of DOSU gives the user information on already existing systems (Systems that have been created with DOSU earlier). It gives the possibility of choosing an old or of creating a new system.

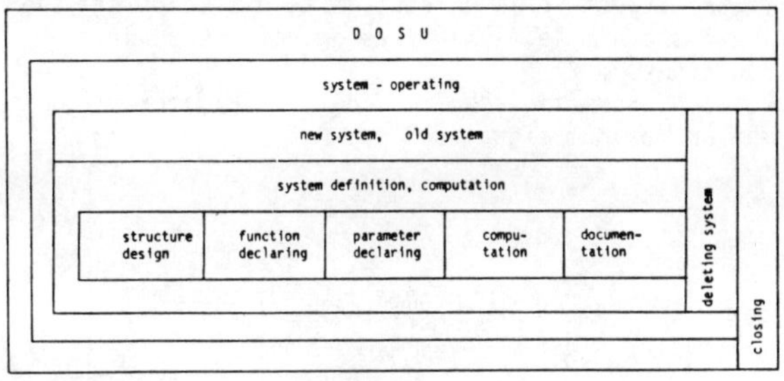

Fig. 2 Principal Structure of DOSU

System Definition and Computation This level, as shown above, is the "heart" of DOSU. The inner segments are only available if necessary (the structure design is always available). Every segment shows on the screen a drawing-plane on which the signal flow diagram is displayed (or parts of it) and a menu-bar at the bottom of the screen.

Every segment has the capability of moving the virtual drawing-plane behind the "screen-window" using a zooming capability.

Normally the user points first to a command and then to a coordinate or to an existing symbol of the signal flow diagram to show where he wants to have the command executed.

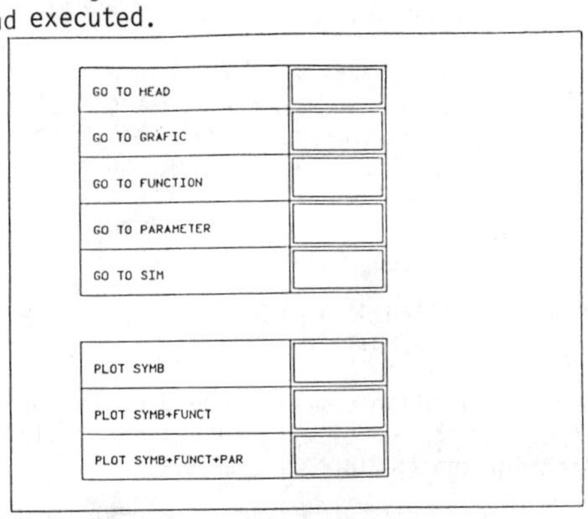

Fig. 3 Main Menu

The steps of developing a system are illustrated with the example of a Helicopter (four-dimensional state space model in constant speed).

a) Structure Design: The structure is defined by a set of the six elements: IN, OUT, SIMPLE ELEMENT, KNOT, CROSSING, CONNECTION. This segment is the only one available when a new system is created. The structure of a signal flow diagram can be defined with these six elements; their meaning or characteristic is defined in the other segments.

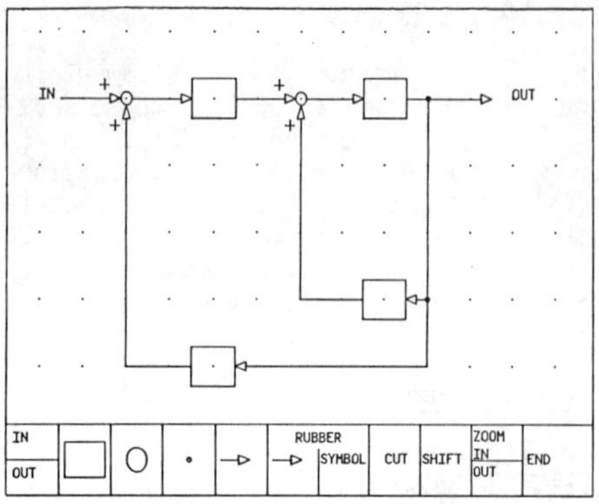

Fig. 4 Design of the System Structure

b) Function Definition: This segment is available, if the structure is correct. It allows to define the knots and simple elements as one of the following functions:
P, I, D, Summation, Multiplication, Limiter, Minimum or Maximum signal switch.

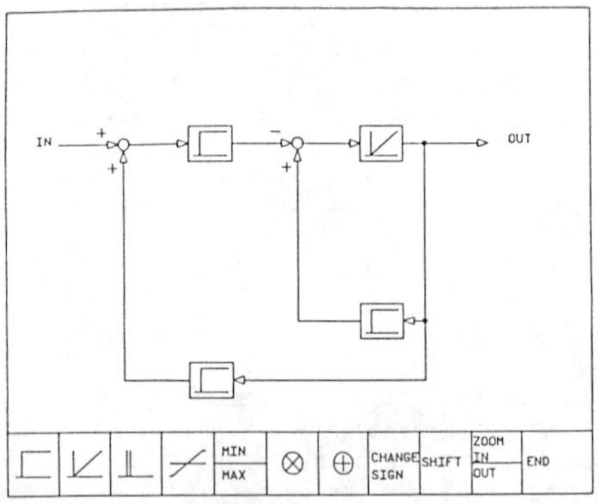

Fig. 5 Function Definition

c) Parameter Declaration: This segment is available, when all functions are declared. Parameters exist for proportional-element and for the signal limiter. For easy definition of parameters in a moredimensional case, a mask appears on the screen that shows the rows and columns to be filled up. This gives also the possibility to point to any value to be modified.

d) Computation: Before a system can be excecuted, there is a last step in completing a signal flow diagram: a segment for declaration of IN and OUT as a process-interface (A/D or D/A converter) or for internal use (test-signal or storage purpose). Also the inital conditions of the integrators are defined in this segment.

(Fig. 6 Definition of Input Signals for Simulation Run, see next column)

Further information concerns the duration of computation and the sampling interval for real time application or for integration steps.

(Fig. 7 Plot of Output Vector, see next column)

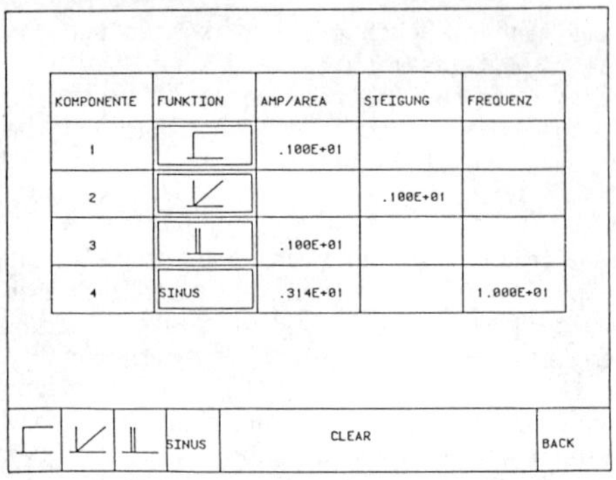

Fig. 6 Definition of Input Signals for Simulation Run

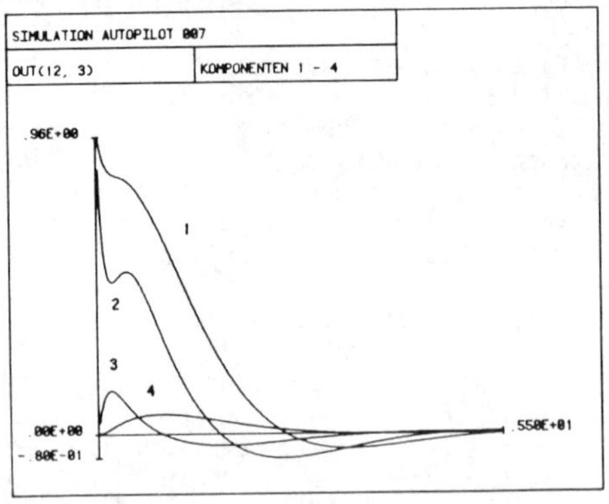

Fig. 7 Plot of Output Vector

3.3 Modification

Each of the inner segments accepts every action of the user either as new definition or as a modification of an existing information. A new compilation is only necessary, when the structure was modified. A function modification causes DOSU to ask for new parameters. A parameter modification has no further effect.

3.4 Real Time Application

DOSU has a built-in timing capability. This feature is activated if either a IN or OUT is defined as process interface. If not, DOSU simulates the system without using the clock.

The shortest possible interval depends on the system (number of functions, type of functions, number of dimensions and others). After the first time the system has been computed, DOSU gives a message and stops the excecution if the intervall which was defined by the user is shorter than the one needed for computation.

3.5 Computation Procedures

Every function mentioned above (e.g. P,I,D..) is associated with a procedure. This procedure can vary according to the needs of the user (e.g. different integration routines). However, the DOSU package includes very simple integration and differentiation routines.

The rules for computing more dimensional signal flow diagrams are those of linear algebra. Besides those rules an integrator is executed on each component separately. Since there are only nxn matrices allowed, the matrix computation is very easy.

3.6 Outlook

The modular structure of DOSU allows the simple adaptation to specific needs.

For example binary signal processing could be added or additional functions may be defined.

Also it is possible to replace calculation routines by procedures from already existing simulation packages.

3.7 Example

The example shows a model of cross flow heat exchanger. The signal flow is one-dimensional with multiple inputs and outputs and nonlinear elements (multiplicators).

REFERENCES

Rimvall M., IMPACT, Interactive Mathematical Program for Automatic Control Theory, Mai 84, Inst. für Automatik, ETH Zürich, Switzerland

Elmquist H., A graphical approach to documentation and implementation of control systems, IFAC Software for Computer Control, Madrid 82, Spain

Aström K.J., Computer aided Modeling, Analysis and Design of control Systems - A perspective, August 82, Departement of Automatic Control, Lund, Sweden

Grimm R., Zur ergonomischen Gestaltung der Mensch-Prozess Schnittstelle: Prozessbeobachtung und Prozessbedienung, Oktober 82, Karlsruhe, BRD

Domeisen H., SIMA, interaktives Simulationspaket für Prozessrechner, Programmbeschreibung, Institut für Mess- u. Regeltechnik, ETH-Zürich, CH, 84

Gray J.O., Nonlinear system design, an integrated approach. Proceedings of CACSD Symposium, Brighton, Great Britain, 1984

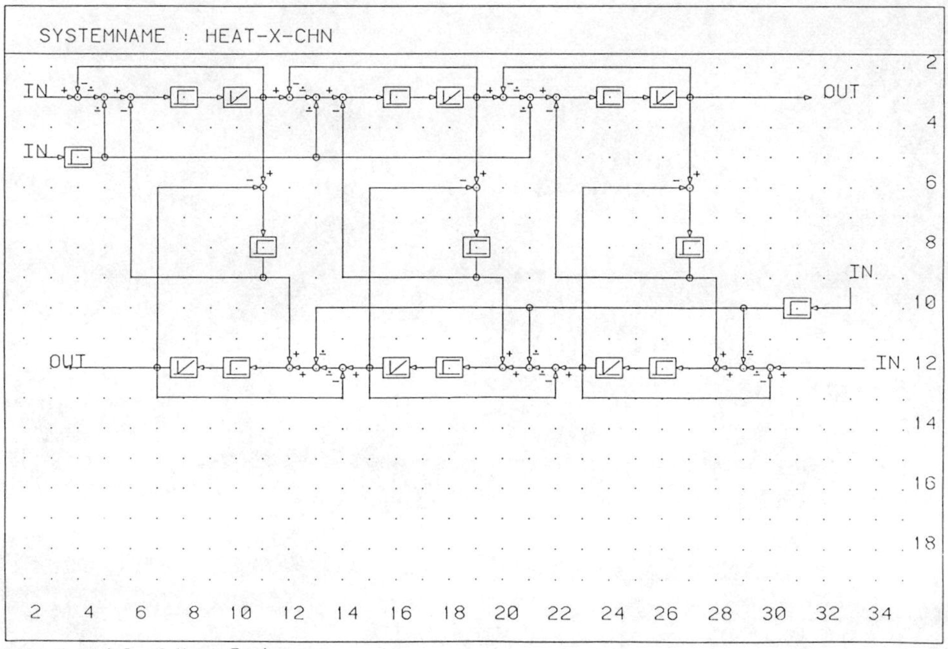

Fig. 8 Model of Heat Exchanger

Input signals: temperatures, massflow
Output signals: temperatures

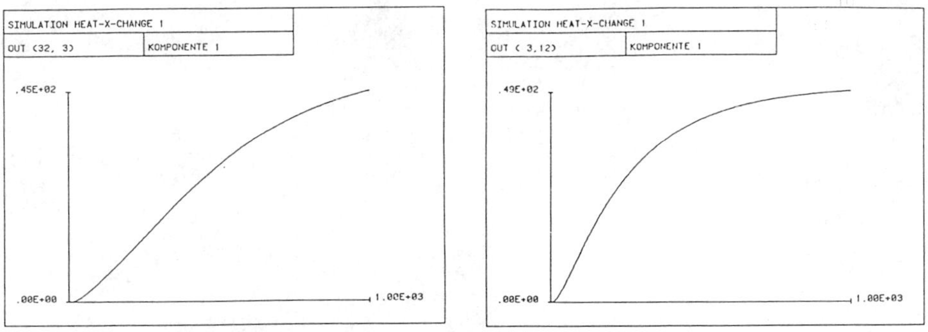

Fig. 9a/ Fig. 9b Plot of Step Responses of medium 1 resp. medium 2

Initial Conditions: medium 1,2: $0^{o}C$
The plots show the responses to an input step of medium 1 (0° to $50^{o}C$)

A DIGITAL SIMULATION SYSTEM OF LINEAR AND BILINEAR PROCESSES WITH OBSERVERS FOR STATE FEEDBACK CONTROL DESIGN

N. Laggoune

Laboratoire d'Automatique, Université de Lyon 1, 43 Boulevard du 11 Novembre 1918, 69622 Villeurbanne, France

Abstract. A conversational Fortran IV software has been built in order to facilitate the design of control systems for linear and bilinear processes. It allows simulation, analysis, observation, and control of those processes. It presents a "menu-driven" structure and a high interactivity with animated bloc-diagrams.Two applications are presented in the field of closed loop linearization of bilinear processes : neutralization pilot plant and D.C. motor. Still in progress, this software is open to introduction of new modules.

Keywords. Bilinear control ; Optimal control ; Time-varying systems ; Observers ; Computer-aided design.

INTRODUCTION

The basic idea of our software (Fig. 1) is implementing a computer-aided design package applied to simulation and state feedback control setting up for both linear and bilinear systems. In case of linear systems, the control is the well known optimal-control which minimizes a quadratic criterion, with optionaly a reference-model following structure including integrators. For bilinear systems, the control law, derived from or researches, is closed loop linearization by non linear state feedback. It has been shown that this control law minimizes a quadratic criterion (Gilles, 1983), slightly different from the one related to the linear case. In each case, if the state-vector is non-completly measurable, the state feedback needs an observer, and the software has to provide a help for designing minimal order ones.

Before a control design, the user can analyze his linear system, e.g. stability, controllability, observability. The Luenberger's observer (1966) computation is followed by the control-law calculation by solving a classical discrete Ricati equation of the appropriate form, depending on the chosen control structure.

The major contribution of the software is in bilinear systems control field. For the closed loop linearization, some mathematical difficulties have been recently solved, and the discrete version of this law is already implemented. Two types of observers can be used :

- a continuous one, proposed by Hara and Furuta (1976). This observer must be discretized and a method for testing its applicability is implemented (Laggoune, 1984a).
- a discrete one, developed by the author (Laggoune and Gilles 1984), and applicable even if the former cannot be used.

The software is specialized, and we have chosen the "menu-driven" structure. This environment is not unconvinient in this case, because of the specialization of the package. In order to increase interactivity, we focused our attention on the use of the terminal screen, by harmonious control of device, cursor and character attributes. For example, an animated block-diagram is a helpful "gadget" for a novice user. Graphics modules permit to the user a following of his manipulations effects on graphical-screen or plotter.

In order to demonstrate the software capabilities, two examples concerning bilinear control are presented.

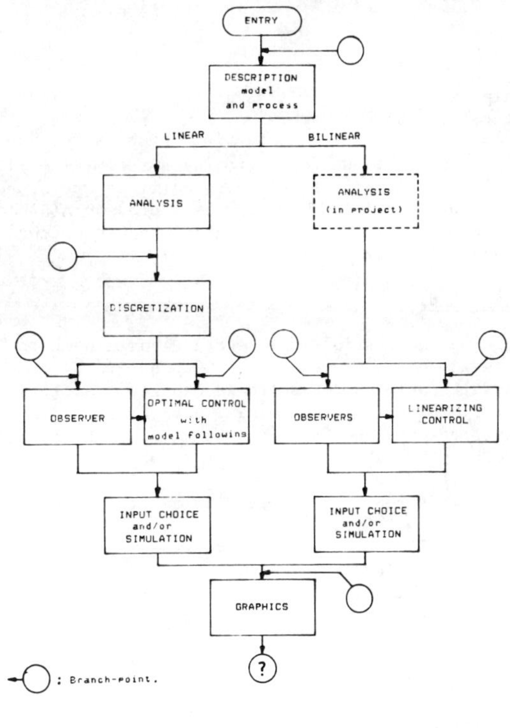

Fig. 1. Software general organization.

SOFTWARE ASPECTS

Organization

The software consists in separate Fortran programs, each one being devoted to a special function (e.g. description, analysis, simulation, etc...). The synchronization between those programs is realized by a "supervisor-program" (Fig. 2). The supervisor, through his menus, performs the major conversation with the user, by means of a "typical-screen" (Fig. 3). The defined three zones of Fig. 3, contain some informations about the current stage (zone 1), the system (zone 2) and the menu (zone 3).

The supervisor performs any action needed by the user, by calling the appropriate program. After the called program execution, the data is stored on mass-storage unit in variable length file. The supervisor takes control after this action and the user can follow his session (Fig. 2).

This modular structure makes the software realy easy to use, and relatively open. Adding new modules only implies a weak modification of the supervisor.

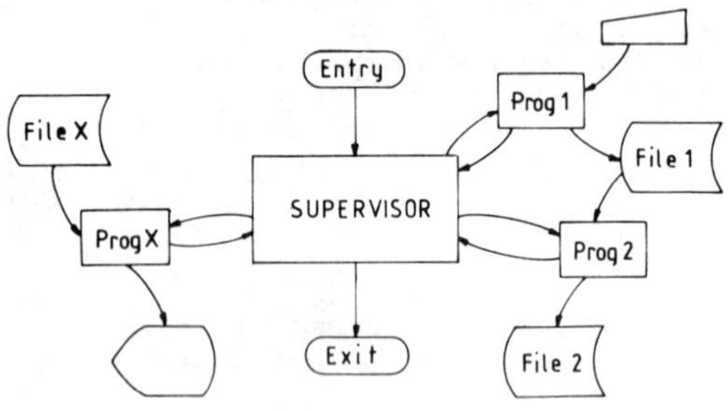

Fig. 2. The supervisor and its function.

Data Structures

Manipulated data are real single precision (in a first time, double precision is not required). These data are stored in matrix, which is the basic data structure. But for operating system (RT - 11 of D.E.C.) compatibility, these matrices are stored as vectors : a A(n,n) matrix is stored as A(n*n) vector. A future extension can be the use of the powerful librairy MATLAB (Moler, 1981).

Files Handling

Begining the session, the user assigns a name to the studied system (example PROCES or SYBIL1). A special module in the software manages a small directory (8 systems) containing data files, with specified name and default extension, as following:

PROCES.COL - COtinuous parameters of Linear system PROCES,
PROCES.SIL - SImulation results ready to display of Linear system PROCES,
SYBIL1.COB - COntinuous parameters of Bilinear system SYBIL1,

and so on. With this ability, the user can easy compare different control structures of the same system, or two systems between them.

CRT Screen Management

One of the very important features of a C.A.D. system is the harmonious use of the sreen. Then we focused our attention on the definition of a typical screen : Fig. 3, tabulated screen, and animated block-diagrams, by using some device attributes (scrolling), cursor attributes (arbitrary positionning) and the character attributes (light, form, size, etc...), for VT100 compatible CRT terminals.

Typical-screen (Fig. 3). Five zones have been defined, for information (zones 1 to 4), or working (zone 5). Zone 1 contains information about the current stage. Zone 2 recalls the system-name on study and some complementary informations (numerical method, block-diagram of control structure). Zone 3 is the current menu. Zone 4, is reserved for error messages. All these zones are controlled by the supervisor, except the fourth, which can also be used by other programs, but for all the scrolling is forbidden. The working space is zone 5, wich is a window in the screen (scroll). This zone is not controlled by the supervisor, it is used by the called-program, for his conversation with the user.

Tabulated-screen. By cursor-position control, the tabulation is a very convenient manner to use the screen as matrix. This feature is used in the description stage, for introducing the state-model of the system.

Animated block-diagrams. In order to make the software easy to learn, and for a better driving of the user during one session, animated block-diagrams have been defined. The user can call these block-diagrams, with any menu. And this simple manipulation can show his location into the software, and he can decide how following, according to the possibilities shown. Calling an animated block-diagram, do not cause any trouble to the current session. This screen represents something like Figure 1, depending on the location inside the software.

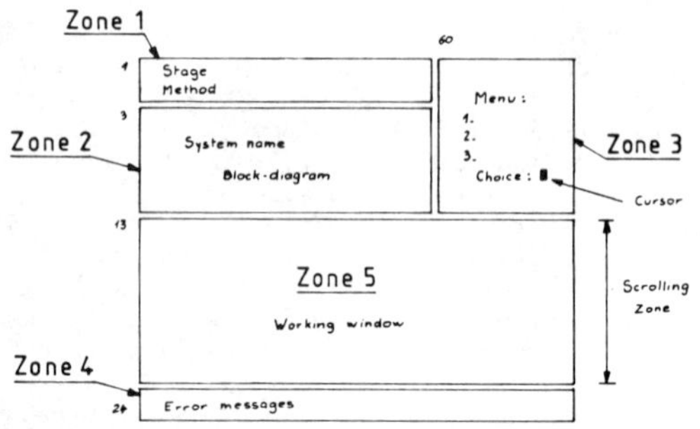

Fig. 3. Typical screen and zones.

LINEAR AND BILINEAR MODELS

In its actual form, the software is not the "state-of-art" in control systems field. Some well known techniques are used, and some original methods are implemented. We will present some capabilities already possible with the actual version of the software, for linear and then for bilinear systems. Before those explanations, we introduce some definitions about the models. For the description stage, only one model is possible : the state-space representation.

In linear case the state-model is :

$$\dot{X}(t) = A.X(t) + B.U(t)$$
$$Y(t) = C.X(t) \qquad (1)$$

where X is the (nx1) state vector, U is the (mx1) input vector and Y is the (rx1) output vector. The matrices are of appropriate dimensions.

For bilinear systems, the state-space model is :

$$\dot{X}(t) = Ao.X(t) + \sum_{i=1}^{m} Ui.Ai.X(t) + B.U(t)$$
$$Y(t) = C.X(t) \qquad (2)$$

with the same definition and dimensions of the vectors. Ui is the ith component of the input vector U(t). In case of discrete modelling, an index "d" is adjoined to the matrices. The matrix C is assumed to be of the form $[I_r : 0]$.

LINEAR SYSTEMS CONTROL

Analysis.

This stage consists in the study of some properties of the system, like stability, controllability or observability. Stability study is performed now through eigenvalues concept (Eigen routines from EISPACK), and very soon Lyapunov's concept will be implemented by using the algorithm for solving A.X+X.B = C (Bartels and Steward, 1972). For controllability and observability the index is derived using the S.V.D. (Singular Value Decomposition), (Lawson and Hanson, 1974 ; Forsythe and coworkers, 1977).

Observation.

The Minimal Order Observer (M.O.O.), is of Luenberger (1966), defined by :

$$\dot{Z}(t) = \hat{F}.Z(t) + \hat{B}1.U(t) + \hat{G}.Y(t)$$

$$\hat{X}(t) = \hat{P}.Z(t) + \hat{V}.Y(t) \quad (3)$$

where the observer's state vector Z(t) is (n-rx1). The observer dynamics is arbitrary, by poles assignment of matrix F, which is under triangular or diagonal from. If necessary, pole-assignment algorithms will be used (Varga, 1982).

Optimal control.

The control structure can include a reference model, or perturbation model, in case of model following technique. In all cases the control law is derived by minimizing a discrete quadratic performance criterion of the form :

$$J = \frac{1}{2} \sum_{k=0}^{N-1} ({}^t\varepsilon_k.Q.\varepsilon_k + {}^tU_k.R.U_k) \quad (4)$$

where $\varepsilon_k = E_k - Y_k$, E_k is the reference vector, Y_k is the output vector, Q and R are defined respectively as non-negative and positive matrices. By applying the optimality principle of Bellman (Athans and Falb, 1966), the solution of the discrete Ricatti equation leads to the control law. The algorithm used for solving this equation involves a factorisation of the Ricatti matrices $K = S^t.S$, which necessarly lead to a symetric positive solution (Barraud, 1974).

BILINEAR SYSTEMS CONTROL

Observation.

Hara and Furuta's M.O.O. (1976). This observer is available for continuous bilinear systems and defined by :

$$\dot{Z}(t) = \hat{A}o.Z(t) + (\hat{B}o + \sum_{i=1}^{m} Ui.\hat{B}i).Y(t) + \hat{B}.U(t)$$

$$\hat{X}(t) = \hat{C}.Z(t) + \hat{D}.Y(t) \quad (5)$$

The applicability test and the calculation of this observer have been recently solved (Laggoune, 1984a). The corresponding algorithm is implemented. Moreover this M.O.O. must be discretized, thus a discrete version of this observer was derived (Laggoune, 1984b).

Discrete M.O.O. Because of the limitations of the former, and in the case of digital control of continuous bilinear systems, an original observer, linear time-varying of minimal order was developed (Laggoune and Gilles, 1984). It is defined by :

$$Z_{k+1} = \hat{A}(k).Z_k + \hat{B}(k).Y_k + \hat{J}(k)$$

$$\hat{X}_k = \hat{P}(k).Z_k + \hat{D}(k).Y_k \quad (6)$$

the time-varying nature of the observer appears in the index "k". This observer is applicable even if the former cannot be used, and his dynamics its arbitrarly fixed.

Linearizing Control.

Let bilinear systems be defined by Eq.(2). If the input vector is piecewise constant (e.g. digital control with zero-hold), we can define a discrete model of those systems by a linear time varying expression :

$$X_{k+1} = Ad(U_k).X_k + Bd(U_k).U_k \quad (7)$$

with : $Ad(U_k) = Exp\{A(U_k).T\}$

$$Bd(U_k) = \int_0^T Exp\{A(U_k).(T-\tau)\}.d\tau \quad (8)$$

$A(U_k) = Ao + \sum_{i=1}^{m} Ui(kT).Ai$ and Ui(kT) is the ith component of U_k at t = kT.

The closed loop linearization by state feedback consists in solving Eq.(7), Eq.(8) and :

$$e_{k+1} = D.e_k \quad (9)$$

where $e_k = Xc(k) - X_k$, and Xc(k) is the reference vector (variable or constant), defined by the user. T is the sampling period and D is the constant closed loop dynamic matrix of the linearized system. Thus the problem is finding the input vector U at time kT according to the vectorial non-linear equation :

$$\left[Ad(U_k) - D\right].X_k + Bd(U_k).U_k + \left[D - I_n\right].Xc(k)=0 \quad (10)$$

The solution is in general non explicit, it is derived by means of numerical techniques (Newton-Raphson, substitution, optimisation, etc...).

The linearization method presents some singularities, in number equal to the system order, and minimizes a quadratic criterion, including a control energy homogeneous term (Gilles, 1983). This kind of singularities must be overcomed by appropriate solution and is the object of our actual interest. In this case a safety control is necessary ; it can be a simple linear control, of appropriate coefficients making the closed loop system stable.

The numerical method used for solving Eq.(10) may also present singularities, or the convergence rate can be too great, thus the software has to give the possibility to switch to another numerical technique.

In the following, in order to demonstrate the achievability of the method two example are presented.

SIMULATION TECHNIQUES

The state-space representation requires the use of the famous exponential matrix in order to solve the state equations (Eq.(1) and Eq.(2)). This exponential matrix is computed by an accurate algorithm using the Padé approximant (Roche, 1980 ; Ward, 1977). The discrete matrices are obtained by computing only one matrix defined as (Barraud, 1980) :

$$\begin{pmatrix} Ad(U_k) & Bd(U_k) \\ 0 & I_m \end{pmatrix} = Exp \begin{pmatrix} A(U_k) & B \\ 0 & 0 \end{pmatrix} .T \quad (11)$$

T is the sampling period and all matrices are constant in the linear case. In the bilinear case the matrices used are defined by Eqs.(7) and (8).

With the help of the software, open loop studies

can be performed, the inputs are in selected form (step, ramp, sinewave, white noise). In order to point out the noise effect on the open or closed loop system, white noise can be added to the state or outputs. Modelling errors can also corrupt the designed control ; thus the user define the system-parameters separatly with the model-parameters and a robustness study can be performed.

SIMULATION EXAMPLES

Neutralization Process

This first example is a mono-input mono-output neutralization pilot plant. The bilinear model avalaible in a wide range (5-20 liters of solution) is :

$$\dot{X} = A.X + A1.U.X + B.U \tag{12}$$

where the state X is the normalized concentration in (H30+), and U is the base (OH-) flow-rate. The identified model is :

$$X = \underset{Am}{9.266}.X + \underset{A1m}{0.1852}.U.X + \underset{Bm}{0.1706}.U \tag{13}$$

In Fig. 4, the model Eq.(13) is assumed to be exact, and the linearizing control is performed from the initial state (Xo = 0,18) to equilibrium state (X = 0.33). Two dynamics are presented $D = -50h^{-1}$ and $D = -25h^{-1}$, the corresponding exponentials (first order closed loop responses)are superposed to the system's responses. In dashed line the response in case of actuator limitation (Umin=-201/h).

DC Motor

The DC motor (Hara and Furuta, 1976), is :

$$\begin{bmatrix} Ia \\ w \end{bmatrix} = \begin{bmatrix} -Ra/La & 0 \\ 0 & -C/J \end{bmatrix} + \begin{bmatrix} Ia \\ w \end{bmatrix} + \begin{bmatrix} 0 & -B/La \\ +B/J & 0 \end{bmatrix} . If . \begin{bmatrix} Ia \\ w \end{bmatrix} + \begin{bmatrix} 0 & +1/La \\ 0 & 0 \end{bmatrix} . \begin{bmatrix} If \\ Va \end{bmatrix} \tag{14}$$

where the state vector has the armature current (Ia) and the angular velocity (w) as components. The input vector is the field current (If) and the armature voltage (Va). The simulated model is :

$$\begin{pmatrix} Ia \\ w \end{pmatrix} = \begin{pmatrix} -10 & 0 \\ 0 & -0.12 \end{pmatrix} + \begin{pmatrix} Ia \\ w \end{pmatrix} + \begin{pmatrix} 0 & -2 \\ 0.33 & 0 \end{pmatrix} . If . \begin{pmatrix} Ia \\ w \end{pmatrix} + \begin{pmatrix} 0 & +1 \\ 0 & 0 \end{pmatrix} . \begin{pmatrix} If \\ Va \end{pmatrix} \tag{15}$$

Figure 5 shows an open loop simulation for two differents inputs vectors, for all examples concerning the DC motor the sampling period is T = 0.1s.

References Changing. In Fig. 6, a goal is to change the second state reference at time 5 seconds, without any change to the first one. In Fig. 7 the iteration number is plotted, as show in the figure, it is not prohibitive. The avearage number is 3, except for the reference changing moment (9 iterations). Thus, for this kind of systems, which accept a sampling period less then one second, a real time processing can be intented.

CONCLUSION AND PROSPECTS

Our conversational software now offers interesting possibilities for helping the user to simulate, analize, observe and design some control systems for linear and bilinear processes. However, this package is still in its beginning phase and its specialized nature offers, with few modifications, a possibility to develop a real-time processing version. This software, still in progress, can be easily extended by adding new modules, such as more efficients algorithms. In project is the introduction of a symbolic calculus module in order to increase the efficiency of the control computation. For the analysis of bilinear systems (controllability, etc...), a special module will be developed in order to bring to the designer a computer aid for the use of new difficult techniques.

ACKNOWLEDGMENTS

The author wishes to thanks Professor A.Y. Barraud for his advices on some numerical aspects of this work and for the routines that he reported us.

REFERENCES

Athans, M., and P.L. Falb (1966). Optimal Control, McGraw-Hill, New-York.

Barraud, A.Y. (1974). Sur la résolution numérique des équations de Ricatti - cas discret - RAIRO Automatique, Oct 1974.

Barraud, A.Y. (1980). A propos du calcul de systèmes linéaires. Equivalents discrets/continus. Internal note, Dec 1980. Laboratoire d'Automatique de Grenoble.

Bartels, R.H., and R.W. Steward (1972). Algorithm 432 : Solution of matrix equation A.X+XB = C, C. Ass. Compu. Machinery, 15, n°9, 481-483.

Forsythe, G.E., M.A. Malcolm, and C.B. Moler (1977). Computer Methods for Mathematical Computations, Prentice Hall, Englewood-Cliffs, New-Jersey.

Gilles, G. (1983). Une loi de commande optimale linéarisante d'une classe de systèmes bilinéaires multivariables. Journées de l'optimisation, 11-13 mai 1983, Montréal.

Hara, S. and K. Furuta (1976). Minimal order observer for bilinear systems. Int. J. of Cont., 24, n° 5, 705-718.

Laggoune, N. (1984a). On the observer of Hara and Furuta for bilinear systems, applicability-test and computation using the singular value decomposition. Submitted to Int. J. Cont.

Laggoune, N. (1984b). Sur la discrétisation de l'observateur de Hara et Furuta. Internal note, Nov. 1984, Laboratoire d'Automatique de Lyon 1.

Laggoune, N., and G. Gilles (1984). Observateur discret d'une classe de systèmes non-linéaires. Workshop on non-linear systems, 11-12 sept. 1984, Grenoble.

Lawson, C.L., and R.J. Hanson (1974). Solving least square problems, Prentice Hall, Englewood-Cliffs. New Jersey.

Luenberger, D.G. (1966). Observers for multivariable systems. I.E.E.E. Trans. on Aut. Cont., AC 11, pp 190-197.

Moler, C.B. (1981). MATLAB user's guide. Computer sciences department, University of New-Mexico.

Roche, J.R. (1980). Application des approximants de Padé au calcul de l'exponentielle matricielle. Third cycle thesis, Grenoble.

Varga, A. (1982). Computer aided design of robust compensator by pole assignment. Preprints of SOCOCO'82 symp, sept. 1982, Madrid.

Ward, R.C. (1977). Numerical computation of the matrix exponential with accurary estimate. SIAM J. Numer. Anal., 14, pp 600-610.

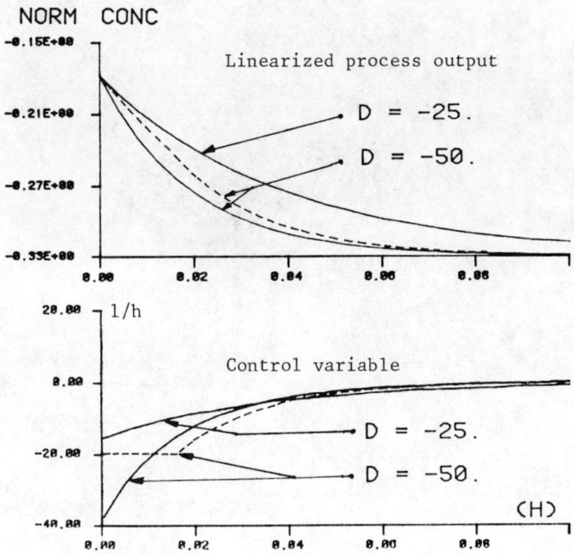

Fig. 4. Neutralization process.
Output and Input variables.

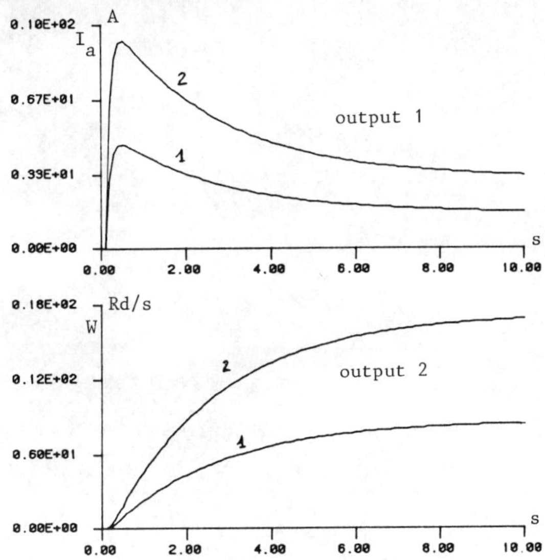

Fig. 5. D.C. motor open loop simulation for different inputs.

Curves 1 : U = $\begin{matrix} 2 & A \\ 50 & V \end{matrix}$

Curves 2 : U = $\begin{matrix} 2 & A \\ 100 & V \end{matrix}$

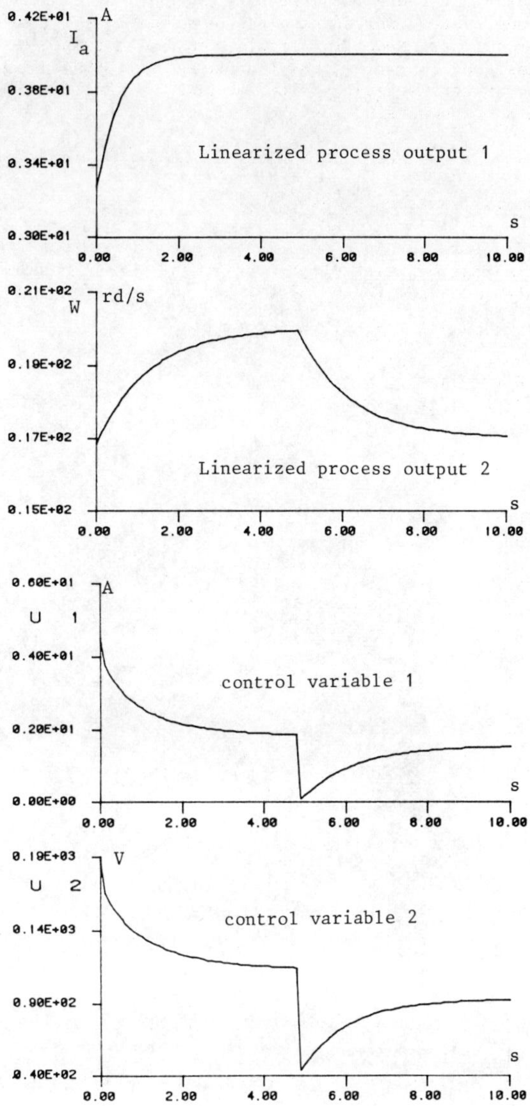

Fig. 6. D.C. motor closed loop linearization.

. first reference $x_c = \begin{pmatrix} 4 \\ 20 \end{pmatrix}$

. second reference $x_c = \begin{pmatrix} 4 \\ 17 \end{pmatrix}$ at t = 5 s.

. initial state $x(o) = \begin{pmatrix} 3{,}274 \\ 16{,}83 \end{pmatrix}$

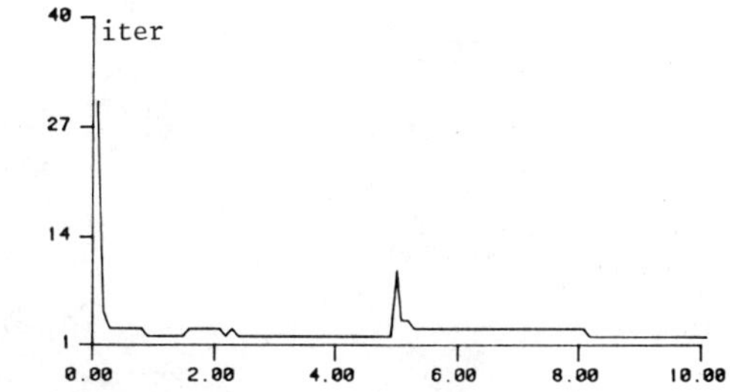

Fig. 7. D.C. motor iteration number.

COMPUTER-AIDED DESIGN OF DECENTRALIZED CONTROL SYSTEMS BY DYNAMIC DECOUPLING

S. K. Lehmann

Institut für Regelungs- und Steuerungssysteme, Universität Karlsruhe, Karlsruhe, FRG

Abstract. A method for the design of decentralized controllers by decoupling is described. The basic concept is the minimization of the influence of the other subsystems on the controlled subsystem by proper choice of controller parameters. Two criteria to measure this influence are proposed. The controller design is carried out by minimization of one of these criteria together with criteria that take into account the performance of the decoupled subsystem. The implementation of the method as part of a program package is sketched and its application demonstrated by means of an realistic example.

Keywords. Computer aided system design; decentralized control; decoupling; eigenvalue sensitivity; linear systems; electric power distribution.

INTRODUCTION

The many methods for the design of decentralized control systems can be grouped into three divisions.
The first division encompasses methods that use the complete model of the overall system for design purposes. All methods for the design of controllers with prescribed structure, as for example PREUSS(1981) and ROPPENECKER(1983), fall into this group. The use of the overall system, however, becomes very costly, if the system order increases.

The second group consists of methods utilizing the subsystem structure for the design by optimal control techniques. The basic concept is a partition of the overall system HAMILTONIAN into subsystem HAMILTONIANs. This leads to subsystem design problems which are coupled. Therefore, they have to be solved simultaneously or - as it is usually done (GUANGQUAN, LEE 1983; HASSAN, SINGH, TITLI, 1979) - recursively. These procedures use the lower order of subsystems, but, due to the recursion, are still extensive, beside the problem of convergence.

The methods of the third group use an aggregate model of the overall system (GEROMEL, BERNUSSOU 1979), thus employing the body of knowledge about connective stability (ŠILJAK, 1979) for the design. The methods can exactly assure stability of the controlled overall system and lead to decoupled subsystem design problems. A setback of these methods so far is their inability to take into account other design specifications than stability.

The method described here is in some way a derivative of the methods in the third group. The exact assurance of overall system stability is sacrificed for the possibility to assign more specifically the behavior of the controlled system. The basic concept is decoupling of the subsystem by its local controller.
If this decoupling approach is at all applicable, as it is for many real systems, it yields completely decoupled subsystem design problems and a good agreement between the performance of the decoupled subsystems and the overall system and therefore a system that is highly robust against failure of subsystem interconnections.

The paper is organized as follows. After system description and problem statement the concept of subsystem sensitivity is introduced. Subsequent sections treat additional design criteria, the used design procedure, and the realization as part of a program package. By means of an example, the application of the method is demonstrated, and the conclusion section summarizes the relevant results.

SYSTEM DESCRIPTION AND PROBLEM STATEMENT

The systems dealt with are assumed to be linear, time-invariant, continous. The system, consisting of r subsystems, is described in the *interaction-oriented notation* (LITZ 1983):

$$\begin{aligned} \dot{\underline{x}}_i &= \underline{A}_{ii}\underline{x}_i + \underline{B}_{Mi}\underline{u}_{Mi} + \underline{B}_{Di}\underline{u}_{Di} + \underline{B}_{Ii}\underline{u}_{Ii} \ , \\ \underline{y}_{Mi} &= \underline{C}_{Mi}\underline{x}_i \ , \\ \underline{y}_{Ii} &= \underline{C}_{Ii}\underline{x}_i \ , \\ i &= 1,\dots,r \ . \end{aligned} \qquad (1)$$

The dimensions of $\underline{x}_i$, $\underline{u}_{Mi}$, $\underline{u}_{Di}$, $\underline{u}_{Ii}$, $\underline{y}_{Mi}$, $\underline{y}_{Ii}$ are n_i, p_{Mi}, p_{Di}, p_{Ii}, q_{Mi}, q_{Ii}, respectively. In (1) $\underline{B}_{Mi}$ and $\underline{C}_{Mi}$ are the usual input and output matrices, M referring to manipulated and measured. $\underline{B}_{Di}$ is the disturbance input matrix. The vectors $\underline{u}_{Ii}$ and $\underline{y}_{Ii}$ describe the coupling of the i-th subsystem to all other subsystems, I referring to interaction.

The aggregation of these other subsystems is referred to as the *complement system* to the i-th subsystem. The effect of those subsystems on the i-th one is described by the coupling equation

$$\underline{u}_{Ii} = \sum_{\substack{j=1 \\ j\neq i}}^{r} \underline{P}_{ij}\underline{y}_{Ij} \quad . \tag{2}$$

The decentralized control-laws are written as

$$\underline{u}_{Mi} = -\underline{R}_i\underline{y}_{Mi} \quad , \quad i = 1,\ldots,r \quad . \tag{3}$$

The problem on stake is the computation of $\underline{R}_i$ for all subsystems such that the performance of the controlled overall system is satisfactory. The desired behavior can be specified via criteria to be minimized by the controller gains or by eigenvalue assignment. For the decoupled subsystem, the controller design is a straightforward task.
However, any nontrivial coupling to other subsystems affects the system behavior in an unpredictable manner, may even evolve in instability.
The degree to which the coupling affects the subsystem behavior is a function of the local controller. Therefore, it is possible to choose a controller that assures

- a satisfactory behavior of the controlled decoupled subsystem,
- a minimal affection of this behavior by the complement system.

Thus, the design specification is "minimal" influence of the complement system on the controlled subsystem. To minimize the influence as much as possible two things are needed: a criterion to measure this influence and a method to minimize it by suitable choice of the controller variables.

SENSITIVITY OF SUBSYSTEMS

The influence of the complement system on the subsystem depends on the subsystem itself and on the complement system.
It is possible to take into account both factors, but this leads to a dependence of the local controller on subsystem and complement system and thus to an interdependence of the subsystem design problems.

In order to avoid such an interdependence, the concept of *subsystem sensitivity* is introduced. Thereby, sensitivity is defined as a property of the subsystem alone. The key assumption is now that the subsystem performance is not substantially influenced by **any** complement system, provided its sensitivity is sufficiently small.

If this assumption holds - it will be shown that it does, if sensitivity is measured by one of the proposed criteria - the controllers for all subsystems can be designed minimizing the sensitivity of **their** subsystem. Thus, for all subsystems analogous design problems are to be solved, and all these problems are completely independent.
Since through the rest of the paper only subsystems are considered, the index i to describe a subsystem is suppressed.

To measure the sensitivity of a subsystem, two criteria are proposed.
The first one utilizes the inverse LAPLACE-transform $\underline{G}(t)$ of the subsystem transfer matrix $\underline{G}(s)$ between interaction input and output and is defined as

$$J_1 = \int_0^\infty \sqrt{\mathrm{tr}[\underline{G}^T(t)\,.\,\underline{G}(t)]dt} \quad ,$$

$$\underline{G}(t) \circ\!\!-\!\!\bullet\ \underline{C}_I\ [s\underline{I} - \underline{A}]^{-1}\underline{B}_I \quad . \tag{4}$$

This criterion is the integral of the EUCLIDEAN norm of the interaction output, when the the interaction input is an impulse function. It was used for example by LUNZE (1981) for stability analysis of complex systems.

The second criterion is based on the concept of *eigenvalue sensitivity* (LITZ 1981). The sensitivity of the m-th eigenvalue to a feedback of the i-th interaction output to the j-th interaction input is given by

$$S_{ij}^{\lambda_m} = -c_{im}^{*}b_{mj}^{*} \quad . \tag{5}$$

with c_{im}^{*} $[b_{mj}^{*}]$ as the (i,m)-th [(m,j)-th] element of the output [input] interaction matrix of the subsystem transformed to JORDAN-canonical form. To obtain a criterion describing the subsystem as a whole, the eigenvalue sensitivities are aggregated as follows:

$$J_2 = \sum_{i=1}^{q_I}\sum_{j=1}^{p_I}\sum_{m=1}^{n} \gamma_m\,(S_{ij}^{\lambda_m})^2 \quad . \tag{6}$$

The positive factors γ_m permit weighting of the contributions of single eigenvalues. With the usual choice

$$\gamma_m = \frac{1}{|\lambda_m|^2} \tag{7}$$

the summands in (6) are the squared relative eigenvalue sensitivities (LITZ 1981).
By aggregation of subsystem and complement system it can be shown that for $J_1 = 0$ or $J_2 = 0$ the subsystem is not influenced by **any** complement system. For reasons of continuity, it follows that the influence of the complement system will be negligible, if one of the sensitivity measures is sufficiently small - independent of the complement system encountered. Thus the sensitivity criteria J_1 and J_2 fullfill the above stated assumption and are therefore both usable for the controller design.
The main advantage of the second criterion is the availability of its gradient to the design parameters (LEHMANN 1984). Therefore, this criterion will be used exclusively.

ADDITIONAL CRITERIA AND MULTIOBJECTIVE OPTIMIZATION

Beside the dynamic decoupling of the subsystem, a satisfactory performance of the decoupled subsytem and - in many practical cases - a specified structure of the controller is to be achieved.

These two requirements can be taken into account by the criteria

$$J_T = \int_0^\infty [\underline{x}^T \underline{Q}_x \underline{x} + \underline{u}_M^T \underline{Q}_u \underline{u}_M]\, dt \qquad (8)$$

and

$$J_N = \sum_{i=1}^{q_M} \sum_{j=1}^{p_M} d_{ij} r_{ij}^2 \quad , \quad d_{ij} > 0 \quad . \qquad (9)$$

Is a controller element r_{ij} to be zero, d_{ij} is set to a large value. The minimization of J_N will then yield a negligibly small r_{ij}. Due to specific design requirements, additional criteria can be defined.
Thus, the sensitivity criterion is just one other design criterion on the subsystem level.

The controller is designed by minimization of these several criteria. To perform this multiobjective optimization, two approaches have been used. For a small number s of criteria say s < 4), the minimization of a weighted sum

$$J = \sum_{i=1}^{s} \alpha_i J_i \quad , \quad \alpha_i > 0 \quad . \qquad (10)$$

of the criteria J_i with interactive variation the weighting factors α_i is completely sufficient. As the number of criteria increases, this handling of the weighting factors, which is crucial for the success of the design, becomes more and more difficult. In these cases, the vector optimization method due to KREISSELMEIER/STEINHAUSER (1979) has been found as an efficient alternative.

The controller design itself relies on a technique developed by ROPPENECKER (1983). The cornerstone of the approach is a parametric expression for a state feedback controller which reads

$$\underline{R} = [\underline{p}_1,\ldots,\underline{p}_n][\underline{v}_1,\ldots,\underline{v}_n]^{-1} \quad ,$$

$$\underline{v}_i = [\underline{A} - \lambda_{Ri}\underline{I}]^{-1}\underline{B}_M\underline{p}_i \quad ,$$

$$i = 1,\ldots,n \quad . \qquad (11)$$

In this expression, λ_{Ri} is the i-th eigenvalue of the controlled (sub)system, $\underline{v}_i$ the i-th eigenvector and $\underline{p}_i$ the i-th *invariant parameter vector* (ROPPENECKER 1981, FAHMY/O'REILLY 1982). The closed loop eigenvalues and invariant parameter vectors are all degrees of freedom of a state feedback controller. In order to minimize the design criteria, the $\underline{p}_i$ and - within certain bounds - the λ_{Ri} are varied.

One advantage of this method is the additional control of the eigenvalues. The problems of instable systems in the course of the optimization are thereby completely avoided.

Moreover, the method is numerically very efficient in connection with the criterion J_2. From (5) it follows

$$S_{ij}^{\lambda_m} = \underline{w}_m^T \underline{b}_{Ij} \underline{c}_{Ii}^T \underline{v}_m \quad . \qquad (12)$$

The $\underline{v}_m$ are the eigenvectors given by (11), the $\underline{w}_m^T$ the row eigenvectors of the controlled subsystem, the rows of the inverse modal matrix. $\underline{b}_{Ij}$ [$\underline{c}_{Ii}^T$] denotes the j-th column [i-th row] of the interaction input [output] matrix. Since the modal matrix and its inverse have to be computed for the controller design [see (11)], the evaluation of the criterion does not require any additional complex calculations.
Even the gradient of the criterion can be expressed as a function of the $\underline{v}_m$ and $\underline{w}_m^T$. Due to (6) and (12) the calculation of the gradient of J_2 requires essientially the evaluation of the derivatives of the $\underline{v}_m$ and $\underline{w}_m^T$ to the design parameters, i. e. to the λ_{Ri} and $\underline{p}_i$. For these derivatives it can be found (LEHMANN 1984):

$$\frac{\partial \underline{v}_m}{\lambda_{Rk}} = [\underline{A} - \lambda_{Rm}\underline{I}]^{-1} \underline{v}_m \, \delta_{mk} \quad ,$$

$$\frac{\partial \underline{v}_m}{\partial \underline{p}_k} = [\underline{A} - \lambda_{Rm}\underline{I}]^{-1} \underline{B}_M \, \delta_{mk} \quad ,$$

$$\frac{\partial \underline{w}_m}{\partial \lambda_{Rk}} = - \underline{w}_k \underline{w}_m^T \frac{\partial \underline{v}_k}{\partial \lambda_{Rk}} \quad ,$$

$$\frac{\partial \underline{w}_m}{\partial \underline{p}_k} = - \underline{w}_k \underline{w}_m^T \frac{\partial \underline{v}_k}{\partial \underline{p}_k} \quad ,$$

$$\delta_{mk} = \begin{cases} 1 & \text{if } m = k \\ 0 & \text{if } m \neq k \end{cases} \quad . \qquad (13)$$

The knowledge of the gradient considerably facalitates the numeric minimization of the second sensitivity criterion. That is why this criterion is appropriate in the most cases.

The design procedure outlined so far, requires a multiobjective optimization, the formulation and evaluation of different design criteria and finally the aggregation and analysis of the controlled overall system. Although the complexity of the problem has been greatly reduced by working on the subsystem level, the use of a digital computer is inevitable. Therefore, the method has been implemented as part of a CACSD-package, which is described in the next section.

IMPLEMENTATION OF THE METHOD

Due to space limitations, only a few aspects related to decentralized systems and multiobjective optimization will be covered here.
The program system consists of a number of programs written in FORTRAN 77 which form three groups:

1. Programs for the handling of system data
2. Programs for system analysis
3. Programs for system design

The programs exchange information in a formalized fashion. For this purpose, the following data types, i. e. unformatted FORTRAN-records of well defined structure, have been defined:

REMAT (Rectangular matrix)
LITIS (Linear, time-invariant (sub)system)
OVSYD (Overall system description)
SIMRE (Simulation results)
PLODA (Plot-data)

As the FORTRAN-records are unformatted, they cannot be handled by usual EDITORS, but only by special programs. This guarantees the compatibility of the records at all instances. The programs to handle the records form group 1. The most important ones are given below:

CONOV1 - computes the controlled overall system.
MATED1 - handles data of type REMAT.
SYSED1 - handles data of type LITIS.
OVSYS1 - computes the overall system using using OVSYD data.

The most frequently used data type is the type LITIS. It is based on the system description (1) and has the following structure:

1. Record: Vector of 12 INTEGER-numbers, containing dimensions (mandatory)
2. Record: Dynamic matrix (mandatory)
3. Record: Manipulation-Input-Matrix (optional)
4. Record: Interaction-Input-Matrix (optional)
5. Record: Measured-Output-Matrix (optional)
6. Record: Interaction-Output-Matrix (optional)
7. Record: Disturbance-Input-Matrix (optional)
8. Record: Documentation-String (optional)

The programs of group 2 provide several options for analysis. Some programs and their purposes are given below.

EIGSY1 - Modaltransformation of a system
EIGMA1 - Eigenvalue/eigenvector computation for a matrix
INZER1 - Computation of invariant zeros of a system
SIMUL1 - Simulation of a linear system

All programs expect input data of type REMAT or LITIS. The program SIMUL1 - as well as other simulation programs - generates output data of form SIMRE which are input to the not mentioned plot programs. The use of these programs is by no means restricted to decentralized systems.

The programs of group 3 are design programs. In the following, only those programs are listed that support the design by dynamic decoupling.
All these programs treat one subsystem at a time. This enables complex computations with acceptable effort due to the low order of subsystems. After design of controllers for all subsystems independently the program CONOV1 can be used to aggregate the controlled overall system for simulation purposes.

The following design programs are available:

DYDEC1
DYDEC2
DYDEC5
DYDEC6 are small, fast programs which only minimize a sensitivity criterion by eigenvalue variation for subsystems with a single manipulation input. They use both mentioned criteria, but don't use gradients. They are mainly used to check, whether design by decoupling is at all feasible.

DYGRA1 minimizes a modal sensitivity criterion, is similar to the DYDECn programs but uses gradients.

DECEN1 is a large, comfortable program which by vector-optimization technique minimizes several criteria, especially all criteria mentioned in the paper.

MINVS4 is designed to handle large subsystems with several manipulation inputs. The eigenvalues of the controlled subsystem are to be specified exactly and a weighted sum of only three criteria is minimized using gradients.

Standard mathematical problems in all programs are - if possible - solved by subroutines of the commercial NAG (National ALgorithms Group) program library.

The computations for the following example were carried out by MINVS4 and DYDEC6.

EXAMPLE

In order to demonstrate the use of the method and the program system described, the design of load-frequency control for an electric power distribution network is shown.
The system is described by CUNO (1981). It consists of three local networks connected by power lines. The order of the overall system is 36, its structure is depicted in Fig. 1.

For the purpose of decentralized control system design the system must be decomposed into subsystems. One obvious decomposition are the local networks. This approach yields three sub-

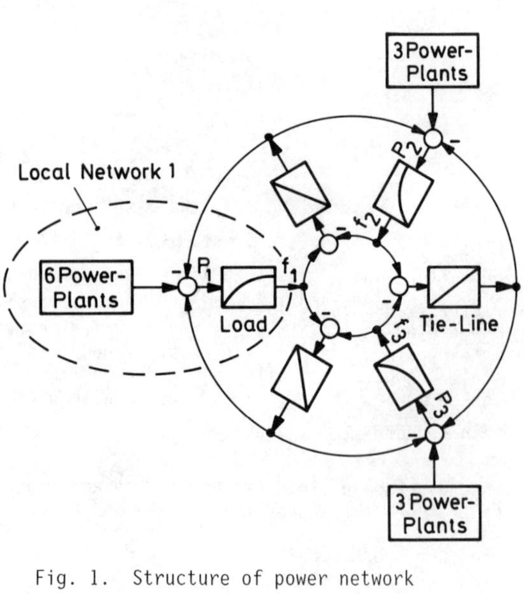

Fig. 1. Structure of power network

systems of order 16, 9 and 8. Using MINVS4, for all three subsystems state feedback controllers were designed. Since all state variables are measurable, no additional structure contraints were placed on the controllers. The performance of the controlled system show Fig. 2 to 5. In all four simulations frequency responses to a load increase of 1000 MW in the local network 3 are depicted. In Fig. 2 the decoupled subsystem 3 and the overall system are compared.

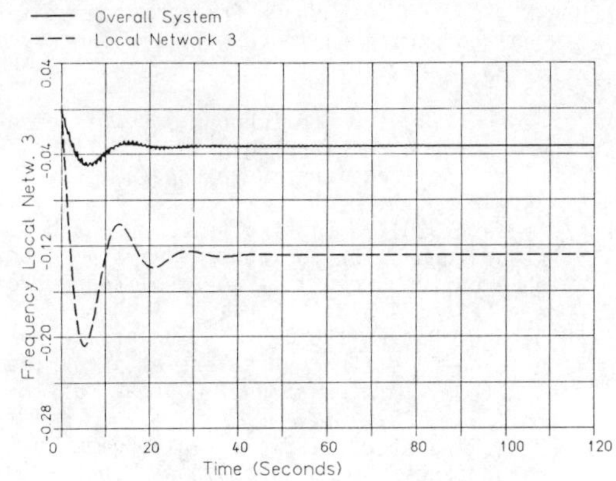

Fig. 2. Decoupled and overall system

It can be seen that the dynamic behavior of the subsystem does not substantially differ from the behavior of the overall system. However, the size of the frequency drop is reduced by the support of the other subsystems, a desired property.

Fig. 3 compares the overall system controlled by state feedback with the same system controlled by conventional load-frequency control.

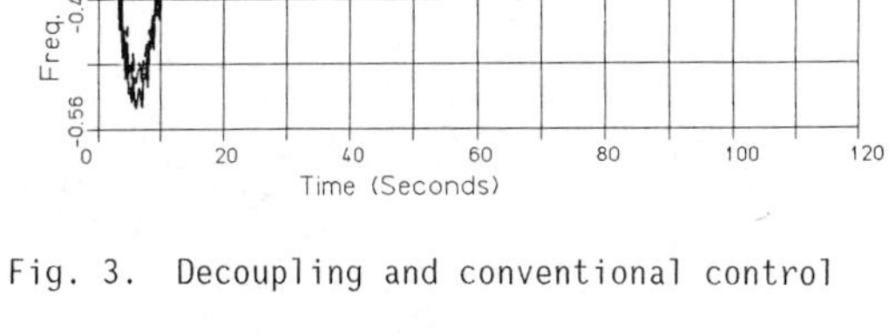

Fig. 3. Decoupling and conventional control

The frequency drop is only minimally reduced by the more refined control system, as this drop is mainly due to the slow dynamics of the power-plants as compared to the fast dynamics of loads and tie-lines. At first sight, the situation concerning the stationary frequency offset appears even worse: the more complex control gives rise to a significantly larger offset. However, as zero offset is required, a secondary control using integral action is necessary in both cases. The design of such control scheme is considerably easier for the the system with the state feedback control. The reason for the problems designing the secondary control for conventionally primary controlled plant is the peak at 18 s. This peak leads very easily to positive frequency deviations, when secondary control is applied, and this is to be avoided at all instances. On the other hand, for the system controlled by the state feedback controller the secondary control is designed quite easily using the classical method (GRANGER 1931) with a slight modification.

Fig. 4 shows the overall system with primary and secondary control.

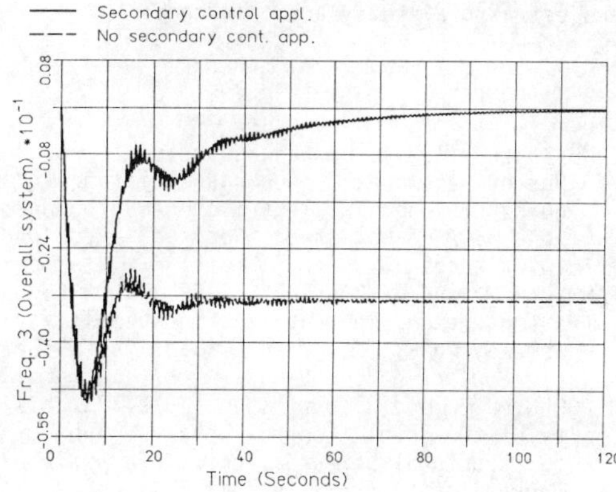

Fig. 4. Primary and secondary control

Another important feature of the decoupling design is the reliability of the controlled system. To demonstrate this, in Fig. 5 a typical "deasaster-type" situation is considered: A sudden load increase in local network 3 and at the same instance failure of all power lines to the other subsystems. As the decoupled subsystem was considered in the design, the now isolated system shows the expected acceptable behaviour, offcourse with a greater offset. The conventional load-frequency control however, completely fails.

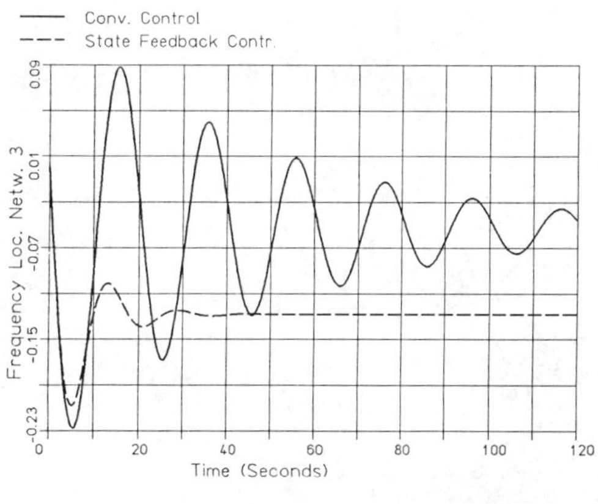

Fig. 5. Behavior of decoupled subsystem 3

CONCLUSION

The design of decentralized control systems via decoupling methods enables design completely on the subsystem level. Moreover, the subsystem design problems are completely decoupled and can be tackled by any method known from centralized control, whereby the decoupling requirement can be taken into account by one additional design criterion. Two additional advantages of the procedure are the reliability to failure of subsystem interconnections and the assignment of eigenvalues specifically to certain subsystems.

ACKNOWLEDGMENT

The support of the Deutsche Forschungsgemeinschaft (Grant Nr. Fo/71-13) for this research project is gratefully acknowledged.

REFERENCES

CUNO, Bernd (1981). Anwendungsbeispiel zur robusten Regelung: Übergabeleistungs-Frequenzregelung in elektrischen Verbundnetzen. *2. Workshop Robuste Regelung.* Interlaken.

FAHMY, M.M. and J. O'REILLY (1982). On Eigenstructure Assignment in Linear Multivariable Systems. *IEEE Transactions on Automatic Control*, *27*, 690-693.

GEROMEL, J.C. and J. Bernussou (1979). Stability of two-level control schemes subjected to structural perturbations. *International Journal of Control*, *29*, 313-324.

GRANGER, H. (1931). Deutsches Reichspatent DRP DRP 634 025 from Feb. 8th,1931.

GUANGQUAN, Li and Gordon K.L. LEE (1983). Decentralized control of large scale systems with dynamic interconnected subsystems. *International Journal of Control*, *37*, 775-786.

HASSAN, M.F., M.G. SINGH and A. TITLI (1979). Near Optimal Decentralized Control with a Prespecified Degree of Stability. *Automatica*, *15*, 483-488.

KREISSELMEIER, G. und R. STEINHAUSER (1979). Systematische Auslegung von Reglern durch Optimierung eines vektoriellen Gütekriteriums, *Regelungstechnik*, *27*, 76-79.

LEHMANN, Stefan K. (1984). Entwurf dezentraler Regelungen mit Hilfe von Entkopplungsverfahren. *Regelungstechnik*, *32*, 345-448 and 373-376.

LITZ, Lothar (1981). Berechnung stabilisierender Ausgangsrückführungen mit Hilfe von Polempfindlichkeiten. *Regelungstechnik*, *29*, 434-440.

LUNZE, Jan (1980). Ein Verfahren zur Stabilitätsprüfung für gekoppelte Systeme bei fehlerhaftem Modell. *Messen-Steuern-Regeln*, *23*, 374-378.

PREUSS, Hans Peter (1981). Successive pole shifting by state controllers of prescribed structure. *International Journal of Control*, *33*, 1059-1071.

ROPPENECKER, Günter (1981). Polvorgabe durch Zustandsrückführung. *Regelungstechnik*, *29*, 228-233.

ROPPENECKER, Günter (1983). Vollständige modale Synthese linearer Systeme und ihre Anwendung zum Entwurf strukturbeschränkter Zustandsrückführungen. *Fortschrittsberichte der VDI Zeitschriften*, *Reihe 8*, Nr. 59.

ŠILJAK, Dragoslav D. (1978). *Large-Scale-Dynamic-Systems*. North-Holland, New York-Oxford-Amsterdam.

INTERACTIONS IN DYNAMICAL SYSTEMS IDENTIFICATION

N. E. Madzarov, I. G. Kalaikov, E. M. Garipov and P. G. Ganev

Department of Automatics, Higher Institute of Mechanical and Electrical Engineering, Sofia, Bulgaria

Abstract. The paper deals with the improvement of system identification procedures by interactions between the user and the computer. The basic identification algorithms are discussed with respect to the necessity of dialog at some places of the procedure and the efficiency of that approach for model and parameter estimation. All refered problems are implemented in the interactive software system named INSOF-I. The conception and consideratons for its hierarchical structure are given. All ideas of the man-machine interaction supported by the protocol and graphical output of INSOF-I are demonstrated by the generalized least squares estimation method.

Keywords. Interactive systems; identification; parameter estimation; computer aided design; linear continuous-time and discrete time systems.

INTRODUCTION

The interactive principle is applied as a modern approach in development of software paskages for control system design. Some authors (e.g. Lemmens, 1979; Astrom,1981) have pointed out the following main advantages of the interactive operation in dialog with the computer in comparison with the traditional computer usage:

- the problems can be solved by users having minimal knowledge in programming, most often by means of natural for each group of specialists language;
- differently qualified persons can be effectively teached not only in computer practice but in their speciality details as well;
- a real-time interactive mode of operation is available, therefore a better studying of computer control problems can be done;
- the terms and the financial ammount for project development can be sihnificantly reduced, etc.

The system identification is one of the problems that can be solved effectively in interactive mode of operation. Many interactive packages for identification (in some cases as subsystems of more complex software systems) are described in the specialized literature (e.g. Denham and Abaza,1975; Wieslander and Gustavsson,1976; Schmid and Unbehauen, 1979; Dymshiz,1979; Furuta and co-workers,1981; Young,1982). This list can be continued by another similar works. It must be noted that the comments on the identification methods and algorithms implementation in the mentioned above references are not sufficient if their rational use in interactive mode is considered. Report on it, a general conclusion of the main steps in the identification procedere, in which the engineer's intervention and making a decision for turning the computation flow to desired direction takes an important place, is made in the present paper. The authors' experience in building up of an interactive program system INSOF-I (INteractive SOFtware for Identification) is shared. Illustrative examples support a concrete assumed but quite general approach of working in interactive mode.

METHODS AND ALGORITHMS FOR INTERACTIVE SYSTEM IDENTIFICATION

The identification procedure includes three main stages;

- separation of the plant from its environment, development of general concept about it and an idea about its interrelations with the environment;
- design of experiments, according to this idea, to acquisite information about measured signals;
- data processing via definite methods to obtain the plant mathematical model and its validation.

In practice this sequence of operations is realized always by repeating it entirely or in stages. Thus, after the engineer's competent intervention in certain moments of the man-computer communication, the degree of uncertainty in the identification problem solving is eliminated considerably. While the last two stages in the identification procedure are based on strong scientific theory and use well mathematized and algorithmized methods, the first one is associated above all with putting empirical rules into practice. Here it is relied on the researcher's experience and intuition for implicit use of the results in the third stage.

The identification, in spite of the tendency towards maximal formalization as a procedure, remains a creative step in the process of control system design. The creative element is developing by implementation of interactive systems, which include not only algorithms based on exact mathematical rules but on expert subsystems with the specialists' experience putted there. This, in the condition of man-computer interaction in the identification problems, it is get over the rupture between the applied engineering approach and the mathematical one, which intil recently was blamed for inecessary theorizing and establishing difficult to use methods for great number of engineers.

The following considerations are to be taken into account in the implementation of interactive software systems for identification:

1. The identification is not a self-purpose problem. The ungrounded model complication with unnecessary details leads usually to "excess" in terms of the problem solved by this model. If the problem is controller synthesis, this "excess" can not only bother it but make worse the control system behavior as well. It is obvious that the model have to represent only these real plant characteristics that correspond with the solving problem. Thus the identification must be considered as a tool to reduce the nature variety and this reduction have to (may) be purposefully controlled by man-computer interaction.

2. The control system design problems solved via the plant model, are normally better adopted to certain model structure. Every software system for identification have to offer some possibilities for choice of an appropriate model structure (e.g. parametric or noparametric model, continuous-time and discrete-time model, model contaning also the noise model and so on). The choice of a structure and the clarification of its advantages over the rest structures is a hard formalized problem that successfully can be solved in interactive mode.

3. The practice (e.g. the controller synthesis) most often requires an examination of the plant local properties. These ones are well expressed by means of linear continuous-time or discrete-time models. This class of models has an essential advantage that their structures are defined unambiguously by the order of the corresponding differential (difference) equation and the time delay at the model inputs. In spite of the possibility of some kind of formalization, the estimation of these model structural parameters is a multivariant procedure and it is desirable to solve it in interactive mode.

4. The most identification problems can be solved successfully in batch mode processing as well. The interaction approach increases significantly the intellectual power of man-computer system and ensures all modern conveniences for operative activity. In such case the specialists solve their problems directly with no need of the programmer's mediation in clear and easy to learn problem-oriented language.

Taking into account the objective trends for computer aided design exclusively in interactive mode of operation and the refered features of the identitication problems the given below main groups of tasks are included in INSOF-I system (Kalaikov and co-workers,1983), made with the participation and under the guidance of the authors.

Time Series Analysis

The effectiveness of the identification methods depends on the statistical properties of the experimental data. Most of the parameter estimation methods require the input-output observations to be zero mean gaussian stationary stochastic processes. Therefore the stage of expetimental data preprocessing includes a normality test and a stationarity test (Bendat and Piersol,1971). If some of these tests gives unsatisfactory resulrs, the user must try to detect and remove the corresponding components in the realizations, e.g. applying low-pass or high-pass filtering, trend removal, centering and so on, which are implemented by appropriate software modules.

Model Order and Time Delay Estimation

The parameter estimation is accomplished after the model structure has been obtained. If linear models are considered, their structure is determined unambiguously by the differential (difference) equation order and the time delay. Subjective estimates of the structural parameters are often prefered since the objective criterias usually lead to unacceptably large values of the model orber. It is possible to reduce it involving time delay. The significant advantages of the interactive approach are clearly delineated if the wide variety of structure identification methods are taken into account.

Two ways to estimate the model structure parameters, which depend on the available information, are provided in INSOF-I:

Input-output observations. A combined method, including subjective and objective procedures, is applied. The subjective procedure is based on visual evaluation of the determinant ratio of information matrices (Unbehauen and Gohring, 1974) under fixed time delay and variable model order. In case of ambiguous estimates the most probable model order value is choosing by the objective procedure of the F-criteria after the standard loss function has been computed. This approach is illustrated in Table 1, where a concrete example of making a decision is outlined. There IDn,r means the determinant ratio of the corresponding information matrices, ß is the output additive noise level, Q is the loss function value, m and r are the model order and time delay, (n^*, r^*) and ($\hat{n},\hat{r}$) are respectively the subjective and objective estimates of (n,r). The graphic logID=f(n) is shown in the middle of the table.

TABLE 1 Structure Identification Based on Input - Output Observations

Model	n	r	ß [%]	lg $ID_{n,r}$	n^*	r^*	Q	($\hat{n},\hat{r}$)
$W(z^{-1})=\frac{-0,102z^{-1}+0,173z^{-2}}{1-1,425z^{-1}+0,496z^{-2}}z^{-1}$	2	1	15	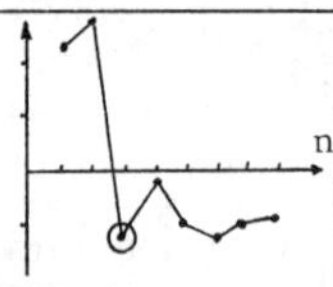	3	0	$0,111.10^7$	(2, 1)
				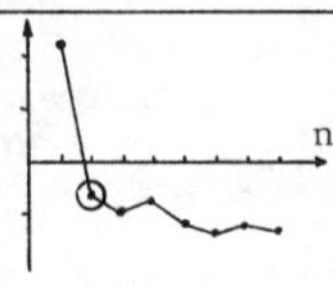	2	1	29,83	
				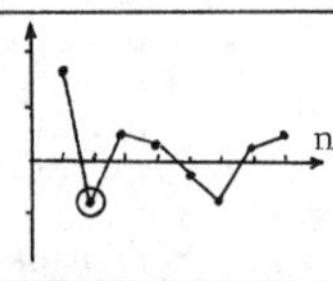	2	2	$0,151.10^7$	

Pulse response. In this case the infotmation matrices are formed by estimates of the pulse response instead of the input-output observations. The couple (n,r),for which the corresponding determinant ratio is approximately constant, is accepted as a valid estimate of n ahd r. The time delay value is determined visually looking at the pulse response curve. By the same example this method is shomn in Table 2, where $d_j^{n,r}$ means the determinant ratio of infotmation matrices, and k_1 and k_2 are the initial and end value of the examined time interval. The graphic d=f(j) with different values of n and r is placed in the middle of the table.

It is evident that both problems are based on efficient man-computer interaction as at this step an input information (data arrays, structural parameters) is formed or selected for the real stage of the identification procedure.

TABLE 2 Structure Identification Based on Pulse Response

Pulse response	$d_j^{n,r}$	Q	(n,r)
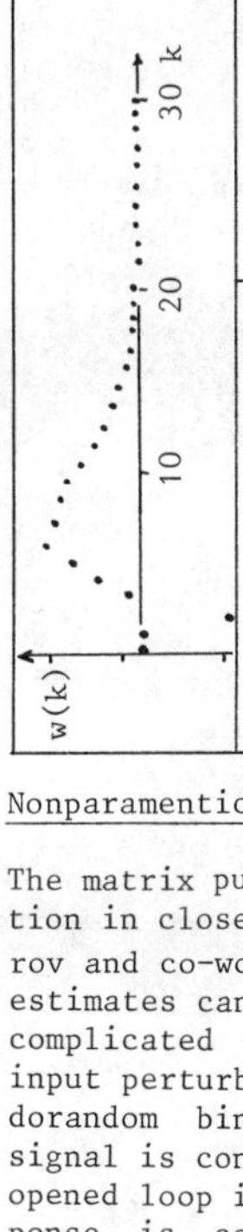 $W(z^{-1})=\frac{-0,102z^{-1}+0,173z^{-2}}{1-1,425z^{-1}+0,496z^{-2}}z^{-1}$	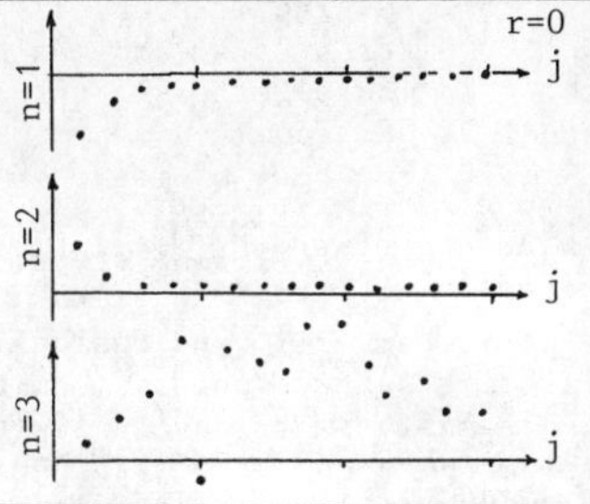 r=0 (n=1, n=2, n=3)	$0,67.10^{4}$	
	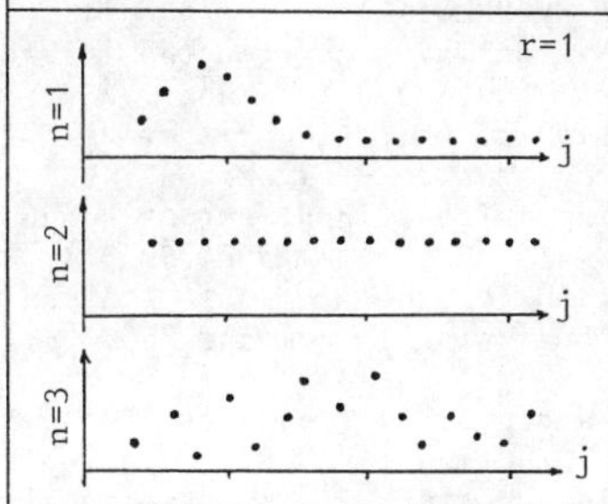 r=1 (n=1, n=2, n=3)	$0,872.10^{-5}$	(2,1)
	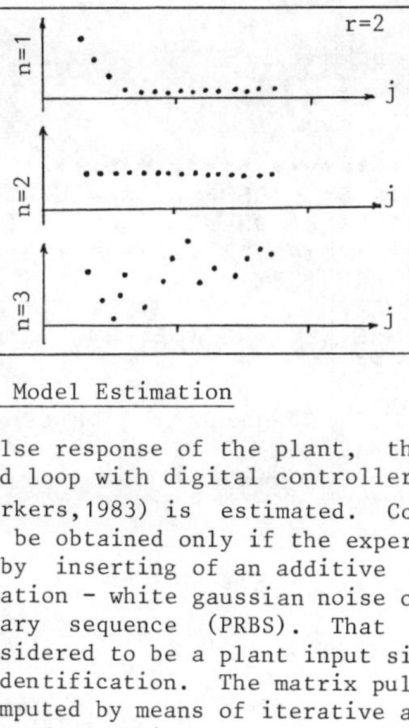 r=2 (n=1, n=2, n=3)	53,61	

Nonparamentic Model Estimation

The matrix pulse response of the plant, that function in closed loop with digital controller (Madzarov and co-workers, 1983) is estimated. Consistent estimates can be obtained only if the experiment is complicated by inserting of an additive measured input perturbation - white gaussian noise or pseudorandom binary sequence (PRBS). That additive signal is considered to be a plant input signal in opened loop identification. The matrix pulse response is computed by means of iterative algorithm using the found recursively or in off-line mode crosscorrelation functions estimates of the corresponding input-output and additive signals. The obtained pulse response estimate can be used as a final nonparamentric model of the plant or as an intermediate product for parameter identification.

Parametric Model Estimation

The parameters of a p input and r input model, represented as a matrix transfer function, are estimated. The choosen form permits the plant to be decomposed into r subplants with single output (the MISO model). After the structural parameter have been obtained the well known estimation methods - least squares, instrumental variable, maximum likelihood - are applied. The tasks are constructed in sich a way that both discrete-time and continuous-time (via disctere orthogonal transformations, Kalaikov and Garipov, 1984) models can be utilized. The numericaly stable procedures of LINPACK (Dongarra and co-workers, 1979) are used for solving the respective systems of algebraic equations.

On-line Estimation Procedures

The on-line estimation is aimed first and foremost with the determination of control signal to the plant when the plant dynamic properties or environment's ones are changed, i.e. when an adaptive control is needed. These procedures can be realized if a computer with a data acquisition subsystem is available. In our case a personal microcomputer (like Apple II) with necessary accessories (A/D and D/A converters, floppy-disk drives, digital plotter and corresponding software modules) is used. The presence of such an intelligent terminal in the interactive system allows the implementation of various operations - beginning with data acquisition from the examined plant and ending with the design and analysis of complex control algorithms in hierarchical systems.

The included in INSOF-I on-line identification algorithms are based on the square root filtering method (Peterka, 1975) and some decomposition methods from LINPACK, which ensure a sufficient precision on computers with limited machine word. All the algorithms are intended for both discrete-time and continuous-time MISO model estimation.

STRUCTURE AND ORGANIZATION OF THE INTERACTIVE SYSTEM

The INSOF-I has a four level hierarchical structure. The modules of the first and second level are programmed as callable subroutines. Those of the third and forth level are main programs that call only the subroutines of the lower levels.

On the first (the lowest) level basic mathematical functions are placed which are an extention to the standard subroutine libraries. Other part of the first level includes functions depending on the operating system (input-output operations, communication modules with the operating system). These modules move the intersection lines with the OS from the higher levels to the lowest one. In this way the programs from the higher levels become portable. The adaptation of INSOF-I to different computer or OS requires changes in the lowest level modules only.

The second level includes modules that realize the numerical methods as the program package LINPACK and form the scientific and engineer library of the system.

On the third level the main programs, which implement the identification and parameter estimation tasks, are placed. They all are executed independently, activated by a monitor program (kernel) of the INSOF-I, placed on the forth (the highest) level. The monitor program organizes and dispatches the system operations, manages the distribution of the computer resourses, alternates the working mode of the lower level programs.

The INSOF-I processed data are organized in disk mass storage files with specified and standartized a priori structure. Each file is a data array simply defined during the program execution. The file content is indicated via parameters included in a common information block of the INSOF-I files:

* file name;
* file structure parameters:
  - data type (e.g. time series,matrices),
  - data word length in bytes,
  - information type (e.g. system representation form - continuous-time or disctere-time model);
* additional parameters:
  - date and time of file creation,
  - file creating task name and comments.

After that title information the real content of the file is stored. Its dimension and structure depends on the actual data.

The INSOF-I gives up the user a set of commands which implement the above mentioned identification and estimation problems as well as some specific utility functions (data entry and manipulation, file copying and transformation etc). Each operation performs a single command carried out by an individual main program. The command string is formed in a standard manner in order to be learned and used easily:

$$\text{Name } OF_1, OF_2, \ldots/KW_1/KW_2 \ldots = IF_1, IF_2, \ldots/KW_3/KW_4 \ldots ,$$

where the attributes are:
- Name - the command name,
- OF_i - output file identifier,
- IF_i - input file identifier,
- KW_i - keyword,

Exampe:

CORF RUY/20 = U,Y/500

(estimation of 20 ordinates of the crosscorrelation function RUY using 500 points of signals U and Y).

The command string interpretation is realized in a way that the interaction details depend on the user's qualification degree. The higher his qualification the shorter the dialog is (or is missing at all) and vice versa. This is specific for each command and depends on how complete and correct the command string is. The list of the basic INSOF-I commands is given below.

TABLE 3 List of INSOF-I Commands

Name	Function
MSTE	Model structure parameter estimation (model order and time delay)
NPME	Nonparametric model estimation (matrix pulse response)
DLSQ	Least squares parameter estimation
DIVA	Instrumental variables parameter estimation
DMXL	Maximum likelihood parameter estimation
RNPE	Recursive nonparametric model estimation
RLSQ	Recursive LS parameter estimation
RIVA	Recursive IV parameter estimation
RMXL	Recursive ML parameter estimation
ARME	Parameter estimation of AR models
MDTR	Model transformation (nonparametric model parametric model)
WPWZ	Discretization of continuous-time systems
SNST	Normal distribution and stationary tests
SPRO	Signal processing
CORF	Correlation function estimation
SPDF	Spectral power density function estimation
READ	Data input
TYPE	Data output (on system printer)
PLOT	Data output (on plotter)
SGEN	Test signals generation
SMOD	Modelling of continuous-time and discrete-time systems
DEFM	Definition of macro
CALL	Execution of macro
RTRN	Exiting from macro
GOTO	Conditional or unconditional branch in macro
ASKI	Input of vatiable, used in macro

In order to give the users to delelop their own problems by means of the established command set defined in advance, command sequences named macros are used. Their application is recommended if preliminary fixed sequences of operations have to be carried out or complex computational methods examined alreaby must be realized. Every macro is created in text editing mode under the control of the INSOF-I monitor program and can be called by all users. The macro text is a set of translated (interpreted) INSOF-I command strings. As an example on Fig. 1 the text of macro "DGLS"is given, which performs the generalized least squares estimation method in the following steps:

- initial parameter "PAR" and residuals "E" estimation via least squares method (command "DLSQ");
- statistical analysis of residuals (observing the correlation function "RE", obtained by "CORF" and plotted by "PLOT");
- making a decision to stop the iterations and to exit from the macro (commands "ASKI" and "RTRN");
- synthesis of whitening filter "FIL" on the base of the resisuals "E" (command "ARME");
- filtering of the input-output sequences "U" and "Y" through "FIL" (command "SPRO");
- repetition of the procedure (unconditional jump to the beginning - command "GOTO").

Protocol and graphical outputs of the INSOF-I work including the macro "DGLS" execution are shown as an example on Fig. 2 (the structural parameters estimation procedures are not illustrated because they are outlined in Table 1 and Table 2).

```
 1 * *************************************
 2 * * "DGLS" - GENERALIZED LEAST SQUARES *
 3 * *  ESTIMATION OF DISCRETE-TIME MODEL *
 4 * *************************************
 5 DLSQ  PAR , E = U , Y /DI
 6 TYPE  PAR
 7 CORF  RE /20 = E , E /@@
 8 PLOT  RE
 9 * MAKE  YOUR  DECISION :
10 *  1 - TO CONTINUE THE PROCEDURE,
11 *  2 - TO EXIT.
12 *$=>
13 ASKI  I1
14 GOTO  I1 /16
15 RTRN
16 ARME  FIL , EE = E
17 SPRO  Y = FIL , Y
18 SPRO  U = FIL , U
19 GOTO  5
```

Fig. 1. Text of macro "DGLS".

CONCLUSION

The paper deals with the problem of the dynamical system identification in interactive mode of operarion. Some considerations about the active user's intervention to obtain an acceptable (from engineer's point of view) plant model in short terms, are disscussed. The main principles of the interactions realized in the software system INSOF-I for parametric and nonparametric identification are described. This package satisfies the most modern CAD system features: modular structure, portability, possibility of extension without rebuilding, multiuser mode of operation, convinient data base, flexible and easy to adjust dialog, error protection, etc. The development and improvement of INSOF-I is connected with the extension of both the classes of models and identification methods.

REFERENCES

Astrom, T., and H. Elmquist (1981). Perspective on interactive software for computer aided modeling and design of control systems. IEEE Conf. on Dec. and Control, San Diego.

Bendat, J. S., and A. G. Piersol (1971). Random data: analysis and measurement procedures. Wiley-Interscience, New York.

Denham, M. J., and B. A. Abaza (1975). System identification package (SYSID). Publ. 75/28, Dept. of Comp. and Control, Imp.College,London.

Dongarra, J. J., J. R. Bunch, C. B. Moler, and G.W. Stewart (1979). LINPACK user's guide. SIAM, Philadelphia.

Dymshiz, E. (1979). A process computer program package for interactive computer aided design of multivariable control systems. Proc. of 2th IFAC/IFIP Symp. SOCOCO, Prague.

Furuta, K., S. Hatakeyama, and H. Kominami (1981) Structural identification and software package for linear multivariable systems. Automatica, vol. 17.

Kalaikov, I. G., and E. M. Garipov (1984). A generalized algorithm for identification of continuous multivariable systems. Prepr. of 3rd Int. Conf. on Syst. Engineering, Dayton, Ohio.

Kalaikov, I. G., E. M. Garipov, S. B. Maleshkov, M. S. Sulemezova, and P. G. Ganev (1983). Inter. active system for identification of dynamical models. 8th Int. Conf. on Syst. Sci., Wroclaw.

Lemmens, W., and A. Van den Boom (1979). Interactive computer programs for education and research: a survey. Automatica, ovl. 15.

Madzarov, N. E., S. B. Maleshkov, N. S. Sulemezova, and P. G. Ganev (1983). Identification of process in closed loop. 8th Int. Conf. on Syst. Sci. Wroclaw.

Peterka, V. (1975). A square root filter for realtime multivariable regression. Kybernetika, vol. 11.

Schmid, C., and H. Unbehauen (1979). KEDDC, a general purpose CAD software system for application in control engineering. Proc. of 2nd IFAC/IFIP Symp. SOCOCO, Prague.

Wieslander, J., and I. Gustavsson (1976). IDPAC - An efficient interactive identification program. Prepr. 4 th IFAC Symp. on Ident.and Syst. Param. Est., Tbilisi.

Unbehauen, H., and B. Gohring (1974). Test for determining model order in parameter estimation. Automatica, vol. 10.

Young, P. (1982). A computer program for general recursive time series analysis. Prepr. of 6th IFAC Symp. on Ident. and Syst. Param. Estimation, Arlington.

```
>DIA

          I N S O F - I   <V 1.1>

DIA>READ U/SQ/DK

[READ] SEQUENCE LENGTH [POINTS] ? => 800
[READ] NUMBER OF SEQUENCES ? => 1
[READ] TIME INTERVAL ? => 2
[READ] COMMENTS FOR "U    ":
=> PLANT INPUT SIGNAL

DIA>PLOT U

[PLOT] X - DIMENSION ? [IN MM] => 90
[PLOT] Y - DIMENSION ? [IN MM] => 40
[PLOT] SYMBOL SIZE ? => 2
[PLOT] READY PLOTTER - PRESS <CR> cr
```

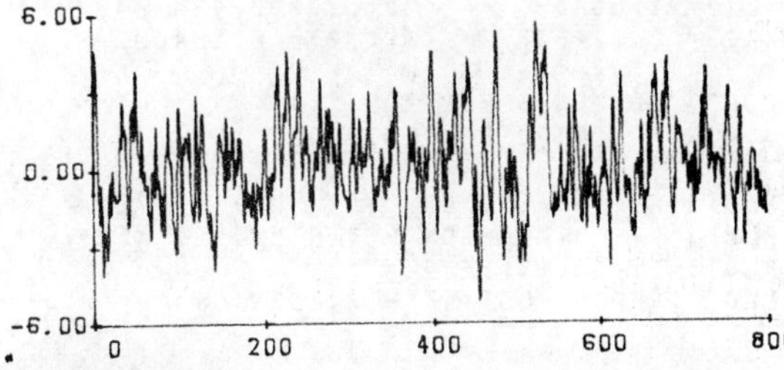


```
[PLOT] DISK OUTPUT FILE [Y/N] ? => N

DIA>READ Y/SQ/DK

[READ] SEQUENCE LENGTH [POINTS] ? => 800
[READ] NUMBER OF SEQUENCES ? => 1
[READ] TIME INTERVAL ? => 2
[READ] COMMENTS FOR "Y    ":
=> PLANT OUTPUT SIGNAL

DIA>PLOT Y

[PLOT] X - DIMENSION ? [IN MM] => 90
[PLOT] Y - DIMENSION ? [IN MM] => 40
[PLOT] SYMBOL SIZE ? => 2
[PLOT READY PLOTTER - PRESS <CR> cr
```

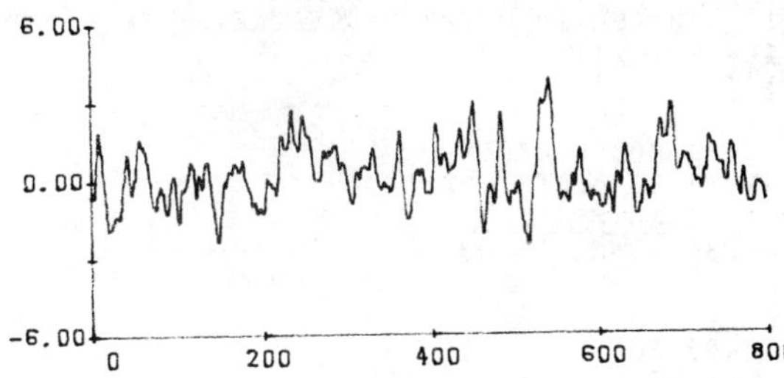


```
[PLOT] DISK OUTPUT FILE [Y/N] ? => N

DIA>SNST U/NO/0.99

[SNST] THE SIGNAL IS STATIONARY AND
       NORMALLY DISTRIBUTED.

DIA>SNST Y/NO/0.99

[SNST] THE SIGNAL IS STATIONARY AND
       NORMALLY DISTRIBUTED.
```

Fig. 2. Protocol and graphical outputs of INSOF-I work (cont. on next page)

```
DIA>CALL DGLS

*************************************
* "DGLS" - GENERALIZED LEAST SQUARES *
*  ESTIMATION OF DISCRETE-TIME MODEL *
*************************************

[DLSQ] MODEL DENUMERATOR ORDER ? => 2
[DLSQ] MODEL NUMERATOR ORDER   ? => 2
[DLSQ] MODEL TIME DELAY ? => 1
[DLSQ] COMMENTS FOR "PAR ":
=> LEAST SQUARES PARAMETER ESTIMATES
[DLSQ] COMMENTS FOR "E   ":
=> RESIDUALS

PAR  REGM.DI "DLSQ" : LEAST SQUARES PARAME

     TIME DELAY = 1

   DENUMERATOR:          NUMERATOR:
A( 0)= 1.00000       B( 0)= 0.000000
A( 1)=-1.56460       B( 1)=-0.484143E-01
A( 2)= 0.620637      B( 2)= 0.104082

[CORF] COMMENTS FOR "RE  ":
=> COR.FUNCTION OF RESIDUALS "E"

[PLOT] X - DIMENSION ? [IN MM] => 70
[PLOT] Y - DIMENSION ? [IN MM] => 45
[PLOT] SYMBOL SIZE ? => 2
[PLOT] READY PLOTTER - PRESS <CR> cr
```

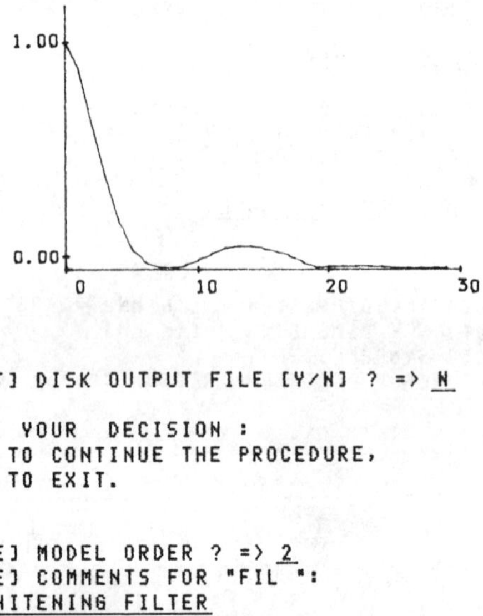


```
[PLOT] DISK OUTPUT FILE [Y/N] ? => N

MAKE  YOUR  DECISION :
 1 - TO CONTINUE THE PROCEDURE,
 2 - TO EXIT.
=> 1

[ARME] MODEL ORDER ? => 2
[ARME] COMMENTS FOR "FIL ":
=> WHITENING FILTER
[ARME] COMMENTS FOR "EE  ":
=>

[SPRO] COMMENTS FOR "Y   ":
=> FILTERED PLANT OUTPUT SIGNAL

[SPRO] COMMENTS FOR "U   ":
=> FILTERED PLANT INPUT SIGNAL

[DLSQ] MODEL DENUMERATOR ORDER ? => 2
[DLSQ] MODEL NUMERATOR ORDER   ? => 2
[DLSQ] MODEL TIME DELAY ? => 1
[DLSQ] COMMENTS FOR "PAR ":
=> PARAMETER ESTIMATES
[DLSQ] COMMENTS FOR "E   ":
=> RESIDUALS

PAR  REGM.DI "DLSQ" : PARAMETER ESTIMATES

     TIME DELAY = 1

   DENUMERATOR:          NUMERATOR:
A( 0)= 1.00000       B( 0)= 0.000000
A( 1)=-1.49146       B( 1)=-0.758688E-01
A( 2)= 0.554671      B( 2)= 0.138345

[CORF] COMMENTS FOR "RE  ":
=> COR.FUNCTION OF RESIDUALS "E"

[PLOT] X - DIMENSION ? [IN MM] => 70
[PLOT] Y - DIMENSION ? [IN MM] => 45
[PLOT] SYMBOL SIZE ? => 2
[PLOT] READY PLOTTER - PRESS <CR> cr
```

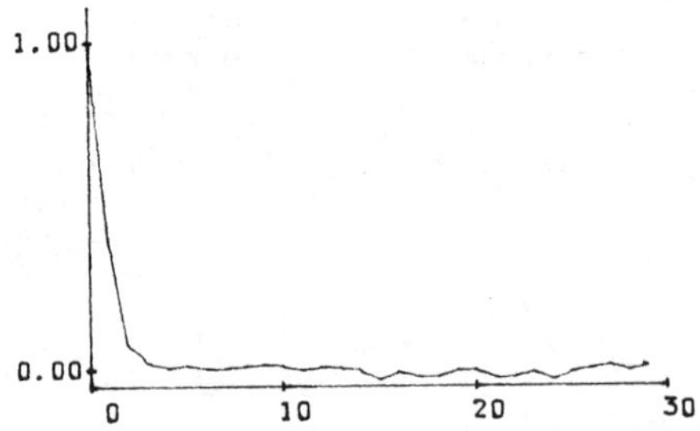


```
[PLOT] DISK OUTPUT FILE ? [Y/N] => N

MAKE  YOUR  DECISION :
 1 - TO CONTINUE THE PROCEDURE,
 2 - TO EXIT.
=> 2

DIA>FINF /FU

I N S O F - I   FILE INFORMATION :

U    SEQ .   SPRO :FILTERED PLANT INPUT SI
Y    SEQ .   SPRO :FILTERED PLANT OUTPUT S
PAR  REGM.DI DLSQ :PARAMETER ESTIMATES
E    SEQ .   DLSQ :RESIDUALS
RE   SEQ .   CORF :COR.FUNCTION OF RESIDUA
FIL  REGM.DI ARME :WHITENING FILTER
EE   SEQ .   ARME :

DIA>^Z

>
```

Fig. 2. Protocol and graphical outputs of INSOF-I work (cont. from previous page)

INTERACTIVE PACKAGE FOR THE IDENTIFICATION OF THE ELEMENTS OF A CONTROL LOOP

M. Messely, R. De Keyser and A. van Cauwenberghe

Laboratory of Automatic Control, University of Gent, Belgium

Abstract. The paper presents an interactive package for the identification of the elements of a control loop. This includes not only the dynamic behaviour of the process, but also the static behaviour of the process and the controller and the static nonlinear behaviour of the sensors. The program contains subpackages for model estimation, parameter estimation and curve fitting. Model evaluation and parameter estimation are necessary for the simulation of processes and controllers in order to develop appropriate control strategies. Although ARMAX models are chosen for most of the identification techniques, one algorithm allows the estimation of continuous system parameters. Curve fitting is needed for the static behaviour of processes and for the non-linear behaviour of industrial sensors. The program is fully interactive and offers many graphical possibilities such as x-y engineering plots and three-dimensional plots.

Keywords. computer-aided system design, continuous systems, control system analysis, identification, iterative methods, least-squares approximations, linear systems, modelling, parameter estimation, sampled-data systems, stochastic systems

INTRODUCTION

All modern CAD programs have two common features :
- they handle and manipulate many data
- they offer the user a graphical representation of these (processed) data.

A good CAD package makes the manipulation of data easy and offers powerful graphical possibilities. This increases the interaction and gives the designer a thorough understanding of his design. Much attention is paid to these two points when developing INTER. In this package for identification the data are generated from input/output observations of the process to be identified. The graphics capability contains x-y engineering plots, three-dimensional plots and pole/zero diagrams. These drawings together with purely numerical results give the user as much information as possible. They can not only be used in research and industry but are also very attractive for didactical purposes such as the training of students in identification.
The paper first presents the possibilities of INTER and then discusses the implementation and design of the package. The use of the program is illustrated on an example. Finally INTER is used for the control of an industrial absorber process.

POSSIBILITIES OF INTER

INTER contains subpackages for parameter estimation, model evaluation and curve fitting.

Parameter estimation.

The package contains a maximum likelihood estimation of the parameters of a continuous process with discrete observations (Ghysels, 1982). The equation of the continuous is given by (1) and the equation of the discrete observations by (2) :

$$\dot{x}(t) = A\,x(t) + B\,u(t) + v(t) \tag{1}$$

$$y(k) = C\,x(k) + D\,u(k) + e(k) \tag{2}$$

$x(t)$ is the state vector, $u(t)$ the input vector and $y(k)$ the observation vector. $v(t)$ and $e(k)$ are gaussian noise inputs with zero mean. The procedure involves the solution of a "mixed Kalman filter". The optimization routine is the BOX-algorithm that uses no gradients and looks for the global extremum of the likelihood function. This routine is used to obtain a continuous-time simulation model of a process unit.

Other routines of INTER compute the parameter estimates for the ARMAX model given by

$$[1 + A(q^{-1})]y(k) = z^{-d}\,B(q^{-1})u(k) + [1 + C(q^{-1})]e(k) \tag{3}$$

$\{y(k),u(k)\}$ are observed outputs and inputs and $\{e(k)\}$ is a white noise sequence. d is the dead-time index and A, B and C are polynomials in the backward shift operator :

$$\begin{aligned} A(q^{-1}) &= a_1 q^{-1} + \dots + a_{n_A} q^{-n_A} \\ B(q^{-1}) &= b_1 q^{-1} + \dots + b_{n_B} q^{-n_B} \\ C(q^{-1}) &= c_1 q^{-1} + \dots + c_{n_C} q^{-n_C} \end{aligned} \tag{4}$$

The orders n_A, n_B and n_C are not necessarily equal to each other.

INTER contains two iterative estimation procedures : a maximum likelihood and an extended least-squares estimation. These methods have in general some drawbacks : they require much computation and offer only a numerical result.

A new interactive procedure called RHYP was developed by Prager (1980) and a version of this procedure is one of the most powerful tools of INTER. The outline of this procedure is given here.
Suppose N observations $\{y(k),u(k)\}$, k=1,...N and suppose that the C-estimates are known. The observation data can be filtered as follows :

$$y_f(k) = y(k) / [1 + C(q^{-1})] \quad (5)$$

$$u_f(k) = u(k) / [1 + C(q^{-1})] \quad (6)$$

Then equation (3) becomes :

$$[1 + A(q^{-1})]y_f(k) = q^{-d} B(q^{-1})u_f(k) + e(k) \quad (7)$$

Define the residuals ϵ as :

$$\epsilon(k) = (1 + \hat{A}(q^{-1}))y_f(k) - q^{-d} \hat{B}(q^{-1})u_f(k) \quad (8)$$

Equation (8) is now linear in the unknown parameters A and B. So minimization of the cost function

$$J = \frac{1}{N}\sum_{k}^{N} \epsilon(k) \quad (9)$$

is obtained by a simple linear least-squares algorithm. The problem is that the C-estimates are not known. But it is possible to construct a priori a set of polynomials 1 + C that have roots outside the unit circle. When the order of C is restricted to two then all the allowable coefficients of C lie in the triangle of fig. 1. $n_C = 2$ is not a severe restriction for practical problems and has the advantage that only a limited region for the C-coefficients has to be examined and that graphical representation of the cost function J is still possible. The user can scan the allowable region of C-coefficients, he can define the resolution of this scanning and he can focus on a smaller region. Three-dimensional plots give him the necessary information for each decision. In Messely (1983) it is shown that this algorithm is numerically robust compared to other estimation routines.
INTER has also some recursive estimation algorithms : the least-squares recursion, the extended least-squares recursion and the maximum likelihood recursion. With the help of these routines the user can examine the convergence and smoothness of convergence of the different parameter estimations.
s-z conversion and z-s conversion routines allow the transition from the continuous model (1) to the discrete model (3) and vice versa.

Model evaluation.

A routine estimates the order of the unknown system prior to parameter estimation. It is a determinant ratio test of the product moment matrix. The first version uses the normal product moment matrix, a second version the instrumental product moment matrix. The latter gives better results in the case of low signal to noise ratios (Wellstead, 1978).

The other model evaluation techniques available in INTER can only be applied after parameter estimation.

If the model structure and the parameter estimation are correct then the residuals should form a white noise sequence and should also be independent of past values of the input signal. Cross-validation computes the autocorrelation function of the residuals and the crosscorrelation between the residuals and the input samples. It warns the user for estimations of the model order that are too low, as it is impossible to get white residuals when the estimated order is lower than the exact one.
Overestimation of the model order can be discovered by calculation of the poles and zeroes. When the estimation of the order is too high, then the poles and zeroes lie closely together so that cancellation may occur.
INTER permits simulation of the process with the estimated parameters. Comparison of the simulated output with the measured output can tell the designer quite a lot about the model and order he used.
Another model evaluation technique checks the behaviour of the cost function J (9). For one specified model J reaches a minimum for the "best" parameters, but that doesn't tell anything about the model itself. Akaike (1974) introduced a loss function that also takes the model complexity into account :

$$AIC = J + 2p \quad (10)$$

where p is the number of parameters in the model. The supplementary term 2p protects the designer against overestimation.

Curve fitting.

Least-squares fitting can be applied on any function that is linear in the unknown parameters. This is useful for identification of the non-linear behaviour of industrial sensors. It is also used for estimation of the static gain of the process when this has reached a steady state. This can serve as a control for the estimation of all the dynamic parameters of the system.
Table 1 gives an overview of the possibilities of INTER.

TABLE 1 Possibilities of INTER

INFO	**list options available at this level**
PAREST	**parameter estimation techniques**
MOD	definition of process and noise order
FIL	specification of the file with observations (y,u)
TSZ/TZS	s-z and z-s conversion
TWZ/TZW	w-z and z-w conversion
ELS	iterative extended least-squares method
MLH	maximum likelihood iterative technique
SEA	display of the path followed by an iteration technique
INT	interactive maximum likelihood technique (RHYP version)
GRA	three-dimensional plot of cost index
REC	recursive algorithms
PAR	display of the time evolution of the estimated parameters
SVM	maximum likelihood estimation of the continuous system parameters

MODEVA **model evaluation techniques**
ORC calculation of the determinant ratios of the product moment matrix
ORG display of the determinant ratios
SIM simulation of the system with estimated parameters
TIM graphical representation of observations, simulated output and estimation error
PZC calculation of poles and zeroes of the estimated process and noise model
PZG display of the pole/zero diagram
COC calculation of autocorrelation and crosscorrelation functions
COG graphical representation of autocorrelation and crosscorrelation
COS calculation of the costindex J and of AIC
CURFIT **curve fitting**
CFI input of the observations
CFC calculation of the curve fitting
CFG graphical representation of the curve fitting
END **return to higher level**

IMPLEMENTATION OF INTER

INTER contains about 6000 FORTRAN lines. It makes full use of the well known numerical libraries LINPACK and BLAS (Dongarra, 1979) and EISPACK (Smith, 1976). INTER runs on a PDP11/34 under the operating system RSX11M. Soft plots can be made on VT100 terminals with graphical extensions, hard plots on a TEKTRONIX plotter.

If there is only 64 K RAM available for each user, INTER can't be entered as a whole. In this case it can be divided into several independent tasks. Connection between the tasks is made by files. This file management is completely hidden for the user. All the installation dependent code for each task is gathered into one or two subroutines. A small supervising program that is loaded into memory during the whole identification session, manages the different tasks.
INTER has a hierarchical structure. Changing from one level to another can be done by commands given in a menu. The choice between the possibilities within one level can be made by commands of a menu or by a question/answer method which has preprogrammed default answers. Error messages inform the user if a task can't get started or does not work properly. Wherever possible suggestions are made to the user to avoid the failure. Additional information is available by an INFO utility.
Because of the modular concept, extension of INTER is easy. Existing tasks don't have to be changed once they are built.

THEORETICAL EXAMPLE

500 samples are generated using the difference equation

$$y(k) = -a_1\, y(k-1) + b_1\, u(k-1-d) + e(k) + c_1\, e(k-1) + c_2\, e(k-2) \qquad (11)$$

- with dead-time index d = 1
$a_1 = -0{,}5$ and $b_1 = 1$
$c_1 = 0$ and $c_2 = 0$

- The input sequence u and noise e are pseudo random noise with zero mean. The signal to noise ratio S/N is 0,924.

This example is nearly the same as Bohlin (1971) used in his paper on the ambiguities in maximum likelihood identification.
As in Bohlin (1971) a first-order model with no transport delay ($\hat{d} = 0$) will be fitted to the data. After many iterations the maximum likelihood search routine finds two different results depending on the start values :

1. $\hat{a}_1 = 0{,}4105$ $\hat{b}_1 = -0{,}4848$
$\hat{c}_1 = 1{,}048$ $\hat{c}_2 = 0{,}4235$
$J = 0{,}1405$

2. $\hat{a}_1 = -0{,}8340$ $\hat{b}_1 = 0{,}4109$
$\hat{c}_1 = -0{,}5875$ $\hat{c}_2 = -0{,}1199$
$J = 0{,}1705$

So at least two local minima exist for the cost function J. Applying the interactive procedure to this problem discovers these two minima in one run. Figure 2 shows the shape of the cost function. The curves are contours where J has a constant value. The two extrema can easily be seen. The paths followed by the maximum likelihood search routine are also visualized on fig. 2. The designer can now change the resolution of the interactive procedure or change the region in the c_1c_2-plane in order to examine the problem somewhat closer. Figure 3 gives a close-up of the critical region.
Fitting the data to the model with dead-time index $\hat{d} = 1$ results in the following maximum likelihood estimations :

$\hat{a}_1 = -0{,}5187$ $\hat{b}_1 = 1{,}043$
$\hat{c}_1 = 0{,}0069$ $\hat{c}_2 = -0{,}0219$
$J = 0{,}0868$

Figure 4 shows the results of the interactive estimation procedure and also the search path of the maximum likelihood routine. There is only one extremum now. The autocorrelation of the residuals is drawn in fig. 5 and ressembles that of a white noise. So it may be concluded that the last model applied to the data is the correct one.

INDUSTRIAL APPLICATION

As a further example we shall now use INTER to identify an industrial absorber-stripper plant for removal of CO_2 from acid gases.
Figure 6 shows the absorber plant. The feed gas flow rate is considered as the input signal and the CO_2 concentration as the output. They are represented in fig. 7. During the observations the steam flow rate is kept as constant as possible, so that the process can be considered as a single-input single-output system. The sampling period is 20 seconds.

After examination of the input and output signals it can be seen that the process has a dead time of about 8 sampling periods. The data are first fitted on a first-order model for A and a second order model for B because of the uncertainties of the value of the dead time :

$\hat{n}_A = 1$
$\hat{n}_B = 2$
$\hat{d} = 8$

The identification results of three different methods are summarized in table 2 :
MLH = maximum likelihood iterative technique
ELS = iterative extended least-squares
INT = interactive maximum likelihood method

TABLE 2 Identification results

	MLH	ELS	INT
$\hat{a}_1$	-0.8797	-0.8818	-0.8842
$\hat{b}_1$	$0{,}390.10^{-3}$	$0{,}358.10^{-3}$	$0{,}235.10^{-3}$
$\hat{b}_2$	$0{,}175.10^{-2}$	$0{,}175.10^{-2}$	$0{,}183.10^{-2}$
$\hat{c}_1$	0,0111	$-0{,}653.10^{-3}$	$0{,}477.10^{-6}$
$\hat{c}_2$	$0{,}458.10^{-2}$	$-0{,}892.10^{-2}$	-0,0400
J	$0{,}528.10^{-4}$	$0{,}528.10^{-4}$	$0{,}527.10^{-4}$

From fig. 8 it can be seen that there is only one strongly pronounced minimum of the cost function J. Hence it is not surprising that the different methods give the same results.
The process is then simulated with the estimated parameters and the same input signal. A comparison of the simulated output signal and the measured output signal can be made with the help of fig. 9. As a final test for the identification of this process loop, the static behaviour is analyzed during periods of steady state. Figure 10 represents the gain obtained by the curve fitting subprogram. A gain of 0.0178 is found. The gain obtained by estimation of the dynamic parameters is

$$(\hat{b}_1 + \hat{b}_2)/(1 + \hat{a}_1) = 0{,}0178$$

Because different methods lead to the same results, the estimated model can be used for simulation of this process.

CONCLUSION

The program package INTER for the identification of the elements of a control loop has been presented. Examples have shown that it can be used for research and for industrial applications.
INTER has not yet reached its final state. Routines can easily be added due to the modular concept. Adaptation of INTER to a 32-bit super minicomputer is planned... and a tracking ball or a "mouse" would increase the interaction with the designer.

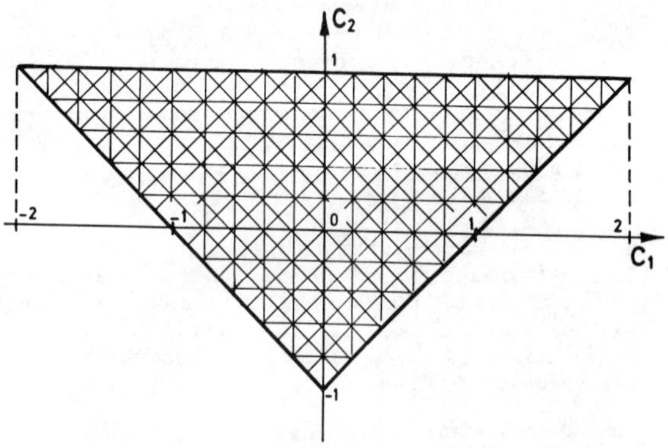

Fig. 1. Allowable region for C_1 and C_2 .

ACKNOWLEDGMENT

This work was partially sponsored by the Foundation of Fundamental Collective Research (FKFO) of the Belgian National Fund for Scientific Research (NFWO).

REFERENCES

Akaike, H. (1974). A new look at the statistical model identification. IEEE Trans. AC-19, 716-722.

Bohlin, T. (1971). On the problem of ambiguities in maximum likelihood identification. Automatica, 7, 199-210.

Dongarra, J.J., J.R. Bunch, C.B. Moler and G.W. Stewart. (1979). Linpack User's Guide. SIAM, Philadelphia.

Ghysels, J. (1982). Parameterschatting met maximum-likelihoodmethode. Internal report, Laboratory of Automatic Control, Gent, Belgium.

Messely, M. and R. De Keyser (1983). Interaktieve identifikatie van ARMAX-processen. Internal report, Laboratory of Automatic Control, Gent, Belgium.

Prager, D.L. and P.E. Wellstead (1980). Interactive recursive maximum likelihood estimation. IJC, 32, 1005-1030.

Smith, B.T., J.M. Boyle, B.S. Garbow, Y. Ikebe, V.C. Klema and C.B. Moler (1976). Matrix Eigensystem Routines - EISPACK Guide, vol. 6, Lect. Notes in Comp. Sci., 2nd ed. Springer, New York.

Wellstead, P.E. (1978). An instrumental product moment test for model order estimation. Automatica, 14, 89-91.

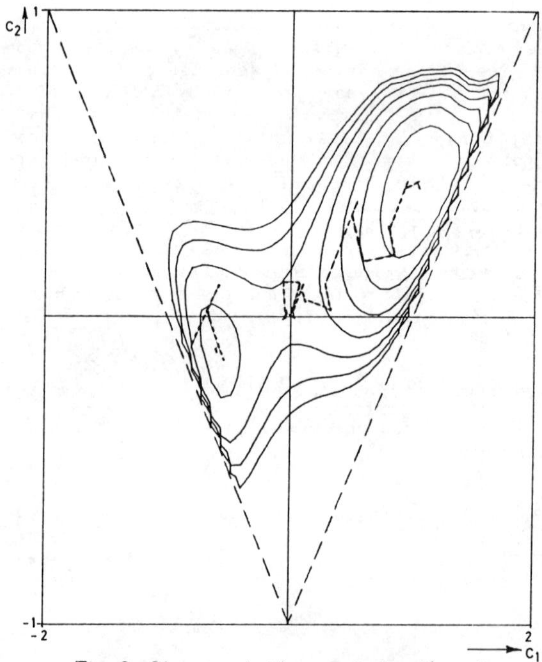

Fig.2. Shape of the cost function J.

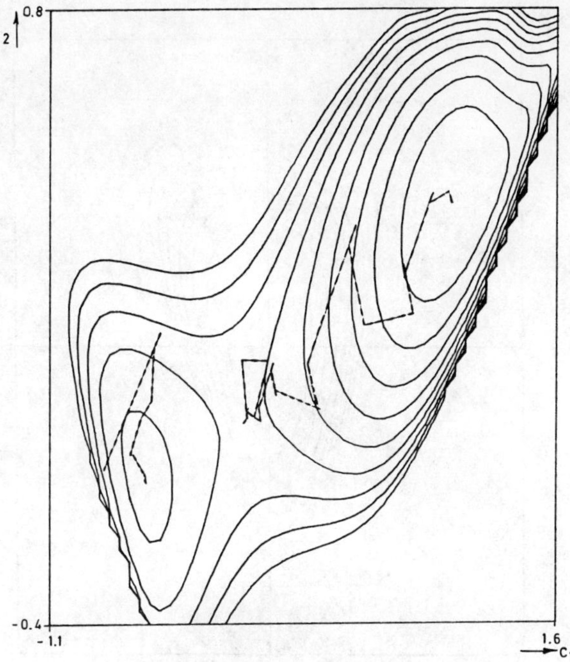

Fig. 3. Close-up of critical region.

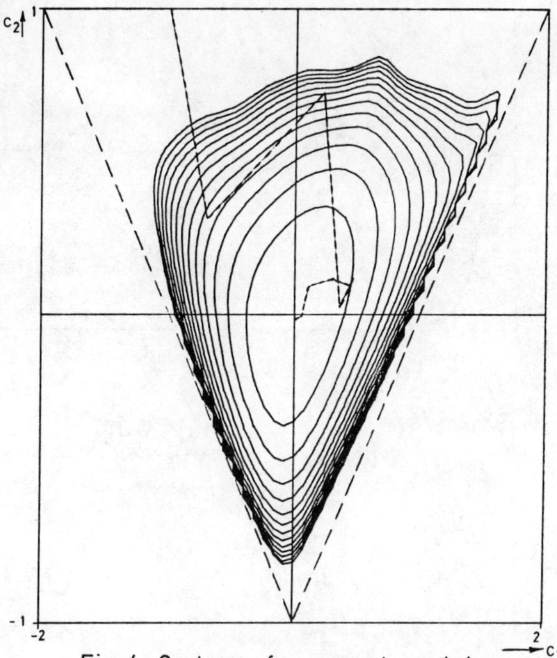

Fig. 4. Contours for correct model.

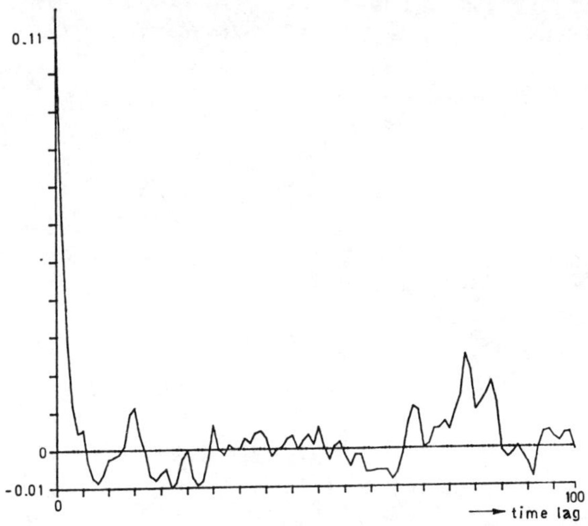

Fig. 5. Autocorrelation of residuals.

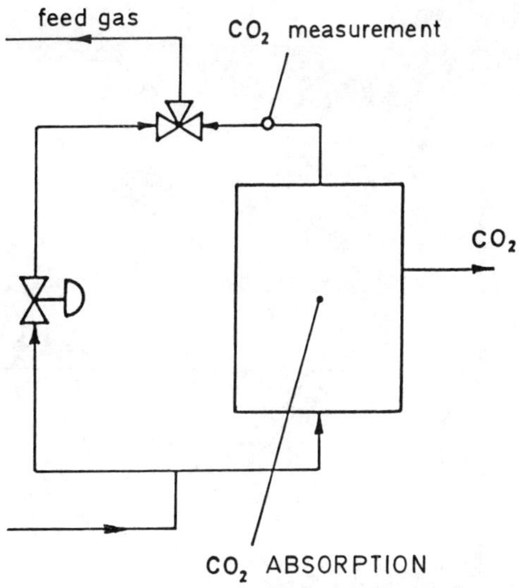

Fig. 6. Industrial absorber-stripper plant.

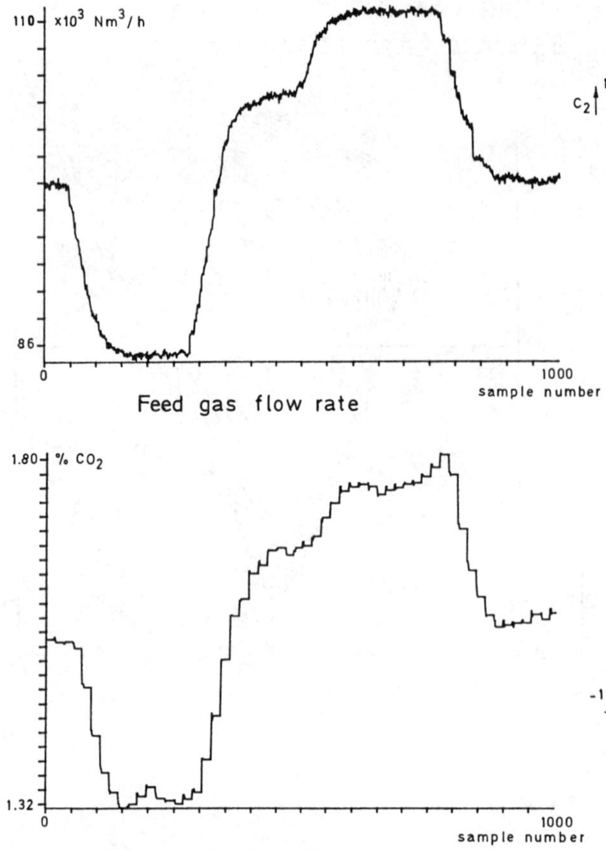

Fig. 7. CO_2 concentration.

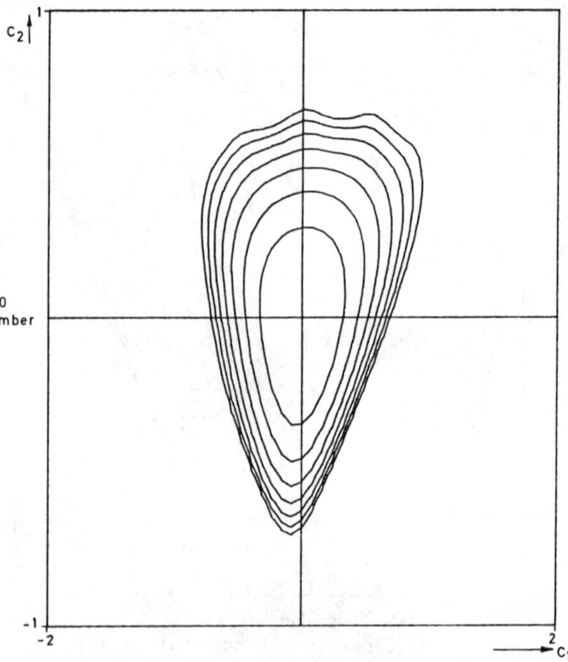

Fig. 8. Cost function J for absorption process.

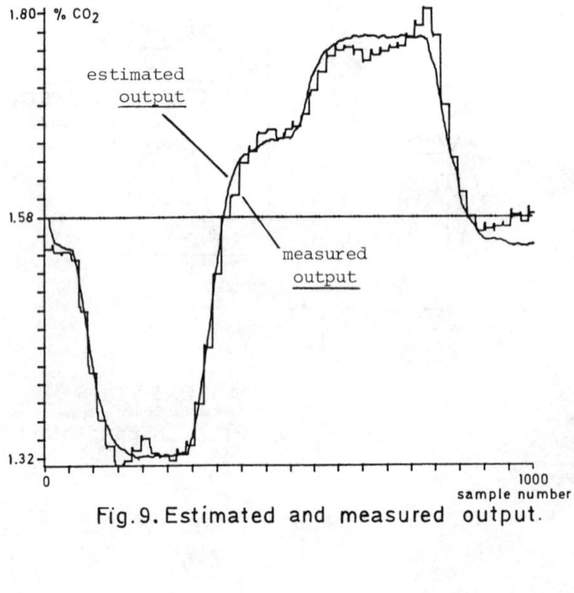

Fig. 9. Estimated and measured output.

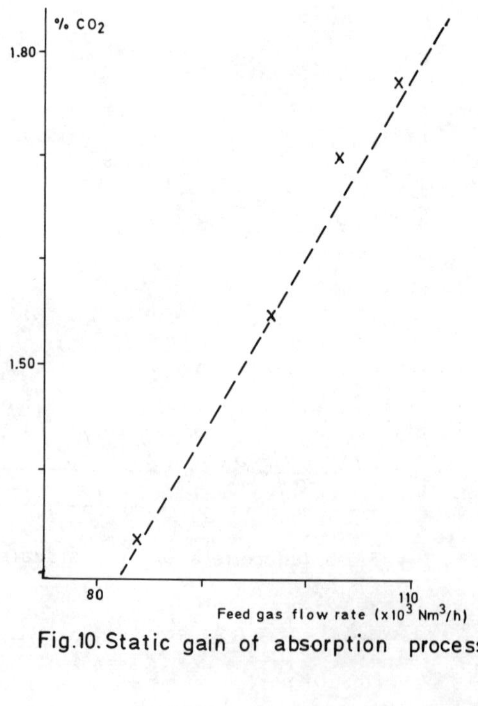

Fig. 10. Static gain of absorption process.

A PROGRAM PACKAGE FOR MULTICRITERIA OPTIMAL DESIGN OF HIERARCHICAL AND DECENTRALIZED LINEAR CONTROL SYSTEMS

A. Ollero and R. Marín

Departamento de Automática y Electrónica, Escuela Técnica Superior de Ingenieros Industriales de Vigo, Universidad de Santiago (Campus de Vigo), Vigo, Spain

Abstract . In this paper we present a program package which includes alternative routines to compute hierarchical and decentralized control laws together with a proposed interactive multicriteria optimal control algorithm to consider appropiately non conmensurable and conflicting criteria at different levels. The package includes an interpreter, based on fuzzy set theory, which permits to perform the designer's interaction in the algorithm in a qualitative linguistic way. The package is written in Pascal and permits a low cost implementation in microcomputers with the CP/M operating system. Moreover we present an example of two area megawatt frecuency control which illustrates the designing procedure.

Keywords. Control system synthesis; computer-aided system design; hierarchical systems; multicriteria optimization; fuzzy set theory; optimal control; man-machine. systems.

INTRODUCTION

A significant number of packages for computer aided design of control systems have been presented during the last years (see for example Åstrom, 1981; Munro, 1981; Furuta and Yamano, 1984; Schafer, 1984). Classical single-input single-output system design techniques (Frecuency domain, root locus) and its recent generalizations (inverse Nyquist array, multivariable root locus) are well suited to be studied by means of digital computer packages with graphical display terminals. However, the application of clasical linear control theory is very restricted by the dimensionality of the mathematical model. This restriction is alleviated, from the point of view of theoretical formulation of the synthesis procedures, when using state space approaches but the computer requeriments increases very rapidly with the number of state variables. For instance, the computing time to solve the linear quadratic regulator problem varies cubically with the system order.

In the context of modern control theory, large scale systems are usually studied by using extensions of optimal control methods. A methodology widely employed is the hierarchical system approach. Indeed, open loop control synthesis can be solved by applying the early works on hierarchical systems (Mesarovic, Macko and Takahara; 1970) and dynamical optimization of large scale systems (Lasdon, 1970; Pearson, 1971) Later this methodology was used for closed loop control system synthesis (Singh, 1980; Bernussou and Titli, 1982). Applications of hierarchical methods always imply decomposition of the system into interacting subsystems, which provide a basis for the formulation of independent optimal control subproblems linked together by a separate coordination unit. Moreover, this designing methodology is suitable to be applied by means of decentralized computing (Tamura,1975). We notice that subproblem decomposition can be considered only as a tool to solve large scale problems. However, most of the practical large scale control systems such us multiarea electrical power systems, road traffic networks or steel rolling consist of interconnected subsystems with local states, outputs and controls. Then, the application of hierarchical system theory can be supported by physical considerations about the process.

Moreover we notice that in the above we consider only the problem of computing control laws but no reference is made to the implementation of these controls. There exists hierarchical control systems configurated with several lower level controllers, acting on the process, coordinated on line by a control unit at a higher level. In this paper we deal with the particular case or multilevel structures, based on the state feed back concept, in which coordination variables are added to the local feedbacks. This structure, has special interest for linear system with quadratic performance criteria. An alternative to the hierarchical control structures is the decentralized control (Wang and Davidson, 1973; Slijac and Vukcevic , 1976). In the simple decentralized control, although each local control can be computed taking into account criteria and models of the whole system, each local controller only uses on-line the local state variables. Thus decentralized control can be viewed as a special case of the above mentioned structures based on the state feedback concept in which the coordination variables are not generated at all; instead of this we implement only local feedbacks.Obviously this is a suboptimal structure when comparing with the hierarchical ones, but may have a great interest from a practical point of view as for example for large scale systems which are widely distributed in space and the main cost of the implementation of the control may will lie in the telemetry links necessary to transmit information. Alternatively we could think of readjusting the local feedbacks gains at some time intervals, which could be looked on as an adaptation which would present a way of on-line coordination.

Finally note that in the above we are concerned with methods to compute and implement hierarchical and decentralized controls from well defined performance criteria to be optimized. However the design of a large scale control systems involves the consideration of a great number of non conmensurable and conflicting local and global performance criteria which adds an additional complexity to the problem. Little attention has been paid in the hierarchical and decentralized control literature to the appropiate consideration of the multicriterion character of the problem of design these control systems. In this paper we propose the use of multicriteria optimization theory to design hierarchical and decentralized linear control systems. We propose an interactive multicriteria algorithm which generalise the results presented in Ollero and Marin (1982,1984) for reduced order linear control system design. A program package written in Pascal is presented. This package includes an interpreter which permits to perform the designer's interactions of the algorithm in a qualitative linguistic way. The interpreter is based on the use of fuzzy set theory as an vehicle to overcome the well known difficulties in the numerical estimates of trade-offs between criteria (Baptistella and Ollero, 1980). Moreover we present an example of two area megawatt-frecuency control which illustrates the main features of the designing procedure.

MULTICRITERIA FORMULATION OF HIERARCHICAL AND DECENTRALIZED LINEAR STATE FEEDBACK CONTROL

Let us consider the problem of the optimum control of a group of N dynamic, linear, interconnected subsystems by minimization of an additive, separable, quadratic criterion in the form:

$$\min \sum_{i=1}^{N} \int_0^{\infty} (\underline{x}_i^T Q_i \underline{x}_i + \underline{u}_i^T R_i \underline{u}_i)\, dt \tag{1}$$

subject to: $\dot{\underline{x}}_i = A_i \underline{x}_i + B_i \underline{u}_i + C_i \underline{z}_i;\ \underline{x}_i(\emptyset) = X_{i0}$

$$\underline{z}_i = \sum_{j=1}^{N} L_{ij}\, \underline{x}_j \quad ;\ i=1 \text{ to } N \tag{2}$$

where $\underline{x}_i \in R^{ni}$ (state), $\underline{u}_i \in R^{mi}$ (control), $\underline{z}_i \in R^{li}$ (linking variable), L_{ij} is a matrix of interconnection on the states, $Q_i \geq \emptyset$ and $R_i > \emptyset$.

It is well known (Bernussou and Titli, 1982; Sing, 1980) that the local control law $\underline{u}_i$ which minimices this problem can be obtained in the form:

$$\underline{u}_i = K_i\, \underline{x}_i + \underline{v}_i \tag{3}$$

where K_i is a local feedback gain matrix obtained by resolving the local Riccati equation and $\underline{v}_i$ is a coordination variable supplied by a coordinating level. This variable can be compused by a linear combination of the states in other subsystems.

We can obtain a Descentralized control structure if we impose as an aditional restriction the local control form:

$$\underline{u}_i = K_i\, \underline{x}_i \tag{4}$$

Procedures to compute and to analyze the stability of the solutions of those two kinds of problems (hierarchical and decentralized control) have been studied by many authors (Singh, 1980, Findeinsen and others, 1980; Bailey and Wang, 1972; Bailey and Davison, 1973).

It is clear that if we change the values of the matrices Q_i and R_i we get a different solution for the design problem. Then, the designer must choose the matrices Q_i and R_i according to relative importance of the cost of all the control and state variable involved in order to obtain a good solution to the design problem. Several guidelines to solve this problem has been proposed. However, difficulties remain and the choice of the matrices Q_i and R_i is, in general, somewhat arbitrary.

Multicriteria optimization methods offers interesting ideas for a systematic studie of the trade offs between designers criteria. A multicriteria optimization formulation can take the form:

$$\begin{aligned} &\text{Min } J_1 = \int_0^{\infty} (x_1^1)^2\, dt \\ &\text{Min } J_{k1} = \int_0^{\infty} (x_i^j)^2\, dt;\ i=1 \text{ to } N;\ j=1 \text{ to } n_i; \quad k1 = j + \sum_{p=1}^{i} n_p \\ &\text{Min } J_{k2} = \int_0^{\infty} (u_i^j)^2\, dt;\ i=1 \text{ to } N;\ j=1 \text{ to } n_i; \quad k2 = j + \sum_{i}^{N} n_i + \sum_{p=1}^{i} m_p \end{aligned} \tag{5}$$

Subject to (2) and (3) or (4). Where x_i^j and u_i^j are state and control variable j of the subsystem i.

To solve this multicriteria or vector valued optimization problem a wide variety of methods can be applied. However, all the proposed techniques of resolution have a common and important characteristic: conversion, from the multicriteria problem, to scalar optimization (one criterion) whose solutions can be found by standard optimization techniques.

In this paper we use an interactive multicriteria method proposed by the authors in a previous paper (A. Ollero and R. Marin, 1982). In this method the scalar optimization formulation takes the form:

$$I = \underline{w}\, J = \sum_{e=1}^{n+m} w_e J_e \ ;\quad w_e > 0 \tag{6}$$

where n+m is the number of criteria and the weighted vector $\underline{w}$ is modified interactively according to the designer's trade-offs between the criteria in each iteration. We notice that by optimizing the index I we obtain a non-inferior (non-dominated, or Pareto optimal) solution of the multicriteria problem (5). The designer may give qualitative diagnostics over each criterion (J_1 is good, J_2 is poor etc) and an interpreter based on the fuzzy set theory obtains a new weighted vector $\underline{w}$ for the next scalar optimization. It is clear that the index of the scalar problem can take the form (1) if the matrices Q_i and R_i are diagonals with the elements:

$$q_1^1 = w_1$$

$$q_i^j = w_{k1} \ ;\quad k1 = j + \sum_{e=1}^{i} n_e$$

$$r_i^j = w_{k2} \ ;\quad k2 = j + \sum_{e=1}^{N} n_e + \sum_{e=1}^{i} m_e$$

where q_i^j and r_i^j are the q_{jj} and r_{jj} elements of the matrices Q_i and R_i.

There are usually many states and control variables in Hierarchical or Decentralized control problems, and the number n+m of criteria in (5) is large. However, we notice that, in many applicationes, we can define a set of NA < n + m quadratic indices having physical meaning. Let JA be an NA-dimensional vector $JA = (JA_1 ......JA_{NA})^1$ thus, it can be writen JA = LA J where LA is an aggregation matrix defined from the designer interest and

$$JA_1 = \sum_{j=1}^{n+m} la_{1j} \; J_j; \; 1 = 1 \text{ to } NA \qquad (7)$$

Note that, for a given NA dimensional weighting vector $\underline{wa}$; the matrices Q_i and R_i can be obtained by:

$$q_i^j = w_p = \sum_{k=1}^{NA} \underline{wa}_k \; la_{kp}; \; p = j + \sum_{e=1}^{i} n_e \qquad (8)$$

$$r_i^j = w_s = \sum_{k=1}^{NA} \underline{wa}_k \; la_{ks}; \; s = j + \sum_{e=1}^{N} n_e + \sum_{e=1}^{i} m_e$$

where $\underline{wa}_k$ is the k element of the vector $\underline{wa}$, w_p and w_s are the p and s element of the correspondent vector w. By this procedure we can get one criterion for each subsystem, or aggregate all the criteria corresponding to the states in one criterion and all the criteria corresponding to the control variables in another criterion. It is possible to use many aggregation matrices with a physical meaning. The solution of the multicriteria problem can be changed depending on the aggregation matrix chosen. Usually, large scale control system involves several hierarchical levels of criteria. Thus, an aggregation matrix for each level must be defined. For each level there is one criterion space for the qualitative diagnostics of the designer. We can define, for each aggregation matrix L^j, one weighting vector $\underline{wa}^j$ and one desaggregated weighting vector WL^j, related by:

$$WL_p^j = \sum_{k=1}^{NA} la_{kp} \; Wa_k \; ; \; p=1 \text{ to } n+m \qquad (9)$$

where WL_p^j is the p element of the WL^j vector. The composed vector W can be computed by:

$$W_i = \prod^{n+m} WL_i^j \; ; \quad i=1 \text{ to } n+m \qquad (10)$$

and the elements of the matrices Q_i and R_i are obtained by using (6).

AN ALGORITHM FOR MULTICRITERIA OPTIMAL DESIGN

In order to assure a systematic investigation of non inferior solutions which lead us to determine the preferred solution of the multicriteria problem (5), we propose the following interactive algorithm:

Step 1: Determine the matrices of the model A_i, B_i, C_i, L and the initial value of the state variable x_{io}.

Step 2: Choose a hierarchical structure for the criteria and a suitable set of matrices LA.

Step 3: Choose a fuzzy set for each qualitative diagnostic and for each criteria level.

Step 4: Fix one qualitative protocol of change in the vector of weights for each criteria level.

Setp 5: Choose an initial vector of weights for each level.

Setp 6: Determine the composed vector W by (10)and the matrices Q_i and R_i by (6).

Setp 7: Determine K_i and v_i (hierarchical control) or K_i (decentralized control)by a suitable method of calculation.

Setp 8: Simulate the close loop system response with the values computed in the step 7.

Setp 9: Interact with the designer. The designer may supply a qualitative diagnostic for each criterion at each level (for the criteria without qualification a default fuzzy set is adopted). If an appropiated stopping criterion is satisfied, then the current non inferior solution is the preferred one, otherwise the interpreter generates a vector of weights for each criteria level and goes to the step 6.

We notice that aggregation matrices LA can be obtained directly from designer's opinions or assisted with suitable procedures for control objectives reduction (Johnston and Barton, 1984).

In the next section we present a program package to carry on this algorithm.

PROGRAM PACKAGE GENERAL STRUCTURE AND MAIN FACILITIES

The program package presented in this paper is a subset of a more general control system design package. This package has a modular structure with divisions in several hierarchical levels. Fig. 2 shows the two upper levels of the hierarchy. The module "manager" (M Ø) load in the main memory modules of the second level following the current design phase and the user's options. The "determination and analysis of the model" module (M 1) let the user introduce the model parameters or alternatively execute identification procedures. Moreover, interconnected systems can be introduced by means of special purpose block definition language and represented in memory in a data structure linked with pointers. Furthermore M2 includes several routines for analysis in the state space and frequency response. The "simulation" module (M2) also has several options. Taking into account the block representation of interconnected systems, modifications in the system interconnection structure can be performed easily by reassigning data pointers. For continuous time modelling of the processes, the designer can use alternative numerical integration routines (Euler, fourth order Runge Kutta, exponential matrix). Moreover (M2) disposes of noise generators and non linearities simulators that can be added to the theoretical model. The module "control law sinthesis" (M3) includes procedures to compute hierarchical and decentralized control laws by minimizing a given weighted quadratic index. In the hierarchical option the user disposes actually of both centralized Newton-Raphson method (Kleinman, 1967) and decentralized computing by using an interaction prediction method (Singh, 1980). Decentralized control laws can be computed by using the suboptimal method presented in Bailey and Ramapriyan (1973).

The interaction module is based on the use of linguistic variables (Zadeh, 1975) in the context of Fuzzy Set Theory. The user performs the interaction step 9 in the algorithm supplying linguistic diagnosis for each criterion on each level in accordance with the grammar shown in Fig. 1. For each criterion level, the diagnosis about a particular criterion can be represented as the assignment of a linguistic value to a linguistic variable. In general, the linguistic values can be composed of several atomic linguistic terms. There are three cathegories of these atomic terms: primary linguistic terms ("good", "acceptable", "poor", "unnaceptable") linguistic hedges ("very", "more or less", "highly" "sort of", "rather"), and complementary ("not") and connective ("and", "or") terms. The primary linguistic terms can be defined as fuzzy subset of the closed interval (0,1) by means of the submodule "gendat". The effect of linguistic hedges can be interpreted by using non linear operators over the primary linguistic terms. Complementary and connective terms are applied by means of the complement, intersection and union of fuzzy sets. The linguistic diagnosis are readed by an interpreter called "interact" which performs its syntactic analysis and form fuzzy sets representing each diagnosis. Then the interpreter produces factors of change in relative weights between criteria at a given level from the protocols defined by the user in the step 3 of the algorithm. These protocols consist of a set of linguistic rules, connected by "or" assigning linguistic values to a linguistic variable, representing the factor of change in relative weight between two criteria, from a pair of linguistic values of two linguistic variables corresponding to a pair of criteria. To facilite the specification and use of linguistic protocols we have defined an special purpose minilanguage. The submodule " genproto " performs the syntactic analysis of the rules supplied by the user and produces a representation of the protocol. Finally we note that the interpreter determines a new normalized weighting vector for a level from the old one and the factors of change obtained as noticed above.

SOFTWARE AND HARDWARE CONSIDERATIONS

The actual version of the package is written in PASCAL MT+ by using its overlay system. The package first level is the "root" of the overlay system containing the overlay manager, the library routines of PASCAL and all the global data of the package. Each of the remain levels configurates one overlay area in the main memory.

We notice that PASCAL MT+ is a compiler of Pascal language for microcomputers with the CP/M operating system. Thus, this package can run in a 8 bits microcomputer system with 64 kbytes RAM. An special hardware for floating point real operations its valuable. The medium storage must be a floppy disk with at least 250 kbytes. However a 500 kbytes of medium storage is recomended. Moreover it is necessary to use a video terminal, a printer, and a graphic terminal or a digital plotter.

Obviously, by using this implementation of the program package, the dimension of the control system mathematical model is very constrained. Actually only systems with order smaller than fifteen have been studied.

EXAMPLE: MEGAWATT-FREQUENCY CONTROL OF A TWO AREA ENERGY SYSTEM

We consider the 9th-order linear system modelled in Fosha and Elgerd (1970). This system can be also represented as two interconnected subsystem one for each area, by means of equations (2) with N = 2, $n_1 = 5$, $n_2 = 4$. The state variables of the first area are respectively the integral of the tie line power deviation of the area x_1^1, the integral of frequency deviation x_2^1, the frequency deviation x_3^1, the variations with respect to prescribed steady state values of the generator output x_4^1 and the governor valve position x_5^1. In the second area the tie line deviation is proportional to the tie-line deviation in the first, and so only four state variables are considered with x_i^2; i = 1, 4 having the same interpretation of x_i^1; i = 2,5. The interconnection variables are $z_1 = x_2^1$, $z_2 = x_1^2$. By using typical values of the system parameters the matrices of the representation are:

$$A_1 = \begin{bmatrix} 0 & 0.545 & 0 & 0 & 0 \\ 0 & 0 & 1 & 0 & 0 \\ 0 & -3.27 & -0.05 & 6 & 0 \\ 0 & 0 & 0 & -3.33 & 3.33 \\ 0 & 0 & -5.208 & 0 & -12.5 \end{bmatrix}$$

$$A_2 = \begin{bmatrix} 0 & 1 & 0 & 0 \\ -3.27 & -0.05 & 6 & 0 \\ 0 & 0 & -3.33 & 3.33 \\ 0 & -5.208 & 0 & -12.5 \end{bmatrix}$$

$$B_1 = \begin{bmatrix} 0 \\ 0 \\ 0 \\ 0 \\ 12.5 \end{bmatrix} ; \quad C_1 = \begin{bmatrix} -0.545 \\ 0 \\ 3.27 \\ 0 \\ 0 \end{bmatrix}$$

$$B_2 = \begin{bmatrix} 0 \\ 0 \\ 0 \\ 12.5 \end{bmatrix} \quad C_2 = \begin{bmatrix} 0 \\ 3.27 \\ 0 \\ 0 \end{bmatrix}$$

Following a step-load change it is intended to minimize the change in tie-line power, the frequency deviation, and the integral of frecuency deviation (time error). To solve this design problem by means of the approach presented in the paper, two criteria levels have been defined. The first level includes three criteria: the tie-line power, the frequency deviations and their integrals in both areas, and the two control variables. In the second level we consider five criteria: the tie-line power, the frequency deviations in both areas, the integrals of frequency deviations in both areas, the control in area 1, and the control in area 2. Table 1 shows the protocols used.

In two interactions of the algorithm, the named "preferred solution" of the hierarchical control system design problem was achieved. Table 2 shows the diagnostics used and the normalized weighting vectors in each interaction. The preferred solution is fiven by:

$$\begin{aligned} K_1 &= (-2.6338\ -2.1517\ -3.5609\ -3.469\ -0.689) \\ v_1 &= (-1.4246\ x_2^1\ -0.2565\ x_2^2\ -0.2433\ x_2^3\ -0.053\ x_2^4) \\ K_2 &= (-1.895\ -3.5652\ -3.4649\ -0.685) \\ v_2 &= (3.1368\ x_1^1\ -1.1005\ x_1^2\ +0.2292\ x_1^3\ +0.309\ x_1^4 \\ &\quad +0.0654\ x_1^5) \end{aligned} \tag{11}$$

By using the same criteria the decentralized control solution is given by

$$\begin{aligned} K_1^o &= (-4.096\ -2.724\ -3.719\ -3.561\ -0.72) \\ K_2^o &= (\ -0.6434\ -3.2687\ -3.293\ -0.660) \end{aligned} \tag{12}$$

In Fig. 3 the regulators (11) and (12) are compared with the proposed by Elgerd and Fosha (1970). In Table 3 we compare the control effort required for the three regulators. Notice that state variable response of the hierarchical solution can be considered the best one, but it is paid with a greater control effort. Decentralized solution has a dynamic state variable response slightly worse than the hierarchical one, but may have the practical advantage of the implementation without transmitting information between local controllers of each area.

CONCLUSIONS

In this paper we present an algorithm for interactive multicriteria optimal design of hierarchical and decentralized control. By using this algorithm it is possible to consider problems with several design criteria at different hierarchical levels. The algorithm is integrated in a program package with alternative routines to compute hierarchical and decentralized control laws. The package includes an interactive module, based on fuzzy set theory,which permits to perform the algorithm interactions in a qualitative linguistic way. The package offers the option of block oriented simulation to represent interconnected systems offering additional facilities to modify the structure by reassigning data pointers. The package is written in Pascal MT+, by using its overlay sistem, and can be implemented at low cost in a microcomputer system with the CP/M operating system. A megawatt frequency control design is used as an example of the designing procedure proposed in the paper.

Finally we note that large scale control system design requires an implementation of the package in a more powerful computer system.

REFERENCES

Åström, K.J. and H. Elmqvist (1981). "Perspective on interactive software for computer aided modeling and design of control system". Proceedings of the 20th IEEE Conference on Decision and Control, San Diego.

Bernussou, J. and A. Titli (1982). "Interconnected Dynamical Systems: Stability, Decomposition and Decentralization".North Holland.

Bailey, F.N. and F.C. Wang (1972). "Decentralized control strategies for linear systems". Proceedings, Asilomar Conference.

Baptistella, L.F.B. and A. Ollero (1980). "Fuzzy Methodogies for Interactive Multicriteria Optization". IEEE Trans. Syst. Man and Cybern. Vol. SMC-10, July 1980, pp. 355-365.

Findeinsen, W., F.M. Bailey and others (1980). "Control and Coordination in Hierarchical systems". Jhon Wiley and Sons, pp. 369-391.

Fosha, Ch. E. and O.L. Elgerd (1970). "The Megawatt-Frequency Control Problem: A New Approach Via Optimal Control Theory". IEEE Trans on Power and Systems, PAS 89, nº 4 pp. 563-577.

Furuta, K. and O. Yamano (1984). "Computer aided control and real time control program organizing system 'CACOS'". Preprints of 9th IFAC World Congress Budapest 1984. Vol. VIII, pp. 173-179

Johnston R.D. and G.W. Barton(1984). "Control objetive reduction in single-input single-output control schemes". Int. J. Control, vol.40, pp. 265-270.

Kleinman, D.L. (1967). "On the interactive technique for Riccati equation computation". IEEE Trans. Autom. Control, AC-13, pp. 114-115.

Lasdon, L.S.(1970). "Optimization Theory of Large Systems". Mcmillan, New York.

Mesarovic, M.D., D. Macko and Y. Takahara (1970). "Theory of hierarchical multilevel systems". Academic Press, New York.

Munro, M. (1981)."Illustration of the applicability of computer aided design packages". Proceedings of the 20th IEEE Conference on Decision and Control, San Diego.

Ollero A. and R. Marin (1982). "Algorithms for Interactive Multicriteria Optimal Design of Control Systems". Proceedings of 3th IFAC/IFIP Symposium on Software for Computer Control. Pergamon Press. pp. 331-337.

Ollero A. and R. Marin (1984). "An Interactive Multicriteria Optimization Approach for Linear Control System Design". Preprints of the 9th IFAC World Congress Budapest 1984.Vol.VIII pp. 244-250.

Pearson J.D. (1971). "Dynamic decomposition techniques" In "Optimization methods for large scale systems". D.A. Wismer (Editor) Mc Graw-Hill.

Schafer R., M. Sain (1984). "A computer aided design package for the total synthesis problem". Preprints of the 9th IFAC World Congress Budapest 1974. Vol. VIII. pp. 179-185.

Siljak, D.D. and M.B. Vukcevic (1976). "Decentralization, stabilization and estimation of large scale linear systems". IEEE Trans. Autom. Control, AC 21, pp. 363-366.

Singh, M.G. (1980). "Dynamical hierarchical control". North Holland, New York.

Tamura, H. (1975). "Decentralized optimization for distributed-lag models of discrete systems". Automatica. Vol. 11, pp. 593-602.

Wang, S.H, and E. Y. Davison (1973). "On the stabilization of decentralized control systems". IEEE Trans. Autom. Control, AC - 18, pp. 473-478.

Zadeh, L.A. (1975). "The Concept of a Linguistic Variable and its applications to Approximate Reasoning". Int-Journ. Of Sciences Vol. 8, pp. 149-199, Vol. 8, pp. 301-357, Vol. 9, pp. 43-80.

If criterion C1 is D1 and criterion C2 is D2, then criterion C3 D3					
level 1					
C1	D1	C2	D2	C3	D3
1	unnacceptable	2	unnacceptable	3	large-decrease
2	good	3	acceptable	1	large-increase
2	good	1	good	3	large-increase
2	poor	1	acceptable	2	small-increase
level 2					
C1	D1	C2	D2	C3	D3
2	acceptable	3	poor	3	small-increase
4	acceptable	5	poor	5	small-increase
3	acceptable	2	poor	2	small-increase
5	acceptable	4	poor	4	small-increase
5	acceptable	4	acceptable	1	small-increase

Table 1. Protocols for an Interaction

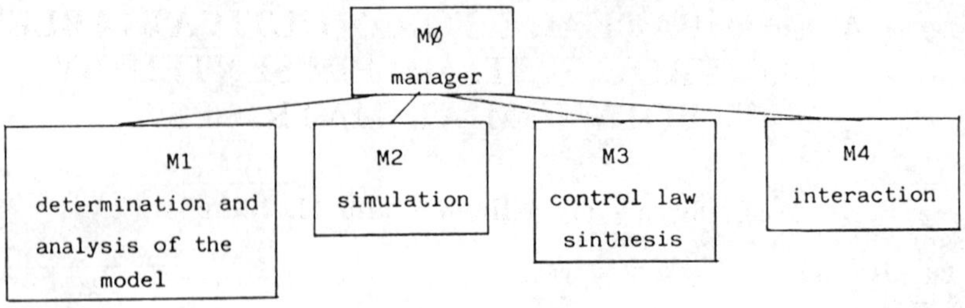

Fig. 2 Structure of the program package

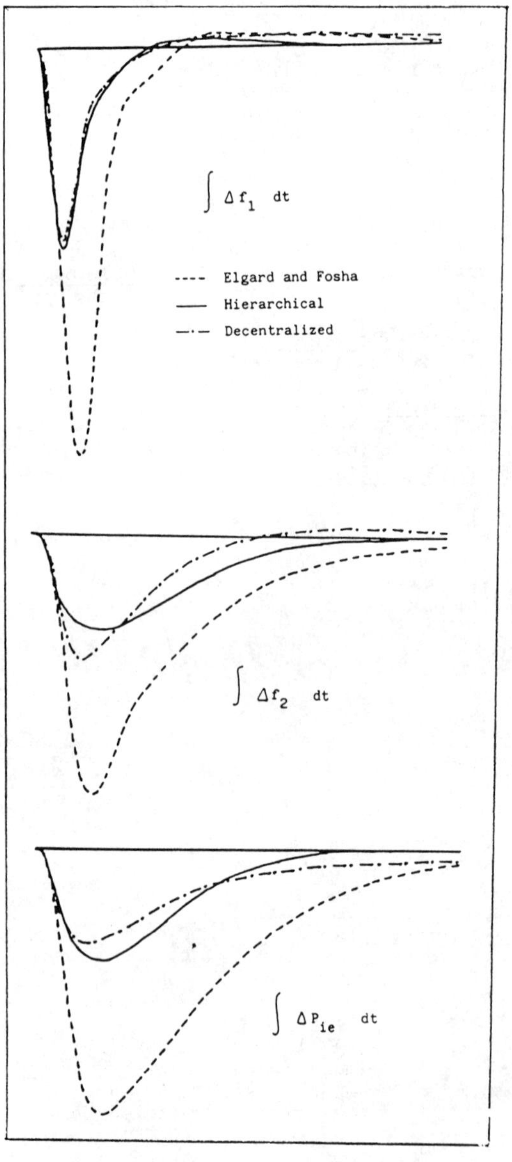

Fig. 3 Dynamic of the solutions

Interaction = "end"/ Diagnostic Interaction
Diagnostic = Definecriterion (modifier term/term)
Definecriterion = "level" number "criterion" number / "criterion" number
Number = digit/digit number
Digit = "1"/"2"/"3"/"4"/"5"/"6"/"7"/"8"/"9"/"0"
Modifier = "highly"/"very"/"more-or-less"/"sort-of"/"rather"
Term = "good"/"acceptable"/"poor"/"unnacceptable"

Fig. 1 grammar for an Interaction

w_1	w_2	w_3	w_4	w_5
0.447	0.447	0.447	0.447	0.447
level 1:	criterion 1	unnacceptable		
	criterion 2	unnacceptable		
	criterion 3	good		
0.612	0.537	0.537	0.082	0.082
level 1:	criterion 1	acceptable		
	criterion 2	sort of poor		
	criterion 3	acceptable		
	criterion 3	sort of poor		
	criterion 4	acceptable		
0.741	0.480	0.464	0.044	0.044

Table 2 Interactions in the example

x Fosha and Elgerd, xx hierarchical, xxx Decentralized

$\int(\dot{u}_1)^2\,dt$	$\int(\dot{u}_2)^2\,dt$	$Max\lvert\dot{u}_1\rvert$	$Max\lvert u_2\rvert$	
0.2136	68.27 E-4	1.28	0.114	x
0.8495	93.13 E-4	3.47	0.309	xx
0.8854	17.01 E-4	3.56	0.0737	xxx

Table 3 Control effort

A CAD PACKAGE FOR MULTIVARIABLE CONTROL SYSTEMS PRESENTED BY POLYNOMIAL MATRICES

P. Sipari*, R. Ylinen and H. Blomberg***

**Helsinki University of Technology, Systems Theory Laboratory, SF-02150 Espoo, Finland*
***Technical Research Centre of Finland, Electrical Engineering Laboratory*

Abstract. The basic theory for analysis and synthesis of differential systems using polynomial matrix representation is considered. Concepts like controllability, observability, stability etc. are defined and methods based on elementary row operations for checking them are given. As an example of synthesis methods an algorithm for synthesis of observers is given. A similar algorithm for synthesis of feedback compensators has been presented elsewhere. Computer aided design of multivariable control systems is performed interactively using a program which enables manipulation of polynomial matrices, and a simulation program that uses polynomial matrices as models of differential systems. The program for manipulation of polynomial matrices performs row operations and other transformations, calculates determinants and their roots etc. Synthesis problems can also be considered by studying different regulator compositions, constructing estimators and feedback compensators. The simulation program communicates with the manipulation program by means of polynomial matrices stored in files. Simulation results are presented by graphic displays or by separate plotter and design decisions can be based on these results. The use of the CAD package is demonstrated by considering an observer synthesis problem.

Keywords. Computer-aided design; control system analysis; control system synthesis; linear systems; multivariable control systems; observers; polynomials.

INTRODUCTION

The polynomial matrix representation introduced by Blomberg and co-workers (1969, 1978), Rosenbrock (1970) and Wolovich (1974) seems to be successful in solving various analysis and synthesis problems concerning multivariable control systems.

The polynomial matrix representation utilizes an operator algebra which is a natural generalization of the Laplace and Z transforms. Its use does not imply zero initial conditions, which in turn makes the calculus more complicated. On the other hand it can be shown that many problems can and should be treated without specified initial conditions. In particular the initial conditions for multivariable systems are so awkward to treat, that it is best to try to avoid them entirely.

In this article we shall give a brief presentation of the operator algebra mentioned above, and of the associated polynomial matrix representation along with a CAD package based on this representation. A more comprehensive presentation of the theory and algorithms for feedback compensation and observer synthesis can be found in the book by Blomberg and Ylinen (1983). The CAD package enables the designer to apply these analysis and synthesis methods to real multivariable control system design.

As an example an observer synthesis problem is considered. An example of a feedback compensator synthesis is presented in a paper by Blomberg, Ylinen and Sipari (1983).

DIFFERENTIAL INPUT-OUTPUT RELATIONS

Let X be a vector space of complex-valued functions defined on an open interval $T \subset \mathbf{R}$. Suppose further that the elements of X are infinitely differentiable, which yields that the differentiation operator p is a linear operator on X. The ring $\mathbf{C}[p]$ of polynomials in p over the field $\mathbf{C}$ of complex numbers is a commutative subring of the ring of endomorphisms of X and X is a (left) module over $\mathbf{C}[p]$.

A system of s linear time-invariant differential equations can be written as

$$A(p)y = B(p)u \tag{1}$$

with $u \in X^r$, $y \in X^q$, and with $A(p)$ and $B(p)$ polynomial matrices of sizes s by q and s by r, respectively. The *differential input-output relation* determined by Eq. (1) is defined as the set S of input-output pairs (u, y) that satisfy Eq. (1). Eq. (1) can also be written as

$$[A(p) \,\vdots\, -B(p)] \begin{bmatrix} y \\ \hdashline u \end{bmatrix} = 0 \tag{2}$$

and we shall say that S is *generated* by the partitioned matrix $[A(p) \,\vdots\, -B(p)]$ and that $[A(p) \,\vdots\, -B(p)]$ is a *generator* for S. S is clearly uniquely determined by its generator but the same S can be generated by infinitely many generators which are said to be *input-output equivalent* to each other.

A generator $[A(p) \,\vdots\, -B(p)]$ such that $\det A(p) \neq 0$ is called *regular* and an input-output relation is regular if it has a regular generator. A regular input-output relation has the property that its every subset of input-output pairs coinciding with a pair of given initial segments up to a given time is a mapping.

COMPOSITIONS OF DIFFERENTIAL INPUT-OUTPUT RELATIONS

Let $\mathcal{B} = \{S_1, \ldots, S_k\}$ be a family of differential input-output relations generated by regular generators $[A_i(p) \,\vdots\, -B_i(p)]$, $i = 1, \ldots, k$. $\mathcal{B}$ can be presented as a regular input-output relation S consisting of pairs $(u, y) = ((u_1,\ldots, u_k), (y_1,\ldots, y_k))$ with $(u_1,y_1) \in S_1,\ldots, (u_k,y_k) \in S_k$ and generated by $[A(p) \,\vdots\, -B(p)]$ with $A(p)$ = block diag $(A_1(p),\ldots, A_k(p))$ and $B(p)$ = block diag $(B_1(p),\ldots, B_k(p))$.

The interconnections between the input-output relations are described by means of an interconnection matrix C, which tells how the inputs $u_o \in X^{r_o}$ and $y_o \in X^{q_o}$ of the overall relation are chosen and how the inputs $u_i \in X^{r_i}$ and the outputs $y_i \in X^{q_i}$ are connected. C is a $(q_o+\Sigma r_i)$ by $(r_o+\Sigma q_i)$ matrix such that

(i) every entry is either 1 or 0
(ii) every row contains one and only on 1
(iii) each of the first r_o columns contains at least one 1
(iv) the first r_o entries of the q_o uppermost rows are zeros.

The interconnection constraints can be expressed via the matrix multiplication

$$\begin{bmatrix} y_o \\ u \end{bmatrix} = \begin{bmatrix} 0 & C_2 \\ C_3 & C_4 \end{bmatrix}\begin{bmatrix} u_o \\ y \end{bmatrix} = C\begin{bmatrix} u_o \\ y \end{bmatrix} \tag{3}$$

Now the pair $(\mathcal{B}, C)$ is called a *composition* of the family $\mathcal{B}$. Obviously, the family $\mathcal{B}$ can be replaced by a singleton set {S} or simply by S or, furthermore, by the generator $[A(p) \,\vdots\, -B(p)]$.

A composition (S, C) always defines an *internal input-output relation* S_i, an *overall input-output relation* S_o and an *internal-overall input-output relation* S_{io} consisting of pairs (u_o, y), (u_o, y_o) and $(u_o,(y, y_o))$ respectively, satisfying Eq. (3) with $(u, y) \in S$.

The internal input-output relation S_i determined by (S,C) is a differential input-output relation generated by the equation

$$\begin{bmatrix} A(p)-B(p)C_4 & 0 \\ C_2 & I \end{bmatrix}\begin{bmatrix} y \\ y_o \end{bmatrix} = \begin{bmatrix} B(p)C_3 \\ 0 \end{bmatrix} u_o \,. \tag{4}$$

If S and S_i are regular we call the composition (S, C) regular, too.

The composition (S,C) is said to be *observable* if for each $(u_o,y_o) \in S_o$ there exists a unique y such that $(u_o,(y,y_o)) \in S_{io}$.

If S_o is the overall input-output relation determined by a composition (S,C), the composition (S,C) is said to be a *decomposition* of S_o. Two compositions are *input-output equivalent* if they determine the same overall input-output relation.

POLYNOMIAL AND RATIONAL MATRICES

Suppose that the signal space X is such that p is an indeterminate over **C**. Then **C**[p] is an integral domain and has the field **C**(p) of quotients. X can be considered as vector space over **C**(p) if and only if every nonzero $a(p) \in \mathbf{C}[p]$ is a bijection.

An s by q-matrix A(p) over **C**[p] is *invertible* as a polynomial matrix if and only if s = q and det A(p) is a nonzero constant. The matrices A(p) and B(p) are *row equivalent* if A(p) = P(p)B(p) with P(p) invertible. Correspondingly, A(p) and B(p) are *column equivalent* if A(p) = B(p)Q(p) with Q(p) invertible. A polynomial matrix A(p) with $\det A(p) \neq 0$ is row equivalent to an upper triangular matrix where every entry above a diagonal entry is of lower degree than the diagonal entry itself and the diagonal entries are monic i.e. their leading coefficients are equal to one. This matrix can be found by means of a *triangularization procedure* using *elementary row operations*.

If the upper triangular matrices A(p), B(p) with $\det A(p) \neq 0$, $\det B(p) \neq 0$ are row equivalent and of the form described above, then A(p) = B(p). Such a form is called a *canonical upper triangular form* (CUT form). The triangularization can be applied to matrices with zero determinants and to nonsquare matrices, too, and the corresponding results are CUT forms. With the aid of *elementary column operations* a polynomial matrix can be transformed to a canonical lower triangular form.

Another important canonical form for the row equivalence of polynomial matrices is the *canonical row proper form* (CRP form) which is defined as follows. The ith *row degree* of a polynomial matrix is the highest degree of the entries in the ith row and the corresponding leading row coefficient is the row matrix consisting of the coefficients of the terms of row degree. The matrix A(p) with $\det A(p) \neq 0$ is *row proper* if its leading row coefficients are linearly independent i.e. the matrix having them as its rows is invertible. A(p) is of CRP form if it is row proper and the matrix consisting of the leading terms of the entries of A(p) is of CUT form. Row properness and CRP form can also be generalized to matrices with zero determinants and to nonsquare matrices. *Column proper* forms can be obtained by elementary column operations.

A q by q-matrix C(p) is a *greatest common left divisor* (GCLD) of a q by q-matrix A(p) and a q by r-matrix B(p) if it is a common left divisor of A(p) and B(p), i.e. $A(p) = C(p)A_1(p)$ and $B(p) = C(p)B_1(p)$, and every other common left divisor D(p) of A(p) and B(p) is a left divisor of C(p), i.e. C(p) = D(p)E(p). If the matrix $[A(p) \,\vdots\, B(p)]$ is transformed to a column equivalent matrix $[C(p) \,\vdots\, 0]$, then C(p) is a GCLD. In particular, if the GCLDs of A(p) and B(p) are invertible then A(p) and B(p) are *left coprime*. For A(p) and B(p) left coprime there are matrices $Q_3(p)$ and $Q_4(p)$ such that

$$Q(p) = \begin{bmatrix} A(p) & B(p) \\ Q_3(p) & Q_4(p) \end{bmatrix}$$

is invertible. Conversely, if

$$P(p) = \begin{bmatrix} P_1(p) & P_2(p) \\ P_3(p) & P_4(p) \end{bmatrix}, \text{ with } P_1(p), P_4(p) \text{ square}$$

is invertible, then $P_1(p)$ and $P_2(p)$ are left coprime, and so are $P_3(p)$ and $P_4(p)$.

A rational matrix G(p) over **C**(p) can be presented in the forms

$$G(p) = A(p)^{-1}B(p) = L(p)M(p)^{-1} \tag{5}$$

with A(p), B(p) left coprime and L(p), M(p) right coprime polynomial matrices. G(p) is *proper* (alternatively *strictly proper*) if its entries are proper (strictly proper), i.e. their numerators are at most of as high degree as (lower degree than) their denominators. If G(p) is proper (strictly proper) then the row degrees of B(p) are at most as high as (lower than) the corresponding row degrees of A(p). If A(p) is row proper this also holds conversely. The corresponding properties hold for the column degrees of L(p) and M(p).

ANALYSIS OF INPUT-OUTPUT RELATIONS

Let S_1 and S_2 be differential input output relations generated by $[A(p) \,\vdots\, -B(p)]$ and $[L(p)A(p) \,\vdots\, -L(p)B(p)]$, respectively. Then $S_1 \subset S_2$, and if L(p) is invertible then $S_1 = S_2$, i.e. row equivalent generators are input output equivalent.

Let X be "sufficiently rich" in the sense that it contains all possible complex-valued solutions y to all equations $a(p)y = 0$, $a(p) \neq 0$, or equivalently, contains all signals of the form $t^n \exp(\lambda t)$, $t \in T$, $n = 0,1,2,\ldots$, $\lambda \in \mathbf{C}$. Then the regular input-output relations S and S_1 generated by $[A(p) \vdots -B(p)]$ and $[A_1(p) \vdots -B_1(p)]$ satisfy $S_1 \subset S$ if and only if the generators satisfy

$$[A(p) \vdots -B(p)] = L(p)[A_1(p) \vdots -B_1(p)] \qquad (6)$$

for some square polynomial matrix L(p). Moreover, $S_1 = S$ if and only if L(p) is inverible. This means that input output equivalent regular generators are also row equivalent, i.e. for them the input-output equivalence and row equivalence coincide. Furthermore, the canonical forms for row equivalence are then canonical forms for input-output equivalence, too.

Let $[A(p) \vdots -B(p)]$ be regular. Then A(p) is invertible as a rational matrix. The rational matrix $G(p) = A(p)^{-1}B(p)$ is called the *transfer matrix* determined by $[A(p) \vdots -B(p)]$. It should be noted that G(p) can be identified with a mapping $X^r \to X^q$ if A(p) is invertible as a polynomial matrix. However, arbitrary transfer matrices can be identified with mappings if and only if X can be considered as a vector space over $\mathbf{C}(p)$.

Two generators as well as the corresponding input-output relations are said to be *transfer equivalent* if they determine the same transfer matrix. For each rational matrix $G(p)$ there is a regular generator $[A_1(p) \vdots -B_1(p)]$, with $A_1(p)$, $B_1(p)$ left coprime such that that every other transfer equivalent generator $[A(p) \vdots -B(p)]$ satisfies Eq. (6) for some L(p).

If X is sufficiently rich, then the transfer matrices determined by all generators of two regular input-output relations S_1 and S_2 such that $S_1 \subset S_2$ are equal. The same holds if X has a $\mathbf{C}$-subspace X' such that p restricted to X' is an operator of X' and an indeterminate over $\mathbf{C}$, and X' can be regarded as a vector space over $\mathbf{C}(p)$.

A regular generator $[A(p) \vdots -B(p)]$ and corresponding input-output relation are called (strictly) proper if the transfer matrix $G(p)=A(p)^{-1}B(p)$ is (strictly) proper.

A regular generator $[A(p) \vdots -B(p)]$ and the corresponding input-output relation are said to be *controllable* if A(p) and B(p) are left coprime. A controllable input-output relation is *minimal* among the input-output relations generated by transfer equivalent generators.

Finally, a regular input-output relation S generated by $[A(p) \vdots -B(p)]$ is *asymptotically stable* if every $y \in S(0)$ asymptotically approaches zero with increasing time. If X is sufficiently rich, S is asymptotically stable if and only if the roots of det A(p), i.e. the roots of the *characteristic polynomial* of S, have negative real parts.

It is important to note that for algebraic reasons we have to use numbers to describe the system parameters. This means that we lose the relationships between the model parameters and the original quantities of the real system. Therefore the methods based on the model have to be robust, i.e. insensitive with respect to variations in parameters, possibly in degrees of polynomials, too.

ANALYSIS OF COMPOSITIONS

Let the internal-overall relation S_{io} determined by a composition (S, C) be generated by Eq. (4). Using elementary row operations this can be brought to the form

$$\begin{bmatrix} \tilde{A}_1(p) & \tilde{A}_2(p) \\ 0 & \tilde{A}_4(p) \end{bmatrix} \begin{bmatrix} y \\ y_o \end{bmatrix} = \begin{bmatrix} \tilde{B}_1(p) \\ \tilde{B}_2(p) \end{bmatrix} u_o \qquad (7)$$

If for each (u_o, y_o) such that

$$\tilde{A}_4(p)y_o = \tilde{B}_2(p)u_o \qquad (8)$$

there exists a corresponding y such that

$$\tilde{A}_1(p)y + \tilde{A}_2(p)y_o = \tilde{B}_1(p)u_o , \qquad (9)$$

then the overall input-output relation S_o is generated by Eq. (8). In particular, this holds if $\tilde{A}_1(p)$ is a surjection. If $\tilde{A}_1(p)$ is in addition a bijection then the composition (S, C) is completely observable. If X is sufficiently rich then $\tilde{A}_1(p)$ is a bijection if and only if it is invertible, and furthermore, $\tilde{A}_1$ is invertible if and only if $A(p)-B(p)C_4$ and C_2 are right coprime.

OBSERVER SYNTHESIS

Let S be a regular differential input-output relation with two outputs y_1, y_2. Suppose that S is generated by the equation

$$\underbrace{\begin{bmatrix} A_1(p) & A_2(p) \\ A_3(p) & A_4(p) \end{bmatrix}}_{A(p)} \begin{bmatrix} y_1 \\ y_2 \end{bmatrix} = \underbrace{\begin{bmatrix} B_1(p) \\ B_2(p) \end{bmatrix}}_{B(p)} u. \qquad (10)$$

Only the output y_2 can be observed and used together with u as input of a regular differential input-output relation $\hat{S}$ called an observer.

Consider the composition shown in Fig. 1 and suppose that S is given and regular. The problem is to find a regular differential input-output relation $\hat{S}$ such that the internal input-output relation S_i is regular and at least the following conditions are fulfilled.

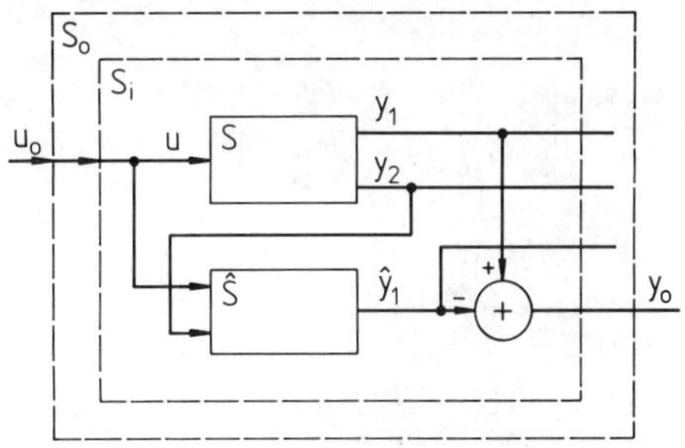

Fig. 1. The observer synthesis problem.

If

$$\hat{y}_1|(-\infty,t) \cap T = y_1|(-\infty,t) \cap T \\ \text{for some } t \in T \text{ then } \hat{y}_1 = y_1. \qquad (11)$$

The overall input-output relation S_o should be asymptotically stable and the roots of its characteristic polynomial should be located in a satisfactory way with regard to the resulting dynamic properties.

The observer should also fulfil certain other conditions. For instance $\hat{S}$ should be realizable as simply as possible, which in practice means that it has to be controllable and of low order. Furthermore, the properties of S_o should be insensitive to small variations in parameter values.

The first step towards a solution is to bring the generator for S to upper triangular form

$$\left[\begin{array}{c|c|c} \tilde{A}_1(p) & \tilde{A}_2(p) & -\tilde{B}_1(p) \\ \hline 0 & \tilde{A}_4(p) & -\tilde{B}_2(p) \end{array}\right] . \qquad (12)$$

Let $[C(p) \,\vdots\vdots\, -D_1(p) \,\vdots\, -D_2(p)]$ be a generator for $\hat{S}$.

It is quite easy to see that in order to fulfil the condition (11) the overall input-output relation of the composition in Fig. 1 has to be generated by a generator of the form $[E(p) \,\vdots\, 0]$. This, furthermore, leads to the result that the observer $\hat{S}$ has to satisfy

$$U(p)[C(p) \,\vdots\vdots\, -D_1(p) \,\vdots\, -D_2(p)] = [T_1(p) \,\vdots\, T_2(p)]\left[\begin{array}{c|c|c} \tilde{A}_1(p) & -\tilde{B}_1(p) & \tilde{A}_2(p) \\ \hline 0 & -\tilde{B}_2(p) & \tilde{A}_4(p) \end{array}\right] \qquad (13)$$

and

$$T_1(p)A_1(p) = E(p), \qquad (14)$$

with $U(p)$ and $[T_1(p) \,\vdots\, T_2(p)]$ left coprime.

Obviously

$$[C(p) \,\vdots\vdots\, -D_1(p) \,\vdots\, -D_2(p)] = [\tilde{A}_1(p) \,\vdots\vdots\, -\tilde{B}_1(p) \,\vdots\, \tilde{A}_2(p)] \qquad (15)$$

satisfy the condition (13) but possibly not the other conditions. If $A_1(p)$ is 'stable', then a 'stable' and proper observer can be constructed with the aid of a number of suitable transformations $T^1(p), T^2(p), \ldots, T^k(p)$ according to

$$U(p)[C(p) \,\vdots\vdots\, -D_1(p) \,\vdots\, -D_2(p)] = [I \,\vdots\, 0]\underbrace{\overbrace{\left[\begin{array}{c|c} T_1^k(p) & T_2^k(p) \\ \hline 0 & I \end{array}\right]}^{T^k(p)} \cdots \overbrace{\left[\begin{array}{c|c} T_1^2(p) & T_2^2(p) \\ \hline 0 & I \end{array}\right]}^{T^2(p)} \overbrace{\left[\begin{array}{c|c} T_1^1(p) & T_2^1(p) \\ \hline 0 & I \end{array}\right]}^{T^1(p)}}_{[T_1(p) \,\vdots\, T_2(p)]}$$

$$\times \left[\begin{array}{c|c|c} \tilde{A}_1(p) & -\tilde{B}_1(p) & \tilde{A}_2(p) \\ \hline 0 & -\tilde{B}_2(p) & \tilde{A}_4(p) \end{array}\right] \qquad (16)$$

with $T_1(p) = T_1^k(p) \ldots T_1^2(p)\, T_1^1(p)$ 'stable' and with $T^1(p), T^2(p), \ldots, T^k(p)$ further so chosen that a proper observer results.

It can be shown that if $[C(p) \,\vdots\, -D_2(p)]$ as obtained from Eq. (16) and S are proper. Then the whole observer is proper.

The transformations needed above can be chosen according to the algorithm developed in the book by Blomberg and Ylinen (1983) (suitably modified).

It should be clear that the observer always can be chosen controllable: suppose that in Eq. (16) a 'maximal' $U(p)$ is extracted ($C(p)$, $D_1(p)$, $D_2(p)$ left coprime). If $U(p)$, $T_1(p)$, $T_2(p)$ are not left coprime, then a GCLD of $U(p)$, $T_1(p)$, $T_2(p)$ is cancelled from Eq. (15). On renaming the remaining left coprime factors as $U(p)$, $T_1(p)$, $T_2(p)$ respectively we obtain a final (13) leading to a controllable observer.

CAD PACKAGE

The CAD package for multivariable control systems represented by polynomial matrices consists of two separate programs; one for algebraic manipulations of real coefficient polynomial matrices (RPMAT) and the other for simulation of differential systems represented by polynomial matrices (PSIM). Both programs were written using Pascal in HP 1000/A 700 computer environment. They are interactive and easy to use for solving different kinds of analysis and synthesis problems. Although the program for manipulation of polynomial matrices and the simulation program are separate they cooperate with each other through polynomial matrices stored in files.

Manipulation Program RPMAT

Using the program for algebraic manipulation of real coefficient polynomial matrices it is possible to handle a specific so called object matrix at a time. The object matrix can be given from the terminal or read from the file and it is analyzed and manipulated by using commands. Commands and possible additional parameters are given interactively after programs prompt. With the aid of help command it is possible to get information about all possible commands and also detailed information about each command and its usage.

Some parameters such as summation error level and parameters for calculation of the roots of a polynomial which are used repeatedly during many different commands are initialized at the beginning of the execution and they stay valid until changed by definition command.

While executing the program it is also possible to handle files for storing the object matrix or its submatrix and for making the matrix stored in file a part of the object matrix.

The program checks the correctness of a given command and its parameters and informs about possible errors without terminating the execution. It is also possible to return to the previous object matrix i.e. to annul the latest command.

The present version of the program uses virtual memory file so that if the program is terminated its execution can be continued using the latest object matrix and parameter definitions.

Commands. Commands consist of three-character command codes and possible parameters such as row or column indices and polynomials etc. Polynomials are normally given by writing the coefficients in decreasing order separated by blanks or commas.

For input and output there are different commands for giving the whole object matrix or a part of it from the terminal, for reading the object matrix from file and for defining a zero or identity matrix of desired size as object matrix. The object matrix or a part of it can be displayed completely or by giving only the highest degree terms of each entry.

For operating on the object matrix via elementary row or column operations there are commands for interchanging the rows or columns and for multiplying a row by a polynomial and adding a row to another multiplayed by a polynomial.

There are commands for transposing the object matrix or its submatrix. There are also commands for transforming the object matrix to upper triangular or row proper form and for removing the greatest common left divisor of the columns of the object matrix.

In addition if the object matrix is considered as a generator a corresponding state-space representation can be constructed.

For using files to manipulate the object matrix there are commands for joining the matrix stored in file to the object matrix in different ways. The object matrix can also be pre- or postmultiplyed by the matrix in file and the matrix in file can be added to the object matrix.

There are commands for solving feedback compensator and estimator design problems via algorithms based on the theory. Different feedback compensator structures can be analyzed and candidates for feedback compensators and estimators can be constructed.

Finally there are commands for defining some parameters, and the names of used files, for returning to the previous object matrix and for terminating the execution of the program.

Numerical considerations. Frequent use of arithmetic operations in elementary row operations results in growing truncation errors.

In the worst case this leads to incorrect or unreliable results in solving analysis or synthesis problems. Actually it is most important that the degree of each entry of the object matrix is correct after each operation. Therefore double precision (about 16 digits) is used in the program. In adding polynomials term by term the following closeness test is used:

$$\left| \frac{c_k + c_2}{c_1 - c_2} \right| < \varepsilon \quad , \tag{17}$$

where c_1 and c_2, $(c_1 \cdot c_2 < 0)$, are coefficients to be added and ε is a summation error level defined by the user. If Eq. (17) holds the corresponding coefficient of the resulting polynomial is supposed to be zero, otherwise equal to $c_1 + c_2$. Some other common methods are also used to reduce the errors caused by truncations.

The effect of the truncation errors depends naturally on the numerical values of the object matrix and on the amount of operations to be done. The test problems considered seem to show that systems of order 10 or less can be analysed and e.g. feedback compensators for them can be constructed.

Simulation Program PSIM

Using the simulation program it is possible to simulate linear time-invariant proper differential systems. The system can be represented by a composition of subsystems which also may have input delays. Simulation results are stored in a file from which they can be plotted out using a graphic display or a plotter.

Subsystems are represented by generators in row proper form. Each subsystem is simulated applying the fourth order Runge-Kutta method and a state-space type representation which is easily consructed from the row proper form. The whole system is simulated simulating the subsystems in the order in which they were defined and taking their interconnections into account.

The simulation program stores the definitions given at the beginning of the simulation in a definition file. Previously stored definitions, possibly appearing, are read as optional settings and each of them can be accepted as such or changed if necessary.

The program asks first the names of the files where the generators of the subsystems have been stored. After that input delays in the subsystems can be defined.

An interconnection is given by defining for every input whether it is equal to some output or equal to an overall input. The values of overall inputs can be given using constants, ramps, sinusoidals and uniformly or normally distributed random functions.

Next the initial and terminal times and the lenghts of integration and printing intervals are given.

Finally the input and output components to be stored are named and the printing file to be used is specified.

After these definitions, the program starts the simulation with zero initial conditions. Our test simulation runs have not revealed any particular numerical problems.

A NUMERICAL EXAMPLE

Let us consider the following observer synthesis problem. The generator of the original system S is

$$\begin{array}{ccccc} y_1 & y_2 & y_3 & y_4 & u \end{array}$$
$$\left[\begin{array}{cccc|c} p+1 & 2p+6 & 0 & 0 & -1 \\ 0 & p^2+5p+6 & 0 & 0 & p+1 \\ -1 & 0 & p+3 & -1 & 0 \\ 0 & -p-3 & 2 & p+1 & 0 \end{array}\right], \tag{18}$$

where the input u and the outputs y_3 and y_4 can be observed. Our problem is to design a strictly proper observer $\hat{S}$ for estimating the outputs y_1 and y_2. S can be interpreted as representing e.g. a series composition of two subsystems.

Using the RPMAT program the generator (18) is first brought to an upper block triangular form (12)

$$\begin{array}{ccccc} y_1 & y_2 & y_3 & y_4 & u \end{array}$$
$$\left[\begin{array}{cc|cc|c} -1 & 0 & p+3 & -1 & 0 \\ 0 & 2p+6 & p^2+4p+3 & -p-1 & -1 \\ \hline 0 & 0 & -p^2-2p-3 & p^2+2p+1 & p+2 \\ 0 & 0 & 0 & p^2+3p+2 & p+1 \end{array}\right] \tag{19}$$

giving $\tilde{A}_1(p) = \begin{bmatrix} -1 & 0 \\ 0 & 2p+6 \end{bmatrix}$.

It follows that (y_1, y_2) cannot be uniquely determined on the basis of y_3, y_4 and u. A stable observer can nevertheless be constructed because the only root of the determinant of $\tilde{A}_1(p)$ above is -3.

The generator (19) is chosen as a first candidate for the construction of the observer. New candidates are then constructed with the aid of the RPMAT program. When a factor is added to characteristic polynomial of the candidate, $(p+5)$ is chosen as satisfactory. After each step the program tests the properness of the candidate. Finally a strictly proper candidate (20) is obtained for the observer. The characteristic polynomial of the resulting observer chosen as given by the generator (20) has one root at -3 and three roots at -5.

$$\begin{array}{c} \\ \\ \end{array}\begin{matrix} \hat{y}_1 & \hat{y}_2 & u & y_3 & y_4 \end{matrix}$$

$$\left[\begin{array}{cc:c:cc} 5p^2+50p+125 & -8p^2-64p-120 & -23p-63 & -80p-52 & 94p+174 \\ 0 & p^2+8p+15 & p+1 & -6 & -3p-3 \end{array}\right] \tag{20}$$

The behaviour of the constructed observer connected to the original system can be studied with the aid of the simulation program. One simulation run obtained is depicted in Fig. 2. The initial values of the outputs and their derivatives of the original system S are here equal to zero. The initial values of the outputs of the observer $\hat{S}$ are equal to -0.5 with zero derivatives. The input u is a unit step at 0.

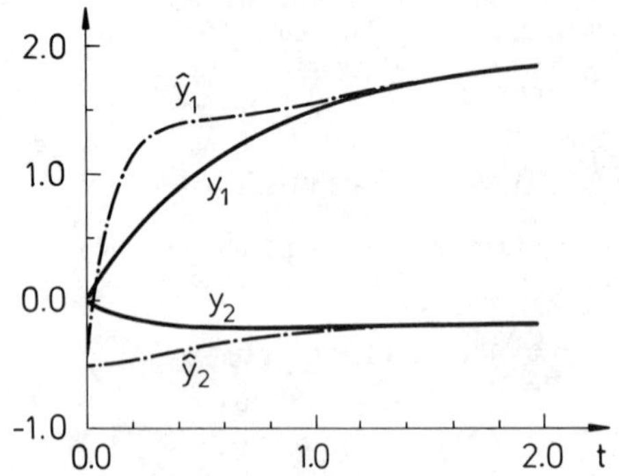

Fig. 2. Simulated behaviour of true and estimated outputs.

Other examples have been considered in theses and in other unpublished papers.

CONCLUSIONS

In this paper a CAD package for multivariable control systems represented by polynomial matrices has been presented along with the theoretical background and an example. The created theoretical machinery seems very promising and it works rather well in treating problems concerning analysis and synthesis of multivariable linear systems. At present the truncation errors in manipulating polynomial matrices limit the order of the systems to be considered to about ten. The simulation program is easy and flexible to use and it has proved very suitable for our purposes. Present research focuses on numerical aspects, in particular on utilizing structural properties of the system to achieve improved versatility.

REFERENCES

Blomberg, H., A. Halme, J. Sinervo, and R. Ylinen (1969). On algebraic methods in systems theory. Acta Polytechnica Scandinavica, Ma 19, Helsinki.

Blomberg, H., and R. Ylinen (1978). Foundations of the polynomial theory for linear systems. Int. J. General Systems, 4, 231-242.

Blomberg, H., and R. Ylinen (1983). Algebraic theory for multivariable linear systems. Academic Press, London.

Blomberg, H., R. Ylinen, and P. Sipari (1983). Polynomial systems theory and multivariable feedback compensator design. In V. Kelhä, M. Luukkala, and T. Tuomi (Ed.), Topics in Technical Physics. Acta Polytechnica Scandinavica, Helsinki. pp. 28-38.

Rosenbrock, H.H. (1970). State-space and multivariable theory. Nelson, London.

Wolovich, W.A. (1974). Linear multivariable systems. Springer, New York.

MAX – A PROGRAM SYSTEM FOR SYMBOLIC MANIPULATION OF POLYNOMIAL MATRICES

J. T. Tanttu and M. Aaltonen

Department of Electrical Engineering, Tampere University of Technology, PO Box 527, SF-33101 Tampere, Finland

Abstract. MAX is an interpreted programming language for polynomial matrix manipulations. It contains especially those operations which are frequently needed in linear control. MAX is written using the C programming language under the VAX/VMS operating system. In this paper the design, implementation and available operations of MAX are described.

Keywords. Computer aided design, symbolic computation.

1. INTRODUCTION

Many design methods of multivariable control systems lead to symbolic manipulation of polynomial or rational matrices (Kucera, 1979, Kailath, 1980, Wolowich, 1983, 1984). These calculations are extremely tedious and time consuming to work out manually. There exist some general purpose program systems for symbolic mathematics - the best-known of which is perhaps MACSYMA (Mathlab, 1983) written in LISP. It is a very large program that requires an effective virtual memory computer. The recent trend is, however, to implement symbolic mathematics even with microcomputers (mikromath and SMP on IBM PC).

MAX, a program for symbolic computation of polynomial matrices, is by no means intended as a rival for these extremely sophisticated systems. On the contrary MAX is a special purpose system intended exclusively for manipulations of polynomial and rational matrices having one free variable. MAX contains the basic operations (+,-,x,/) for scalars, polynomials, polynomial and rational matrices. More importantly MAX contains algorithms that are often needed in control system design and synthesis:

- greatest common divisor (gcd) and least common multiple (lcm) of polynomials and polynomial matrices

- Smith form of a polynomial matrix

- extraction of matrix divisors

- coprime matrix fraction representation of a rational matrix.

Most of the algorithms are from Kucera (1979). Other algorithms can be found in Pernebo (1980).

In section 2 the design and implementation of MAX is described. Section 3 describes the available functions. In section 4 an example is presented how MAX can be used in controller design. Finally in section 5 some conclusions are given and the effect of symbolic computation methods in future CAD-systems for control engineering is briefly discussed.

2. DESIGN AND STRUCTURE OF MAX

MAX is an interpreter like ("command driven") program especially designed for computations with polynomial matrices. Actually MAX is a very high level interpreted programming language. Its data types are scalars (floating point numbers and integers) (s), polynomials (p), polynomial matrices (pm), rational expressions (polynomial fractions) (r) and rational matrices (rm) each having one free variable and lists (l) consisting of objects of the other data types.

The command language includes besides the computational functions (see section 3), some control structures for user programming and a possibility to define new functions.

MAX manipulates named objects that reside in its working space. When the program is started the working space and nametable are both empty. The program allocates memory for the new objects dynamically and there is also a garbage collection system in the system to prevent excessive use of memory. When the program is terminated all data in the working space including the nametable and the objects are lost. The user has however a possibility to store objects into external files using a specific save command.

MAX is written using the C-language under VAX/VMS operating system. Its command language in implemented using YACC (Johnson, 1978) a compiler compiler. This makes changes of the syntax quite easy.

Because most of the algorithms in MAX are done in floating point form it would be more accurate to call MAX a semi symbolic program system in contrast to systems that use accurate arithmetic with infinite precision" numbers. This is not believed to be any serious deficiency since usually the data used in control system design is not accurate. MAX contains an interface to MACSYMA where more accurate calculations may be performed if needed.

3. SYNTAX AND AVAILABLE OPERATIONS

The Data Types of MAX

Integers and reals. Integers and reals are in the

sequel called scalars to distinguish them from more complex data types. They are presented according to the syntax of the C programming language.

Polynomials. Polynomials are expressed almost as in the normal mathematical notation. However the multiplication is denoted with * symbol and the power with a ↑-sign.

ex. 1.0 + 5.0*z↑3 is the MAX notation for polynomial $1 + 5z^3$.

Rational functions (polynomial fractions). Rational functions are expressed as a ratio of two polynomials

ex. (z + 2.0*z↑2)/(7.0 + z) is the MAX notation for $z + 2z^2/7 + z$.

In the notation the paranthesis are necessary.

Matrices. Matrices can have as their elements any of the above mentioned datatypes. They are defined as

```
(mat m,n) L
```

where m and n are the column and row dimensions of the matrix and L is a list of its elements.

Lists. List is an indexed array of elements. Its elements may be of any allowable MAX datatype even other lists or matrices.

ex. [A, (1 + z↑2)↑3, 2.71]

Lists are extensively utilized in functions that return many values. Returning these values in a single list saves considerably the user's typing efforts.

The Syntax of MAX

Instead of the complete syntax of the MAX programming language only some of its features are presented with examples.

The variables of MAX are character strings starting with a letter. The length of the names of variables is not restricted and both lower and uppercase letters may be used. In the subsequent examples the variables are written in capital letters and the names of the MAX functions with lowercase letters.

The assignment operator. The basic use of the assignment operator is to give values to named variables. No type checking is performed and the old value of the variable is lost if the same name is used twice.

Ex. let us give values to 4 objects

```
A = 1.0*z + 2.5*z↑2;
B = A↑10;
L = [A,B];              notice that L is a list
M = (mat 2,2) [A,B,1];
```

In the last example the matrix element M(2,2) is assigned the value zero as the list contains only three elements. An alternate way to perform the same assignment is:

```
M = (mat 2,2) [ ];    definition of a zero matrix
M(1,1) = A;
M(1,2) = B;
M(2,1) = 1;
```

In the examples above another MAX operator, the semi-colon (;) operator, is used. It indicates to the command interpreter the termination of a command.

Basic arithmetic operations. An essential feature of MAX is its ability to do arithmetics with all its datatypes excluding lists. Addition, subtraction and multiplication are defined with all the datatypes (except lists) with the requirement that matrices be of the proper dimensions. Division of polynomials or rationals may be done with the solidus (/) operator resulting to a rational value but there is a separate function for polynomial division called div returning both the quotient and remainder polynomials in a list. Tables 1, 2 and 3 illustrate the allowable types in addition, multiplication, and division operations.

The power operation. The power operation of MAX has the syntax

```
A↑n
```

where A is an expression of any of the data types (s, p, r, pm, rm) and the exponent n has an integer value. Negative exponents are not allowed with matrices.

Special Operators

The special feature of MAX are its functions for polynomial and rational matrix manipulations some of which are described in this section.

1. Division of polynomials has the syntax

```
[U, V] = div(A,B);
```

where A,B,U,V are polynomials. Notice that the quotient U and the remainder V are returned in a list. Thus the statement

```
X = div(z↑2+z+1,z);
```

would yield X(1) = z + 1 and X(2) = 1.

2. Division of polynomial matrices. As matrices are non-commutative there is a separate function for left and right division of polynomial matrices each returning a two element list consisting of the quotient and the remainder matrices

```
[U1, V1] = ldiv(A, B);  where A = B*U1 + V1
[U2, V2] = rdiv(A, B);  where A = U2*B + V2
```

3. Extraction of a divisor from a polynomial matrix. There is a separate function for extracting a left or a right divisor from a polynomial matrix:

```
C1 = rextract(G1, C);   where C = C1*G1
C2 = lextract(G2, C);   where C = G2*C2
```

4. The euclidean algorithm for polynomials. The operator euc does the euclidean algorithm for two polynomials

```
[G, P, Q, R, S] = euc(A, B);
```

Thus the function euc returns a list of 5 polynomials satisfying the equations

```
AP + BQ = G     G = gcd(A,B)
AR + BS = 0     L = AR = lcm(A,B)
```

There are also separate functions gcd and lcm which return the greatest common divisor and least common multiple polynomials of the two argument polynomials respectively.

5. Zeroes of a polynomial. The function zeroes returns the zero points of a polynomial in a list.

6. Smith form of a polynomial matrix. Function smith returns a list of three matrices that are the smith decomposition of its argument polynomial matrix:

```
L = smith(A);
```

where L = [U, S, V] is a list of three polynomial matrices satisfying S = UAV, U and V being unimodular and S having non-zero elements only in the diagonal.

7. Determinant of a polynomial matrix. One of the most often used functions is det which computes the determinant of a polynomial matrix

```
R = det (A);
```

As an example the following statement yields the zeroes of the determinant of a polynomial matrix:

```
X = zeroes(det(A));
```

where A is a polynomial matrix.

8. The euclidean algorithm for polynomials matrices. The euclidean algorithm, gcd and lcm functions for polynomial matrices are similar to those of polynomials except that separate algorithms for left and right operations are needed. This yields 6 different functions reuc, rgcd, rlcm, leuc, lgcd and llcm having the obvious definitions.

9. The coprime matrix fraction description. The coprime left and right matrix fraction descriptions for a rational matrix are obtained with functions lmfd and rmfd respectively. As an example the following staments

```
L    = lmfd(S); A = L(1); B = L(2);
char = det(A);
```

compute the characteristic polynomial of the transfer function matrix S. First the lmfd function returns in a list two left coprime polynomial matrices A and B satisfying

$$S = A^{-1}B$$

then the determinant of the square polynomial matrix A is computed.

10. The inverse of a polynomial matrix. The inv function returns the inverse of a polynomial matrix. The result is a rational matrix which may be used in other expressions. The statement

```
G = C * inv(z*I - A)*B;
```

yields the transfer function matrix of the state space representation S(A, B, C) assuming that I is an identity matrix of proper dimensions.

There are also other computational functions which are not described here. It is believed that using the basic functions of MAX several controller design and analysis algorithms that are based on the polynomial equation approach may be implemented quite easily.

Input operations. In general input operations are done using the assignment operator as was illustrated in the previous examples. Especially for user created functions there is a separate read operation having the following syntax:

```
read (type definition) NAME
```

where the type definition is needed only for matrices and the NAME is the variable whose value is to be read. The read command will cause the NAME followed by an equal sign to be printed on the next line of the terminal output and the program to wait for user input.

Output operations. The basic output operator of MAX is the question mark (?) having the effect that MAX prints on the terminal the value of that variable or expression which immediately preceded the ?-operator. Suppose that symbol X is a polynomial ($X = 7 + 2z + z^2$) then the command

```
X?;        yields
% 7.0 + 2.0*z + z↑2
```

To get a slightly more attractive output also a print function is available. It prints both the name of the variable and its value. For example:

```
print X;
X = 7.0 + 2.0 z + z↑2
```

Save and load operations. The save command is used to put named objects, i.e variables having a value, into an external file. The file can be then brought into the working space of MAX with the load command. Only those variables that are saved can be used when MAX is run again after being terminated.

The load command is used to fetch any excutable MAX staments from an external file. The staments are executed immediately after the loading operation. Using the load facility new functions can be effectively built into MAX.

4. AN EXAMPLE

Let us solve as an example the unilateral polynomial matrix equation

$$AX + BY = C \qquad (1)$$

where A (lxp), B(lxq) and C (lxm) are known polynomial matrices and X (pxm) and Y (qxm) unknown polynomial matrices. The algorithm of Kucera (1979) will be used. The solution of the problem consists of two main steps. First the left euclidean algorithm is performed for the matrices A and B. Then the polynomial matrix C is decomposed so that C = Gl * Cl, where Gl is the greatest common left divisor of C. If this decomposition is not possible (1) has no solution. As a MAX program the algorithm is as follows:

```
print "Solution of the Diophantine equation";
read l; read p; read q; read m;
read (mat l,p) A;
read (mat l,q) B;
read (mat l,m) C;
[Gl,Pl,Ql,Rl,Sl] = leuc(A,B);
Cl = lextract(Gl,C);
XO = Pl*Cl; YO = Ql*Cl;
print "the particular solution is";
print XO; print YO;
print "the general solution is";
print "X = XO + Rl*T  Y = YO + Sl*T";
print "where";
print Rl; print Sl;
print "and T is a polynomial matrix"
```

The complete function should also include some kind of checking whether Gl really is a left divisor of C. This could be done with the if then else construction and comparison operator of MAX but it is not included in the example for the sake of simplicity.

5. CONCLUSIONS

A program system MAX for polynomial matrix computations is presented. Besides the basic arithmetic operations it contains some special function for polynomial and rational matrices which are needed especially in discrete linear controller analysis and design. Being implemented as an interpreted programming language MAX offers considerable freedom and flexibility for the user to create new algorithms.

MAX is a very specialized program which has been used only for research purposes. However it is our belief that very high level programming languages will be used in the implementation of future CAD packages for control system design. As the computers dedicated for symbolic programming (e.g the newly introduced LISP machines) become cheaper the best way to build a system like MAX is to use a general purpose program for symbolic mathematics - MACSYMA or SMP for example - as a building tool. The same trend will probably affect the design and implementation of all future CAD packages for simulation and control. They will be built using high level simulation languages combined with extensive use of high level graphics functions. This will also free the CAD system designers to concentrate more on control engineering aspects instead of low level details of programming and what is even more important this will result in flexible CAD systems which are easily expanded by the users.

Acknowledgement - This work has been supported by the Academy of Finland.

REFERENCES

Johnson, S.C. (1978). Yacc: yet another compiler-compiler. Program report, Bell Laboratories, New Jersey.
Kailath, T. (1980). Linear Systems. Prentice-Hall, New York.
Kucera, V. (1979). Discrete Linear Control: The Polynomial Equation Approach. Wihey, New York.
Pernebo, L. (1980). Numerical algorithms for polynomial matrix systems. Report TFRT-7205/1-048. Department of Automatic Control, Lund Institute of Technology, Lund Sweden.
Mathlab Group (1983). A brief overview of MACSYMA. MIT, Cambridge, Massachusetts.
Wolowich, W.A. (1983). Dead-beat error control of discrete multivariable systems. Int. J. Control, 37, 567.
Wolowich, W.A. and P.J. Antsaklis (1984). The canonical diophantine equations with applications. SIAM J. Control Optimiz, 22, 777.

TABLE 1 Allowable datatypes in addition (subtraction) operation.

±	s	p	pm	r	rm
s	s	p	-	r	-
p	p	p	-	r	-
pm	-	-	pm	-	rm
r	r	r	-	r	-
rm	-	-	rm	-	rm

S = F1 ± F2

TABLE 2 Allowable datatypes in multiplication operation.

*	s	p	pm	r	rm
s	s	p	pm	r	rm
p	p	p	pm	r	rm
pm	pm	pm	pm†	rm	rm†
r	r	r	pm	r	rm
rm	rm	rm	rm†	rm	rm†

S = F1 * F2

† matrices must be of proper dimensions

TABLE 3 Allowable datatypes in division operation.

/	s	p	pm	r	rm
s	s	p	-	r	-
p	r	r	-	r	-
pm	-	-	-	-	-
r	r	r	-	r	-
rm	-	-	-	-	-

S = F1 / F2

THE FREQUENCY DOMAIN DESIGN OF APPROXIMATELY DECOUPLING AND NORMALIZING COMPENSATORS FOR MULTIVARIABLE SYSTEM

Wu Zhi-Ming and Hsu Hsiao-Ming

Department of Electrical Engineering and Computer Science, Shanghai Jiao Tong University, Shanghai 200030, People's Republic of China

Abstract The method presented here illustrates how to specify a set of diagonal elements $q_{ii}(s)$ $i=1,2,\cdots m$ of the open loop transfer function matrix (TFM) Q(s) to obtain such a precompensator K(s) that satisfies the requirements of stability, decoupling, robustness and dynamic behavior of output. This method is suitable for computer-aided design (CAD) since the whole design procedure are accomplished by computer except specifying $q_{ii}(s)$. A 4x4 examples are included.

In recent years, the design techniques for multivariable systems have been well developed and many efficient methods appeared one after other. Each of them may be suitable to different cases. Even in today, they are still interesting topics to many researchers. This paper introduce the ideas of approximate decoupling (Rosenbrock,1974) and pesudo-diagonaling (Hawkins,1972) into RFN method, then we give a new approach of design Approximetely Decoupling and Normalizing (ADN) compensators. Compare with the previous methods, less number in the trial and error method is required. Furthermore, the compensated system is decoupled and a better dynamic behavior is achieved.

We list symbols and notations which used in this paper as following

R,C	: field of read and complex number
$R^{m\times n}$, $C^{m\times n}$	: set of mxn matrices with elements in R and C respectively
R(s)	: field of rational function in s with coefficients in R
$R(s)^{m\times n}$	: set of mxn matrices with elements in R(s)
A^H	: conjugate transpose of A
$\bar{\sigma}(A)$	: maximum singular value of A
$\underline{\sigma}(A)$	: minimum singular value of A
$\Vert A\Vert$	: Frobenius norm of A, [tr(A A)]
$\Vert A\Vert_2$	: spectral norm of A, (A)
MS(A)	: measure of skewness of A
#Z[f(s),C]	: number of zeros (multiplicities counted) of f(s) in the right half plane (RHP), f(s) R(s).
#E[f(s),a]	: number of enclosement (positive for anticlockwise) of closed contour which maps f(s) when s goes alone the Nyquist D-curve around the point of (a,j0.0).

(1) Preliminary

Theorem 1.1 (from Stewart, 1973)

Let $A\in C^{m\times n}$. Then there are unitrary matrices U and V such that

$$V^H AU \doteq \begin{bmatrix}\Sigma & 0\\ 0 & 0\end{bmatrix},\quad \Sigma = \mathrm{diag}[\sigma_1, \sigma_2 \cdots\cdots, \sigma_r]$$

where $\sigma_1 > \sigma_2 > \cdots > \sigma_r > 0$

Definition 1.1 (see Wilkinson, 1965)

If $Q\in C^{m\times n}$ commute with its conjugate transpose, i.e. $Q\cdot Q^H = Q^H Q$, then Q is called a normal matrix.

Theorem 1.2 (from Wilkinson, 1965)

$Q\in C^{m\times n}$ is normal iff there is a unitrary matrix W such that Q has the eigenvalue decomposition $Q = W\Lambda W$, where $\Lambda = \mathrm{diag}[\lambda_1, \lambda_2 \cdots\cdots \lambda_m]$

Definition 1.2 (normal system)

If the TFM $G(s)\in R(s)^{m\times n}$ and the behavior of G(s) is normal at every working frequency in $S = (s_1, s_2, \cdots\cdots, s_N)$, then the system is said to be a normal system on S.

From the viewpoint of design, normal systems have the following advantages

(i) The principle gain and principle phase of G(s) are very near to the magnitude and phase of characteristic gain of G(s)

(ii) The characteristic loci of a normal systems are insensitive to the disturbances.

Theorem 1.3

Let the normal matrix of a given system be $Q\in C$ and its disturbed form be $\tilde{Q} = Q(I + \Delta)$. Suppose λ_i is the ith eigenvalue of A and λ is any eigenvalue of $\tilde{Q}$, then

$$\min_i|\lambda - \lambda_i| < \Vert Q\Vert_2 (\Vert\Delta\Vert_2 + MS(Q)) \qquad (1\text{-}1)$$

Proof: See Appendix A

When Q is a normal matrix, MS(Q)=0, then (1-1) deforms as $\min_i|\lambda - \lambda_i| < \Vert Q\Vert_2 \Vert\Delta\Vert_2$.

When Q is normal approximately, MS(Q) is so small that (1-1) may keeps the spectral insensitivity of matrix Q.

(2) Decoupling of Normal Matrix

Until now, decoupling control is still required to make each component of output Y(t) satisfies the specified transient behavior indepently and accurately. So we come to discuss the condition of decoupling a normal matrices.

To design a compensator K(s) that Q(s)= G(s)K(s) is fully decoupled would usually leads to the difficulties such as K(s) is physical unrealizable or its structure is too complicated to form. Thus we can only obtain the approximately decoupled systems in some senses.

In many previous design methods, diagonal dominant (DD) systems were considered as decoupled systems approximately (see Rosenbrock,1974). It is, however, lack of accuracy in some cases (especially when number of input or output < 3). For example:

$$A = \begin{bmatrix}1.0 & 0.99\\ 0.99 & 1.0\end{bmatrix}$$

satisfies the definition of DD, but its coupling is very strong clearly. So it is necessary to give a more reasonable definition for approximate decoupling.

Definition 2.1 (α-DD):

Let $A \in C^{m \times n}$, if $|a_{ii}| > sa_{ij} = \alpha \sum_{j \neq i}^{m} |a_{ij}|$ (or $|a_{jj}| > \alpha \sum_{i \neq j}^{m} |a_{ij}|$)

then ith row of A (or jth column of A) is α-DD. If all the rows (or columns) of A are α-DD, then A is α-row (or column) DD. When A is both α-row DD and α-column DD, then A is α-whole DD, where $0 < \alpha < \infty$. Generally speaking, should be selected according to requirement of decoupling and dimension of the matrix. Since $\eta = (m-1)|a_{ii}|/saij$ gives the average ratio of diagonal element to off-diagonal one, we can write α as a fraction $\alpha = \eta/(m-1)$, where η reflect requirement of decoupling and m is the dimension of matrix. For instance, when $\eta=5$, m=3, $\alpha=5/(3-1)=2.5$

Definition 2.2 (α-approximate decoupling)

If $Q(s) \in R(s)^{m \times n}$ is an open-loop TFM with unity feedback and Q(s) is α-whole DD on the set of working frequencies ($s_1, s_2, \ldots \ldots s_N$), then the closeloop system is said to be α-approximately decoupled, and α is called dominant margin. Considering the situations of $\alpha \to \infty$, which correspond to fully decoupling and $\alpha = 1.0$ to approximate decoupling in the sense of conventional DD. When Q is a normal matrix and its eigenvectors are known, the condition that Q is α-DD is given out by followings.

Theorem 2.1

Let the normal matrix $Q = Y \Lambda Y^H \in C^{m \times m}$, where Y is a unitrary matrix, rank(Q) = r, and $\Lambda = \mathrm{diag}(\lambda_1, \lambda_2, \ldots\ldots, \lambda_r, 0, 0, \cdots 0)$.

Suppose

$$\det \begin{bmatrix} |y_{11}|^2 & \cdots & |y_{1r}|^2 \\ \vdots & & \vdots \\ |y_{r1}|^2 & \cdots & |y_{rr}|^2 \end{bmatrix} \neq 0$$

and

$$X=[x_{ij}]_{r \times r} = \begin{bmatrix} |y_{11}|^2 & \cdots & |y_{1r}|^2 \\ \vdots & & \vdots \\ |y_{r1}|^2 & \cdots & |y_{rr}|^2 \end{bmatrix}$$

exists, assume

$$qa=(\sum_{i=1}^{r} q_{ii})/r, \quad sml= \sum_{K=1}^{r} y_{iK}\, \bar{y}_{jK}\, x_{Kl},$$

$$smj= \sum_{l=1}^{r} sml, \quad SMi1= \sum_{j \neq i}^{m} \sum_{l=1}^{r} sml,$$

$$SMi2 = \sum_{j \neq i}^{m} \sum_{l=1}^{r} sml$$

where $\bar{y}_{jK}$ is the complex conjugate of y_{jK}, if

$$\max_{i \in r} \left| \frac{q_{ii} - qa}{qa} \right| < \min_{i \in r} \frac{1 - \alpha\, SMi1}{1 + \alpha\, SMi2}$$

then the upper r rows and left r columns of Q are α-DD.

Proof: see Appendix B.

Corollary 2.1

Let all assumptions in Theorem 2.1 be same but rank(Q) = m, and if

$$\max_{i \in m} \left| \frac{q_{ii} - qa}{qa} \right| < \frac{1}{1 + \alpha \max_i (SMi2)}$$

then Q is α-whole DD.

Proof: see Appendix B.

These two relations of approximate decoupling are concerned with diagodnal elements of Q(s). When what we know are eigenvalues of Q(s), we may get the similar results in the same way.

Theorem 2.2

Let the normal matrix $Q = Y \Lambda Y^H \in C$, rank(Q)=r,

if $\lambda a = (\sum_{i=1}^{r} \lambda_i)/r$, $Ysi = \sum_{K=1}^{r} |Y_{iK}|^2$,

$$Yi^*_1 = \sum_{j \neq i}^{m} \left| \sum_{K=1}^{r} y_{iK}\, \bar{y}_{jK} \right|, \quad Yi^*_2 = \sum_{j \neq i}^{m} \sum_{K=1}^{r} |y_{iK}\, \bar{y}_{jK}|$$

and

$$\max_{i \in r} \left| \frac{\lambda_i - \lambda a}{\lambda a} \right| < \min_{i \in m} \frac{Ysi - \alpha Yi^*_1}{Ysi + \alpha Yi^*_2}$$

then Q is α-whole DD.

Proof: omitted.

Corollary 2.2

Let all assumption in theorem 2.2 be the same, but rank(Q) = m. If

$$\max_{i \in r} \left| \frac{\lambda_i - \lambda a}{\lambda a} \right| < \frac{1}{1 + \alpha \max_{i \in m} (Yi^*_2)}$$

then Q is α-whole DD.

Proof: omitted.

(3) The Analysis of Approximately Decoupled and normalized system

1. Stability.

When the openloop TFM Q(s) is normal, stability of the closed-loop system can be determined from the diagonal elements of Q(s).

Theorem 3.1:

Generalized Nyquist Criterion when Q(s) normal

Let Q(s) = G(s)K(s), where $Q(s) \in R(s)^{m \times m}$, $G(s) \in R(s)^{m \times m}$ and $K(s) \in R(s)^{m \times m}$ are TFM's of open-loop system, plant and compensator respectively. Suppose clcp(s) and olcp(s) are characteristic polynomials of closed- and open-loop system, if Q(s) is normal at any s on the Nyquist D-contour and $\#Z[clcp(s), C_+]=p$, then closed-loop system with unity feedback is stable iff

$$\sum \#E[q_{ii}(s), -1] = p \qquad (3\text{-}1)$$

Proof:

For $\dfrac{clcp(s)}{olcp(s)} = \dfrac{\det F(s)}{\det F(\infty)}$, where F(s)=I+Q(s)

then $\#Z[clcp(s), C_+] - p = -\#E[\det F(s), 0]$

So closed-loop system is stable iff

$$\#E[\det F(s), 0] = p \qquad (3\text{-}2)$$

If Q(s) is normal, then $Q(s)=W(s)\Lambda(s)W^H(s)$, from theorem 1.2, and F(s)=I+Q(s) is still a normal matrix.

Nwokah (1983) has proved that

$\sum_{i=1}^{m} \#E[(1+\lambda_i(s)), 0] = \sum_{i=1}^{m} \#E[f_{ii}(s), 0]$, so we may have $\#E[\det F(s), 0] = \#E[\prod_{i=1}^{m} (1+\lambda_i(s)), 0] =$

$$\sum_{i=1}^{m} \#E[(1+\lambda_i(s)), 0] = \sum_{i=1}^{m} \#E[(1+q_{ii}(s)), 0] = \sum_{i=1}^{m} \#E[q_{ii}(s), -1] \qquad (3\text{-}3)$$

Combining (3-2) and (3-3), results in (3-1).

In designing, specify Q(s) first and then find out the K(s). Thus we always choose p as the number of Smith-MacMillan zeros of Q(s) in RHP. We keep K(s) has no RHP poles and don't allow K(s) to cancel the RHP poles of G(s).

2. Transient Behavior of Output

Since the system is approximately decoupled, the transfer function between $y_i(s)$ and $u_i(s)$ is discribed by $q_{ii}(s)$ and the influence of $u_j(s)$ to $y_i(s)$ is negligible. Thus, we may use classical approaches to find the transient behaviors.

3. Robustness.

Following Doyle and Stein (1981), let nominal open-loop system be Q(s) and a real one contain uncertainty be $\tilde{Q}(s)$, losing no generality, we suppose that uncertainties appear in output, as $\tilde{Q}(s)=[I+L(s)]Q(s)$. If $\|L(s)\|_2 \leq lm(s)$, we may have

Theorem 3.2 (Robustness for Normal System)

Let $Q(s) \in R(s)^{m \times m}$ be a TFM of normal system and $Q(s)[I+Q(s)]^{-1}$ is stable, the closed-loop system $\tilde{Q}(s)[I+\tilde{Q}(s)]^{-1}$ preserves stability iff

$$\min_K |\lambda_K + 1/(1.0 - l\hat{m}^2)| > lm/(1.0 - l\hat{m}^2), \quad lm(s) < 1.0 \qquad (3\text{-}4)$$

$$\min \mathrm{Re}(\lambda_K) > -1.0/2.0, \quad lm(s) = 1.0 \qquad (3\text{-}5)$$

$$\max_K |\lambda_K - 1/(l\hat{m}^2 - 1.0)| < lm/(l\hat{m}^2 - 1.0), \quad lm(s) > 1.0 \qquad (3\text{-}6)$$

Proof: see Appendix C

If we know the value of lm(s), for instance, from the measurement introduced by Owens(1983), we may analyse the robustness of ADN system accurately. For a practical system, Doyle(1981) pointed out the

the robustness condition as: $l_m(s) \ll 1$ in low frequencies, $l_m(s)=1$ in middle frequencies and $l_m(s) \gg 1$ in high frequencies. Thus, by Theorem 3.2 following requirements to $[\lambda_1(s), \lambda_2(s), \ldots, \lambda_m(s)]$ are expected:

(a) loci of $\lambda_i(s)$'s depart from critical point (-1,j0) as far as possible in low frequencies.

(b) loci of $\lambda_i(s)$'s are kept to the right side of line Re(s)=-1/2 in middle frequencies.

(c) loci of $\lambda_i(s)$'s tend to original point at (0,j0) in high frequencies.

In fact, (3-5) is also an iff condition that preserves integrity of the system. It is easy to verify that when some of actuators or feedback devices are in failure, the system is just correspond to $l_m(s)=1$. Though (3-4) & (3-6) are conditions concerned with eigenvalues of Q(s), if we take notice of

$$q_{ii}(s) = \sum_{K=1}^{m} |Y_{iK}|^2 \cdot \lambda_K(s) \quad \text{and} \quad \sum_{K=1}^{m} |Y_{iK}|^2 = 1,$$

we may consider $q_{ii}(s)$ is a qeighted average of $[\lambda_1(s), \ldots, \lambda_m(s)]$'s, then the relation like (3-4) (3-5) and (3-6) still hold, when $\lambda_i(s)$ is replaced by $q_{ii}(s)$.

Furthermore, we may deduce robustness condition concerned with $q_{ii}(s)$'s in the range of low, middle and high frequencies.

Theorem 3.3

If all the assumptions in Theorem 3.2 be fixed, $[\lambda_1(s), \ldots\ldots \lambda_m(s)]$ satisfy Corollary 2.2 (α-DD), then $\widetilde{Q}(s)[I+\widetilde{Q}(s)]$ preserves stability if

$$|qs+1/(1-l_m^2)| - c|qs| > l_m/(1-l_m^2), \quad l_m(s)<1 \tag{3-7}$$

$$\mathrm{Re}(qs) - c|qs| > -1/2, \quad l_m(s)=1 \tag{3-8}$$

$$|qs-1/(l_m^2-1)| + c|qs| < l_m/(l_m^2-1), \quad l_m(s)>1 \tag{3-9}$$

where $qs=(\sum_{i=1}^{m} q_{ii}(s))/m$, $c=1/(1+\alpha Yi^*)$

Proof: the eigenvalue of Q(s) satisfy Corollary 2.2, then $\max_K |\lambda_K-\lambda a| < c|\lambda a|$.

$$\begin{bmatrix} q_{11} \\ \vdots \\ q_{mm} \end{bmatrix} = \begin{bmatrix} |Y_{11}|^2 & |Y_{12}|^2 & \cdots & |Y_{1m}|^2 \\ \vdots & \vdots & & \vdots \\ |Y_{m1}|^2 & |Y_{m2}|^2 & \cdots & |Y_{mm}|^2 \end{bmatrix} \begin{bmatrix} \lambda_1 \\ \vdots \\ \lambda_m \end{bmatrix}$$

Multiply both side by [1,1,...1]/m and note that Y is an unitrary matrix, we may have $qs = \lambda a$.

From (3-4), (3-5) and (3-6), we may get

$$\begin{aligned} \min_K |\lambda_K+1/(1-l_m^2)| &> \min_K [|\lambda a=1/(1-l_m^2)| - |\lambda_K - \lambda a|] \\ &= |\lambda a+1/(1-l_m^2)| - \max_K |\lambda_K-\lambda a| \\ &> |qs+1/(1-l_m^2)| - c|qs| \end{aligned} \tag{3-10}$$

$$\begin{aligned} \min_K \mathrm{Re}(\lambda_K) &= \mathrm{Re}(\lambda a)+\min_K \mathrm{Re}(\lambda_K-\lambda a) \\ &> \mathrm{Re}(\lambda a)-\max_K \mathrm{Re}(\lambda_K-\lambda a) > \mathrm{Re}(\lambda a) - \max_K |\lambda_K-\lambda a| > \mathrm{Re}(qs)-c|qs| \end{aligned} \tag{3-11}$$

$$\begin{aligned} \max_K |\lambda_K -1/(l_m^2-1)| &< \max_K [\lambda a-1/(l_m^2-1)| + |\lambda_K-\lambda a|] < |qs-1/(l_m^2-1)| + c|qs| \end{aligned} \tag{3-12}$$

In a similar manner, we can prove the necessary condition which preserve integrity for normal systems is

$$\mathrm{Re}(qs)-c|qs| > -1/2$$

(4) Design of ADN Systems

For the purpose of simplicity in real system, we restrict the compensator to a precompensator feedback to unity feedback. Then the closed-loop system can be depicted as follow.

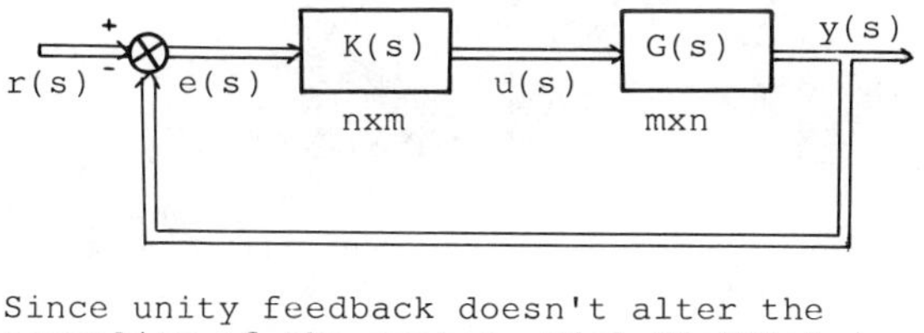

Since unity feedback doesn't alter the normality of the system, Q(s)=G(s)K(s) is also approximate normal. From the method of Reverse Frame Normalizing (REN) presented by Hung (1982), if the singular value decomposition (SVD) of G(s) is $G(s)=Y(s)\Sigma(s)U^H(s)$ and let $K(s)=U(s)\Omega(s)Y^H(s)$, then $Q(s)=G(s)K(s)=Y(s)\Lambda(s)Y^H(s)$ is normal, where $\Lambda(s)=\Sigma(s)\Omega(s)=\mathrm{diag}(\lambda_1(s), \lambda_2(s), \ldots, \lambda_r(s), 0, 0, \ldots 0)$ where r=rang(Q).

Generally, the design of compensator K(s) can be separated into two steps:

(i) given an openloop TFM Q(s) with satisfactory behavior.

(ii) Seeking the solution of K(s) which make the performance index J = f(G(s), Q(s), K(s)) as small as possible.

In RFN design method, Q(s) is determined by choosing ($\lambda_i(s)$, i=1,2,...r) since the eigenvectors of Q(s) are fixed by G(s). Unforturnately, it is difficult to choose approprite (s)'s to meet the performance specifications which are given in real time form as rise time, overshoot and settling time of the output y (s), i=1,2,...m. On the other hand, if an index of

$$J = \sum_{K=1}^{N} \| Q(jw_K) - G(jw_K)K(jw_K) \|^2 \quad \text{is adopted,}$$

then a problem of solving nonlinear optimisation is required when the commen denominator of K(s) not being affirmed. Furthermore, since the parameter number in K(s) is considerable, the amount of computation is quite large.

We improve the RFN method in the following

(i) Taking diagonal elements of Q(s), rather than its eigenvalues $\lambda_i(s)$'s.

(ii) Taking

$$J = \sum_{K=1}^{N} \beta_K^2 \frac{\|G(jw_K)\|^2 \, \|K(jw_K)-K(jw_K)\|^2}{\|Q(jw_K)\|^2}$$

as performance index, where K(s) satisfies that Q(s)=G(s)K(s) is approximately normal.

The main advantages of this method lie in:

(i) Applying the conclusions in section II & III (as (2-2),(3-1),(3-4),(3-5),(3-6),(3-12) etc) we consider stability, robustness, decoupling & transient behaviors from q 's simultaneously.

(ii) Performance index (4-1) changes the problem of finding a polynomial matrix K9s) into mxn separated rational polynomial of $k_{ij}(s)$. Such an improvement not only save computational time but also arise accuracy.

Based on the above method, a CAD package of ADN compensator calculation with Fortran 77 are provided. The programs are implemented in Buroughs 1955 computer in an interactive way. We are sorry for there is no graph terminal can be used. Though the results of simulation are inspiring, it is from the lineprinter. Only data can be obtained from it.

In some cases, the coupling between elements in G(s) may be strong that the right side of (2-1) would vanish. Then, it is hard to decide $q_{ii}(s)$, i=1,2,...r independently except $q_{11}(s)=q_{22}(s)\ldots=q_{rr}(s)$ is taken. At that time, we can use a real matrix K_0 for pre-decoupling of G(s). K_0 could be obtained by pseudo-diagonal algorithm from Hawkins (1972) or alignment algorithm from MacFarlane (1977). The function of K_0 is to make $G(s)K_0=I$, where s is either a single value or a set of weighted frequencies. Example

$$G(s) = \begin{bmatrix} \frac{6}{s+3} & \frac{4}{s+2} & \frac{-2}{s+1} & \frac{7}{s+2} \\ \frac{2}{s+3} & \frac{5}{s+1} & \frac{3}{s+3} & \frac{4}{s+2} \\ \frac{3}{s+1} & \frac{7}{s+3} & \frac{8}{s+2} & \frac{5}{s+4} \\ \frac{2}{s+2} & \frac{4}{s+1} & \frac{1}{s+4} & \frac{1}{s+3} \end{bmatrix}$$

The requirements of compensated system are stable, robust and decoupling. If the transient response of system are specified as follow:

Y (t): $\sigma_{p1} <$ 10%, t < 1 sec. ess = 0.0
y (t): $\sigma_{p2} <$ 8%, t < 1 sec. ess = 0.0
y (t): $\sigma_{p3} <$ 5%, t < 0.8 sec. ess = 0.01
y (t): $\sigma_{p4} <$ 3%, t < 0.8 sec. ess = 0.05

where σ_{pi}- overshoot, t -settling time, ess -steady state error, then we may choose

$$q_{11}(s) = \frac{48}{s(s+8.5)} \qquad q_{22}(s) = \frac{50}{s(s+9)}$$

$$q_{33}(s) = \frac{53}{s(s+10)} \qquad q_{44}(s) = \frac{55}{s(s+11)}$$

take s logarithmically, s= 0,j0.1,j0.301, ...,j10 . Suppose the approximate decoupling coefficient = 1.4, checking all kthe conditions presented in section II & III, and q (s)s' are thought of satisfactory, we can fix K(s) as

$$K(s) = \frac{1}{s}\begin{bmatrix} \frac{4.26s+8.44}{s+6.56} & \frac{1.89s+3.52}{s+3.47} & \frac{1.32s+3.71}{s+12.5} & \frac{9.61s+17.7}{s+10.0} \\ \frac{0.29s+0.06}{s+0.85} & \frac{4.74s+5.13}{s+8.60} & \frac{-1.2s+0.72}{s+8.19} & \frac{-3.9s-4.04}{s+8.02} \\ \frac{0.06s-1.44}{s+2.47} & \frac{-0.7s-2.60}{s+3.44} & \frac{8.13s+14.8}{s+10.4} & \frac{-16.5s-30}{s+15.6} \\ \frac{-2.6s-4.20}{s+7.63} & \frac{5.21s+12.6}{s+9.50} & \frac{-4.3s-9.74}{s+14.4} & \frac{11.5s+23.9}{s+14.7} \end{bmatrix}$$

The simulation result of closed-loop system shows that the design is very successful.

REFERENCES

Doyle, J.C. & G.Stein(1981) Multivariable feed-back design concept for a modern synthesis IEEE Trans. Vol AC-26 pp4-16

Hawkins,D.J.(1972) Pseudo-diagonalization and the inverse Nyquist array method Proc.IEE Vol 119 pp337-342

Hung,Y.S.& A.G.J.MacFarlane(1982) Multivariable Feedback: A Quasi Classical Approach. Lecture Notes in Control and Information Science, Springer-Verlag, Berlin.

Klema, V.C.& A.J.Laub(1980) The singular value decomposition and some applications IEEE Trans. Vol AC-25 pp164-176

MacFarlane,A.G.J.& I.Postlethwaite (1977) Characteristic frequency and characteristic gain function Inter. J. of Control Vol 26 pp265-278

MacFarlane,A.G.J. & I.Postlethwaite (1977) The generalized Nyquist stability criterion and multivariable root loci Inter. J. of Control Vol 25 pp81-127

Macfarlane,A.G.J. & D.F.A.Scott-Janes(1979) Vector gains Inter.J. of Control Vol 29 pp65-91

Nwokah,O.D.I.(1983) Stability of linear multivariable systems Inter.J. of Control Vol 37 pp623-629

Owens,D.H. & A.Chotai (1983) Robust controller design for linear dynamic systems using approximate models Proc.IEE part D,Vol 130 May p45

Postlethwaite,I.,J.M.Edmunds & A.G.J.MacFarlane (1981) Principle gains and principle phases in the analysis of multivariable feedback systems. IEEE Trans. Vol AC-26 pp32-46

Rosenbrock,H.H.(1974) Computer Aided Control System Design Academic Press, London

Stewert,G.W.(1973) Introduction to Matrix Computation Academic Press, New York

Wilkinson,J.K.(1965) The Algebraic Eigenvalue Problem. Oxford, Clarendon Press

Appendix

A. Proof of Theorem 1.3

Let $\tilde{Q} = Q(I+\Delta) \in C^{n\times n}$. It has been shown in Wilkinson(1965) that if Schur triangular decomposition (STD) of Q is $Q=U(D+T)V^H$, then

$$(1+ab+ \quad +a^{n-1}b^{n-1})a\|Q\Delta\|_2 > 1 \qquad \text{(A-1)}$$

where U and V are both unitrary matrices. $D = \text{diag}[\lambda_1, \lambda_2, \ldots\ldots, \lambda_n]$, T is a strictly upper triangular matrix, and
$a = 1/[\min_i|\lambda_i-\lambda|]$, $b = \|T\|_2$

Since the left side of (A-1) is a monotonic function, it is still true when $a^*=\inf[a]$.

$$(1+a^*b+\ldots+(a^*)^{n-1}b^{n-1})a^*\|Q\Delta\|_2 = 1 \qquad \text{(A-2)}$$

Denote $x=\sup_\lambda \min_i|\lambda_i-\lambda|$, then (A-2) can be in $f(x)= [(1+b/x+\ldots+b^{n-1}/x^{n-1})\|Q\Delta\|_2]/x-1 =0$ (A-3)

It is easy to versify that f(s) is a monotonic decrease function and $f(\|Q\Delta\|_2)>0$, $f(\|Q\Delta\|_2+b)<0$ hence the unique real solution of (A-3) must lie between $\|Q\Delta\|_2 < x < b+\|Q\Delta\|_2$. Then, we may have

$$\min_i|\lambda_i-\lambda| < \sup_\lambda[\min_i|\lambda_i-\lambda|] < \|T\|_2 + \|Q\Delta\|_2 < \|Q\|_2\cdot(\|\Delta\| + MS(Q))$$

where $MS(Q)= \|T\|_2 / \|Q\|_2$

B. Proof of Theorem 2.1 and Corollary 2.1

(i) Proof of Theorem 2.1

$$|q_{ii}|=qa+(q_{ii}-qa) > |qa| - |q_{ii}-qa| > |qa| - \max_i|q_{ii}-qa| \quad i=1,2,\ldots\ldots r. \qquad \text{(A-4)}$$

for

$$\begin{bmatrix} q_{11} \\ \vdots \\ q_{rr} \end{bmatrix} = \begin{bmatrix} |y_{11}|^2 & \cdots & |y_{1r}|^2 \\ \vdots & & \vdots \\ |y_{r1}|^2 & \cdots & |y_{rr}|^2 \end{bmatrix} \begin{bmatrix} \lambda_1 \\ \vdots \\ \lambda_r \end{bmatrix} \qquad \begin{bmatrix} \lambda_1 \\ \vdots \\ \lambda_r \end{bmatrix} = \begin{bmatrix} x_{11} & \cdots & x_{1r} \\ \vdots & & \vdots \\ x_{r1} & \cdots & x_{rr} \end{bmatrix} \begin{bmatrix} q_{11} \\ \vdots \\ q_{rr} \end{bmatrix}$$

or $\lambda_i = \sum_{k=1}^{r} x_{ik} q_{kk}$, for i=1,2,......r.

$$\begin{aligned} \sum_{j\neq i}^{m}|q_{ij}| &= \sum_{j\neq i}^{m}\Big|\sum_{k=1}^{r} y_{ik}\bar{y}_{jk}\Big(\sum_{l=1}^{r} x_{kl}q_{ll}\Big)\Big| \\ &= \sum_{j\neq i}^{m}\Big|\sum_{l=1}^{r}\Big(\sum_{k=1}^{r} y_{ik}\bar{y}_{jk}x_{kl}\Big)(q_{ll}-qa) + qa\sum_{l=1}^{r}\sum_{k=1}^{r} y_{ik}\bar{y}_{jk}x_{kl}\Big| \\ &\leqslant \max_l|q_{ll}-qa|\,SMi1 + |qa|\cdot smj \qquad \text{(A-5)} \end{aligned}$$

where SMi and smj are given in the paper.

Let

$$|qa| - \max_i|q_{ii}-qa| > \alpha(\max_i|q_{ii}-qa|SMi2 + |qa|SMi1)$$

$$\max_{i\in r}\left|\frac{q_{ii}-qa}{qa}\right| < \frac{1-\alpha SMi1}{1+\alpha SMi2}$$

$$\max_{i\in r}\left|\frac{q_{ii}-qa}{qa}\right| < \min_{i\in r}\frac{1-\alpha SMi1}{1+\alpha SMi2}$$

Combining (A-4)(A-5)(A-7), according to definition 2.1, we know that the upper r rows of Q are α-DD. Since α-DD of columns of Q can be studied by α-DD of rows of Q, then the left r columns of Q are α-DD if

$$\max_{i\in r}\left|\frac{q_{ii}-qa}{qa}\right| < \min_{i\ r}\frac{1-\alpha SMi1}{1+\alpha SMi2}$$

Obviously, (A-7) & (A-8) are same because magnitudes of two complex conjugate values are equal. Thus, Theorem 2.1 is proved.

(2) Proof of Corollary 2.1

Since

$$\begin{bmatrix} x_{11} & \cdots & x_{1m} \\ \vdots & & \vdots \\ x_{m1} & \cdots & x_{mm} \end{bmatrix} \begin{bmatrix} |y_{11}|^2 & \cdots & |y_{1m}|^2 \\ \vdots & & \vdots \\ |y_{m1}|^2 & \cdots & |y_{mm}|^2 \end{bmatrix} = I_m \qquad \text{(A-9)}$$

if we post multiply both sides of (A-9) by vector [1,1,......1] and note that Y is an unitrary matrix then we can get

$\sum_{k=1}^{m} x_{ik}=1, \quad i=1,2,\ldots m.$

when $i\neq j$, we may have

$$\left|\sum_{l=1}^{m}\sum_{k=1}^{m} y_{ik}\overline{y}_{jk}x_{kl}\right| = \left|\sum_{k=1}^{m} y_{ik}\overline{y}_{jk}\left(\sum_{l=1}^{m} x_{kl}\right)\right| = \left|\sum_{k=1}^{m} y_{ik}\overline{y}_{jk}\right| = 0 \quad \text{(A-10)}$$

Substituting r=m and (A-10) into (2-1), the corollary is proved immediately.

C. Proof of Theorem 3.2

Let $\widetilde{G}(s)=[I+L(s)]G(s)$, $\|L(s)\|_2 \leq lm(s)$. Suppose K(s) is fixed and $G(s)K(s)[I+G(s)K(s)]$ is stable, Wilkinson (1965) has proved: $G(s)K(s)[I+G(s)K(s)]$ preserves stability iff

$$[GK(I+GK)^{-1}]<1/lm(s) \quad \text{(A-11)}$$

For $Q(s)=G(s)K(s)$, we have

$\widetilde{Q}(s)=\widetilde{G}(s)K(s)=[I+L(s)]G(s)K(s)$
$=[I+L(s)]Q(s)$

then (A-11) becomes

$\overline{\sigma}[Q(I+Q)^{-1}]<1/lm(s)$

When Q(s) is a normal matrix,

$\overline{\sigma}[Q(I+Q)^{-1}]=\max_i|\lambda_i/(1+\lambda_i)| =1/\min_i|1+1/\lambda_i|$

Hence,

$$\text{(A-12)}\Leftrightarrow \min_i|1+1/\lambda_i| >lm(s)\Leftrightarrow|1+1/\lambda_i|> lm. \quad \text{(A-13)}$$

Substituting $\lambda_i=a+jb$ into (A-13), we get

$$(a^2+b^2)(1-lm^2)+2a+1>0, i=1,2,\ldots m \quad \text{(A-14)}$$

Considering all the cases of: 1) $lm(s)<1$, 2) $lm(s)=1$, 3) $lm(s)>1$, we can prove that (A-14) is equivalent to (3-4)~(3-6).

A DATABASE FOR PROCESS AUTOMATION DESIGN

R. Heinonen and J. Haarla

Technical Research Centre of Finland, Electrical Engineering Laboratory, Otakaari 5 I, 02150 Espoo 15, Finland

Abstract. Process and automation system design sets great demands on the design database. A relational database system is a general and quite natural way to represent design data. However, a relational system is not necessarily an optimal way because a large design system contains information which has complex structures and dependancies and which is modified continuously during the design process. An experimental database system for process automation design has been implemented using VAX11/750/VMS. The programming language has been Pascal. The database system has, instead of relations, six basic entities: objects, object types, connections, connection types, procedures and requirements, and certain relationships between these. The system has been applied to some power plant processes. Descriptions of processes, subprocesses, components and control algorithms have been stored into the database. Also various functional and other requirements on the subsystems of the plant are stored into the database so that, for example, following kinds of questions can be answered: "which requirements should component xx satisfy", "to which subsystems of the plant is requirement R123 connected", "in which plant functions is component C234 used". The purpose of the developed system is twofold. First, the plant and automation system descriptions are in such a form that the requirements can be connected to the right places in the descriptions. Second, the system is a framework by which various computerized design tools can be integrated and the information transfer between then managed.

Keywords. Automation design; computer aided design; modelling; database.

INTRODUCTION

Computer aided design usually deals with the lowest level descriptions of industrial processes and automation systems (diagrams, algorithms etc.). In large design projects it is, however, quite important to manage also the higher level descriptions of the plant (requirements, functional descriptions) and their connections to the implementation level (Heinonen, 1984).

To realize some ideas of the Nordic co-operation project NKA/LIT (Human reliability..., 1981; Heinonen, 1983) a quite general database management system has been implemented in the form of a System Modelling Language (SML). The SML is based on the three level model (Ross, 1977; Haase, 1982; Heinonen, 1983) of system descriptions:

- requirement specification (why-level)
- functional specification (what-level)
- implementation specification (how-level)

Each level represents a certain view into the entity descriptions. There are six separate entity types in the SML database. Every entity type has its own attribute types. Different attribute types of the SML-entity answer to different questions of the three level model. Figure 1 visualizes this.

There are many requirements set on the modelling language (Heinonen, 1984). The main requirement is that the formal description has to be able to help in handling the three level model. The following requirements were also considered in the implementation (Eberlain, 1982):

- Classification of entities by defining different entity types.
- A possibility to make compositions. Composition is the reference of constituents to a composite entity.
- A possibility to make abstractions. Abstraction is the reference of 'concrete' to 'abstract' responsive that of individuals to type.
- Attribute definitions must be carried out by assignment of predicates to basic value sets.
- Semantic integrity rules and operational specifications must complete the entity type definitions.

FUNCTIONAL DESCRIPTION

There are six different types of entities in the SML database:

- Object type (name, subtype of, purpose, requirement, parameter, port, algorithm, components, structure).
- Object (name, type, purpose, requirement, parameter, port, algorithm, components, structure).
- Requirement (name, description, parameter, satisfied by).
- Procedure (name, purpose, requirement, attached to, algorithm).
- Connection type (name, purpose, parameter, from, to).
- Connection (name, type, purpose, requirement, parameter, from, to).

These entities can be connected to each other with many different relations. The SML has several properties that are useful from the point of view of computer aided design.

Special relationships between entities.

1. Entity types

   Types define classes of entities which have same kind of properties. The entity inherits the parameters from its type. The values for the parameters are given separately to each entity. If the type of the entity has no parameters defined, then they can separately be defined to each entity of that type. In addition to the parameters inherited from the type of the entity the user can add separate parameters to each entity of the same type.

2. Connections.

   The connections of an entity to other entities are defined with ports that are attached to the entity. The definition of the connection is made by defining an entity of type connection to the database. That connection can then be attached between two other entities. Also the ports are inherited from the type of the entity in the same way as parameters.

3. Components.

   For each entity can be defined components. The entity is composed of them. This definition is made to the 'composition'-field of the entity. The components are separate entities that exist in the database. The existence of them is checked when the component is created. For example a display of a control systems can be built up from the displays of its' components.

Requirements.

Requirements are special kinds of entities which can be connected to other entities. To each entity can be attached requirements that the entity must satisfy. The definition of the requirement can happen in two ways:

1. The requirement can be attached directly to the entity. This is possible when the entity is created or when it is modified. In both cases the program asks the user the name of the requirement. If the addition is wanted, then the name is given, otherwise only the carriage-return is pressed.

2. The requirement can be attached to the entity by attaching the name of the entity to the 'satisfied-by'-field of the requirement.

DATABASE OPERATIONS

In the following it is described how the entities and the relations between them can be manipulated.

Object-type.

1. Adding.
   - If the type is already defined in the database it can't be recreated.
   - All references to the requirements are checked. If the requirement created doesn't exist in the database, it is added there. If it does exist, the satisfied-by field of it is updated.
   - The possible existence of higher types is checked.
   - The object-type created inherits requirements, parameters, ports, components and structure from its higher type if the higher type exists.

2. Deleting
   - The type is marked "deleted" but it is not deleted from the database before all references to it are deleted.
   - The requirements are updated.
   - As a default all the subtypes and all the objects of the type are deleted, but when necessary they can also be remained in the database.

3. Modifying
   - The name of the object-type can't be modified.
   - The purpose-, parameter-, port- and algorithm-fields can be modified if they are not inherited from the higher type.
   - The structure- and components-fields can't be modified if there are objects in the database that are of this type.

Object.

1. Adding.
   - If the object is already defined in the database the creation can't be done.
   - The type of the object must be defined in the database.
   - Parameters, ports, algorithms, components and structures are inherited from the type of the object if they are defined there, otherwise they can be defined separately, in which case they are attached only to the object now created.
   - The requirements are handled in the same way as in the case of object-types.

2. Deleting
   - The requirements are updated.
   - The connections are updated.

3. Modifying
   - The name of the object can't be modified.
   - The requirements inherited from the type of the object can't be modified.
   - The parameter- and port-names inherited from the type of the object can't be modified, the values them can.
   - The algorithm inherited from the type of the object can't be modified.

Requirement.

1. Adding
   - If the requirement is defined in the database it can't be added there as a separate entity.
   - The requirement can be added in two ways:
     - As a separate entity
     - In connection to another entity, when the other entity is added or modified.
   - If the entity is added as a separate entity, to those entities, whose name is in the satisfied-by field of the requirement, is attached the name of the requirement.

2. Deleting
   - The requirement can be deleted only as a separate entity.
   - If the requirement is deleted from another entity then only the references to it from other entities are deleted.
   - If the requirement is deleted as a separate entity then all the references to it from other entities are deleted.

3. Modifying
   - The name of the requirement can't be modified.
   - The description of the requirement can be changed.
   - The "satisfied-by"-field of the requirement can be modified but then must the references to other entities be checked.

Procedure.

The same procedure takes care of all the attachment types to one attribute. The attachment types are 'if-inserted', 'if-deleted' and 'if-needed'. The procedures can be attached the attributes of other entities in the following ways:

Object-type	name,purpose,parameters, ports,algorithm,structure
Object	same as object-type
Requirement	name,description
Procedure	-
Connection-type	name,purpose,parameters, from,to
Connection	same as connection-type

1. Adding
   - If the procedure is defined in the database then the addition can't be done.
   - The procedure doesn't have to be attached to any attribute when it is added. If it is attached to any attribute then the existence of the attribute are checked.

2. Deleting
   - The possible connections to other attributes are deleted.

3. Modifying
   - If the "attached-to"-field of the procedure is modified then the references to other attributes are checked.

Same kind of rules exist also for the connections and connection types.

USE OF THE SML

The SML system can be used to store the design information in the following ways:

- Process and automation system components can be defined:
  - Purpose of the components
  - Requirements which the component must satisfy
  - A set of attribute values
- Plant components can be connected to each others with various relationships.
- A requirement list can be created for the plant. Requirements can also have a set of attributes (type, verification method, status).
- Composition relation can also represent abstract relationships i.e. we can, for example, define a plant function as an object. The plant components that are necessary for that function are defined as the components of it.
- Connections can also be more abstract, for example, a connection could be a redundancy relationship between two plant components.
- Quite complex consistency rules for the database can be created by using attached procedures.
- One attribute of objects is ALGORITHM. This attribute has no meaning in the SML context, so it can be used quite freely for various purposes (for example to define control algorithms, to describe functions etc.).

To create a design database a set of object types, connection types and attached procedures must be defined. These form a framework for the design.

IMPLEMENTATION

The SML software was implemented with VAX11/750 computer with VMS operating system. The programming language was Pascal. The file manipulations were realized with the RMS software of the Digital Equipment Corporation. The user can manipulate the SML database either through the SML user interface or through application programs. The application program interface is realized by an entity manipulation subroutine package.

The Database Structure.

The database of the SML consists of many separate files so that every entity of the same entity type are stored in the same file. The files are keyed access files. In addition to these files the database consists of three text files, which are needed for the texts and parameters of the entities. This way the size of the database can be remained as small as possible, because the text files need only as much memory as the texts need. The text files are also made direct access, otherwise the speed of the database search would suffer too much, expecially if the database is large. In SML all the procedures which manipulate the database files are placed into the same module. All other modules use standard Pascal as much as possible. This way the transportation of the program is made easy. The structure of the database is visualized in fig. 2.

Interactive User Interface.

The users point of view to the SML is shown in fig. 3. There can be six separate entities simultaneously in the working area, each of them of different entity type. The entity existing in the working area can be listed in many different ways and with the help of it can also new database searches be made. After the database files are opened in the beginning of the user session, they are kept open until the end of the session. It is also possible to use command files and macro commands.

EXAMPLES

In the following some entities of the SML are listed:

1. Requirement:

```
r025 REQUIREMENT
  TEXT
    The collection tanks of subprocesses
    1, 2 and 3 are emptied through the
    lowest points. However, there must
    also be an outlet at the height of
    0.5 m from the bottom.
  PARAMETER
    type                  :  mandatory
    verifying_method      :  manual
    status                :  verified
  SATISFIED BY
    t11
    t16

  AUTHOR:  jh  26-NOV-1984
END      r025
```

As we can see the requirement is attached to tanks t11 and t16.

2. Object type:

```
tank  OBJECT-TYPE
  PURPOSE
    Tank definition
  PARAMETER
    name                  :
    type                  :
    volume                :
    location              :
    calculationdata       :

  AUTHOR:  jh  26-NOV-1984
END  tank
```

3. Object:

```
t11  OBJECT OF TYPE  tank
  PURPOSE
    Collects the systemdrainage
  REQUIREMENT
    r025:
      The collection tanks of subprocesses
      1, 2 and 3 are emptied through the
      lowest points. However, there must
      also be an outlet at the height of
      0.5 m from the bottom.
    r026:
      The outlet of the collectiontanks
      is selected manually (lowest or
      0.5m point).
  PARAMETER
    name                  : t11
    type                  : cylinder
    volume                : 225m3
    location              : level9
    calculationdata       : waterfilling/60C

  AUTHOR:  jh  26-NOV-1984
END  t11
```

The object has inherited the parameter names from the type. The requirement that was attached to this object can be seen also in the listing.

Example Session.

With the help of the SML it is possible to create, delete and modify entities and also to navigate in the database from an entity to other entities which are realted to it. All listings can be made so that the last updating date and the initials of the updater of every attribute can be seen. In the following SML session some of these features are shown. (All the comments are marked with (*- and *)-marks).

```
(* a key value to the object-type
   is created *)
. create-key object-type
THE OBJECT-TYPE KEY NAME ?  valve
(* the object-type valve is read
   from the database *)
. get o-t
(* it is listed *)
. list o-t

valve  OBJECT-TYPE
  PURPOSE
    Valve definition.
  PARAMETER
    name                 :
    type                 :
    loc.                 :

  AUTHOR:  jh  18-DEC-1984
END  valve
(* the object v113 is read from the
   database with a macro command.
   The command file get.com includes
   the commands create-key and get *)
.  get object v113
```

```
. li o

vl13  OBJECT OF TYPE  valve
  PURPOSE
    The filter fl1 outlet valve.
  PARAMETER
    name                   :  vl13
    type                   :  valve
    loc.                   :  level7

  AUTHOR:  jh  18-DEC-1984
END  vl13
(* the owners of the object vl13
   are listed *)
. list-owners obj
THE OWNERS OF THE OBJECT vl13 :
    fl1-connection : filter-connection
    fl1-function : filter-function
    parallel-connection : filter-connection
    serial-connection : filter-connection
    vl13-display : valve-display
(* the second owner of the object vl13
   is read from the database *)
. get-owner object
. get-owner object next
(* the second owner is listed so that
   the last updating dates and initials
   can be seen *)
. list object author

fl1-function  OBJECT OF TYPE  filter-function
  PURPOSE
    jh  18-DEC-1984
  PARAMETER
    name :  fl1-function      jh  18-DEC-1984
    type :  filter-function   jh  18-DEC-1984
  COMPONENTS
    component1= vl13          jh  18-DEC-1984
    component2= fl1           jh  18-DEC-1984

  AUTHOR:  jh  18-DEC-1984
END  fl1-function
(* the first component of the fl1-function
   is read from the database *)
. get-component object
(* it is listed *)
. li o aut

vl13  OBJECT OF TYPE  valve
  PURPOSE
    jh  18-DEC-1984
  PARAMETER
    name :  vl13              jh  18-DEC-1984
    type :  valve             jh  18-DEC-1984
    loc. :                    jh  18-DEC-1984

  AUTHOR:  jh  18-DEC-1984
END  vl13
(* so we got back to the original
   object vl13 *)
```

The shortened forms of the instructions are also legal as long as the shortened form is unique.

CONCLUSIONS

An important feature of the SML database is that the the requirements and goals can be scattered everywhere in the plant description so that they are available when needed. Other properties of the SML-system are:

- attached procedures,
- property inheritance of objects,
- variable length data fields which give the possibility to store qualitative descriptions of the system into the right points in the database.
- logging of the updates of the data at the data item level,

Future work will pay attention to the development of requirement management and to the use of attached procedures.

The software has been applied to a few small subprocesses and automation systems in power plants with good results.

REFERENCES

Eberlain,W., Wedekind,H. (1982). A methodology for embedding design databases into integrated engineering systems. In Encarnacao,J., Krause,F.-L., File structures and databases for CAD.

Haase,V.H., Koch,C.R. (eds.) (1982). A special issue on application oriented specifications. Computer, 15, nr 5, 10-59.

Heinonen,R., Ranta,J., Wahlstrom,B. (1984). A conceptual framework for the design of complex automation systems. IFAC'84 preprints vol. VI. pp. 246-250.

Heinonen,R., Ranta,J., Wahlstrom,B. (eds.) (1983). Design methods and computer aided design of process automation. NKA/LIT -project workshop, Espoo, 1-2 November 1982, Technical Research Centre of Finland, Symposium series 31/83.

Human reliability in complicated energy systems (1981). Nordic Liaison Committee for Atomic Energy, project description NKA/LIT(81)9.

Ross,D. (ed.) (1977). The special collection on requirement analysis. IEEE Trans. Software Engineering, SE-3, nr 1.

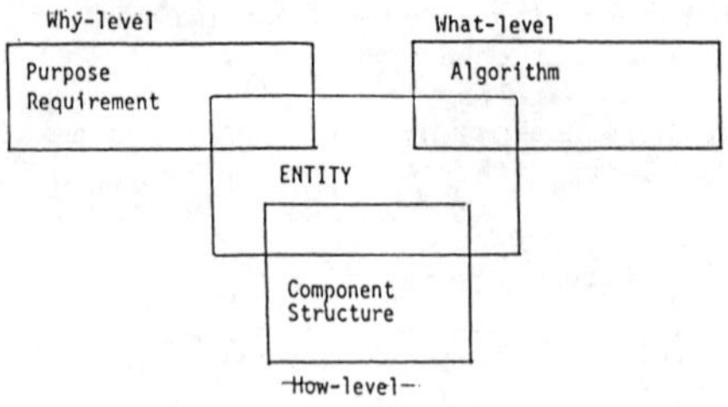

Fig. 1. The three level model in the SML.

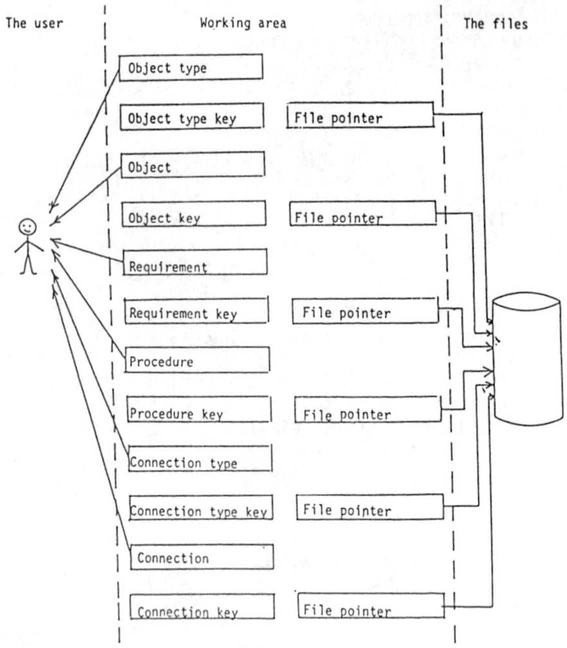

Fig. 3. SML from the users point of view.

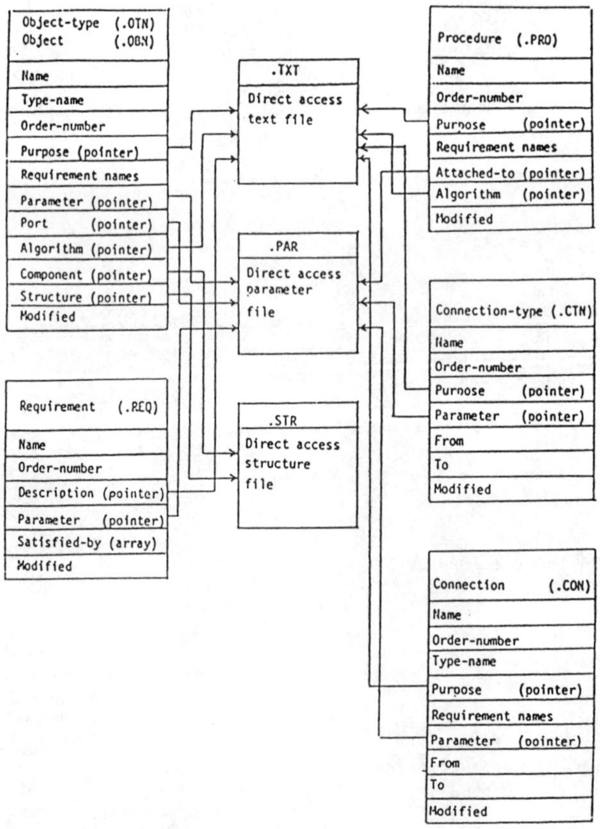

Fig. 2. The structure of the SML database.

COMPUTER AIDED CONTROL SYSTEM DESIGN FOR POWER GENERATION PLANT

B. W. Hogg, E. Swidenbank and D. G. Morrell

*Department of Electrical and Electronic Engineering,
The Queen's University of Belfast, Northern Ireland, UK*

Abstract. The paper describes the development of a computer-aided modelling and control-system-design facility for power generation plant. This has been fully implemented on a laboratory model turbogenerator, and extensive tests have been conducted on full-scale plant. The system is based on a dedicated PDP 11-23 microcomputer, with background processing on a VAX 11-780 minicomputer, using specially-written software packages. Results obtained both in the laboratory and from site tests are presented.

Keywords. Computer aided system design, data acquisition, identification, computer control.

INTRODUCTION

The rapid developments in computer technology, and advent of inexpensive microcomputers, have radically altered the approach to the control of industrial plant and processes. In power stations, the trend is clearly towards integrated control of the plant, encompassing the various aspects of boiler, turbine and generator control. The development of suitable control strategies requires accurate representation of the plant dynamics, together with a specialised control systems design, analysis and simulation facility.

Modelling fulfils an important role in investigations of the performance of electric power systems and the development of improved controllers for turbogenerators. Numerous mathematical models have been used to describe these large and complex systems, and representations have become increasingly detailed, keeping pace with advances in technology and new problems in power system operation and control. The modelling of plant is particularly important, with two main objectives. Firstly, to determine the simplest form of model which will describe the unit satisfactorily, and secondly, to develop test procedures and methods of deriving models from test data.

The models are used in various ways to examine and predict system responses and performance, and in design studies. The main difficulties which arise (other than in computation) are in determining the structure of a model, providing an accurate database, and establishing the validity of the model. Scaled laboratory models using small generators are sometimes used, with analogue or digital real-time simulation of thermal plant. These are useful in the development of modelling techniques[1], controllers[2], test instrumentation[3], etc. However, the accuracy of models can only be relied upon if they can reproduce results obtained by tests on the generators or systems they represent. Numerous tests have been conducted by power companies, and a substantial effort continues internationally in this area, particularly in seeking to determine values of plant parameters from test data[4,5]. Linear models are adequate for dynamic stability and response studies, and control system design. These may be obtained with minimum disturbance while the plant is operating normally. The methods of system identification may be applied to provide models in transfer function[6], state variable[1,7] or input-output[8] format.

The paper describes a hardware/software system based on a dedicated PDP 11-23 microcomputer, with background processing on an interconnected VAX 11-780. It presents results obtained in tests on full-scale plant and in the laboratory. The microcomputer has a range of interface modules, peripherals and software which enable it to be readily connected to real plant, monitor large quantities of data, and provide rapid graphical display of the results of analysis, performance assessments, simulation, or plant tests. It has been used in tests on 200 MW units in power stations, monitoring the responses of up to 48 quantities to disturbances input from the computer. Recorded plant responses and models obtained by identification techniques are transferred to the VAX to validate the non-linear models used for simulation in the assessment of performance, and the linear models which form the basis of control system design.

The controllers are designed interactively, using modern control theory, and a number of multivariable and adaptive techniques have been implemented. After evaluation on simulated models, the controllers are transferred to the PDP-11, and extensively tested in real time on a laboratory model turbogenerator.

DESCRIPTION

The general arrangement and philosophy is illustrated by the schematic in Fig. 1. It can be used with a laboratory model turbogenerator[1], or on full-scale plant.

The laboratory model consists of a micromachine with extensive instrumentation and control facilities, which has been designed to represent a large turbogenerator unit in a power station. A PDP 11-23 microcomputer monitors and controls the system, collecting data for modelling and identification, and implementing controllers in real time. It also checks and validates data for transfer to the VAX 11-780 minicomputer, which contains the main modelling, CAD and simulation packages. This does not operate in real time, and updating of controllers with changing conditions is performed by the microcomputer.

Tests are also conducted in power stations, when up to 48 quantities on a single unit are monitored simultaneously over periods of up to half an hour[9]. The data is then brought to the laboratory and transferred to the VAX for analysis and design studies.

At present, the complete procedure has been implemented on the laboratory system, and the resulting controllers evaluated. Extensive modelling of real plant has also been undertaken, together with the design of prototype controllers. It now remains to complete the modelling/design loop by testing controllers on the plant. Initially, this will be in the form of "on-top" controllers superimposed on the existing control systems.

DATA ACQUISITION

The initial goal in control system design is to obtain an accurate model of the system to be controlled. Previous research on a physical simulation of a generator has shown that controllers designed from models generated by identification methods perform better than those designed from linearised analytical models. The identification process requires data measured directly from the system under consideration. Small-amplitude deterministic signals are injected into selected set-points of the system, and the responses logged. This operation requires a computer system which is robust, transportable and versatile.

Hardware

The data acquisition system is illustrated in Fig. 2. A Modular Instrument Computer (MINC) has been employed to generate and inject deterministic signals into the set-points of systems under test. Up to 48 signals from the system can be logged. The MINC consists of an LSI 11-23 processor with 256 K bytes of RAM, two 512 K byte floppy disk drives, an IEEE bus and various hardware interface modules which slot into the LSI backplane. The system includes analogue-to-digital (A/D), digital-to-analogue (D/A), digital input and output modules the timing of which is controlled by a programmable clock. Many configurations of the system can be realised to meet continuously varying interface requirements. Analogue input through the single A/D chip is by three 16 channel multiplexers. To sample all signals at one instant, a sample-and-hold unit is used which receives a holding pulse from the clock every sample period. An interrupt is simultaneously generated to initiate an assembly language ISR which converts and stores in memory the sampled data. The D/A conversion is required to provide disturbance inputs. This consists of 8 memory mapped 12 bit converters. The system VDU is a Newbury terminal which is capable of emulating either DIGITAL VT100 or TEKTRONIX T4010 displays.

Signals to be monitored on the generating plant have been mostly available from existing current-loop instrumentation, but some extra transducers have been installed on request by power station personnel. All signals are cabled to one location, and the current loop signals terminated by suitable resistors to provide signals in the range ±5V. Signals from these "soft" sources are fed through various signal processing analogue circuits before being presented to the sample-and-hold unit. High input impedance differential amplifiers prevent loading or earthing of plant transducers, and thus ensure safety. Signals which exhibit a small deviation about a large offset are processed by fixed gain subtractor/amplifiers. This action considerably reduces measurement errors caused by quantisation. Compensation for the subtracted voltage is accounted for in the software by scaling.

Software

The structure of the data acquisition program is that of low priority code, performing tasks such as operator interaction and graph plotting, interrupted at each sample interval by a high priority subroutine which controls the interface hardware and processes the incoming data. Through question-and-answer dialogue, the operator can determine sampling rates, data scaling, and disturbance signal generation. Data may also be examined either numerically as a screen print, or in graphical form. Eight graphs on each of five displays can be drawn. This feature is essential as plant responses must be constantly monitored during tests for safety. The subroutines which execute these low priority jobs are written in Fortran, and are overlayed in memory to reduce space. High priority code and system variables reside permanently in lower memory while the top 192 K bytes are used for fast data storage. Sampling rates are controlled by the programmable clock which is synchronised to mains frequency. At each interval, an interrupt is generated which initiates the assembly language ISR. The processor registers are immediately stored, and the priority increased to maximum thus preventing further interrupts. Using floating point instructions the ISR reads, scales, filters and stores incoming and outgoing data. Deterministic signals are generated and output through the D/A modules. Software limits may be imposed on the levels of these outputs. Running at maximum capacity, i.e., 48 inputs and 8 outputs, the ISR executes in under 17ms.

Once the system has been installed at a test site, the operator can control the data acquisition process entirely from the keyboard. As the system is operating in an industrial environment, many safety features have been included. Operator responses are constantly checked to catch errors which may result in a system crash.

The system has been used to gather data from three power stations in Ireland. In Poolbeg Power Station, Dublin, disturbances were applied to field voltage and governor setpoints. The responses of the 270 MW alternator and turbines were logged for four minutes per test. Tests at Ballylumford Power Station in Northern Ireland were to model the dynamics of the boiler/turbine system of a 200 MW set. Due to the longer time constants of the system, signals were injected for up to one hour, and responses logged.

DATA CHECKING

The initial data verification is done locally, using the microcomputer off-line. When setting up the system, the first task is to ensure that all the signals are correctly phased, both to each other and to the existing power station control room instrumentation. This is initially achieved by obtaining a hard copy of the data logged on a digital plotter and comparing it with the plant chart recorders. For signals not logged on plant chart recorders, auto and cross correlation routines are applied to check these against the initial ones. Unexpected poor quality signals can then be further examined using Fast Fourier transformations, and a digital second order notch filter can be applied to remove any undesired frequencies.

The data is then transferred via floppy disks to the VAX 11-780 minicomputer. The logged signals are manipulated and combined to produce new signals, the aim being to derive a new signal from a combination of the others previously logged. These are then compared, and provide an extra check on the transducers involved with respect to tracking, offsets and sensitivity.

MODELLING

The object is to produce a minimal model that accurately describes the turbogenerator, and which can be updated on-line. This would permit changes of the turbogenerator performance to be accommodated over a period of time. This requirement precludes the use of time series analysis which has been applied to a power station boiler , where the dynamics are very slow in comparison to the turbo-alternator. Recursive modelling identification algorithms were then selected, and a software package has been written comprising of three such recursive algorithms (least squares (RLS), unbiased least squares (URLS), and instrumental variables (RIV)).

A subset of the signals available is selected, the choice of which is determined by the requirement that the model will be used as the basis of a design for a controller, and so the ill-conditioning of the state variable set must be minimal.

The modified data file is then processed by the modelling programs. An example of the comparison of a 15th order reference model, 5th order reduced model and the original logged data is shown in Fig. 3. It can be seen that as long as the quality of the logged signals is high, the 5th order reduced model provides an accurate description of the turboalternator, e.g., Field Volts. The 15th order model performs well even with poor signal quality, due to its ill-conditioning or redundancy, e.g., Active Power.

MODEL VERIFICATION

A number of procedures are used to check accuracy, offsets and cross coupling effects.

Accuracy

Two models are normally determined, a high order ill-conditioned state-variable set to give a base reference, and a well-conditioned low-order model. A numerical performance index (Q) is calculated, based on the signal to error power ratio. This calculation does not take signal noise into consideration, therefore it is confined to comparing different model performances on the same set of logged data.

The best determination of the model accuracy is obtained by studying the plots of the plant and model responses.

Another useful method of verification is, at a given set of plant operating conditions, to perform two tests, using Pseudo Random Binary Sequence (PRBS) and Random Amplitude (RA) disturbance inputs respectively. The models are then determined and the data from the PRBS test used with the RA model and vice versa.

Offsets

It is important with regard to subsequent controller design that the models be not only accurate but also unbiased by the effects of noise on the logged signals. These effects are screened by comparing the RLS and RIV models to the URLS model.

RANGE OF MODEL VALIDITY

The linear model produced is only accurate for a small range of operating conditions, being the model of a non-linear system. A turbogenerator operating point can be defined by the Real (P) and Reactive (Q) power output, where P and Q are normalised into per unit (pu) values. A model was first obtained at the base operating point from the micromachine. Data was then gathered over a selected range of P and Q values. These sets of data were then applied to the model and the values of Q drawn on a three dimensional map as shown in Fig. 4. A range of base operating points was investigated with similar results.

From these results it can be seen that the range of model validity is in the order of 0.1 pu for P and Q.

ADAPTIVE CONTROLLER

An example of control systems designed by this method is a series-parallel adaptive model-following controller. This provides inputs to both exciter and governor, and has been derived as an output-variable controller. It is shown schematically in Fig. 5.

This controller has been fully tested on the micro-machine, and shown to perform very well. A typical response to a three-phase short circuit applied at the generator transformer for 120ms is shown in Fig. 6, showing that there is close agreement between the system and model responses.

CONCLUSIONS

The paper describes a CAD facility for control systems associated with power generation plant. It may be used to evaluate or update controllers on individual units of existing plant, or examine options available for new installations.

ACKNOWLEDGEMENTS

The authors are grateful for the support and co-operation of staff at the Northern Ireland Electricity Service, and the Electrcitiy Supply Board, Dublin.

REFERENCES

1. Sharaf, M.M., and B.W. Hogg (1980). Identification and control of a laboratory model turbogenerator. International Journal of Control, Vol. 31, no. 4, 723-739.
2. Pullman, R.T., and B.W. Hogg. Discrete state-space controller for a turbogenerator, Proc. I.E.E., 15, 31-46.
3. Pullman, R.T., and B.W. Hogg (1978). Instrumentation for testing and control of a laboratory model turbogenerator. I.E.E.E. Trans., IM-27, 188-192.
4. De Mello, F.P., and L.H. Hannett (1981). Validation of synchronous machine models and derivation of model parameters from tests. I.E.E.E. Trans., PAS-100, 662-672.
5. Sugiyama, T., T. Nishiwaki, S. Takeda, and S. Abe (1982). Measurements of synchronous machine parameters under operating conditions. I.E.E.E. Trans., PAS-101, 895-904.
6. Hope, G.W., S.T. Nichols, and J. Carr (1977). Measurement of transfer functions of power system components under operating conditions. I.E.E.E. Trans., PAS-96, 1789-1808.
7. Sharaf, M.M., and B.W. Hogg (1981). Evaluation of on-line identification methods for optimal control of a laboratory model turbogenerator. Proc. I.E.E., 128, Pt. D, 65-73.
8. Abdalla, O.H., and P.A.W. Walker (1979). Optimal control of a laboratory power system model with output feedback. I.E.E.E. Summer Power Meeting, Paper no. A79, 451-456.
9. Swidenbank, E., I. Boyd, and B.W. Hogg (1984). On-line identification of large turbogenerators. Proc. 6th European Conf. on Electrotechnics, Brighton, 74-78.
10. Sharaf, S.M., B.W. Hogg, and O.H. Abdalla (1985). Model-following adaptive control of a turbo-

generator system. Internation Conf. "Control 85", University of Cambridge, UK.

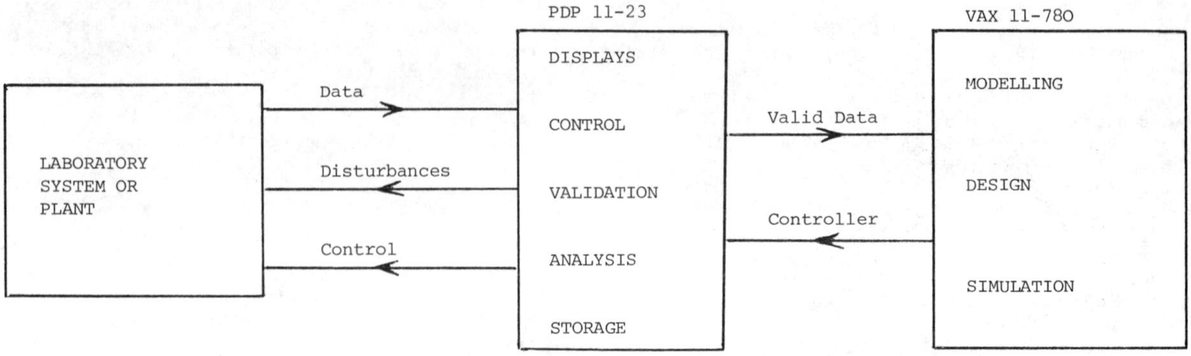

Fig. 1. Schematic of CAD System

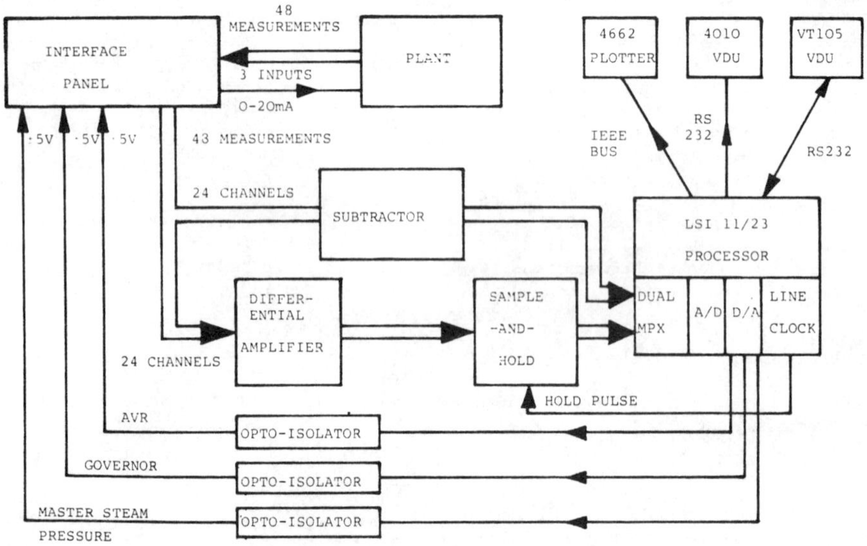

Fig. 2 Data Acquisition System

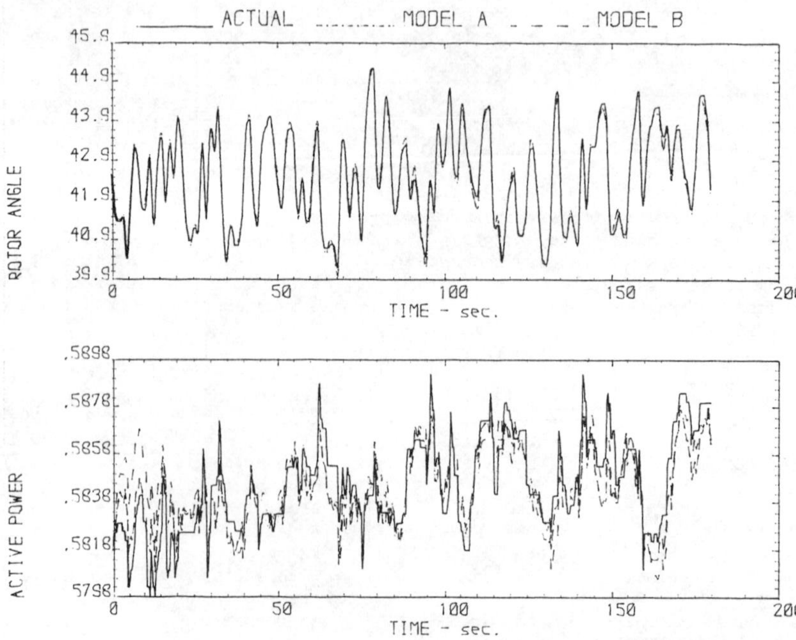

Fig. 3 Comparison of Plant and Model Responses

Recursive instrumental variables identification algorithm
Model A = 15th order, Model B = 5th order

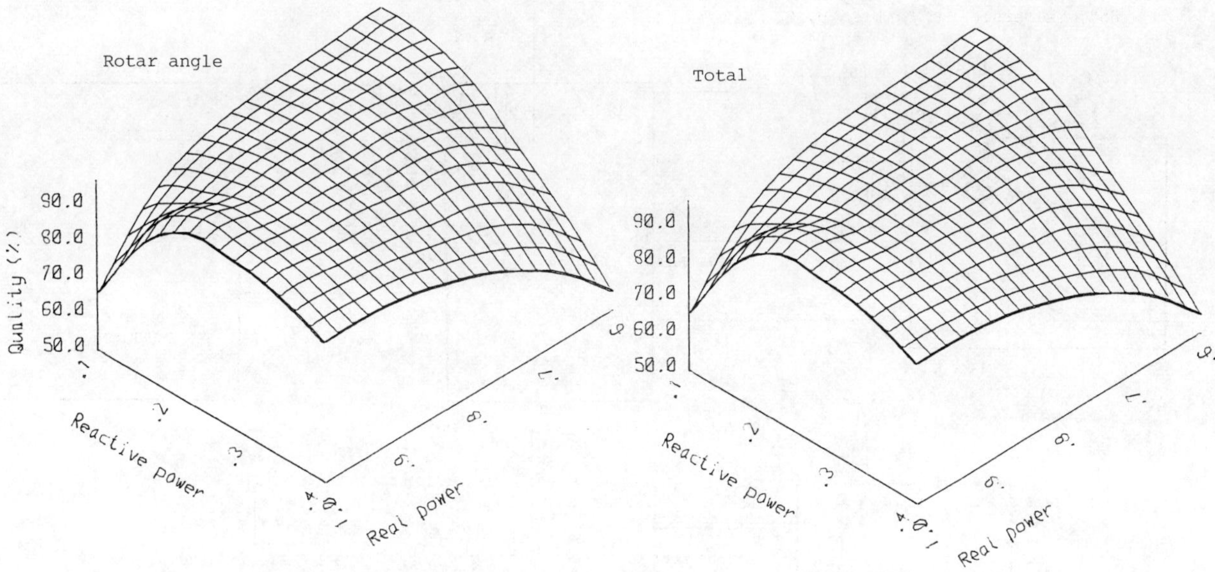

Fig. 4 Micromachine model quality over range of operating points

Model operating point: P.= 0.8, Q = 0.2
Date range : P = 0.6 to 1.0, Q = 0.1 to 0.4
Data increment : P = 0.1, Q = 0.1

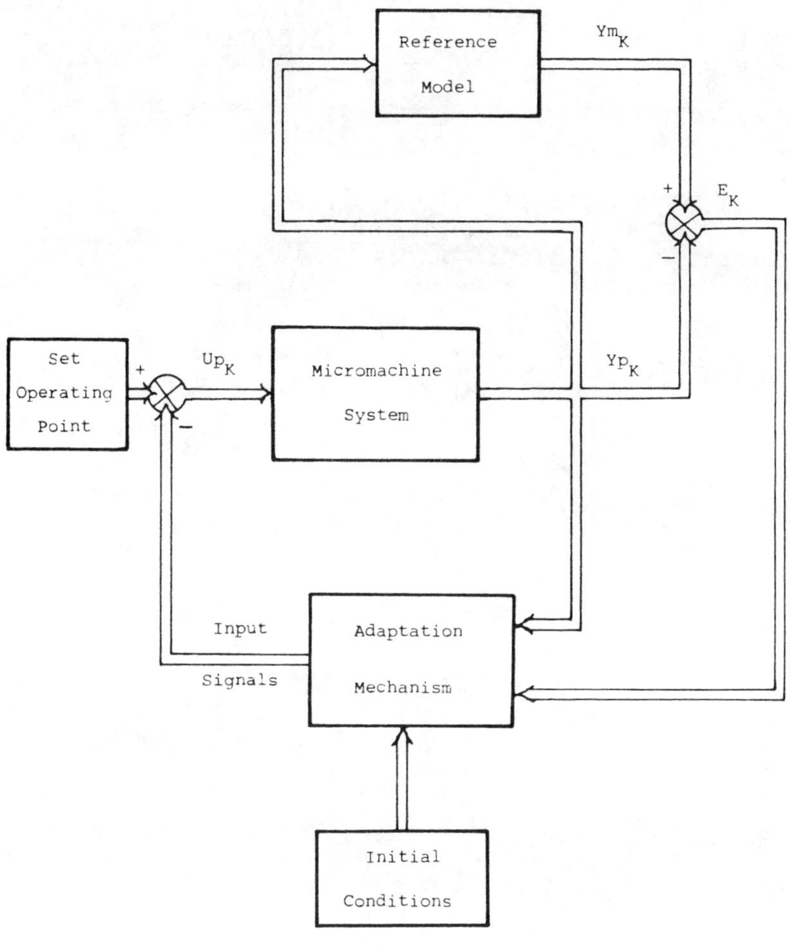

Fig. 5 Series-parallel Adaptive Model-following Controller (AMFC)

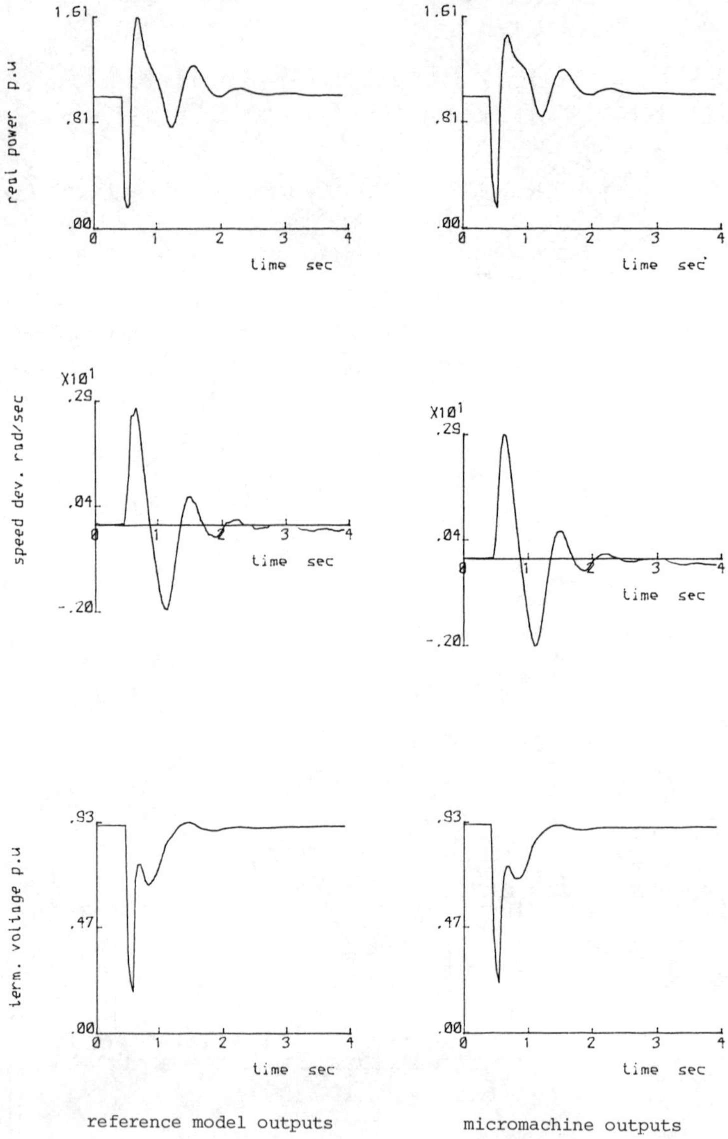

Fig. 6 Response of Micromachine with AMFC

Three-phase short circuit applied at generator transformer for 120ms

STABO – A PROGRAM PACKAGE FOR DESIGN AND SIMULATION OF STEAM BOILERS

D. Marganović, D. Petrovački and I. Pešenjanski

Faculty of Technical Sciences, P.fah 104, Novi Sad, Yugoslavia

Abstract. This paper deals with the program package STABO which is intended for the calculation and simulation of static behaviour of various types of steam boilers. Backgrounds for calculation are given. The concept of the overall package and the description of its organization are presented. Some simulation results for an industrial steam boiler are also given.

Keywords. Steam generators; computer-aided design; simulation.

INTRODUCTION

Computer aided design of steam boilers, as complex technical systems, is of an increasing importance nowadays. Apart from design itself, it is of interest to have such a program which will make possible simulation of static behaviour of the boiler with a view to analyzing the behaviour of steam power plant and to obtain basic parameters for control decisions.

In practical work the need has arisen for a parallel analysis for various steam boiler constructions for the same purpose as well as for the analysis (through simulation) of behaviour of various constructions at various loads (to satisfy control requirements). Therefore, it would be suitable to have such a program which relatively quickly and as simply as possible (from the user's point of view) can perform necessary simulations for various constructions and various (static) work regimens on the basis of the unique methodology, so that the results of simulation could be mutually compared.

For example, it can be interesting to compare the bahaviour of the steam boiler with different fuels (because of unstable fuel costs nowadays).

These were the basic motives for the development of the program package STABO (statics of the steam boiler) which makes design and simulation of various types of steam power plant possible.

BACKGROUNDS FOR CALCULATION

Majority of functional relations of physical processes in steam power plant which are applied in the program package STABO fall under the category of normative materials being generally acknowledged in contemporary technical practice. CKTI methodology has been accepted for the basic concept |1|, although there are no obstacles to some other recommendations to be applied - for example |2|.

By such a choice, application of the program has been attained to the overall class of steam power plant in which combustion of gas, oil and coal-fired fuel takes place.

Apart from its application to a large number of various constructions, advantages of the accepted conception are also seen in reliability of the results, particularly in the vicinity of nominal (designed) loads which are checked by numerous experimental investigations. Some of these investigations the authors have performed as well (for example |9|). Besides, calculation procedures after CKTI yields methodologically completely defined and encircled approach.

As a matter of fact, CKTI methodology is adapted to the manual way of calculation - a series of relations is given in tables, diagrams, nomograms and similar ways of presentation. Manifold checks of the results and possibilities of additional interventions are envisaged in the calculation procedures, tolerance of the calculation errors is from the calculation aspect extremely "rough", etc. In addition, recommended calculations for the states which considerably deviate from the nominal ones are rather intended for checking the construction than they are intended for the calculation of a real process.

In view of the fact that the goal of the program package STAKO is simulation of all stationary steam power plant situations which could be really expected (a wide interval of load changes, even the regimens at the edge of damage) it has been necessary to change and work out the original CKTI methodology.

Firstly, control calculations, i.e. programs which calculate heat state for a known construction are established. This need not be a disadvantage in case of design calculation, because construction is assumed at first, its behaviour is calculated and the procedure is repeated until a satisfactory result is obtained.

The overall steam boiler is divided into a series of elementary volumes with different mathematical expressions of heat transfer mechanism: Furnace - the basic form is radiation, banks of tubes - the form is convection with radiation in parallel flow, counterflow, crossflow, etc.

Different tube arrangements (in-line or staggered arrangement with different transversal and longitudinal pitches) and any kind of change in construction is treated in the framework of the separate elementary volumes.

The number of elementary volumes, depending upon capacity of computer resources, can be increased, but considerable partitioning does not lead to a remarkable increase in accuracy of calculation.

Calculation of heat exchange, i.e. resulting temperature, is sequentially performed for each individual element and for heat exchangers, for a group of exchangers with the same phase and for a steam boiler as a whole. Boundaries of phase change within elementary volume (beginning and completion of evaporating) are automatically determined.

In fact, this is a finite element method applied to a problem of solving temperature field in a heat exchanger which is descretized in space along the gas tract.

The above cited procedure eliminates the necessity of independent programming of a large number of combinations of flow schemes appearing in practice.

Heat exchange in the furnace is calculated according to a semiempirical equation after A.M.Gurvitch |1|.

$$\theta_o = \frac{(B_o/A)^{0,6}}{M + (B_o/A)^{0,6}} \tag{1}$$

which links undimensional gas temperature at the furnace exit θ_o with Boltzmann's criteria B_o, furnace darkness degree A and the parameter of the flame position M. It is well known that the cited equation is not valid for the poorly cooled furnaces, for the furnaces with significant convection, and probably for extremely "cold" furnaces. In this connection, the problem of the application of equation (1) appears at low loads with large air excesses. Reliable results are attained in the diapason of $0.5<\theta_o<0.85$ |3|.

Heat transfer in superheaters, economizer, air heater is performed predominantly by convection, so that for these exchangers equations of heat transfer and heat balance are (for simbols - see appendix 1):

$$Q=KF\Delta T_m=\phi B(H_g'-H_g''+\Delta\lambda H_o)=D(H_f''-H_f') \tag{2}$$

Mean overall heat-transfer coefficient K is based on the surface of the gas side and is calculated from the expression

$$\frac{1}{K} = \frac{1}{\alpha_{cg} + \alpha_{rg}} + W + \frac{F}{F_f}\frac{1}{\alpha_f} \tag{3}$$

Heat transfer coefficient by convection on the surface of the gas side α_{cg}, heat transfer coefficient by radiation α_{rg}, heat resistance of the tubes with deposits W, as well as heat transfer coefficient by convection on the side of the cooler fluid α_f (in cases of water boiling and streaming $\alpha_f \approx \infty$ is adopted) are calculated in the way which is a routine method in technical calculations for arithmetic mean temperature difference of combustion gases. Mean heat transfer coefficient is in effect a function of the local temperature difference between gas and cooler fluid (T_g-T_f), and enthalpies are also dependent on temperature and space (lenght) coordinate (because of entrance of air in the gas canal $\Delta\lambda H_o$, and heat losses in the environment being expressed by the coefficient ϕ). On account of that, equations (2) do not yield a possibility of an immediate calculation of unknown temperature of output gases T''_g if mass flow rates B and D are known, and one temperature of cooler fluid is known (T'_f or T''_f) in addition to known functional dependences $H_g=H_g(T_g,\lambda)$ and $H_f=H_f(T_f)$ (for more information about the solution of similar problems, see |5|,|6|).

On the other hand, procedures of the mean temperature difference ΔT_m encompass work with four temperatures (instead of two) i.e. work with corresponding bulky subroutines for an explicit, and particularly for an implicit calculation of thermophysical features of working medias, which in cases with limited operative memory may present a limiting factor.

For cited reasons, an iterative procedure of the calculation of output gas temperature is implemented in STABO, whereby the value $K(T_g-T_f)$ is iteratively improved, as well as

$$\phi B(H'_g-H''_g+\Delta\lambda H_o)/(T'_g-T''_g)$$

and

$$D(H''_f - H'_f)/(T''_f - T'_f).$$

For slight temperature changes ($T'_g - T''_g$), which we provided by the choice of relatively small elementary volumes, nonlinearities are not so much striking in order to interfere the reliable convergence of solution. Investigations of convergence are given in |7| and in most cases they finish in 2-8 iterations.

For the calculation of the values of water and steam states, programs have been developed according to basic approximate equations IFC 1967 |8| which thoroughly reproduce values of previously adopted international sceleton tables. Equations from |3| and |8| have been used for the calculation of dynamic viscosity of water and steam and for the coefficient of heat transfer, respectively. They also well reproduce values of international sceleton tables.

Calculation of enthalpies of combustion gases and air is performed as in the case of the mixture of semiideal gases. Enthalpies of gas, oxygen, nitrogen, carbon dioxide, steam from |8| and flying ashes from |1| are approximated by polinomial of the eight degree, i.e. by the method of least squares. Thermpohysical properties of flue gases (coefficient of heat transfer, kinematic viscosity and Prandtl's number) are performed according to |5|.

BASIC NUMERICAL CALCULATIONS

For solving nonlinear balance equations for individual steam boiler elements as well as for the balance of the overall steam boilers, three numerical methods have been tested: method of successive approximations, secant method and a combined method of bisection and inverse quadratic interpolation according to |7|. Contrary to expectations, the method of successive approximation has proved to be very efficient |10|. Explanation for this efficiency lies in the fact that the most curves not only satisfies the necessary convergence condition, but their slopes are almost ideal for the attainment of fast convergence in the wide neighbourhood around the solution.

When calculating the zeroes of nonlinear functions the greatest problem is the fact that the majority of these functions have vertical asimptotes (poles) in the interval in which the zero of the function may be expected. Therefore, in the program package STABO, a subroutine for finding zero of functions on the basis of the method of bisection and inverse quadratic interpolation is used with a simple method for avoidance of the field in which there is an asimptote.

ORGANIZATION OF THE PROGRAM PACKAGE

The principal scheme of the organization of the program package STABO is presented in Fig.1. The concept of the package is made in such a way that it satisfies the following requirements: high modularity, small demands in the main memory, simplicity of the use and high software portability.

For this purpose, "segmentation" of bound units is carried out. Bound units consist of a resident part ("root") and more changable overlays.

Communication between individual overlays (data exchange) is partly performed through the common zone in the main memory and partly through the dedicated files on external memory (disc).

In order to simplify the use of STABO, this program package is organized in three bound units: preprocessor processor and postprocessor. Processor is the most important part because it actually performs the calculation and simulation of the given steam power plant.

Processor consists of a resident root and the bound unit contains all necessary common zones in the main memory in which exchange of data between individual overlays takes place. In the first overlay, modules for stoichiometric calculation, properties of combustion gases and the calculation of some initial values before starting the main (external) loop are situated. In the second overlay there are both modules within the external loop (furnace, tube network, superheater, evaporator,economizer, air heater) which depends on the previously selected steam boiler as a whole . Criterion of the exit from the loop is the combination of relative error according to independent variable (temperature of flue gases at the chimney entrance) and the absolute error of the function (satisfaction of the heat balance equation of the steam boiler). This is the change in relation to |1| where the criterion is the absolute error according to the independent variable.

In the third overlay there are program modules for the determination of the necessary water and steam values, obtained on the basis of |8| and |3|.

When making a decision about the way of processor formation, two alternatives related to the problem of solution of the choice of the corresponding module within the selected steam boiler configuration are considered. Namely, the problem is seen in the fact that from a large number of modules for the same function of steam boiler (for example, air heater) one should automatically select a corresponding module without making a previous intervention on original source programs (which would complicate the use of the program package, increase the possibility of error and require recompilation). According to the first alternative, bound unit of the processor would have all possible various modules linked to the various types of steam boiler elements (furnace, tube network, superheater, etc.), but in such a way that they would be situated in various overlays and the processor itself would, in dependence of input data which define steam boiler configuration, recall only those overlays which are the part of steam boiler construction. Disadvantage of this alternative is seen both in an extremely large number of overlays linked to a bound unit and in time loss when changing corresponding overlays in the main memory for each passage through the main itterative loop. Because of that, the authors have decided to take another alternative which assume the existence of utility programs, or a complete preprocessor which links the bound unit of the processor on the basis of the data on steam boiler configuration before starting the execution of the processor. Advantage of this alternative is seen in the fact that the main loop of the algorythm is contained in a unique overlay, so that the program executes faster and the total number of processor's overlays is reduced to three. Disadvantage in respect to the first alternative is in decreased portability of the overall program package. In the second alternative which is implemented in STABO, there are utility programs which make possible "alternative" linking of the processor only to those modules for individual steam boiler elements which are given in configuration. This linking process is an automatic one on the basis of the input data.

Pre- and post-processors are designed in such a way that they represent a suitable interface between the user and the machine in the sense of simple-interactive data input (steam boiler configuring, and others) and control of necessary output documents- results. The most important function of preprocessor is to ensure a proper linking of the processor on the basis of automatic editing of linker directive files according to the input data.

For example, if we would like to replace one type of air heater (SUB1A) and the economizer (SUB2A) with another constructive types (SUB1B and SUB2B) then it is necessary only to replace the following instructions in a linker file

```
LDEF   SUB1,  SUB1A
LINKN  SUB1A
LDEF   SUB2,  SUB2A
LINKN  SUB2A
```

by means of automatic editing (for Honeywell level 6 minicomputer) with the following instructions

```
SDEF   SUB1,  SUB1B
LINKN  SUB1B
LDEF   SUB2,
LINKN  SUB2B.
```

In source programs there is a sequence of FORTRAN instructions which should not be altered (i.e. there is no need for re-compilation:

```
EXTERNAL   SUB1
EXTERNAL   SUB2
---
CALL  ZERO (SUB1,...)
---
CALL  ZERO (SUB2,...)
---
END
SUBROUTINE  ZERO (FUN,...)
---
CALL FUN (...)
---
END
```

where ZERO is the name of the subroutine for finding zero of the function FUN (which is actually SUB1A or SUB1B, SUB2A or SUB2B).

ILLUSTRATION OF THE USE OF THE PROGRAM PACKAGE STABO

The construction of a steam boiler, which has been analized by the use of STABO, is presented in Fig.2. The boiler is considered with or without the air heater. Analysis is performed for two kinds of fuels: oil and coal (pulverized coal). The technical data of the boiler are: the steam production D_m=40 000 kg/h, steam pressure p_s=45 bar, steam temperature t_s=450 ^oC, feedwater temperature t_w=130 ^oC. This type of a small boiler can be often found in the process industry.

Some of the results obtained by simulation with STABO are presented in Figs. 2-10 where dash lines are used for coal, full lines for oil and dot lines for overload. For the different values of the coefficient λ and load D/D_m, fuel consumption per kg of the steam B/D (for oil and coal) is given in Fig.3 and in Fig.4, respectively. These results can be used both for the control of steam boilers and in various economic analyses. Dependence of η on λ and D/D_m as well as the kinds of fuels are given in Fig.5.The analysis of the steam boiler construction with or without the air heater (economic justification of its implantation) is obviously in Fig.6. Change of η for two values λ, and in dependence of coal moisture is presented in Fig.7. (Nominal values for coal moisture in calculations are 40%). In Figs. 8 and 9 dependence of the combustion gas temperature at the exit of the boiler, TG", as a function of λ and load D/D_m for coal and oil are presented. These results are of interest when considering an additional utilization of rest heat and for avoidance of certain steam boiler work regimes due to low temperature corrosion. In Fig.10 heat amount QPHL is given which should be considered in the cooler in order to control steam temperature at the output of the steam boiler.

REFERENCES

|1| "Teplovoij rasčet kotelnyh agregatov",Normativnij metod, Energija, Moskva 1973.

|2| VDI-Wärmeatlas, VDI-Verlag, Düsseldorf, 1974.

|3| Rikvin,S.L., Aleksandrov,A.A.: "Teplofizičeskie svoistva vody i vodjanogo para", Energija, Moskva, 1980.

|4| Bloh A.G.: "Teplovoij izlučenije", Energija, 1967.

|5| Hansen,H.: "Wärmeübertragung im Gegenstrom, Gleichstrom und Kreuzstrom", Springer, Berlin, 1976.

|6| Roetzel,W.: "Heat Exchanger Design with Variable Transfer Coefficients for Crossflow and Mixed Arrangements", Int.J.of Heat and Mass Transfer, Vol.17, 1974, pp.1037-1049.

|7| Forsythe,G., Malcolm,M., Moler,C.: "Computer Methods for Mathematical Computations", Prentice Hall, Englewood Cliffs, 1977.

|8| Schmidt,E.: "Properties of Water and Steam in SI-Units", Springer,Berlin, 1969.

|9| Gulič,M., Pešenjanski, I., Perunović,P.: "Heat Distribution in a Gas-Fired Furnace" (in Serbocroation), Strojarstvo,17(1975),No.6,pp.273-277, Zagreb

|10| Marganović,D., Petrovački,D., Pešenjanski,I.: "A Program Package for Design and Simulation of Various Steam Boilers", EES›84, Nice, 1984.

APPENDIX-Symbols used

Q	total heat-transfer rate W
K	mean overall heat-transfer coefficient based on the surface area of the tube from gas side W/m^2
F	heat-transfer area m^2
ΔT_m	mean temperature difference ^oK
ϕ	nondimensional coefficient including heat loss in the environment
B	fuel mass flow rate kg/s
D	mass flow rate of cold fluid
H	specific enthalpy kJ/kg
λ	coefficient of air excess in combustion gases
T	temperature ^oK
α	heat-transfer coefficient
W	wall resistence of separating surfaces of fluids including deposits
Δ	finite difference

Subscripts

r	radiative
C	convective
g	refers to fuel gas
f	refers to cold stream (steam, wather,air)
′	at the inlet
″	at the outlet
o	refers to environment state

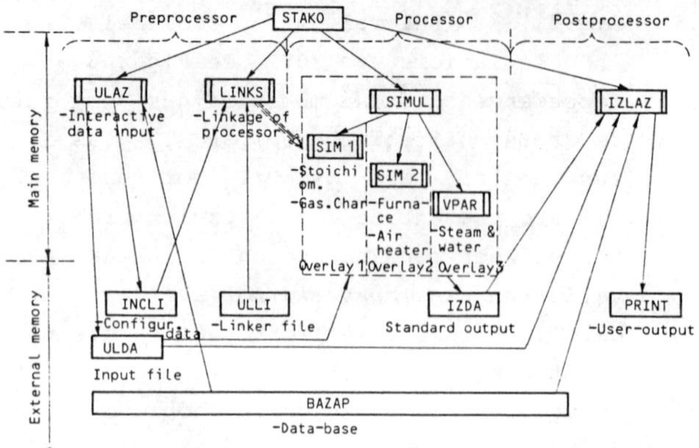

Fig.1

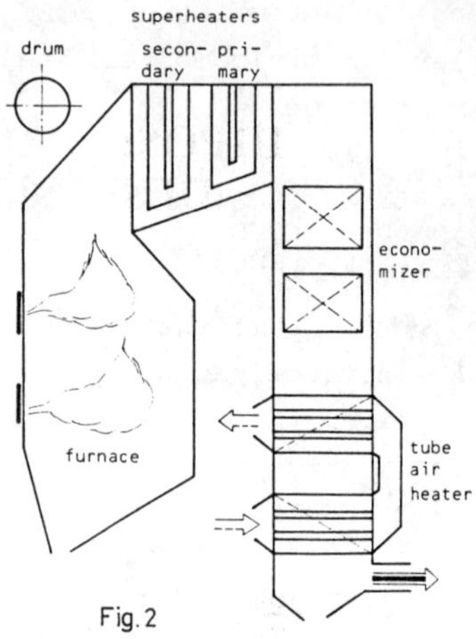

Fig.2

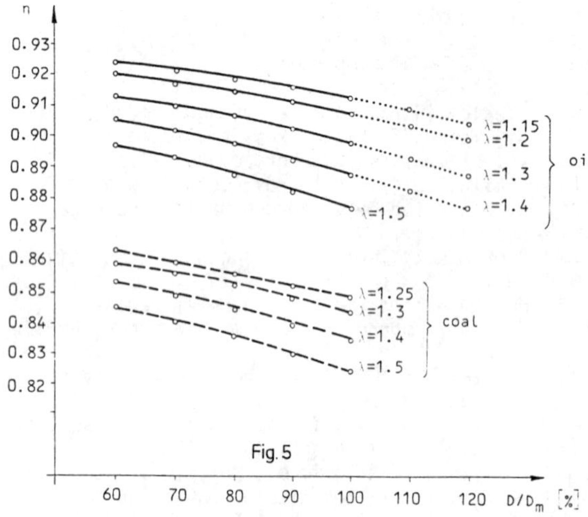

Fig.5

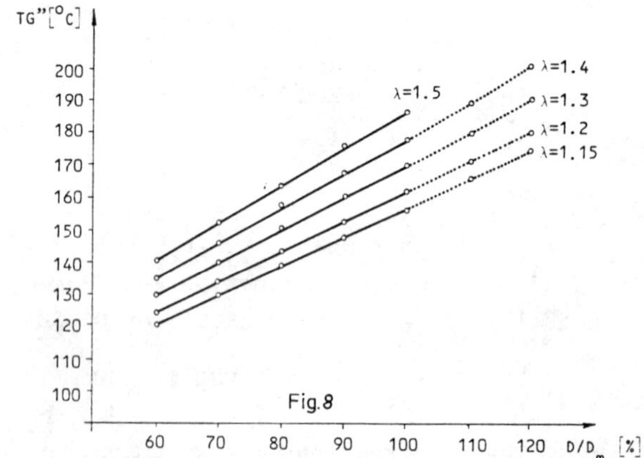

Fig.8

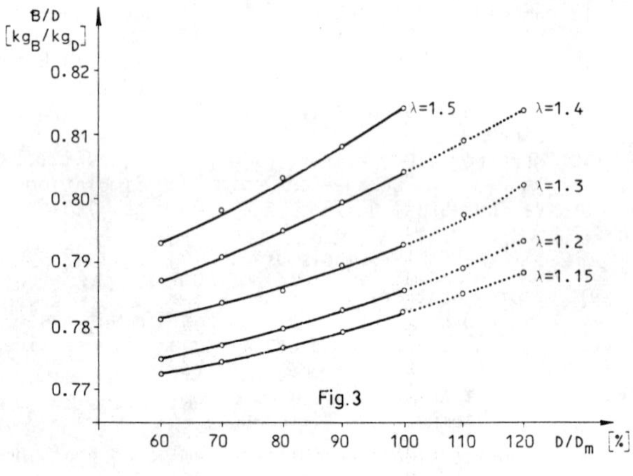

Fig.3

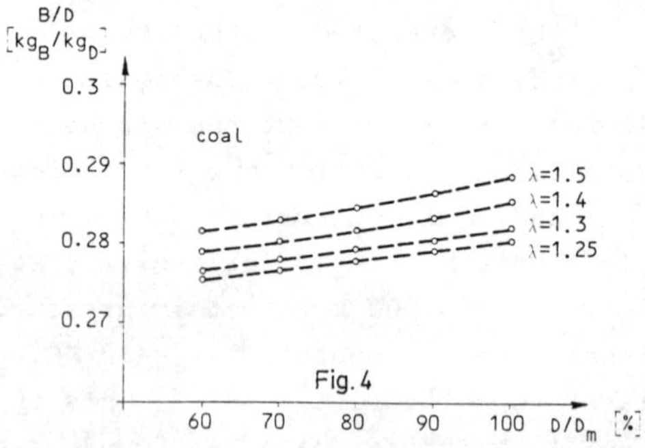

Fig.4

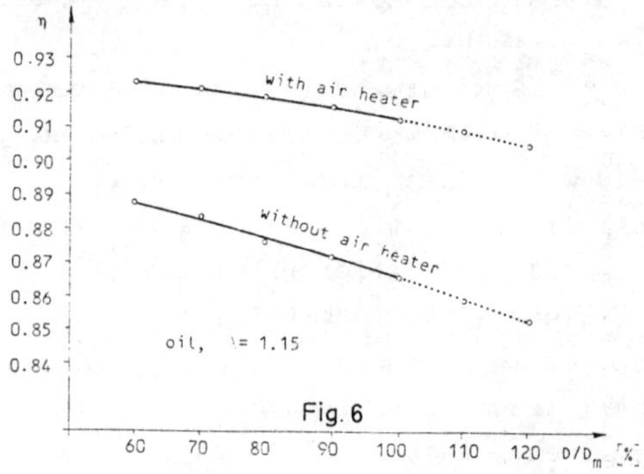

Fig.6

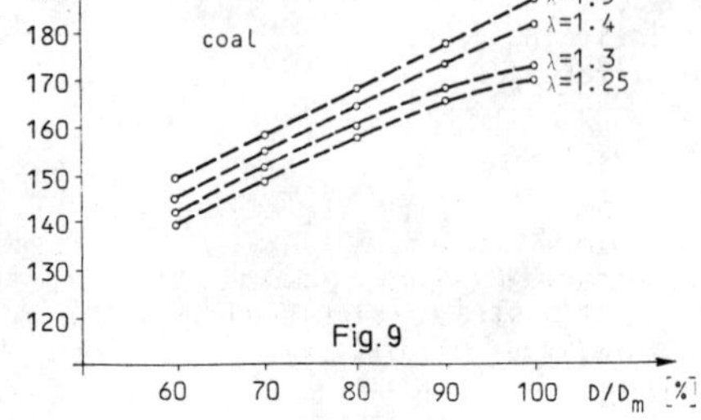

Fig.9

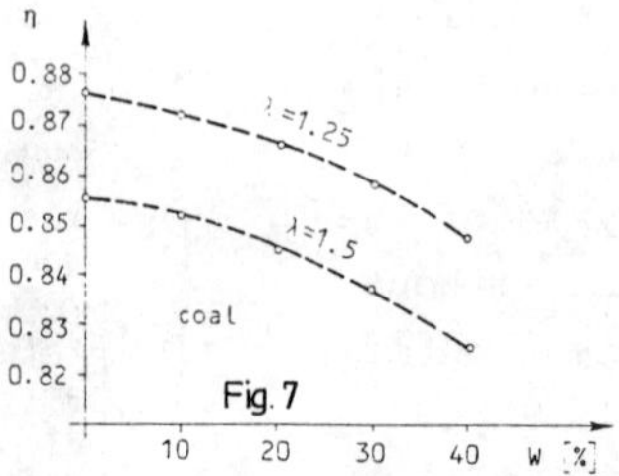

Fig.7

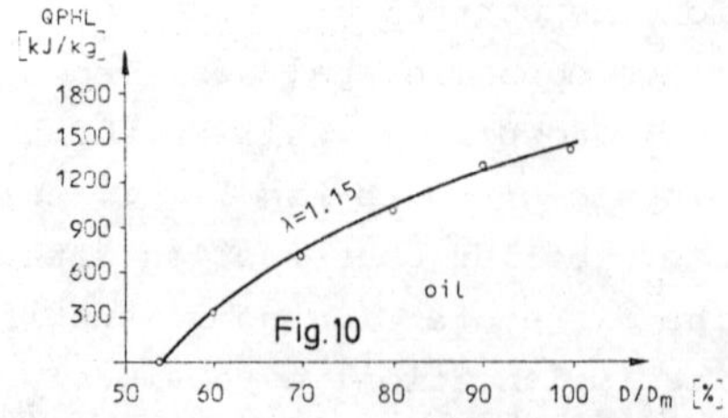

Fig.10

DIPOS: A DATABASE FOR COMPUTER AIDED DESIGN IN ELECTRIC POWER SYSTEMS

L. Borzemski and S. Lebiediewa

Institute of Control and Systems Engineering, Technical University of Wroclaw, 50-370 Wroclaw, Poland

Abstract. The database system named DIPOS intended for use in computer-aided electric power system expansion planning is presented. The CAD system containing the DIPOS database is briefly explained in the context of the transmission network expansion planning. There are many reasons that in CAD applications we should rather think about specialized or problem-oriented databases, which would give unique solutions in particular situations, than general purpose databases with maintenance and operation overheads. In the paper the user's view of data, user interface, database data model and physical database organization are discussed. The system is the relational database from the user's point of view but it employs the network database data model at the system level. This allows the user to have well known natural relational data model with its correctness and ability to perform profitable relational operations, whereas the DIPOS database network data model gives additional advantages in the pre-definition of possible use patterns of the data. An example of database application is also presented.

Keywords. Computer-aided system design; data handling; databases; power systems planning.

INTRODUCTION

The electric power system design can be considered in the context of system expansion planning in different time ranges. The expansion planning deals with the generation, transmission and distribution systems in the electric power network. The methods used in the planning of the power system expansion require a great number of data necessary for calculations or obtained as the results of complex calculations. This gives rise to the need for a computer-aided planning system containing a database.

As part of a comprehensive program to support the electric power system design, the CAD system with the integrated database has been developed at the Technical University of Wroclaw /Borzemski and co-workers, 1983; Kinsner and Nawojska, 1982/.

The database named DIPOS /Database In Power Systems/ making use of the problem-specificity, is characterized by access and storage structures which guarantee a fast data access and the minimization of space utilization, as well as by the data manipulation language taking into account the user's needs.

The relational database approach has been chosen because it allows the user to operate within database profiting by a unified and simple data viewpoint. Some non-relational data operations are also provided.

The database extends the problem-solving capabilities of the computer-aided planning system through providing data handling support for all available problem-solving programs.

In the paper the main principles of the DIPOS database are discussed. The CAD system is also outlined, as well as the connection between the electric power system and the database is described.

ELECTRIC POWER SYSTEM EXPANSION PLANNING - THE CAD ENVIRONMENT

Electric power system expansion planning can be viewed as a complex decision making process dealing with the generation, transmission and distribution systems planning for different time ranges. Because of the high investment costs involved and the long life-time of electric supply, transmission and distribution equipment, the decisions have far - reaching consequences. Thus the most precise and productive planning methods must be used. Application of such complex and sophisticated methods raises a difficulty in handling the data. This can be overcome through the development of suitable computing environment such as a CAD system.

The electric power system expansion planning has been extensively studied and several systematic heuristic and mathematical approaches have been proposed, for example by Backlund and Bubenko /1978/ and Indurawa /1980/.

The group from the Technical University of Wrocław has developed the CAD system for supporting transmission network expansion planning. It is built from two main functional layers: the lower layer referring to data handling and the upper layer dealing with problem - solving programs.

Sample expansion planning procedure is shown in Fig.1. The aim is to find the expansion variant of network configuration at the minimum cost for a given time range. The procedure starts from the redundant maximum variant covering all existing and planning nodes and connections /that is, lines and transformers/. The least-cost expansion variant is achieved through successful eliminations of connections from admissible succeeding variants. Technical constraints, network consistency and reliability of power flow are checked for given power demand, for which network expansion is to be done.

Characteristics and Use Patterns of Transmission Expansion Planning Data

The data processed in the algorithms of transmission system expansion planning can be classified as either

/a/ catalogue data - data specifying technical, constructional and economical features of possible transmission network component types, or

/b/ parameter data - data concerning the characteristics of components of both existing and planning transmission systems. Parameters define the components as they are included in the configuration of the system. Localization conditions and planning features are also taken into account.

The catalogue data are provided by the component supplier or constructor whereas the parameter data are individually determined for each component. Both data types provide full spectrum of information needed /or obtained/ in the expansion planning.

The catalogue data are provided for the following transmission network components:

- transmission lines,
- conductors used for transmission lines and bus-bars,
- bus-bar sections,
- switch fields equipment,
- circuit-breakers,
- transformers.

To define a transmission system /now operating, under construction, considered in a plan of expansion etc./ and its state, the parameter data are considered for:

- transmission lines,
- bus - bar sections,
- switch fields equipment,
- substations,
- transformers,
- switching stations,
- variants of expansion,
- strategies of expansion,
- power demand.

The contents of individual data sets of both kinds has been presented by Borzemski and co-workers /1983/. For example, the transmission line catalogue data set includes twenty nine data items specifying each line type, e.g. type of line, rated voltage, number of circuits, average span length, catalogue index number of ground wire, ground area covered by the line and by the steel tower /or wood pole/, length of insulators, their type and mounting co-ordinates, mass of the supporting towers and intermediate poles. On the other hand, each element of the line parameters set is described by twenty four data items, e.g. name of the line, commission year, planning feature, rated voltage,localization code, climate zone code, type of line, line length, average ground resistivity, admissible power.

It is obvious that there are one-to-many associations between the components of the transmission system, which must be modelled to represent directly the network configuration, e.g. transmission lines connected to a certain bus-bar. The associations may exist between components being of different or the same type. These associations are also reflected in the use pattern of data. The user may ask to get the component data either together with information on associated components or without it.

The planning methods are characterized by a great number of data necessary for calculations and used by a series of succeding algorithms in the proceeding steps of the planning procedure.

The users are rather non-computer oriented and the total task coding in high level language is divided between them. Tasks often involve the iterative execution of several computer programs, each requiring data generated by the others in addition to user input data.

As a possible solution to achieve consistent network expansion planning in which all input, processed and output data will be immediately accessible to all the users, is the development of computer-aided design system with integrated database, providing analytic and data handling support.

Databases in CAD Systems

There are many reasons that in CAD applications we should rather think about specialized or problem-oriented databases, which would give unique solutions in particular situations, than general purpose databases with maintenance and operation overheads. The industry developed databases are mainly concentrated on fulfilling the requirements from some general user with a very wide data use pattern. This data management concept cannot be fully applied in scientific databases /Borzemski, 1983; Borzemski and Lebiediewa, 1982/.

As the problem environment is well defined the problem - oriented database can be developed. Here it means that database is oriented to the problems supported by the CAD system. Such database, making use of the problem specificity, provides the users with high - level data access relevant to the needs. This allows the users to concentrate on the problem, without necessity to carry on /of course, in some extent they must do it/ data during design activity.

THE DIPOS DATABASE

In general, three levels of data models in

database are considered: external, conceptual and internal. An external model /known also as a user's view/ shows how information within the database is viewed by a user. A conceptual model /a database model/ describes which information may enter and be stored in the database. This model is used by the database management system /DBMS/ in handling all access to the database. The user's view is mapped into the database model. An internal model describes how and where the information is represented in storage structures. The logical database organization deals with the user's view and database model, whereas the physical database organization concerns the internal model.

Data Structures and Data Manipulation

The user data comprises two data sets: the catalogue data set CATALOGUES and the parameter data set PARAMETERS. Each of them consists of separate homogeneous data subsets. These data subsets are in the form of tables of records of the same type, where each record contains data items values specifying certain transmission network component. Thus there are six tables of records in the CATALOGUES and nine tables of records in the PARAMETERS. All records pertaining to the same table are different because at least one data item has distinct value.

The associations are now defined between network components as data relationships between records from particular tables in the data set PARAMETERS. Due to the lack of the circuit-breaker table in the PARAMETERS, the data relationships also exist between records from the switch fields equipment parameter table and records in the circuit-breaker table in the CATALOGUES. The circuit-breaker table is not included in the PARAMETERS because no information about these components is relevant to network configuration or operating conditions.

It is obvious that the list of associated components /now we can say - associated records/ with a given component is not of a fixed length. For the purposes of this study, the maximum numbers of associations /i.e. their sizes/ have been estimated. For example, in the case of a medium size regional power system, a record of the table of substation parameters may be associated with up to ten records from the table of transformer parameters and with up to thirty records from the table of line parameters. A record of the table of variant parameters may be associated with up to 20 500 records from the following three tables, viz.: substation parameters /max. 500 records/, transformer parameters /max. 5000 records/ and line parameters /max. 15 000 records/. All associations are illustrated in the figure related to the database model discussed in the next section.

In transmission expansion planning study the data are treated by the users as some generalized arrays of numeric or character data items. Users manipulate on different data aggregates /e.g. single numeric data, vectors, arrays/.

Available data may be interrogated in different ways and the users may wish:

- to read and write a table,
- to select certain colums of a table and to remove from the resulting table any duplications in the records,
- to select certain records of a table satisfying a given condition,
- to window /to display a sub-table/ certain part of a table,
- to select certain record of a table together with associated records.

The tables of records are very comprehensive and the users want to have data manipulation possibilities allowing them rather to narrow the contents of the tables than to enlarge them.

Logical Database Organization and Data Operations

As the user's view the relational data model /Date, 1977/ has been chosen because it corresponds to the tabular data view. The basic data structure in the model is a relation which can be viewed as a two-dimensional non-homogeneous data array consisting of rows and colums, with the rows corresponding to records and the columns representing data items within the records. Columns are called attributes and take the values from the domains containing the same type data item values.

The analysis shows that all attributes of each relation /also the primary keys/ take values from simple domains. The relations do not possess candidate keys and functional dependencies between any attributes do not exist. Hence the relations are set up in fourth normal form.

The relational database approach, as used in DIPOS, assumes the development of a data manipulation language based on relational algebra. In the DIPOS the retrieval language is extended by some user-dependent operations.

The system provides the following operations: PROJECT, SELECT, MASK, LIST, READ and WRITE. Two of them, i.e. PROJECT and SELECT are fundamental operations in the relational approach, allowing to select, respectively, certain colums or rows from a relation. The third operation - MASK - means that the resulting relation contains selected rows with selected colums; this is the way to obtain a "window" from a relation.

In the relational model all access paths are determined by row content alone, order is determined /by content/ at the time a collection of rows is retrieved. Thus the user can establish any relationship to access the data, e.g. to represent discussed before associations. However, the size of associations in not a priori known and adding new attributes or building new "linking" relations may be space-consuming. Moreover, the names of associated rows and relations must be known in advance in retrieval process. This may results in some complications in data manipulation. Therefore, the users are concerned with an operation allowing direct retrieval of the set of associated rows. The DIPOS database system deals with this problem, providing the LIST operation. This operation selects the associated rows along the predefined access paths and the list of qualified rows is presented to the user.

Apart from the above operations, the system provides a means to read a relation as well as to write /update/ a relation, row, column or attribute value.

Of course, LIST operation is non-relational and in that form it cannot be realized within the relational database model. The following approach seems to be convenient. The user is aware of well-known natural relational data view, whereas the DIPOS database management system manipulates on the network view of data. The DIPOS database logical organization is shown in Fig.2. There are all possible one-to-many associations as shown by the double arrows.

Consider an example of a database which includes substation, switching station, transformer and line parameters relations as it is illustrated in Fig.3. The contents of the relations can be displayed by READ operations /for space reasons only a few attributes per relation are assumed/. The user may ask if there are substations in the network, commissioned before 1981. Then SELECT operation specifies SUBSTATION -P relation as argument and "CY<1981" as condition. The result contains only one row named S2. The figure also shows the list of rows obtained after issuing LIST operation concerning all components associated with substation S1. The order of rows in the list is defined by the user. The user is able to put some selection condition in retrieving the associated records.

Physical Database Organization

The DIPOS physical database consists of two types of files, viz.: the database hierarchical index and the main data files. The database index is used to search data in the main data files containing data described in the CATALOGUES and PARAMETERS data sets. The physical database organization is remarkably simple and provides unique advantages for operational efficiency as well as for ease - of - definition and database evolution. There is complete separation of data and access structures. This separation allows DIPOS to manage the data and the access structures independently. The main data files contain relations with no embedded data for logical relationships between rows. This provides the data independency.

The relations are stored as two-dimensional Fortran - like arrays of equal - length elements. There is also an advantage of physical data organization consisting in representing within database only three types of data items, that is 8-character data strings, integer and floating-point numbers. All these data items are encoded into two words, each of four 6-bit character in length. The relations are sequenced rows by rows. The space of the main file is logically continuous but the physical space distribution accross disk depends on operating system.

SYSTEM STRUCTURE AND IMPLEMENTATION

The system consists of Data Definition Subsystem /DDS/, Data Manipulation Subsystem /DMS/ and Data Utility Subsystem /DUS/.

The DDS clearly defines /creates/ relations, their descriptions, associations, levels of protection and maintains the database indexes and dictionary. The role of the database dictionary is to provide with information about each object and process within DIPOS, for example, the dictionary stores full descriptions of relations.

The means that are used to interface the user with a database mostly depend on user's needs. The CAD algorithms are written in Fortran, so that all data operations have the general form of CALL statements to callable routines under control of DMS.

Utility subsystem DUS provides with functions including programs for initial loading, unloading and reloading as well as creating a back-up copy of the database. The problem-solving program library is maintained separately by a means of operating system.

The database is implemented on ODRA 1300 series /ICL 1900 - like/ computers. The GEORGE-3 operating system is preferred although DIPOS can run under other available operating systems.

CONCLUSIONS

The DIPOS database intended to support the computer-aided transmission expansion planning has been presented. Using the database the users may develop their algorithms without focusing on problems related to data handling. The system supports two different data model types. The user operates within relational database profiting by a unified viewpoint. But due to the network database model maintained at the system level, the user is able to find the data along predefined access paths, as well.

REFERENCES

Backlund,Y. and J.A.Bubenko Sr. /1978/. Computer-aided distribution system planning. Proc.of the Six Power Systems Computation Conference, Vol.I. Darmstadt, IPC Science and Technology Press.

Borzemski,L. /1983/. Data structures and storage structures in scientific data base for multistage experiment. Acta Cybernetica, 6,1,13-21.

Borzemski,L. and S.Lebiediewa /1982/. Problem-oriented data base design. Proc.8th Int. CODATA Conference, Jachranka.

Borzemski,L.,A.Jaroszewska,S.Lebiediewa and A.Mały /1983/. Database DIPOS /in Polish/ Rep. 59/83, Inst. of Control and Systems Engg.,Tech. University of Wroclaw.

Date,C.J./1977/. An Introduction to Database Systems. Addison-Wesley Pub.Co.,Reading.

Indurawa,A.S./1980/. Power system generation and transmission planning using probability methods. Ph.D.Thesis, Imperial College, London.

Kinsner,B. and L.Nowojska /1982/. Computer system for transmission system planning /in Polish/. Rep. 214/82, Inst. of Electric Power Engg.,Tech. Univ. Wrocław

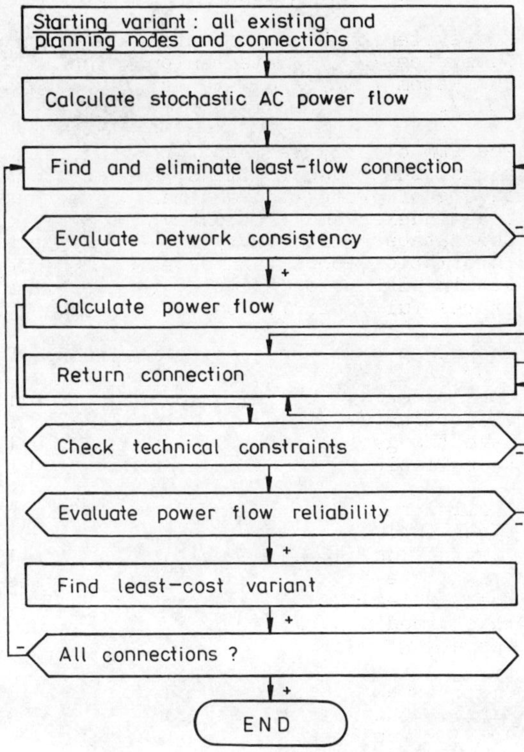

Fig. 1. Basic structure of the transmission system expansion planning

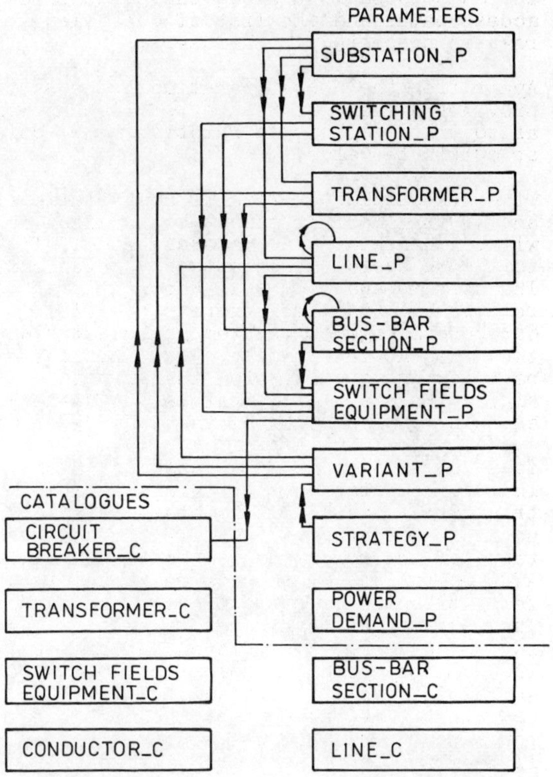

Fig. 2. The DIPOS database logical organization

a

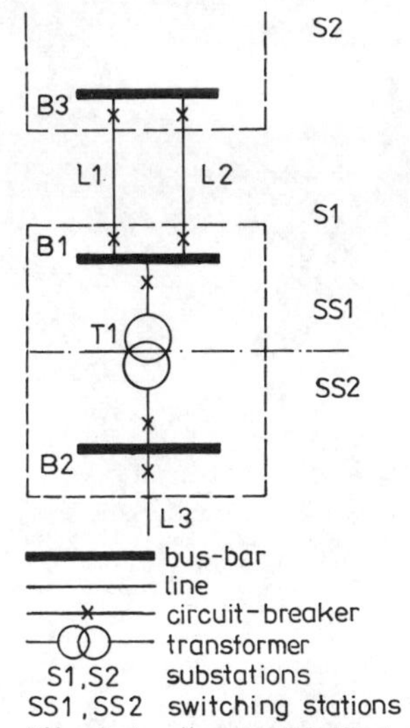

b

LINE_P

L_NAME	UN	LOC_CD	TYPE
L1	110	10	OVD
L2	110	11	OVD
L3	35	20	OVD

SUBSTATION_P

S_NAME	PF	CY	UN_MX	S_AREA	SS_NO
S1	0	1982	110	150	2
S2	0	1980	110	100	1

TRANSFORMER_P

T_NAME	CY	PF	TYPE	UN
T1	1982	0	HV/MV	110/35

SWITCHING_STATION_P

SS_NAME	PF	CY	UN	F_NO	SS_AREA
SS1	0	1982	110	3	50
SS2	0	1982	35	2	75
SS3	0	1980	110	2	100

UN-rated voltage, LOC_CD-localization code, PF-planning feature, CY-commission year, *_P-parameter relation

c

Resulting list after issuing LIST SUBSTATION_P.S1

S_NAME	PF	CY	UN_MX	S_AREA	SS_NO	SS_NAME	PF	CY	UN	F_NO	SS_AREA
S1	0	1982	110	150	2	SS1	0	1982	110	3	50

SS_NAME	PF	CY	UN	F_NO	SS_NAME	T_NAME	CY	PF	TYPE	UN
SS2	0	1982	35	2	75	T1	1982	0	HV/MV	110/35

L_NAME	UN	LOC_CD	TYPE	L_NAME	UN	LOC_CD	TYPE	L_NAME	UN	LOC_CD	TYPE
L1	110	10	OVD	L2	110	11	OVD	L3	35	20	OVD

Fig. 3. Network diagram (a), database (b), LIST operation (c) samples

COMPUTER AIDED DESIGN OF ELECTRIC DISTRIBUTION NETWORK

S. Kujszczyk

Department of Electrical Engineering, Warsaw Technical University, Warszawa, Plac Jedności Robotniczej 1, Poland

Abstract

Electric distribution networks are and essential part of power systems, In Poland system of computer - aided design of HV and MV networks have already been elaborated and put into operation.
The system is based on the data base. It is composed of the following data sets:
- catalogues, containing the technical and economic parameters of the equipment,
- catalogues classifying customers,
- data catalogues for reliability calculations,
- data for technical and economic calculations,
- parameters essential for the computer drawing of designs of electric networks,
- numerical representation of the existing networks.

The system has a block structure. It has following blocks of programs:
- block for writing and maintenance of data sets,
- blocks for generating new variants of the network,
- blocks for data processing,
- blocks for technical and economic calculations,
- blocks for drawing,
- auxiliary blocks.

The system operating in a conversational way enables the user to guide the calculating process as it takes place.
The user determines the sequence of performing the respective elements of the calculating process and, after every stage of the calculations, is able to return to any previous stage he chouses. The various calculations are performed in accordance with the user´s instructions. These instructions are in the form of instructions in so called problem oriented language, adapted to the needs of the system. Aknowledge of programming and the principles of the operation of computers is not required in using instructions in the language.

Keywords. Computer aided design. Electric distribution networks.

INTRODUCTION

Electric distribution networks are an essential part of power systems. In Poland, distribution networks comprise networks from low voltage /LV/380/220 V through middle voltage /MV/6 kV, 10 kV, 15 kV, 20 kV, and 30 kV to high voltage /HV/,110 kV.

These distribution networks constitute about 85% of the value of the whole power system.

System of computer-aided design of HV and MV distribution networks have already been elaborated and put into operation. It based on the data base.

The systems has a block structure.Because of the calculations which require controlling indirect results up dating the data and adjusting them, the elements of the system constitute autonomous programmes, that carrying out the following functions:
- formation and administration of data base of existing networks,
- the generation of alternative networks,
- the technical and economic calculations necessary in the process of desing.

The system is equipped with a problem oriented language of an interactive type, particularly suited to "computer-aided" work.

The system of design of electric distribution networks presented by Backlund,Bubenko 1979 , Hindi Brameller 1977 and Kujszczyk 1974, 1984 are computer-aided system. In the future, the system will be gradually transformed into a full automatic one.

THE DATA BASE

The system is based on the data base of a universal character. It enables the carrying out of both design calculation and operational calculations. The base in composed of the following data sets:
- catalogues, containing the technical and economic parameters of the equipment,

- catalogues classifying customers and informing about their daily load curves,
- data catalogues for reliability calculations, containing per-missible loss of load periods and the costs of the energy not supplied,
- data for technical and economic calculations,
- parameters essential for the computer drawing designs of electric networks,
- numerical representation of electric networks,
- loads in the existing networks.

For networks generated during the process of design, only a list of changes in relation to the existing network is given.

THE STRUCTURE OF THE SYSTEM

The structure of the system comes from its functions and interactive method of work. The best-suited is a block structure, in which every elementary task is conducted by a separate block. The work of the computing and organizational blocks is guided by a special block, called the executive programme.

The source programme, written in a problem oriented language, is translated into an usual programme by the translator block. After translating part of the source programme, translator transfers guidance to the executive programme utilizing the respective blocks. After implementing the translated part of the source programme, the executive programme transfers guidance to translator, etc.

The flow diagram of the system is shown in Fig. 1. Using the system of computer design and operation of electric networks, one can do several kinds of calculations charing appropriate blocks of programs. They are:

1. block for the writing and maintenance of data sets AKTU ,
2. blocks for generating new variants of the network,
3. blocks for data processing,
4. blocks for technical and economic calculations,
5. blocks for drawing,
6. auxiliary blocks.

Selected blocks are now discussed. The largest group of blocks consist of blocks for technical and economic calculations.

The main calculations done by group 4 are:
- calculation of the load flow and current flow in radial networks,
- calculation of load flow and current flow in mesh networks,
- calculation of voltages and particularly: voltage drops in network elements, nodal voltages, the cost caused by voltage deviations,
- calculation of short circuits, and particularly: the characteristic parameters of short circuits in the nodes of the network, the flow of short-circuit currents in radial networks, the flow of short-circuit currents in mesh networks, ground faults currents in MV network,
- calculation of power and energy losses: load and iron losses in transformers, load losses in electric lines,
- calculation of reliability: loss of load, probability parameters the cost

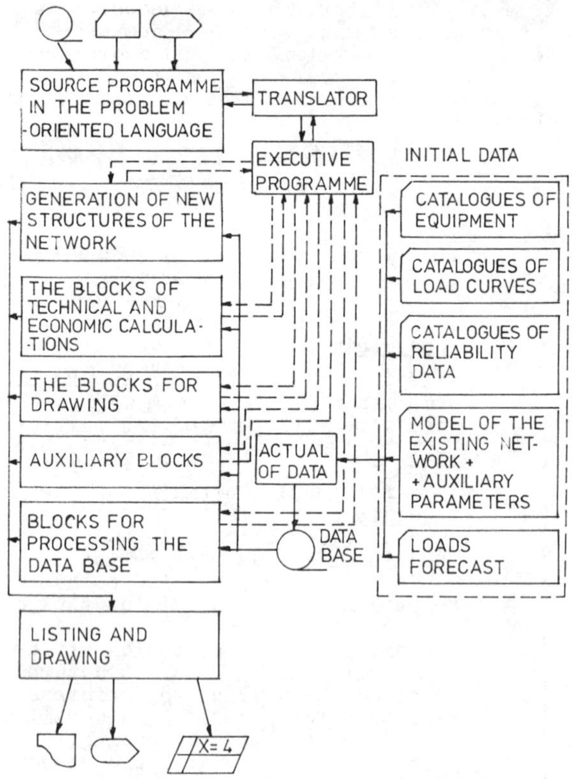

Fig. 1. Block diagram of the integrated system of designing and operating electric distribution networks

of customers losses caused by outs in the power supply,
- supervisory calculations: the supervision of permissible load flow, checking the level of short-circuit currents, checking voltage levels,
- economic calculations: calculating annual running costs, the costs annual power and energy losses, discounted costs, the synthetic cost index and the unmortgaged parts of the network costs,
- data base processing: listing of equipment according to: catalogue classification, belonging to various substations, and date of installation, manufacturers, etc.

Group 6 comprises auxiliary blocks,dealing with the network graph and the transformation and decomposition of the graph. This group includes the blocks which:
- generate the network tree,
- generate the tree with a given root,
- determine the path between nodes on a graph,
- change branches in tree.

The group of auxiliary blocks includes the programs which produce certain additional results, requested by the user.

The integrated system has modular structure. It can be extanded to include new calculating blocks and functions.

The printing of results can be done either in alphanumerical form or in the form of a drawing. The graphical printing of the network diagram can be accomplished with the use of a plotter. The data necessary for this purpose must contain a numerical model of the network diagram and coordinates of nodes, the scale of the drawing and the contents of the descriptions names, numbers .

CONVERSATION IN THE INTEGRATED SYSTEM

The computer system operating in a conversational way enables the user tu guide the calculating process as it takes place. The user determines the sequence of performing the respective elements of the calculating process and, after every stage of the calculations, is able to return to any previous stage he chooses. The various calculations are performed in accordance with the user's instructions. These instructions are in the form of instructions in so-called problem-oriented language, adapted to the needs of the system. A knowledge of programming and the principles of the operation of computers is not required in using instructions in the language. The language enables the user:

- forming, bringing up to date and processing the data base,
- solving various problems connected with the operation and design of the network,
- intervention in the calculating process, depending on intermediate results,
- printing and drawing the results.

The main parts of the language are its set of instructions and grammar.

The alphabet is a finite substantive set of symbols the vocabulary which can be divided into five parts.
Instruction heading PROCESS;
- key words of types of instructions,e.g. INPUT, OUTPUT, CHANGE, FLOW OF POWER, SHORT-CIRCUIT,...
- key words of objects, e.g. KZ, PNO, TZW, L, NR,...,
- limiters , ; = - .

The grammar comprises:
- syntax, which analyse and construct sets of correct expressions,
- semantics, which involves semantic laws of the language.

The performance of an operating design calculation of a network constitutes one process. This process is composed of instructions written by the designer and the relevant print-outs which represent the system's reply.

The construction of the instructions depends on its type, and the most developed one is in the following form: type of sentence function object qualifiers. The key words of the type of sentences define the scope of the sentences operations. For each type of sentence a list of functions is defined. For each functions a list of objects is defined. For each object a list of qualifiers is defined.

In operating and design calculations three kinds of functions are carried out with the aid of the language: describing the diagram and state of the network, calculations of the network, and printing the results.

Below are examples of sentences:
- CHANGE ERASE x_i-x_j;...;x_l-x_k; x_m;...;x_n; the branches x_i-x_j;...;x_l-x_k and nodes x_{mj};...;x_n referred to are eliminated from the network,
- TENSION x_j;...;x_j; calculation of the tension in the nodes referred to,
- ACTIVE FLOW LIVSCHITZ x_i; determing the flow of active power in the tree of roof x_i by the LIVSCHITZ method,
- THREE-PHASE SHORT-CIRCUIT x_i; x_j; calculating the short-circuit parameters for the indicated nodes of a mesh network in the case of a three-phase short-circuit,
- VOLTAGE REGULATION HV/MV-MV/LV x_i; calculating the optimal settings of high and middle-voltage and middle and low--voltage, transformes ratios in a network tree of roof x_i in the case of voltage regulation,
- PRINT PATH x_i,-x_j; the print-out of the branchs and nodes of the path x_i-x_j.

For the purposes of the language, TRANSLATOR has been written using a mixed technique. The COMPILER of the language produces on end programme in an intermediate language specially created for this purpose, which is a set of macroinstructions for performing technical, economic and auxiliary calculations. The macroinstructions constitute a sort of library of programmes which closely cooperate with each other by using common sets of data and results.

This structure permits great flexibility in performing calculations which can be arbitrary interrupted, and them re-started at any point in the previously performed process. This solution is more efficient than a strictly compilatory technique.

The end programme is then prepared by a simple INTERPRETER of the intermediate language. It calls forth and implements the successive macroinstructions.

The task of the TRANSLATOR is not only to compile and verify the formal correctness of the instructions but also to check the sequence in which the respective instructions occur in the and programme. An example of an incorrect sequence in the instruction to calculate power losses without previously calculating the load flow. It should be emphasised that the COMPILER and INTERPRETER have been written in FORTRAN IV. This makes it relatively easy to introduce changes to the TRANSLATOR. It also enables existing programmes to be utilized.

SUMMARY

A computer system of designing HV and MV electric distribution networks has been introduced in many cities in Poland, including the two largest, Warsaw and Łódź, It is currently being introduced in rural networks. Of course, before the systems were introduced, the data base had to be prepared. This was a particularly difficult and labour-consuming process. However, the systematic bringing up to date of the data base is incomparably simpler.

REFERENCES

Backlund, Y., J.A. Bubenko 1979 . Computer-aided distribution system planning. Electrical power and energy system, Nr 1

Hindi, K.S., A. Brameller, 1977 . Design of low-voltage distribution networks; a mathematical programming method. Proceedings IEE Power Record, January

Kujszczyk, S. 1974 . Automatic design of housing estate distribution networks. Proceedings IEE Power Record, June

Kujszczyk, S. 1984 . Nowoczesne metody obliczeń elektroenergetycznych sieci rozdzielczych. WNT, Warszawa

COMPUTER-AIDED DESIGN OF INDUSTRIAL ELECTRICAL DISTRIBUTION NETWORKS

J. Kulczycki* and M. Zabicki**

**Institute of Power Systems Engineering, Academy of Mining and Metallurgy, B-1, Al. Mickiewicza 30, Krakow, Poland*
***ALPHA, ul. Balicka 176, 30-149 Krakow, Poland*

Abstract. The paper presents general-purpose computer-aided design system for the determination of the optimal design of the industrial, electrical distribution system. A cost-effective structure for such a network is often a multilevel, hierarchical structure with one or more voltage levels and several kinds of equipment. It consists of main supply point, transformer substations and/or switchboards together with supplying lines (which all together form a "backbone network") and cable feeders connecting demand points to backbone network (local networks). A distribution planning model is formulated using graph theory and discrete programming. Decisions can be taken on two levels - the first concerns backbone network, the second specifies assignments of demand points to particular elements of the backbone. The model is rather general - backbone can be any of loopless structure. The problem structure typically leads to heuristic solutions of an iterative nature. A kind of such method is implemented in a general-purpose software package letting on automatic design of industrial electrical distribution systems.

Keywords. Computer-aided design; modelling; heuristic programming; optimisation; power distribution.

INTRODUCTION

The designing of electrical networks is a complex process in which the application of computing techniques has grown steadily, from the well established areas of load flow and security analysis for high-voltage transmission systems into the more recent computer-aided procedures for planning electrical distribution systems.

In practical distribution system design problems, the number of possible line routes, transformer and switchboards locations is very large. Furthermore, equipment capacities must usually be chosen from a finite set of available options. The planner is confrontated with a multitude of technical and economical constraints. The equations relating construction and operating costs to system configuration are usually nonlinear. In summary, the planning problem is very complex, even without considering the difficulties posed by uncertainties in load forecasts, and sometimes beyond the resolving power of the unaided human mind. The planner needs to have access to appropriate computational technique; this would give him a unified approach to the many different distribution system design problems, and thus would enable him to systematically check a great number of project variants. In other words, the planning of electrical distribution system should be formulated as a problem of mathematical optimization, which could serve as a basis for computer-aided design procedures.

The selection of a suitable computational technique for the outlined problem poses severe difficulties, and several attempts have been made in the application of mathematical programming to the solution of distribution planning models. Mathematical synthesis was first attempted by Knight (1960), who proposed a spatial model of the network and solved it using linear programming. Boardmanand Hogg (1972), later Adams and Laughton (1974), as well as Kulczycki (1976) presented alternative approaches to the problem using dynamic programming and branch-and-bound method. More recently Wall, Thompson and Northcote -Green (1979) proposed very effective transshipment formulation of the problem, and Backlund (1978) brought together many modern techniques to provide a model package for electric distribution systems synthesis.

Heuristic methods (Meckiff and co-workers, 1979; Zabicki,1982) have received much less attention than mathematical ones, which is somewhat surprising taking into account model simplification needs. Very often the implementation of the heuristic algorithms in an interactive system employing computer graphics enable a study of many aspects of a problem and intuitive deductions of alternative designs leading to computational speed and simplicity.

This paper outlines the principles of

a heuristic method for optimization of any loopless distribution system and presents based on it computer-aided design package which can serve as a tool in many realistic problems of searching for acceptable design of industrial, electrical distribution networks.

THE PROBLEM

Many of industrial, electrical distribution systems are low voltage or medium voltage/low voltage networks where power flows in one direction only, from the substation (or switchboard) to demand points. The unidirectionality of power flow, in addition to the low statistical failure rate of equipment at low voltage levels, led various authorities to adopt loopless arrangement for these networks.

The distribution network is characterized by a number of loads, dispersed within the planning area at known locations, and assumed to be assigned to one or more energy sources (supply points). Moreover, the potential sites for locating intermediate transformer substation and/or switchboards are specified. This enables to use electric lines more efficiently - more cost-effective lines can be employed and the number of long lines can be reduced. From this basis one outlines all feasible routes for connection of the switchboards, substations and supply points, as well as demand points. Eventually existing system facilities can also be accounted for. The resulting set of feasible routes, potential sites, loads and their locations, together with specification of standard equipment and technical and economical data, form the data base for optimization. The network structure can be conceived as hierarchical tree - loads connected by cable feeders to switchboards, switchboards to substations, substations to supply points. In describing the problem further, we confine ourselves to a three-level hierarchy: demand points transformer substations (or switchboards) and one supply point. Extensions to more levels and more supply points are straightforward. The objective is to select:
(1) transformer substations with respect to number, sites and sizes;
(2) cable routes and sizes,
to supply given loads at the lowest possible cost, and subject to capacity limitations on network elements, thermal rating and voltage drop constraints, as well as topology and reliability constraints.

We can treat the supply point and transformer substations with supplying lines as a "backbone network", and cable feeders connecting demand points to "backbone" as low level, local networks. Thus the global design problem consists of two interdependent subproblems: the design of the "backbone network" and the design of the local distribution networks.

THE MODEL

The network model is a mathematical formalisation of the described above network design problem. For the purpose of the research covered in this paper, the model consists of a number of loads, substation and one supply point, which are to be interconnected so that the complete network satisfies all load and reliability conditions and minimizes the chosen objective function. A set of mathematical constraints which define network topology, reliability and other important features are derived from the engineering considerations which apply to industrial, electrical distribution networks.

The model to be described requires three kinds of data as inputs: the demand point locations and their load, the feeder network segment description and their possible routes, the description of substations and their supplying lines. Thus:
(1) we assume that the total demand for electrical power in a small area is represented by a single demand point; the power demand of that area is a load of the demand point;
(2) the feeder network is described by means of two data structures - the first contains descriptions of the types of cables which can exist in the local networks, the second gives the lengths of line segments between each pair substation - demand point;
(3) the data required for each substation are its capacity and its location, actual or potential. The substation capacity represents existing or proposed maximal transformer rating. Two data structures, similar to that of feeder networks, are specified for each supply line.

System Costs

The total cost of distribution network includes the cost of transformers, cost of civil work at substations, cost of supplying lines, cost of feeder cables, cost of trenching and cables lying and cost of electric losses in the system.

It is generally accepted that the cost of a substation or cable can be represented by cost-capacity functions. Such a function comprises all the cost components associated with the establishment of power flow distribution capacities, inclusive of capitalized cost of losses. All investments are converted to annual costs considering the economic life of equipment. Also, annual costs of losses during the period of analysis are discounted to present worth. The cost-capacity functions are obtained by superimposing these cost items. Fig. 1 illustrates cost--capacity curves for transformer different

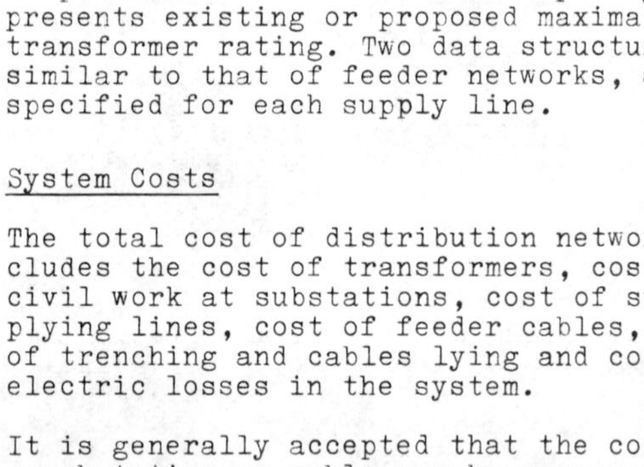

Fig. 1. The Cost-Capacity Curves.

ratings. Similar cost functions can be obtained for cables.

We assume that the cost of supplying lines as a function of substation capacity, is superimposed on the cost of transformers and the resultant cost function is then linearised to yield a two-stage cost function of the form represented by Eq. (1).

$$f = \alpha + \beta S \tag{1}$$

Network Representation

The representation of the design problem is by a network of nodes and links. The nodes represent the supply point, the potential and existing substation sites and locations of demand points. The links represent the feasible cable and supplying line routes. The schematic and network representation of the discussed electrical distribution system are exemplified in Fig. 2 and Fig. 3, respectively. With each node representing substation site associated are two quantities, first, an upper bound on carrying power, which in fact is the maximum capacity for the transformer type of that substation (or the capacity of the number of transformer units that can be installed), second, a cost coefficient which appears in the objective function of the design problem and represents a linear approximation to the cost of power losses and installation costs for that site. With each feasible

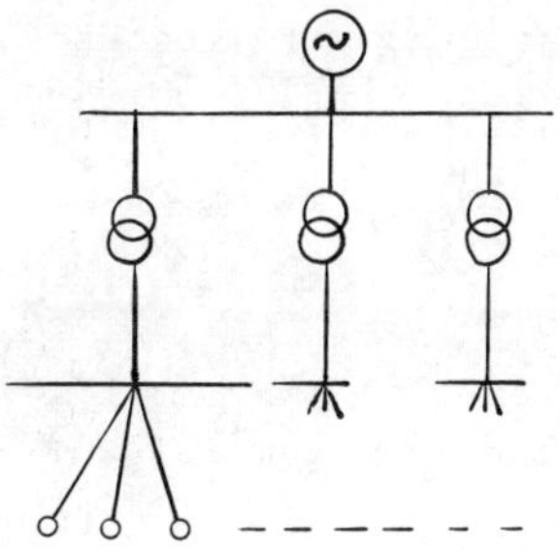

Fig. 2. Schematic Representation of the Three-Level Distribution Network.

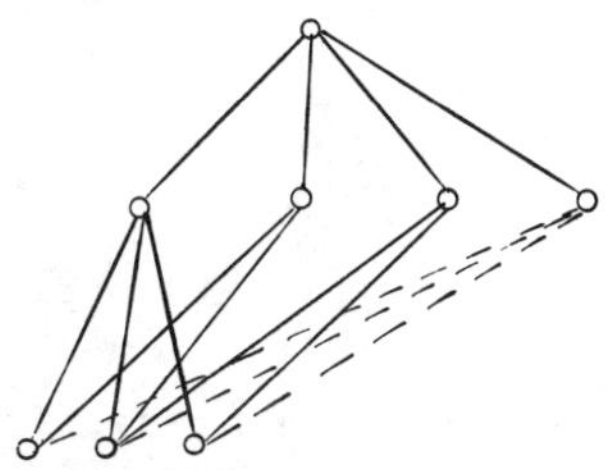

Fig. 3. Network Representation of Possible Cable Routes and Substation Sites.

connection substation-demand point we associate connection costs, which can be computed in advance, based on the cost--capacity curves for cables, load of each demand point and distance between that pair.

Mathematical Formulation

We are now in position to state the problem in a mathematical fashion. The problem to solve is:

$$\text{Min} \sum_{i=1}^{m} (\alpha_i + \beta_i S_i) y_i + \sum_{i=1}^{m} \sum_{j=1}^{n} b_{ij} z_{ij} \tag{2}$$

Subject to

$$R \ni R_t = \{ i \; y_i = 1 \} \tag{3}$$

$$\sum_{i=1}^{m} z_{ij} = k \; ; \quad j = 1,\dots,n \tag{4}$$

$$S_i = \sum_{j=1}^{n} S_j z_{ij} \leqslant S_i^{max} \; ; \quad i = 1,\dots,m \tag{5}$$

$$z_{ij} \leqslant y_i \; ; \quad i = 1,\dots,m; \quad j = 1,\dots,n \tag{6}$$

$$y_i \in (0,1) \; ; \; z_{ij} \in (0,1) \; ; \quad i = 1,\dots,m; \; j = 1,\dots,n \tag{7}$$

The notation used are:

- m number of feasible substations
- n number of loads
- α_i fixed charge for substation i (including supplying line)
- β_i incremental cost of substation i
- S_i load of substation i
- b_{ij} cost of connection between i-th substation and j-th demand point
- $y_i = \begin{cases} 1, & \text{if substation i is opened} \\ 0, & \text{otherwise} \end{cases}$
- $z_{ij} = \begin{cases} 1, & \text{if j-th load is connected to substation i} \\ 0, & \text{otherwise} \end{cases}$
- R set of feasible configurations for backbone network
- k number of substations to which each demand point have to be connected
- S_j load of j-th demand point
- S_i^{max} maximum capacity of the i-th substation

The inclusion of the integer variables in the objective function (2) yields a correct representation of the system costs. Constraint (3) is a "backbone" topology constraint; constraint (4) is local networks topology constraint - for k=1 they have standard, radial configuration, and k=2 means double radial configuration for these networks. Thus constraint (4) can be treated also as reliability constraint.

Condition (5) ensures that the load of each substation is zero when it is not chosen, and less than the maximum rating of the largest available transformer or switchboard (or group of them) when it is chosen. The configuration constraints (3 - 4) give the model a flexibility to incorporate many of special prerequisities which can be found in real applications. For instance, if substations are the main elements of the backbone network that constraint can force some specified substations to be a part of any solution, to ensure or suppress certain combinations etc. It can also incorporate tree topology of the backbone network (f.e. if the distribution system is to be low-voltage, cable-busbar network, or if we allow a tree connections of switchboards to supply point.

The formulation as given above is similar to plant, warehouse and facility location problem found in operations research literature. Most proposed methods for solving it employ branch-and-bound technique or heuristic procedures. The problem as stated above is a hard integer programming problem running times which probably grow exponentially with the size of the problem. Thus for large problems, the optimal algorithms are generally found to be too expensive in terms of storage space and computation time. Furthermore, they do not easily accomodate extensions to layouts other than standard (i.e. each element connected directly to the next highest element in the hierarchy). The choice of solution technique for the electrical distribution network design problem has therefore been concentrated on heuristic procedures.

OUTLINE OF THE METHOD

The global design problem consists of two subproblems: the design of the backbone network and design of local networks. The two subproblems interact with each other through the following paramters:

(1) backbone elements number, their layout and sizes
(2) demand points assignment to backbone elements
(3) capacity constraints for backbone and local networks
(4) reliability requirements.

Once the above variables are specified, the subproblems can be solved independently. This naturally leads to heuristic solution process of an iterative nature as follows:

(i) determine backbone network configuration
(ii) assign loads to backbone
(iii) check if solution is improved.

To implement step (i) one can employ well known ADD generating scheme (Kuehn and Hamburger, 1963), often used in the ordinary plant location problem. The algorithm starts with no substation installed. It then investigates, step by step, all allowable sites. The selection of the added site is based on the greatest savings in the cost function. The procedure thus involves a sequence of demand point - substation assignment problems (step ii) for different sets of specified substation locations, each with a distinct new substation site, and selecting the best new , which gives the minimum cost. The procedure terminates when no more savings in cost can be accomplished by adding substations.

The need for solving a sequence of demand points - substation assignment problems in each step of the algorithm demands the very effective solution procedure for that problem. However, it can be easily shown that for a fixed set of installed substations, problem (2 - 7) reduces to the generalized assignment problem, which belongs to the class NP-complete (Fisher, 1981). The best solution methods (exact) available today for that kind of problems are based on implicit enumeration or branch-and-bound, and thus are inadequate for our purposes. What we need is a fast heuristic algorithm. Under some conditions such an algorithm can be constructed. It proceeds as follows: initially all loads are unassigned. Then at each step least cost assignment is found, and if is feasible (i.e. the chosen substation is not overloaded) the step is terminated. Otherwise, the second cheapest, then the third and so on, is tried. The detailed discussion of that algorithm and conditions under which it works is presented in the work of Zabicki (1982).

On the first sight, the assignments could be found based on l.v. connection costs. But this do not accomplish obvious interdependence between backbone and local networks costs. More exact decisions can be made if we would also take into account increase in substation costs due to connection of some demand point. This can be achieved by constructing mxn, so-called decision matrix c_{ij} , where

$$c_{ij} = b_{ij} + \beta_i S_j \tag{8}$$

Now, the cost of the system consists of the fixed cost of substations, given by

$$\sum_{i=1}^{m} \alpha_i y_i \tag{9}$$

and assignments cost, given by

$$\sum_{i=1}^{m} \sum_{j=1}^{n} c_{ij} z_{ij} \tag{10}$$

It is evident, that the formulation given in (2 - 7) is equivalent to the formulation with cost function given by the sum of formulas (9) and (10). The attractiveness of this approach lies in the fact that decision matrix is to be constructed only once, at the start of the solution process, while cost calculations during the solution process are very simple and effective.

Although the mathematical model is based on the linearisation of the cost functions (cost vs. power) the true discrete costs can be made to govern the search for a "good" solution. Once the network configuration is fixed, the solution can be thought as a schedule of power flows in a network. Hence an appropriate cable, transformer or switchboard size can be nominated for a cable route or substation site, according to the power flow indicated by the solution. Consequently the true discrete costs can be calculated.

The proposed solution method is not optimum. The sites found earlier should not necessarily be retained as more substations are entered into backbone. Similarly, on lower level, demand point assignments to substations can be not optimal. To alleviate problems like that, the design system incorporates a variation of the algorithms, allowing a decision at each step to drop one of the sites already selected, or to interchange a site already in solution with the site actually out of it.

COMPUTER-AIDED DESIGN SYSTEM

The purpose of the system is to enable designers to study industrial, electrical distribution networks within the framework of the mathematical model. It gives them possibility to efficiently check great number of project variants and thus encourages a more consistent and comprehensive system planning.

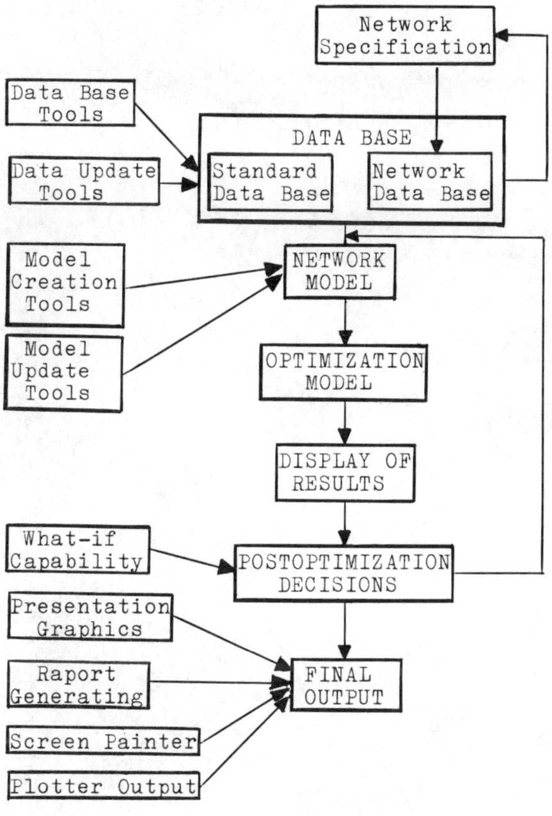

Fig. 4. Block Scheme of the Computer-Aided Design System

The design process is consisted of three major phases. The first one is preparing the Network Data Base. As shown in Fig. 4 the computer analyzes the user's input (i.e. network specification) and determines what additional information is required to perform the model building phase. Some of this information may be supplied automatically by the computer through the use of intelligent defaults from the Standard Data Base (i.e. network equipment catalogue and specifications, cost data). The remaining information is provided by the user in response to prompts via a simple interactive display interface. The specification of the problem is progressively refined until the Network Model can be generated automatically.

The Data Base creation and update employ a variety of support tools, among them Data Dictionary is essential. It represents a data model maintained by the Data Base. Support facilities such as Data Modelling Tool and Data Analysis Tool are needed for synthesizing user informations into a fully normalized data model. Subsets of the Standard Data Base are then extracted by a user and together with the Network Data Base go under integrity studies. These includes range and data type checks, checks for permissible values, and checks involving cross-reference to other data.

In the second phase of computer-aided design process, the network model of the of the specific system under study, is created. It requires as input: (1) locations of supply point, substations sites, and demand points, (2) matrix of minimum feasible distances between pairs substation site - demand point, (3) vector of minimum feasible distances between all substation sites and supply source, (4) demand points loads and maximum load for each substation site, (5) technical and economical indices of the equipment to be utilized. The network model is then transformed into Optimization Model which states a base for a configuration search. The decisions are performed using an optimization package, which employs heuristic algorithms outlined in the previous section. However, other methods can be applied at this stage, f.e. branch-and-bound technique for problems of smaller size. The output from that phase gives the total load for each substation and its optimum service area, i.e. a set of demand points connected to that substation. After accepting the results postoptimization decisions can be performed and model can be modified.

Finally, at the third stage of the design process, the optimum substations capacities are determined from their loads. Cables sizes are determined by economic considerations involving line lengths, load, losses and cost. Voltage drops can be also considered. The user gets from the design system tabular summaries by demand location, substation and supply point (if more than one), construction and losses costs and voltage drops in percent. In addition, a printer plot of the substation service areas is provided and a graphical display of the network layout is given.

CASE STUDIES

The described, computer-aided design system is actually under development and will finally run on Digital PRO 350 microcomputer. However, the "heart" of the system, an optimization model has been already implemented and heavily tested on CDC6600 for several different network structures, among them medium voltage/low voltage distribution network with radial configuration of low voltage feeders and reliability constraints, the low voltage radial network with switchboard substations as an intermediate link between supply sources and demand points, and the cable--busbar low voltage distribution network often used in industrial plants. In the last case the backbone (busbar) part of the network is to have tree topology, while the demand points are radially connected to busbars. Running times have been on the order of seconds for networks containing about hundred of loads and fourty substations (or busbar segments). The quality of the obtained solutions has been generally in the range 5 - 10% of the optimum value reached using branch-and-bound package (where it was possible to find optimum). The results suggests that the model is well suited for the design of loopless electrical distribution networks. It seems very likely that using the full, computer-aided design system solutions will be even closer to optimum.

CONCLUSIONS

We have presented in this paper an approach for optimizing loopless, electrical distribution networks. The distribution planning model has been formulated based on graph theory and discrete programming, and with simplicity as an important objective. The proposed solution method has been typically heuristic, however results of using it have been very pleasing. The method does not give an exact solution with one run of the package, but instead gives the planner a tool with which to evaluate different network options. The full computer-aided design system is currently under development. It will allow a review of distribution projects to be done in a fraction of time and with far more confidence than previously possible, moreover, an interactive capability of the system will give a planner a large amount of control at each stage of the design process.

REFERENCES

Adams, R. N., and M. A. Laughton (1974). Optimal planning of power networks using mixed-integer programming. Proc. IEE, 121, 139-147.

Backlund, Y. (1978). Hierarchical Modelling of Regional Electric System. Ph. D. Thesis, The Royal Institute of Technology, Stockholm.

Boardman, J. T., and B. W. Hogg (1972). Computer method for design of electricity supply networks. Proc. IEE, 119, 851-856.

Fisher, M. L. (1981). The lagrangian relaxation method for solving integer programming problems. Management Sci., 27, 1-18.

Knight, U. G. (1960). The logical design of electrical networks using linear programming methods. Proc. IEE, 107 A, 306-314.

Kuehn, A. A., and M. J. Hamburger (1963), A heuristic program for locating warehouses. Management Sci., 9, 643-666.

Kulczycki, J. (1976). Optimization of Industrial Electrical Distribution Networks. Zn AGH, Krakow, Poland.

Meckiff, C., J. T. Boardman, I. Richards, and J. R. Green (1979). Comparative analysis of heuristic synthesis methods for electricity transmission networks. IEEE Trans. on Power Apparatus and Systems, PAS-98, 187-196.

Wall, D. L., G. L. Thompson, and J. E. D. Northcote-Green (1979). An optimization model for planning radial distribution networks. IEEE Trans. on Power Apparatus and Systems, PAS-98, 1061-1068.

Zabicki, M. (1982). Combinatorial Optimization of Electrical Distribution Networks. Ph. D. Thesis, Academy of Mining and Metallurgy, Krakow, Poland.

DESIGN AND VERIFICATION OF SEQUENCE CONTROL SYSTEMS WITH TIMED PETRI NETS

Y. Nakamura, S. Abe and S. Takeda

Central Research Laboratory, Mitsubishi Electric Corporation, Amagasaki, Hyogo, Japan

Abstract. An interactive design support system "INDEED" for industrial sequence control is developed at an engineering work-station. The main features of the system are a user-friendly man-machine interface, a timed Petri nets based simulator and an intelligent verifier for the design modification. An integrated graphic editor/simulator is operated by a menu and a mouse. After a controller is designed as a logic diagram with the editor, a simulation can be carried out on the same logic diagram and a timing chart appears upon the diagram by means of a multi-window system. The timed Petri nets based simulator converts the logic diagram into a timed Petri net. The whole system (the controller and the controlled devices) is simulated. The properties of Petri nets, such as liveness, conflict and reachability, can be used for analyzing and improving the design. As a design is often carried out by the modification of an old design, a verifier for the design modification is developed. The verifier extracts different portions of the timed Petri nets of these designs and converts them into automata, and the detected difference is shown as state-transition maps.

Keyword. Sequence control; sequential machine; Petri nets; automata theory; controllers; computer-aided design; control system analysis; digital system.

INTRODUCTION

With the development of microprocessors, the implementation of an industrial sequence controller has changed from a wired relay logic to a microprocessor based programmable digital controller (PC). Due to the flexibility of PC, not only small scale systems but also large systems such as electric power plants have been implemented in PC. PC decreases the controller's hardware cost. However, the efficiency of the control logic design has not been improved.

In the logic design of LSI, many CAD systems, such as a logic diagram drawer, a logic simulator, a test tool and so on, have been developed (Parker, 1981; McWilliams, 1978; Saito, 1981; Werner, 1983). These tools can not be introduced directly in the field of the sequence control:

(1) In a drawer, the shapes and the representing functions of symbols are different,
(2) a simulation or verification method for a controller with controlled devices must be developed,
(3) as a test and an adjustment are carried out at the power plant's location, a tool for use at that location must be developed.

The design aids for sequence controllers have hardly been considered, except for a description of the controller by programming languages (Biel, 1981; Takahashi, 1980). The languages are based on those developed for general-purpose computers in the scientific field (e.g. Pascal, PL/1). Another approach deals with the sequence control system as an asynchronous and parallel machine. Several methods were proposed for modeling and designing using Petri nets (Peterson, 1977). But these methods (Masuda, 1980; Famchandani, 1974; Azema, 1976) have not been widely used in practical applications. Designing by these methods not only requires a change of design style, but also leaves out a logic diagram of a control logic, which is necessary either for simulation or for testing and modification at the location of the plant. Therefore, many designers use a logic diagram instead of these methods, but reliability and productivity do not improve.

A sequence control system consists of two parts: a controller and controlled devices. In this paper, the controller is designed as a logic diagram. But the controlled devices are modeled with timed Petri nets, because it is easier to represent the devices' concurrent behavior by Petri nets than by logic diagrams or state machines. The designed logic is converted into a timed Petri net (TPN), because

(1) a TPN is easy to combine with other TPNs, thus making a simulation of a whole system with devices possible,
(2) a simulation is performed by a simple firing rule, and
(3) analysis methods, such as a reachability tree (Peterson, 1977), are applicable.

In the following sections, an interactive aspect of sequence design system "INDEED" and a use of timed Petri nets for simulation and verification are presented as an example of a design of an electric power plant.

TIMED PETRI NETS

Petri nets are developed for graph models of information flow. We use this modeling tool to model the state transitions that occur in a sequence control system. A transition firing time is introduced that assigns firing times to the transition of the Petri nets. These Petri nets are called "timed Petri nets".

Figure 1 is a basic building block of a Petri net (PN). The vertical bar is called "transition", while the circles are called "places". The dots on the places are called "token". A Petri net is the interconnection of such modules. Thus a Petri net is considered as a directed bipartite graph which

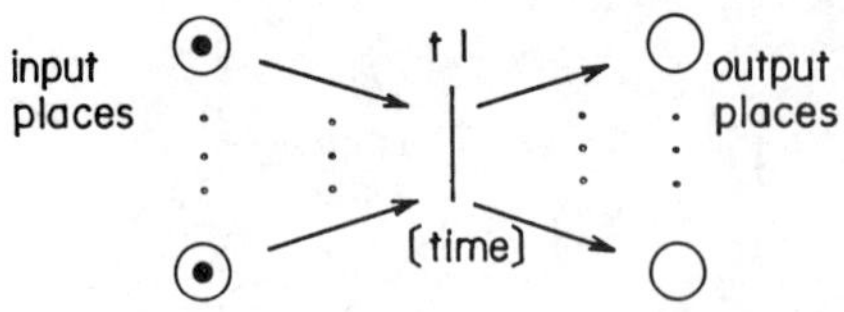

Fig. 1. A Petri net with a transition time.

allows a directed arc from a "place" to a "transition" or vice versa. In our model, the transition represents an event taking place in the device and in the control logic. The event is a change of state from one state to another caused by input signals. The place represents a condition of the devices and signals.

To simulate the state change flow in a Petri net, a "token" is put on each place. A token represents that some condition associated with its place is satisfied, or that a device holds that state. The token distribution is called the "marking". The marking can be thought of as a state of the modeled system. A transition is said to be "firable" or "enable" if each of its input places contains at least one token. The "firing" of transition is performed by removing one token from the transition's input places and adding one token to each of its output places. The transition time is added to the firing rule. If a transition becomes firable, the transition must fire after the transition time is spent.

If the number of tokens is any places never exceeds one, Petri nets are called "safe nets" or "safe". Timed Petri nets with the safe condition are used in this work.

SYSTEM CONFIGURATION OF "INDEED"

INDEED supports the whole design process of the controller which is designed as a logic diagram and is implemented in PC. Figure 2 shows a job flow of a logic design and the corresponding support tools. The software structure of INDEED at a work-station is shown in Figure 3. The design policies of INDEED are:

(1) to design a controller using a logic diagram in accordance with the current design style;
(2) to use timed Petri nets as basic tools (timed Petri nets are suitable for representing the asynchronous and concurrent behavior of the devices);
(3) to support the whole process of the design, i.e. drawing a logic diagram, simulation, verification and formatting;
(4) to provide a user friendly interface with a mouse in order to avoid complex keyboard commands.

The software tools of INDEED are described as follows.

DDB (a device database). The components of a power plant are modeled in DDB. DDB contains TPN representation and input/output (I/O) of the devices. As an example of the contents of DDB, Figure 4 shows an entrance valve of a hydro-electric power plant.

SEE (a schematic entry and editing system). A controller is not designed as a timed Petri net (TPN) but as a logic diagram. SEE is used for the logic design. First of all, a user selects the devices of his plant. SEE extracts their models and I/O signals from DDB. A control logic of the devices is designed on a screen. I/O signals of the devices are registered in SEE. If a designer uses other signals except these signals as I/O signals, SEE confirms whether a new signal is really required or not. This avoids the misuse of I/O signals.

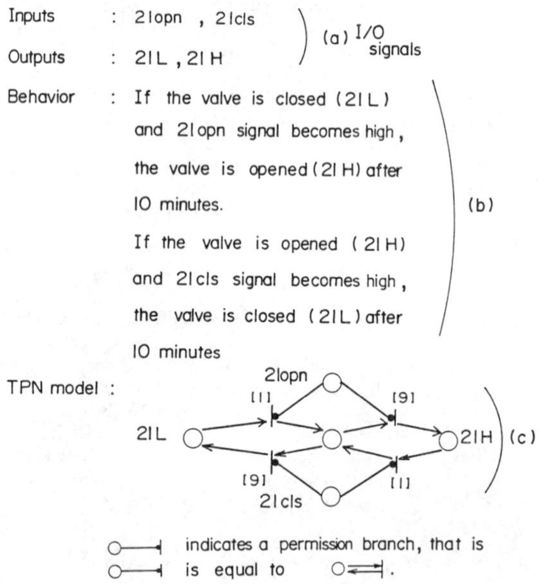

Fig. 4. An example of the contents of DDB. A part (a) is a definition of input/output signals, (b) is a description of the valve's behavior in a text, and (c) is a TPN representation of a state's transition.

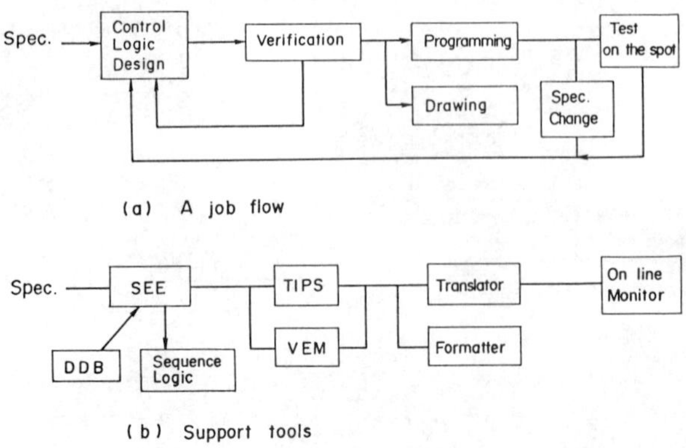

Fig. 2. A job flow of the sequence control system design and the corresponding support tools.

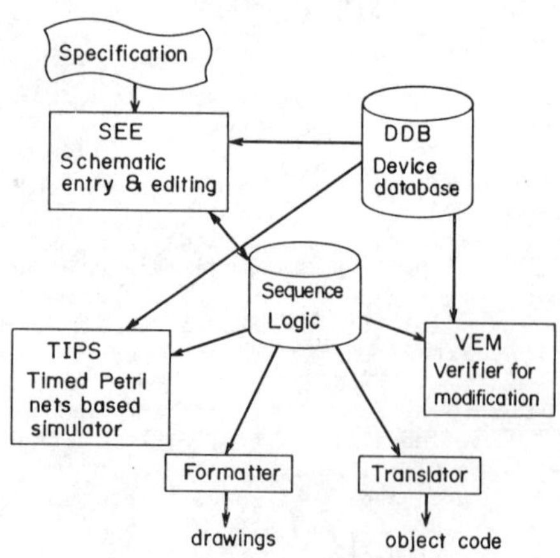

Fig. 3. The software structure of INDEED.

Figure 5 shows the screen of SEE, on which a start-up sequence logic of the hydro-electric power plant is designed. The right-lower window of Figure 5 shows a whole logic in a smaller scale. The editing area is indicated by the reversed color in the upper left corner of the right-lower window. This area is enlarged in the left-upper window. In this area, a logic diagram is composed or modified. The interactive facilities and features are summarized in Table 1.

Table 1 The features of SEE

(1)	A mouse based operation	Menu selection, symbol arrangement, deletion, etc. can be done without complex keyboard operations.
(2)	Menu driven command	Device insertion, symbol selection, etc. are indicated on a menu by a mouse.
(3)	Wiring	There are two alternatives. One uses a self-searching algorithm, when a source and a destination points are indicated. The other is a rubber band.
(4)	Database	Behavior of devices and their input/output signals are easily referred from DDB. A designed logic is entered in DDB as a sub-logic.
(5)	Interface to other systems	A network list of the logic diagram is available for other systems; for example, other simulators in a main frame or other CAD systems. A printed form (hardcopy) of the logic diagram is available. The logic is translated into object codes for a programmable controller.

TIPS (a timed Petri nets based simulator). TIPS converts the designed control logic into a TPN automatically. A TPN of a control system is composed by a combination of the TPNs of the logic and the devices. TIPS makes a TPN work according to the firing rule of TPN.

VEM (a verifier for design modification). In many cases of the design of a power plant, the new design is performed by the modification of an old similar system. VEM verifies the modified function of the system.

TIPS and VEM are the major tools, called from SEE. The interactive facilities and the methods used on TIPS and VEM are presented in the following sections.

TIMED PETRI NETS BASED SIMULATOR

We use timed Petri nets as a simulation tool, because timed Petri nets have the following advantages:

(1) a simulation is performed by the firing rule only;

(2) machines represented as timed Petri nets are easy to combine;

(3) analysis methods such as the liveness check, the conflict check, and the reachability tree are available.

The timed Petri nets based simulator (TIPS) converts a designed logic into a TPN. Then a TPN which represents a control system with devices is composed by a combination of the TPNs of the logic part and the devices. This composition is simply the overlapping of the output places with the input places which have the same names. A conversion example of the AND gate into TPN is shown in Figure 6.

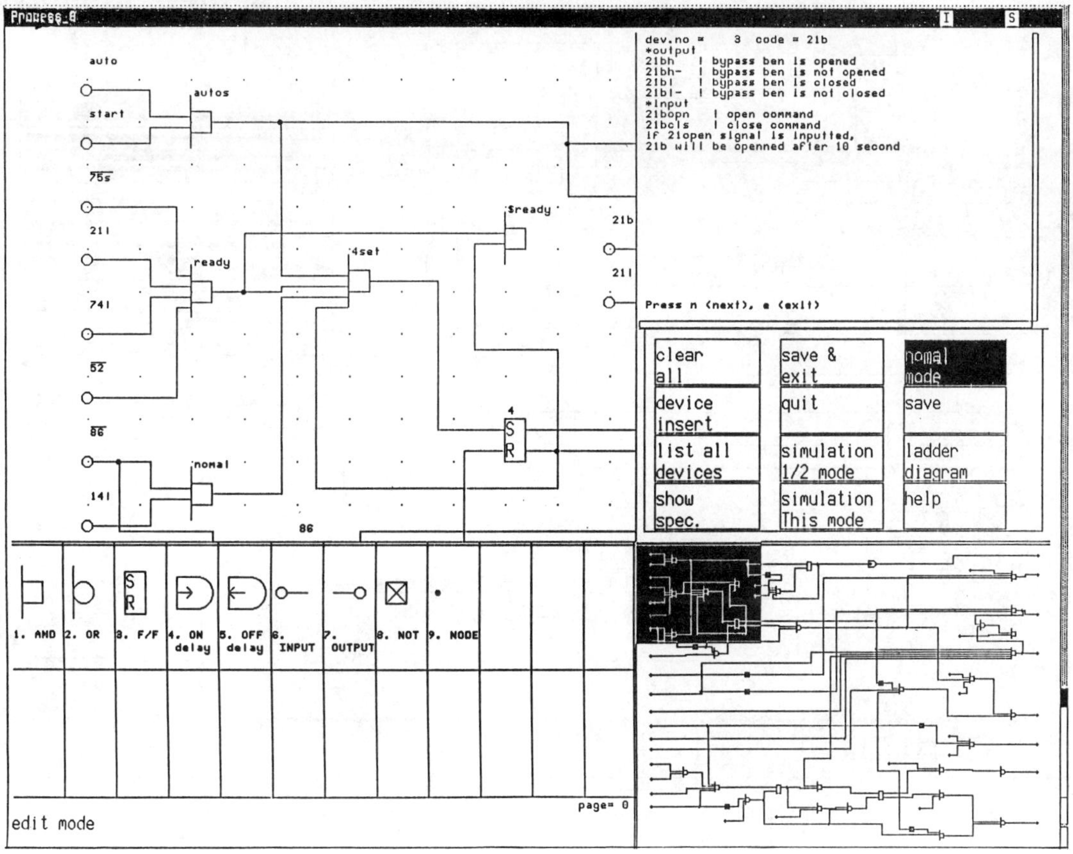

Fig. 5. The display screen of SEE.

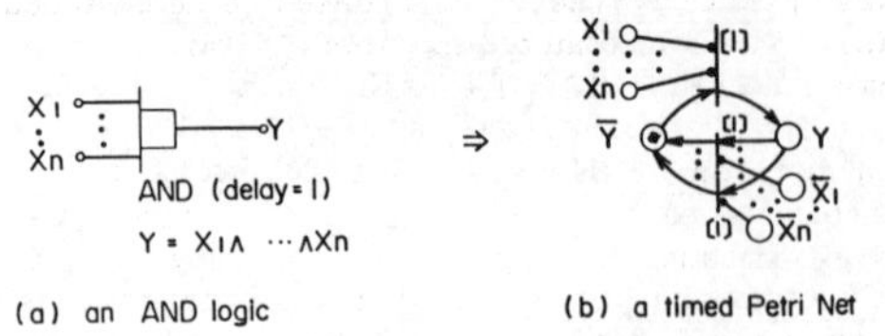

Fig. 6 A conversion of an AND logic into a TPN.

The TPN of a whole sequence control system is simulated to examine its behavior. To start a simulation on TIPS, an initial marking of the TPN that represents a state of the devices and the logic gates must be given. From any state of the system, a simulation can be begun. A state change takes place by firing the transition associated with that state according to the firing rule of TPN. During the simulation, TIPS checks whether conflicts, deadlocks or endless loops occur. A non-deterministic state transition is detected as a conflict. TIPS can compose a reachability tree which is used to inspect the system.

Interactive Operations

A mouse device is used to set or change an initial state of flip-flops and input gates. To assign an initial marking of devices, TIPS lists all states of a device (e.g. valve is opened, or closed), then a user selects one of them. If the start command is entered, a simulation is begun. The outputs of a logic are sent to the TPN models of the devices. If any transitions of the TPN become "enable" and they have no conflicts, they are fired by TIPS. If the devices have feedback inputs to the logic, TIPS makes the inputs of the logic change. This causes a successive change of the state. Then the simulation goes on.

If deadlocks or conflicts are detected during a simulation, TIPS stops to ask for a user's instruction, because they indicate design errors. The user can select whether to change a state locally and continue, or can ask TIPS to display a timing chart in order to inspect the state transition process.

Figure 7 shows a timing chart and a state of a logic. The state at the time indicated by the vertical line on the timing chart is displayed on the logic diagram in a background window. Once the vertical line is moved left or right by the mouse, the state at that time is also visible. A graphics management program which changes the color of logic symbols on the background window and a simulator are running simultaneously. Any timing of simulation can be viewed on the screen as both a timing chart and a state of a logic. This makes it easy to detect design errors.

The features of TIPS are summarized as follows.

(1) A sequence control system with controlled devices is simulated.
 - A simulation is performed as feedback loops are closed.

(2) Graphic and interactive facilities.
 - A timing chart and the state of each logic symbol are visible at the same time. Interactive control of simulation is indicated by the mouse or commands.

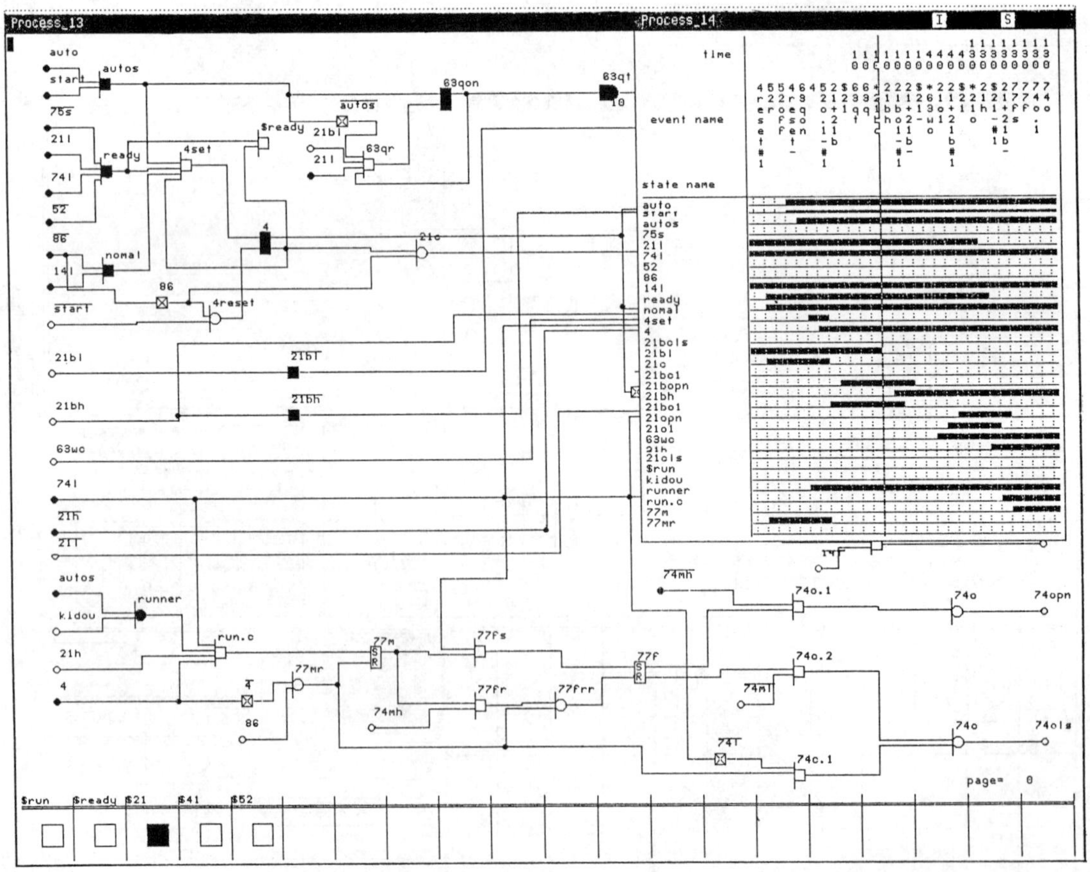

Fig. 7. The screen of TIPS. A state of the logic is shown as the color of the gates on the background; black symbols show "1" and white symbols show "0". A timing chart is displayed on the diagram. On the timing chart, the horizontal axis represents time, and the names of the signals are tabulated along the vertical axis.

(3) Event driven simulation.
- Only the firing rule makes a simulation continue.

(4) Check facilities.
- Deadlocks, conflicts or endless loops are detected during a simulation.

VERIFICATION FOR DESIGN MODIFICATION

The design of a large scale sequence control system is so often accomplished by the modification of an old similar system that a verification method for these cases is strongly required. We propose a verification method for a design modification (VEM) that detects different functions between an old system and a new one by comparison. If the different functions do not affect the remaining portions, the change is correct.

To verify the difference, we use the automaton rather than the TPN, because the automaton has more deciding capability that the TPN.

Extraction of Difference

A sequence control system (SC) is devided into a logic part (L) and a device part (D).

$$SC = (L\ ,\ D) \tag{1}$$

We represent an old system and a modified new one as

$$SCo = (Lo\ ,\ Do) \tag{2}$$

and

$$SCn = (Ln\ ,\ Dn) \tag{3}$$

respectively. The different portion between SCo and SCn is expressed as follows:

$$SCo' = (Lo - Lo \cap Ln\ ,\ Do - Do \cap Dn)$$
$$= (Lo'\ ,\ Do') \tag{4}$$

$$SCn' = (Ln - Lo \cap Ln\ ,\ Dn - Do \cap Dn)$$
$$= (Ln'\ ,\ Dn') \tag{5}$$

The different parts of the logic, Lo' and Ln' have the same input and output signals. The inputs are the signals used in a modified portion which come from an un-modified portion. The outputs are the signals used in an un-modified portion. Lo' and Ln' are converted into TPNs and interconnected TPN models of Do' and Dn', respectively. They are expressed as TPNo and TPNn.

Conversion into Automata

A modified reachability tree algorithm of PN is applied to the two TPNs: TPNo and TPNn. The well-known reachability tree algorithm makes all markings of PN as nodes of the tree (Peterson, 1977). But our algorithm makes only stable markings of TPN as nodes of the tree, so that the tree becomes smaller. A stable marking means a state in which no transition occurs at the state if none of the inputs are changed. Additionally, the changed inputs and outputs during the state transition are added to a branch from one node to another node, because the difference between the two systems is detected in the succession of input signals and related outputs. If and only if a new system generates outputs the same as an old one for any inputs, both systems are equivalent.

These algorithms are applied to sequence control systems. Figure 8 shows different portions of the logic and devices as an example of a hydro-electric power plant. These portions are converted into TPNs. The automata obtained from Figure 8 are shown in Figure 9.

Equalization of State Transitions

The minimization algorithm (Dietmeyer, 1978) is applied to the automata. When the state transitions of the simplified automata are not the same, an equalization algorithm is applied. The algorithm is described briefly as follows.

The initial states of both automata are equivalent.

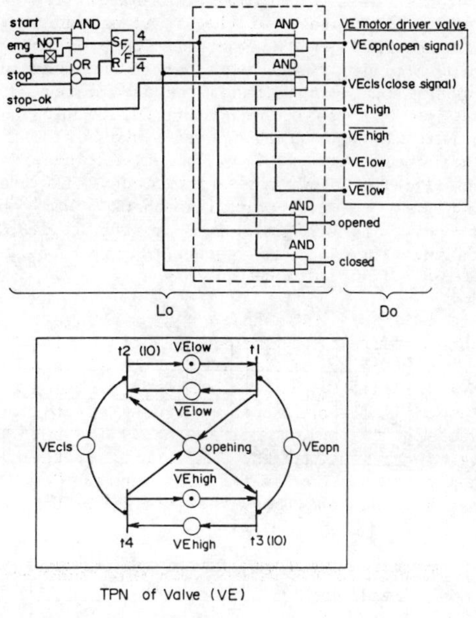

(a) an old sequence control system

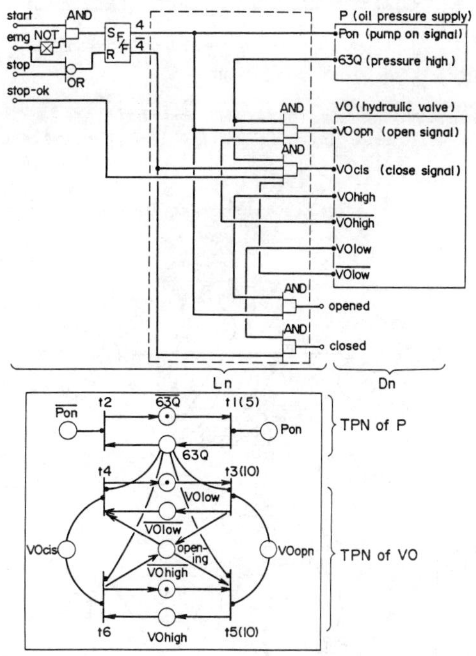

(b) a new sequence control system

Fig. 8. An example of different portions between an old sequence control system and a new one. (a) is an original control logic (Lo) and a TPN of the controlled device (Do). (b) is a modified part of a control logic (Ln) and a TPN of the controlled devices (Dn).

Thus the states which are reached from the initial states with the same input signals are also considered equivalent. The same number assigns such states. New automata are obtained, and they have the same state transitions for any inputs, but they have different outputs.

Figure 10 is the result of this algorithm when applied to the automata of Figure 9. The different functions, which may have an unexpected effect on a portion not modified, are caused by different outputs. The outputs of the automata are compared to find such different functions. When the difference is found, the change of inputs and outputs is presented as a state-transition map, and inputs are sent to TIPS for checking. A user can examine whether it is an unexpected function or not by using a timing chart and a state display, which is the same as Figure 7. If it is correct, the user offers VEM to search for other different functions. If not, the user corrects the errors.

The different functions detected by VEM are shown in Figure 10. A user will check them and will correct some design errors in the logic of a new design. Then, VEM is again applied to the corrected logic for confirmation.

state \ input	I_1	$\bar{I}_1$	I_2	$\bar{I}_2$
S_0	S_1	S_0	S_0	S_0
S_1	S_1	S_2	S_0	S_1
S_2	S_1	S_2	S_0	S_2

(1) next state function

state \ input	I_1	$\bar{I}_1$	I_2	$\bar{I}_2$
S_0	$\bar{O}_2 \to O_1$	–	–	–
S_1	–	$\bar{O}_1$	$\bar{O}_1 \to O_2$	–
S_2	O_1	–	O_2	–

(2) output function

(a) An old system

state \ input	I_1	$\bar{I}_1$	I_2	$\bar{I}_2$
S_0	S_1	S_0	S_0	S_0
S_1	S_1	S_2	S_2	S_1
S_2	S_1	S_2	S_2	S_2

(1) next state function

state \ input	I_1	$\bar{I}_1$	I_2	$\bar{I}_2$
S_0	$\bar{O}_2 \to O_1$	–	–	–
S_1	–	$\bar{O}_1$	$\bar{O}_1$	–
S_2	O_1	–	–	–

(2) output function

(b) A new system

Fig. 9. Automata obtained by modified reachability tree algorithm. (a) is associated with Fig. 8 (a). (b) is associated with Fig. 8 (b).
$I_1 = 4$, $I_2 = \bar{4} \wedge$stop-OK, O_1 = open, O_2 = close

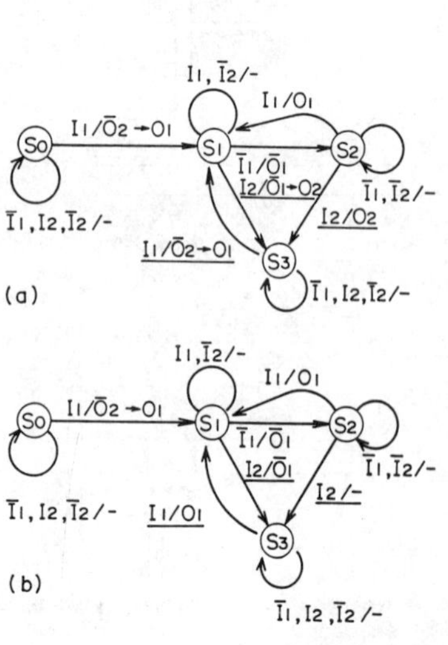

Fig. 10. Automata with the same state transition. "_" indicates difference. "01 → 02" means that 01 is output, then 02 is output.

CONCLUSION

The interactive man-machine interface of INDEED, and the use of timed Petri nets for simulation and verification have been presented. It is easy to make these tools work on other computer systems. Figure 11 shows a personal computer implementation used at the plant site. SEE provides a menu driven and user-friendly interface. The graphic interface of TIPS makes a check easier. TIPS detects design errors as conflicts or deadlocks. If a design is performed by modification of an old system, VEM is applicable.

We think that these tools increase the reliability, the productivity and the efficiency of a design and a test. In the future, an automatic verification method based on general check rules, such as a fail-safe rule, or a fault detection method will be introduced to INDEE.

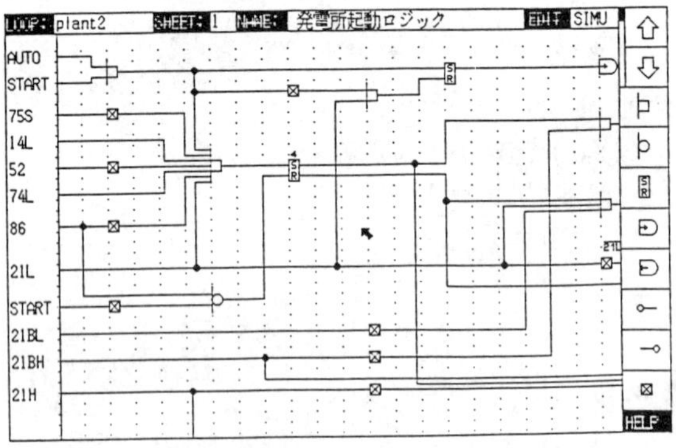

Fig. 11. A personal computer implementation of INDEED.

REFERENCES

Werner, J., G. Robson, and R. Harris (1983). Comparing the computer-adied engineering systems in action, VLSI DESIGN, 11, 24-35.

Saito, T., N. Kawato, and S. Uehara (1982). A CAD system for logic design based on frames and demons, Proc. 18th D. A. Conf., 451-456.

McWilliams, T.M. and L.C. Widdoes (1978). SCALD: Structured computer aided logic design, Proc. 15th D. A. Conf., 318-327.

Parker, A.C., D.P. Siewiorek, and D.E. Thomas (1981). A design methodology and computer aids for digital VLSI system, IBM J. Res. & Dev., 25, 4, 272-280.

Takahashi, H. (1980). An automatic-controller description language, IEEE Trans. Software Engineering, SE. 6, 1, 53-64.

Biel, G., and A. Ditzinger (1981). Computer-aided design of microprocessor-based digital controllers, Microproc. and Microprog., 7, 5, 326-333.

Peterson, J. (1977). Petri nets, Computing Survey, 9, 3, 223-252.

Famchandani, C. J. (1974). Analysis of a synchronous concurrent system by timed Petri nets, MAC-TR-20, MIT.

Masuda, R., and K. Hasegawa (1980). Mark flow graph and its application to complex sequential control system, Proc. 13th Hawaii Int. Conf. on Sys. Sci., 1, 194-203.

Azema, P., R. Valette, and M. Diaz (1976). Petri nets as a common tool for design verification and hardware simulation, Proc. 13th D. A. Conf., 109-116.

Dietmeyer, D. (1978). Logic design of digital systems, Allyn and Bacon Inc., Chap. 12.

Copyright © IFAC Computer Aided Design in Control and Engineering Systems, Lyngby, Denmark, 1985

COMPUTER AIDED DESIGN FOR ELECTRICAL SYSTEMS

M. R. Heller

Simulation Technology, Control Data GmbH, Berg-am-Laim Str. 47, 8000 München 80, FRG

Abstract. Engineering simulation is a design tool for evaluation of advanced electrical concepts, control systems and system integration.

The entire simulation system is designed to be highly interactive and flexible, tailored for engineering development.

Using digital computers requires to survey the numerical techniques in continuous simulation together with major software systems which exist.

This paper shall basically consist of the current state-of-the-art giving some prospectives for future development.

INTRODUCTION

Electronic design automation minimizes the routine, noncreative and time consuming tasks necessary to effect a hardware design, thus allowing the designers more opportunity to exercise the more creative, rewarding and valuable aspects of their jobs. At the same time, design automation provides a data interface between engineering and manufacturing to enable the automating of repetitive production process. It is a goal to evaluate, design, develop, implement, improve and support computer-aided design (CAD) and computer-aided manufacturing (CAM) systems for the primary purpose of decreasing electronic hardware development, production and maintenance costs.

ELECTRICAL ENGINEERING-CIRCUIT DESIGN

The interpretation of electronic design automation is a set of computer application tools that are available to engineering and manufacturing to assist in:

- Functional system-level design
- Logic design
- Verifying the functional and/or logic design using computer simulation
- Transforming the logic design into hardware
- Generating documentation including wire lists, logic prints and bill of materials
- Producting manufacturing tooling including printed circuit board artwork, integrated circuit masks, numerical control tapes (printed circuit board drilling, component insertion, or wirewrap), and test routines for IC's and modules

All of these functions are drawn together into a comprehensive Modular Integrated Design Automation System called MIDAS.[1]

THE MODULAR INTEGRATED DESIGN AUTOMATION SYSTEM - MIDAS

Modular

MIDAS achieves a level of modularity not hereto foreseen in electronic design systems. First, the system is available in compatible modules for logic simulation, fault simulation, and layout. Second, user defined modules can be developed and added quickly through the MIDAS data base, such that they are automatically compatible with all other system modules.

Integrated

MIDAS is an integrated system, not a disjointed set of separate programs. The programs are modular and communicate through standardized formats. Data generated by one application program can be directly utilized by another program through a common data base without the need for manual reformatting.

[1] MIDAS is a trademark for Control Data

Design

MIDAS is a design tool for engineers and logic designers. It provides work amplification to the designers during specification, implementation, verification and continuation phases of the project. Design results are automatically transformed from engineering to manufacturing via the MIDAS data base.

Automation

MIDAS is a computerized set of application tools that automates much of the routine, tedious and detailed work. MIDAS utilizes the decision making capability of the user while making use of the computer for data calculations and information storage to automatically do such functions as on-the-fly rules checking, vendor-specific tooling, inclusion of tooling data in final design verification, configuration management, and many more.

System

MIDAS is a true system with all parts interacting in a compatible manner. Since all information is contained in a common data base and all data transactions with this base are controlled by the system, all aspects of a project are linked together. Application programs and project data are tied together in a systematic manner.

MIDAS tools have been used during the design phase of many digital products, including the Control Data supercomputers, where prototype checkout time was dramatically reduced from several years on a similar product to only three months. The primary reason for this success is threefold:

1. The design was simulated until it was 95 percent operational.

2. The tooling of PC boards, wiring and LSI circuits was done from the central data base already simulated.

3. The design was resimulated using the exact tooling to ensure a working reproduction of the original design.

Consequently, 95 percent of the normally expected timing and logic problems were eliminated before the hardware was built.

Management advantages of simulation have also been found. Since it is now possible to prove designs prior to actually building the hardware products are brought to market quicker and with far less risk of schedule slippage. Project costs are also more easily controlled and reduced since the largest costs occur when the equipment is fabricated and put on the test floor. Rework cycles are reduced or eliminated. Finally, and perhaps most significantly, product quality is improved. Product quality is greatly enhanced by doing "worst case" timing analysis.

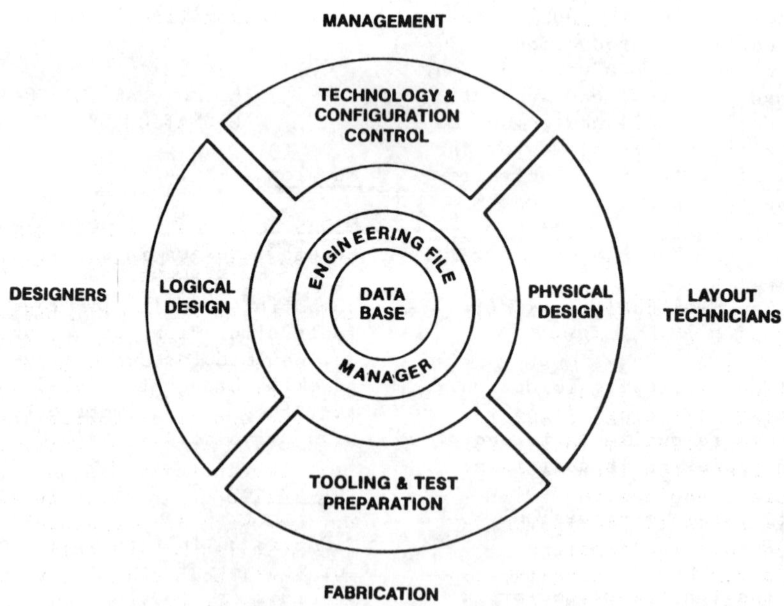

Fig. 1. Modular Integrated Design Automation System

This assures that, in a production run requiring many components, all fabricated items will work with any component mix. Marginal timing situations are revealed and corrected before they are built into the product. Reproduction of field repair costs in this manner is very important in reduction of overall product costs.

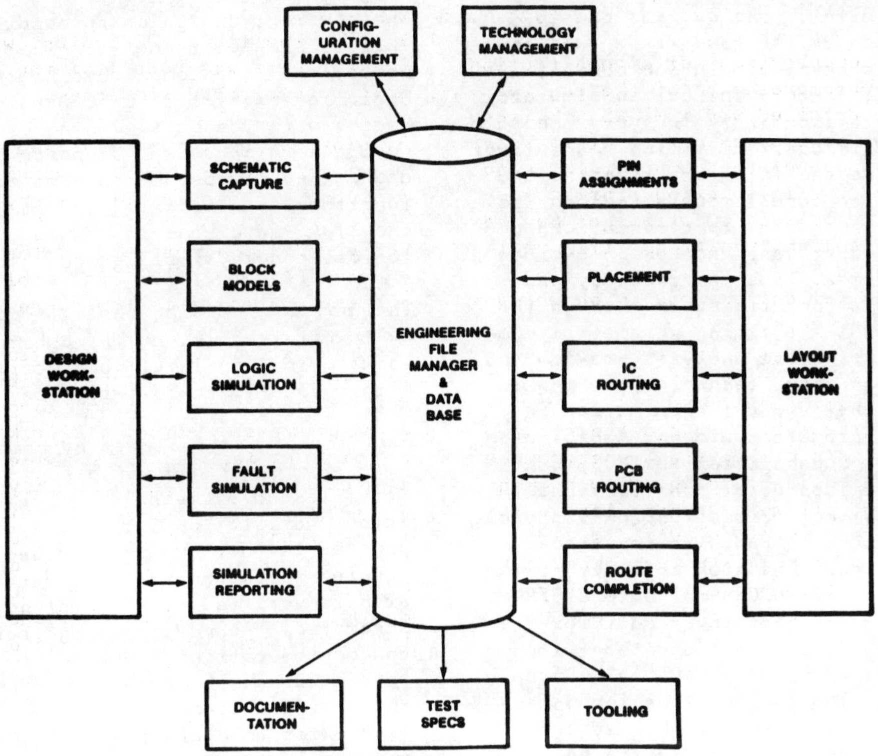

Fig. 2. Computer Automated Design System

FUNCTIONAL DESIGN

The design process begins with the functional design of the desired electronic system. With MIDAS, this process can truly be managed using a "Top Down" design methodology where a conceptual model is simulated, including clocked timing, in its clock diagram form until optimized. Each block can then be parceled into specific engineering tasks whereupon a gate level or "Bottom Up" design is merged into the functional design. This "Bottom Up" merging is necessary since the actual utilization of some selected chip technology has significant influence on design tradeoffs.

As merging takes place, selected block models are replaced with gate models and the simulation process continues in a mixed mode. Gate model performance can be correlated with that of the corresponding block model, and design discrepancies resolved. At this point the completed gate design may be set aside, and the equivalent block model reused as design simulation activities move on to other areas of the system. This selective use of gate and block models for system level simulation continues until the entire gate-level system design is completed and simulation at the gate level verifies the design. With this approach, design errors are detected at the earliest possible time, eliminating potentially vast amounts of re-design downstream.

Functional and logic design is performed by the designer using a design workstation. MIDAS is compatible with a wide variety of workstation configurations, including nonintelligent interactive terminals, intelligent graphics "turn-key" systems, and distributed processing systems built around 32-bit CPU's. Functions performed at the workstation will vary with the configuration to be used.

Dynamic digital logic simulation capability in MIDAS is provided by the ASSIST simulator. ASSIST is believed to be the most comprehensive simulator available for the actual design and production of MOS and bi-polar digital logic systems due to features such as its true mixed functional and gate level simulation capabilities. ASSIST is an event driven simulator with complete capability for very accurate (picosecond if desired) timing resolution. The rise time and fall time (leading and trailing pulse edges) models include fast representations, slow representations and specified "worst case" ranges on average switching times. Thus the accuracy and design resolution attainable through the matrix of ASSIST's timing parameters combined with its state analysis provide the capability to track technology as the IC geometries shrink and the variation in electrical parameters widens. ASSIST also has the full capabilities for MOS, ECL and other semiconductor constructions. MOS features include bidirectional signal flow on data buses and high impedance states on nodes. The high capacity of the ASSIST simulator on Control Data CYBER mainframes means that lower cost more reliable products are produced than with lower capacity systems. MIDAS also provides static logic simulation for selective path delay analysis. This system, called TRACE, operating from the engineering data base, uses Set-and-Hold timing models to determine the critical timing path effecting any given output path(s). These models are a part of the component library and contain best, nominal and worst timing for both the "0" and "1" logic states for each component in the library. TRACE takes into account not only all data paths, but also all clock and control paths as well. Selected paths may be disabled by the designer to drop from consideration any that are not of immediate interest. Flexible reporting allows him to analyze overall system performance quickly and efficiently, finding and correcting timing problems that otherwise would take many times longer using conventional logic evaluation methods.

In addition to logic and timing verification, Control Data believes that testability has to be a design consideration as well. Each new design has to be analyzed for testability and in MIDAS this is done with an Automatic Fault Simulator known as AFS. An integral part of the system, AFS generates a dictionary of all possible faults, eliminating redundancies, and then automatically generates test operands (vectors) for fault simulation and subsequent grading of the fault dictionary for coverage. This system provides a complete set of test data to the designer, as well as neutrally formatted files of test operands that will be used later to test the facricated IC's, PC boards, etc.

PHYSICAL DESIGN

The physical design phase parallels and follows the functional (logical) design phase, and is accomplished by the MIDAS Logic Layout System, LLS. Physical design begins with the partitioning of the system design into chassis, PC boards and IC's, and can easily become an extension of the functional partitioning process, since the MIDAS data base is common to both the logical and physical design phases.

The physical design phase is accomplished by layout technicians through the layout workstation, an intelligent terminal with high resolution color graphics.

Partitioning is followed by component/macro placement and pin assignment. Accomplished by technicians through the layout workstation inter-actively, this process is fast and user friendly. Tightly coupled to the engineering data base, each step is verified for consistency with the schematic, technology design rules (DRC), and configuration ground rules, ensuring that errors will not be introduced into the designs and carried forward. Facilities exist to evaluate placements prior to routing, not only for route-ability, but also for induced timing problems due to long and short interconnect paths. The latter evaluation is accomplished through automatic estimation of path delays by LLS, which are fed back to the data base for subsequent logic simulation and/or path delay analysis. Following component/macro placement, layout technicians generate the inter-connect routing patterns using LLS. A variety of routing software packages exist in MIDAS, including back panel, PC board and gate array routers. Route completion (if any) is performed interactively by the technician using the color graphics workstation. These routers, though functionally different, complete the physical design and calculate actual path delays, including gate array metalization paths, printed wiring paths on boards, and back panel wiring paths. These delays automatically replace estimates in the data base calculated earlier, and provide the facility for performing design verification, including all physical factors, of IC's, boards, back panels, and indeed the entire system design. At Control Data, we routinely perform system level simulations following completion of the physical design steps for system complexities up to 300,000 gates.

IC TECHNOLOGY

Even though MIDAS has been developed to support design activities utilizing a wide range of IC technologies, recent development and enhancement efforts have been mainly aimed at ECL and CMOS gate arrays.

Using MIDAS, a standard cell approach is provided for array personalization. A library containing at least 80 logic macro elements has been developed and contains both the logical and physical design of each, including metalization patterns and full electrical characterization. The library includes both static and dynamic storage elements, as well as tri-state bus devices.

The designer simply references the macros by part number and specifies the inter-connects to build his logic network. The layout technician places the macros within the coordinate grid of the array, and needs only to route the inter-connects, since metal paths within each macro are pre-defined. Furthermore, pre-characterization of entire macros gives excellent initial simulation results, followed later by refinements as placement and routing activities make inter-connect path delays available to the simulation process.

SUMMARY

Release of MIDAS presents an integrated approach to electronic product design. The MIDAS system of programs allows an integrated design process coupled with manufacturing outputs. From capture of design, through partitioning, component organization and placement, board design and layout simulation to provide correct operation, and fault simulation to generate test vectors, the MIDAS system employs a common data base providing extensive configuration management and technology control. This allows control of the design process, while simultaneously improving product quality and decreasing the development schedule. After the initial release of MIDAS, continual enhancements can be expected as advanced products become available. Sucessful advanced products will be sought from high quality products developed by software companies.

The use of MIDAS produces lower project design costs and a more field-reliable product. In turn, this means products which generate additional revenue since they are brought to market quicker, with a longer life cycle, and at the lowest possible cost.

DESIGN AND TUNING OF CONTROLLERS IN DISTRIBUTED COMPUTER CONTROL

M. J. Shah

*IBM Corporation, National Accounts Division, Industrial Automation Center,
2077 Gateway Place, San Jose, CA 95110, USA*

INTRODUCTION

Distributed control has been implemented for some time in the process industries such as refineries, chemical plants and paper mills. Manufacturing industries, which employ a large number of distributed discrete and continuous processes are attempting to formulate an architecture for distributed computer control. In many cases, a computer integrated manufacturing (CIM) concept is being planned and proposed. The interpretation of distributed control is generally dependent on the potential supplier or the user of CIM.

In semicondutor industry, a variety of manufacturing processes are employed, including continuous, discrete, batch, and continuous-batch. The fluctuating volume requirements and competitive pressures as well as rapid change in product specification and product obsolesence, create manufacturing nightmares in this industry.

The semiconductor industry is unique because the cycle time from research to production is very short, giving rise to a need for dramatic increase in production volumes as well as large oscillations in these volumes. The industry must also prepare to switch from one product to another in a short time. While some of the oscillations are manifested by chaotic large scale employment and layoffs, some companies have implemented flexible robots to resolve at least the bottleneck on the assembly lines.

To the semiconductor industry, then, process control, or designing controllers for higher quality products is relatively new. While in the sixties, chip yields of 10% was not an uncommon phenomena, the scale of quality control achieved in process industry is unheard of in the semiconductor industry. However, the potential payback from better quality control and improved prodution is large, so that the computer component industry itself is now embarking on computer control on a large scale.

DISTRIBUTED COMPUTER CONTROL

Because of the variety of processes involved in manufacture of semiconductor components, distributed computer control is ideally suited. In Figure 1, we show a hierarchical distributed control system, which allows a combination of monitoring and control of a variety of discrete and continuous processes. The system also interfaces to different vendor instrument controllers, ladder logic controllers and DDC Controllers. In addition, the system provides a centralized table driven architecture for application software and the capability of a common display base at upper levels in hierarchy. The system monitors multiple distributed units (processes distributed around the plant) with information display in a relatively central location. The system also provides a common link to the management information system, with preprocessed data sent to a large data base host, for production information purposes. The details of this system and description of the distributed software architecture may be found in one of our earlier publications (1).

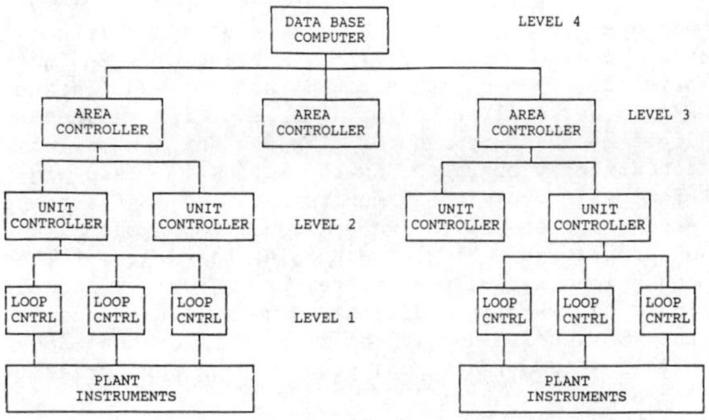

Figure 1. A Multilevel Hierarchical Control System

CONTROL REQUIREMENTS

In this paper, we will concentrate on the design and tuning of controllers in this hierarchy of distributed computer system. We recognize the fact that computer control in this industry is still relatively primitive, compared to, for example, the chemical industry. A major obstacle to proper controller design and subsequent tuning in the semiconductor industry is that often the process is new. The process engineer in some cases may not be aware of what to measure in order to perform proper control. In electrochemical operations, often one encounters key control of process done by periodic titration in a laboratory and manual recording. The judgement of control action is totally left to manual practices. With the increase in concern of pollution from chemicals, clearly the manual methods are inadequate by today's technology standards.

Another obstacle in controller design is that once a new process is started up, considerable production scheduling pressure requires immediate production even at lower yield. Little time is left to study the process for controller design, whenever control is possible. Any attempts to characterize process dynamics, for example, by step responses, meets major resistance from operational staff. The result is, that a distributed Control Computer often ends up performing only the monitoring function, collecting a great deal of data, often not too meaningful for anything other than statistical analysis.

What is required in this industry then, is a scheme which provides a method of on-line controller design and tuning, based on selective data gathering from computer monitored units. Since distributed control also distributes intelligence among the various processors, successful on line controller design should take full advantage of this process. The scheme should be employed for both continuous and semicontinuous processes in the production system.

We will illustrate this scheme, by examining the control of a multizoned oven used in processing ceramic substrates. These substrates have prepared circuit modules (chips) mounted on them requiring careful processing. In this furnace, as shown in Figure 2, wet hydrogen is continuously fed from the bottom to maintain a reducing atmosphere and to provide uniformity of heat by adding forced convection to radiant heating from the walls. This is a semibatch process, in that a large batch of the substrates is loaded and heated in a preprogrammed stages. The stages consist of heat startup to bring the temperature grdully to a desired value, followed by holding the temperature for several hours and subsequent cool down. Other similar processing of substrates employ travelling beam furnace with graduated temperature zones.

The hierarchical control system of Figure 1 was employed for this furnace. The area and unit control functions in the figure are logical and not physical, so that they could be combined into a single Series/1 processor. In addition, a ladder logic based state sequential processor program was concurrently operated on the Series/1. This program performed digital monitoring and fast action processing of critical alarm conditions, much like a programmable logic controller hardware. The heavy loading of the central processor by ladder logic software in addition to normal functions on the Series/1 made it difficult to execute any off-line (non-real time) programs on the Series/1 processor.

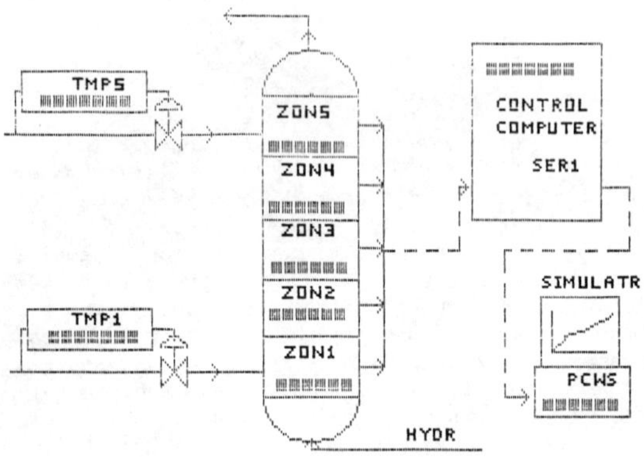

Figure 2. A sintering furnace under distributed computer control

DISTRIBUTED SOFTWARE FUNCTIONS

In Figure 3, the application software functions of the various distributed processors of the control system are shown. The area controller system is where the time stamped data collection takes place. The data consists of both output measurements such as temperatures in feedback loops as well as measurements of control variables such as the electrical heat input in each of the furnace heat zones. In normal operation, the temperature setpoint is accurately ramped with the control computer in a preprogrammed fashion. The electrical heat input is manipulated by a direct digital controller. The problem at hand is not only to design a feedback controller for each of the multiple zones but also to consider, at the same time, the effects of heat input changes in one zone on the temperatures in the adjacent zones. Furthermore, the effect of disturbance of hydrogen flow rates on the temperature must also be compensated.

In our case, the process was available for a limited amount of experimentation at startup, when open loop dynamic behavior of the individual zone temperature feedback loop could be examined. While the electrical input was ramped in steps during heat start up, responses to temperatures in the various parts of the zone were recorded. The same experiments were repeated (with deliberate step disturbances) in the period when the temperature reached the ultimate level. The experiments were repeated for different hydrogen flow rates. Since the temperature measurements in the adjacent zones were also being collected in the control computer data base, together with the zone under study, all the required data for dynamic analysis was available for dynamic analysis of the process.

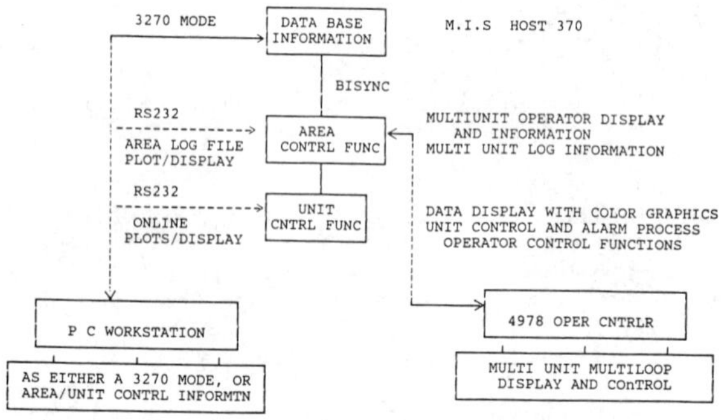

Figure 3. Software Functions in Distributed Control System

In figure 3, we see that the temperature measurements, during the experiment, were available for observation not only at the operator station in the furnace room but also to an engineer in the office via a telecommunications line on an IBM personal computer. The personal computer operates in this case as an intelligent work station. This on line tracking then provides a means to an approximate characterization of the dynamic behavior of the process in each zone, by visual observation of data during a ramp stage. It is possible to observe up to four plots of different measurements concurrently on the same screen. This allows visual observation of any effects of heat input changes in one zone on the temperatures in adjacent zones.

The collected data was also available on the area controller Series/1 log files, which could be selectively retrieved later by the personal computer work station. A time period could be specified to observe the heat zone dynamic behavior, during the various stages of furnace operation.

Thus with a personal computer work station, an engineer can observe either offline or online, the relationship between the input and output of a control system, at any time interval. A first approximation of the dynamic relationship between the input and output, in this case the electrical energy and the temperature, can be ascertained from this data.

PROCESS SIMULATION AND CHARACTERIZATION

Having obtained the crude approximation of the process dynamic behavior, the next step is to characterize the process in some combination of known transient behaviors, so that a controller design could be arrived at.

There are two possible ways of implementing this analysis. A simulation program, similar to the IBM 1130/1800 digital simulation language (3), was available both on the IBM Series/1 and the personal computer. Depending upon the available processor time, the program could be run on the Series/1. However, because of the heavy CPU load on the Series/1, it was decided to transmit the selected data from the area controller Series/1 computer to the PC work station file for this analysis. The simulation program on the PC can subsequently operate in a stand alone mode for process characterization, and multiple trial and error runs may be performed to obtain a good process model fit.

The simulation program on the PC provides a library of functions and subroutines that represent known dynamic systems. It also includes a digitized generated function capability. The actual plant measurements represent the generated function in simulation. With this facility then, the simulation program is employed to make several runs to prepare plots for comparison of a process dynamic model with the plant data collected during a process transient. If one desires, mean square error criterion can also be used to arrive at a good fit of the dynamic model by adjusting the process time constants, till a minimum error criterion can be satisfied. We found in general, that a second order model with a transport delay gave a reasonable fit to the transients. As expected, the time constants are not only different during the warm up temperature stage and the stage when the desired sintering temperature is reached (effect of T^4 due to radiant heating) but also the hydrogen flow has an effect on the time constants. During the first pass of controller design, the interaction effects on the adjacent zones can be ignored.

CONTROLLER DESIGN

Once a process model is obtained, the next step is the controller design. The personal computer simulation program allows multiple parameter runs with different controller settings of a proposed PI controller. To select a good starting design, the "CC" program from California Institute of Technology (2), can be utilized for stability analysis by the standard Root Locus and Frequency response methods. The personal computer simulation program sets up a file for CC program to perform the root locus and Bode Plots. With a quick examination of Bode Plots, one arrives at initial values of controller parameters, then returning back to the PC simulation program, a closed loop step response analysis is performed using the controller design. Tuning is performed on the simulator till a satisfactory response is obtained. Because of set point ramping in the heating stage, it is important to ensure quiescence of the loop sufficiently ahead of the next ramp step time. The controller design is then implemented on the individual zone temperature controller.

During a set point ramp step, the set of temperature data collected after closed loop control, is retransmitted to the PC workstation. The closed loop simulation response is then compared on the simulator with the actual response and the controller is retuned for improving the performance. Figure 4 shows a simplified flowchart of the controller design process for an individual zone. The same procedure is repeated for each zone and also for the two or three stages of furnace operation.

The control computer software allows replacement of new values of controller coefficients (K_R and K_C) automatically in the DDC table as the furnace is moved from one heating stage to another. Furthermore, if the hydrogen flow rate is altered for some specific reason, the controller coefficient sets could also be altered to compensate for the change.

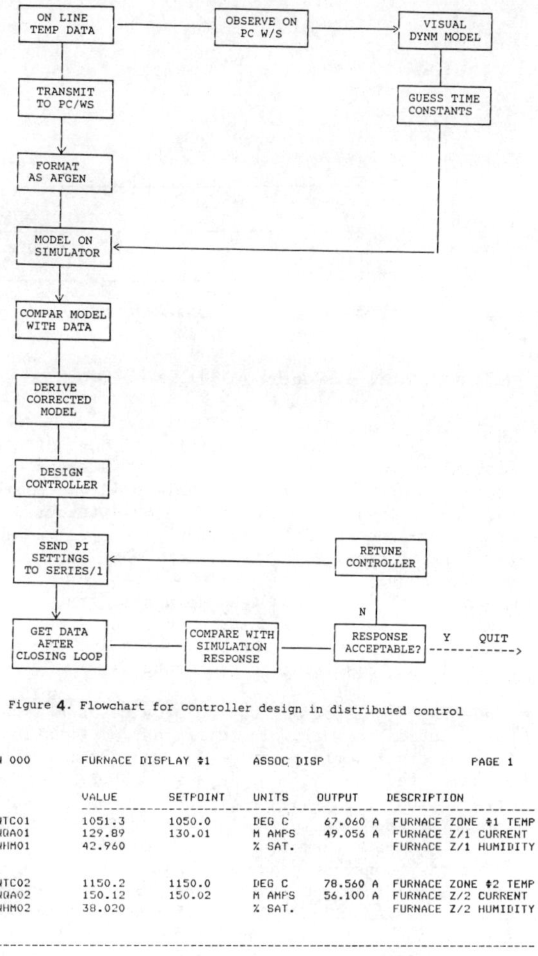

Figure 4. Flowchart for controller design in distributed control

FURN 000 FURNACE DISPLAY #1 ASSOC DISP PAGE 1

	VALUE	SETPOINT	UNITS	OUTPUT	DESCRIPTION
FURNTC01	1051.3	1050.0	DEG C	67.060 A	FURNACE ZONE #1 TEMP
FURNQA01	129.89	130.01	M AMPS	49.056 A	FURNACE Z/1 CURRENT
FURNHM01	42.960		% SAT.		FURNACE Z/1 HUMIDITY
FURNTC02	1150.2	1150.0	DEG C	78.560 A	FURNACE ZONE #2 TEMP
FURNQA02	150.12	150.02	M AMPS	56.100 A	FURNACE Z/2 CURRENT
FURNHM02	38.020		% SAT.		FURNACE Z/2 HUMIDITY

:ITEM REQ 2:ASSOC DISP 3:PVS REQ 4:ALT PAGE 5:SERVICE 7:
:OUTPUT 9: 10: 11: 12:SVD 13:CSF PRINT

Figure 5. Showing a multivariable display from area controller

Figure 5 shows the on line tabular displays of the furnace control system. It provides on line measurement and control information on the operator display station, updated every 8 seconds. In addition the PC workstation also provides trend plots of upto four measurements concurrently. These plots, after implementation of closed loop control, provide the controller performance information at the control station near the furnace. The same information is available to the engineer on a

PC connected remotely to the Series/1 control computer. Figure 6 shows a display on the PC with the simulator program for the open loop process dynamics characterization for an individual zone. Again, as with open loop, the actual measurements during closed control, can be represented by the generated function capability of the simulation system, allowing controller retuning based on the comparison between actual and simulated controller responses.

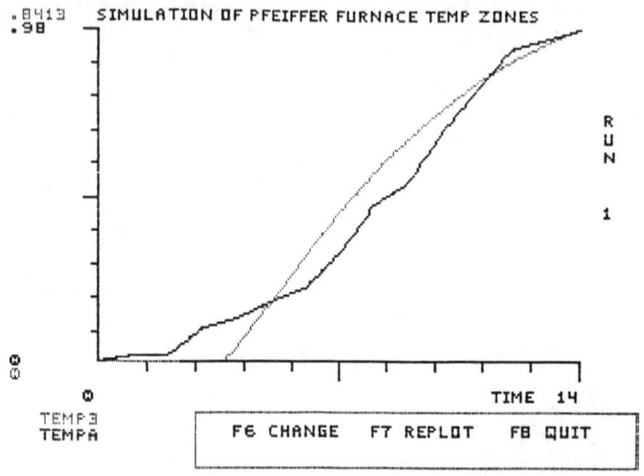

Figure 6. Comparison of actual and model temperature in Furnace

INTERACTION BETWEEN ADJACENT ZONES

This is perhaps the most difficult part of the controller tuning, especially, since a great deal of open loop data is not available prior to a process going into production. The only recourse is to perform the final tuning, on line, using some trial and error adjustments of the controller coefficients. Also, in general, set point ramping may be performed concurrently in all zones, thus making even more difficult, experimental observation of the effect of step changes in one zone heat on the adjacent zone temperatures.

Since the simulator on the PC provides simultaneous solution of multiple differential equations, the process dynamic behavior for all the zones can be simulated in one model pass. The effect of interaction from adjacent zones can thus be approximated using interaction terms first by adding a seperate dynamic term including a first order time constant and gain. Since the simulator permits multiple generated functions, all the zone temperature on line data (with and without control) can be represented in comparing the open loop and closed loop behavior of the entire furnace. The open loop data can be used for process characterization and the closed loop data are useful for controller tuning.

The relationship between the temperature in the center of each zone and the heat inputs can be given by

$$T_i = Q_i \frac{K_{p_i} e^{-s\tau_{di}}}{(s\tau_{1i}+1)(s\tau_{2i}+1)} + Q_j \frac{K_{p_{ij}}}{(s\tau_{ij}+1)}$$

with the assumption that the interaction is only in one direction, namely the direction of the hydrogen flow. It is obvious that even the process characterization can become a fairly complex problem since six parameters have to be fitted to the model in each zone. One has to use an engineering judgement rather than sophisticated mathematical techniques of control theory to be of any acceptable solution on the plant floor.

Figure 7 shows a block diagram representation of closed loop system with interaction from adjacent zone as a load disturbance. The PC simulator provides simultaneous closed loop response of all the furnace zones. Starting with the values of the PI controller settings designed for individual zones, fine tuning can be done based on preliminary estimates of interaction Kp_{ij} and T_{ij} from process characterization. The ability of looking at multiple plots concurrently on the PC as well as the facility to replot with different variables, quickly allows one to zero in on a satisfactory controller parameters.

With sufficient practice and training of the plant engineers, it becomes fairly routine to implement this "practical" adaptive control system, with which the plant personnel feel comfortable.

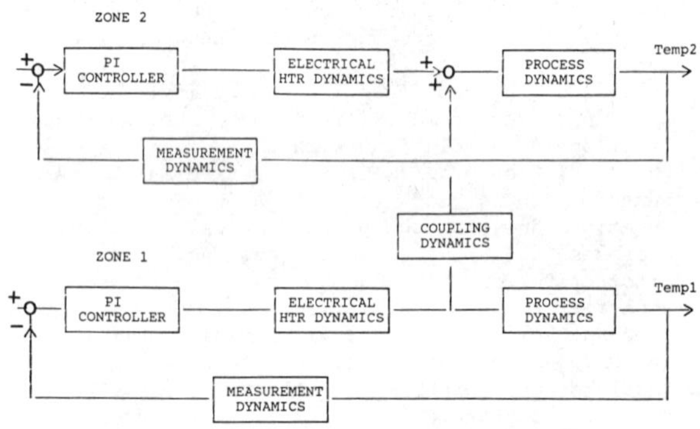

Figure 7. Closed loop block diagram with interzone coupling

TIMING CONSIDERATIONS

In real time computer control applications, one always wonders whether any sophisticated control calculations can be performed in computing time considerably shorter than the process time constants. Though this problem is somewhat simplified in distributed control by delegating workload to the intelligent work stations, the computing power of the microprocessors on the PC may still raise the same question.

We found that for both the open loop and closed loop simulation, the PC/XT performed the simulation in about 20 seconds whereas the PC/AT reduced that time to about six seconds. Several runs could thus be made with the simulator and a satisfactory retuning of parameters could be performed in a matter of about two minutes, once an engineer gains familiarity with the program.

The time constants for the process turn out to be of the order of several seconds to minutes, so that the simulation speed is more than adequate. The control computer has very little load in the controller design calculation, since it only transmits measurement data to the PC in short bursts.

For on line controller design and tuning the distributed computer control system thus provides an advantage over the system where the simulation is performed on a single control computer.

SUMMARY AND CONCLUSIONS

We have presented in this paper a method of controller design of an industrial process unit, where distributed monitoring and control is implemented using a variety of control computer processors. A simple method of model building and model fitting for process characterization is provided. The method uses an on-line software control simulator on a PC workstation for the controller design. The same simulator program can be used for further on-line tuning of controllers during the operation of the plant. This yields a practical, easy to implement, adaptive controller in the distributed control environment.

REFERENCES

1. Shah M.J. and Brecher V. H.,"Distributed Control in Automation of Semiconductor Manufacture", paper presented at the International Industrial Controls Conference, Philadelphia, September, 1984.

2. Thompson P., "A Computer Aided Control Sysytem Design", Prentice Hall Inc., 1985.

3. SHAH M., "Engineering Simulation Using the IBM Personal Computer", Prentice Hall, 1985.

DESIGN OF A CONTROLLER IN AN UNDERGROUND GASIFICATION PLANT

R. Gorez

Department of Mechanical Engineering, University of Louvain, Bâtiment Stevin, Place du Levant 2, B-1348 Louvain-la-Neuve, Belgium

Abstract. A new underground gasification plant is described and a low order dynamic model of the thermal state in the exhaust well is set up. The process being highly nonlinear, the design of the control system, including selection of output variables, sensors locations and control laws and parameters, is achieved via computer simulation and trial and error techniques. The computer program for this model will also be implemented in the process computer on-site and it is shown how this program can be used by the process operators, for training or as an aid for controlling the process.

Keywords. Computer aided design ; control engineering ; computer applications ; energy control ; heat systems ; modelling ; non linear systems.

INTRODUCTION

Underground gasification experiments are conducted in Belgium right now. Hot gas is produced by burning coal layers approximately 1000 m beneath the ground level. Gas is collected through an exhaust well which consists of two concentric pipes, as sketched in Fig. 1. The gas flows upwards through the inner pipe and cooling water is injected through the annular space between the two pipes, in view of keeping the wall temperatures within some safety limits.

In this kind of well, the cooling water must transform into superheated vapor before reaching the bottom of the well (if liquid water came to the bottom of the well, the gas generator might be flooded ; it would disturb the gas production, and possibly, this one should be stopped during the time which is needed to exhaust water from the well through an auxiliary pipe acting as an air-lift system). Superheated vapor will be carried away by the gas stream flowing upwards through the inner pipe. Clearly, the mixing of gas and superheated vapor in the bottom of the well induces a positive feedback on the thermal state of fluids and walls in this part of the well, resulting in a thermal process which is not self-regulatory. Then, this process must be controlled by the process operators or by automatic controllers manipulating the cooling water flow rate.

Gasification plants are already in operation elsewhere in the world, especially in U.S.A. In all reported experiments, the depth of the well is much smaller than in this case ; then, the process can be controlled without any special difficulties. In the case which is considered here, there will be a long transportation time delay between the operation of the control valve which is located at the top of the well, and the subsequent variations of the flow rate of coolant in the bottom of the well. This time delay is due to the film type flow of the cooling water along the walls in the upper part of the well ; it is roughly equal to the time interval inside which the volume of superheated vapor might collapse in case of large disturbances at high production rates. Then, the control of the process will be very tight, and a model of the thermal state dynamics in the well is needed in order to determine safe operating conditions, select locations of thermocouples for sensing temperatures inside the well, and investigate control strategies.

As the process to be controlled is highly non linear, all those objectives have been achieved by trial and error techniques, using a computer for simulation of the process with or without controller. Simulation of a complete distributed parameter model would require excessive computer times ; thanks to some simplifying assumptions based on the analysis of previous simulation results, it has been possible to set up a low order lumped parameter model, which consists of three ordinary differential equations only, but still holds the main non linear features of the process behavior. This model has been simulated on a minicomputer and is used for investigations on the process dynamics, selection of sensors locations and experiments with various automatic controllers or manual control strategies. Besides, this model has been translated into a FORTRAN program which will be implemented in the process computer on-site ; this model will be the core of a non linear observer which can be used by the plant operators for reconstructing the global thermal state of the process and monitoring some variables which cannot be measured, such as the position of the transition point between the lower part of the well where vapor is superheated and the upper part where water is vaporizing. Moreover, this model can be used by operators for training and for improving their control strategies or their knowledge of the process dynamics by investigating the effects of various simulated disturbances ; it can be used as well as an aid for controlling the process, the operators being able to predict the responses of the process to the manipulation of the control variable.

The following section of the paper contains a short description of the plant and introduces its mathematical model. Another section is dedicated to the design of manual control strategies or automatic controllers. The last section outlines other uses of the model on-site and/or on-line.

PLANT DESCRIPTION AND MODELLING

Figure 1 shows a sketch of the gas exhaust well. Basically, the well consists of two concentric pipes, called tubing and casing, the diameters of which are 5 and 7 inches respectively. Cooling water is injected through the annular space between the two pipes and transforms into superheated vapor in the lower part of the well. Vapor is carried away by the upstream flow of gas resulting from the combustion of deep coal layers located around the well; the mixture of gas and superheated vapor flows upwards through the tubing.

The depth of the well is 865 m. Then, from a thermal point of view, the well can be considered as a heat exchanger of infinite length. Heat and mass transfers have been described by Gorez and Duqué (1983), using a distributed parameter model taking into account the axial and radial distributions of the state variables : temperatures of the pipe walls, temperatures, pressures and compositions of the fluids. They have found that the simulation of this model would require a big computer facility and excessive computer times. Then, low order lumped parameter models consisting of a few ordinary differential equations have been set up by Gorez and Paillet (1983) and by Gorez and Passau (1984), thanks to some simplifying assumptions based on the analysis of simulation results from the complete distributed parameter model :

- the dynamics of the heat exchanger depends only upon the temperatures deviations in the superheated zone (lower part of the well, where vapor is superheated) ; elsewhere, the temperatures of walls and water can be considered as equal to the water saturation temperature corresponding to the pressure in the well ;
- the pressure in the well is constant (pressure losses in the upstream flow are very low, except at the highest production rates ; pressure losses in the downstream flow are compensated by gravity effects) ;
- the heat exchanges can be calculated from the average temperatures of walls and fluids in the superheated zone.

These assumptions result in the following equations expressing heat and mass balances in the well :

1) in the mixing zone (bottom of the well) :

$$F_m = F_g + F_w \quad , \quad F_m c_m = F_g c_g + F_w c_w \quad ,$$

$$G_{mo} = G_{vo} + H_g \; ,$$

$$G_{vo} = F_w c_v \theta_{vo} \quad , \quad H_g = F_g c_g \theta_g \quad ,$$

2) in the superheated zone :

$$H_t = Lc'_t \theta_t \quad , \quad H_c = Lc'_c \theta_c \quad ,$$

$$\Phi_{mt} = LK_{mt}(\theta_m - \theta_t), \; \Phi_{vc} = LK_{vc}(\theta_v - \theta_c), \; \Phi_{cr} = LK_{cr}\theta c,$$

$$\theta_m = \frac{2G_{mo} - F_m gL + LK_{mt}\theta}{2F_m c_m + LK_{mt}} = \frac{\theta_{mo} + \theta_{mL}}{2} \quad ,$$

$$\theta_v = \frac{LK_{vt}\theta_t + LK_{vc}\theta_c + F_w gL}{2F_w c_v + LK_{tv} + LK_{vc}} = \frac{\theta_{vo}}{2} \quad ,$$

$$L = \exp(\lambda) \; ,$$

$$\dot{H}_c = \Phi_{vc} - \Phi_{cr} \quad ,$$

$$\dot{H}_t = -\Phi_{vc} + H_g - F_g g L_t - F_w r \quad ,$$

$$H_t \dot{\lambda} = -\Phi_{mt} + H_g - F_g g L_t - F_w(r + gL) + G_{vo} \; ,$$

where subscripts t, c, r, g, w, v, m denote tubing, casing, surrounding rocks, gas, water, vapor (superheated), mixture of gas and vapor, respectively ; if two subscripts are used, they refer to exchanges between the two corresponding entities, except if the second subscript is o or L, denoting respectively the bottom of the well or the transition point between the superheated zone and the upper part of the well. The meaning of the other symbols is as follows :

c	specific heat (per mass unit)
c'	specific heat (per length unit)
F	mass flow
g, r	gravity, vaporization heat
G, H	enthalpy
K	heat transfer coefficient (per length unit)
L	length of the superheated zone
L_t	total length of the well
T	temperature
T_w	temperature in the upper part of the well (liquid-vapor equilibrium at given pressure)
$\theta = T - T_w$	relative temperature
$\lambda = \log_e L$	transformed length
Φ	heat flux.

As for the temperature sensed in the superheated zone of the annulus, it can be represented by the following relationship :

$$\theta_s = \theta_{vo}(1 - Ls/L) \text{ if } L > L_s \; ,$$

$$0 \qquad \text{if } L \leq L_s \; ,$$

where L_s is the distance between the sensor location and the bottom of the tubing.

This low order model has been validated by comparing simulation results obtained when using this model and the complete distributed parameter model in similar conditions. For instance, Fig. 2 shows the time responses of the superheated zone length L and the sensed temperature θ, for a step change of + 200° of the gas temperature.

DESIGN OF A CONTROLLER

As it can be seen from Fig. 2, the process is not self-regulatory. Then, the process must be controlled in order to keep the superheated zone length L and the casing temperature in the bottom of the well within given safety limits : if L is too low, there is a risk of flooding the gas generator in case where the superheated zone would collapse ; on the opposite, excessive pipes temperatures would result in dangerous thermal expansions of the pipes, with risks of damages to the cement coat around the casing, to the fixing device of the tubing or to the pipes themselves.

Either the flow rate of the gas-vapor mixture or the cooling water flow rate could be used for controlling the process. Manipulating the first variable would yield better performances, as a result of shorter response times, but it would disturb the operation of the gas generator. Therefore, manipulating the flow rate of the gas vapor mixture must be restricted to cases of emergency, and in normal operation, control must be achieved by manipulating the cooling water flow rate. However, this control system must be checked carefully, because of the long time delay inside the control loop. This time delay is due to the film type flow of the cooling water along the walls in the upper part of the well, which results in a velocity :

$$V = 1.64 + 1.44 \text{ Log } F_w \; ;$$

for a well depth of 765 m and water flow rates from 0.2 to 1.2 kg/s, the resulting time-delay ranges from 20 to 5 min.

Besides, the following questions must be answered :

- locations of temperature sensors
- choice of manual or automatic control
- in case of manual control, selection of appropriate sampling rate and control strategy
- in case of automatic control, calculations of the control law and parameters.

The process being highly nonlinear, answers to those questions have been obtained by trial and error techniques, involving many experiments on the model of the process which is presented in the previous section and has been simulated on a digital computer.

In those experiments, the time history of the superheated-zone length and the temperatures at the possible sensors locations or at other critical points were recorded. The experiments included :

- determination of responses of the process itself to step variations in the cooling water flow rate, the gas flow rate or the gas temperature ;
- investigation of the behavior of the process when it is controlled by various control laws or strategies and is disturbed by step or random variations of the gas temperature.

Analysis of the simulation results lead to the following decisions :

- 6 sensors should be located at distances of 10 m from each other, the first one being located at 15 m from the tubing end ;
- the first sensor should be used for control and the reference value of the temperature to be regulated should be the same at all production rates, resulting in longer superheated zone lenghts for higher production rates ;
- in case of manual control, if the process is disturbed, the sensors should be sampled at least each 5 minutes, and a control strategy very similar to a PI control law should be used :

  $$\Delta F_w = K(T_s - T_r) + K'\Delta T_s$$

  where ΔF_w is the flow rate variation which should be made by the operator, T_r and T_s are respectively the reference and actual values of the sensor temperature and ΔT_s is the variation of T_s from the previous to the current sampling time (Fig. 3)
- in case of automatic control, the best results are obtained when using a conventional PI controller (Fig. 4) ; other control laws have been tried: deadbeat control, weighted minimum variance control with on-line parameter identification . The last one may give good results (Fig. 5) but the sampling period must be carefully selected and cautions must be taken in view of the convergence of the identification algorithm.

Therefore, by means of computer simulation analysis, it has been possible to design a satisfactory control system.

OTHER USES OF THE MODEL. CONCLUSIONS

The FORTRAN program corresponding to the model presented in this paper can also be implemented in the process computer on-site, requiring a core memory size of 6.5 Kbytes. The process operator will be allowed to initialize the model for selected operating conditions, to introduce step variations of the cooling water flow rate, the gas flow rate or the gas temperature as well as random variations of the gas temperature, and to track or record the time history of the superheated zone length and the temperature at the sensor location. A run corresponding to a real time interval of 1 hour would take only a few seconds. Moreover, the program can be interrupted at prespecified time intervals, allowing the operators to modify water and/or gas flow rates during interruptions.

Then, this program can be used off-line by the process operators for training and for improvements of their control strategy or their knowledge of the process dynamics. It can also be used on-line as an aid for controlling the process, the operators being able to predict in a few seconds the effects of selected changes in the control variable or to observe via the model some variables which cannot be measured, such as the superheated zone length.

This study confirms that for highly non linear processes, appropriate models are valuable tools for analysis and design of a satisfactory control system. In the case considered here, models have been used off-line for design of the control system, including selection of sensors locations, and reference values of the output variables as well as selection of appropriate control laws. It is also intended to use the same models on-line as an aid for controlling the process.

REFERENCES

Gorez, R. and C. Duqué (1983). Modelling and Control of an Underground Gasification Well. Proc. IMACS Symposium on Simulation in Engineering Sciences, Nantes, pp. 179-184.

Gorez, R. and D. Paillet (1983). Simulation of the thermal state of an underground gasification well. Proc. First European Simulation Conference , Aachen, pp. 460-465.

Gorez, R. and V. Passau (1984). Models of the thermal state in an underground gasification well. Tenth Annual Underground Coal Gasification Symposium, Williamsburg.

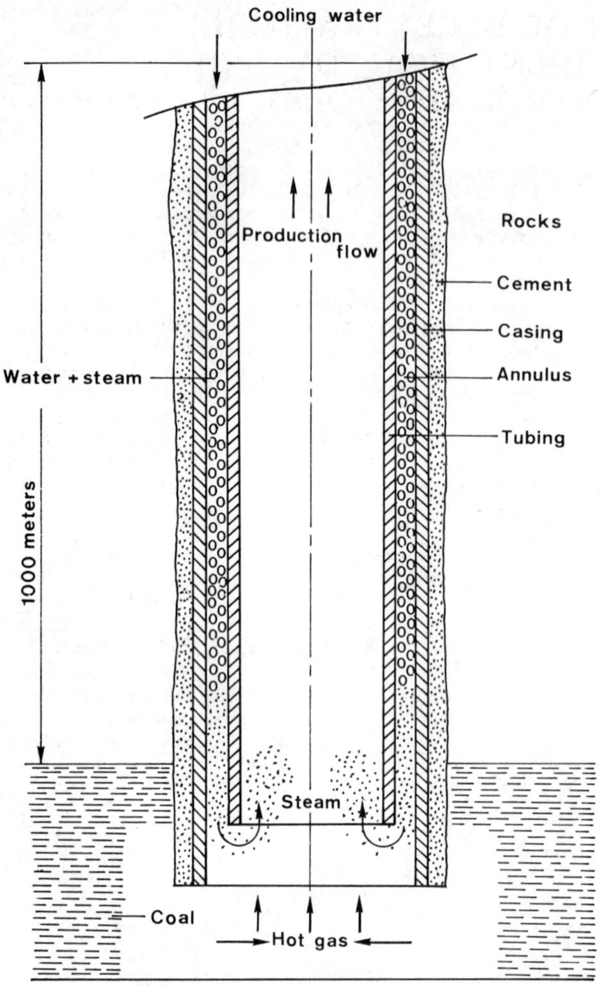

Fig. 1. Sketch of an exhaust well

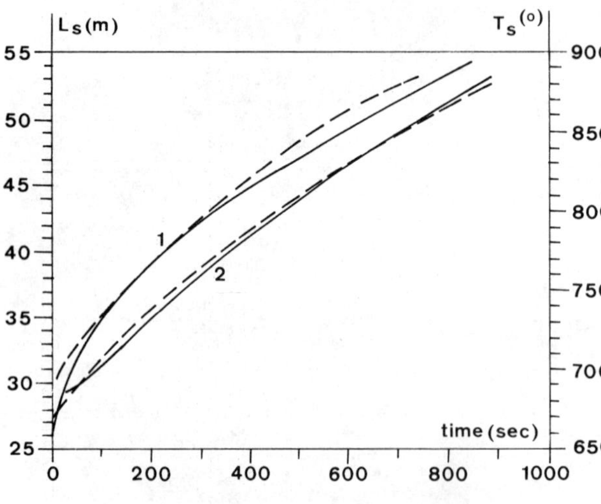

Fig. 2. Transient responses at a +200° step in the gas temperature
1 : distributed parameter model
2 : lumped parameter model
solid lines : superheated zone length
dash lines : temperature at the sensor location

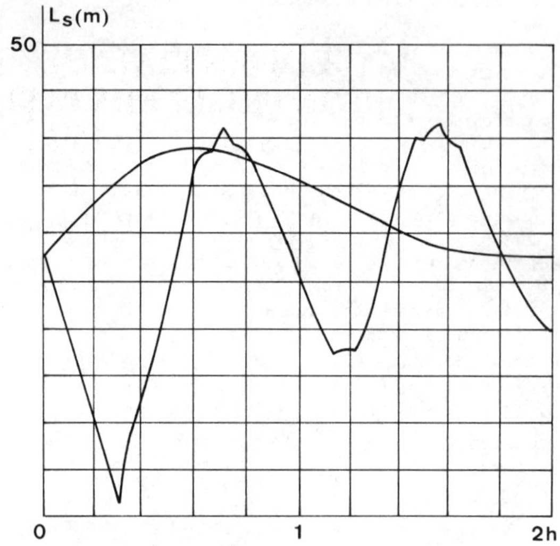

Fig. 3. Transient response to ±50° steps in the gas temperature ; PI type control strategy

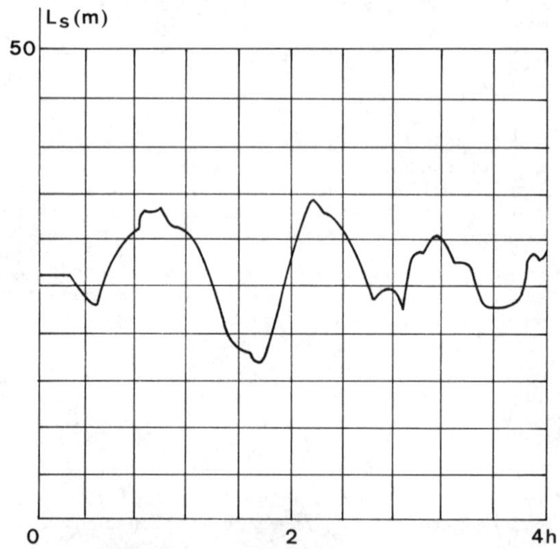

Fig. 4. System controlled by a PI controller Response to randoms variations of the gas temperature.

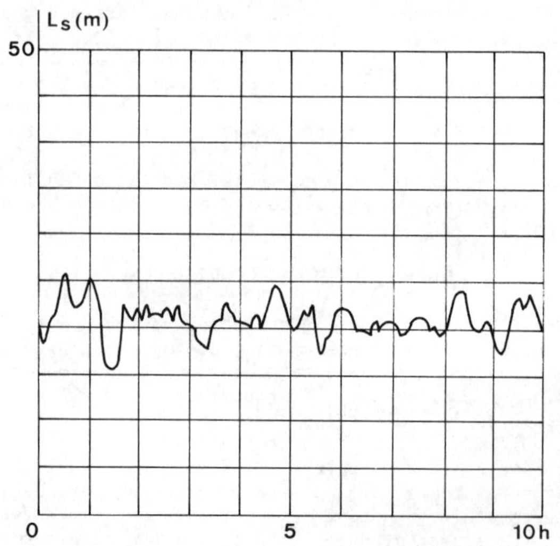

Fig. 5 .System with weighted minimum variance adaptive controller.

SEQUENTIAL DESIGN OF DECENTRALIZED CONTROLLERS WITH IDENTIFICATION AND SELFTUNING CONTROL METHODS

R. Isermann and H. Hensel

Institut für Regelungstechnik, Technische Hochschule Darmstadt, Schlossgraben 1, D-6100 Darmstadt, FRG

Abstract. One of the problems with the design of decentralized controllers for complex processes is that the interactions of the process parts are not known and therefore design methods for centralized multivariable controllers cannot be used. By a sequential (successive) procedure for process identification and computer aided controller design it is possible to design decentralized controllers where the interactions are automatically taken into account. It is shown by simulations that this approach converges rapidly. The sequential design can be applied as well for multivariable, as for cascaded and hierarchical control systems. Furtheron the sequential procedure is especially attractive if selftuning controllers are used.

1. INTRODUCTION

During the last decade several methods and software packages were developed for

- on-line identification with process computers
- computer aided design of control algorithms with process computers
- self-tuning control algorithms for process and micro computers.

They were used in the meantime for many pilot and real industrial processes and have proven to be useful, especially if there were little knowledge on the internal behaviour, difficult dynamical behaviour and long settling times, see e.g. Isermann (1984a). However, the application was mostly limited to single-input single-output control, despite the methods were also developed for multivariable processes. For these multivariable processes generally "central" models are used, consisting of the complete information(structure and parameters) on the transfer behaviour for the inputs and outputs including all important cross interactions. These identified models then led in a straightforward way to "centralized" controllers, e.g. state controllers and parameter optimized main and coupling controllers, with a unique sampling time. Some of the difficulties with this "central" multi-input multi-output approach are:

- it is frequently not possible to operate industrial multivariable processes for longer time periods in open loop
- it is difficult to find suitable model structures
- it is difficult to design fixed multivariable controllers for all operation modes (manual or closed loop operation for each control variable separately)
- the "central" design does not agree with the sequential closing of loops during commissioning
- it is frequently not possible to use a unique sampling time even for the main loops
- complex multivariable control algorithms are mostly not accepted in the process industries
- decentralized micro computers were developed, which replaced the centralized process computers and have no access to the inputs and outputs of neighboured control loops.

These difficulties and developments forced the design of "decentralized" controllers where it is assumed that there is only knowledge on the directly connected process parts, which includes that the interactions between the process parts are unknown.

If then models of the process parts are known the influence of the interactions can for example tried to be taken into account by properly extended observers and state controllers, Kuhn (1983).

In the following a special way of identification and decentralized controller design is considered where both, the models of the process parts and the interactions are assumed to be unknown.

2. SEQUENTIAL IDENTIFICATION AND CONTROLLER DESIGN

2.1 Multivariable control systems

As first example *two process parts* P_1 and P_2 are considered with one manipulated variable and one control variable each, u_1 and y_1 respectively u_2 and y_2, Fig.1. It is assumed that the transfer behaviour of the process parts $y_1(z)=G_{11}(z)\ u_1(z)$ and $y_2(z)=G_{22}(z)\ u_2(z)$ and the interactions between the process parts can be described by linear difference equations of finite order. The task now consists in designing two decentralized digital controllers (regulators) R_1 and R_2 which are only connected with their process parts P_1 and P_2, Fig.2. For the microcomputer controllers it is further assumed that they may carry software packages for process identification and controller design or that they may be connected with a central main station or a commissioning computer with implemented required software packages. These software packages may be developed for discrete-time parametric process models and consist for example of:

- process identification: single-input single-output parameter estimation methods like least squares (LS), instrumental variables (IV), correlation and least squares (COR-LS) or their recursive versions; order and dead time search procedures; signal generators (PRBS) and signal filters (e.g. drift elemination); model verification.
- controller design: single-input single-output controllers like e.g. PID-controllers (parameter optimization), state-controllers with observers (matrix Riccati design), minimum variance controllers (direct calculations).

Corresponding methods and software-packages are described elsewhere, e.g. Isermann (1981), (1984a,b)

The procedure then is as follows, compare Fig.3:

(1) Overall behaviour $G_1^1(z) = y_1(z)/u_1(z)$ is identi-

fied and controller $R_1^1(z)$ is designed.

(The interactions between P_1 and P_2, relevant for R_1 are automatically taken into account).

R_1^1 is then connected to process part P_1.

(2) Overall behaviour $G_2^1(z) = y_2(z)/u_2(z)$ is identified and controller $R_2^1(z)$ is designed.

(The interactions between P_2 and P_1, now with connected controller R_1^1, are taken into account).
R_2^1 is then connected to process part P_2.

(3) Overall behaviour $G_1^2(z) = y_1(z)/u_1(z)$ is identified with disconnected controller R_1 and controller $R_1^2(z)$ is redesigned.

(The interactions between P_1 and P_2, now with connected controller R_2^1, are taken into account).
R_1^2 is then connected to process part P_1.

(4) As (2) to redesign controller R_2^2.

This procedure continues until both controllers R_1^i and R_2^j , i=1,2,...,j=1,2,..., converge to fixed controllers R_1 and R_2. Before an example is shown, some remarks are given:

- The sequential approach is similar to iterative tuning rules for multivariable processes, but now based on actual process models and computer aided design.
- Only single-input single-output (SISO) methods are used, in order to keep the approach simple and to end up in decentralized SISO-controllers.
- The (automatic) order and dead time search procedures within the process identification packages allow models with changing orders/dead times, especially during the first steps.
- Necessary conditions for the sequential procedure are that the corresponding systems are stable and identifiable. The last condition means that the systems must be controllable and observable with regard to the very input and output.
- The sampling times may be arbitrarily different for the decentralized controllers, which is a significant advantage.

Example 1. Sequential identification and controller design for a two-input two-output system.

Fig. 4 shows a block diagram of a two-variable process with p-canonical structure and strong interactions. This process was simulated on an analog computer and connected to a process computer. For the identification program package OLID was used, for the controller design the package CADCA. PID controllers were used for this example. Their parameters result from minimization of the quadratic loss function

$$I = \sum_{k=0}^{n} y^2(k) + rK_p^2 \Delta u^2(k).$$

In Fig.5 the control behaviour is shown for some design steps. After design step (1) for R_1^1 a good behaviour for u_1, y_1 results. During step (2) the system for controller R_2 is identified, with connected controller R_1^1. The response to a setpoint change of controller R_2^1 shows a satisfactory behaviour. However, setpoint change of controller R_1^1 results in too strong oszillations. A redesigned R_1^2 in step (3) gives much better damped behaviour.

After 6 identification and redesign steps the controller parameters converge to nearly constant values. The signals for step (3) and (6) show only minor differences. Hence, a fast convergence can be observed.

This also becomes obvious, if after step (1) R_1^1 is disconnected and R_2 is designed in the same way. After closing both loops a weakly damped behaviour results. Hence step (2) in the procedure described above is already a good improvement.
If all signals are available in one (central) computer, the model of the two-variable process is known after step 2 and the ongoing sequential design can be performed by simulations of the system only. □

The sequential design can be extended to more than two process parts. It is then proposed to begin with process parts with fast dynamics and with little dependence on the remaining system and to proceed to process parts with slow dynamics and strong interactions.

2.2 Cascaded control systems

The sequential procedure is straight forwardly applied to cascaded control systems, compare Fig.6. One begins with the identification and controller design of the inner loop, connects the (slave) controller R_1' and repeats the same for the outer loop. If the inner loop is much faster than the outer one a redesign of its controller in general is unnecessary. Otherwise redesign procedures as for multivariable systems may follow.

2.3 Feedforward control systems

Usually parameter optimized feedforward and feedback controllers are designed separately. This may lead to an overcompensation after a change of a disturbance variable. Therefore it may be better to first design the feedback controller and then to design the feedforward controller for the closed loop. This can be performed by sequential identification and controller design steps.

2.4 Hierarchical control systems

The sequential procedure described for multivariable control systems moved in a horizontal way, the procedure for cascaded control systems in a vertical way. In vertical direction more controllers may be added, see Fig.6, e.g. for coordination or for reaching a certain block behaviour. The sequential identification and controller design can then be applied in vertical direction, making use of model reduction of the underlying systems through identification methods.

In general it is recommended to proceed sequentially as follows:

- from lower levels to higher levels
- from fast dynamics to slow dynamics
- from weak interactions to strong interactions.

Some properties of the sequential procedure are:

- It is rather adaptable to the systems structure.
- With increasing design progress more and more process parts and their interactions are taken into account.
- Repeated identification of system parts includes a progressive model reduction (order reduction).
- Sampling times may be very different.

- Fast convergence can usually be expected, because of the intermediate identification phases.
- It seems to be difficult to proof the stability of the resulting system. However, the sequential design should be more reliable than other design procedures for decentralized controllers which do not get information on the interactions.

The described procedure was applied for the control system design of a *Benson-boiler steam-turbine unit*, which was simulated on an analog computer (30 integrators), Bayer (1984). The sequential procedure was especially suited to study various control system structures for steam temperature, steam pressure and mass balance (evaporator end point) control including several feedforward controllers. One starts with the design of the less coupled cascaded steam temperature control systems of the 3 superheaters and continues with the design of the turbine power control. Finally the design of the steam pressure controller follows. The turbine power control has a strong influence on the pressure control. However, the turbine power control is much faster than the pressure control, such that the interactions to the pressure loop can be neglected. Hence, two sequential steps are sufficient in this case.

3. SEQUENTIAL SELFTUNING CONTROL

The sequential identification and controller design may need a considerable amount of time. In Isermann (1984b) it was shown that under noisy conditions about 50% of time can be saved by using selftuning controllers. These controllers result, if recursive parameter estimation methods and controller design methods are combined properly, see Fig.7. Parameter estimates $\underline{\hat{\theta}}(k)$ of the process are obtained after each sampling interval which are used to design the controller parameters $\underline{\beta}(k)$. Following procedure has demonstrated to be useful in the first step:

a) short *pre-identification* in open loop by introducing a perturbation signal to obtain a rough process model and therefore good starting conditions for b)

b) *selftuning control* until reaching a sufficient control performance. The controller parameters are then fixed and the control continues with a fixed controller.

The automatic design of two decentralized controllers for *two process parts* P_1 and P_2 corresponds to the sequential approach described in section 2.1; compare Fig.8:

(1) After a short pre-identification phase controller R_1 is switched to selftuning mode. A probing signal may be introduced to accelerate the controller parameter adaptation. Then controller R_1^1 is obtained and its parameters are fixed.

(2) The same procedure as (1) is now applied to controller R_2. The selftuning results in controller R_2^1 which is also fixed.

(3) R_1 is again set to selftuning and results in R_1^2, which is fixed.

(4) R_2 is retuned as (3).

The sequential selftuning continues until both controllers converge to fixed controllers R_1 and R_2. At least for the first steps it is proposed to include an automatic search of model order and deadtime, see Mäncher and Hensel (1985). Also in this case different sampling times can be used for both controllers.

Example 2. Sequential selftuning of a two-input two-output process.

An analog simulated two-variable process according to Fig.9 was connected to the (fault tolerant) 16-bit double microcomputer system FIPS, see Mänc her (1984). The selftuning controller consists of the recursive least squares parameter estimation method in discrete square root filtering form and a state controller with observer. The state controller is designed by solving the recursive matrix Riccati difference equations after each sampling which minimizes the quadratic performance criterion

$$I = \sum_{k=0}^{\infty} \underline{x}^T(k)\underline{Q}\,\underline{x}(k) + ru^2(k).$$

An integrator in the state controller was introduced by adding an integrator to the estimated process model.

Fig.10 shows the time history of the signals for the first six steps. During step (1) the pre-identification of process P_1 is performed by using a PRBS-signal for u_1. Then the loop 1 is closed with R_1 in selftuning mode and the control variable y_1 reaches rapidly its setpoint $w_1=0$. A step function of the reference variable w_1 from 0 to 1 and back from 1 to 0 introduces an external excitation of the selftuning controller and shows the resulting transient behaviour of loop 1. During step (2) the same is repeated for controller R_2, but with fixed R_1^1 connected to process part 1 so that the process interactions are taken into account in selftuning of R_2. A well damped behaviour for y_2 is obtained for the second setpoint step w_2. Because of the interactions the signals for step (3) show a different behaviour as for step (1). R_1^2 now results with a longer settling time for y_1. The change of R_1 during step (3) is clearly shown by comparing y_1 and u_1 from step (2) with step (4). During step (4) R_2 is slightly changed and during step (5) the control behaviour of y_1, u_1 is improved only a little bit. The following steps indicate no changes of controller parameters anymore.

This example shows that with sequential selftuning a fast convergence is obtained and that a few steps are enough to reach a good control behaviour. It was demonstrated for a noiseless system. However, the described procedure of selftuning is already determined for a noisy system. In Isermann (1984a), (1984b) it was shown, that in approximately the same time selftuning controllers converge also for significantly disturbed processes. □

The procedure of sequential selftuning control can, accordingly to sequential identification and controller design described in section 2, also be applied to

- *cascaded control systems*, by selftuning first the inner controller and then the outer controller. For an example see Isermann (1984a).
- *feedforward control systems* by using selftuning feedforward controllers, see Isermann (1982).
- *hierarchical control systems*, by proceeding with sequential selftuning in a vertical way, as described in section 2.4.

The same properties given at the end of section 2 in principle also hold for the sequential selftuning control. But in comparison to sequential identification and controller design considerable time can be saved and the resulting control behaviour can be observed immediately. Therefore this approach is suitable also for on-line changing of controller design parameters, like the sampling time T_0 or the weighting factors r_i for the manipulated variables, which usually are found iteratively.

4. CONCLUSION

In this paper a sequential (successive) application of identification and controller design and a sequential application of selftuning control is proposed for the design of decentralized controllers. It can be applied for multivariable, cascaded and hierarchical control systems by moving horizontally and vertically through the system. There are several advantages which make the procedure attractive for complex industrial processes:

- simple single-input single-output models and controllers are used
- the procedure is rather adaptable to the systems structure
- the design and closing of loops may follow the usual commissioning procedure
- the procedure includes a progressive model reduction
- very different sampling times can be used
- the expense of required software is acceptable for decentralized microcomputers used for process industries, at least in the version of selftuning controllers.

Simulations have shown that the sequential design converges rapidly for the investigated cases. However, a theoretical analysis has to show general convergence conditions.

REFERENCES

Bayer, Th. (1984). Sukzessiver rechnergestützter Entwurf von dezentralen Regelungen für Großanlagen komplexer Struktur. Internal Report, Inst. für Regelungstechnik, TH Darmstadt.

Isermann, R. (1981). Digital control systems. Springer, Berlin.

Isermann, R. (1982). Parameter adaptive control algorithms - a tutorial. Automatica 18, 513-528.

Isermann, R. (1984a). Rechnerunterstützter Entwurf digitaler Regelungen mit Prozeßidentifikation. Regelungstechnik 32 (1984), 179-189, 227-234.

Isermann, R. (1984b). Computer aided design and selftuning of digital control systems. ACC-American Control Conference, San Diego, Juni 1984.

Kuhn, V. (1983). Verfahren zur dezentralen Zustandsbeobachtung linearer Systeme mit komplexer Struktur. Regelungstechnik 31, 44-50.

Litz, L. (1983). Dezentrale Regelung. Oldenbourg, München.

Mäncher, H. and H. Hensel (1985). Determination of order and deadtime for multivariable discrete-time parameter estimation methods. 7th IFAC Symposium on Identification and System Parameter Estimation, York.

Mäncher, H. (1984). Synchronization tools and restart method in a fault tolerant distributed automation system FIPS. FI/NTG/GMR-Fachtagung 'Fehlertolerierende Rechnersysteme'. Informatik Fachberichte Nr. 84, Springer-Verlag,1984.

Singh, M.G. (1981). Decentralized Control, Elsevier, Amsterdam.

Acknowledgement: This research was supported by the Deutsche Forschungsgemeinschaft.

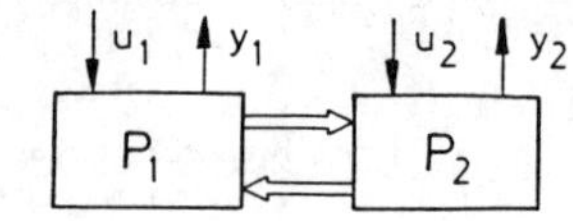

Fig.1 Two process parts P_1 and P_2 with (unknown) interactions.

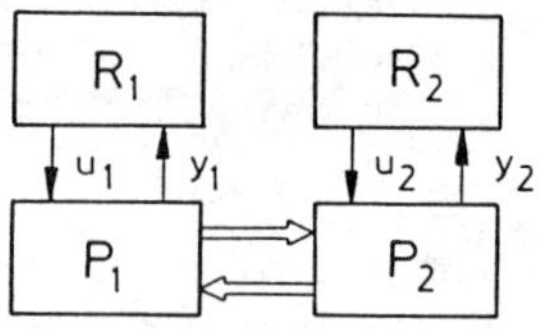

Fig.2 Two process parts P_1 and P_2 with two decentralized controllers R_1 and R_2.

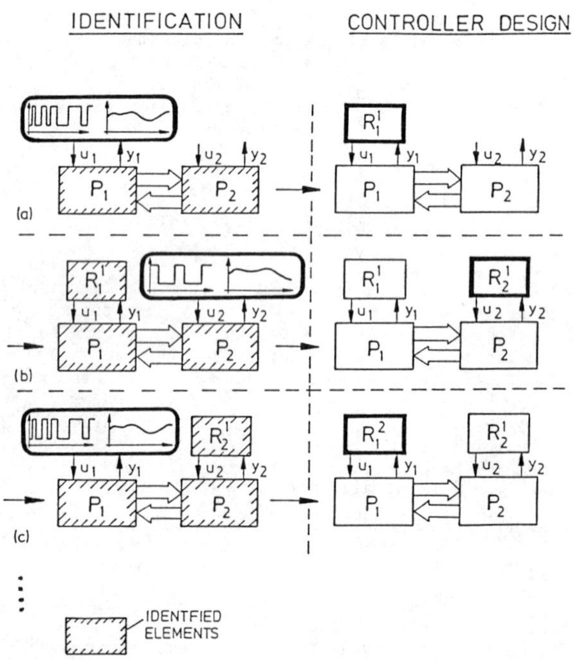

Fig.3 Procedure for sequential identification and controller design of a two-input two-output process.

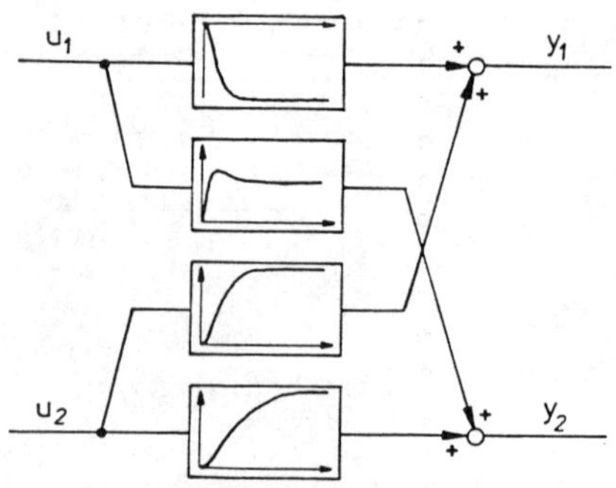

Fig.4 Blockdiagram of a two-input two-output process with strong interactions.

$$G_{11}(s) = \frac{-1}{(1+2s)^3}$$

$$G_{12}(s) = \frac{0,5+4s}{(1+s)(1+2s)(1+4s)}$$

$$G_{21}(s) = \frac{1}{(1+s)^2(1+3s)(1+2s)}$$

$$G_{22}(s) = \frac{1}{(1+s)^2(1+3s)(1+10s)}$$

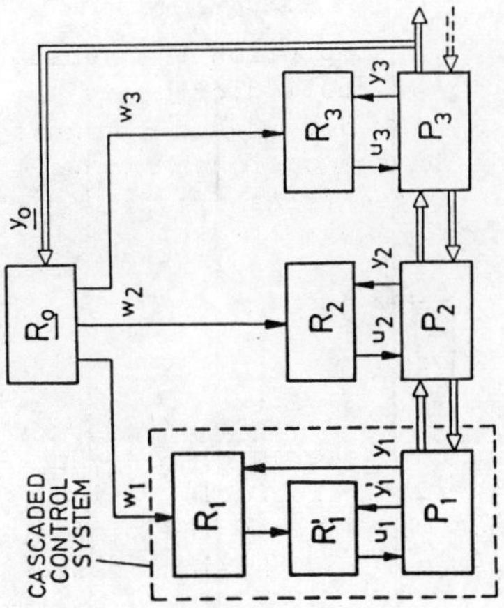

Fig.6 Multivariable process with decentralized controllers R_1', R_1, R_2, R_3 and an overall controller R_o (hierarchical control system)

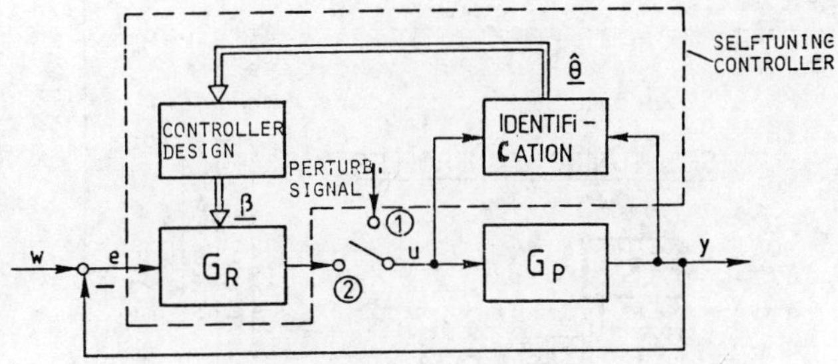

Fig.7 Blockdiagram of a selftuning controller

① pre-identification

② selftuning control

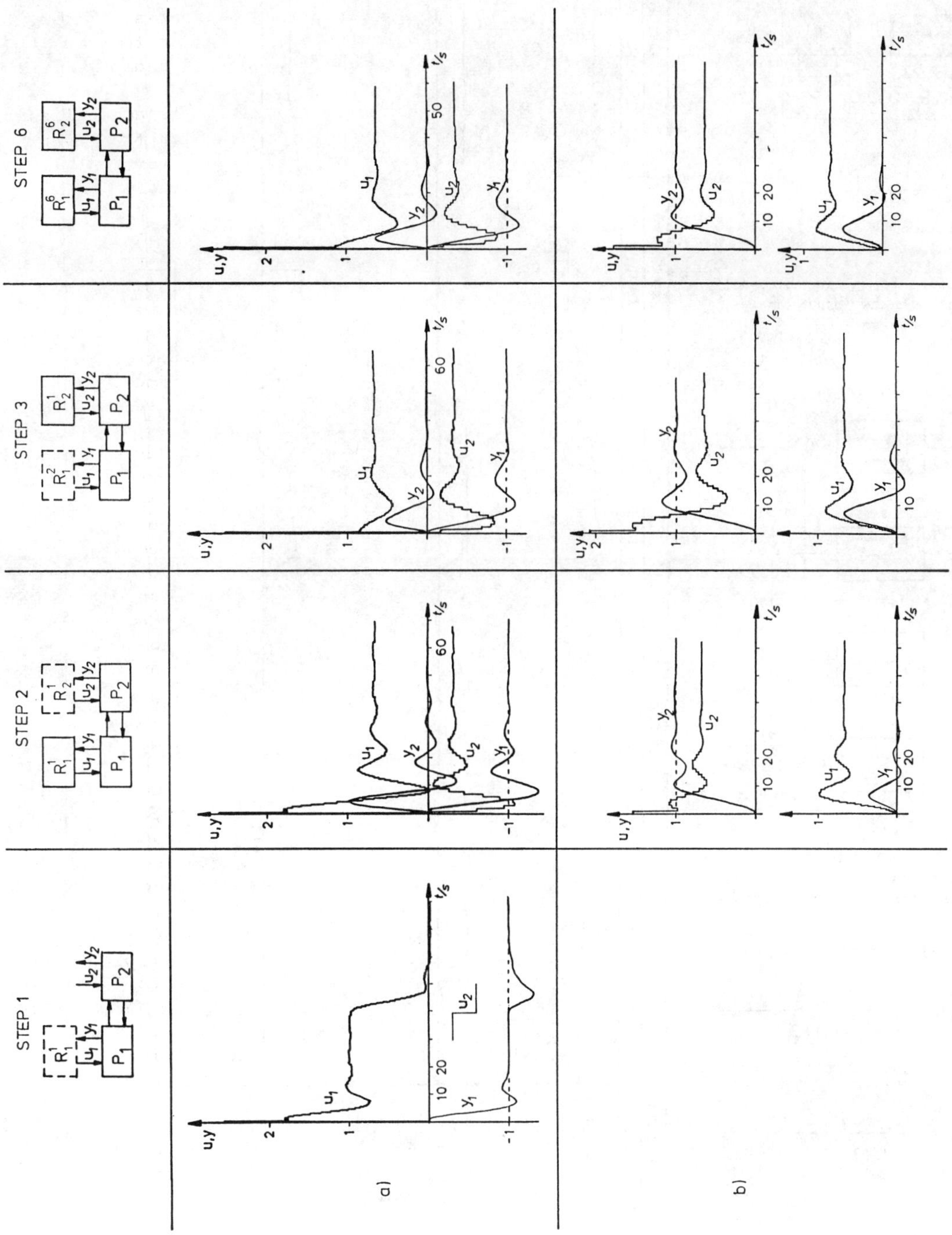

Fig.5 Signal records for the sequential design of two decentralized PID-controllers

a) setpoint change $w_1 = -1$

b) setpoint change $w_2 = 1$

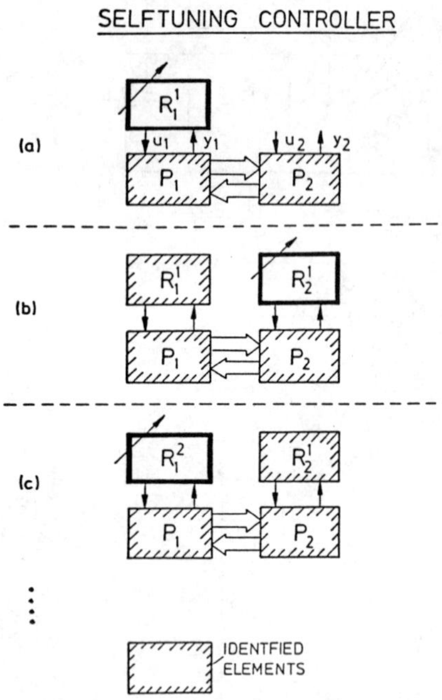

Fig.8 Procedure for sequential self-tuning of a two-input two-output process.

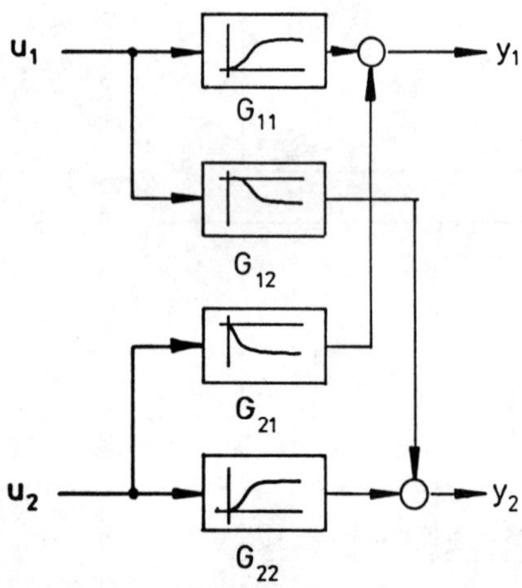

Fig.9 Block diagram of a two-input two-output process. (Rough model for an airconditioning plant).

$$G_{11}(s) = \frac{1}{(1+3s)^2}$$

$$G_{12}(s) = \frac{-0.67}{(1+3s)(1+\frac{5}{3}s)(1+s)}$$

$$G_{21}(s) = \frac{-0.38}{1+3s}$$

$$G_{22}(s) = \frac{1}{(1+3s)(1+\frac{5}{3}s)}$$

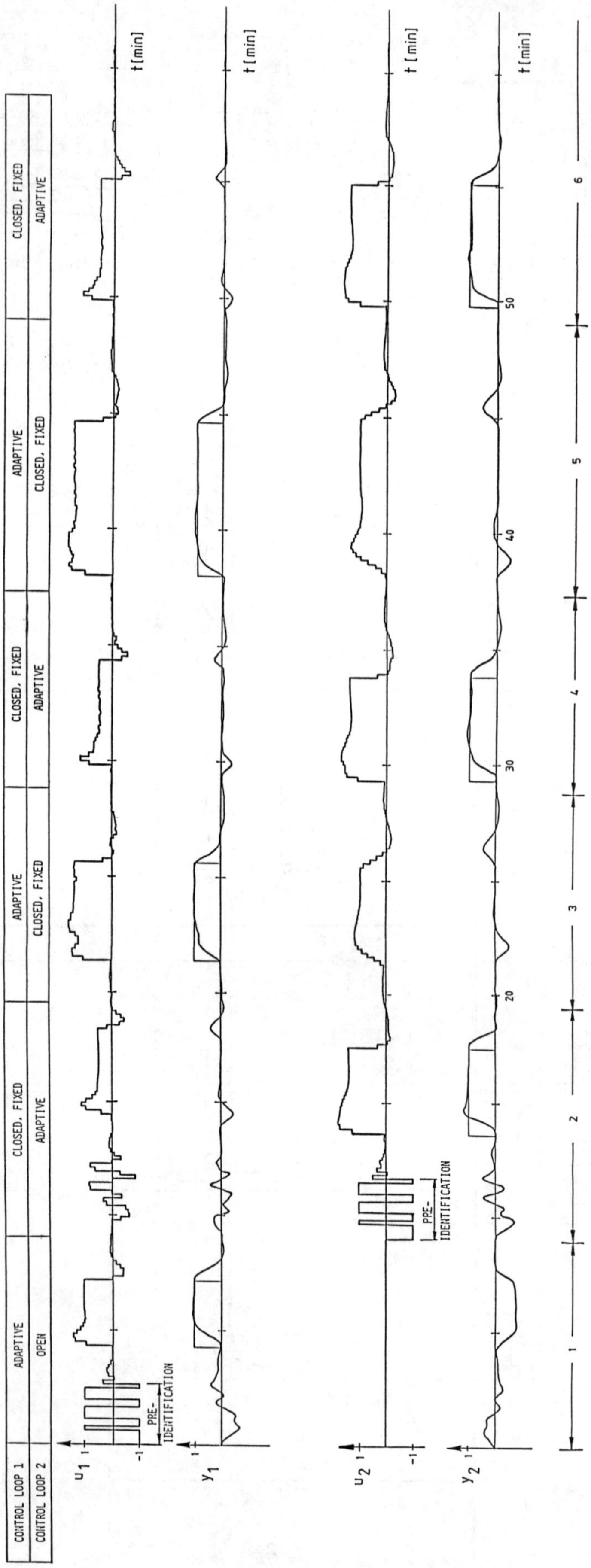

Fig. 10 Signal records for sequential selftuning of two decentralized state controllers with observers. T_0 = 2 sec. q_{ii} = 1; r = 0.5.

COMPUTER – AIDED DESIGN OF INDUSTRIAL AUTOMATION

L. A. Jacić

College of Engineering, Požarevac, Yugoslavia

Abstract. In recent years, computers with their great computational power have assumed very important role in the logic design of industrial digital control systems. The implementation of large-scale digital systems is an example of one design problem which can be impractical to solve by hand. How to enable production engineer to improve the existing control network, or to design the absolutely new one, presuming, as it is very often, that he is not a control design specialist. On the basis of new technological demands, an engineer should form a desired control specification which represent the input format to available computer. Computer should produce desired logic equations as a basis for the immediate realization, since the system is reliable and minimal in sense of number of required components. Software for complete design of digital control networks will be presented in the paper. It is applicable for treating deterministic as well as stochastic systems, both combinational and sequential etc. The concept of use of interactive computer graphics and other available computer aids in synthesis process is given. A relevant example is included in order to illustrate the program package.

Keywords. Digital systems; digital control; Computer-aided logic design; finite automata; sequentials machines; industrial control; hazards and race conditions; sequential control systems.

INTRODUCTION

A total machine-the aggregate of the power system, the control system and the external system is given, due to Stecki (1980), at fig. 1.

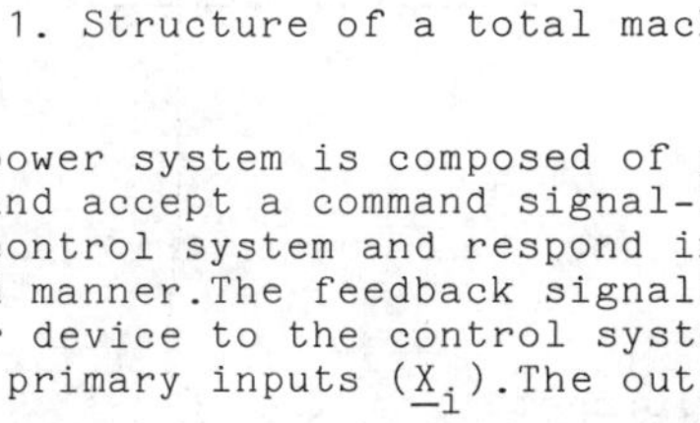

Fig. 1. Structure of a total machine

The power system is composed of power devices, and accept a command signal-inputs from the control system and respond in a prescribed manner. The feedback signals from the power device to the control system are called primary inputs ($\underline{X}_i$). The output signals from the control system to the power devices are termed primary outputs. The external system is composed of all the devices that are not included in the power system or in control system (emergency stops and reset, anti-tie down two-hand control, etc.). The control system represent the interface between the external system and a machine.

The design of external system and its own control system is outside of the design problem, and shall not be considered here.

The control system is set of components which will accept primary inputs and external inputs (from external system), combine them together with information stored in its memory, and make a decision. The decisions are output signals from the control system that are communicated to the power system and the external system. The output signals can be primary ($\underline{Y}_j$) and secondary outputs ($\underline{Z}_k$). So, the control system to be designed is considered as an information processing device, subjected to a string of input information and producing a string of output information. Furthermore, as the processed information is in digital binary form, the control system is called a digital control system.

The most general form of a digital control system to be synthesized is given by following eqs. and fig. 2.

$$\underline{Z}_k` = F_z(\underline{X}_i, \underline{Z}_k)$$

$$\underline{Y}_j = F_y(\underline{X}_i, \underline{Z}_k) \qquad (1)$$

It is evident, due to Huffman (1954), that we consider digital sequential control system as a system consist of input and output subsystems, both of combinational behaviour.

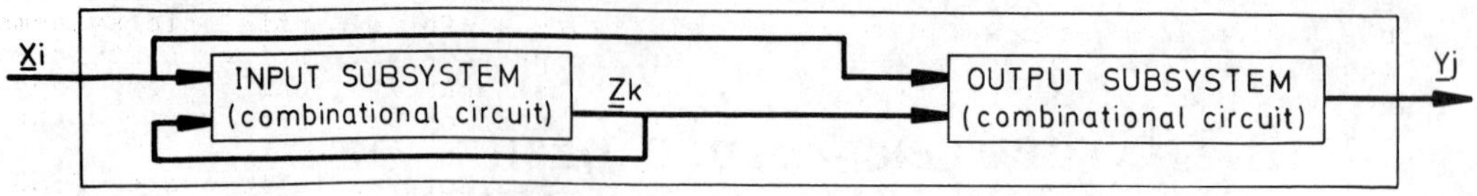

Fig. 2. The general concept of digital control system

According to the conventional clasification there are two types of control systems: combinational and sequential. In combinational systems, the output signals depend only on the input signals simultaneously presented (one to one correspondence between sets of input and output signals). So, the main characteristic of this type of a system is uniqueness of output signals. In a sequential logic system, the output are related not only to the present inputs of the system, but also to the past history of the inputs. Hence, sequential network must incorporate memory elements. This type of networks can be classified into a two groups: synchronous and asynchronous.

The synchronous type sequential system must be externally timed in order to produce the correct operations. An asynchronous circuits are designed to be internally "self-timed". This type of circuits, because of self-timing feature and faster overall response time than the synchronous circuits, is desirable for sequential logic systems. In other words, concerning industrial digital control systems, each control step commences immediately after the previous step is completed. Asynchronous type circuitry can be divided into a two classes: Regularly Activated Networks (RAN) and Irregularly Activated Networks (IAN).

The main characteristic of RAN is predictable input sequence. On the other hand side, when the schedule of input signals is indefinable, the network is IAN. Maybe the better way is to give the following clasification: deterministic and stochastic systems, which is identical to previous classification. So, we can say that deterministic circuit is logic circuit in which only predetermined inputs are allowed, while stochastic means that it is logic circuit in which predetermined, random or both types of inputs are permissible. We shall treat sequential asynchronous systems, both deterministic and stochastic, fig. 2.

So, in order to produce the desired digital control system, we must derive the logic equations of type (1), namely for $\underline{Y}_j$ and $\underline{Z}_k$. The optimality criteria J_{opt} which has to be respected during the synthesis procedure, is given by eq. (2):

$$J_{opt} = J_{min} \wedge J_{rel} , \qquad (2)$$

where J_{min} is a component of J_{opt} with respect to a minimal number of components, which represent a cost function. J_{rel} represent a component of J_{opt} with respect to reliable operation (hazards and races to be elimminated during the synthesis process).

THE SYNTHESIS PROCEDURE

The choice of synthesis method is dictated by the type and complexity of the problem. The method used here represent the modification of well-known Huffman method (1954) This method is applicable for synthesis of asynchronous digital control systems without respect to the size of a problem. As we mentioned, the deterministic and stochastic systems can be handled by this method. The general algorithm of synthesis procedure (Jacić, 1984) is described below, fig. 3.

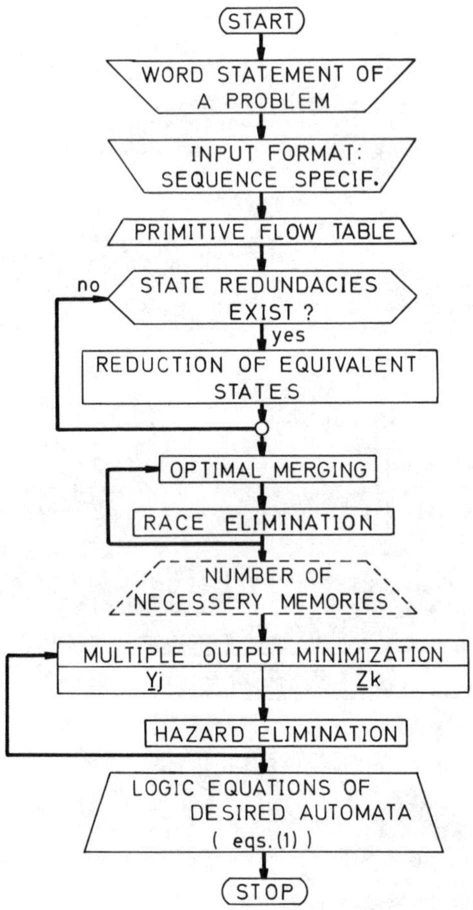

Fig. 3. Synthesis algorithm

The disadvantages of classical Huffman`s method are described in (Jacić, 1983a, 1983 b), and served as a basis for further improvement and modifications of this method.

The logic description of a problem is based on word statement which describe the expected machine operation. Input data obtained from the word statement can be a step by step specification, truth table, primitive flow table, timing chart etc. Adequate formatting technique must be used in order that, regardless of the type of network, all possible combinations of input states affecting the operation are considered in prescribing the output conditions. The efficient format that is used here is Primitive Flow Table (PFT). It is produced on the basis of Step by Step Sequence Specification (SSSS). PFT is very powerful format since we employ it later in phase of minimization of automata.

The relation between SSSS and PFT is shown at figure 4., for deterministic systems and for stochastic systems, fig. 5. It is suggested to use GRAFCET-technique for proper description of operation of desired automata.

For a given deterministic system, SSSS is given by a string: A+,B+,B-,A-. The comma divides the sequence into a steps. It is obvious from the graph of this sequence that it is pure deterministic system, fig. 4.

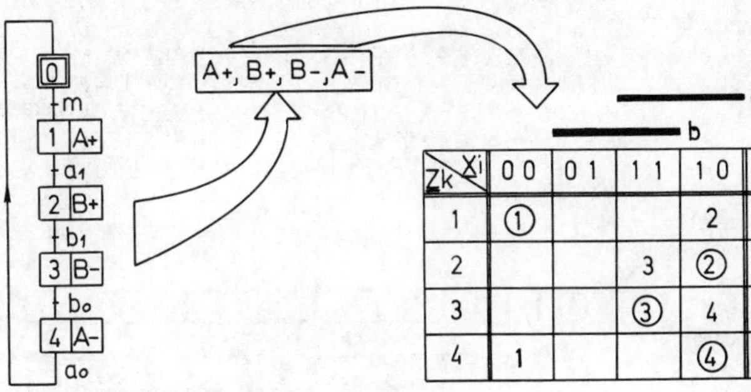

Zk \ Xi	00	01	11	10	YA	YB
1	①			2	1	0
2			3	②	1	1
3			③	4	1	0
4	1			④	0	0

DETERMINISTIC CIRCUIT

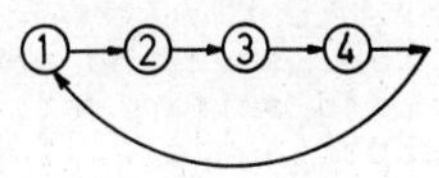

Fig. 4. GRAFCET-SSSS-PFT correlation for deterministic systems

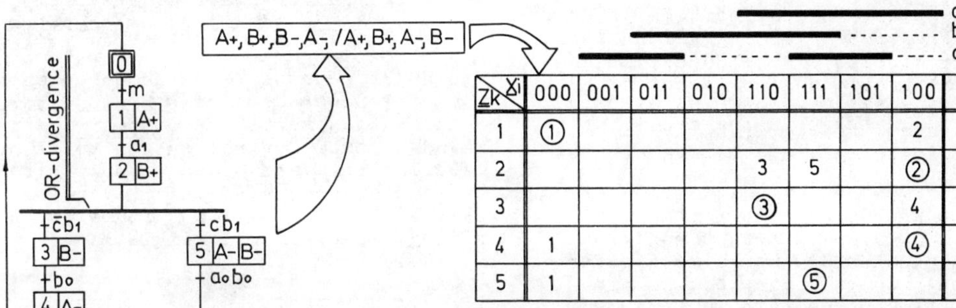

Zk \ Xi	000	001	011	010	110	111	101	100	YA	YB
1	①							2	1	0
2					3	5		②	1	1
3					③			4	1	0
4	1							④	0	0
5	1					⑤			0	0

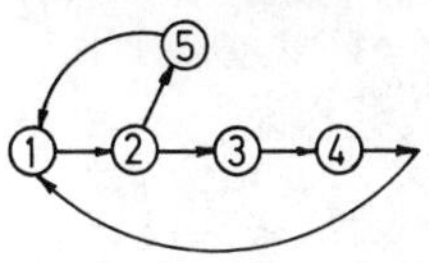

STOCHASTIC CIRCUIT

Fig. 5. GRAFCET-SSSS-PFT correlation for stochastic systems

Looking into a GRAFCET-representation at fig. 5., we do recognize the concept of alternative paths. The system will follow the sequence under the OR divergence depending on external input "c". The SSSS is given by a string: A+,B+,B-,A-/A+,B+,A-B-. Slash (/) represent the OR divergence. For the case of simultaneous paths, two slashes (//) have been used (the AND divergence). How many external inputs are needed depend on number of alternative paths. The system, for this stochastic case will make transition 2→3 for c=0, and 2→5 for c=1, fig. 5. A and B denote power devices, ie. pneumatic cylinders, "+" and "-" denote the type of movements, a_0, a_1, b_0 and b_1 are signals from limit switches, and c is external signal, according to the fig. 6. The power system for B is identical as this of fig. 6. Of course, the method is applicable to any type of automata, without respect to informational and working media (electronic, pneumatic, hydraulic, electrical etc.).

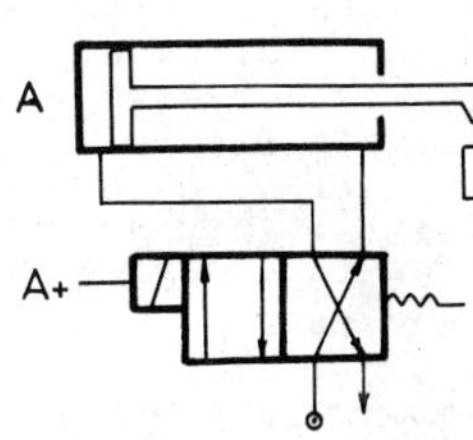

Fig. 6. Power system concept

Each type of a sequence that is common in industrial automation, can be handled by this method (ordinary sequence, simultaneous sequences, alternative sequences, partial override cycle and partial repeat cycle).

The primitive flow table (PFT) is produced in obvious manner,mannualy or by computer, and serve as a basis for further synthesis.

Next step in synthesis is reduction of equivalent states according to the well known rules in digital automata theory (Paull,Unger).Given is powerful algorithm for systematic identification and elimination of redundant states.This step represent a preliminary stage of reduction of necessary memory elements.

The final stage of reduction of number of necessary memory elements is achieved by optimal merging.Disadvantages of classical merging technique (Jacić, 1984) are: immensity,great possibility of error,nonunique solution,impossibility of CAD merging,nonexistence of algorithm for the choice of optimal subgroups,impracticity of application for large-scale systems etc.A long term research in this area (Jacić,1983a,1983 b,1984) resulted in optimal merging method which has been computerized.Introducing the definitions of "state-image" and "minimal-image",the theorem on "minimal-state-image" is established and proved,giving the neccessary and sufficient conditions for the existence and identification of minimal memory digital control system.

Next step,race elimination is connected to a previous step.The races are eliminated by itterative checking for race all the possible combinations between the raws of reduced flow table.Very often,the desired sequence of rows of reduced flow table is obtained directly by the application of the proposed algorithm,in first itteration. Races due to primary inputs are eliminated in the early beginning by using cyclic Grey code.

In order to derive the logic equations of desired automata with respect to the optimality criteria (eq. 2),a slight modification of well known Mc Clyskey`s method for multiple output Boolean functions is used for the first time,opposite to Huffman`s approach.Hazard elimination is done at the same time during the minimization process, based on the eq. 3.:

$$XY + \overline{X}Z = XY + \overline{X}Z + YZ \qquad (3)$$

In this expression term YZ represent a hazard elimination term.

The logic equations obtained at the end of briefly described synthesis processs represent the behaviour of desired digital control system.Later,they serve as a basis for immediate realization.It is good to point out that we do realize a control system in the Boole basis (NOT,AND,OR),where memory functions are achieved implicitly in combinational subsystem.

COMPUTER-AIDED-LOGIC-DESIGN

The manual synthesis is almost impractical when large-scale networks are encountered. In order to perform the system synthesis, the user need to supply the SSSS information to computer,describing the desired logic.The program package will then perform the necessary steps to obtain the digital equations by the synthesis procedure already programmed.With this capability,the designer do not need to know a formal synthesis procedure.He need to know only how to write SSSS,call the program and then implement the resulting equations.

The program package LJUBA consists of main program and 19 subroutines (Jacić,1984).The computerization of proposed method is done in part,at George Washington University,Washington DC,and has been completed in Computer Center in Požarevac,on Honeywell L6/53 Computer.Operative system is GCOS 6-MOD 400 and the program language is Fortran,while the compiler is Fortrana.The package is easy to apply on each universal computer,with slight modifications if necessary.Some ideas have been used from Surjaatmadja (1973), mainly the output format specification.

The SSSS is fed into a computer via videoterminal in proposed format.In order to illustrate the software package,consider a following example.It is necessary to design digital control system described by bar-chart,fig. 7.

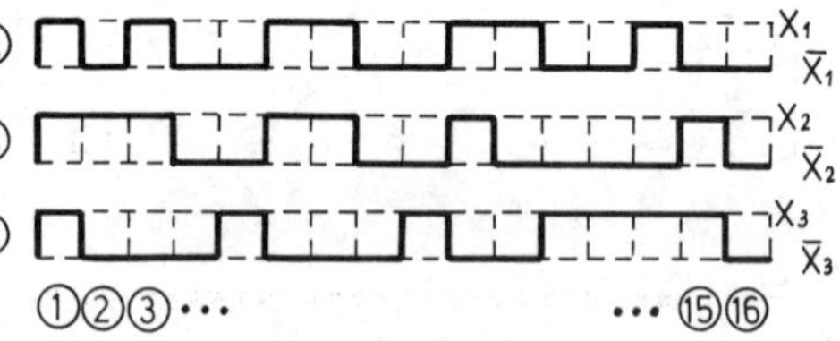

Fig. 7. Bar-chart of a system to be designed by computer

This chart represent the sequence of operation of 3 power devices A,B and C,for which activation serve the output signals Y1, Y2 and Y3 from control system to be synthesized.The relevant output-input triples for this system are: $A \div (Y1,(\overline{x}1,x1))$,$B \div (Y2,(\overline{x}2, x2))$,$C \div (Y3,(\overline{x}3,x3))$.Input format to computer is given by a following string:

A+B+C+,A-C-,A+,A-B-,C+,A+B+C-,A+B+C-,A-B-,

C+,A+B+C-,B-,A-C+,A-C+,A+,A-B+,B-C- .

This test problem involving 3 input and 3 output variables resulted in processing time of 9 CPU seconds.The computer printout is given in table 1.

The horizontal lines in the printout represent conjunction terms(AND`s) while the relation between them is of dinsjuctive type (OR`s).

After,the control network has been implemented by constructing a physical equivalent of an abstract model of the control system-set of Boolean equations.The logic elements NOT,AND,OR (and their combinations) which perform basic functions can be identified and used to construct the control system.

The program limitations are: number of feedback inputs is 16, number of outputs is 16 and number of internal states (memory) is 32.This limitations are just formal in its nature.

FUTURE TRENDS IN CAD

We are just about to rebuild and improve the existing CAD concept for digital control systems.We plan to achieve it by making a program package which would do the following,based on the outputs of previous software package LJUBA :

Table 1 Computer printout

	Z1	Z2	Z3	Z4	Z5	X1	X2	X3
Z1 =								
	1	-	-	-	-	-	-	-
	-	-	-	-	-	-	0	1
	-	-	-	-	-	0	1	0
Z2 =								
	1	1	-	-	-	-	-	-
	-	1	-	-	-	0	-	-
	1	-	-	-	-	0	0	-
Z3 =								
	-	-	1	-	-	-	0	-
	-	-	1	-	-	-	-	1
Y1 =								
	-	1	-	-	-	1	-	-
	0	0	-	-	-	-	0	-
	0	0	-	-	-	0	-	-
	-	-	-	-	-	0	1	0
	1	1	-	-	-	-	1	-
	0	-	-	-	-	1	0	-
	-	1	-	-	-	-	1	0
	0	0	-	-	-	0	-	0
Y2 =								
	-	1	-	-	-	1	-	-
	-	-	-	-	-	0	1	0
	-	-	-	-	-	1	1	1
	1	1	-	-	-	-	1	-
	0	0	0	-	-	-	0	0
	-	1	-	-	-	-	1	0
	0	-	0	-	-	1	0	0
	0	0	0	-	-	0	-	0
Y3 =								
	-	-	1	-	-	-	-	-
	-	-	-	-	-	0	0	-
	1	0	-	-	-	-	0	-

- forming a list of necessary logic components,
- optimal interconnection between components (connection matrix) and logic diagram,
- functional diagram.

Also, we plan to establish the data base of existing logic components at control market with relevant parameters such as: producer, type, price, technical characteristics etc. The integral concept of CAD of digital systems is given at fig. 8.

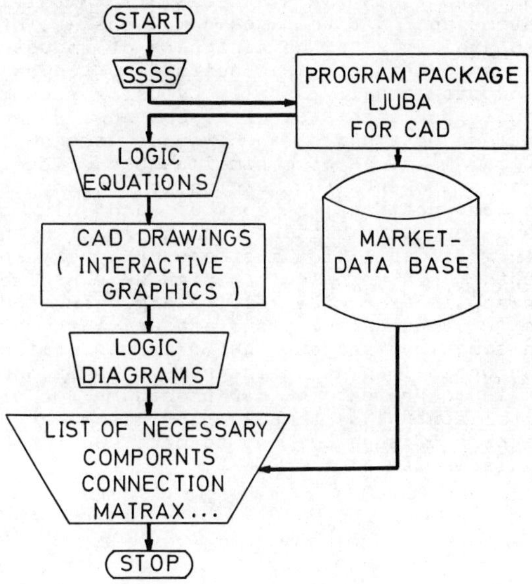

Fig. 8. The integral CAD concept

A CAD working station is configured due to fig. 9.

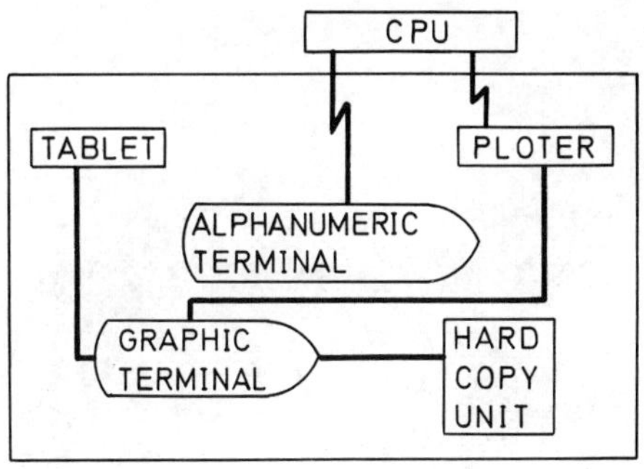

Fig. 9. CAD working station

The tablet used is Tektronix 4954, interactive graphic terminal Tektronix 4014 and interactive plotter Tektronix 4663.

An integral CAD concept for digital control networks reduce the time spent on manual synthesis. This allow the designer to experiment with different design approaches, and test them for hardware cost, feasibility etc. Beside, this approach to the logic synthesis enable to keep track of the components used in the design so that if the specification of a device used in many places is changed,the change will carry throughout the design.Also,the integration of graphics with text documentation represent a powerful basis for easy maintenance and troubleshooting.There are numerous advantages of CAD of digital control systems over the classical-manual synthesis approaches. Sometimes,the CAD approach is the only possible tool,especially when large systems are treated.

CONCLUSION

In spite of existing and availability of many CAD tools for synthesis of digital systems,and fact that the computers are often available to the designers of industrial automation,the synthesis is often handled out manually,usually by using step modules for each step of a sequence.The solutions obtained on such manner were not optimal in sense of minimal number of required components for realization,the systems were not fast,the cost of such solutions were rather high etc.Computers with their great abilities must occupy a very important place in logic design process.Their application will releive engineers from the routine calculations and drawings,which sometimes prevent them from performing the optimal solution.The role of this paper is to present one CAD procedure for digital control systems and to induce the designers of such systems to more use the available CAD aids.

REFERENCES

Huffman, D.A. (1954). Synthesis of sequential switching circuits. J. Franklin Inst., 257/3, 161-191.

Jacić, A.L. (1983a). A generalized methodology for minimizing fluid logic hardware.In R.Henke (Ed.), The Fluid Power Controls and Systems Conference, University of Wisconsin,Milwaukee. pp. 1-47.

Jacić, L.A. (1983b). Computer-aided-logic-design of digital systems.In S. Tzafestas (Ed.), MECO`83 Symposium,Vol.2, Acta Press, Athens. pp. 135-139.

Jacić, L.A. (1984). A method for synthesis of optimal digital systems of sequential type and CAD. Ph.D. Thesis,University of Belgrade.

Paul, M.C.,and Unger, S.H. (1959). Minimizing the number of states in incompletely specified sequential switching functions. IRE Trans. on Elec.Comput., EC-8, 245-249.

Stecki, J.S.,and Reddecliffe, O.A. (1980). Designing complex control systems with the computer. Hydr.and Pneum.,1, 107-109.

Surjaatmadja, J.B.,Bench, L.E., and Chan, C . (1973). Program and user`s guide for the classical synthesis of fluid logic networks. Report No. R73-1, University of Oklahoma.

COMPUTER AIDED DESIGN SYSTEM FOR INSTRUMENTATION ENGINEERING

K. Koskinen

Department of Automation and Electrical Engineering, EKONO Oy, Helsinki, Finland

Abstract. The utilization of computer in the design work is particularly based on three facts: fastness, errorfree processing and integration of data. Great advantages can be achieved by integrating the data management system and the graphic system into a functional unit.
The paper describes a computer-aided instrumentation design system developed by EKONO Consulting Engineers. In step by step progressing design work, the data entered earlier are utilized in all phases of planning. The work advances by adding and revising the data stored in the system. The same piece of information can appear in several forms. Any data is, however, fed and updated only once. The systems support engineers throughout the design process and produce most of the documents needed.

Keywords. Computer-aided design; data handling; computer-aided circuit design; computer graphics; integrated planning programs; instrumentation engineering.

INTRODUCTION

EKONO is a Finnish Consulting Engineering Company the main fields of which are energy, forest and mining industry, urban development and buildings.

The company provides a complete range of consultancy and engineering services such as business development, project management and complete engineering services as well as start-up, operation and maintenance services of plants. EKONO uses computer in the feasibility-study phase for calculation programs to find out the rentability of different process alternatives. Engineering phase of the plants computer is used for process dimensioning, piping-, electricity- and automation planning. The use of the engineering systems of different fields has not been integrated, they have no common database, but the data transfer is manual for the time being.

In the automation planning we have been using computer since 1976. As there was no economically realistic, suitable and flexible enough system at that time we started to develop it ourselves. The data management system developed supports basic engineering. The system was expanded with graphical programs in 1980 that supports detail engineering. The systems were realized first with the DEC system 10 time sharing computer of Digital Equipment Corporation. Since 1982 the systems have been used in the VAX 11/780 of the same manufacturer. The systems have been developed and modified in the past few years so as to better meet the changing demands of the automation technics planning and on the other hand to facilitate the man-machine communication.

DESCRIPTION OF THE SYSTEM

The system consists of about 300 program modules and 45 data files of different sizes. System comprises two highly integrated phases: data management system and graphic programs system. In basic engineering the system is used for data management, calculation tasks and reporting. The user communication is carried out by menu technics and the form management system (FMS).

Graphics programs system is used in detail engineering for drawings, documentation routines and management of drawing information and is based on interactive graphics data processing.

Programs

There are totally 45 programs with 170 integrated subprograms at the user's disposal. Different types of programs are available. From the users point of view, the most important programs are the data input programs and listing programs. The data input programs feed the data into the computer memory and change single data. The listing programs create reports from the data stored in files assorting them according to required criteria.

In addition, the users have several subprograms at their disposal that expedite the different routines. They serve for transferring data from one file to another so that the same data can be used in different displays and reports or e.g. for changing the items in all files at any stage of the project. It is also possible to sort the information on specific equipment, count of equipment functions or the vendors. There are practically no limitations on the selection of the sorting criteria.

Data Files

The files are divided into two groups basic data files and project data files.

Basic data files. The basic data files contain e.g.

- translations of codes in clear nomenclature
- calculation equations
- list- and form models, drawing symbols
- information on alternative methods of cable connections and various terminal numbering systems, cable types and sizes
- symbol library for standard equipment of major manufacturers
- dictionary for multilingual listing

Part of the data of project data files are fed in codes that refer to translations in clear nomenclature stored in the basic data files. Thus the user need not write the same text over and over again. In reports and listing the codes are translated in clear nomenclature. Nowadays the reports are obtainable in English, German, Swedish and Finnish.

The basic data files contain alternative methods of cable connections practices and various terminal numbering systems. In each project the user must specify which of the systems is applied for field junction boxes, terminal strips and control room enclosures. Further the user defines the kind of cable to be used between each location and how the cables are to be numbered.

Symbols for every component in the system are necessary for creating wiring diagrams. These have been prepared with the computer graphics station and stored in the symbol library according to manufacturers. The data of the basic files can be changed with the updating programs.

Project data files. The project data files are created and updated with data input programs. They contain specific data on the function of the loops, the equipment and the cables. Project data files are e.g.

- functional register
- equipment specifications
- specifications for flow elements and control valves
- alarm and interlocking point lists
- purchase specifications
- equipment type file
- junction box file
- cable number file

The functional register is the basic document of the project comprising initial data of the circuit e.g. process and piping data, initial equipment specifications and functions of loops. The object of the functional register is to serve as instrument index and to summarize technical features of all the measurement and control loops of the project.

The equipment specification is another main document of the basic engineering. The specifications define technical features of each equipment by loop numbers. The additional data and precisions accumulated in the progress of the basic engineering are fed directly into the equipment specifications.

The flow elements and control valves specifications show the process and piping data relating to the said equipment as well as other detailed data needed for the calculation and dimensioning the equipment and in connection of requests for bids and orders.

The requests for bids are obtainable either as turn-key delivery or by each equipment group according to the way instrumentation is purchased.

The turn-key purchase specification is in the form of equipment specification. The loops are listed in growing loop number order and all the equipment of each loop with functional demands are specified in clear nomenclature.

If the equipment is purchased by the equipment groups from different vendors, the equipment to be purchased is arranged to functionally similar groups such as control valves, temperature transmitters and recorders.
The equipment of each group is listed in loop number order on the purchase specifications by the equipment groups.

The equipment type file is an auxiliary file that defines the type data on the equipment for equipment specification and wiring diagrams.The junction box and cable number files are auxiliary for graphic system.

There is a specific project data file for each document to be printed so that the same piece of information can be in different places in the computer memory. It is significant that any data are fed into the system only once, generally into the functional register and that the data are transfered automatically to other files.

PROGRESS OF THE COMPUTER AIDED DESIGN

Basic Engineering

When the flowsheets (PI-diagrams) of the plant are completed the initial data needed for the functional register are collected in co-operation with the process designers.

In the functional register are recorded all the initial data on measurement and control loops, besides process and piping data also all the necessary equipment in process piping, process connection data, hook-up drawing for installation, function of loop and service of alarm and interlocking limits to such an extent as it is possible at this stage of planning.Of the technical features of the equipment are defined e.g. operating principle, input and output signals, supply energy, etc.

The data fed into the functional register are utilized in other listings, Fig. 1,

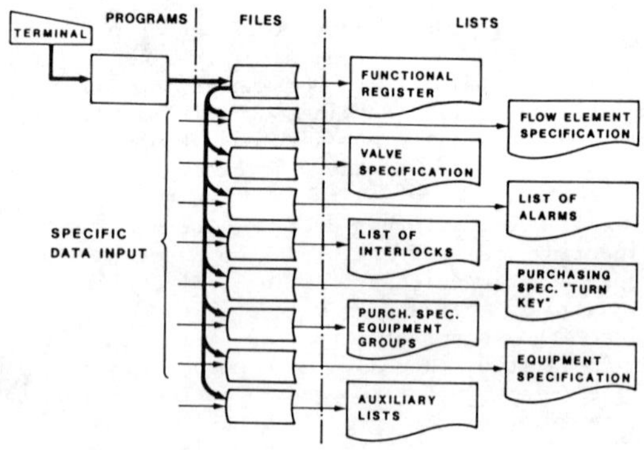

Fig. 1 Function of data management system

The listing programs transfer automatically the data requested from the functional register into

- flow element specification
- control valve specification
- purchase specification
- equipment specification
- alarm and interlocking points lists

The necessary data are transferred in the same manner into other supplementary auxiliary lists. The files are revised with additional data by the specifications as the planning progresses. Of each file a report can be produced when necessary, the form of which is defined separately.

Request for proposals can be listed in two different ways. The user can choose to list all transmitters, valves, pressure switches, etc., separately on their own lists either to go to separate vendors for bids on each equipment group or to print one complete list of all the equipment arranged according to control loops if a turnkey contract with everything supplied by a single source is desired.

After vendors receive the list, their proposal preparations are very rational. Rather than having to prepare thick, impressive proposal documents of their own, the vendors simply fill out the forms with their equipment model numbers and prices, make a copy for their records, and mail the completed forms to the client as proposals. In a turnkey project including installation and material, a vendor also gets a complete list of installation material that makes pricing quick and easy.

When the proposals are opened, their comparison is very simple because they are all alike except for equipment model numbers and prices. First of all, it is very easy to check whether a proposal is complete or if something has been omitted. If a supplier wants to suggest an alternative to a specification, space is reserved for comments on each page.
When the vendors have been selected, the equipment specification is completed with final equipment types and models and other supplier specific data.

In the equipment specification reference is made to flow element and control valve specifications. Any data previously entered for the flow element and the control valve in the functional register appears automatically on the form. The user fills out the missing lines with process data for calculation for manufacturing and the results are transferred into the flow element specification.

Detail Engineering

The second phase of the system, graphic programs, Fig. 2,

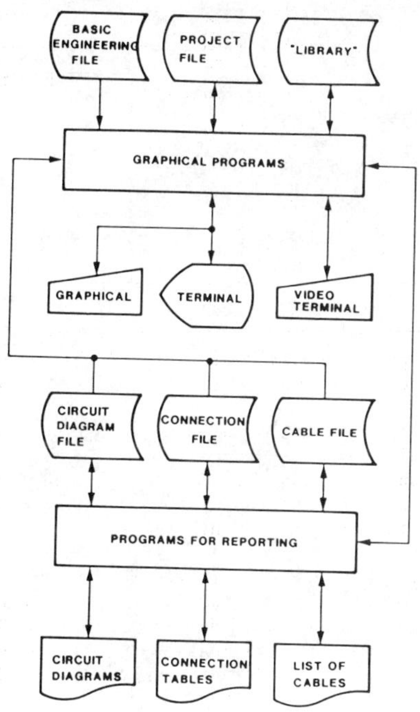

Fig. 2 Graphics programs functions

produces documents for final instrumentation implementation. These documents are mainly control and measuring-loop wiring diagrams, terminal connection tables, and cable lists. The list of installation material can also be included here if not already prepared during the basic engineering. In order to be able to produce an accurate list of installation material, the program needs information for each loop separately; i.e. number of the hook-up drawing, lengths of impulse pipe and cable runs. The rest is produced by the program totalling the details for every type of material into a comprehensive list.

The graphics programs read the location codes from the functional register files. The equipment model numbers are read from the equipment file. The symbol library provides terminal symbols for each piece of equipment and places the symbols in appropriate locations on the CRT. Provided that a circuit-wiring diagram with the same equipment configuration has been prepared earlier and stored, the program automatically proceeds by connecting the familiar terminals together on the CRT "drawing" asking only for project-specific data such as numbers of the junction boxes. If the equipment appears for the first time, the user draws the connecting lines between the terminals using a cursor on the CRT.

Terminal connection tables. After all the wiring diagrams connected with any one junction box or terminal strip have been graphically completed, the terminal numbering program can be started. This program assigns numbers to terminals and establishes the terminal connection tables and cable lists in a corresponding file based on project-specific guidelines and the graphical connections on the CRT drawing. The program finishes by assigning the drawing a number and by updating the list of drawings. Any changes and corrections are made on the CRT, and the program automatically does corresponding updates in the connection tables and cable lists.

Because the whole system has been designed as an integrated and interactive set of programs, each piece of information has to be dealt with only once, and the program automatically distributes the information throughout the system. For example, if an alarm point is deleted, the user deletes the point from the functional register and it automatically disappears from the equipment specification, list of alarms, list of hook-up drawings, installation materials, wiring diagrams, and connection tables. Furthermore, the corresponding pair of cores in the cable list will appear as spears when the next update of documentation is printed. A list of deletions is also available.

USER EXPERIENCES

In developing the system special attention has been paid to the needs of the user. Therefore, the user experiences have been mostly positive, the advantages being

- filling out of the functional register carefully guarantees the accuracy of other lists and specifications
- typing work is deleted and thus also the typing errors
- revisions and corrections can be made quickly and easily. Data is entered into the functional register or wiring diagram only once and is after that usable where needed
- detailed engineering can be realized with a few persons only compared with manual planning
- as the system is easy to use the user needs no actual training in data processing

As a disadvantage could be mentioned that the use of the computer requires a new kind of thinking from the user and the difficulty in adopting it may arise objection in some designers and reduce motivation of work.

In developing or purchasing an engineering system, negative surprises can be avoided by careful analyzing of the requirements and adaption possibilities of the computer based system. As for the engineering methods and routines it should be emphazised that poor engineering routines are poor even if carried out by a computer.

CONCLUSION

Design work can be characterized as iterating: data are completed and they change along with the progress of the project. To ensure the management of the changes and to guarantee flexible data transfer it is imperative that the data be fed into the computer only once whereafter the data are available where needed and in the requested form.

The design system is to be built for the designer, for him to use as an auxiliary tool. The structure of the system is to be flexible enough to meet with future and changing requirements.

After almost ten years' experience from the development and use of computer-aided design, we at EKONO are convinced that the only way to succeed in the future is the automation of the routines of the design work.

REFERENCES

Rusanen, H. (1984). Carrying and Planning Programmes for Electrification and Instrumentation at Tampella. CAMP' 84 Computer Graphics Congress Berlin, pp. 405-407.

Scherf, K. H., V. Koskinen and S. Aarnio (1983). Computer-aided design reduces labor on instrumentation projects. Pulp & Paper, 66-69.

COMPUTER-AIDED CONFIGURATION AND START-UP OF PROCESS CONTROL SYSTEMS

R. Schumann

VDO Mess- und Regeltechnik GmbH, Hackethalstrasse 7, D-3000 Hannover 1, FRG

Abstract. A systematic approach for the computer aided planning, configuration and operation of a process control system is outlined which supports systematic collection of all required informations, comfortable symbolic configuration of the system function and automatic compilation of the collected information to various documentation files. It is indicated, how the functional configuration and start-up of the process control system can be supported by computer simulations, process identification and selftuning features being part of the process control system itself.

Keywords. Process control; industrial control; computer aided design.

INTRODUCTION

Before a process control system can be set in operation, different preparatory phases must have been passed in which a great amount of information about process and system function is collected, converted to the hardware and software configuration of the process control system and compiled to system documentation files which have to support installation, start-up, operation and maintenance of system and process.

Fig. 1 shows the simplified structure of a decentralized process control system.

Three types of system devices can be found on the three basic hierarchy levels: The process I/O devices represent the interface between process and system and may consist of sensors, actuators, transducers, A/D and D/A converters. These I/O devices are clustered to process stations which

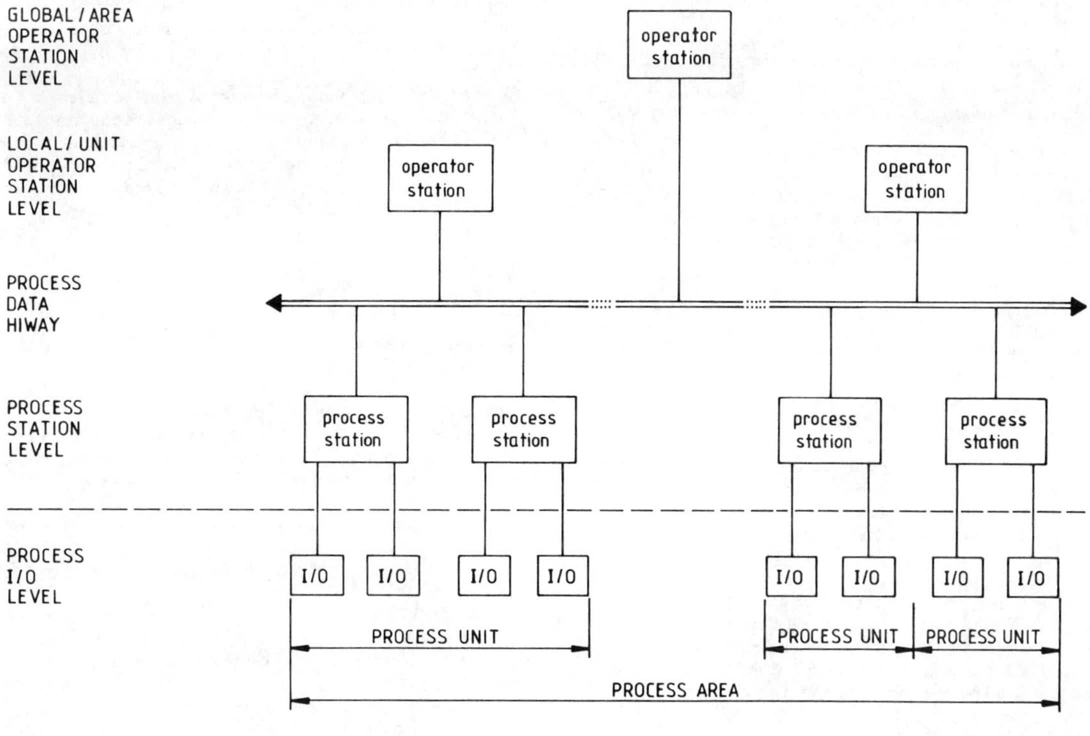

Fig. 1: Principle Structure of Decentralized Process Control System

have to handle the main part of the process control functions like continuous control, PLC and batch control. The interface to the human operator is realized by the operator stations and has to support system and process monitoring and operator interactions. When the process control system is applied to a process the system specific structure is complemented by a logical system organization according to the functional modules of the process, i.e. subsystems including process and operator stations are assigned to process areas, units, subunits etc..

Design procedure and operation of a process control system can be divided in 5 different phases, see Fig. 2. During the planning phase, the process I/O points and the functional requirements for the system are defined, such that in the configuration phase the system hardware and software can be designed.

I PLANNING	definition of process I/0 points functional specifications reliability demands
II CONFIGURATION	system organisation hardware configuration function configuration
III TEST	debugging of function configuration optimization by simulation operator training by simulation
IV REALIZATION	system installation system start-up on-line optimization
V OPERATION	process control set-up / alarm management maintenance

Fig. 2: System Design and Operation Phases

During the test phase the system configuration is analyzed with respect to system and process operation.

The realization phase comprises the ordering of system parts, system installation and system start-up. Tasks of the operation phase are besides the general process control and monitoring functions also system and process set-up management and on-line diagnosis functions.

Although modern process control systems are based on digital process computers they do not support sufficiently the above described preparatory stages, i.e. most informations are collected in planning sheets, often repeatedly, the functional configuration of the process control system is prepared by filling out configuration sheets and in a third step the information is compiled by hand for the different documentation files necessary for system installation and operation. Test of the system configuration can be done by a test installation of the configured system, however, in general without simulations of the process based on realistic models. During the start-up phase the finetuning of the system functions is done with specially trained personal without major computer support besides the test functions already used during the system test phase.

In the paper a systematic approach for the computer aided planning, configuration and realization of a process control system is outlined which supports systematic collection of all necessary information, symbolic configuration of the process control function and automatic compilation of the collected information to various documentation files. This procedure may be realized on the process control system itself or on special configuration units. The following prerequisites for the realization of the outlined computerized design procedure are

- a system data base containing all informations about all hardware and software modules required for system configuration
- a configuration data base in which all configuration information is collected
- a CAD systematic which utilizes interactive graphic and text editors allowing efficient input of ASCII and symbolic informations

An important general feature of the computerized design procedure must be an efficient user guidance through the configuration path utilizing extensively help information assistance thus making sure that all informations necessary for the realization of the system function are collected.

In addition it will be indicated, how the functional configuration, test and start-up of the process control system can be supported by computer simulations, process identification and self-tuning features being part of the process control system itself.

SYSTEM PLANNING

In the planning phase for the process control system all requirements for process control and operation have to be collected as basis for the system configuration. These informations are obtained from the designers of plant and processes for which the process control system is to be applied. The information interface between process and process control system designers consists in general of

- P&I (piping and instrumentation) diagrams in which the principle interconnections between different process units, measurement and control points can be found with process specific names as defined by the process designers which are named in the following "tags", see the example in Fig. 3
- functional specifications of the process I/O points
- specification of process control and operation functions.

The collection and use of these specifications can be supported efficiently by computer aided procedures during which the following informations are entered in the configuration data base

- P&I diagram: this is constructed by means of a CAD systematic using standard symbol libraries for process units like valves, pipes, boilers and enabling the efficient manipulation of all graphic objects, i.e. placement, copying etc. By entering I/O point tags in the P&I diagram in a standardized way basic informations on the I/O point related functions may be transferred automatically, i.e. informations on the physical character of the signal like temperature, pressure or flow, the kind of use for indication or recording, and the kind of use for process control.

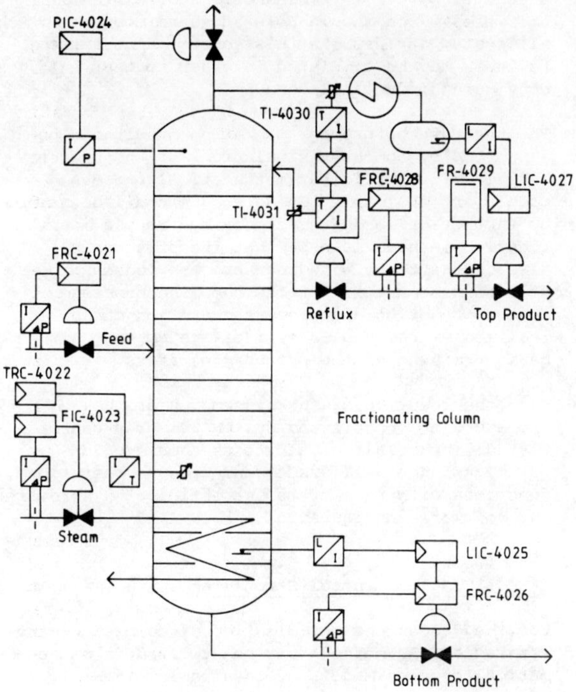

Fig. 3: P&I Diagram

- process I/O specification lists: the I/O point specific informations which are included in the P&I diagram are automatically selected from the graphic information and entered into process I/O specification lists. The user has to complement these lists by further functional specifications of the I/O points and the process control functions directly related to them
- formalized process control and operation function specification lists: additional specifications of the process control and operation functions which cannot be related directly to process I/O points must be collected in formalized requirement lists concerning
  - hierarchical organization of process modules defining process areas, units, etc.
  - complex continuous control strategies
  - sequence control functions
  - batch control strategies
  - operator displays and interactions
  - process and system set-up management
  - safety and reliability requirements

All these informations must be transferred from the designers of the process to the process control system designers. This can be accomplished either manually, i.e. during the planning phase the information is selected from planning sheets and P&I diagrams and entered into the computer, or more or less automatically provided that the informations have already been collected during the process design in a general process data base from which the above described specification informations can be generated and entered automatically in the configuration data base.

SYSTEM CONFIGURATION

During the system configuration phase hardware and software components of the system are configured to meet the requirements established in the planning phase. Using the computer all configuration can be done interactively making use of the specification information collected in the configuration data base and entering the new configuration information into it; furthermore the computer can provide from the system data base all required information on I/O and system devices possibly required for system hardware configuration and also on the function modules available for the system software configuration.

The first step of system configuration is the organization of the different system devices on the three system levels:

- process I/O level: assignment of process I/O points to process stations
- process station level: assignment of process stations to process modules, i.e. definition of standard/supervisory stations for process units/areas etc.
- operator station level: assignment of operator stations to process modules, i.e. definition of local/central operator stations for process units/areas etc.

The result of this step are assignment lists and hierarchy trees which are stored in the configuration data base.

The second step of system configuration is the definition of the required system hardware. This will be done making use of the system data base providing all informations on hardware devices available:

- process I/O hardware list: assignment of sensors, actuators, transducers to process I/O points, definition of back-up/redundant devices based on the process I/O specification lists
- process station hardware list: configuration of process station components like processor boards, local operator interfaces, termination panels etc, definition of back-up/redundant devices
- operator station hardware lists: configuration of operator station components like processor boards, monitors, keyboards, mass storage devices, interfaces etc., definition of back-up/redundant devices

From the information collected in this configuration step in the configuration data base and from the system data base, I/O and system part lists and also I/O and system installation instructions can be generated.

The third step of system configuration is the transformation of the formalized description of the process control and operation functions into the software configuration of the system. In many process control systems this is done by connecting function blocks in configuration lists and defining their parameters. Using CAD methods this can be accomplished by connecting symbols for function blocks on the screen e.g. for PID-controllers or logic functions, compare Fig. 4, and combining this with fill-in-the-blanks procedures for the definition of direct function connections, thus generating all functional interconnections defining as a whole the software configuration of the system.

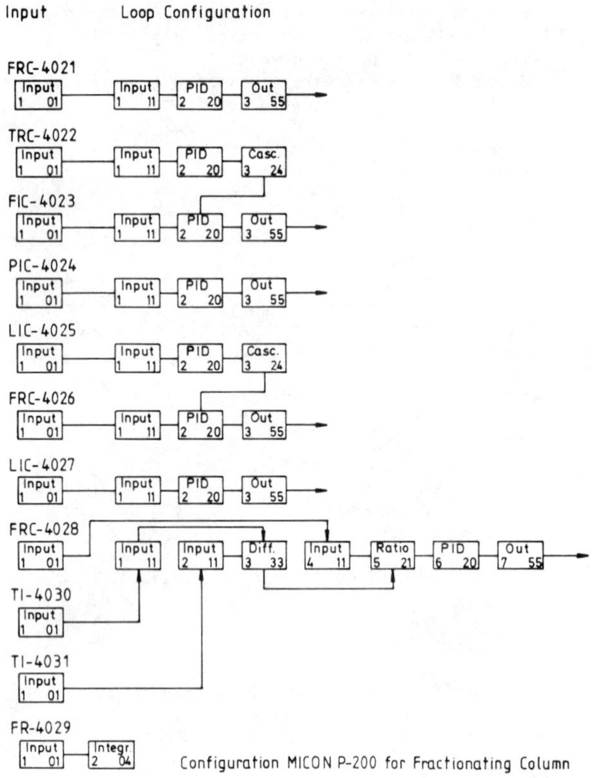

Fig. 4: Symbolic Control Function Configuration

The operator displays are generated in a CAD like fashion with function elements for dynamic displays parts, virtual keys etc.

On the different system levels the following tasks have to be accomplished for the system function configuration:

- configuration of process I/O related functions: signal conditioning like filtering, linearization, compensation; alarming by limits and alarm reporting mechanism; preprocessing like position control, local logics
- configuration of process station functions: continuous control; sequence and batch control; alarming and alarm management; local operator interfaces
- configuration of operator station functions: process displays for monitoring and operator interactions; reports and report formats; process specific help assistance displays; process/system set-up and alarm management.

The results of this step are the symbolic configuration displays and the process control and operation programs derived hereof for process and operator stations which are stored in the configuration data base.

SYSTEM TEST AND SIMULATION

During the system test phase the correct operation of the system itself and in connection with the process to be controlled must be tested. For the test of the system itself this includes the simulation of system operation which will enable for the different system devices the analysis of memory and processing power requirements and of timing problems, especially with respect to the communication lines. This can be efficiently supported by built-in diagnosis and selftest utilities which should be a standard feature of all system devices. In this context also system failures may be simulated in order to test configured redundancy concept.

The test of the process operation itself can be supported by computer simulations of the process behaviour if sufficient mathematical process models are available. These models could be gained by theoretical modeling and/or by process identification methods in case that the process is already existing. With such process models a realistic simulation of the complete process control function and process operation can be realized which enables besides system test also a basic training of the process operators.

Also the reaction of the system and/or the operators to process and system failures can be tested. The result of all tests should be an optimization of all system and process operation functions with respect to the anticipated hardware and software configuration.

CAE of Complex Control Strategies

For the design and optimization of complex control strategies CAE methods can be utilized if mathematical process models are available. These process models can be gained by theoretical modelling of the process dynamics or by computer aided process identification, compare Fig. 5.

Process identification is based on the analysis of the measured process input and output signals using correlation or parameter estimation methods; this means that process models can be obtained from measurements of process signals provided that the process is available. A variety of process identification procedures have been developed and applied during the last decade for the experimental modelling of single- and multivariable processes, compare e.g. Eickhoff (1974), Isermann (1971, 1981), such that they are now ready for industrial application e.g. in process control systems.

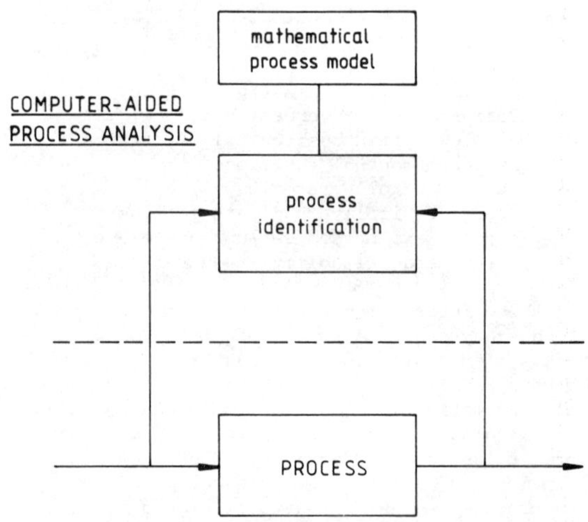

Fig. 5: Process Identification

Based on the computer simulation of process models control strategies for single loops and complex control problems can be designed and optimized with CAE methods by the engineer working interactively with the computer. A typical example of such a computerized process analysis and design procedure is described in Mann (1983) for a large rotary dryer in the sugar industry.

SYSTEM INSTALLATION AND START-UP

The first step in this phase is the ordering of all parts and devices which are required for system installation. The required part lists can now be gained from the system data base and the configuration data base in which during the system planning and configuration phase all necessary information have been stored.

For the system installation all necessary documentation can be gained from these data bases too, i.e. the stored informations are rearranged to system installation instructions containing all necessary informations about wiring, power requirements etc..

CAE Functions for System Start-up

The system start-up phase can be supported efficiently by the built-in test utilities and CAE functions already used during the system test phase. On-line the CAE functions as already described for the system test and simulation phase can be applied for controller tuning and process optimization.

Another CAE function, which may be utilized during the system start-up phase is the selftuning function, which can be characterized as the on-line combination of process identification and computerized controller design, compare Fig. 6.a. The selftuning function will automatically tune control parameters like the PID parameters thus simplifying and shortening the tuning of control systems considerably. An example for the tuning of a PID-controller as realized on a small industrial operator station (VDO MDC-60) is shown in Fig. 6: The tuning procedure starts with the on-line identification of the process dynamics in open or closed loop applying a defined test signal to the process input. After the process model has been gained the step response of the identified process model is shown on the screen, see Fig. 6c, thus giving a rough information on the quality of the identified model. The next step is the determination of the PID parameters by an optimization

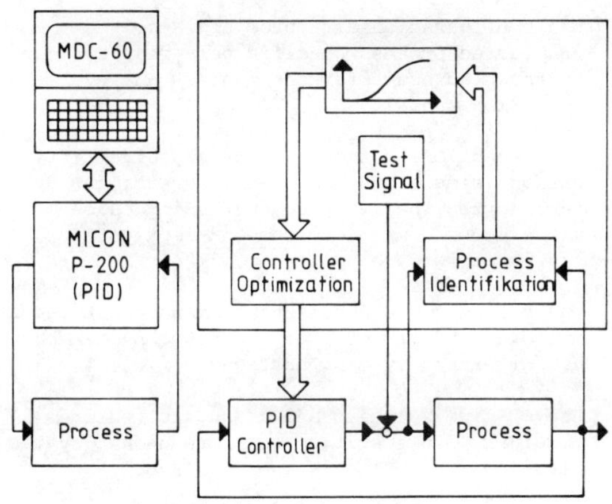

Fig. 6a: Selftuning Control Principle

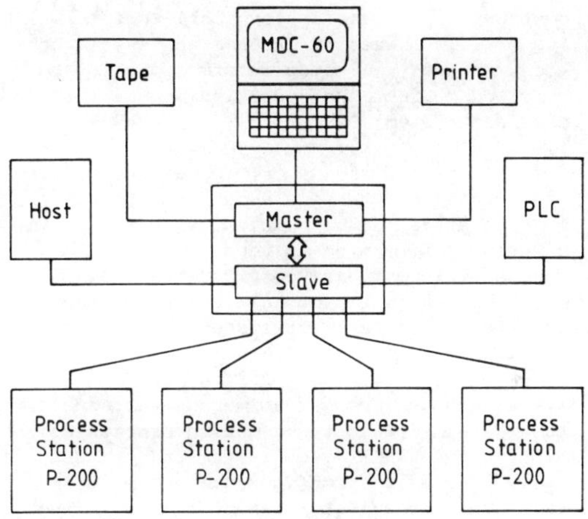

Fig. 6b: Basic MDC-60 Configuration

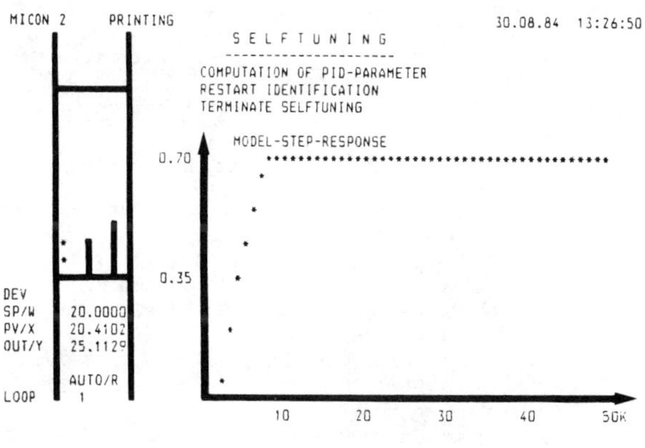

Fig. 6c: Process Model Step Response

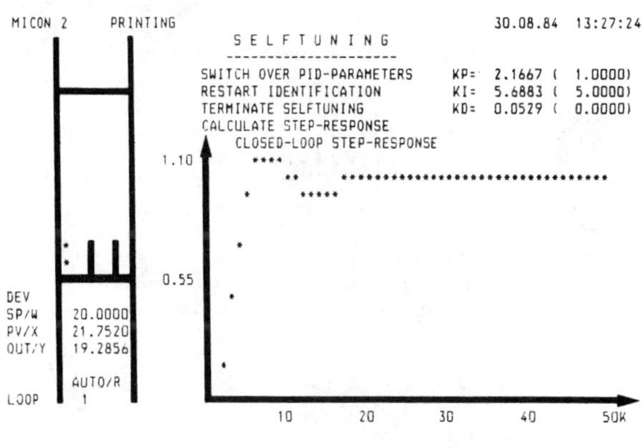

Fig. 6d: Simulated Closed Loop Step Response

Fig. 6: Selftuning of PID-Controller with MDC-60

algorithm, resulting in a proposal for these parameters which is illustrated by the simulation of the resulting closed loop step response, see Fig. 6d. After the result of the selftuning procedure is considered sufficient the controller parameters can be transferred to the on-line controller, i.e. to the process station P-200.

The selftuning functions based on process identification can be applied in open and closed loop which makes them in principle also applicable as on-line adaptive controllers, see e.g. Schumann (1982), Radke (1984). In this form they may also be applied during the process operation phase for the control of processes with nonlinear or time-varying dynamic behaviour.

SYSTEM OPERATION

During the system and process operation phase the process control system can support all operations by using the information in the configuration data base and the system data base. During normal process operation the following aids can be provided

- built-in system and process operation assistance: This function can be looked at as a generalized help assistance function, i.e. all help informations provided for the system and all help displays which have been generated for system and process operation during the system configuration phase are available now to support all phases of process and system operation. In addition these informations may be compiled to form the operation manual for process and system

- flexible report and report format generation: all actual informations about system and process state can be combined with the configuration information to reports including journal informations, configurable logs and protocols which may be initialized by the operator, time marks or system and process events

- system and process set-up management: one of the most important tasks during process operation is the handling of system and process set-ups. By using the configured alarm management functions the operator can be guided through process set-ups by appropriate alarming paths, alarm help information and emergency shut-down strategies. Also expert system support can be made available for the set-up management which allows both, the systematic collection of set-up management strategies as applied by experienced operators, and the utilization of these informations for set-up management by the computer. The storage of sequence-of-event informations allows in addition the analysis of causes for system set-ups

- on-line system diagnosis: By using the built-in selftest and on-line diagnostics, failures of system functions can be readily discovered. If redundant devices have been configured, the faulty devices may be replaced automatically with respect to their functions. In any case the system can provide maintenance recommendations to the operator thus allowing easy replacement of the respective system parts.

Furthermore all informations collected during system configuration can be utilized now as system and process reference information related to plant organization (area/unit/group), system organization (local/global stations), I/O point references (assignment lists, functional interconnections) and function configuration (symbolic displays, function lists). Together these reference informations represent an on-line operation manual including all assistance information required during operation.

CONCLUSIONS

Design and operation of a process control system can be considerably simplified by making use of the computation facilities of the systems in the above described ways. The main advantages for design and operation of the system are summarized in the following statements:

- the general handling of system and process is simplified by extensive use of preconfigured and configured help assistance

- the user guidance makes sure that all informations required for the system configuration is collected completely and only once

- in the system the complete information about system devices and functions is stored and can be accessed during all phases of system design and operation

- the collected information can be utilized to provide all kinds of documentation files including part lists, installation instructions, assignment lists, reference lists etc.

- the configuration data base and also the configured process specific help displays can be taken as basis for the process and system operation manual

- the use of CAE methods for process identification, simulation and control system design can shorten the start-up and tuning phase of the process control system and improve the system performance

- the system and process set-up management can be improved and may even be supported by making use of expert system functions

- the on-line selftest and diagnosis features together with specific use of redundant system parts can dramatically increase system availabilty.

Looking at the features of process control systems available now it can be stated that a new phase of computerization of process control is just beginning.

REFERENCES

Eyckhoff, P. (1971). System Identification. John Wiley, Londen.

Isermann, R. (1971). Prozessidentifikation. Springer, Berlin.

Isermann, R. (1981). Digital Control Systems. Springer, Berlin.

Mann, W. (1983). Digital Control of a Rotary Dryer in the Sugar Industry. Automatica, Vol. 19, No. 2, pp. 131-148.

Radke, F. (1984). Ein Mikrorechnersystem zur Erprobung parameteradaptiver Regelverfahren. Dissertation, TH Darmstadt. VDI Fortschrittsberichte, Reihe 8, Nr. 77.

Schumann, R. (1982). Digitale parameteradaptive Mehrgroessenregelung. Dissertation, TH Darmstadt. PDV-Bericht KfK-PDV 217, Kernforschungszentrum Karlsruhe GmbH.

INTERACTIVE TOOL WITH TRANSPARENT REAL TIME SYSTEM SOFTWARE FOR PROCESS CONTROL

E. Bouzou

Servolaboratory, Building 327, Technical University of Denmark, DK-2800 Lyngby, Denmark

Abstract. An interactive graphical tool to define real time applications for process control is presented. Interactivity and graphical representation make the work of the user of the tool easy and the real time system software transparent to him. The task of the programmer is also simplified, roughly reduced to the writing of the controllers, all real time and communication aspects being solved.

Keywords. Computer application, computer control, computer software, industrial control, process control.

INTRODUCTION

The purpose of this paper is to present an Interactive Tool (IT) for process control. The IT provides an easy way to define, modify real time applications, and display data to help the designer of the controller. A real time application is essentially composed of processes (part of programs which must be executed sequentially) and connections between processes to allow those to communicate. Interactivity is helped by a graphical representation that gives an updated image of the real time application. The target computer is a Hewlett Packard 1000 A 600 + (real time software: RTE-A). The IT is written in FORTRAN 77 and is itself composed of two processes: one for the graphic (independent of the computer's software) and one for the interface between the graphic and the real time software which is dependent on this software.

DESCRIPTION OF THE IT(user's point of view)

Description of a process control real time application. A simple process control real time application is represented in figure 0. The process CONTROL in communication with the real process to be controlled provides the control command. If the controller is to be tested, it is required, for safety reasons, to dispose of a backup command provided by a backup controller and used whenever the controller to be tested command is not suitable. The real time application structure is changed like in figure 0.b. There is no reason to suppose that the controller to be tested is only composed of one process, but of several processes communicating between each other (the same remark applies for the backup controller). The communication (exchange of data or information between processes (boxes) is represented by connections between processes. As two commands for the real process are now available, a supervisor process is needed to choose which one has to be used. It is now up to the supervisor to communicate with the real process. In the following, the chain containing the backup controller will be called 'backup chain' and the chain containing the controller to be tested will be called the 'controller chain'. As a lot of elaborated controllers need it, it has been provided the possibility to define a parallel reference model for the real process: REFMO shown in figure 1. At the same time data is sent or got to or from the real process by the supervisor, data is sent or got to or from the reference model. All this data is then available for treatment in the backup and controller chains.

The last element of the structure (fig. 1) SERVICE is a service process to display, in different forms, data sent by other processes.

The complete structure (fig. 1) defines then five kinds of processes:

- backup processes
- controller processes
- supervisor processes
- reference model processes
- service processes

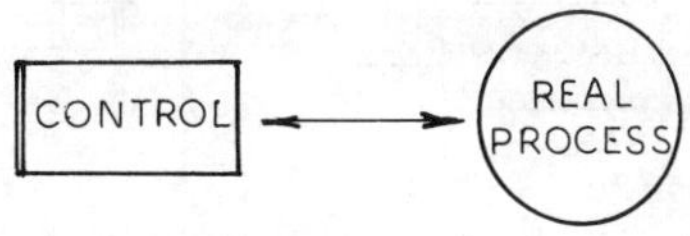

Fig. 0. Simple process control structure.

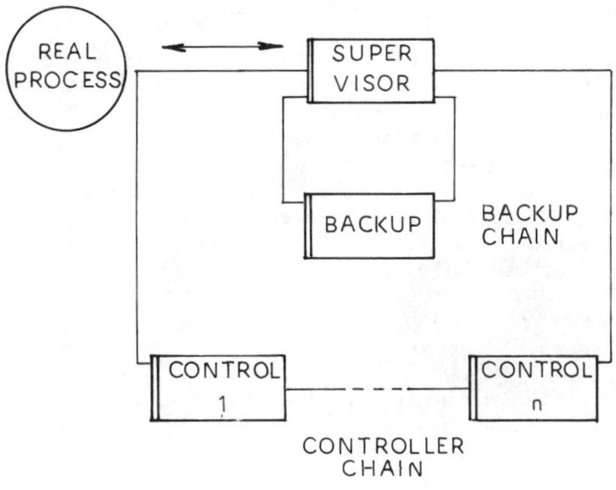

Fig. 0.b. Basic real time application structure.

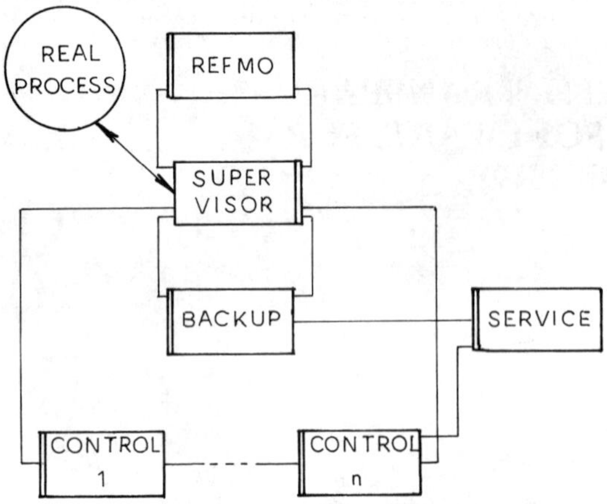

Fig. 1. Complete real time application structure.

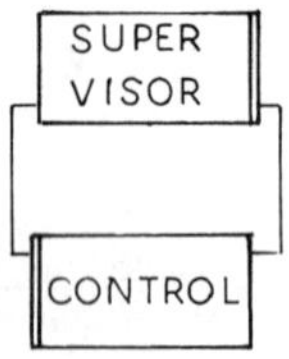

Fig. 1.b.Smallest real time application structure.

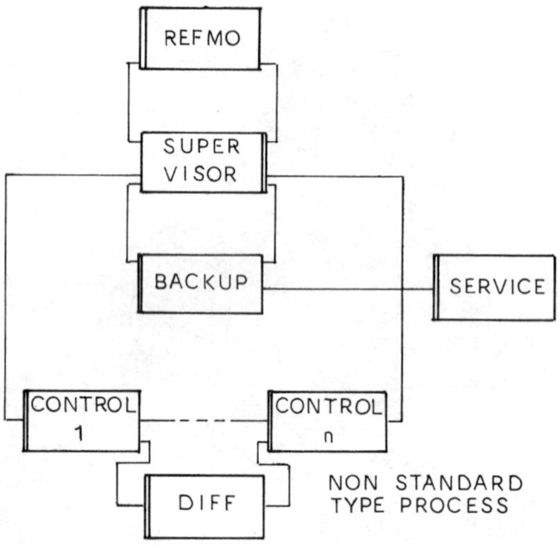

Fig. 2. Extended real time application structure.

The smallest possible structure supported by the IT is represented in figure 1.b. One supervisor and one controller process (or backup). Before running a real time application, the IT checks that the structure agrees with the one represented in figure 1. This structure can be extended, but no supplementary checks of structure are added (cf. fig.2). The structure in figure 1, can be used for simulation purposes. The reference model process can be replaced by a process simulating an existing real process not available.

Graphical Representation. The graphical representation (on graphical screen) of the real time application is like the one of figure 1. A process is represented by a box (cf. fig. 3). The name in the box is the name of the program in memory associated with the process.

The screen is split in two parts: on for the graphic and one for data other than graphic used by the IT to display commands questions and error messages.

Graphical Part. Small dots represent a matrix on the screen (cf. fig. 4). Each dot of the matrix is an available place to draw a box.

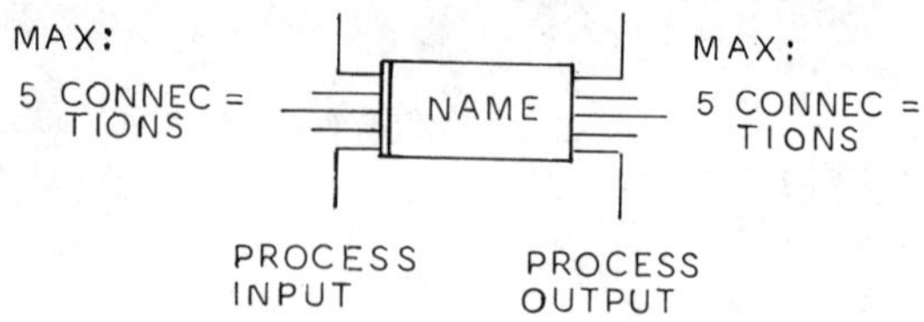

Fig. 3. Graphical process representation.

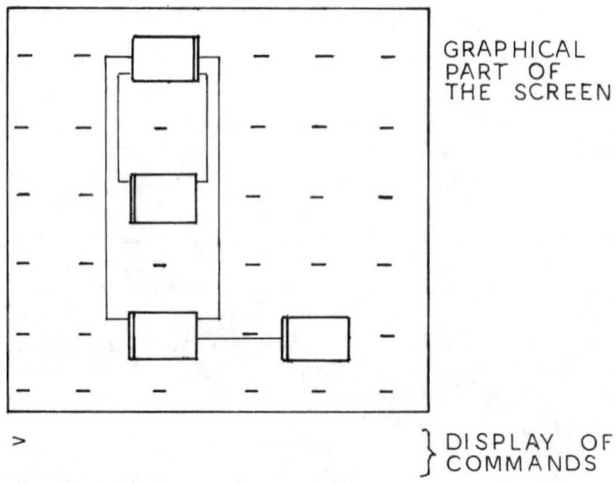

Fig. 4. Graphical screen definition.

Part to display commands, questions, error messages. The graphical commands available to the user are:

- create element:
  To create a box on the screen.
- delete element:
  To delete a box on the screen and all the connections attached to this box.
- connect elements:
  to connect two boxes.
- disconnect elements:
  to disconnect two boxes.
- name element:
  to give or change the name of the box. The name can be given when creating the element.
- displace element:
  to displace a box and relevant connections on the screen, to make the drawing more readable.
- clear all:
  to reinitialize screen and memory.
- use "A" to abort and "?" for help.

A graphical pointeris used for the graphical commands.

Besides the previously defined graphical commands, it is required to add the following commands:

- start chain:
  To start backup or controller chain.
- stop chain:
  To stop backup or controller chain.

- initialization:
  To initialize process parameters, clocks or sharable extended memory area.

Each command makes the interface with the real time software transparent to the user.

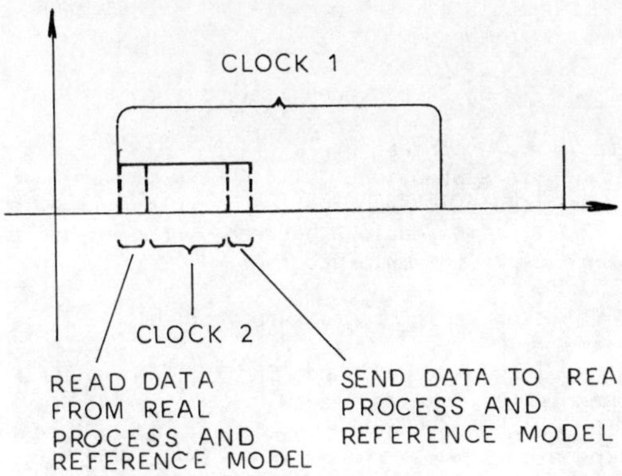

Fig. 5. Clocks description.

Running a real time application. When a suitable structure for the real time application has been set up, the user starts execution of either one or two of the backup and controller chain, using the command "start chain". After checking the structure, the IT asks the user to define the clocks associated with the supervisor of the real time application. There are two clocks to be defined. One precising the sample time period: clock 1; the second being an estimation of the computation time (upper limit) in the backup chain and/or the controller chain: clock 2 (cf. fig. 5). Clock 2 is used by the IT, as a time parameter to wait answers from processes when dialoging with them, and the supervisor: It is a way for the supervisor to know if the controller or backup chain command arrives in due time. When the clocks are defined, the IT sends a message to the supervisor to precise that the backup or controller chain is to be activated. Then the IT schedules, according to clock 1, a clock process, in charge to send a clock buffer to the supervisor every due time.

Flow chart during one sample time period for the real time application. When the supervisor receives a clock process buffer (clock 1), it gets data from the real process (and the reference model process if any), then sends a buffer containing this data to each active chain (backup and/or controller) and puts itself in a suspend time state, according to clock 2. After this waiting time has elapsed, the supervisor is rescheduled and tries to get the buffers at the end of the active chains, then chooses which buffer has to be used and sends relevant data to the real process (and the reference model process if any). It then waits for an other clock process buffer. To stop a chain, use the command "stop chain".

Modifications on line.
Modifications of structure. Any modification (changing names of processes, disconnection of two processes etc..) online of the real time application structure (cf. fig. 1), deactivates automatically the chains involved by the modification, if active. This does not apply for service processes, reference model processes and processes of a different kind than those cited.

Modifications of parameters. Each process has usually some parameters to be intialized. These parameters are initialized when creating the element or on line when chains are active. The initialization parameters buffer is 256 words. When communicating between each other, processes exchanged buffers: size 256 words. The size of these two kinds of buffer (initialization parameters buffer, communication buffer) can be a too big limitation, if processes need to initialize big matrices or pass them to other processes. Therefore a shareable extended memory area (SEMA) has been defined accessible by all processes. As synchronisation problems could arise a resource number (equivalent to a semaphore), the same for all processes, is passed to each created process. This resource number gives exclusive access to SEMA.

Display of data. As said before, it is possible to connect or disconnect processes with a service process without stopping any chain. The period for display of data is a multiple of the sample time period clock 1, display period that can be initialized during process initialization. When connected to a service process, a process sends a buffer containing the data to be displayed to the service process every due time.

DECRIPTION OF THE IT(programmer's point of view).

The IT has to exchange different kinds of information with each process. This is possible only if each process has a predefined structure allowing this dialogue with the IT: father process of all other processes.

A suitable process is then composed of the following elements:

- a prefix file: it is a symbolic file.
- a standard main program
- programmer's subroutines

The prefix file: The first information in the prefix is the type of process. Two characters are used to precise the process type. Standard types are:

IP : for reference model process (or simulation process)
BP : for backup process
CP : for controller process
SV : for supervisor process
SP : for service process

Standard process types are used by the IT to check the structure of the real time application. Other process types can be defined by the programmer.

The second information in the prefix concerns possible connections for the process. A connection between two processes is the possibility for these two processes to exchange a buffer using a mail box. To differentiate the connections (from and to the process), the IT expects to find a type of connection associated to each of them. Two characters are used to precise the connection type. The standard types are (cf. fig. 6):

PC : connection to or from reference model process (or simulation process)
BC : connection to or from backup process
CC : connection to or from controller process
SC : connection to service process

Standard connection types are used by the IT to check the structure of the real time application. Other connection types can be defined by the programmer. Two processes can be connected by the IT only if they have a same connection type in common (and only one).

The third information in the prefix concerns the questions, asked to the user, to initialize process parameters. This part of the prefix is used whenever the user uses the command "initialization" to initialize a process.

The standard main programs. There are two sorts of standard main programs: one for the supervisor process (to be used whenever a supervisor process is built) and one for any other kind of process: ordinary process. The ordinary standard main program supports a process structure like in fig. 7. The main input and output connection types should be precised in the prefix file. The supervisor standard main program supports a process structure like a fig. 8. If other connections than those provided by the standard structures are desired, the IT will establish them, when ordered to, but it is up to the programmer in his own subroutines to use the relevant real time software subroutine calls for the exchanges of buffers using these connections. Part of the data defined in the standard main programs has to be available in the programmer's subroutines. It is therefore stored in named COMMONS available in these subroutines.

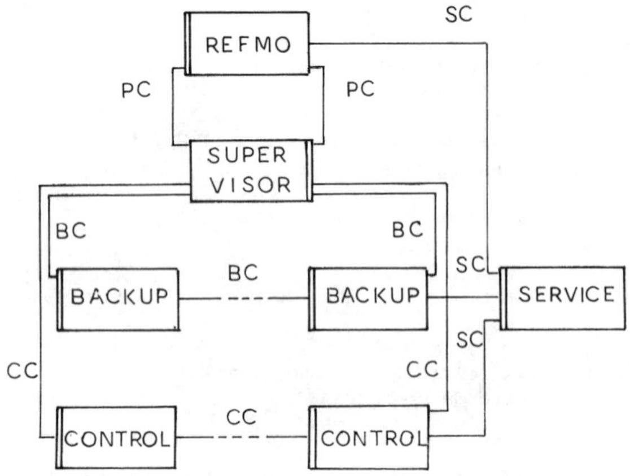

Fig. 6. Standard communication types.

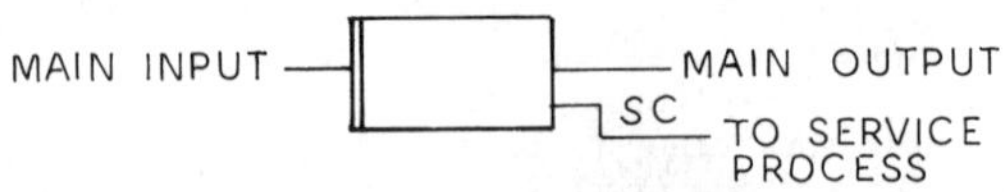

Fig. 7. Ordinary process standard structure.

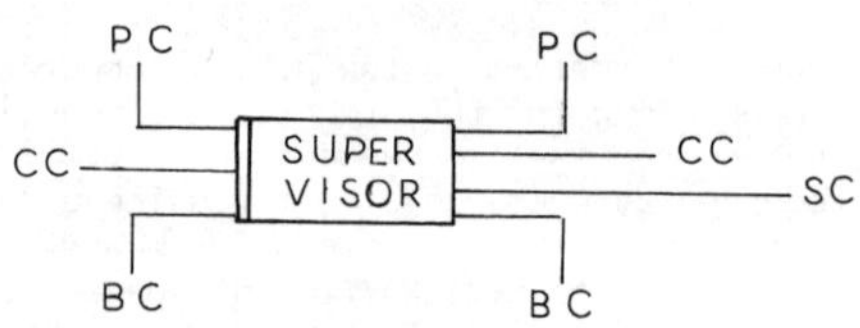

Fig. 8. Supervisor standard structure.

Programmer's subroutines. There are at least two programmer's subroutines: one called INIT and the other one called APPLIC. When updated, the process parameters are stored in a buffer. The programmer should update his own variables using this buffer in INIT. APPLIC contains the intented task of the process. In the supervisor case, this task is to choose between the backup and the controller chain. INIT and APPLIC can access useful data through the COMMONS defined in the standard main programs. For a supervisor process, two extra subroutines have to be defined: one to initialise data communication with the real process INIIO and one to execute the communication with the real process DOIO. In fact two standard subroutines have been written which should cover all applications, so that even for a supervisor process, only INIT and APPLIC have to be written.

The shareable extended memory ares: SEMA (specific RTE-A software feature). As stated a SEMA has been defined to allow large data exchange. This SEMA can be used in each process, provided to load as first module, during linking, a predefined BLOCK DATA containing the array in common, and to specify in due places, the compiler's directives needed when using the SEMA.

Sum up. To built a process, it is required:

- to write a process prefix file
- to write subroutines INIT, APPLIC....
  for a real time application structure like in fig. 1,communications between processes are taken care of in the main programs.

In the executable program are loaded:

- the block data corresponding to SEMA as first module (if SEMA is needed)
- the relevant standard main program
- the programmer's subroutines: INIT, APPLIC, ...

During the linking sequence, each process should be given a priority.

CONCLUSION

The interactive software for process control presented in this report is partly depending on the real time software implemented on the HP 1000 A: RTE-A. Nevertheless, keeping the same structure for the program, it could be adapted to an other real time software using its corresponding features. The safety of the real process being the highest priority, a real time software providing a high degree of security (for example during exchange of information between processes) is needed. From this point of view, the RTE-A software is not ideal, but the interest of the IT remains. Using the IT presupposes that process control programs are available. The construction of these programs has been standardized as much as possible. Nevertheless, the existence of a library of programs compatible with the IT would increase its interest (existing software could be used and modified).

REFERENCES

Blanke, M., Holt, H., Levin, M.L., Spiegelhauer, A.V.: Multiprocessor System for Automatic Control Purposes, System-description. Servolaboratory, Technical University of Denmark. 1979.

Spiegelhauer, Arne, V.: System Software for Process Computers. Servolaboratory, Technical University of Denmark. 1982.

Farsøe, Henrik, F., Andersen, Torben, R.: A regulator Management System at the Mulreg System. Servolaboratory, Technical University of Denmark. 1982.

Hewlett Packard Manuals: RTE-A User's Manual, RTE-A Programmer's Reference Manual, RTE-A System Design Manual etc..

APPLICATION OF A MULTIVARIABLE ROBUST CONTROLLER DESIGN METHOD TO HARD-COAL PREPARATION

B. Cuno* and W. Neddermeyer**

**AEG, Frankfurt/M., FRG*
***Technical University of Darmstadt, Institute of Control Engineering, Darmstadt, FRG*

Abstract. The paper describes an application of a robust controller design method for calculating the parameters of a controller for a multivariable hard-coal preparation process. The parameters are optimized via a sequential design method in the frequency domain, being based on a combination of the Horowitz/Sidi design (1972), Mayne's (1973) sequential design and Steinhauser/Kreisselmeier's (1979) vector performance criterion method. The design was carried out using a relatively sophisticated CAD-program. The paper shows how the parameters are calculated with this CAD-method and discusses the simulated control results. The simulation is compared with the measured results at the plant after implementation of the designed controller on a process computer AEG 80/30 with a real time software control package ARSI*.

Keywords. Multivariable robust controller design, parameter uncertainty, vector performance criterion, CAD.

I. INTRODUCTION

The plant at hand is given by two coal bunkers in a coal preparation process, which follows the underground operation of a coal mine. In order to reduce the production changes in crude coal the bunkers are used to keep an uniform coal flow. The investigated part of the whole plant is sketched in fig. 1.

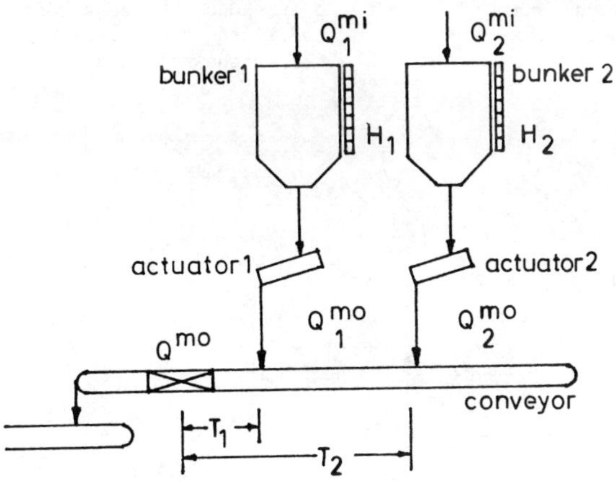

Fig. 1. Sketch of the process

The control task considered here is complicated by some special effects of the dynamic plant behaviour; such as

i. The bunkers input mass flows are not measured.

ii. The bunker volume is small compared with the mass flow.

iii. The measurement of the coal level in the bunkers is strongly disturbed (up to 50%) by several effects.

iv. The actuators behaviour depends on the grain size, the coal surface moisture and the raw coal level in the bunkers.

v. The output mass flow measurement has a considerable delay time compared to the remaining dynamic parameters.

The system can be denoted as multivariable, nonlinear and parameter uncertain. Potential solutions of such a control problem may be given by the use of adaptive algorithm and the robust controller design. As the bunker behaviour sometimes results in abruptly changing parameters adaptive control must be excluded as a control alternative:

- self adaptive control cannot provide an acceptable system performance.

- gain scheduling needs switching conditions which cannot be derived from available measurements.

That is why a robust control structure based on a systematic robust multivariable design method has been chosen.

II. THE MATHEMATICAL DESCRIPTION OF THE PROCESS AND BASIC CONTROL REQUIREMENTS

The mathematical description of the process behaviour in general can be given by a non-linear differential equation

$$\dot{\underline{X}} = \underline{f}(\underline{X},\underline{Y},\underline{U},\underline{P}) \tag{1}$$

with
$\underline{U} \in R^{n_U}$ input vector
$\underline{Y} \in R^{n_Y}$ output vector
$\underline{X} \in R^{n_X}$ state space vector
$\underline{P} \in R^{n_P}$ parameter vector

where $\underline{P}$ represents possible changes in system parameters e.g. ρ the specific gravity of the raw coal.

Defining the workspace of the system by

$$W: = \{\underline{X}/\underline{X}_{lb} \le \underline{X} \le \underline{X}_{ub}\} \qquad \text{lb: lower bound, ub: upper bound} \tag{2}$$

* ARSI is a trademark of AEG

one can assume that the system can be represented in the workspace W by a set of linearized system equations. They can also be noted as transfer functions with varying parameters $\underline{a}$

$$P(s,\underline{a}):=\{P(s,\underline{a})/P(s,\underline{a}_{ij})=L\{S_{ij}\}\ \underline{X}^j \epsilon W\ \underline{p}^i \epsilon R^n P\ ; \quad j=1...k\ ;\ i=1...l\} \tag{3}$$

with
- $L\{\ \}$ the Laplace operator
- S_{ij} the linearized system equations in state space representation at the operating point $\underline{X}^j$ and the parameter set $\underline{p}^i$
- $P(s,\underline{a})$ the transfer function or transfer function matrix of the linearized system in the workspace W
- $\underline{a}_{ij}$ vector of parameters of the transfer function

The scalar k is the number of different fixed parameter set points $\underline{X}^j$, and its choice depends on the admissible error ϵ between the real non-linear plant and the linearized one.

The plant as shown in fig. 1 has an input vector $\underline{U}$ containing the actuator set points U_1,U_2 and the unmeasureable input mass flows Q_1^{mi} , Q_2^{mi}. The output vector $\underline{Y}$ contains the elements Q^{mo} the measured output mass flow and the coal level of each bunker H^1 and H^2. The differential equations describing the process behaviour are

$$\frac{dM_j^s}{dt} = Q_j^{mi} - Q_j^{mo}$$

$$\frac{dM_j^s}{dt} = \rho_j \cdot A(H_j) \cdot \frac{dH_j}{dt} \tag{4}$$

$$\frac{dQ_j^{mo}}{dt} = \psi_j(H_j,\alpha_j,\gamma_j)Q_j^{mo} + \phi_j(H_j,\alpha_j,\gamma_j)U_j \quad j=1,2$$

with ψ_j and ϕ_j as non-linear functions of H_j the coal level, α_j the grain size and γ_j the surface moisture of the raw coal.

Fig. 2 shows the bar chart of the linearized process model with the above mentioned input/output parameters.

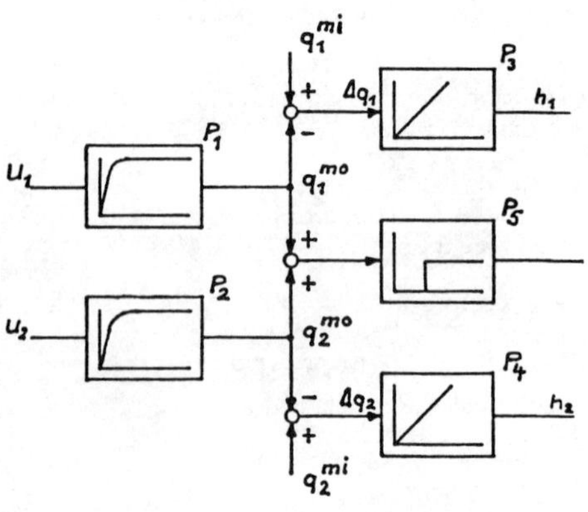

Fig. 2. Linearized process model

The transfer functions P_i are given by

$$P_i(s,\underline{a}) = \frac{K_i^A}{1+T_i s} = \frac{q_i^{mo}}{u_i} \quad ; \quad i=1,2$$

$$P_i(s,\underline{a}) = \frac{K_i^A}{s} = \frac{h_{i-2}}{\Delta q_{i-2}} \quad ; \quad i=3,4$$

$$P_5(s) = e^{-T_T s} \quad ; \quad T_1 \approx T_2 = T_T \tag{5}$$

The varying parameters $\underline{a}$ of the model are summarized as:

$$K_i = K_i^A/_j$$

$$T_i = T_i^A/_j \quad ; \quad i=1,2$$

$$K_i = 1/(\rho_i \cdot A_i/_j) \quad ; \ i=3,4 \tag{6}$$

with $/_j$: parameter depending on different linearization points

In particular the parameter A_i represents the changing of the bunkers cross sections depending on the coal level in the bunkers. The parameter ρ_i describes the varying specific gravity of the bunkers contents. Due to coal level, grain size and coal surface moisture the actuator parameters K_i^A and T_i^A change. The variation limits of the above mentioned parameters were measured (K_i^A, A^i) or estimated (T_i^A, ρ_i). They are listed in the following table:

$20 \le K_1^A \le 33$ [t/hV]

$23 \le K_2^A \le 36$ [t/hV]

$5.36 \le A_i \le 24.97$ [m^2] with $1.5 \le h_i \le 4.8$[m]

$1.2 \cdot 10^3 \le \rho_i \le 1.4 \cdot 10^3$ [kg/m^3]

The model in fig. 2 represents the actuator and bunker behaviour. In our practical application we have two sensors for measuring the coal level in the bunkers. Due to a lot of effects we get uncertainties in the sensor signals. One can observe errors up to 50%. A controller design has to take these effects into consideration. Therefore the model in fig. 2 has to be extended by a sensoring part. The complete model including the disturbance inputs d_i and the filters F is shown in fig. 3.

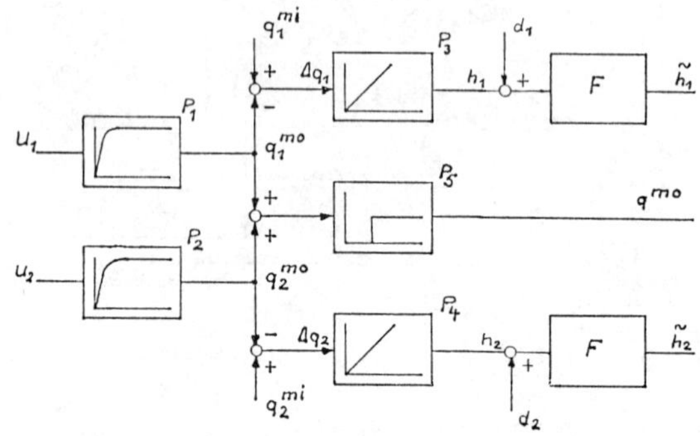

Fig. 3. Linearized process model including measurement units

The basic requirements for the design are:

i. A nearly constant mass flow q^{mo} should be reached, because the process units following the bunkers work in a much better way when the mass flow is constant over a time intervall which is as long as possible.

ii. The control loop should work in such a way that the measurement disturbancies of all the h_i-values do not affect the mass flow q^{mo}.

iii. The control loop should prevent an overflow of the bunkers.

iv. The bunkers should not become empty.

III. STRUCTURE OF THE MULTIVARIABLE CONTROL LOOP

The variables which could be used for feedback are the coal level $\tilde{h}_1$, $\tilde{h}_2$ and the output mass flow. The actuator set points are u_1 and u_2. In a preanalysis some controller structures using the above mentioned input/output variables had been tested and compared on the real process. As a result of the preanalysis we chose the structure which is depicted in fig. 4.

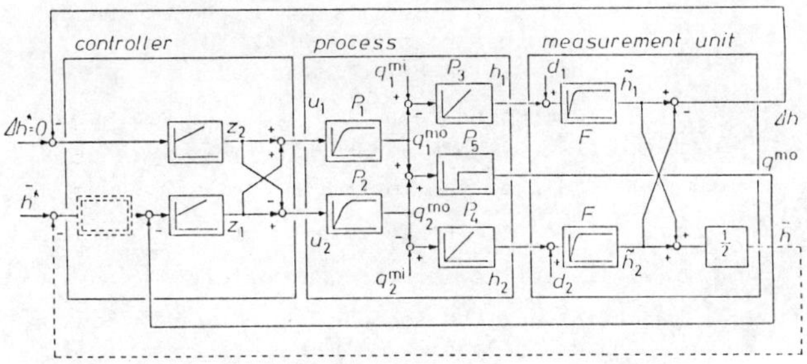

Fig. 4. Control structure

The control structure contains PI-algorithms as sufficient in most applications in process industries. The measured difference of $\tilde{h}_1$ and $\tilde{h}_2$ is one of the controlled variables. The other two are given by q^{mo} and the average coal level $\overline{h}$ which is controlled via a so called master loop. The $\Delta\tilde{h}$-loop comes into consideration because both bunker volumes should be used symmetrically. The q^{mo}-loop controlls the main variable q^{mo} in order to produce a nearly constant mass flow. However, the output mass flow must be connected with the input mass flows otherwise the bunkers might become empty or full. Therefore a third control loop has to be designed. This master regulator which is not considered in this paper sets the commanded output mass flow. However, to fulfill the controll demands one has to design the multivariable inner loop in order to control $\Delta\tilde{h}$ and q^{mo} in a robust way at first. The required multivariable design was carried out by a CAD method described in section IV.

IV. ROBUST MULTIVARIABLE DESIGN METHOD

The plant description indicates that the applied design method must allow to deal with the following restrictions:

i. The plant is multivariable and parameter uncertain.

ii. One loop contains a considerable delay time compared to the remaining dynamic parameters.

iii. The process has very strong limitations for the actuator signal range.

In order to handle the multivariable aspects we used the sequential design method proposed by Mayne (1973). As in most multivariable design methods using the frequency domain the m-input, m-output multivariable system is split down by this method into m-single input/single output systems for design. Frequently used and well-known design procedures can now be applied for the SISO design. To handle the parameter variations and especially to guarantee stability in spite of them the quantitative design method proposed by Horowitz/Sidi (1972) for SISO systems is combined with Mayne's method, as suggested by Shaked/MacFarlane (1977). For the iii-th design restriction and a comfortable way of work with the Horowitz/Sidi method (Gräser, Neddermeyer, Tolle 1982) a vector performace criterion in analogy to the Kreisselmeier/Steinhauser's (1979) design method was defined and optimized.

The whole procedure which makes use of sequentiell design, robust design and a vector performance criterion can only work in a computer aided way.

The following sections introduce to Mayne's method (IV.1) and the robust design very briefly. The use of the vector performance criterion and the CAD package which were discussed already by Gräser, Neddermeyer, Tolle (1982) will be explained in section IV.2 and IV.3.

IV.1 MULTIVARIABLE DESIGN PROCESS

The structure of the considered control loop with two degrees of freedom is shown in fig. 5.

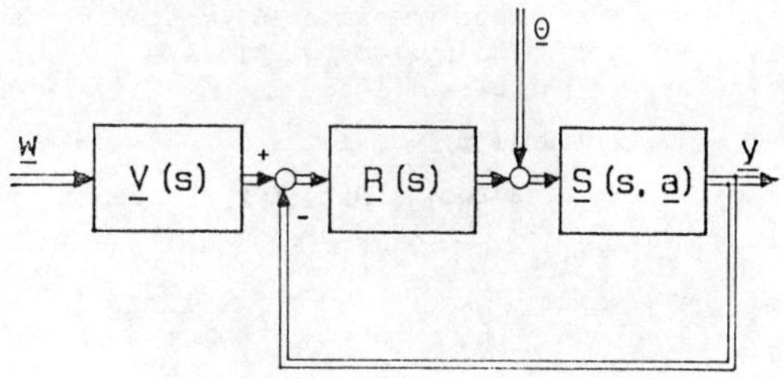

Fig. 5. Multivariable control loop

In our special application a one degree of freedom structure is sufficient because a prescribed disturbance behaviour is required only. According to Mayne an additional input vector $\underline{\theta}$ is used for the derivation of the equations (see fig. 5). The dynamic behaviour of the closed control loop depending on the two inputs is given by

$$\underline{y} = \underline{\tilde{T}}\,\underline{V}\,\underline{w} \quad \text{with} \qquad (7)$$

$$\underline{\tilde{T}} = (\underline{I} + \underline{S}\,\underline{R})^{-1}\,\underline{S}\,\underline{R}$$

$$\underline{y} = \underline{T}\,\underline{\theta} \quad \text{with} \qquad (8)$$

$$\underline{T} = (\underline{I} + \underline{S}\,\underline{R})^{-1}\,\underline{S}.$$

For the sequential design in connection with a non-diagonal dominant MIMO system, the i-th design step depends on all the controllers determined previously and it is necessary to mark the different transfer function matrices with a superscript. According to Mayne, the following notation is used

$$\underline{S}^{o}(s) = \underline{S}(s), \quad \underline{R}^{o}(s) = \underline{O}, \quad \underline{R}^{m}(s) = \underline{R}(s)$$

$$\underline{S}^{q}(s) = (\underline{\Delta}^{q})^{-1}\cdot\underline{S} = (\underline{I} + \underline{S}\,\underline{R}^{q})^{-1}\underline{S}$$

$$\underline{\Delta}^{q}(s) = (\underline{I} + \underline{S}(s)\,\underline{R}^{q}(s))$$

$$\underline{R}^{q}(s) = \mathrm{diag}(r_{11},\ldots,r_{qq},0\ldots0) \qquad (9)$$

In general, after closing the q-th control loop we find for the elements of the system transfer matrix

$$S^{q}_{ab} = \frac{S^{q-1}_{ab} + R_{qq}\,(S^{q-1}_{qq}S^{q-1}_{ab} - S^{q-1}_{aq}S^{q-1}_{qb})}{\delta_q} \qquad (10)$$

δ_q denotes the characteristic equation of the q-th-SISO-control loop.

$$\delta_q = 1 + S^{q-1}_{qq}\,R_{qq} \qquad (11)$$

For the sequential method, there are only m stable controller designs necessary for S^{q-1}_{qq}; q = 1,2...m and all methods used for SISO design can be taken for that.

IV.1.1 ROBUST CONTROLLER DESIGN

The Horowitz/Sidi design method is based upon the plant transfer function of the process (here $S^{q-1}_{qq}(s,\underline{a})$) and the Nichols chart for controller design.

For SISO systems the Horowitz/Sidi-design is normally divided into three design areas according to growing frequencies.

In area one, high accuracy and small effects of the parameter variations are demanded. The relative sensitivity of the closed loop compared with the one of the open loop has to be less than one. For control systems with an excess $e \geq 2$ of poles over zeros the Bodeintegral

$$\int_0^\infty \log |E(j\omega)|\, d\omega = 0 \tag{12}$$

is valid. As a consequence $|E|$ can not be less than one over the whole range of frequencies. So in the second design area $|E|$ can only be limited to an magnitude $1 \leq |E| \leq \gamma$. In area three, the controller gain is very small and the sensitivity of open and closed loop are almost the same $|E| \approx 1$. For quantitative design of SISO-systems within area one, the sensitivity of the closed loop

$$|\Delta T_{qq}(j\omega, \underline{a})| \leq \alpha(j\omega) \qquad 0 \leq \omega \leq \omega_w \tag{13}$$

is limited. In area two, limits for the sensitivity

$$|E_{qq}(j\omega)| \leq \gamma_1 \tag{14}$$

or of the maximum peak of

$$|T|, \quad |T_{qq}(j\omega)| \leq \gamma_2 \tag{15}$$

with $\gamma_1, \gamma_2 < 1$ are given, respectively, guaranteeing in this way also stability in spite of parameter variations. With the Nichols chart and parameter areas for different values $s = j\omega_i$; $i = 1...n$ of the transfer function $S_{qq}^{q-1}(j\omega,\underline{a})$, Horowitz/Sidi transforms the requirements as formulated in Eq. 13, 14 or Eq. 13, 15 into boundaries for the nominal open loop

$$L_{qq}(j\omega, \underline{a}_o) = R_{qq}(j\omega)\, S_{qq}^{q-1}(j\omega, \underline{a}_o)$$

frequency response.

For the design of a multivariable control system two problems have to be distinguished:

- robustness of the stability under the presence of parameter variations,
- quantitative design.

The first case demands stability of the characteristic equations of the m-SISO-systems sequentially designed. A SISO-system is stable for all parameter sets $\underline{a}$ when the open loop nominal frequency response plot does not penetrate into the closed curves (boundaries in the Nichols chart) which are derived from Eq. 14 or Eq. 15. Additionally the Nyquist criterion has to be fulfilled.

For quantitative design of the multivariable system the sensitivity $|\Delta T_{qq}(j\omega, \underline{a})|$ should be limited within the frequency range $0 \leq \omega \leq \omega_w$. In the SISO-case one holds the design requirements Eg. 13 exactly if the controller brings the magnitude of the nominal open loop on the boundaries (Horowitz 1973) in the Nichols chart. For the multivariable case these relations are not easy to describe. In the sequential design method the q-th design step may also influence all transfer functions designed previously

$$S_{aa}^{q} \neq S_{aa}^{q-1}; \; a < q.$$

Up to now, it is not possible to formulate an exact interrelation between the design requirements Eq. 13 and the sequential design steps. Only for the last row of the transfer function matrix T(s) a direct correlation can be derived (Gräser, Tolle 1982). But due to the relatively high gains normally used in region 1, the already designed closed loops are very often changed only marginally by closing additional loops so that one can live with this design method in general also in the multivariable case.

IV.2 OPTIMIZATION OF A VECTOR PERFORMANCE CRITERION

Section IV.1 reduces the multivariable robust design problem to the following task: Find a regulator which holds the boundaries for the nominal open loop frequency responses.

A most convenient way to find the denominator and nominater parameters $\underline{r}$ of $R_{qq}(s, \underline{r})$ is to make use of a vector performance criterion $\underline{G}(\underline{r})$ which was very succesfully applied bei G. Kreisselmeier and R. Steinhauser (1979, 1984) for some other problems. Therefore the constraints for $L_{qq}(j\omega, \underline{a}_o, \underline{r})$ and/or the controller have been formulated as such a vector performance criterion $\underline{G}(\underline{r})$. The robust design requirements derived from the boundaries in the Nichols chart are normally representable by the following six elements of the vector criterion:

- $g_1(\underline{r})$: Sum of the deviations of the frequency response points $L_{qq}(j\omega_i, \underline{a}_o)$ from the boundaries B_i to higher gains;
- $g_2(\underline{r})$: Sum of the deviations of the frequency response points $L(j\omega_i, \underline{a}_o)$ from the boundaries B_i to lower gains;
- $g_3(\underline{r})$: Component resulting from the violation of the high frequency boundaries;
- $g_4(\underline{r})$: Gain of the controller $(j\omega \to 0)$;
- $g_5(\underline{r})$: High frequency gain K^∞ of the controller;
- $g_6(\underline{r})$: Stability of the controller.

In a practical application one has not only requirements which can be easily represented by the frequency domain methods. Such demands are time domain spezifications as noted under section IV (strong limits for the actuator signal range).

With the reduction of the design problem to an optimization of a vector performance criterion such constraints can easily be considered by defining additional vector elements. In our practical application such elements can be given by

$$g_{6+i} = \int_0^{T_E} (y(t,\underline{a}_i) - y_{max})^2\, dt$$

$$\forall\, t \text{ with } y(t,\underline{a}_i) \geq y_{max} \qquad i = 1,2...n. \tag{16}$$

IV.2.1 OPTIMIZATION TECHNIQUE

As the above performance vector elements indicates each design aspect is to be rated quantitatively by means of a suitable positive design criterion in such a way, that reducing the value of a criterion always means improvement of a corresponding design aspect. In a second step the design is achieved iteratively as follows. In the γ-th design step a target vector $\underline{c}^\gamma$ is chosen within the range given by

$$\underline{G}(\underline{r}^{\gamma-1}) < \underline{c}^\gamma < \underline{c}^{\gamma-1} \tag{17}$$

Here $\underline{r}^{\gamma-1}$ denotes the controller poles and zeros which resulted from the previous design step, and $\underline{c}^{\gamma-1}$ denotes the associated target vector. In the above choice of $\underline{c}^\gamma$, it is advisable to take $c_j^\gamma = c_j^{\gamma-1}$ for those design criteria, which have been made sufficiently small in earlier design steps, whereas $c_i^\gamma = g_i(\underline{r}^{\gamma-1})$ may be chosen for design criteria, which are to be reduced in value further. This determines the direction in which the design proceeds. Then $\underline{r}^\gamma$ is defined as the solution of a scalar optimization problem

$$\min_{\underline{r}} \{-\frac{1}{\beta} \ln \sum_{i=1}^{L} \exp(\beta g_i(\underline{r})/c_i^\gamma)\}. \tag{18}$$

Using the definition

$$\alpha(\underline{r}) := \max_{1 \leq i \leq L} (g_i(\underline{r})/c_i^\gamma) \tag{19}$$

the optimization problem can be rewritten in the form

$$\min_{\underline{r}} \{\alpha(\underline{r}) + \frac{1}{\beta} \ln \sum_{i=1}^{L} \exp[\beta(\frac{g_i(\underline{r})}{c_i} - \alpha(\underline{r}))]\} \tag{20}$$

and the minimization is essentially reduced to the minimization of the value of $\alpha(\underline{r})$. With Eq. 19 we have

$$g_i(\underline{r}) \leq \alpha(\underline{r}) \cdot c_i^{\gamma} \qquad i = 1,2...L \tag{21}$$

that means minimizing $\alpha(\underline{r})$ is the same task as reducing the value of all design criteria simultaneously. In the program package described below the optimization task is realized with the Hook-Jeeves method.

IV.3 CAD PACKAGE

The program package used here is written in FORTRAN 77 as a multitasking dialog system which runs on a VAX 11/780. Some detailed program descriptions are given by Ersü, Neddermeyer, Tews (1983). However, the global structure is shown in fig. 6.

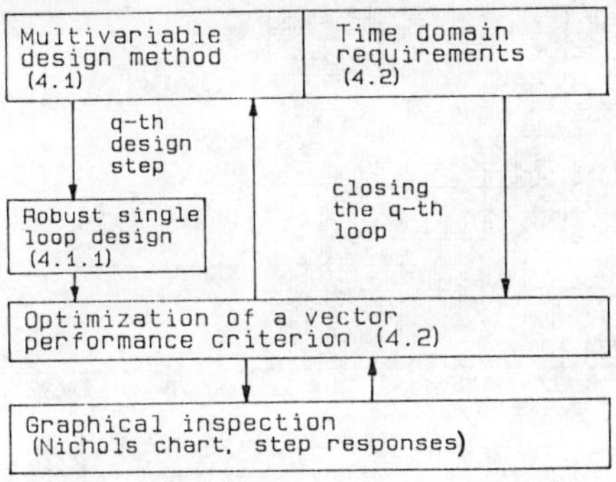

Fig. 6. Global software structure

On the higest level we have the multivariable frequency domain design and time domain requirements for the multivariable system. As shown in IV.1 and IV.2 the multivariable design problem reduces to m-single loop designs which are carried out by an optimization of a vector performance criterion. During this process a grafical output supports the user choosing the target vectors for the optimization and for inspection of the design results.

V. APPLICATION OF THE INTERACTIVE COMPUTER AIDED CONTROLLER DESIGN

At first the plant is rewritten in a P-kanonical system representation. Fig. 7 shows the input/output variables and the location of the transfer functions $S_{ij}(s, \underline{a})$.

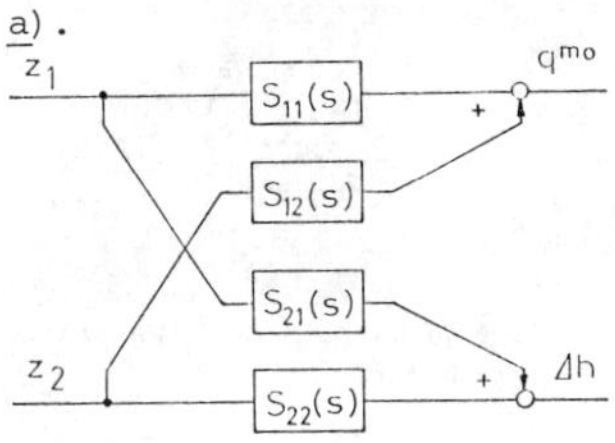

Fig. 7. P-kanonical plant model

They are given by

$$\begin{aligned} S_{11}(s,\underline{a}) &= P_5(s) \cdot (P_1(s,\underline{a}) + P_2(s,\underline{a})) \\ S_{12}(s,\underline{a}) &= P_5(s) \cdot (P_1(s,\underline{a}) - P_2(s,\underline{a})) \\ S_{21}(s,\underline{a}) &= F(s) \cdot (P_3(s,\underline{a})P_1(s,\underline{a}) - P_4(s,\underline{a})P_2(s,\underline{a})) \\ S_{22}(s,\underline{a}) &= F(s) \cdot (P_3(s,\underline{a})P_1(s,\underline{a}) + P_4(s,\underline{a})P_2(s,\underline{a})) \end{aligned} \tag{22}$$

Fig. 8 contains the corresponding Bode plots of the parameter uncertain transfer functions S_{ij}.

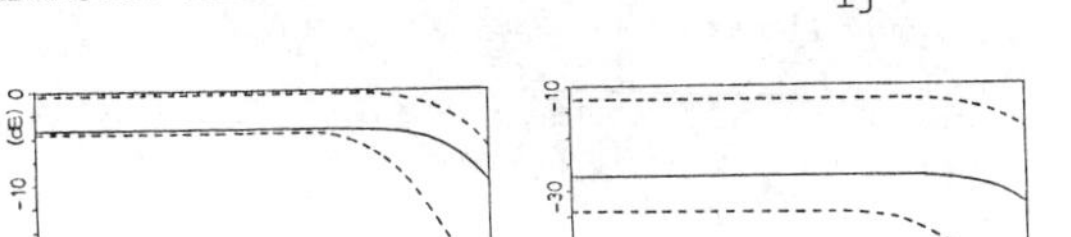

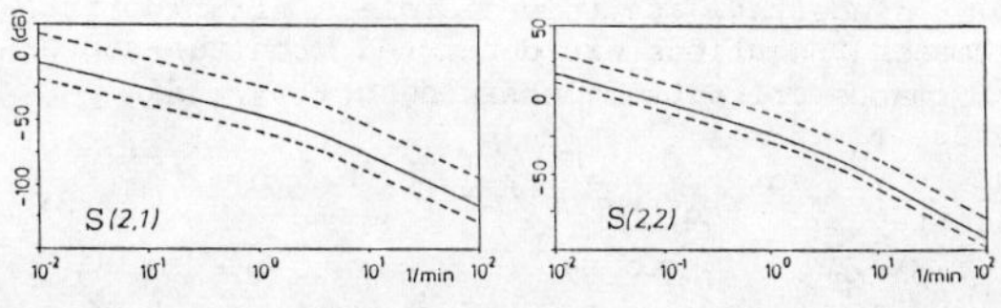

Fig. 8. Bode plots of the uncontrolled plant

The design requirements for the MIMO problem can be summarized as:

i. $|\Delta T_{qq}(j\omega,\underline{a})|$ as small as possible for $0 \leq \omega \leq \omega_W$

ii. $|T_{qq}(j\omega,\underline{a})| \leq 2\,dB \quad \forall \underline{a},\omega$

iii. The overshooting of the step response of the q^{mo}-loop should be limited to 10%.

iv. An abrupt measurement disturbance of the coal level (0.5 m is a typical value) should not disturbe the bunkers output mass flow.

For the first item i. it makes no sense to require absolute bounds on $|\Delta T|$. Due to the non-minimum phase behaviour of the plant the bandwidth and as a consequence the realizeable robustness are limited. However, the Horowitz/Sidi-boundaries for the first open loop $R_{11}(S,\underline{r})\, S_{11}^{o}(s,\underline{a}_o)$ are noted in fig. 9.

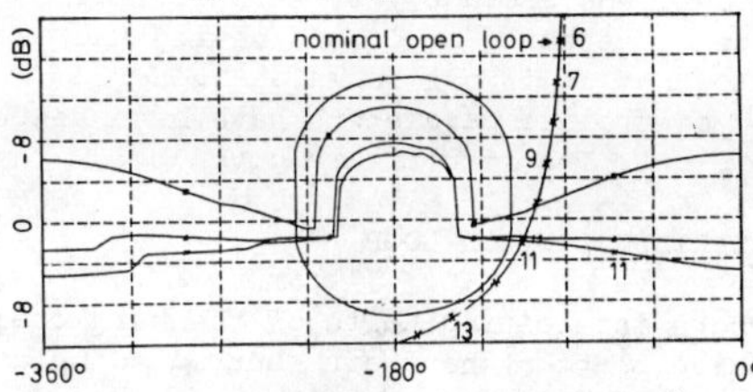

Fig. 9. Nichols chart of the nominal open loop frequency response

The chosen vector performance index contains the terms $g_1 ... g_6$ mentioned under IV.2 and the time domain requirements

$$g_{6+i} = \int_0^{T_E} (q^{mo}(t,\underline{a}_i) - w^{q^{mo}})^2 dt$$

$$\forall\ t \text{ with } q^{mo}(t,\underline{a}_i) \geq w^{q^{mo}} \qquad i=1,2...n \tag{23}$$

The optimized nominal open loop (see fig. 9) does not penetrate the closed boundaries and frequency point 11 (ω=0.6 [1/min]) is on a line of constant $|\Delta T|$ with $|\Delta T|$=2.7dB. Additionally in fig. 12d the response of a q^{mo} set point step for the linearization point L_1 is shown. The q^{mo}-trajectory fulfills the overshooting restriction.

The controller parameters are computed as

$$R_{11}(s) = 0.228\left|1 + \frac{1}{11.4s}\right| \tag{24}$$

With this regulator we obtain for the frequency response of the first closed loop $\tilde{T}_{11}^{1} = S_{11}^{1} R_{11}$ the Bode plot noted in fig. 10.

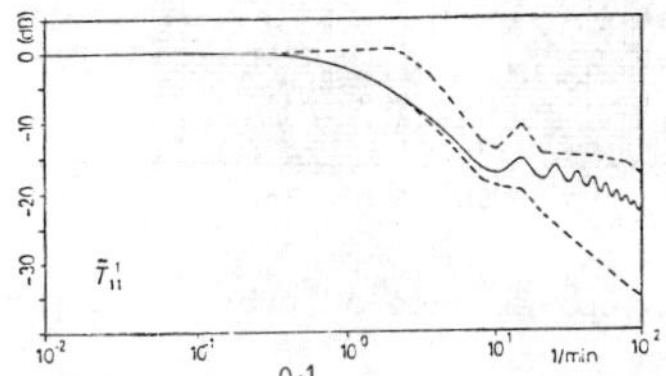

Fig. 10. Bode plot of $\tilde{T}_{11}^{1}$

The frequency plot of S_{22}^{1} with the boundaries according to item (ii.) and the frequency plot of the nominal open loop $R_{22}(s,\underline{r})S_{22}^{1}(s,\underline{a}_o)$ are not shown, since they are in principle similar to fig.9. The second regulator was optimzied with the vector performance criterion containing $g_1 \ldots g_6$ and

$$g_{6+i} = \int_{o}^{T_E} (q^{mo}(t,\underline{a}_i) - w^{q^{mo}})^2 dt$$

$$\forall \; t \text{ with } w^{q^{mo}} = \text{const.} \qquad i=1,2\ldots n$$

The resulting regulator parameters are

$$R_{22}(s,\underline{r}) = 1.46(1 + \frac{1}{460s})$$

Fig. 13d shows a simulated disturbance of 0.5 m at the linearized plant (L_2). One can see that the output mass flow is not disturbed.

In fig. 11 the frequency response of the closed loop $\tilde{T}^2_{22} = S^2_{22} R_{22}$ is given.

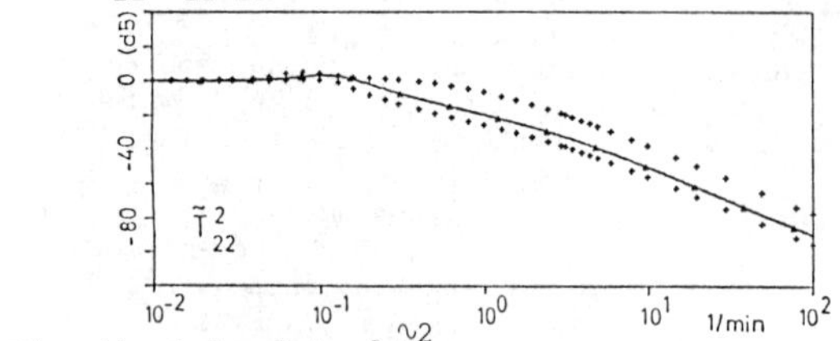

Fig. 11. Bode plot of $\tilde{T}^2_{22}$

The couplings are not very strong. That is why the frequency response of $\tilde{T}^1_{11}$ is nearly the same as $\tilde{T}^2_{11}$. Due to that it is not shown again here. The optimized controller parameters have been tested at the real plant. The results are given below.

VI. REALIZED CONTROL LOOP

The here interesting units of the hard coal preparation plant at hand are controlled by a AEG 80-30 process computer system.

The application software used for the realization of the feedback control is ARSI80 and AS80 for the logic/sequence control. Both systems provide operating procedures to run the plant automatically from start up to normal operation and to manage emergency and shutdown operations. Custom tailoring of the modular program packages has been made mainly off-line while necessary modifications of control structures could be done during on-line operation.

The control structure and the control algorithm has been realized in a quasi continuous way.

Fig. 12,13,14 show the simulation results together with the measured traces.

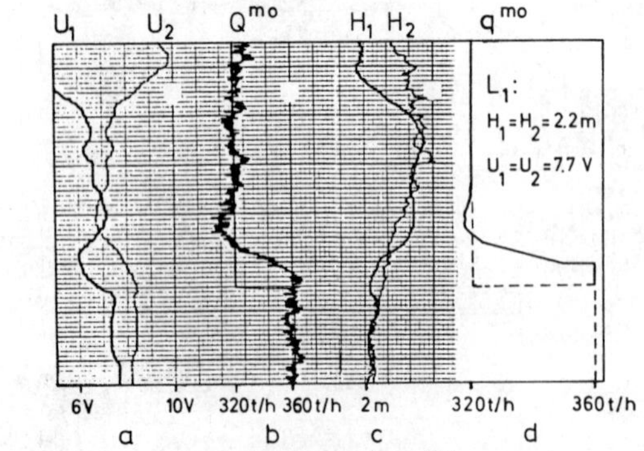

Fig. 12. Step response of the q^{mo}-loop

Fig. 12 compares the simulated to the real step responses of the q^{mo} loop.

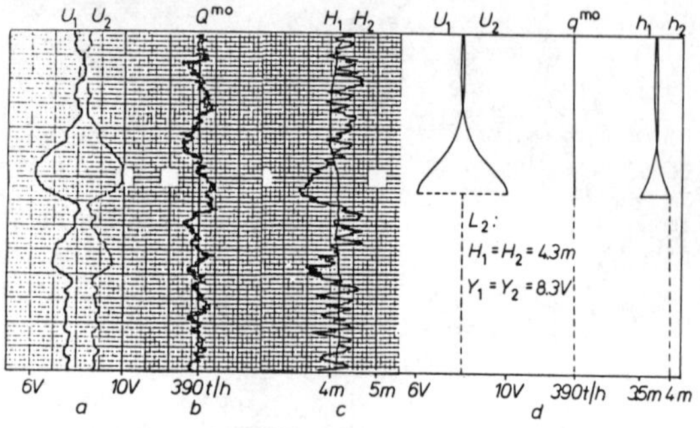

Fig. 13. Disturbance response (0.5 m)

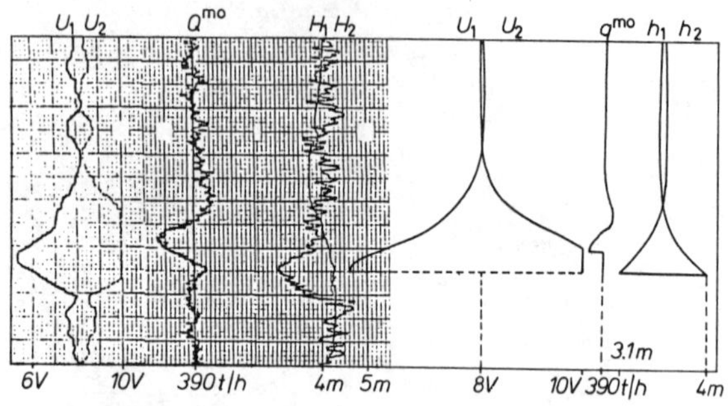

Fig. 14. Disturbance response (0.9 m)

Fig. 13 and 14 show the influence of the disturbed coal level measurement. Looking at the measured and simulated trajectories one finds that they are very similar.

In a more general way one can say that the controller fulfills the design requirements in the simulation as well as on the real plant.

VII. CONCLUSIONS

The paper describes a multivariable, robust design process. The design was carried out with an interactive CAD-system. The optimized controller parameters are applied to the real process. The process behaviour fulfills the required system performance in an impressive way.

VIII. REFERENCES

Ersü, E./Neddermeyer, W./Tews, V. (1982). REDECK - A CAD-program package for the control engineer, using only mini-computers. IASTED, Applied modelling and simulation - AMS'82, Paris, France

Gräser, A./Neddermeyer, W./Tolle, H. (1982). CAD of the Horowitz/Sidi-design for feedback systems with large parameter uncertainty. IFAC, CAD in Control and Engineering Systems, Purdue, USA

Horowitz, I.M. (1973). Optimum loop transfer function in single-loop minimum-phase feedback systems. Int. J. Control, Vol. 18, pp. 97-113

Horowitz, I.M./Sidi, M. (1972). Synthesis of feedback systems with large plant ignorance for prescribed time domain tolerances. Int. J. Control

Kreisselmeier, G./Steinhauser, R. (1979). Systematic control design by optimizing a vector performance index. IFAC, CAD of Contr. Systems, Zürich, Switzerland

Kreisselmeier, G./Steinhauser, R. (1984). Application of vector performance optimization to a robust control loop design for a Fighter Aircraft. Int. J. Control, Vol. 37

Mayne, D.Q. (1973). The design of linear multivariable systems. Automatica

Shaked, U./MacFarlane, A.G.J. (1977). Design of linear multivariable systems for stability under large parameter uncertainty. IFAC, Multivaribale technological systems

TOWARDS SUPPORT SYSTEM FOR ELEMENTARY TECHNOLOGICAL UNIT DESIGN

P. Domański*, P. Marczyński* and J. Kopytowski**

*Institute for Control, Systems Engineering and Telecommunication, University of Mining and Metallurgy, Krakow, Poland
**Institute of Industrial Chemistry, Warszawa, Poland

Abstract. The framework of the support system for elementary technological unit (ETU) design is presented. The ETU is a part of chemical installation which works on the base of one reactor. The problem of working out the concept of the ETU is discussed. The artificial intelligence production system and the branch and bound method as a base of system's algorithm were chosen. The system will be hierarchical and composed of: structure generator (which works on higher level on generality) and process unit evaluation module. The first one says "what is to be done", the second one "how it can be done". The alternate work of those parts is shown. Two-dimensional quality coefficient energy/investements coming down from cost engineering was chosen. The simple example as well as working with the system are also discussed. The use of the system as a part of economic decision support system is proposed.

Keywords. Computer aided design; chemical industry; structure generation; heuristic programming; artificial intelligence; economics.

PROBLEM STATEMENT

The methods of computer aided design (CAD) which are used in chemical industry enable:
- the computation of process's parameters for assumed installation concept (flowsheeting methods),
- installation optimization related to chosen parameters which are decisive variables.

If it occurs after optimization procedures that the project of the installation does not satisfy designer's demands, then he must do some changes in it and has to start the flowsheeting and optimization procedures once more. This way of working out the design does not enable optimization on the level of concept. It is not only time-wasting, but also the best one may be omitted.

The problem is to find a method, which enables consideration of structure alternatives in elementary technological unit design, in order that the designer will be able to focus his attention on the problem of reactions and reactors, which is frequently the crucial point of the chemical installation design.

ELEMENTARY TECHNOLOGICAL UNIT

The elementary technological unit (ETU) is a part of chemical installation which works basing on one reactor only. The concept of ETU contains the most important data implying the structure of ETU. It is defined as directed graph:

$$\text{CONCEPT} = \langle \text{PU}, \text{F} \rangle, \qquad (1)$$

where PU is a set of "process units" and F is a set of "flows". Process units correspond to nodes of the graph, flows to directed arcs.

What the Process Units and Flows Are

The elementary technological unit is composed of unified units described in catalogues, e.g. distillation columns, cyclones, heat exchangers, etc. They are called "process units" and can be divided into three main groups:
I -which are used for changing the physical parameters of the mixtures flows ,
M- -which are used for separations,
M+ -which are used for connections.

I, M- and M+ will be also called "operations" and described as: M-(F1,F2,F3) , if it separates F1 into F2 and F3 , and: M+(F1,F2,F3) , if it connects F1 and F2 into F3.

Process units are connected by flows.

Flow is the homogenous mixture of the elements called components of the flow. Flows correspond to arcs of the graph (1). It must be noticed that designer can divide given flow in any way he wants.

Example. Air can be divided in following ways:
1/ Components: N_2, O_2, the remaining ones are not important.
2/ Components: N_2, O_2, Ar, CO_2, SO_2, O_3 (the greater number of components of the same flow).
3/ Components: clean air (N_2+O_2), pollutions (some components can be joined).

DESIGN TASK

The design task is described by:
1/ Desired products of the ETU
2/ Available resources
3/ Reactor's description
4/ Physico-chemical data concerning considered components and flows (mixtures)
5/ Set of available process units with their descriptions
6/ Quality coefficient

The 4/-6/ will be treated as a given data base of the system. The remaining ones can be described as sets of flows:
1/ ETU's output
2/ ETU's input
3/ Reactor's input and output

The Design Task Formalism

The design task is formalised as artificial intelligence "production system" described by Nilsson (1982). The production system is composed of global database, production rules and control system.

The global database is a main data structure used by the system, which describes state of design task solving. Do not identify it with database 4/-6/ above. Here it can be defined as following:

$$\text{DATABASE} = \langle \text{DATA, GOAL} \rangle \tag{2}$$

where DATA is ETU's input and reactor's output, GOAL is ETU's output and reactor's input. It will be useful to distinguish the initial DATA from DATAs which will occur during design task solving. It will be called INIDATA.

Production rules are operations on global database. In this case they are operations M-,M+. It is assumed that GOAL structure does not change during problem solving. Rules will be described as transforming the DATA only.

Example. Let DATA = $\{$ABC, D$\}$, then operation M_1 = M-(ABC, AC, B) transforms it into: DATA' = M_1(DATA) = $\{$AC, B, D$\}$.
ABC means flow of three components: an A, a B and a C.

Applicable rules. The rule can be unapplicable to a given database. It is assumed that M+/M- operations cannot use FLOW as their argument if:

$$\exists \text{FLOW}' \in \text{GOAL: FLOW} = \text{FLOW}' \tag{3}$$

It can be derived from: do not spoil what is already good.

Control system chooses rules (operations) which are to be applied in a given moment of problem solving. It is in fact an algorithm of the system.

Solution of the Design Task

Solution of the design task is defined as a sequence of applied rules operations which satisfy the condition:

$$\text{FULL SOLUTION (INIDATA)} = \text{GOAL} \tag{4}$$

The partial solution is a sequence of rules which does not satisfy (4).

Design task is defined then as a pair:

$$\text{TASK} = \langle \text{INIDATA, GOAL} \rangle \tag{5}$$

It can be noticed that M operations change initial TASK into the new one:

$$\text{TASK}' = \text{M(TASK)} = \langle \text{M(INIDATA), GOAL} \rangle = \langle \text{INIDATA}', \text{GOAL} \rangle .$$

Every state of problem solving can be called "current design task".

In fact it would be more "elegant" to define the solution of the design task as a graph, but this definition is more simple for computer applications.

It must be stressed that such a formalisation assumes that all data 4/-6/ are stored in database of the system. It is in fact too strong assumption, but for the sake of simplicity let us say it holds.

ASSUMPTIONS

Let us summarize assumptions done:
1/ Design of the elementary technological unit can be developped from so called concept of ETU containing (a few) most important data only
2/ The working out the concept of the ETU is a hierarchical process
3/ The structure of ETU is determined by M+ and M- operations, the structure of I operations will be proposed after working out the M+/M- structure
4/ The physico-chemical data concerning considered mixtures and substances are available as well as data concerning process units
5/ The solutions, partial ones also, can be described by approximate quality coefficient
6/ Reactor and its parameters are known
7/ Design task has a solution

The most important request to the system is not to lose the optimal structure (the best solution).

Those assumptions imply that it can occur a situation when it is impossible to find the best solution. The reasons are:
- the quality coefficient may be multidimensional, then only Pareto set can be obtained,
- finite accuracy of the quality coefficient may cause that two, or more, solutions have overlapping intervals of its estimation.
The system produces a set of equivalent, from the point of view of quality coefficient, solutions.

BASE ALGORITHM - CONTROL SYSTEM

As the system has to find the best solution of the design task the backtracking control strategy (Nilsson, 1982) is applied. As it is important to system not to consider so many alternatives the known from operations research branch and bound method is taken as a base of the algorithm.

Decision / Solutions Tree

All possible solutions of the design task can be shown in form of decision tree

(solutions tree). The nodes of the tree correspond to the solutions, also partial, the arcs to chosen elements of the solution, i.e. units or operations. If arc M_i connects the solutions S_2 and S_1, than $S_1 = S_2 \cup \{M_i\}$. One can say: S_1 is developped from S_2 by adding M_i operation.
The element of the tree which is not developped from any other solution is called "an empty solution".

It is assumed that quality coefficient for developped solution is not less than for the original one:

$$S_1 \supset S_2 \Rightarrow Q(S_1) \geqslant Q(S_2). \qquad (6)$$

The solutions tree is to be checked, but it is impossible to check all of nodes. Some of nodes correspond to solutions, also partial, which cannot be accepted nor developped because of too high value of quality coefficient. Those solutions are too expensive. Every solution for which lower estimation of quality coefficient is greater than upper estimation of unless one accepted solutions is rejected. When two solutions have common part, i.e. $S_1 = S_{1a} \cup S_{1b}$, $S_2 = S_{2a} \cup S_{2b}$ and $S_{1a} = S_{2a}$, then the different parts, S_{1b} and S_{2b} will be compared.

Short Description of Algorithm's Acting

The last considered partial solution is developped by applying the rule (operation) which has not been used yet. The rules are settled in order by the system. If there is not such a rule (operation) the last element of the solution is rejected and this reduced solution is considered.

Now the estimation of the quality coefficient of received solution is computed.

If it is acceptable then:
- if it is a partial solution: it is developped,
- if it is a full solution: the set of accepted solutions is updated, the last element of the solution is rejected, and received partial solution is to be developped.

If it is not acceptable then the solution will not be developped, it is reduced and the reduced solution is to be developped.

The algorithm stops when the empty solution cannot be developped in any unchecked way.

The algorithm produces the set of equivalent solutions, concepts, which can be put through flowsheeting and optimization procedures.

MAIN PARTS OF THE SYSTEM

The hierarchical structure of the system was chosen. The working out of the concept of ETU will be divided into two levels:
1/ Level of structure generator (strategy)
2/ Level of process unit evaluation module, PUEM (verification of the strategy), the quality coefficient is its important part.

STRUCTURE GENERATOR

The structure generator considers the most important data only. It does not use the explicit knowledge about process units, but every decision it makes basing on the set theory and physico-chemical data concerning considered substances will be tested on the second level, where this knowledge is used. There is a difference between solution of the design on those levels. For structure generator it is a sequence of M+/M- operations, for process unit evaluation module it is a sequence of process units. The quality coefficient can be calculated for the second one only. The second one is regarded as a final product of the system.

What Does the Structure Generator Do?

The structure generator solves the design task on the level of M+/M- operations. Choice of the new element of the solution consists of four steps:
1/ Choice of applicable rules (operations)
2/ Setting them in order from probably the best one to probably the worst one
3/ Getting first applicable and unchecked one and trying it (with the help of PUEM), when PUEM stops, the estimation of the quality coefficient is calculated:
4/ Depending on results of 3/ it goes to step 1/ or 3/.
Detailed description of structure generator and process unit evaluation module will be presented later.

Applicable Rules (Operations)

Following preconditions of applicability are proposed:

M-(FLOW, F1, F2) can be applied if and only if:

$$\forall A,B \in FLOW: [\exists FLOW' \in GOAL: A,B \in FLOW' \Rightarrow (A \in F1 \Leftrightarrow B \in F1)]. \qquad (7)$$

M+(F1, F2, FLOW) can be applied if and only if:

$$\exists A \in F1\ \exists B \in F2\ \exists FLOW' \in GOAL: A,B \in FLOW' . \qquad (8)$$

It means: M- operation can be applied if it breaks at least one connection of components which is not requested in the GOAL structure and M+ operation can be applied if it creates at least one connection requested in the GOAL structure.

Additional Memory vs. Special Limitations

The limitations which are imposed, (7) and (8), are very weak in order not to lose the best solution. On the other hand it leads to inconvenience of overlooped structure generation. S is an overlooped solution if:

$$\exists S' \subset S: S' \neq \emptyset \quad S'(DATA) = DATA \qquad (9)$$

The trouble expressed by (9) can be eliminated in following ways:
1/ The system must check whether current design task is equal to any previously obtained one. In this case the current solution is rejected, it is not developped. It needs additional memory.

2/ Special limitations for M+/M- operations can be settled, owing to these the overloops will never arrive. Here the question arises: will the best solution be omitted?

Setting Rules in Order

Before one can start with choice of process unit it must be known what this unit ought to do. As separations, M- operations, are regarded as most difficult tasks, let us focus attention on this problem. If, for example, flow ABCD is to be separated into four flows: A, B, C and D , then which of the components will be separated first? It is said (Beddow, 1981; Roberts and co-workers, 1971) that for given separation a given method is good if it differentiates between components on the basis of the most appropriate property. The answer is: the component which is the most different of the others, which sticks the less to the others, is to be considered as a first alternative, because probably good unit for such separation can be found.

The rules (alternative M operations) can be settled in order with respect to certain function which is "quality of M operation". The problem is how to measure this quality, how to measure the difference between chosen component and the others for M- operation.

Example of Heuristic Procedure for M-Operations Ordering

The simple proposal is presented as an example:

Let $\{x_i\}$ be a set of parameters such as: size, density, surface wetability, magnetic permeability, electrical conductivity, temperature and heat of cristallization and condensation, water and other solubilities, etc.

Let $x_i(A)$ be a value of x_i for A expressed in any chosen way: linear, logarithmic, polynomial, ... The value x_{iA} is defined:

$$x_{iA} = \min_{B \in FLOW,\ B \neq A} \left\{ |x_i(B) - x_i(A)| \right\}$$

$$\text{if: } x_i(A) = \min_{B \in FLOW} \{x_i(B)\} \text{ or } \max_{B \in FLOW} \{x_i(B)\}$$

and:

$$x_{iA} = 0 \text{ , elsewhere; } A \in FLOW \qquad (10)$$

As x_{iA} are expressed in different units (m, kg, K, ...) there is no way to compare them. Let $\{c_i\}$ be a set of coefficients for which it holds: $\dim(c_i \cdot x_{iA}) = [1]$.

Values of c_i say which of x_{iA} are relatively small or relatively huge. The maximum of $\{c_i \cdot x_{iA}\}$ points out which property is the most appropriate for separation. It is a "measure" of difference between component A and the others. But it is not a "measure of quality" good enough for real applications.

It is possible such a situation when the "difference" for a given separation is relatively great, but despite it the operation is not so good, because, for instance, explosivity of the products. Let $Y = \{y_i\}$ be a set of such crucial parameters as: explosivity, toxicity, corrosivity, etc.

Let F(Y, A) be a correction function, which basing on the values of those crucial parameters changes "measure of differance" into "measure of quality":

$$V(A) = \max_i \left\{ c_i \cdot x_{iA} \right\} \cdot F(Y, A) \qquad (11)$$

F can be a step function, which tolerates small values of y parameters and rejects operations which give products with exceeded limits of those parameters.

Then the M- alternatives will be ordered with respect to the values of V(A) described in (11).

It is a difficult task to make adequate setting procedures. A lot of heuristic knowledge is to be applied here. It will be possible for the designer to put into the system a part of heuristic knowledge of his own.

PROCESS UNIT EVALUATION MODULE

Process unit evaluation module (PUEM) has to choose the process unit or configuration of process units for a given M+/M- operation chosen by structure generator.

The analysis of technological literature points out that it is most difficult to choose the process unit which is adequate to certain M- operation. For example in order to realise the gas-solid separation one must choose one process unit among minimum ca 20 units proposed. But the analysis of four basic parametres of the flow enables to limit this list up to a few process units (Sargent, 1971).

Considering the speed of the design task solving and properties of the branch and bound method it would be useful to order this subset as to begin from the process unit which is probably the best one to probably the worst one among feasible units.

The Way PUEM Chooses the Process Units

Let M be an operation chosen by structure generator. PUEM has to assign a set of alternative process units to this operation. It has to compute which units will be best for this operation. The module acts as following:

1/ Finds available units

2/ Proves technological feasibility of those units

3/ Sets them (feasible ones) basing on approximate quality coefficient in order from probably the best to worst one.

4/ Worst ones are eliminated

5/ The remaining ones are taken under more detailed consideration (cost engineering).

Approximate Quality Coefficient for PUEM

For every feasible unit an approximate quality coefficient is calculated. It is based on energy consumption (calculated from heuristic formulae too) and roughly estimated investement costs. The accuracy is small. The calculations should be made with rather great reserve for avoiding the influence of "contextual costs".

For example: the distillation column is expensive and needs a lot of energy. But it gives mixtures with high temperatures as an output, so part of the energy can be recovered - and it must be taken into account.

If there are units which cannot be the best under any condition, they are eliminated.

QUALITY COEFFICIENT

Structure generator gives the strategy for solution's development, process unit evaluation module shows the way the strategy can be realised, but final verification of decision made so far is quality coefficient computation. Here the two-dimensional quality coefficient is proposed. It is composed of the following:
1/ Energy consumption
2/ Estimated investement costs
This quality coefficient cannot be identified with the one used by process unit evaluation module for preliminary evaluation of feasible units.

Energy Consumption

It is regarded as most important factor from manufacturing costs, the factor which strongly depends on the structure of the installation. It is not reasonable to consider here the factors which are not depend or weakly depend on the structure, e.g. fixed charges.

Estimated Investement Costs

It is based on estimation of Total Capital Investement on Battery Limits. Battery Limits is the cost of basic devices, installation, in main production line.

The Problem of Accuracy

Planned accuracy of the quality coefficient is about 20-25% (Rejewski and Kumur, 1983), but in some cases it can be greater. It must be noticed that for the stage of concept, defined as above, available methods do not enable estimation with accuracy greater than 12-15% (Viola, 1981).

The role of estimation accuracy must be here strongly emphasized: it influences the time which must be spent to find the whole set of equivalent full solutions. The second: the accuracy is better the number of set's elements is smaller (remember that those elements must be finnaly examined by flowsheeting procedures).

The problem of I operations is to be discussed when the problem of accuracy is considered. It is already known how M+/M- operations are chosen.But where do I operations come from? The most difficult is to find the unit which changes the flow's temperature. It results from the fact that known methods have global character: they consider the set of heat exchangers in all installation whereas the considered problem is to do the partial estimation. The estimation accuracy is so important that it is proposed to apply these global methods to every partial structure (solution).

First Estimation

Before system starts the first estimation is to be made. It may base on any existing installation (counted using Marshall-Stevens Index), or on any method "order of magnitude" (Rejewski and Kumur, 1983).

REMARKS ON COOPERATION OF SYSTEM'S PARTS

Described above main parts work alternately. After every M+/M- operation has been chosen by structure generator PUEM attemps to find adequate process unit (or group of process units). This choice enables to modify the value of quality coefficient. It is the base to decide whether the search in chosen direction should be continued or not. If PUEM has chosen more than one adequate unit, only for the first one quality coefficient is computed immediately, and then structure generator starts again. The remaining possibilities will be considered after backtracking.

EXAMPLE

Simple example is presented:
Gaseous propylene $CH_2=CHCH_3$ is reacting with chlorine Cl_2 giving the mixture of allylchloride $CH_2=CHCH_2Cl$ and hydrochloric acid HCl. The mixture, composed of 28-29% of HCl and of 72-71% of allylchloride by weight, is cooled down after reaction to about 318-323K. It is to be separated. Select the unit (M- operation) using PUEM.

Choice of available units. Let there be six available units / operations: crystallization, adsorbtion, extraction, distillation, water absorbtion and liquid absorbtion.

Proving feasibility of the units.

TABLE 1 Availability and Feasibility

Unit/operation	Av.	Feas.	Why unfeasible?
Crystallization	+	-	low melting point
Adsorbtion	+	-	chemical reaction
Extraction	+	-	no agent available
Distillation	+	+	-
Water Absorb.	+	+	-
Liquid Absorb.	+	+	-

There are three units/operations available and feasible. All of them have to be taken under consideration - approximate quality coefficient must be calculated (AQC).

TABLE 2 Relative AQC

Unit/operation	Relative AQC:EC	IC
I Distillation	2.5	1.0
II Water Absorbtion	1.0	4.0
III Liquid Absorbtion	3.5	3.0

Where EC and IC are two components of AQC (approximate quality coefficient) expressed in relative way, EC means "energy consumption" and IC means "estimated investement costs".

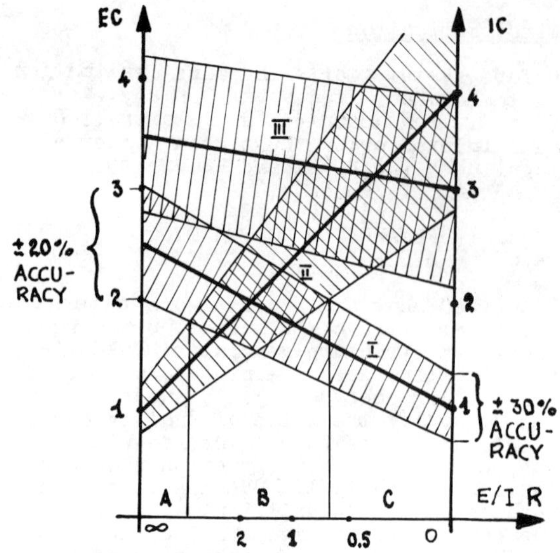

Fig. 1. The choice of the unit.

The choice depends on energy/investement rate E/I R. III cannot be chosen, one unit is chosen in cases A,C, two in case B. If E/I R = 1, the first one will be I. Notice: the example does not show all factors which are to be considered in such a choice.

WORKING WITH THE SYSTEM

It is possible to designer to put the design task into system and receive the solutions only, but the authors find this way unreasonable. It will be much better to control system's actions. The interventions may occur in the following moments:
1/ Updating, completing database with data concerning specific substances, equipment
2/ Changes of quality coefficient
3/ Imposing conditions on solutions
4/ Defining part of the solution
5/ Changing the sets of alternatives
6/ Changing the order of alternatives in both structure generator and PUEM
7/ Changing parameters of heuristic procedures which set alternatives in order

1/-4/ can be treated as a part of design task formulation, the designer uses the knowledge he had before. 5/-6/ are arbitral direct changes in system's acting, e.g. it may occur that some units are unfeasible because of unforeseen reasons. 7/ is a form of heuristic knowledge coming down.

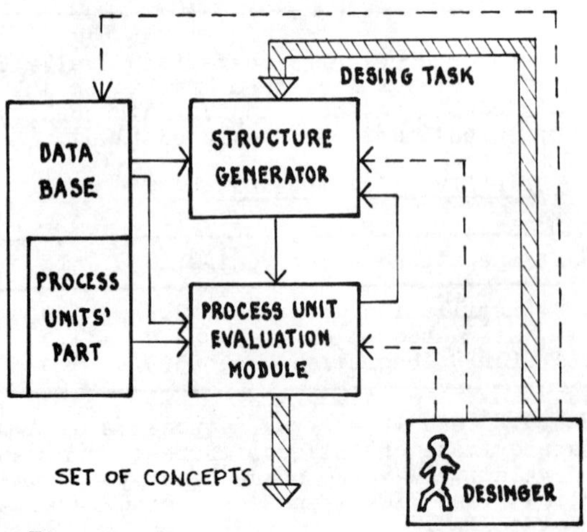

Fig. 2. Working with the system.

Working with the system is shown at Fig. 2, continuous lines show internal connections of the system, dashed lines possibilities of designer's intervences.

Last but not least: the solutions of the design task depend strongly on task formulation, e.g. on the way the designer defines and divides the flows.

CONCLUSION

There is a trend in CAD software to stabilize mathematical methods (Sterling, 1983). Presented framework was based on known formalism of production system and branch and bound method. The new is to apply them to the problem of working out a concept of chemical ETU in such a hierarchical way.

The use of the future system is planned in preliminary works on ETU design, when it can help not to forget about the structure alternatives and avoid mistakes done because of too strong and directed routinism. There is a second possible its application, the idea is compact with that proposed by Górecki and co-workers (1983) in problem of investigations of possible trends in chemical industry development. Such a system can be used as a decision support tool estimating costs of new investements with accuracy better than "order of magnitude".

It is worth saying that for introduced formalisms the assumption "ETU is based on one reactor only" is not necessary. The final shape of the system depends on solution of some important problems. The main one is how to represent the heuristic knowledge about designing, e.g. that one which forms the F(Y, A) functions, see (11). It is planned the system will have selflearning abilities (will learn from the designer).

REFERENCES

Beddow, J.K. (1981). Dry separation techniques. Chemical Engineering, 88, no.17, 70-84.

Górecki, H., J. Kopytowski, T. Ryś and M. Żebrowski (1983). A multiobjective procedure for project formulation - design of a chemical installation. In Lecture Notes on Economics and Mathematical Systems, vol.229, Springer Verlag, Berlin. pp. 252-259.

Nilsson, N.J. (1982). Principles of Artificial Intelligence. Springer Verlag, Berlin.

Rejewski, P. and A. Kumur (1983). Podstawy inżynierii kosztów w przedprojektowym stadium technologii chemicznej. Internal report of Institute of Industrial Chemistry, Warszawa.

Roberts, E.J., P. Stavenger, J.P. Bowersox, A.K. Walton and M. Metha (1971). Solid/solid separations. Chemical Engineering, 78, no.4, 89-98.

Sargent, G.D. (1971). Gas/solid separations. Chemical Engineering, 78, no.4, 11-22.

Sterling, M.W. (1983). New hardware and software trends in CAD/CAM. In Autofact 5 - Conference Proceedings, Computer and Automated Systems Association of SME, Detroit. pp.1.8-1.23.

Viola, J.L.Jr. (1981). Estimate capital costs via new, shortcut method. Chemical Engineering, 88, no.7, 80-86.

A FLEXIBLE COMPUTER PROGRAMME FOR CALCULATION OF ENERGY AND MATERIAL BALANCES IN A SUGAR FACTORY

L. B. Jørgensen*, N. Fonnesbech*, R. F. Madsen*, M. Kümmel and H. Søeberg****

**The Danish Sugar Corporation, Research and Development Laboratories, Nakskov, Denmark*
***Technical University of Denmark, Department of Chemical Engineering, Lyngby, Denmark*

Abstract. A programme system dedicated to the simulation and balancing of energy and material balances in a sugar factory has been developed. Special concern has been put in the development of a flexible and comprehensible user inferface which can be adjusted according to the needs of the individual user. The calculation part of the programme uses a modified version of the sequential modular approach where the calculation path, within the process calculation modules, can be changed. This approach has shown to facilitate the determination of calculation sequence and iteration parameters in the flowsheet. Further, the sugar factory programme includes a method for balancing material balances with redundant plant data and a method for graphical presentation of the flowsheet and calculation results.

Keywords. Sugar industry; computer-aided design; man-machine systems; computational methods; computer graphics.

INTRODUCTION

The calculation of stationary material and energy balances in a sugar factory is vital for design and operation of a plant. Due to the complexity of the plant, first of all due to the many process recycling streams, this is not a trivial task.

The sugar manufacturing process can briefly be described as follows (Schneider, 1968; McGinnis, 1982). Sugar beets are washed and sliced into small oblong pieces which are fed to a diffuser, where the sugar is extracted with countercurrent hot water. Hereby a thin sugar juice (about 15% dry matter) is obtained, consisting of sugar, water and several impurities. These impurities are named nonsugar and treated as one component in the material balances. Most of the nonsugar is then separated from the sugar juice by repeated treatments with slaked lime and carbon dioxide, respectively, succeeded by filtrations. After the purification, most of the water is removed from the sugar juice in a five-effect evaporator, resulting in a thick sugar juice, containing about 70% dry matter. Finally, the thick juice enters the sugar house, where a 3-product crystallization is carried out. The crystals are separated from the mother liquor by centrifuging. In the Danish Sugar Corporation the first two products are mixed to form a white sugar. The mother liquor of the third product forms the molasses, and its crystals are remelted and recycled. Thus the sugar factory consists of series of unit operations (extraction, evaporation, crystallization, centrifuging, etc.), some being batch processes, other continuous.

In the first part of the factory (washing, slicing, extraction) the mass and energy balances can be calculated relatively easily, but the final part, the evaporator and the sugar house, is much more complicated regarding what goes on in the single process as well as the interaction between the processes. This is mainly due to the fact that the five-effect evaporator functions as a steam supplier to other processes, especially the crystallization processes in the sugar house. Furthermore, the simulation of the evaporator and the sugar house is made difficult by the lack of general mathematical models for the heat transmission in the evaporators and the separation processes in the centrifuges. Consequently, empirical values and expressions, dedicated to the individual plant, have to be used in the calculations.

A broad spectrum of the commercially available programme packages (Leesley, 1982) for process simulation was examined with respect to a set of requirements concerning sugar factory calculations, covering the purely technical requirements as well as demands to the interface between the user and the programme system. This examination clearly showed that very advanced programmes, such as Concept (1982) and Flowpack II (Perris, 1982), had to be used in order to fulfil the major part of the requirements. However, even when using such advanced programmes, it would still be necessary to do a lot of additional programming to fulfil all the given requirements. On the other hand, many of the facilities provided by the commercial programmes were not required in the sugar factory calculations. Thus it was decided to develop a material and energy balancing programme, dedicated to the sugar industry.

REQUIREMENTS TO THE PROGRAMME

General

The programme structure and the user interface must have a form which enables the user to spend the major part of his effort and creativity on a critical evaluation of the flowsheet configuration and the results of the calculations instead of a time consuming work with the entering and editing of data.

At his terminal, the user should be able to work in an iterative process, consisting of the sequences

a) Definition of flowsheet configuration
b) Definition of process and stream data
c) Calculation
d) Evaluation of results
e) Editing flowsheet and/or process and stream data
f) New calculation.

The points d)-f) will often be executed several times during the problem solving.

The calculation problems can roughly be divided into two groups. The first group consists of routine solutions of material and energy balances in an existing production plant. The second group requires possibilities for doing experiments with the flowsheet configuration and the process parameters in order to optimize existing and to design new plants. The first group requires only a very simple interface, while the second group implies much more detail and thus requires a more complex interface to the programme. Common for the potential users of the programme will be that they are experts with regard to the sugar plants, but often unfamiliar with computers, programme language and terminals. Therefore, the programme should be able to "speak" the language of the sugar factory staff.

Specific Requirements

The general requirements can be expressed in a number of specific items divided into two parts, one part describing the purely technical aspects and one part concerning the human factor. Although presented separately, these two parts are highly dependent.

Technical Requirements

Calculation of sugar factory processes. The programme must contain basic models (exact or empirical) for the behaviour of the processes in the sugar factory, and the user must have possibility for contributing his own models.

Flexibility regarding the specification of flowsheet and data. Based on the process calculation models, the user must be able to build and store his own flowsheet and specify process and stream parameters. The programme should especially facilitate corrections in existing flowsheets and data.

Flexibility regarding the calculation path in the flowsheet. Once a flowsheet is defined, the user should be able to solve 3 different kinds of problems, which are simulation, balancing of measured plant data and design of new plants. Each of these problems requires different sets of input parameters, meaning that in one problem a parameter will be a given input parameter and in another problem the same parameter will be a calculated output parameter.

Balancing of redundant plant data. Especially in the calculation of material balances of existing plant data there will be redundant data to consider. These data will result in an over-specified set of equations, meaning that the number of linear independent equations will be greater than the number of unknowns. The programme must therefore provide a method by which all data are included in the solution of the balance equations. The method must treat the single data with a weight corresponding to its standard deviation.

Graphical presentation of results. The user should be able to get his flowsheet and calculated values in a graphical form, either by means of a graphics routine implemented in the sugar factory programme or by means of an independent graphics system. If an independent system is chosen, the calculation programme must be able to provide the data concerning the flowsheet configuration and the process parameters in a form which minimizes the necessary additional interface programming.

Requirements Concerning the User Interface

Low response time. The iterative working process at the terminal requires that the response time of the programme is kept as short as possible. This does not only concern the entering and editing of flowsheet and data, but also the specific calculation procedure.

Few operations at the terminal. Getting through the work with defining and editing data in a quick and clear way requires that the number of necessary operations at the terminal are as few as possible.

A flexible interface. The number of details and the style of the user interface should be variable according to the specific needs of the individual user. This means that the number and nature of prompts for process and stream parameters should be variable.

Clear and simple screen pictures on the terminal. When prompting for and presenting data on the visual display unit, it is very important that not too much information is presented at one time on a single screen picture.

Error checking. The programme should contain routines for control of syntax and logical errors made during the specification of data. The user should get his feed-back during the input session, so that a minimum of time is wasted when an error occurs.

THE BASIC STRUCTURE OF THE SUGAR FACTORY PROGRAMME

The basic idea in the chosen programme structure is shown in Fig. 1. By means of a master menu the user is enabled to choose between three main functions, which are

NETWORK. The flowsheeting part, specifying new and editing existing flowsheet configurations.

DATAWORK. The data entering part, specifying and editing process and stream data.

EXECUTE. Execution and presentation of calculations on the process plant defined in NETWORK and DATAWORK.

The three programme parts are able to work separately, as they have all free access to the data files. Data describing the configuration of the flowsheet and physical/chemical data concerning the individual processes and streams are stored on separate data files. Hereby, the user is able to create and store several sets of plant data without having to redefine the flowsheet configuration.

All the programmes work in such a way that all relevant data from the files are read into the memory at the beginning of the execution and written back at the end. This approach requires a certain amount of accessible memory in the computer, but in return the number of disk operations is kept at a minimum, resulting in a low execution time. In the following sections a more detailed description of the three parts of the programme is presented.

THE FLOWSHEETING PART

The user specifies the flowsheet configuration by means of the NETWORK programme, defining how the individual process units and streams are connected. This is done interactively at the terminal by entering names of selected types of processes and streams. A specification of a heat exchanger working on thin sugar juice will typically have the following form.

```
PROCESS NAME   :  THINJUICEHEATEX
PROCESS TYPE   :  HEATEXCHANGER
INPUT STREAMS  :  STEAM1 , THINJUICE1
OUTPUT STREAMS :  CONDENSATE1 , THINJUICE2
```

Figure 2 shows how a simple flowsheet consisting of a single evaporator and a heat exchanger is defined using NETWORK. The programme solely uses the names given by the user for the description of the connections between processes and streams and, consequently, during the definition session the programme continuously controls that logical errors are caught and corrected. There are no limitations to what the processes and streams may be named. The only condition is that the selected type of process must be available in the executing part.

Related to the flowsheet specification procedure an editor is available, by which the user by means of simple commands is enabled to correct already specified configurations. Some of the important editing facilities are

- changing the type specification of a process,
- changing names of processes and streams,
- deleting processes and streams,
- inserting new streams,
- moving streams.

The basic structure of the NETWORK programme is presented in Fig. 3. The programme works in such a way that entered commands, such as CREATE PROCESS, MOVE STREAM, etc., result in calls of specific subroutines. This module structure makes it easy to maintain and expand the programme.

THE DATA SPECIFICATION PART

After the specification of the general flowsheet configuration, the physical and chemical parameters of the processes and streams must be defined. By use of the programme DATAWORK the individual processes and streams are selected by entering the names specified in NETWORK. These names are the only parameters connecting the flowsheet and the data. After data is specified, it is still possible to return to NETWORK for editing purposes, because the identification of the single data refers to a name and not to a specific location on the flowsheet. It is thus possible to move a stream around in the flowsheet without losing its already specified data. During renaming of the elements in the flowsheet a trace-back nametable is created and used in DATAWORK.

If a process or stream is without any data specifications, the user is prompted for some, otherwise a data edition session is prepared. Figure 4 gives an example of how a heat exchanger parameter is edited. Besides the specification of physical/chemical data, a more detailed specification of the flowsheet is given in DATAWORK. The connections between specific positions in the individual processes and the input/output streams, given in NETWORK, are specified where necessary. Furthermore, all the streams are assigned a label, such as STEAM, CONDENSATE, JUICE, etc. These stream labels, like the process types, are used by DATAWORK when it selects what parameters to prompt for and how they should be presented on the terminal.

THE FLEXIBLE USER/PROGRAMME INTERFACE

The terminal screen pictures, presented by NETWORK and DATAWORK, can be adjusted according to the needs of the individual user. As indicated in Fig. 2, the programmes read the interface from a file before contact to the user is established. The flexible interface consists of a selection of character strings, which are written to the terminal screen in a fixed format during the presentation of and prompt for data. Figure 5 shows the contents of records describing the interface for a simple heat exchanger compared with the corresponding screen picture presented to the user. This interface specification method is used both for processes and for streams, and the user is able to modify the interface records, creating his own personal screen pictures.

The interface determines the storage sequence of the process and stream data on the files, as the individual data are stored in the order they are prompted for. This gives a clear coherence between the user interface and the parameter list communicating with the process calculation modules in the executing part of the programme.

THE EXECUTING PART

The calculation of mass and energy balances in the sugar factory can be divided into three basically different problems, being

- simulation of existing process plants,
- balancing of production data,
- design of new process plants.

The simulation problem is a conventional task, which in most programme packages for process simulation is solved by using the sequential modular approach (Westerberg and colleagues, 1979). The sequential modular approach is a method which treats the individual processes in the flowsheet as specific calculation modules connected by the process streams. The entire flowsheet is then calculated as a sequence of process module calculations until convergence of selected parameters are reached. Although this approach works well solving the simulation problem, it can often be a heavy and complicated task using it on the balancing and design problem.

The executing part in the sugar factory programme is based on a modified version of the sequential modular approach combined with dedicated modules for solution of special tasks, such as balancing, design and graphics. The structure of the executing part is shown in Fig. 6, The main programme, EXECUTE, contains a menu from which the user can select among the calculation methods and related options. All the executing modules are able to communicate with the datafiles created by NETWORK and DATAWORK. The programme structure facilitates the implementation of additional dedicated executing modules as the need arises, the only condition being that they make use of the standard routines for communicating with the datafiles.

The Modified Sequential Modular Method

The sequential modular method is a separate executing module communicating with a library of process calculation modules and a selection of mathematical and thermodynamic standard routines. The thermodynamics needed in a sugar factory calculation can in fact be expressed in relatively few equations. In the general and traditional sequential modular method the process calculation modules work in such a way that specified or earlier calculated parameters of inlet streams result in a calculation of the outlet streams of the process. Using this approach, the flowsheet configuration has to be only slightly complicated before the determination of calculation path and iteration parameters is made so difficult that automatic methods have to be used. Often the flowsheet must be recalculated several times before convergence is reached, resulting in rather long response times.

In the sugar factory programme the construction of the process calculation modules makes it possible to alter the calculation path within the module. By means of switches in the parameter list of the module it is possible to decide which parameters will be input and which will be output. The switch-option makes it possible to modify the calculation path in the module according to its position in the flowsheet. This approach does not only facilitate the determination of the calculation path in the flowsheet, but does also minimize the number of iteration parameters and necessary process module calculations, resulting in a short response time. Using the switch-option it is possible to solve the three basic calculation problems mentioned above, although the balancing of redundant data still needs a dedicated approach.

Balancing Redundant Data

The dedicated balancing module solves material balances based on redundant production data. The method solves the fundamental material balances simultaneously with a quadratic minimization of the ratio between the differences of measured and calculated process parameters and the standard deviations on the measured parameters. This approach ensures that the flow of material is exactly in balance and that the measured plant data carries a weight in the set of equations corresponding to their respective standard deviations.

Control of Output Parameters

Controlling the value of calculated output parameters is a problem strongly related to the sequential modular approach. The sugar factory programme includes a simple method by which the value of a calculated parameter can be kept at a given setpoint controlling selected input parameters. The controller can be selected from a collection of different iteration procedures. An ordinary proportional controller has shown to work excellently on most of the sugar factory flowsheets. Contrary to integrating the controller as a part of the flowsheet, as described by Westerberg and colleagues (1979), it is here designed as a separate executing module able to communicate with the sequential modular module. Thus the experimenting user is enabled to specify the controller without having to work on the flowsheet configuration and the process and stream data.

GRAPHICAL PRESENTATION OF THE CALCULATIONS

The best way to present the results of the process calculations was found to be a hardcopy drawing of the flowsheet, containing the relevant specified and calculated data, selected by the user himself.

A simplified approach to solve this problem has been to use the 2-dimensional drafting system DOGS (1983) from PAFEC, England. By means of an interface module connecting the calculation part and the DOGS system it is possible to transfer data to an already existing drawing of the flowsheet. The parametric symbol facilities of DOGS are used for transferring the information between the programme systems. A parametric symbol in effect is a macro consisting of graphical instructions, which can be stored in a text file. The results from the calculation modules are given the format of a parametric symbol, which can be retrieved as any other DOGS symbol and combined with the appropriate master drawing. This master drawing is selected from a library containing a collection of typical flowsheet configurations in the sugar factory. The transferred process and stream data are then automatically written on to the flowsheet in the positions where the actual process or stream appear on the drawing.

Figure 7 gives an example of a DOGS drawing of a five-effect evaporator containing calculated process and stream data.

APPLICATION EXPERIENCE

The large effort applied in the development of the user interface has shown to be justified by the positive acceptance by the users not familiar with computer programmes.

The modified sequential modular approach combined with the editing facilities in the flowsheeting and data specification parts have resulted in a very quick and uncomplicated execution of the iterative working process taken up by the experimenting and optimizing user.

The graphical presentation of flowsheets and calculation results provides the user with a well structured and overall view of all the relevant process and stream parameters, which can be difficult to obtain through search in an often extensive amount of printer output.

Despite the availability of advanced commercial programme packages, the development of our programme has provided the sugar factory staff and the engineering division with a number of vital facilities that are not available in any other programme packages.

REFERENCES

CONCEPT (1982). Process Flowsheet Simulator: User Manual. CAD-CENTRE, Cambridge, England.

DOGS (1983). User Manual. Program Level Number 2.5. PAFEC Ltd., Nottingham, England.

Leesley, M.E. (1982). Computer-Aided Process Plant Design. Gulf Publishing Company, Houston, Texas.

McGinnis, R.A. (1982). Beet-Sugar Technology, 3rd ed. Beet Sugar Development Foundation, Colorado, U.S.A.

Perris, F.A. (1982). FLOWPACK II: A Third Generation Flowsheeting System. ICI Corporate Laboratory, Runcorn, England. In M.E. Leesley (Ed.) Computer-Aided Process Plant Design. Gulf Publishing Company, Houston, Texas.

Schneider, F. (1968). Technologie des Zuckers. M. and H. Schaper, Hannover, Germany.
Westerberg, A.W., H.P. Hutchison, R.L. Motard, and P. Winter (1979). Process Flowsheeting. Cambridge University Press, Cambridge, England.

FIGURES

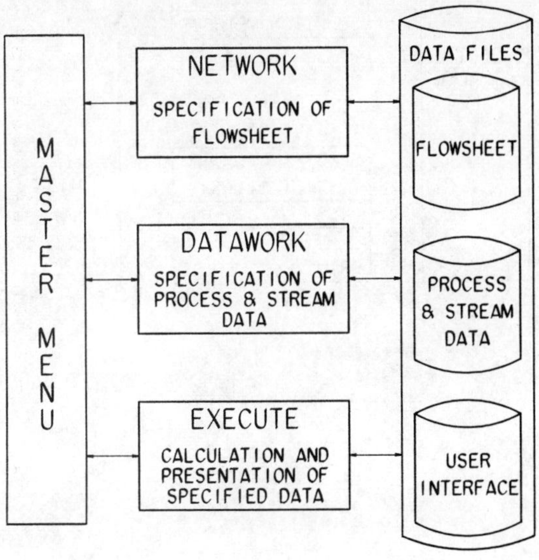

Fig. 1. The basic structure of the sugar plant programme.

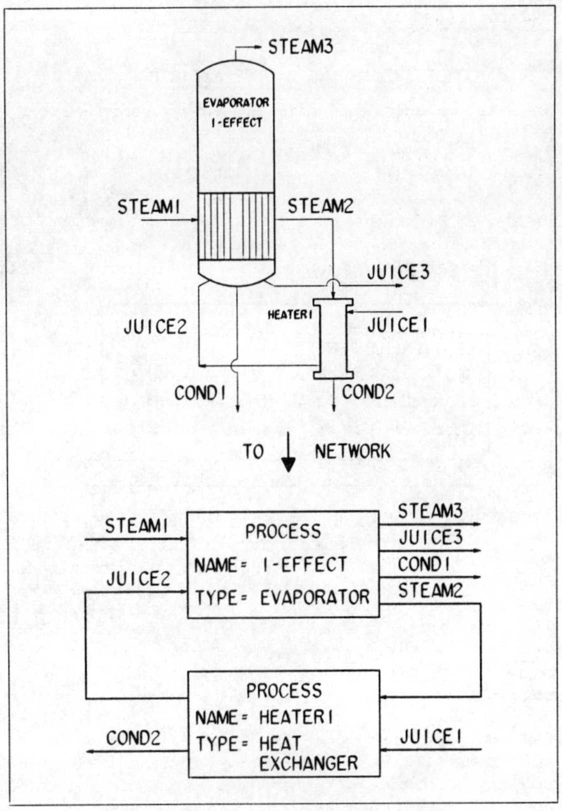

Fig. 2. Specification of a flowsheet using NETWORK. Only names of streams, processes and process types are used.

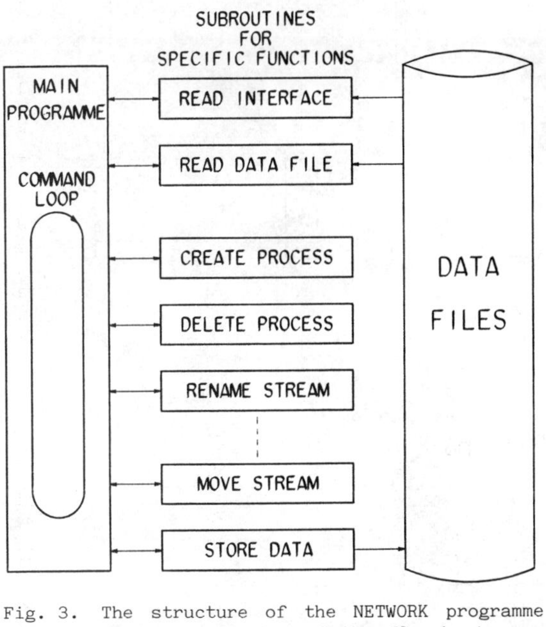

Fig. 3. The structure of the NETWORK programme for specification of the flowsheet configuration.

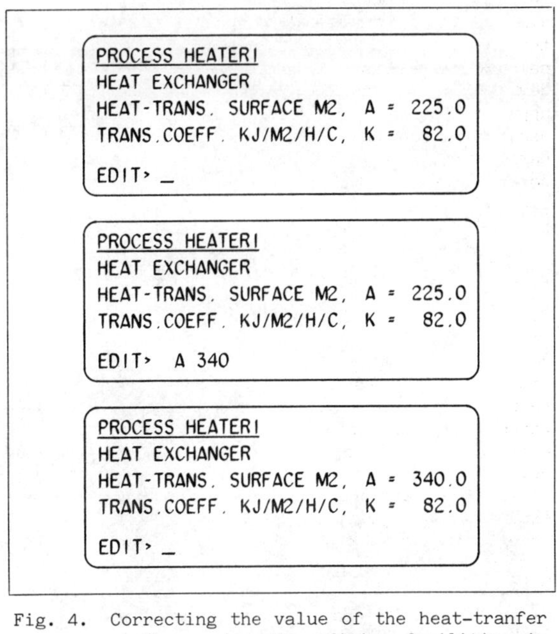

Fig. 4. Correcting the value of the heat-tranfer surface using the editing facilities in DATAWORK.

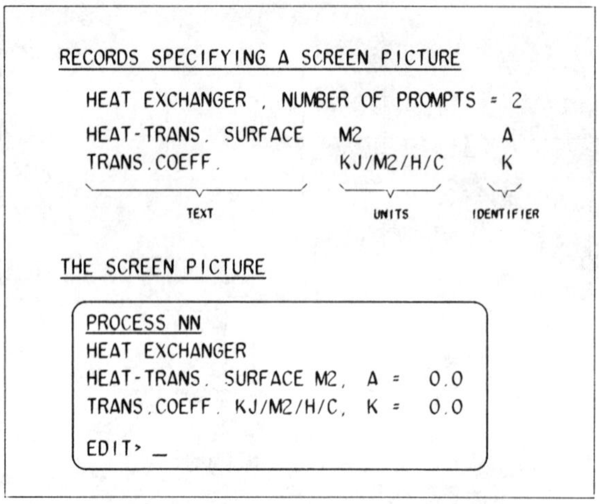

Fig. 5. The flexible user interface. The number of prompts and the contents of the text, unit and identifier character strings can be adjusted to the individual user.

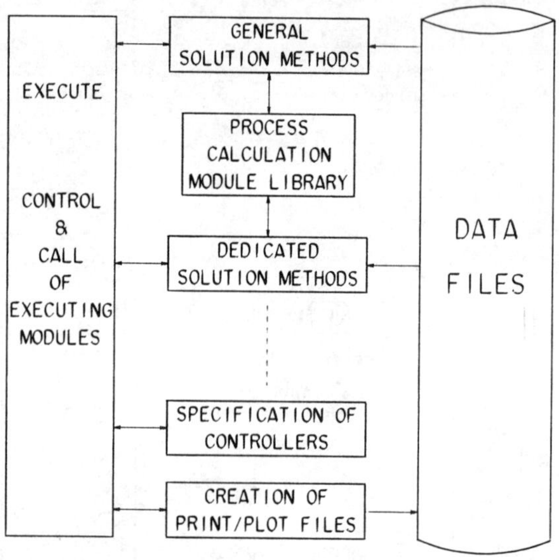

Fig. 6. The basic structure of the executing programme part, EXECUTE.

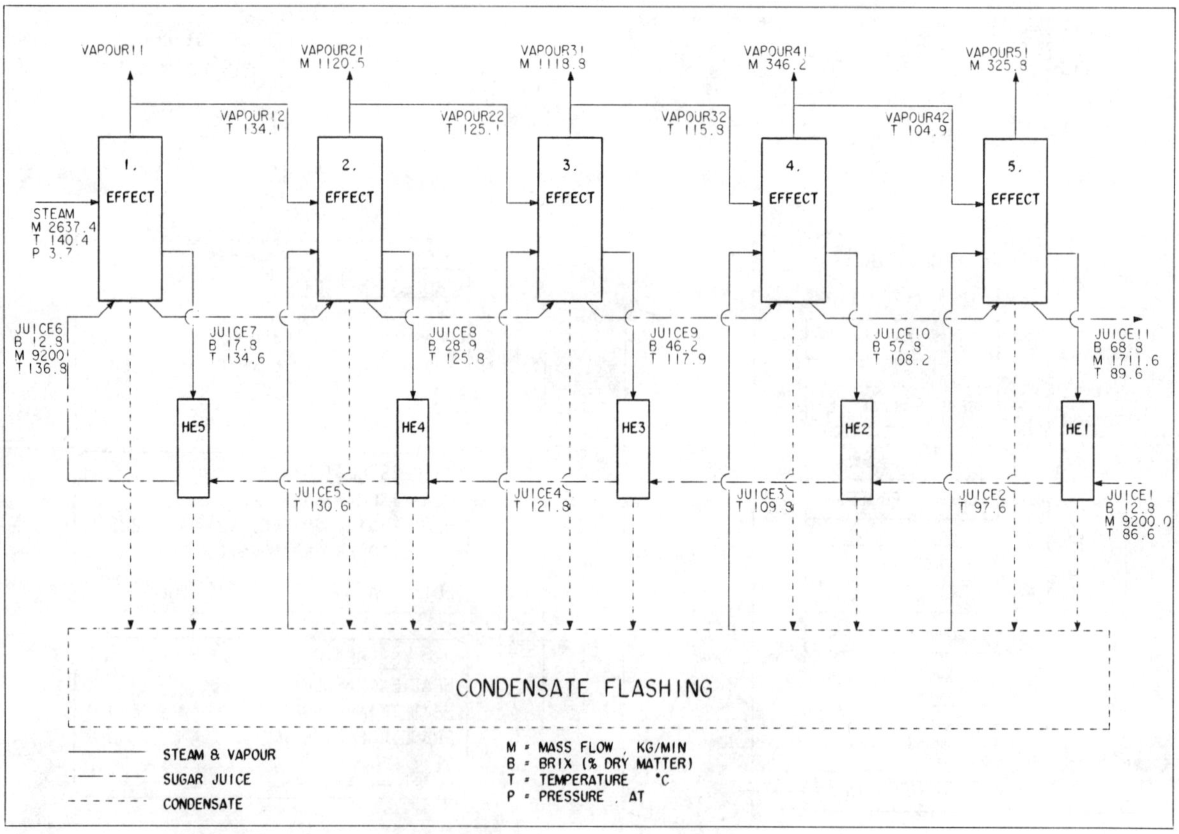

Fig. 7. A DOGS drawing of a five-effect evaporator including results from the calculation programme.

PRACTICAL CAD TOOLS FOR MATHEMATICAL MODELLING

A. Tyssø* and L. Kolbeinsen**

*CAMO A/S, Trondheim, Norway
**Division of Metallurgy, The Norwegian Institute of Technology, University of Trondheim, N-7034 Trondheim, Norway

ABSTRACT

Mathematical modelling of dynamic processes is often considered an intricate and timeconsuming task. Programpackages for simulation, time series analysis and identification in combination with a modern data logging equipment allow the task to be handled in a simpler and more efficient way.

This paper gives a short description of the different tools, and several examples illustrate how they can be utilized in modelling and design of control systems for metallurgic processes.

Keywords

Metallurgical Industries, Data Handling, Computer-aided Design, Parameter Estimation, Modelling, Data Processing, Metallurgy.

Introduction

During the last couple of decades considerable advances have been made in the development of general, mathematical and computational techniques for process analysis, optimization and control. The application of these techniques to many chemical and petroleum engineering operations has met with considerable success, and these techniques are becoming standard tools in chemical and engineering practice. However, in the available textbook literature emphasis appear to be on theoretical development and ideal systems, and rather less attention is paid to the actual practical application of these methodologies to real, complex systems.

The major difference between chemical and metallurgical processing operations is that in general, the former are very much better understood, and the availability or ready development of suitable process models may be taken for granted in the majority of cases. In contrast we are only in the beginning to attain a quantitative understanding of many metallurgical operations. Therefore, in metallurgical systems, practical considerations may force us either to settle for less sophisticated process models or to exert a considerable amount of model-building effort before any work on the actual optimization or control of the process at hand may commence.

Mathematical models and dynamic simulation have long been considered indispensable for the understanding of process dynamics and for control system design. This is particularly true when it comes to plants of a realistic complexity with all process equipment and instrumentation systems included. In such a situation simulation is the only workable way for comparing alternative control design techniques and training of process operators.

The heart of every simulation system is the model of the process, and very often the designer does not have the tools available which make the modelling procedure practicable. One way to improve the situation is to combine interactive programs for simulation, identification and time series analysis with modern programmable datalogging equipment.

Mathematical Models

A mathematical model can be designed using the "physical approach" or the "black box approach". With the "physical approach" the engineer applies balance equations (energy, material, etc.) and fundamental physical laws to generate a set of differential and algebraic equations. Certain assumptions and simplifications are necessary to obtain a model in reasonable form with few unknown parameters.

This is not a straightforward task and a "cut and try" procedure must often be applied. This is particularly the case for metallurgical processes where partial differential equations often must be manipulated.

The modelling procedure is drastically simplified if an interactive simulation program is available. It is then easy to check different model structures, and see how the choice of assumptions and parameter values influences the model behaviour.

Although the derived model reacts reasonably in simulated experiments, it is usually necessary to verify the model by comparing model responses with responses from the process plant. This implies that process recordings must be available and in a form that allows straightforward comparison.

Alternatively to the "physical approach" one may apply the "black-box approach" where, by means of experiments on the real process followed by a system identification or a model adjustment, one derives a suitable model representation. The "black-box approach" thus requires a data-logger and computer programs for time series analysis and identification. A severe limitation though is that only linear models can be considered, and most identification programs cannot handle multivariable models.

Use of Interactive CAD-Program Packages

The use of interactive programs for simulation, identification and time series analysis has to a great extent reduced the number of problems involved in mathematical model building. By combining the intuition and experiences of the user and the data handling and computing capacity of the computer, the modelling procedure can be more efficiently taken care of.

CYPROS *) (Cybernetic program packages) is an interactive program system which is designed to handle a wide range of engineering problems, not only modelling but also analysis and design of control systems. The program system has been developed over a period of almost ten years, with emphasis on applications to industrial processes (1)(2). The programming language is standard Fortran 77. Although Fortran 77 is not a very effective language, Fortran 77 compilers are available on almost all computers. This makes exchange and transfer of programs attractive. One should be aware, however, of all the different Fortran 77 dialects that exist. This is particularly a problem when it comes to implementation on microcomputers because most compilers on this level apply a subset of the standard Fortran 77.

The CYPROS system is interactive and program control is obtained by use of numeric terminals. Output is rapidly examined by extensive use of video colour graphics. The interaction organization is made flexible to meet demands from the experienced user as well as the temporary user and the fresh beginner.

CYPROS consists of a number of special purpose packages for simulation, identification and control. In addition there are packages for matrix handling and time series analysis and manipulation. Figure 1 illustrates the modelling procedures and the capabilities of the CYPROS system.

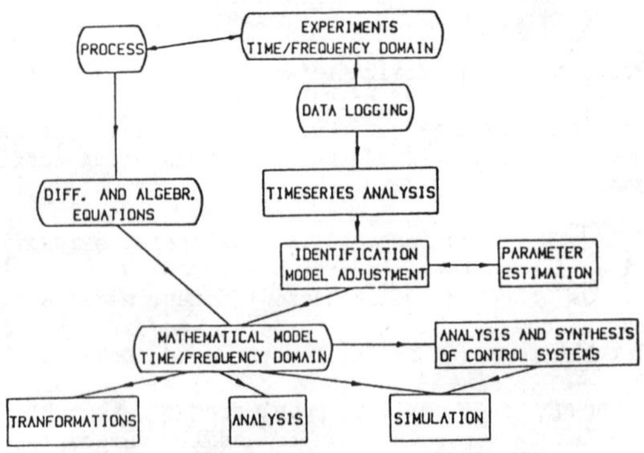

FIGURE 1

*)Developed at the Norwegian Institute of Technology div. of Engineering Cybernetics and SINTEF, the foundation of Scientific and Industrial Research, Dep. of Automatic Control at the University of Trondheim, Norway (see appendix). The system can be purchased from CAMO A/S, Trondheim, Norway.

In the simulation package different methods are available for solving differential and difference equations. The user specifies his/her models by means of Fortran subroutines, and the model parameters are stored in the data base and are easily changed. The simulation results are presented on a colour graphic display and hard copies are easily produced. The most commonly used integration methods are implemented, methods such as Euler and Kutta Merson. A method for solving stiff differential equations is also included. SIM may be used in off-line simulation as well as real-time simulation.

The identification package contains programs for identification and parameter estimation. Many powerful techniques are implemented and single input - single output and multivariable, linear and nonlinear models are considered. Methods such as the Maximum Likelihood method and the Extended Kalman filter method are available. By means of the time series analysis program the user can easily operate on data files (edit, scale, filter) and perform standard statistical analysis (compute correlation functions and power spectra).

There is also available a powerful programmable matrix calculator which can handle a great number of different matrix manipulations and operations.

The control package contains a wide variety of design and analysis methods. Adaptive controllers based on the generalized minimum variance algorithm and a pole placement technique are implemented.

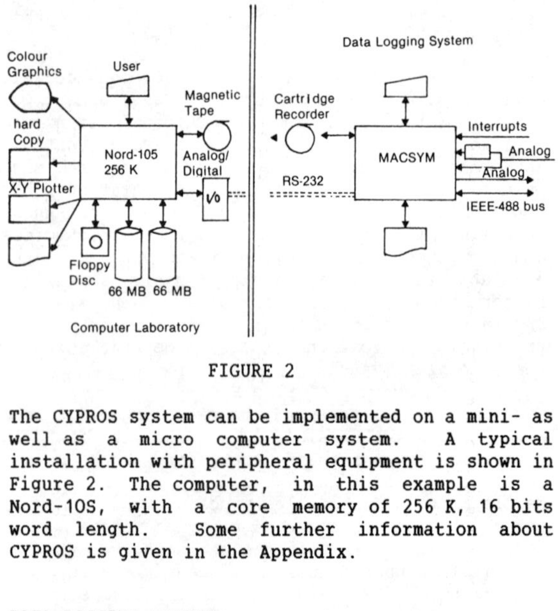

FIGURE 2

The CYPROS system can be implemented on a mini- as well as a micro computer system. A typical installation with peripheral equipment is shown in Figure 2. The computer, in this example is a Nord-10S, with a core memory of 256 K, 16 bits word length. Some further information about CYPROS is given in the Appendix.

DATA LOGGING SYSTEM

Plant recordings are essential in modelling, either to be used for model evaluation or as input data for identification programs. Data logging is known to involve many practical problems such as scaling, analog/digital converting, and noise filtering and theoretical problems such as choice of sampling rate and experiment length. New developments in computer technology and hardware have led to programmable data logging equipment with capabilities that radically simplify the logging procedure. Also the transfer of data from the logging system to the computer laboratory is simplified by the use of a digital cartridge system.

MACSYM (Measurement and Control System) is a series of measurement and control products from

Analog Devices. We have used the so-called Macsym 2 and Macsym 350, of which Macsym 2 is the oldest system. However, the analog and digital i/o-cards are interchangable between the two systems. Macsym 2, although being smaller (64 K RAM) and slower than the 350, is the more rugged (military specifications) of the two systems, and is also easier to transport.

A wide range of i/o-cards are available for the Macsyms and the capacity of any of our systems can be anything up to 96 analog inputs, 24 analog outputs, 8 digital inputs for process interrupts with selectable priority, 16 isolated AC or DC inputs or outputs when signal level and type is hardware selectable. There is also an IEEE-488 bus interface and five serial i/o ports (20 mA current loop or RS-232C communication standard). All channels (ADIO, bus and serial) are easily accessed with BASIC-statements.

Data storage is either on cartridge recorder (Tandberg, TDL-3000) with capacity of 960 K bytes on one cartridge. The 350 is in addition equipped with dual disk drives with 320 K bytes capacity on each. An optional 10M byte Winchester disk is also a possibility.

The Macsym 350 contains two units: Macsym 150 workstation and Macsym 200 intelligent front-end. The 150 has a 8086 5 MHz 16-bit CPU with a 8087 math co-processor in tandem, 256 K bytes RAM (up to 1 M bytes optional in steps of 128 K).

A special communication port (307.2 K baud, synchronous, full duplex RS-422) is reserved for use as the data link to the Macsym 200 or a local area network (LAN) that may service up to 11 front-ends.

The Macsym 200 front-end consists of a single control board, a sixteen slot card cage with back-plane for the ADIO cards and a power supply. Processing power is derived from a 5MHz 16-bit 8088 microprocessor with 64 K bytes of RAM. The analog functions include a precision system reference for calibration purposes, a sixteen channel multiplexer, a programmable gain amplifier, a sample and hold amplifier and an analog to digital converter.

The A/D-converter provides software selectable modes of 12 or 16 bit resolution, integrating or non integrating and differential or oneended input channels. Calibration may be performed periodically throughout system operation in order to maintain a high level of accuracy over time and changing environmental conditions. Standard programming language is Macbasic 3 which is an expanded version of Basic, that has been optimized for real time measurement and control. Special keywords designate the most common measurement and control functions graphics commands. Macbasic 3 uses a line by line compiler which translates program statements directly into machine code as they are entered. By using this approach the ease of interpretive languages is combined with the execution speed of compiled languages.

An important requirement in measurement and control applications is the ability to perform several operations or "tasks" independently of each other. Macbasic 3 is structured as a multitasking language and contains the necessary keywords for managing up to 18 tasks at a given time.

Macbasic 3 uses the MP/M-86 operating system which is a real time, multitasking superset of CP/M-86. CP/M-86 compatible programs will run under MP/M-86 with little or no modification.

The third data logging system is based on the NORD 10S minicomputer with a process interface based on the CAMAC-standard. This is a Norwegian computer with its own operating system, SINTRAN III, which provides full real-time support. Programming language used here is RT-Fortran with i/o routines programmed in assembly code. These routines are a part of the CYPROS system.

As stand alone units, using the simplest form of data logging programs all these systems can handle quite high sampling rates. Maximum is given by the Macsym 350 with 33 K samples/sec for logging one channel using the fastest type of input card and no external storing. For our purposes a sampling time shorter than 5 seconds is seldom necessary.

A typical data logging program package may include the following tasks: Read process signals, scale, linearize, compute derived variables, store variables on the recording unit, display the recorded data, check process responses against alarm conditions and so on. These tasks are easily implemented by combining the BASIC interpreter and the real-time operating system.

The analog and digital output process interface can be utilized for control and identification purposes. Process control algorithms, i.e. PID or self-tuning controllers, may be programmed and evaluated without the need for a large process control computer.

The recorded data are transferred to the CYPROS system as indicated in Figure 2 and stored on a special time series file. All the CYPROS programs may now utilize the recorded data. First comes data editing and then system identification and parameter estimation. The recorded data may also be used for a model verification by comparing simulated and recorded responses.

APPLICATIONS

Three examples of applications illustrating the combined use of the data logging systems and the CYPROS program system are summarized in Table I.

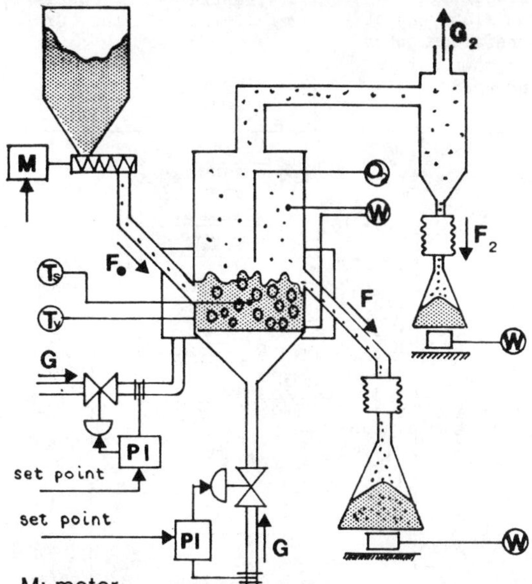

FIGURE 3

Table I Use of data logging - and the CYPROS-system in analysing metallurgical processes.

	Roasting of pyrites in fluidized bed ref (3)	Basic oxygen steelmaking ref (4)	Reduction of iron ore in shaft furnace ref (5).
Process/model-characteristics	Distributed residence time - reduced to lumped parameter model. The process is continous and unstable due to strong exotermic gas/solid reaction	Lumped model with inter-action between foamslag and molten metal. Batch process - gas plus two liquids	Spatial distribution of parameters for shaft. Lumped parameter model used to describe reduction of single pellet.
Data logging and/or control system	NORD 10S with Camac process-interface	Macsym 2	NORD 10S - CAMAC Macsym 2 Macsym 350
Process scale	Bench (100 kg/day)	Full (70 tons/batch)	Pilot (1500 kg/day) batch-lab (0.1 kg/batch)
Model use	- Verification of model reduction (lumping) - Model identification (Maximum likelihood method) - Process simulation - Testing of control strategies - State predictor (Kalman filter) integrated in multivariable control system for the bench scale reactor	- Dynamic process-analysis using the extended Kalman filter program (EXKALM) - On-line state estimation to provide continous "measurements of carbon content, bath temperature and decarburization rate	- Combining experimental data from pilot and lab.-scale to give a total process description. - Testing different con-trol strategies for the pilot plant - Process simulation - Analysing pilot-plant output-data.
Model size	~ 50 state variables reduced to 5	~ 10 state variables	~ 225 state variables
Project size (developement of general computer programs not included	~ 6 man-years	~ 2 man-years	~ 6 man-years

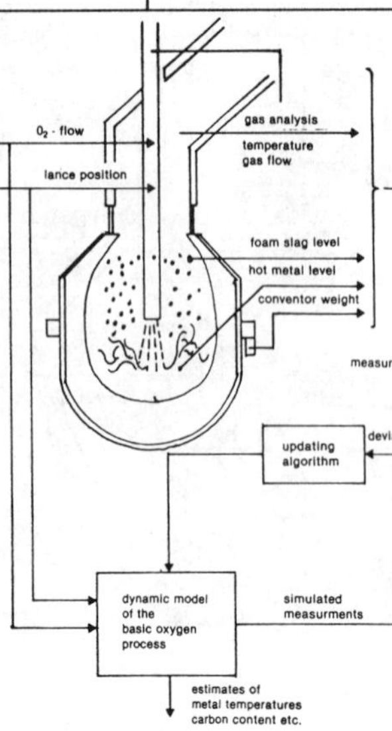

FIGURE 4

As it can be seen from the table all of the chosen examples are taken from the metallurgic industry. The purpose of these examples is to show that the proposed modelling approach is capable of handling a wide variety of process types ranging from laboratory scale to full scale porcesses. Both batch and continous processes and concentrated and distributed parameter models are considered.

An overview of the processes is given in the Figures 3 to 5 and shall not be commented upon further. For further information see references (3) (4) and (5). We would, however, stress two factors: Dynamic models are used both to analyse input/output correlations for a single process, and to combine the results of more than one piece of process equipment into one total model.

In metallurgic industry mathematical models can be utilized in different ways and of particular interest in this context is to use dynamic models on-line for supervisory control. Such uses for dynamic models have been proposed before, but only in recent years has there been enough computing power available to make brute force methods practical. In such a system the computer, through on-line simulation of the process, supplies the plant operation and control system with the necessary process information not available by direct measurement.

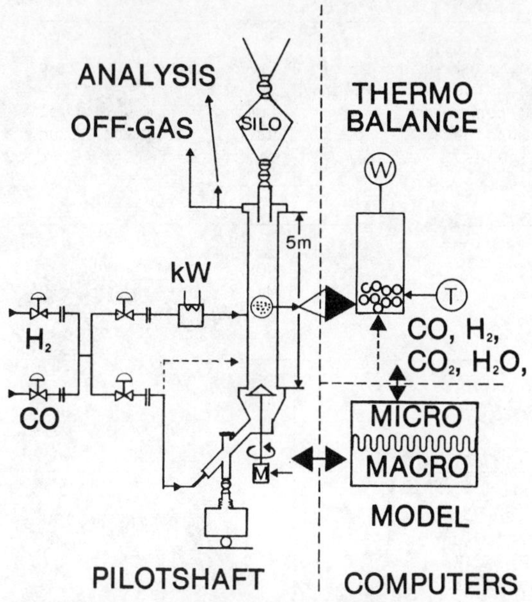

FIGURE 5

In the case of the basic oxygen furnace as well as the sponge iron process, the complexity of the interior of the furnace defy description. The simulation could be run at a time scale exactly matching the process itself or faster in parallel with the actual plant to answer the question "what if". This also would permit a "look ahead" simulation of the system.

The available identification algorithms from the CYPROS system could compare real operating data with corresponding outputs from the on-line model and continually adjust the operating parameters of the model to keep the simulation tracking the actual operation of the process.

Conclusions

Tools for mathematical modelling are discussed. It is shown how a programmable data logging system can be combined with software program packages to reduce the amount of work involved in the modelling activity.

The applications referred to clearly indicate that it is a must to have available efficient CAD-systems and a flexible data logging system in order to solve modelling problems in complex, industrial processes. Through modelling we achieve a better understanding of process behaviour and thereby can design better control strategies and improve process design and operation. The gain in increased productivity is expected to be much higher than the cost of programpackages and data logging equipment.

References

(1) Tyssø A.: CYPROS - Cybernetic program packages
MIC, Vol.1 No.4 October 1980

(2) Tyssø A., Balchen, I: Application of CAD in Modelling Identification and Control of Industrial and Large Scale, Nontechnical Processes.
Paper presented at the 20th IEEE Conference on Decision and Control, San Diego, USA, 1981.

(3) Sælid, S, Jenssen, N.A., Lindstad, T. and Kolbeinsen, L.:
Modelling, Identification and Control of a Fluidized Bed Reactor, Preprints of 5th IFAC Symposium on Identification and System Parameter Estimation, Vol.2,pp 907-916, Pergamon Press, 1979.

(4) Stuan, V,: Control of the L-D Converter.
Report STF48 F81039, 1981

(5) Kolbeinsen, L,: Reduction of Iron Ore, Mathematical Model for Process Control. The Norwegian Institute of Technology 1982 (in Norwegian)

APPENDIX

CYPROS, Additional Software Characteristics

Interactive capabilities: Command dialog as well as question-answer dialog with the HELP facility.

Computers and terminals on which available: Computers: Nord-10, Nord-100, VAX, IBM PC/XT (partial). Terminals: Marconi DTI/Genius, Tandberg TDV 2215, Digital VT100, INFOTON 200, Lear Sigler ADM 3A, Data Terminal Incorporation.

Documentation: User's guides and commented source code.

Memory and disc requirements: Memory: 64 k, 16 bits exclusive operation system. Disc: 2000 pages, 1024 (16 bits).

Availability of code: Can be purchased as a total system or as packages. Typical program price, minicomputers US$ 3000, microcomputers US$ 400.

Contact for details: CAMO A/S, P.O. Box 2893 Elgeseter, N-7001 TRONDHEIM, Norway.

Interactive graphics: Software exists for displaying on Tektronix, Versatec matrix plotter, HP-plotters and alphanumeric terminals of different types (see above).

COMPUTER AIDED DESIGN OF SHIP CONTROL SYSTEMS

J. van Amerongen and P. P. J. van den Bosch

Control Group, Electrical Engineering Department, Delft University of Technology, The Netherlands

Abstract. At the Control Laboratory of the Electrical Engineering Department of Delft University of Technology much research has been carried out during the last fifteen years in the field of ship control systems, with the main emphasis on automatic steering of ships. This work has resulted in the design of an adaptive autopilot (Van Amerongen, 1984). At the moment research is concentrated on the design of an autopilot which uses the rudder not only for course keeping but for stabilization of the roll angle as well. This paper demonstrates how modern computer-aided design methods may help in designing such a system, especially when analytical methods fail because of non-linearities of the process.

Keywords. Computer-aided design, Computer-aided system design, Control engineering computer applications, Ships, Steering machine.

1. INTRODUCTION

When a controller has to be designed for a process it is generally assumed that the process is more or less linear. In that case many design methods are available and a systematical design is possible. This approach is only valid when the non-linearities in the process can indeed be linearized, for instance when small variations around a certain operating condition are considered. The validity of such assumptions has to be demonstrated by means of simulation or by experiments with the real system. In this paper it will be shown how simulation, combined with experimental optimization, can be used to design a controller for a process which has essential, non-linear properties. This design method will be illustrated with a simple example in the field of automatic steering of ships. In Section 2 the mathematical models which describe the ship's steering process will be given. Section 3 describes the design of a controller based on an LQG-approach. It will also be illustrated where this controller design fails. Section 4 summarizes the properties of the interactive simulation program PSI. This program is used in Section 5 to further investigate the controller design for a process with a non-linear behavior.

2. MATHEMATICAL MODELS

2.1 The ship's dynamics

The dynamics of a ship may be described by various models. For the purpose of designing a controller for course keeping, a simple first-order model (Nomoto, 1957) is appropriate to describe the transfer between the rudder angle and the rate of turn:

$$\tau\ddot{\psi} + \dot{\psi} = K\delta \qquad (2.1)$$

where

δ = the rudder angle
$\dot{\psi}$ = $d\psi/dt$ = the rate of turn
and ψ = the heading

Consequently, the transfer between the rudder angle and the heading is of the second order. This model is given in figure 1.

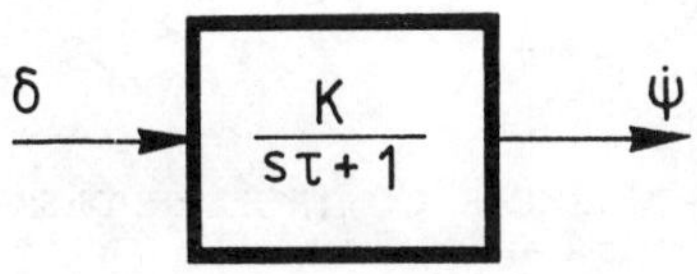

Figure 1 First-order Nomoto model

2.2 Basic controller algorithm

A state-feedback controller for this system can be made by using the control algorithm:

$$\delta = K_p\varepsilon + K_d\dot{\psi} \qquad (2.1)$$

where
ε = the heading error

2.3 The steering machine

The rudder angle computed by the controller is not directly available as an actual rudder angle. The latter is realized by a hydraulic steering machine. As long as the rudder motions are small and the frequency is low the steering-machine dynamics may be disregarded. In that case the controller (2.2) may be computed, based on the assumption that the system is linear. However, for large and high-frequency rudder motions it is essential that the steering-machine dynamics be taken into account. The most simple model of a steering machine for this situation is given in figure 2. The rudder angle is limited to a maximum value of 35 degrees, while the rudder speed is limited to a value of about 3 degrees per second. The maximum rudder angle may be further limited by adding a limiter to the controller, but this will not be considered in this paper. The maximum rudder speed of 3 degrees per second is about the minimum speed required by the classification (insurance) companies. It will be shown below that the limited rudder speed may have a deteriorating effect on the system's performance.

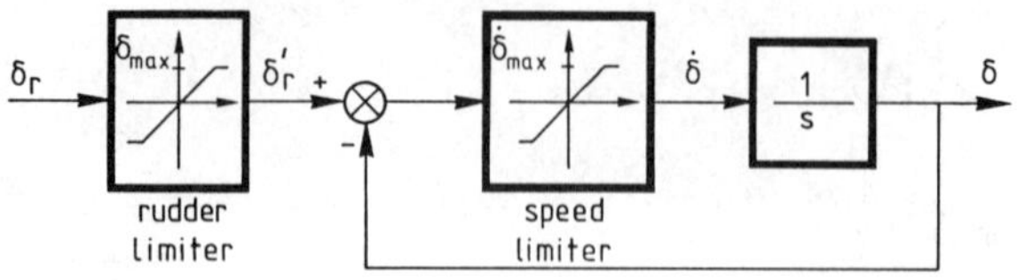

Figure 2. The steering machine

2.4 Disturbances

At sea a ship is influenced by various disturbances, the wind and waves being the most important ones with respect to control of the heading. The wind can be modeled by a low-frequency, almost constant disturbance which adds to the rudder angle. The waves are more high frequency and are of a stochastic character. In simplified form they can be modeled as white noise colored by a second-order filter with low relative damping ($z = 0.1$) and $\omega_n = 0.7$. Of course both values depend upon the sea state and upon the ship's speed and the angle of incidence of the waves. The output of this filter can be added to the rudder signal. The transfer function of the coloring filter is given by:

$$H_{waves} = \frac{\omega_n^2 \, s}{s^2 + 2z\omega_n s + \omega_n^2} \qquad (2.3)$$

3. LINEAR CONTROLLER DESIGN

As long as the steering-machine dynamics are disregarded the design of a controller for the process described by figure 1 is straightforward. In the literature many discussions can be found on the criterion which should be used in order to minimize the speed losses due to imperfect steering (Koyama, 1967, Norrbin, 1972, Van Amerongen, 1982). From a control-engineering point of view the following criterion is a simple one and it has, under certain conditions, a direct relation to the speed losses:

$$J = \int (\varepsilon^2 + \lambda\delta^2)\, dt \qquad (3.1)$$

The optimal controller for the process of figure 1, which minimizes criterion (3.1), can be found by applying a standard LQG approach. This yields the expressions for the proportional gain K_p and the rate-feedback gain K_d.

$$K_p = 1/\sqrt{\lambda} \qquad (3.2)$$

$$K_d = \frac{1}{K} \{\sqrt{1 + \frac{2K}{\sqrt{\lambda}}} - 1\} \qquad (3.3)$$

In literature there is a lot of confusion about the value of the weighting factor λ , which indeed minimizes the loss of speed. This is also due to phenomena which will be demonstrated later in this paper. At the moment a value of λ of 0.1 will be chosen. This yields with:

$$K = 0.1$$
$$\tau = 10$$

optimal controller gains:

$$K_p = 3.16$$
$$K_d = 16.5$$

The behavior of the control system for an initial heading error of 2 degrees is shown in figure 3.

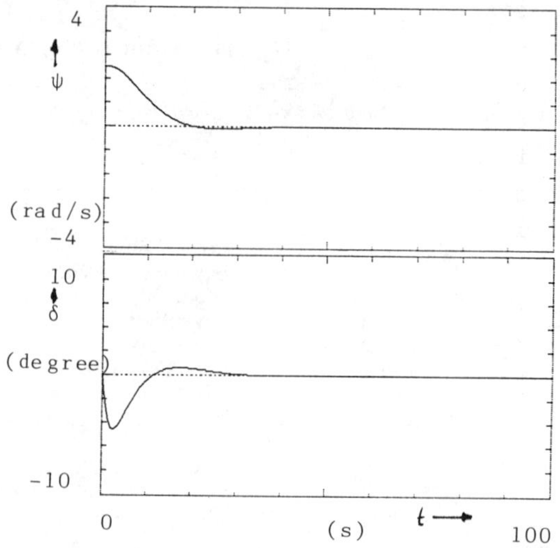

Figure 3. System performance for an initial heading error of 2 degrees.

However, when not only the initial heading error but also a stochastic disturbance is added to the system, it can be noticed in figure 4 that the rudder now almost constantly moves with its maximum speed. In this case it is no longer allowable to disregard the limited rudder speed.

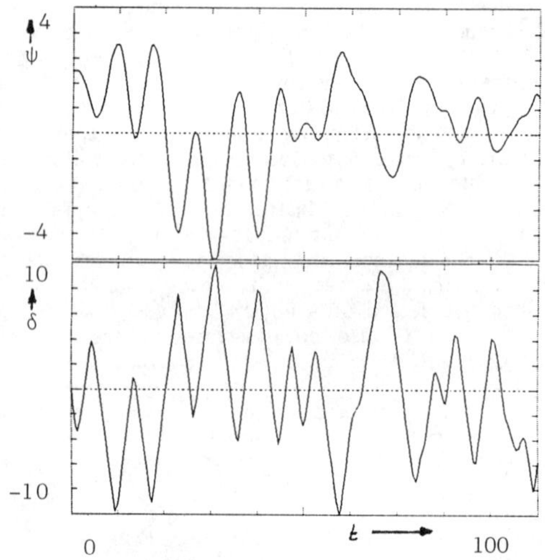

Figure 4. System performance with an additional stochastic disturbance.

In Van den Bosch (1983) a comparison has been made between a mathematically-oriented design method (linear output feedback) and simulation and optimization for designing controllers for linear systems. It turns out that the flexibility in selecting a criterion or a controller structure is much larger with the simulation and optimization approach than with linear output feedback. This advantage has to be paid for with an increase in calculation time ranging from about 25 times for low-order systems to about 5 times for high-order systems. In using advanced optimization techniques during the simulation these numbers can be halved. Consequently, even in designing linear controllers for linear systems, the simulation and optimization approach is a powerful and flexible design tool.

Dealing with nonlinear systems or when designing discrete controllers for continuous systems, mathemathically-oriented design methods cannot be used. Then, simulation and optimization still continue to be powerful and flexible design tools.

In the next section the program PSI will be introduced. The optimization facility of this program enables the design of a suitable controller for linear as well as non-linear systems, so that it is well suited to solving this problem.

4. PSI

The interactive block-oriented simulation program PSI (Van den Bosch 1981, 1982) can be used for simulation purposes, but due to its optimization possibilities it can also successfully be applied to the design of a controller for linear as well as for non-linear systems. The combination of a powerful simulation package with the optimization facility makes PSI especially attractive for control systems which cannot easily be designed with analytical methods. A survey of the facilities of PSI is listed below:

Facilities

- About 90 commands support the user in realizing his design objectives.

- Five numerical integration methods are available, namely four fixed-step methods (Euler, Adams Bashfort 2, Runge Kutta 2 and 4) and one variable-step-size method (Runge Kutta 4).

- Algebraic equations are solved by a fast Newton-Raphson algorithm. If this procedure fails, a more reliable, although slower, optimization algorithm is used.

- Optimization with scaling and constraints is supported. In PSI the user can define the output of an arbitrary block as the criterion and up to eight arbitrary parameters of the simulation model as parameters of the optimization. The parameters that offer the smallest value of the criterion will be accepted as the solution of the optimization procedure. Pattern search (Hooke and Jeeves (1961), equipped with a premature stop of the simulation run, has been selected as the non-linear optimization procedure, due to its robustness and lack of a line-minimization procedure.

Although Pattern Search adjusts its search step size according to the "shape" of the criterion, improvement of the speed of convergence can be obtained by using scaling. Scaling can make each parameter about equally important for the optimization algorithm. Not only is scaling supported by PSI, but constraints are also allowed. Each parameter may have an upper and a lower limit. The optimization algorithm will only search for an optimum in the feasible region of the parameter space.

- Multi-run facilities are available. For example, run-control blocks, comparison of signals between several runs, etc. With the aid of storage variables PSI offers the initial-dynamic-terminal facility of CSMP III. At the end of a simulation run, this run can be continued without resetting the time and the initial conditions.

- Extensive tests on all user-supplied information is implemented. Each error is indicated by a meaningful error message, of which there are about 60.

- About 50 powerful block types are available, among which integrators (limited, mode-controlled, converts), continuous and discrete PI- and PD-controllers, Pulse-Width Modulation, etc. Fortran programming in a non-interactive mode is required to define new block types. The user only needs to write a subroutine in which the output is defined as a function of the input(s) and parameter(s), compile it and after a link step, his block is available.

- There are memories to store signals during a simulation run. These signals can be studied after the simulation run, can be saved on disk or can be used as inputs for future runs. These signals can be redrawn on the screen, as responses or as phase trajectories, after which a cursor, controlled by keyboard commands, can "walk along" these responses. The numerical values appear directly on the screen, so that overshoot, rise time or accuracy can be determined both quantitatively and qualitatively.

- Symbolic block names can be used. Instead of numbers each block or variable can be assigned a user-selected name of up to eight characters. So blocks can get meaningful names like HEADING, RUDDER or INPUT instead of abstract numbers like block 13, 91 or 512, etc.

This section has described a number of facilities which make programs, such as PSI, highly suited to the analysis and design of control systems. PSI is able to solve (non-linear) differential, difference, algebraic and logical, Boolean equations or any mixture of them. Moreover, an attractive and powerful interaction is realized between the user and the program.

5. CONTROLLER DESIGN WITH PSI

With the aid of the program PSI the controller found earlier can be redesigned for the situation with the non-linear steering machine. The system is simulated in PSI. The controller structure and the criterion may be left unchanged. With the optimization facility of PSI the controller gains K_p and K_d are computed for a situation similar to figure 4. The following values are found:

$$K_p = 1.2$$
$$K_d = 1.3$$

The behavior of the system is shown in figure 5.

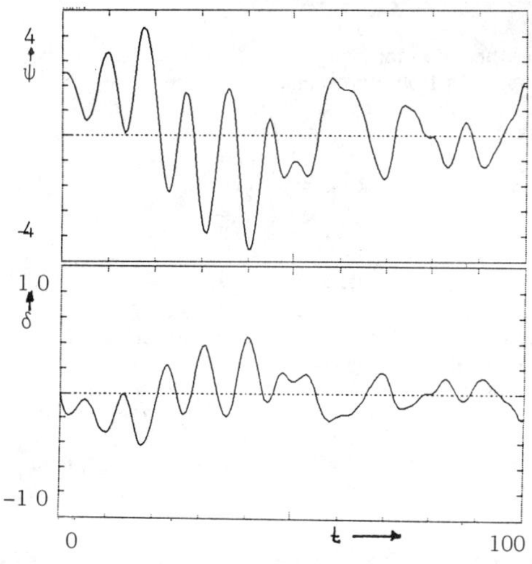

Figure 5. System performance after re-optimization.

These values are much lower than those found before. The difference in the controller behavior can clearly be seen in the performance of the rudder signal, which is much smoother with the redesigned controller. The course-keeping behavior is hardly affected.

Because of the non-linear process which is considered here it is not possible to find a controller which gives an optimal performance for all situations. However, playing with the simulation program and carrying out the optimization for various types of disturbances and initial heading errors gives an indication about the parameters which may be used for a sub-optimal solution in a wide range of situations. The results may also be used for designing an adaptive controller.

6. CONCLUSIONS AND DISCUSSION OF THE RESULTS

In this paper it has been demonstrated how a controller for a non-linear system can be designed by using a simulation program with optimization facilities. The resulting controller differs considerably from the controller which is found when the system's non-linearities are disregarded. A similar result can be obtained when the weighting factor λ in criterion 3.1 is increased. This is a well-known solution for preventing a system from saturating. However, in that case the parameters which are found differ from the parameters found with PSI. This implies that by increasing the value of λ , a sub-optimal solution is found. On the other hand, it appears that the experimental optimization applied in PSI finds the (analytical) optimal solution in the linear case as well.

Of course there are other possibilities for diminishing the influence of stochastic disturbances on the rudder signal (Van Amerongen, 1984). The solution presented in this paper is not intended for designing an optimal autopilot, but for demonstrating the application of computer-aided design facilities for designing appropriate controllers for ships. A similar approach can, for instance, be applied to determine optimum Kalman-filter gains for situations where a straightforward Kalman-filter design would be difficult.

A disadvantage of the proposed method is that it only gives an optimal solution for the investigated type of signals. For other disturbances the optimization should be done again.

The method is not very fast, because each step of the optimization procedure requires a complete simulation run. If there is more than one minimum for the cost function it cannot be guaranteed that the optimum solution will be found. However, this can easily be circumvented by choosing a few different starting points.

REFERENCES

Amerongen, J. van (1982). Adaptive steering of ships - a model-reference approach to improved manoeuvring and economical course keeping, Ph.D. Thesis Control Laboratory, Delft University of Technology, p. 195

Amerongen, J. van (1984). Adaptive steering of ships - a model reference approach, Automatica, vol. 20, no 1, (3-14)

Bosch, P.P.J. van den (1981). PSI - A Software Tool for Control System Design, Journal A, vol. 22, no 2 (55-62)

Bosch, P.P.J. van den (1982). PSI - Interactive Simulation Program. IEEE Control Systems Magazine, vol. 2, no 4, December 1982 (42)

Bosch, P.P.J. van den (1983). Interactive System Analysis and System Design Using Simulation and Optimization. Proceedings IFAC Symposium on Computer Aided Design of Multivariable Technological Systems 1982, Lafayette, Pergamon Press London 1983 (225-232)

Hooke, R. and T.A. Jeeves (1961). Direct search solutions of numerical and statistical problems, JACM, vol. 8 (212-229)

Koyama, T. (1967). On the optimum automatic steering system of ships at sea, J.S.N.A., vol. 122, Dec.

Nomoto, K., T. Taguchi, K. Honda and S. Hirano (1957). On the steering qualities of ships, Int. Shipbuilding Progress, vol. 4

Norrbin, N.H. (1972). On the added resistance due to steering on a straight course, 13th ITTC, Berlin, Hamburg

COMPUTER AIDED DESIGN OF ROBUST MULTIVARIABLE STRUCTURE CONTROL SYSTEMS

C. M. Dorling and A. S. I. Zinober

Department of Applied and Computational Mathematics, University of Sheffield, Sheffield S10 2TN, UK

Abstract. Variable structure control systems employ discontinuous non-linear control functions linked with a manifold in the state space in order to obtain desirable robustness and invariance properties. A CAD package to aid the design of such multivariable control systems has been developed by the authors. The theory and scope of a robust eigenvalue assignment approach is described with reference to a multivariable CAD package and an example of its use is given.

Keywords. Variable structure systems; robust pole-placement; multivariable control systems; non-linear control systems; sliding states.

INTRODUCTION

The essential feature of a variable structure control system (VSCS) is that the (generally non-linear) feedback control has a discontinuity on one or more manifolds in the state space. The structure of the feedback system is altered or (switched) as its state crosses each discontinuity surface. Thus, VSCS may be regarded as a combination of sub-systems, each with a fixed structure and each operating in a specified region of the state space. The central feature is the sliding mode which occurs when the system state repeatedly crosses and immediately re-crosses a switching manifold, because all motion in the neighbourhood of the manifold is directed inwards towards the manifold. In the sliding mode the system is effectively constrained to lie within a certain subspace of the full state space, and the system is thus formally equivalent to a system of lower order, termed the equivalent system. The dynamic behaviour of this equivalent system is different from that of each of the constituent sub-systems.

The transient motion of a VSCS therefore consists of two independent stages: a (preferably rapid) mode bringing the state of the system to the attractive manifold in which sliding occurs; and a slower sliding mode towards the state space origin in the sliding subspace. This two stage behaviour can help to resolve the conflict between the opposing requirements of static and dynamic accuracy which are encountered when designing a linear control system, because a VSCS may be designed to give: a rapid response with no loss of stability; asympotic state regulation; insensitivity to parameter variations and invariance to certain external disturbances.

The theory of VSCS has been developed over the past 25 years. Until recently this development was carried out almost exclusively in the USSR; consequently, the majority of the literature is in Russian, with relatively few papers appearing in the Western literature. Of the few English language monographs in the area, those by Itkis (1976) and Utkin (1978) provide a reasonable survey of the theory up to that time. A survey paper by Utkin (1977) references many of the early contributions available in translation.

As will be shown in the next section, the design of the sliding mode in a nominally linear control system may be completed independently of the form of the actual control functions selected. After defining the nominally linear time-varying and uncertain system to be controlled, some of the options available in the design of a stable sliding mode will be considered. A canonical form is employed for the design of the hyperplane matrix and the control structure. The two main approaches for designing the sliding mode are mentioned and in this paper attention is focussed on the robust eigenvalue assignment problem. The design of the controller which ensures that the two stages of the VSCS operate correctly, is then outlined. The control chatter may be eliminated without serious deterioration of the dynamic closed-loop response.

SYSTEM DEFINITION

The system is expressed in its most general form as

$$\dot{x}(t) = [A+\Delta A(t)]x(t)+[B+\Delta B(t)]u(t) \qquad (1)$$

where x is the state n-vector and u is the control m-vector. It is assumed that $n>m$, B is of full rank m and that the pair (A,B) is completely controllable. The matrix ΔA represents the variations and uncertainties in the plant parameters, while ΔB is the plant/control interface uncertainty. The overall aim of a variable structure regulator control design is to regulate the system state from an arbitrary initial condition $x(0) = x^o$ to the state space origin asymptotically as $t\to\infty$. The components u_j $(j=1,2,\ldots,m)$ of the control vector u are state dependent (feedback) functions, $u_j = u_j(x)$. The switching surfaces are intersecting hyperplanes M_j passing through the state space origin and defined by $M_j = \{x:c_j x = 0\}$ $(j = 1,2,\ldots,m)$ where c_j is a row n-vector. The (ideal) sliding mode occurs when the state lies simultaneously in each of the hyperplanes M_j for $j = 1,\ldots,m$; it is attained when the state reaches and remains in the manifold M which is the intersection of the m hyperplanes: $M = Cx = 0$. In geometrical terms the subspace M is the null space (or kernel) of C, denoted N(C). CB is assumed to be non-singular in this paper to ensure a unique controller. (The singular case has been discussed by Utkin (1978)).

Some workers have replaced the individually discontinuous control functions with a unit vector controller (Balestrino, De Maria and Sciaviccio, 1982; Ryan and Corless, 1984). This is an m-vector state feedback function of the form $s(x) = Cx/||Cx||$, which has a discontinuity on M but is continuous on each of the individual hyperplanes M_j $(j = 1,2,...,m)$. However, once the sliding mode is reached, the behaviour of the system under this form of control is identical to the sliding motion in the case of the individually discontinuous control functions.

In the case of an ideal system ($\Delta A = 0$, $\Delta B = 0$) the sliding mode may be determined from the defining condition $Cx(t) = 0$ for $t \geq t_s$ where t_s is the time when the sliding mode is reached. Differentiating with respect to time and inserting (1) gives

$$C\dot{x}(t) = CAx(t) + CBu(t) = 0 \quad (t > t_s). \qquad (2)$$

An equivalent control u_{eq} may be determined in the linear feedback form $u_{eq}(t) = -Kx(t)$ where the m×n feedback matrix K is given by $K = (CB)^{-1}CA$. The system dynamics are then described by the system equation

$$\dot{x}(t) = A_{eq}x(t) = [I_n - B(CB)^{-1}C]Ax(t) \quad (t > t_s). \qquad (3)$$

Note that this motion is independent of the actual control u during the sliding mode and depends only on the choice of C. In other words, the function of the control u in a VSCS is to drive the state into sliding subspace M, and thereafter to maintain it within M. The convergence of the state vector to the origin is ensured by suitable choice of the feedback matrix K. The determination of the matrix K (or, more precisely, the determination of the matrix C defining the subspace M) may thus be completed with no prior knowledge of the form of the control vector u (although the reverse is not true). The null space of C, N(C), and the range space of B, R(B), are complementary subspaces; i.e. $N(C) \cap R(B) = \{0\}$. Since motion lies entirely within N(C) during the sliding mode, the behaviour of the ideal system during sliding is unaffected by the controls since they act only within R(B).

SYSTEM TRANSFORMATION

For simplicity of presentation we initially restrict attention to the ideal (or nominal) system

$$\dot{x}(t) = Ax(t) + Bu(t). \qquad (4)$$

Our first task is to specify a particular canonical form for the system in order to simplify the development of the design scheme. This form, which is closely related to the controllability canonical form for a multivariable linear system (Kwakernaak and Sivan, 1972), is similar to that used by Utkin and Yang (1978) in an earlier paper on hyperplane design. By assumption the matrix B has full rank m, so that there exists an orthogonal n×n transformation matrix T such that

$$TB = \begin{bmatrix} 0 \\ B_2 \end{bmatrix} \qquad (5)$$

where B_2 is m×m and non-singular. The orthogonality restriction is imposed on T for reasons of numerical stability and to remove the problem of inverting T when transforming back to the original system in the CAD package.

The transformed state is $y = Tx$ and the state equation (4) becomes

$$\dot{y}(t) = TAT^Ty(t) + TBu(t) \qquad (6)$$

and the sliding condition is $CT^Ty(t) = 0$. If the transformed state y is now partitioned as

$$y^T = [y_1^T \; y_2^T]; \quad y_1 \in \mathbb{R}^{n-m}, \; y_2 \in \mathbb{R}^m \qquad (7)$$

and the matrices TAT^T, TB and CT^T are partitioned accordingly, then

$$\dot{y}_1(t) = A_{11}y_1(t) + A_{12}y_2(t)$$
$$\dot{y}_2(t) = A_{21}y_1(t) + A_{22}y_z(t) + B_2u(t) \qquad (8)$$

and

$$C_1y_1(t) + C_2y_2(t) = 0 \qquad (9)$$

where

$$TAT^T = \begin{pmatrix} A_{11} & A_{12} \\ A_{21} & A_{22} \end{pmatrix}; \quad CT^T = [C_1 \; C_2] \qquad (10)$$

and C_2 is non-singular from CB non-singular. The more general system (1) can also be similarly expressed, if $T(\Delta A)T^T$ and $T(\Delta B)$ are also partitioned compatibly with y_1 and y_2.

It will be assumed throughout the remainder of this paper that the uncertainties in the plant control interface occur only on the input channels, i.e. rank(B) = rank(B ΔB). Under this assumption, $\Delta B_1(t) = 0$, and y_1 again becomes independent of the control u.

The canonical form is central to the hyperplane design methods to be described. It also plays a significant role in the solution of the reachability problem, i.e. the determination of the control form ensuring the reaching of the sliding mode. Equation (9) defining the sliding mode is equivalent to

$$y_2(t) = -Fy_1(t) \qquad (11)$$

where the m×(n-m) matrix F is defined by

$$F = C_2^{-1}C_1 \qquad (12)$$

so that the evolution of y_2 in the sliding mode is related linearly to that of y_1. The ideal sliding mode is therefore governed by the equations

$$\dot{y}_1 = A_{11}y_1(t) + A_{12}y_2(t);$$

$$y_2(t) = -Fy_1(t): \quad (13)$$

an (n-m)th order system in which y_2 plays the role of a state feedback control. Closing the loop gives

$$\dot{y}_1(t) = (A_{11} - A_{12}F)y_1(t); \quad (14)$$

so that the design of a stable sliding mode (ie, so that $y \to 0$ as $t \to \infty$) requires the determination of the gain matrix F such that $A_{11}-A_{12}F$ has n-m left-hand half-plane eigenvalues. This may be achieved by using a modified form of any standard design method giving a linear feedback controller for a linear dynamical system. The main methods are those based on the minimisation of a quadratic cost functional and eigenvalue assignment. However, before discussing these methods in more detail, we note that whichever scheme is chosen for the design, fixing F does not uniquely determine C, since m^2 degrees of freedom (d.o.f.) remain in the relationship $C_2F = C_1$. The simplest method of determining C from F is that employed by Utkin and Yang (1978): namely, letting $C_2 = I_m$. This has the merit of minimising the amount of calculation in proceeding from F to C, and hence reduces the possibility of numerical errors.

HYPERPLANE DESIGN BY DIRECT EIGENVALUE ASSIGNMENT

The majority of VSCS designs appearing in the literature to date have suggested eigenvalue assignment methods in the design of the sliding mode. Since these problems have generally been scalar-controlled examples with a single switching hyperplane, specification of the n-1 eigenvalues to be associated with the sliding mode completely determines the feedback matrix F (in this case, an n-1 element row vector).

For the multiple input case, Utkin and Yang (1978) show that the pair (A_{11}, A_{12}) is controllable and that the eigenvalue assignment in (14) is therefore feasible. It is well known, however, that the assignment of eigenvalues of an nth order m-input system requires only n of the nm d.o.f. available in choosing the feedback gain matrix (Srinathkumar and Rhoten, 1975; Shah, Fisher and Seborg, 1975). The remaining n(m-1) d.o.f. may be utilised in some other way; and, in particular, in partially assigning the eigenvectors. Two complementary approaches to the specification of the eigenvectors have been studied. Shah and coworkers show that at most m elements of an eigenvector may be assigned arbitrarily; the remaining n-m elements are then fully determined by the assigned elements. Thus one approach to eigenvector assignment would be to pick m elements according to some scheme and accept the remaining elements as determined. This might allow a degree of adjustment to be carried out by inspection. The alternative method of eigenvector assignment is by consideration of the assignable subspace corresponding to a given eigenvalue (Moore, 1976; Klein and Moore, 1977). This assignable subspace for the eigenvalue λ_i may be characterised as the null space of the (n-m)×n matrix $H(\lambda_i)$ defined by

$$H(\lambda_i) = [A_{11}-\lambda_i I_{n-m} \quad A_{12}] \quad (15)$$

This capability of assigning the eigenvectors may also be used in two complementary ways: to shape the response of the closed loop system; or to maximise the robustness of the eigenvalue placement. Here we consider only the second approach. For simplicity we assume throughout that the non-zero sliding mode eigenvalues are distinct from each other and from the eigenvalues of A_{11}.

ROBUST EIGENVALUE ASSIGNMENT

The freedom to assign (at least partially) the eigenvectors of the equivalent system may be used to ensure that the eigenvalues associated with the sliding mode are maximally insensitive to perturbations in the system parameter matrix A. Including the parameter variation terms in the closed loop form (14) gives

$$\dot{y}_1 = (A_{11}-A_{12}F)y_1+(\Delta A_{11}-\Delta A_{12}F)y_1 \quad (16)$$

where the explicit time dependence of y_1, ΔA_{11} and ΔA_{12} has been dropped. The second term on the r.h.s. of (16) represents the perturbation of the nominal system (14) due to parameter variations. This term is, in general, time-varying and arbitrary; the robust assignment approach aims to minimise its effects.

If the parameter variations are matched, i.e. rank (B) = rank ([B ΔB]), then the robust control designs of Ryan (1983) and Ryan and Corless (1984) may be employed to ensure that the state reaches and remains within the sliding manifold N(C). This is equivalent to maintaining the m zero-valued eigenvalues of A_{eq} corresponding to the dynamics in R(B) at zero. However, if the parameter variations are unmatched, then the ideal sliding mode will be disturbed. Our aim is therefore to design the feedback matrix F such that the n-m assigned non-zero eigenvalues are maximally robust, thereby minimising the effects of the parameter variations in the N(C) on the sliding mode dynamics.

Now the sliding mode system as described by (16) represents an (n-m)th order system with m control inputs. The interface matrix A_{12} has dimension (n-m)xm and rank p which satisfies

$$1 \leq p \leq \min\{n-m, m\}. \quad (17)$$

Note that p = 0 would imply (A_{11}, A_{12}) uncontrollable, contradicting the controllability of the original system. If p<m, the sliding mode system has redundant inputs which may be removed by reordering the control inputs and setting part of F to zero.

We now describe the hyperplane design scheme for robust eigenvalue assignment under the assumption that p = m. Let $\lambda_i \in \sigma(A_{11}-A_{12}F)$, with corresponding right eigenvector $v_i \in \mathbb{R}^{n-m}$ and left eigenvector $r_i \in \mathbb{R}^{n-m}$. Then it is well-known (see Wilkinson, 1965) that the sensitivity of λ_i to perturbations in A_{11}, A_{12} and F depends approximately linearly on the quantity $1/c_i$, the secant of the angle between the vectors v_i and r_i, so that $(1/c_i) \geq 1$. In the case of an equal number of states and inputs for the sliding mode system (n-m = m), we have complete freedom in assigning the eigenvectors of $A_{11}-A_{12}F$, and can do no better than selecting an orthogonal eigenvector matrix $V = [V_1 V_2 \dots V_{n-m}]$ which clearly gives $(1/c_i) = 1$, and in particular we may take $V = I_{n-m}$. More generally, $1 < m < n-m$ and we should aim to make $1/c_i$ as small as possible for each $i = 1,\dots,n-m$. Now, it is also well-known (Wilkinson, 1965) that an upper bound on each of the sensitivities is the spectral condition number $\kappa(V)$ defined by

$$\kappa(V) = ||V|| \cdot ||V||^{-1} \qquad (18)$$

where the spectral matrix norm is employed. Thus minimising $\kappa(V)$ should ensure that the sensitivities of all the assigned eigenvalues are acceptable; and, additionally ensure that V is well-conditioned w.r.t. inversion, so that the solution process for the feedback matrix F is stable and the bounds on the magnitudes of $||F||$ and the transient response are minimised. It should be noted, however, that we are not free to choose our eigenvectors v_i arbitrarily if $m<n-m$, and that the minimum attainable condition number $\kappa(V)$ is therefore not necessarily unity for a given set of eigenvalues.

Four methods for determining a suitable set of eigenvectors for a specified spectrum in the usual linear feedback system are described by Kautsky and Nichols (1983) and Kautsky, Nichols and Van Dooren (1984). The four methods each consist of three steps, labelled A, X and F; steps A and F are identical in all cases, with the crucial step X differing from methods to method. Here we adapt one of these methods to the restricted problem of assigning the eigenvalues of the sliding mode system. This particular method was selected as being generally the most efficient. Fuller details of the algorithm may be found in the reports of Kautsky and colleagues.

Step A: Construct a basis for the null space of $H(\lambda_i)$ $(i = 1,\dots,n-m)$ as the columns of an $n\times m$ matrix $\begin{bmatrix} V_i \\ S_i \end{bmatrix}$ such that $S_i^T S_i = I_{n-m}$.

Step X: Determine vectors v_i in $R(S_i)$ such that $||v_i|| = 1$ $(i = 1,\dots,n-m)$; and set $V = [v_1 \dots v_{n-m}]$.

Step F: Define $w_i = S_i^T v_i$ $(i = 1,\dots,n-m)$, set $W = [V_1 w_1 \dots V_{n-m} w_{n-m}]$, and solve $FV = W$ for the feedback gain matrix F.

In the chosen method the step X is further divided: a set of orthonormal vectors $\tilde{x}_i$ $(i = 1,\dots,n-m)$ which minimise some measure of distance between $\tilde{x}_i$ and $R(S_i)$ is determined; and then for each $i = 1,\dots,n-m$, the vector $\tilde{x}_i$ is projected into $R(S_i)$ to give $\tilde{x}_i$. Since we wish to make each vector as close as possible to $R(S_i)$, we wish to make the angle between each vector and the orthogonal complement of $R(S_i)$ as large as possible. The measure of distance employed is therefore taken to be the sum of the angles between each vector $\tilde{x}_i$ and the orthogonal complement of the corresponding space $R(S_i)$. This measure is maximised by rotating the vectors iteratively, in such a way that the increase in the measure of distance is maximised at each step. Clearly there is a maximum value that this measure may attain, so the iterative procedure will converge. The reason for choosing to maximise this sum (rather than minimising the sum of distances between the vectors and the spaces themselves) is that it is more efficent for systems in which the number of states is very large and the number of controls small.

The method described does not, in fact, produce a minimum for $\kappa(V)$, but it does minimise a conditioning measure giving a good approximation to the optimal conditioning. The minimum attainable conditioning may be determined from the matrices S_i calculated in Step A. We have found the above algorithm to be reasonably successful in practice although we have not yet attempted the case of complex eigenvalues because of the additional difficulties involved (see Kautsky and Nichols, 1983).

EXAMPLE

In order to demonstrate the above procedure we consider a simplified flight control design problem discussed by Harvey and Stein (1978). The state and control vectors x and u are given by

$$x^T = [p_s\, r_s\, \beta\, \phi\, \delta_r\, \delta_a]; \; u^T = [\delta_{rc}\, \delta_{ac}]$$

where p_s and r_s are the stability axis roll and yaw rates, respectively; β and ϕ the sideslip and bank angles; δ_r and δ_a the deflections of the rudder and ailerons; and δ_{rc} and δ_{ac} are the rudder and aileron command signals. The linearised (ideal) model is described by (4) with

$$A = \begin{bmatrix} -.746 & .387 & -12.9 & 0 & .952 & 6.05 \\ .024 & -.174 & 4.31 & 0 & -1.76 & -.416 \\ .006 & -.999 & -.0578 & .0369 & .0092 & -.0012 \\ 1 & 0 & 0 & 0 & 0 & 0 \\ 0 & 0 & 0 & 0 & -10 & 0 \\ 0 & 0 & 0 & 0 & 0 & -5 \end{bmatrix}$$

and B having zero entries everywhere except for $b_{51} = 20$, $b_{62} = 10$.

The transformation matrix reorders the states and switches the signs of the angles β and ϕ. The assignment problem is to find the 2×4 matrix F such that the system (13) with

$$A_{11} = \begin{bmatrix} -.174 & .024 & 0 & -4.31 \\ .387 & -.7460 & 0 & 12.9 \\ 0 & -1 & 0 & 0 \\ .9990 & -.006 & .0369 & -.0578 \end{bmatrix}$$

$$A_{12}^T = \begin{bmatrix} -.416 & 6.05 & 0 & .0012 \\ -1.76 & .952 & 0 & -.0092 \end{bmatrix}$$

has the spectrum $\Lambda = \{-4, -.65, -.6, -.05\}$. The process described above leads to the matrix

$$F = \begin{bmatrix} .627 & .148 & -.082 & 2.107 \\ -2.447 & -.644 & .313 & .799 \end{bmatrix}$$

giving $\kappa(V) = 3.719$ compared with a minimum conditioning of 1.075. The sensitivities of the individual eigenvalues of Λ are respectively 1.365, 2, 1.365 and 2. Setting $C_2 = I_2$ gives the hyperplane matrix

$$C = \begin{bmatrix} .148 & .627 & -2.107 & .0817 & 0 & 1 \\ -.644 & -2.447 & -.799 & -.313 & 1 & 0 \end{bmatrix}$$

By comparison, a design based upon the approach of Harvey and Stein (1978) using eigenvector assignment directly gives

$$F = \begin{bmatrix} .820 & .564 & -.006 & 2.221 \\ -.788 & -.131 & -.024 & 1.727 \end{bmatrix}$$

with the corresponding sensivities 1.717, 26.877, 26.757 and 2.191. The robust design has consistently lower sensitivity indices and the second and third values show a marked improvement over the alternative approach.

CONTROLLER DESIGN

Once the hyperplane matrix C specifying the sliding manifold M has been fixed, the problem remaining is to select a suitable state feedback control function u which will drive the state x to M and maintain it thereafter within this space. The variable structure control law generally has two additive parts: a linear feedback $u_L = Lx$ and a non-linear part u_N. The non-linear section u_N incorporates the discontinuous elements of the control law. The non-linearities may include relays with constant gains, relays with state-dependent gains, linear feedback with switched gains and the scaled unit-vector form

$$u_N(x) = \rho\ Nx/||Px|| \qquad \rho>0 \qquad (19)$$

which yields control components which are discontinuous only on the final manifold Cx = 0. The null spaces of N, P and C are coincident. This control structure has been studied by Ryan and Corless (1984) using a Lyapunov analysis. When uncertainties (and disturbances) are present, the scalar ρ is a time-varying, state dependent function incorporating design parameters based upon the expected magnitude range of the parameter uncertainties.

This control structure drives the system state into the ideal sliding mode. In order to achieve this, it relies on the theoretically valid but practically unobtainable assumption, that instantaneous switching between exact control values may occur. In practice the behaviour of the system will not be ideal, because of time delays in switching, hysteresis and other dynamic non-idealities. Thus the state will remain only within a neighbourhood of N(C), continually passing backwards and forwards through the sliding manifold as the control switches repeatedly. Chattering is obviously highly undesirable in any system having actuator mechanisms which may be damaged by rapid switching. It is necessary in practice to "soften" the non-linearity by substituting a continuous high-gain feedback approximation to the discontinuous part of the control. A straightforward solution is to replace the unit-vector non-linearity (19) by

$$u_N(x) = \rho\ Nx/||Px|| + \delta \qquad \delta>0 \qquad (20)$$

where $\delta>0$ is a small constant (see Balestrino, De Maria and Sciavicco, 1982). This method eliminates chatter motion and simulation studies demonstrate that the resulting transient response can be maintained close to the sliding manifold.

CONCLUSION

This paper has considered the design of the switching hyperplanes of a variable structure control system by means of a robust pole-placement algorithm. It has been shown that the hyperplane design problem reduces to a linear feedback design problem for a lower order system. A particular control form, which ensures that the intersection of the hyperplanes is attained, has been outlined. Other control schemes are presented by Ryan (1983) and Young and Kwatny (1982) while some other aspects of VSCS design are considered by Zinober (1981) and El-Ghezawi, Zinober and Billings (1983).

A CAD package incorporating the ideas and written algorithms presented here has been developed by the authors (Dorling and Zinober, 1984). This Fortran package, named VASSYD (Variable Structure System Design), provides facilities for on-line VSCS design, simulation of transient response and the storage of system data. It incorporates an extensive 'help' module providing the user at all stages with detailed instructions regarding the applicable theory and the mode of operation of the package. A detailed manual is available.

REFERENCES

Balestrino, A., G. De Maria and L. Sciavicco (1982). Hyperstable adaptive model following control of nonlinear plants. System and Control Letters, 1, 232-236.

Dorling, C.M., and A.S.I. Zinober (1984). Computer aided design of variable structure control systems. Proc. IMC Workshop on Computer Aided Control System Design, Brighton, 67-71.

ElGhezawi, O.M.E., A.S.I. Zinober and S.A. Billings (1983). Analysis and design of variable structure systems using a geometric approach. Int. J. Control, 38, 657-671.

Harvey, C.A., and G. Stein (1978). Quadratic weights for asymptotic regulator properties. IEEE Trans. Automat. Control, AC-23, 378-387.

Itkis, U. (1976). Control Systems of Variable Structure. Wiley, New York.

Kautsky, J., and N.K. Nichols (1983). Robust eigenstructure assignment in state feedback controls. Numerical Analysis Rpt NA/2/83, University of Reading, Dept. of Mathematics.

Kautsky, J., N.K. Nichols and P. Van Dooren (1984). Robust pole assignment in linear state feedback. Numerical Analysis Rpt 5/84, University of Reading, Dept. of Mathematics.

Klein, G., and B.C. Moore (1977). Eigenvalue-generalised eigenvector assignment with state feedback. IEEE Trans. Automat. Control, AC-22, 140-141.

Kwakernaak, H., and R. Sivan (1972). Linear Optimal Control Systems. Wiley, New York.

Moore, B.C. (1976). On the flexibility offered by state feedback in multivariable systems beyond closed loop eigenvalue assignment. IEEE Trans. Automat. Control, AC-21, 689-692.

Ryan, E.P. (1983). A variable structure approach to feedback regulation of uncertain dynamical systems. Int. J. Control, 38, 1121-1134.

Ryan, E.P., and M. Corless (1984). Ultimate boundedness and asymptotic stability of a class of uncertain dynamical systems via continuous and discontinuous feedback control. To appear.

Shah, S.L., D.G. Fisher and D.L. Seborg (1975). Eigenvalue/eigenvector assignment for multivariable systems and further results for output feedback control. Electron. Lett., 11, 388-389.

Srinathkumar, S., and R.P. Rhoten (1975). Eigenvalue/eigenvector assignment for multivariable systems. Electron. Lett., 11, 124-125.

Utkin, V.I. (1977). Variable structure systems with sliding modes. IEEE Trans. Automat. Control, AC-22, 212-222.

Utkin, V.I. (1978). Sliding Modes and Their Application in Variable Structure Systems. MIR, Moscow.

Utkin, V.I. and K.D. Yang (1978). Methods for constructing discontinuity planes in multidimensional variable structure systems. Automn. Remote Control, 39, 1466-1470.

Wilkinson, J.H. (1965). The Algebraic Eigenvalue Problem. Clarendon Press, Oxford.
Young, K.-K.D., and H.G. Kwatny (1982). Variable structure servomechanism design and applications to overspeed protection control. Automatica, 18, 385-400.
Zinober, A.S.I. (1981). Controller design using the theory of variable structure systems. In S.A. Billings and C. Harris (Eds.), Self Tuning and Adaptive Control, Peter Peregrinus, London. Chap. 9, pp. 204-229.

COMPUTER-AIDED DESIGN OF SPACECRAFT CONTROL SYSTEMS

Chr. Schmid* and M. E. Stieber**

**Department of Electrical Engineering, Ruhr-University, Bochum, D-4630 Bochum 1, FRG*
***Communications Research Centre, Ottawa, Canada*

Abstract. The KEDDC software package for computer-aided control system design, developed at Ruhr-University, Bochum, F. R. G. is presently applied by the Communications Research Centre (CRC), Ottawa, Canada to the development of control techniques for future large flexible spacecraft in a joint industry, university and government research and development project. The Operational Mobile Communications Satellite (OMSAT) proposed for the mid 1990's featuring a 44m offset-fed antenna, is used as target application. A reduced version of a 146th order high fidelity dynamics model of the satellite serves as a benchmark system. This application is very demanding and forms an important performance benchmark for any software system because the high order of structural models requires the use of highly stable and efficient numerical algorithms. KEDDC has been tested on this problem for more than one year and has performed flawlessly, improving quite considerably the design efficiency as well as the scope of system analysis.

Keywords. Computer-aided system design; program package; control techniques for large flexible spacecrafts.

INTRODUCTION

Future spacecraft for communications, earth resource surveillance and scientific applications will be very large in size and require precise pointing to achieve the mission objectives. An example is the 'Operational Mobile Communications Satellite' (OMSAT) under consideration by the Candian Department of Communications (DOC). The OMSAT spacecraft is illustrated in Fig. 1. This geosynchronous satellite would provide communications services in the 800MHz band to small ground terminals, which may be mounted on cars, trucks or boats throughout Canada and the US. The size of the spacecraft is related to the electromagnetic performance requirements of the multi-beam antenna system, and the power required for the transponders. OMSAT features a 44m diameter reflector, a 70m L-shaped support tower and a 38m x 4m solar array generating 8.2kW of power.

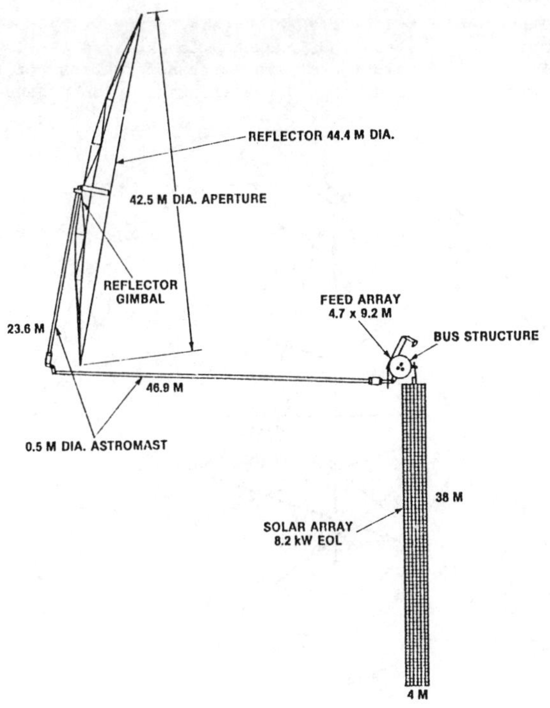

Fig. 1. OMSAT conceptual layout

The spacecraft mass is about 3500kg, with the center of mass located outside the physical structure. The communications beam pointing accuracy required is 0.05 degrees. The reflector, solar array and both parts of the tower are lightweight deployable structures. Their structural flexibility affects the design of the Attitude and Orbit Control System (AOCS). With increasing size of the spacecraft, the fundamental structural frequencies decrease and approach or overlap the control system bandwidth. Control system design for this class of system has been an active area of research for over a decade. Some of the difficulties encountered are the uncertainty in prediction and measurement of structural parameters, the high order of structural systems, and the nonlinear behaviour of deployable structures.

The engineering challenges to design, build and test such a spacecraft are quite significant and control has been identified as a critical area. A joint industry, university and government research project led by the Communications Research Centre (CRC - a DOC laboratory) undertakes to develop the AOCS techniques required for this generic class of spacecraft. A comprehensive CAD system, used for development, evaluation and verification of control techniques has been implemented at CRC in Ottawa. This system essentially is the KEDDC package developed at Ruhr-University, Bochum, F. R. G. by Schmid and Unbehauen (1979). It serves not only as an efficient tool for control system analysis conducted at CRC, also is intended to play a role as an interface and stimulus for the communication between the various research groups within indus-

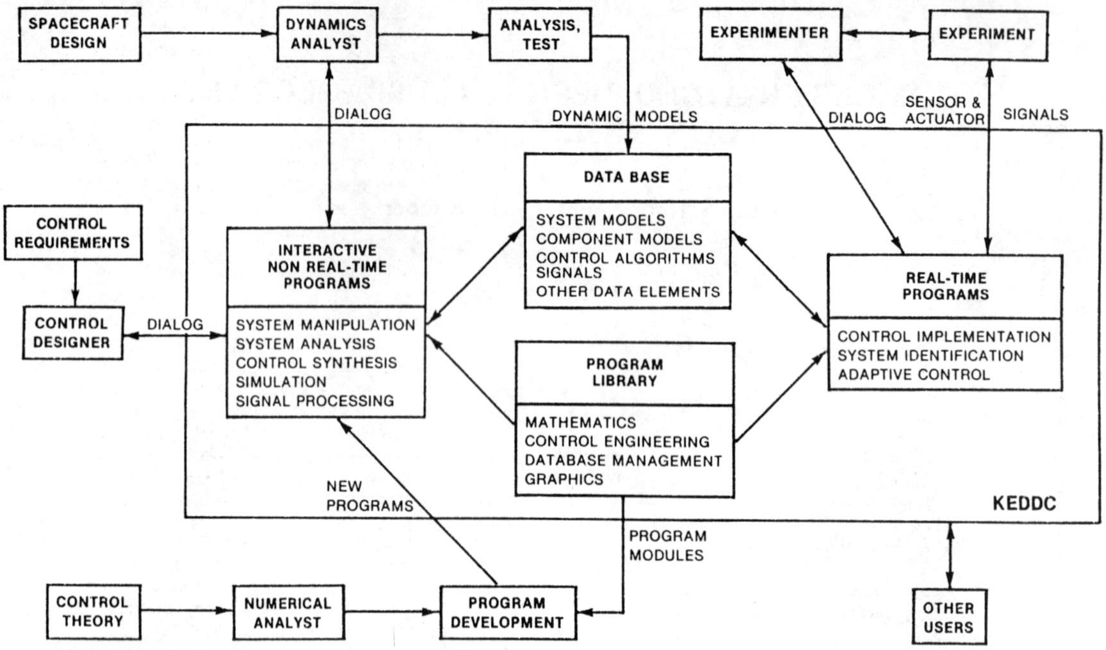

Fig. 2. KEDDC main components and applications environment

tries, universities and CRC. This paper summarizes some aspects of this joint effort.

At first a general description of the most significant capabilities of the KEDDC system is given. The system representation forms and some features of system handling are discussed. Details of the software and interaction, graphics and documentation are outlined here. An introduction into the spacecraft dynamics of the OMSAT vehicle is given. A benchmark spacecraft dynamics model and benchmark control requirements are defined. The experiences in using KEDDC for this spacecraft control problem are presented and finally the extensions to use it in ground-based experiments are discussed.

THE KEDDC SYSTEM

Components of KEDDC

Fig. 2 provides a block diagram of the main components of the CAD system and their interfaces with the applications environment to research and development. The major components are a set of interactive non real-time programs for system and signal analysis, a database, an extensive program library and real-time programs interfacing with experimental hardware. Access to a common database improves the communication between research groups working on a common project. In particular, access by all users to modelling and design procedures can help to improve the interfacing between system analysts and control designers. KEDDC also forms a common software base reducing duplication of program development. It is organized as an open system, parts of which may be added, updated or removed at any time. This unlimited extendability is of particular importance in an applied research and development environment with on-going new programs and methods. Program development is supported by the program library which contains about 1000 routines. Implementation of existing real-time programs will allow immediate implementation of control algorithms, on-line identification and adaptive control.

Capabilities

The package supports a wide variety of operations for system manipulation and analysis, control system synthesis, simulation and signal processing. Systems may be represented in any of the following 10 forms:

1 series of deterministic or stochastic input and/ or output signals,
2 series of auto- or crosscorrelation values,
3 discrete values of the pulse response,
4 discrete values of the step response,
5 transfer function (matrix) in s-domain,
6 transfer function (matrix) in z-domain,
7 discrete values of frequency responses,
8 matrices for continuous state space,
9 matrices for discrete state space,
10 polynomial matrices for MFD.

They can be transformed with ease between time domain, state space and frequency domain as shown in Fig. 3. The solution of system transformations or controller design procedures are usually supported

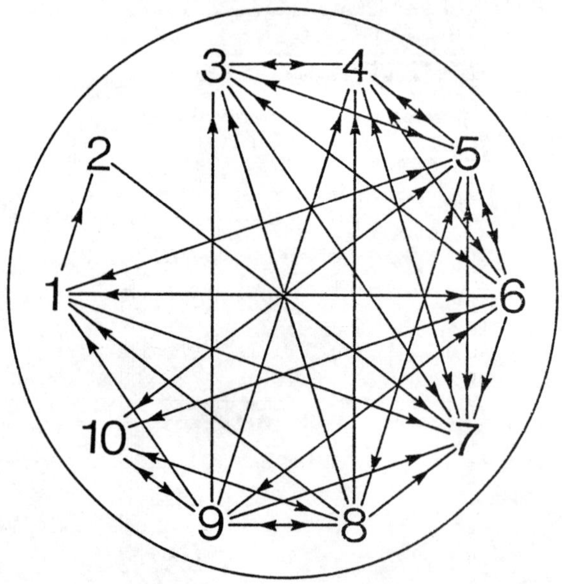

Fig. 3. System representations and transformations. Explanation see text above

by a number of numerical options in order to ensure efficient and accurate solution of high-order problems.

Frequently used tasks, like transformations to different representation forms, data base definitions or basic calculations, are separated from other tasks and are concentrated in main interactive system handlers, the so-called Managers. The following Managers are available:

- Signal Manager for handling signal sequences,
- System Manager for handling systems described by transfer functions or transfer matrices in s- or z-domain,
- Frequency Manager for handling frequency responses or spectra in tabular form,
- Matrix Manager for handling coefficient matrices or systems in state space,
- Polynomial Manager for handling polynomial matrices and matrix-fraction descriptions (MFD),
- Graphics Manager for all graphics operations.

Standard structures for linear control systems are simulated by a standard simulator. The overall system to be simulated may be composed of different subsystems of different representation forms. Each block in this structure may be of the form 5, 6, 8, 9 or 10 given above. For direct digital control operation different special simulators are available. They simulate the environment for testing algorithms in real-time applications. A more detailled discussion of the implemented methods is given by Schmid (1985).

Software Structure and Interaction

KEDDC provides a high degree of portability, which results from the interface of KEDDC to the operating system and to hardware-dependent functions. The strict separation of the core system and the interactive code makes this possible. Program development may take place on other than the target machine as the core system is portable.

The user interacts with KEDDC through a command-driven dialogue combined with a question - and-answer dialogue. He can perform operations in arbitrary sequence, access the implemented methods and algorithms like tools in a toolbox. Each interactive program has its own set of commands, which can be enabled by supervisory scheduling of the programs using the monitor. Menu, status and help facilities normally enable casual users to master the system without reference to manuals. A KEDDC session does not require any pre-planning or programming as forgotten steps can be interlaced in the dialogue at any time, such that a 'hang-up' cannot occur.

Graphics

Graphics operations are performed by a central Graphics Manager, which itself is an autonomous package. Data including attributes (e. g. for a diagram type) are put into a mailbox. The Graphics Manager is a very comfortable specialized task for formatting these mailbox data into graphical form readable by the control engineer. The Graphics Manager can be implemented using a graphics processor, or on an intelligent graphics terminal or as an internal graphics task on the same computer. This concept shows a great deal of flexibility in interfacing to different graphics device configurations and ties diverse parts of KEDDC to an unified system. It incorporates state-of-the-art device independent graphics which is based on the ACM/SIGGRAPH core system standard. All diagrams relevant to control engineering are implemented. The user has the choice of using default diagrams or tailor his plots using a keyword-based descriptive language.

Documentation

The comprehensive documentation consists of a user's manual (1200 pages) and a programmer's reference manual (1400 pages). Self-explanatory code (40% of the source code is comments) supports the maintainability of the package. The programmer's documentation is contained in the source code. Each library routine is documented by a description of the purpose, complete specification of input parameters, output parameters, data types, workspaces, remarks on methods, hints, references to literature, list of subroutine calls and of the revision level. Using these informations, the subroutines can be easily integrated into any software system. In additon, self-explanatory code supports the documentation. For most subroutines and sublibraries, extended documentation in form of reports is available. A KEDDC interface guide, tree diagrams, module crossreference lists, and reports related to special control engineering methods and topics are available. Detailed runtime test documentation listings with many examples including ill-conditioned cases are available.

SPACECRAFT DYNAMICS

Spacecraft Model

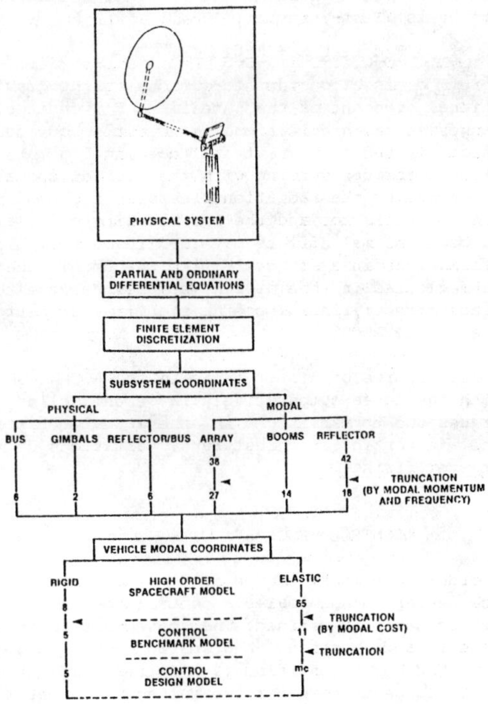

Fig. 4. OMSAT dynamics modelling

Fig. 4 schematically outines the different steps of the spacecraft modelling and model order reduction process. linearisation and spatial discretisation of the system results in a finite element description of the spacecraft, with 29000 degrees of freedom. At this level, the modelling is handled by specialized structural analysis software. The model is reduced to an order of 146 before it is transferred to KEDDC for further reduction, analysis and control synthesis. This model is called the 'high order spacecraft model' and has the following form in modal coordinates (Hughes, 1982):

$$\underline{\dot{x}} = \underline{A}\,\underline{x} + \underline{f}_c + \underline{f}_d \qquad (1)$$

with

$$\underline{A} = \begin{bmatrix} \underline{O} & \underline{I} \\ -\underline{\Omega}^2 & -(\underline{D}+\underline{G}) \end{bmatrix} \qquad (2)$$

$\underline{f}_c$ forcing function due to control inputs
$\underline{f}_d$ forcing function due to disturbances
$\underline{I}$ identity matrix
$\underline{\Omega} = \text{diag}(\omega_i)$, ω_i = frequency of the ith overall vehicle natural undamped mode
$\underline{D}$ nondiagonal damping matrix
$\underline{G}$ nondiagonal gyroscopic matrix.

This model has 3 translational, 3 rotational rigid body modes, 2 rigid body modes associated with the gimbals at the reflector hub and 65 flexible body modes associated with the elasticity of the spacecraft. The lowest flexible model frequency is at about 0.2Hz and its damping ratio is less than 1%. In consideration of the uncertainty of prediction for modal parameters, which has been confirmed by recent space experiments (Lips and Vigneron, 1984), a tolerance of ±20% for modal frequencies, and -70%/+300% for damping is assumed. Conventional laboratory measurement of structural properties of a completely assembled spacecraft of the size of OMSAT is impossible due to gravitational loads.

Disturbances

At geosynchronous altitude, the dominant environmental influences on a spacecraft are solar radiation pressure and gravity gradient torque. The forcing function $\underline{f}_d$ in Eq. (1) has the following components (West-Vukovich, 1984):

$$\underline{f}_D = \underline{F}_O \underline{d}_G + \underline{F}_G \underline{x} + \underline{F}_S \underline{d}_S \tag{3}$$

where $\underline{F}_O \underline{d}_G$ is a constant torque due to the gravitational gradient of the satellite, $\underline{F}_G \underline{x}$ denotes changes in the gravitational gradient torque due to changes in the spacecraft attitude and $\underline{F}_S \underline{d}_S$ are periodic torques varying with the orbital anomaly and caused by the radiation pressure on the earth tracking reflector and the sun tracking solar array. Although thermal effects may lead to depreciable deformations in a structure like the OMSAT, they are neglected in the present studies because they do not directly influence the stability or performance of the AOCS.

A major source of disturbances is the AOCS itself which may cause comparatively large unmodelled torques due to variations in the actuator performance, in particular thrust level variations during thruster firing.

CONTROL BENCHMARK

In order to compare the performance of various control design methodologies a control design benchmark scenario is defined. The spacecraft is assumed to be in a conventional N-S stationkeeping manoeuvre which has been identified as a critical operation due to large unknown torques caused by thrust level uncertainties. The control objective is to maintain attitude control and beam pointing during the stationkeeping manoeuvre. Stationkeeping itself is assumed to be controlled by a strategy independent of attitude control and hence the three translational spacecraft rigid body modes, describing the orbital kinematics, are deleted from the model. The analysis of the communications beam geometry leads to the definition of the performance output vector

$$\underline{z} = \underline{P}\,\underline{x} \quad , \tag{4}$$

where $\underline{P}$ is a performance distribution matrix describing how much the excitation of each state contributes to the elements of the beam pointing error. The control objective is expressed by the cost functional

$$V = \int_0^\infty \underline{z}^T \underline{Q}\, \underline{z}\, dt \quad , \tag{5}$$

where $\underline{Q}$ is a positive definite weighting matrix for the error components (e. g., roll, pitch and yaw error).

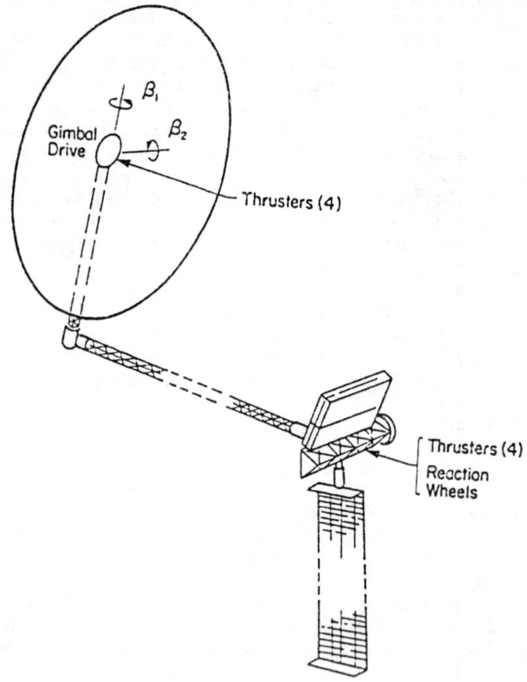

Fig. 5. OMSAT actuator configuration

Fig. 5 shows the baseline actuator configuration with 3 reaction wheels and 4 thrusters located at the spacecraft bus, 4 thrusters at the tower tip and 2 reflector gimbal torquers. During stationkeeping a thruster based control system is assumed with the reaction wheels locked at zero speed. Further the following simplifying assumptions are made: the 8 one-way thrusters are combined to form 5 equivalent independent, linear actuators. An approximately linear thruster characteristic may be obtained by pulse width or pulse frequency modulation although some excitation of high frequency modes will be concomitant with this technique.

A gyropack, encoders and an optical sensor provide the following measurents: spacecraft bus attitude and angular rate about 3 axes, 2 reflector gimbal angles, and the relative displacement between spacecraft bus and tower tip in two directions.

For this input configuration, and a certain power spectrum expected for the disturbances, the model is further reduced by modal cost analysis according to Skelton and Hughes (1980) using Eq. (5) to a control benchmark model of order 32, having 7 control inputs and 10 measurement outputs. For simplicity, the same notation as in Eq. (1) is used to describe the system as it retains the same structure:

$$\dot{\underline{x}} = \underline{A}\,\underline{x} + \underline{B}\,\underline{u}_c + \underline{B}\,\underline{u}_d + \underline{f}_d \tag{6}$$

$$\underline{y} = \underline{C}\,\underline{x} + \underline{\omega} \tag{7}$$

where

$\underline{A}$ systemmatrix as in Eq. (1)
$\underline{B}$ input matrix for actuators
$\underline{u}_c$ control input
$\underline{u}_d$ unknown disturbances caused by actuator variations (assumed stepwise constant)
$\underline{y}$ measurement vector
$\underline{C}$ sensor matrix
$\underline{\omega}$ measurement noise.

This benchmark model can be conveniently manipulated by KEDDC and is used to assess the performance and robustness of control systems. These controllers are usually based on a further reduced 'control design model' (Fig. 2) of the spacecraft to explore the effects of modal truncation.

The benchmark requirements for the attitude control system are:

- stabilization of attitude and communications beam in presence of uncertain and unmodelled plant dynamics,
- regulation of attitude and beam against environmental disturbances and thruster inbalances, and
- spacecraft rigid body response time constants in the order of magnitude of the lowest spacecraft flexible mode. The absolute value of the closed-loop eigenvalues associated with one or more rigid vehicle modes shall be larger than the frequency of the lowest frequency elastic mode.

No tracking of reference signals is required as the spacecraft remains continuously earth pointing after initial earth acquisition. Full state feedback and disturbance feedforward on the nominal benchmark model is the performance reference case.

KEDDC APPLICATION

The large spacecraft control problem is currently being investigated using a non-realtime KEDDC implementation on a mainframe computer system. This centralized facility can be accessed by all members of the research team and is used as a database for the project. The availability of the high (146) order spacecraft model and system order reduction methods to all users help to bridge the gap between the structural dynamics and control disciplines.

KEDDC aided in the definition of the benchmark problem by modal cost analysis and analyzing sensor and actuator configurations for conditioning, controllability/observability of flexible modes and transmission zeros. Presently a number of different model reduction and control system design techniques are put to test on the benchmark problem. The reduction methods considered so far include various versions of modal cost analysis, cost-decoupling transformation and balanced realizations (West-Vukovich, 1984). The following methods are used for control design on the open-loop system reduced to an order between 12 and 32: Variations of the LQG method such as frequency shaping, PI-control, disturbance observers, reduced-order observer design by pole placement, various spill-over suppression techniques such as selective spill-over suppression, model error sensitivity suppression etc. The closed-loop systems of order 32 to 70 are evaluated on the basis of the required model accuracy, system performance to pointing and controller robustness. The methods used include eigenvalue plots, root-locus diagrams, simulations and singular value type robustness tests.

Typically a design case can be analyzed and documented in an afternoon without any pre-planning or following any standard pattern during the session. Computations are carried out with 18 significant digits and have not posed any numerical difficulties. KEDDC offers several numerical choices for critical computations such as eigenvalue/eigenvector calculation, Riccati equation solving (13 options) or polynomial root calculations.

Only minor extensions to the original KEDDC command set was required in order to solve all above analysis and synthesis problems, with the exeption of system-order reduction for which a new module was created.

During new program development, the extensive KEDDC library frees the analyst from routine jobs, like disk I/O, user interfacing, error handling, graphics etc. It offers ready-to-use processing modules at all levels of integration, from basic mathematical functions to complete control engineering modules. For the development of large space structure control technology, KEDDC not only serves as a tool, but also as a software base which plays a role in transferring the technology for future application by industry.

REAL-TIME WORKSTATION

In the next phase of the OMSAT project, a laboratory facility will be developed whose primary purpose is to study the stabilization and control of structurally flexible communications satellites.

Fundamental concepts in the control of flexible space structures must be investigated and evaluated. Realistic sensors, actuators, structures, and control electronics are used in order to ensure the practical aspects of the proposed control techniques. The results of this research must be of direct relevance to the AOCS problem discussed above. For on-ground testing a mechanical model of flexible structures will be built as illustrated in Fig. 6 (Hughes, 1984). This model shows the typical dynamic characteristics of a flexible space structure.

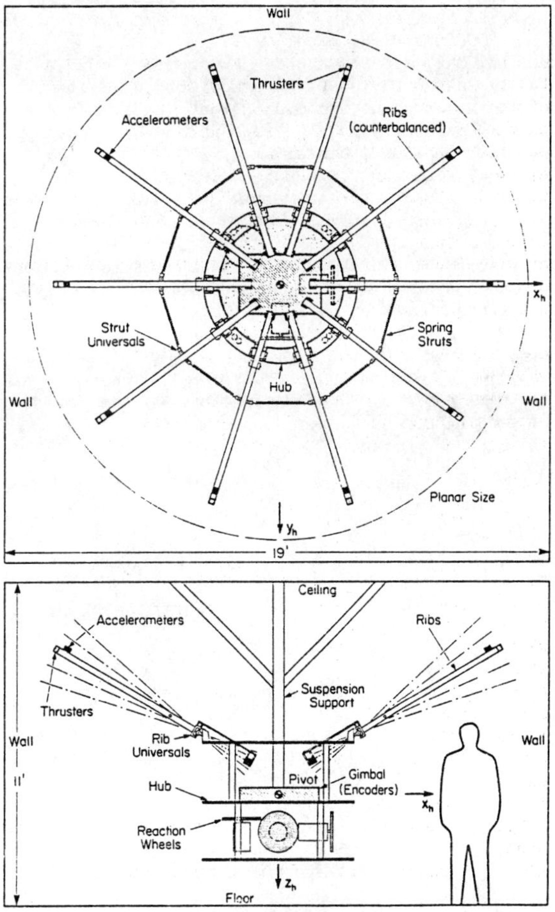

Fig. 6. Design of the ground-based experimental flexible structure model

A Real-Time Workstation will be used as an end-to-end control tool on this flexible structure experiment. It consists of an INTEL-300 series microcomputer system. This workstation will support system identification, control analysis and design as well as the implementation of controllers. Some of the Managers of KEDDC and the real-time suite including adaptive control will be installed.

CONCLUSIONS

The CAD system KEDDC has been found a very efficient and versatile, almost indispensible tool for the reliable analysis of high-order systems. Its architecture and user interface are particularly well suited for an applied research and development environment.

REFERENCES

Hughes, P.C. (1982). *Development of Dynamics Models and Control System Design for Third Generation Spacecraft - Executive Summary*. Report DOC-CR-SP-82-056, Department of Communications, Ottawa.

Hughes, P.C. (1984). *Laboratory Demonstration of Control Techniques for Third Generation Spacecraft-Executive Summary*. Report DOC-CR-84-011, Department of Communications, Ottawa.

Lips, K.W. and F.R. Vigneron (1984). *Damping Synthesis for a Spacecraft using Substructures and Component Data*. Report 1365, Department of Communications, Ottawa.

Schmid, Chr. and H. Unbehauen (1979). KEDDC, a general purpose CAD software system for application in control engineering. Prep. *2nd IFAC/IFIP Symposium on Software for Computer Control (SOCOCO)*, Prag, paper C-V.

Schmid, Chr. (1985). KEDDC - A Computer-Aided Analysis and Design Package for Control Systems. In M. Jamshidi and C.J. Herget (Ed.). *Adcances in Computer-Aided Control Systems Engineering*. North-Holland, Amsterdam.

Skelton, R.E. and P.C. Hughes (1980). Modal Cost Analysis for Linear Matrix-Second-Order Systems. *J. Dynamic Systems, Measurement and Control*, *102*, 151-158.

Stieber, M.E. (1984). Design and Evaluation of Control Systems for Large Communications Satellites. *Workshop on Identification and Control of Large Space Structures*, San Diego.

West-Vukovich, G. and P.C. Hughes (1984). *Development of System Order Reduction Techniques Applicable to Third Generation Spacecraft*. Report DOC-CR-84-044, Department of Communications, Ottawa.

COMPUTER AIDED DESIGN OF VARIABLE STRUCTURE (VSS) CONTROL SYSTEMS

R. Świniarski and A. Knafel

Institute of Control and Industrial Electronics, Electrical Department, Warsaw Technical University, ul. Koszykowa 75 316, Warszawa, Poland

Abstract. The paper presents the the description of computer programme package for interactive designing and simulation of multidimensional linear and nonlinear variable structure systems. The special attention has been paid for utilising of variable structure control to model - folloving control systems. A recursive reconstruction of unaccesible process states in model - following control systems has been discussed as well.

Keywords. Adaptive control, computer - aided design, state estimation.

INTRODUCTION

Analysing of computer - aided design of control systems (CADCS) packages we can notice unsatisfactory passibilities of using a variable structure systems (VSS) technique for control of process with varying or uncertain parameters. Observing a substantial growth of interest in application of VSS technique to model - following control systems the special programme package has been designed for interactive design and simulation of these systems.

Variable structure systems have been described primary in literature by Soviet authors. We can specially pointed out works of Emel'yanov (1964) and Utkin (1971,1977). Recently several scientists from the western countries have successfully develeped some new ideas concerning the variable structure system application to model-following control problems (Young, 1977,1978; Zinober, and all, 1982) and tracking control of non-linear systems (Slotine, 1983).

The main features of variable structure control as insensitivitiness to the process parameter variations and disturbances and simplicity of control still atract the scientist working in discontinuous control technique application field. In the paper the main points concerning the variable structure system application to the model-following systems are pointed-out. The two interactive computer programme package for designing and simulation of adaptive model-following variable structure control system for linear and nonlinear (stationary and nonstationary) systems are presented as well.

VARIABLE STRUCTURE CONTROL

At the begining let us consider the linear system

$$\dot{x}(t) = A(t)x(t) + B(t)u(t) \quad (1)$$

where $x \in R^n$, $u \in R^m$ and A and B are generally a time-varying state and control matrices of appriopriate dimensions. Variable structure system can be characterised by discontinous control which changes structure of system when a reaching a set of switching surfaces (manifolds). The discontinuous control strategy has for this case form (Utkin, 1977)

$$u_i = \begin{cases} u_i^+(x,t) & s_i(x) > 0 \\ u_i^-(x,t) & s_i(x) < 0 \end{cases} \quad (2)$$

where u_i represents the ith component of control vector u, and $s_i(x) = 0$ is the ith component of the m switching hyperplanes

$$s(x) = Cx = 0 \quad (3)$$

in the state space.
In the variable structure system motion on switching hyperplanes alter the structure of feedback control system.
An appriopriate selection of matrix C can quarantee for linear and certain class of nonlinear systems (Slotine, 1983) reaching of switching surfaces $s_i(x)=0$ in finite time, and therefore remaining on this surface approaching the state origin or next switching surface $s_j(x)=0$ in finite time.
Decaying transient of motion on switching surfaces is governed by the eigenvalues which characterized equation $s_i(x) = 0$ (Utkin, 1977). The sliding (chatter for realistic systems) motion when at the point on this surface the "dicrection" of motion along the state trajectories on both sides of surface are toward the switching surfaces. Thus for sliding on switching surface $s_i(x) = 0$ we need fullfilled of relation

$$s_j(x)\dot{s}_i(x) \leqslant 0 \quad (4)$$

in the neighbourhood of $s_i(x) = 0$.
During sliding motion on switching surface $s_i(x)$ the system is described by the equations

$$s_i(x) = 0 \text{ and } \dot{s}_i(x) = 0 \quad (5)$$

and has an invariance properties (Drażenowic, 1969) causing state motion which is independent of some process parameters and disturbances. These feature of variable structure systems is a main "attraction" of this kind of control especialy in application to process with uncertain or varying parameters and unknown disturbances.
Assuming that the sliding motion exist on all switching surfaces, then we can write

$$s(x) = Cx = 0 \text{ and } \dot{s}(x) = C\dot{x} = 0 \quad (6)$$

The so called "equivalent control" u_{eq} (Utkin, 1977) can be found from the equation s(x) = 0

$$u_{eq} = -(CB)^{-1}\, CAx \quad (7)$$

as an linear function of x. The "equivalent control u_{eq} is a smoothed version of the discentinuous control (2).
Thus the sliding dinamics can be reached by applying the discentinuous control (2) or the linear equivalent control (7) (for systems with perfect knowledge about the constant parameters and disturbances).

MODEL - FOLLOWING CONTROL SYSTEMS

The model - following control systems (MFC) (Landau, 1974) based on an idea that the process is controlled in such way that its dynamic behaviour approximates the dynamic of specified (established during designing phase) model.
The extending idea of MFC to the plants with varying parameters or disturbances leads to the so called adaptive model following control (AMFC). The two major attifuded to the AMFC based on Lyapunov function and hyperstability idea (Landau, 1974) quarantee the error between the model and the process goes to the zero as $t \to \infty$ ($\lim_{t\to\infty} e(t) = 0$) but but neither suggest any stricte control

over the error transient.

Young (1977,1978) proposed new design me thod for AMFC systems which utilises the concept of variable structure systems (VSS) for obtaining asymptotically stable systems. In this proposal an effective procedure for specifing the transient response of the error between model and process has ben given. In the AMFC the specified model is the organic part of the system.

Let us assume that the controlled process and the specified model are described by the multidimensional linear equations

$$\dot{x}_p(t)=A_p(t)x_p(t)+B_p(t)u(t) \qquad (8)$$

$$\dot{x}_m(t)=A_m x_m(t)+B_m u_m(t) \qquad (9)$$

where $x_p(t),x_m(t) \in R^n$, $u(t) \in R^m$ and $u_m \in R^L$. In the above equations $U_m(t)$ is the system input and $u(t)$ is the control vector. In the clasical AMFC systems it is assumed that the pairs (A_p, B_p) and (A_m, B_m) are stabilisable, but varying in time of the plant matrices is allowed, however the upper and lower bounds of the component of these matrices are assumed to be known during designing plase.

Defining the error vector as

$$e(t) = x_m(t) - x_p(t) \qquad (10)$$

it can be easily found that

$$\dot{e}(t)=A_m e(t)+(A_m-A_p(t))x_p(t)+B_m u_m(t)-B_p(t)u(t) \qquad (11)$$

The necessary conditions (Erzberger,1968) for perfect model following behaviour can by written as follows

$$\text{rank } B_p=\text{rank}(B_p,B_m)=\text{rank}(B_p,A_m-A_p) \qquad (12)$$

VARIABLE STRUCTURE SYSTEM APPROACH TO AMFC

In Young's (1978) work the variable structure approach to AMFC system has been proposed. In this attitude the designer may specify the desired transient response of system error by the proper choice of the switching surfaces. Assuming that the process state $x_p(t)$ is directly accessible we can consider the linear control law

$$u(t)=K_p x_p(t)+K_e e(t)+K_m u_m(t) \qquad (13)$$

where the matrices K_p, K_e, K_m exist when the conditions (12) hold, and $e(t) = x_m(t) - x_p(t)$.

In VSS application to AMFC (VSAMFC) (Young, 1977) we can substitute in the equ.(13) the matrices K_p,K_o,K_m by the Ψ_p, Ψ_e, Ψ_m defined by

$$\Psi_p=\tilde{\Psi}_p \text{ diag}(\text{sgn } x_p,\dots,\text{sgn } x_p^n), x_p^T = (x_p,\dots x_p^n)$$

$$\Psi_e=\tilde{\Psi}_e \text{ diag}(\text{sgn } e^1,\dots \text{sgn } e^n), e^T = (e^1,\dots e^n) \qquad (14)$$

$$\Psi_m=\tilde{\Psi}_m \text{ diag}(\text{sgn } u_m^1,\dots \text{sgn } u_m^L), u_m^T = (u_m^1,\dots u_m^L)$$

with

$$\tilde{\Psi}_p^T=(\Psi_p^1,\dots,\Psi_p^m) \text{ and } \Psi_p^i=\begin{cases} k_p^i & \text{for } s_i(e)>0 \\ -k_p^i & \text{for } s_i(e)<0 \end{cases} \qquad (15)$$

$$\tilde{\Psi}_e^T=(\Psi_e^1,\dots,\Psi_e^m) \text{ and } \Psi_e^i=\begin{cases} k_e^i & \text{for } s_i(e)>0 \\ -k_e^i & \text{for } s_i(e)<0 \end{cases}$$

$$\tilde{\Psi}_m^T=(\Psi_m^1,\dots,\Psi_m^m) \text{ and } \Psi_m^i=\begin{cases} k_m^i & \text{for } s_i(e)>0 \\ -k_m^i & \text{for } s_i(e)<0 \end{cases}$$

$i = 1,\dots m$

where Ψ_p^i, k_p^i,Ψ_e^i, k_e^i are nx1 vectors, and Ψ_m^1, k_m^i are Lx1 vectors.

The switching hypersurface can be choose to be set of switching hyperplanes

$$s(e(t)) = Ce(t) = 0 \qquad (16)$$

$$s^T = (s_1,\dots s_m)$$

in the n-dimensional error space, where $s_i(e) = 0$ is the ith of m switching hyperplanes.

When the sliding mode exists on all the the hyperplanes then we can write

$$s = C\, e(t) = 0 \text{ and } \dot{s} = C\dot{e}(t)=0 \qquad (17)$$

From the above conditions we have

$$u_{eq}=-(CB_p)^{-1}C\; A_m e(t)+(A_m-A_p)x_p(t)+B_m u_m(t) \qquad (18)$$

and for perfect model - following contro

$$\dot{e}(t)= I-B_p(CB_p)^{-1}C\; A_m e(t) \qquad (19)$$

The above consideration can be extended for specific nonlinear processes (Slotine, 1983) described by the equation

$$\dot{x}_p(t)=f(x_p(t),t) + bu(t) \qquad (20)$$

with the assumption that the nonlinear function of state $f(x_p(t))$ are polynomially bounded

$$|f_j(x_p(t),t)| \leqslant p_j(x_p(t),t) \quad (21)$$
$$j = 1,2,\ldots,n$$

In these case the control vector equation can be medified (Slotine, 1983) by adding to the (13) the term $-K_n$ sqns(t) (and to the (15) the term k_n^i).

DESIGN PROCEDURE

The design procedure of model-following variable structure system is similar to that for multivariable structure systems when the system error vector is considered. The design objective of VSAMFC is to choose the C and to find the vectors k_p^i, k_e^i, k_m^i (equ.(14)) so the system error goes to the zero, as time tends to infinity.

From the above shows that the design procedure consist the two main steps. First it is necessary to select C, i.e design the switching hyperplanes, which guarantee the specified stable eigenvalues of the closed-loop error system and causing decaying of error to the zero with given dynamic. In the second step the appriopriate control law (i.e. k_p^i, k_e^i, k_m^i vectors) generally described by the equ.(15) must be found to ensure reaching and exsistance of sliding motion.

For the system with multivariable control u(t) one can apply the method of control hierarchy described by Utkin (1978) and Young (1978).

The design of the switching hyperplane (selecting of C) is proposed in the several papers (Utkin, 1977; Young, 1978) but in the programme described in the next section also it is possible to use technique of projectors which has been proposed by El-Ghezawi (1984). These methods are usually based on an idea of pole assignment of system error dynamic describing the methion on switching surfaus. The designing of control coefficient must take into consideration fast (for example time - optimal) reaching of switching planes and quarantee existence of sliding motion which tends system error to the origin (Utkin, 1977; Young 1978; Zinober, 1982; Slotine, 1983).

THE COMPUTER PROGRAMME PACKAGES FOR DESIGNING OF VSAMFC

The two computer programme package in FORTRAN Language has been developed for interactive designing and simulation of variable structure systems. First package offers possibility of designing and simulating linear VSS system with special possibilities of considering the VSAMFC systems. The programme provides the Kalman filter utilisation for recursive reconstruction of unaccessible process states. The second interactive programme provides designing and simulation both linear and nonlinear VSAMFC systems. In this programme for reconstruction of unaccessible states for nonlinear plant the extended Kalman filter is provided as well as modelling of process disturbances. In the programme the method of designing which has been mentioned in the previous sections has been applied. Generally the programme offers: designing of various kind of linear and nonlinear plants, specifying the linear model reference, simulating of behaviour of plant and model for deterministic and stachastic cases, designing of switching planes and control vector for VSAMFC systems, simulating of whole VSAMFC systems, reconstruction of unaccesible state of plant by the using of extended Kalman filter. It is possible to consider also the nonstationary plants.

The wide input and interactive possibilities can be used during designing and simulation of VSAMFC system. Among them we can pointed out possibility of changing the plant parameters during simulation, filtering of disceuntinuous control sequences, and modeling the his terese region during the realistic switchies. Furthemore another feature of programme allows designing the so called nonstationary switching surfaces, which can guarantee better system error transient property.

In the cases when the plant disturbance statistic are not known the programme provides using of the adaptive Kalman filter. The programme gives wide possibility of graphic monitoring of obtained design and simulation results.

From the numerous examples which has been tested out by the using of the programme we can present the consideration of VSAMFC of nonlinear field controlled dc motor. The model of this plant can be described by the nonlinear equations (Eichler, 1976)

$$\dot{x}_1 = a_1 x_1 + a_2 u \qquad (22)$$
$$\dot{x}_2 = a_3 x_1 + a_4 x_1^2 x_2 + a_5 x_2$$

where the values of nominal parameters are : $a_1 = -6.6667$, $a_2 = 0.27778$, $a_3 = 381.84$, $a_4 = -11.402$, $a_5 = -1.6667$. The reference model has been selected as

$$\dot{x}_m = \begin{bmatrix} -6.6657 & 0.0 \\ 2000 & -68.96 \end{bmatrix} x_m + \begin{bmatrix} 0.277 \\ 0 \end{bmatrix} u \quad (23)$$

The satisfactory behaviour of system error has been reached for C = (30,1) and the following control coefficients (egus.(13),(14) extending by the term k_n)

$$k_p=(21.8,0.2),\ k_e=(30.0,12.0),$$
$$k_m=(0,0),\ k_n=1.4 \qquad (24)$$

CONSLUSION

The designed programme packages give certain step towards constructing the computeraided design tools for VSAMFC systems. This is only small step, since the methodologies of systematic designing of VSAMFC system are still in evolution phase. The special meeds is for designing of CAD tools for nonlinear and time -varying systems.

REFERENCES

Drażenovic,B. (1969). The invariance conditions in variable structure systems. Automatica,5, 287-295.

Emelyanov,S.V. (1964). Design of variable structure control system with discontinuous switching functions. Eng.Cybern.1,156-160.

Erzberger,H. (1968). Analysis and design of model-following control systems by state space techniques. Proc. of JACC, Ann Arbor, 572-581.

Eichler,J., Mansour,M. (1976). A suboptimal control scheme of nonlinear systems besed on the stored response modeling technique. Regelungstechnik,27

Knafel, A. (1984). Mikrokomputerowe systemy sterowania adaptacyjnego wykorzystujące technikę algorytmów o zmiennej strukturze. MSci. Thesis. Warsaw Technical University. Warsaw.

Slotine,J.J., Sastry,S.S. (1983). Tracking control of non-linear systems using sliding surfaces, with application to robot manipulators. Int.J.Control, 38,465-492.

Świniarski,R. (1980). Optimal control and microcomputer systems. Warsaw Technical University Press.Warszawa

Utkin,V. (1971). Equations of sliding mode in discontinuous systems - 1. Automat.and Remote Control,21,1897-1907.

Utkin,V.I. (1977). Variable structure systems with sliding modes. IEEE Trans.,AC-22,212-222.

Young,K,K.D. (1977). Asymptotic stability of model reference systems with variable structure control. IEEE Trans.,AC-22,279-281.

Young,K.K.D. (1978). Design of variable structure model-following control systems, ibid., AC-23,1079-1085.

Zinober,A.S.I., El-Ghezawi,O.M.E., Billings,S.A., (1982). Multivariable variable - structure adaptive model - following control systems. IEE Proc., 129, Pt.D,1,6-12.

El-Ghezawi,O.M.E., Zinober,A.S.I., Billings,S.A., (1984). Analysis and design of variable structure systems using a geometric approach. Int.J.Control, 38,68-79.

Landau,I.D. (1974). A survey of model reference adaptive techniques-theory and applications. Automatica, 10,353-379.

COMPUTER AIDED DESIGN OF AIRCRAFT FLIGHT CONTROL SYSTEM

M. Vukobratović* and R. Stojić**

*"Mihailo Pupin" Institute, Belgrade, Yugoslavia
**Aeronautical Institute, Belgrade, Yugoslavia

Abstract. A computer implementation of flight control system design method, based on decomposition-aggregation principle and decentralized control of large-scale mechanical systems, is described. Numerical aspects of using the complete nonlinear aircraft model in the synthesis process are considered. Some data about software organization and user interaction are given.

Keywords. Computer-aided design; control system design; flight control; nonlinear control systems; decentralized control; suboptimality; decomposition-aggregation principle.

INTRODUCTION

The capabilities offered by powerful minicomputers have opened up new possibilities in control system design. Interactive computing allowed engineers' intuition to be combined with computer power, so most of existing synthesis techniques were implemented on digital computer to automatize design and save designers' time.

A flight control system is usually designed using a linear decoupled aircraft model valid for a perturbed stationarry horizontal flight. Correctness of control laws obtained in this way is sometimes checked by simulation of the complete nonlinear model. Most of the existing design methods were adopted for approximative and simplified models of real object, with a linear control method as a basic engineering tool (Shah, 1979; Kubbat, 1979). But, advances in air vehicle design and flight control system implementation have begun to overwhelm traditional design and analysis techniques which are most effective in aircraft applications with few surfaces, little dynamic coupling, separation between flight control and other avionic subsystem bandwidths and analog system implementations. So, the need for better and more accurate control system design and analysis techniques has become evident.

Because of recent advances in computer technology, a more sophisticated design method became a real possibility for practical applications. Hence, there appeared, quite naturally, a tendency toward taking into account all relevant information about object dynamics, i.e. apllying as accurate object models as possible. One of such approaches is decomposition-aggregation method, first developed for control of robotic systems (Vukobratović and Stokić 1980; Vukobratović, 1983) and later applied to flight control synthesis (Vukobratović, Stokić and Stojić, 1982; Stojić, 1985). In the paper the attention will be focussed on implementation of the control law synthesis procedure and some numerical algorithms.

The basic idea is that control synthesis should be based on a complete dynamic model, by which we mean a model accurate enough to allow simulation of real object with desired accuracy. But model complexity causes some difficulties during implementation, or in use, which had to be overcome by proper numerical and conceptual interventions. So, we have tried to develop suitable algorithms which, for every step in the synthesis, get all possible informations about object dynamics from simulation program modules.

The plan of the paper is as follows: First, the general approach to control synthesis and stability analysis, whose detailed explanation is given by (Stojić, 1985; Vukobratović and Stokić, 1980), is briefly presented. Then, some new numerical algorithms, which made possible implementation of the synthesis procedure, are given. Implementation of an experimental package on DEC VAX-11/780 with user interaction is explained.

OUTLINE OF THE SYNTHESIS PROCEDURE

The full-scale nonlinear model of the aircraft can be described as

$$\dot{X} = F(t, X) + G(t, X)U \tag{1}$$

$$Y = H(t, X) \tag{2}$$

where $X \in R^n$, $U \in R^m$, $Y \in R^m$ are state, input and output vectors, respectively. The control task is to realize the desired output vector

$$Y = Y_0(t) \qquad t \in T \tag{3}$$

with tolerances within given limits, in the presence of disturbances in the type of initial conditions. In (3) T is a finite interval of time in which the system is observed.

In the first step, nominal control $U_0(t)$ and state space trajectory $X_0(t)$ are calculated suing equations (1, 2) to satisfy (3), i.e. triplet ($X_0(t)$, $U_0(t)$, $Y_0(t)$) makes (1-3) identities over the time interval T. This is, actually, solving of the inverse dynamic problem or construction of nonlinear system inverse.

An the control $U_0(t)$ realizes the desired motion in an ideal case, in the second step the control u(t) is introduced to achieve tracking of nominal trajectory $X_0(t)$ with control

$$U = U_0 + u \tag{4}$$

in the presence of disturbances acting on the aircraft. For this purpose, the model of aircraft deviations from the nominal state space trajectory is used:

$$\dot{x} = f(t, x) + b(t, x)u \qquad (5)$$

where $x = X - X_o$ is the state deviation, $f(t, x) = F(t, X) - F(t, X_o) + G(t, X)U - G(t, X_o)U_o$. Practical stability conditions are imposed: for $\forall x(t_o) \in D_o$ it should be ensured that $x(t) \in D_\tau$ for $\forall t > t_o + \tau$ where $D_o \subset R^n$, $D_\tau \subset R^n$ and τ=const are specified in advance.

To synthesize control u(t) the system (5) is considered as a set of s subsystems, according to the adopted input and state space decompositions

$$R^m = R^{m_1} \times \cdots \times R^{m_s}, \quad R^n = R^{n_1} \times \cdots \times R^{n_s} \qquad (6)$$

with $x_i \in R^{n_i}$ and $u_i \in R^{m_i}$ as the i-th subsystem state and input vectors. Every isolated subsystem

$$\dot{x}_i = f_i(t, \hat{x}_i) + b_i(t, \hat{x}_i)\hat{u}_i \quad i=1,\ldots,s \qquad (7)$$

where $\hat{x}_i$ and $\hat{u}_i$ are projections of x and u on subspaces R^{n_i} and R^{m_i}, respectively, is stabilized by the local control

$$u_i = u_i^\ell(t, x_i) \quad i = 1,\ldots,s \qquad (8)$$

which is synthesized using the approximative subsystem model

$$\dot{x}_i = \bar{f}_i(t, x_i) + \bar{b}_i(t, x_i)u_i \quad i = 1,\ldots,s \qquad (9)$$

The u_i^ℓ is chosen so as to achieve such a subsystem exponential stability degree α_i which will satisfy the stability requirements for isolated subsystem (7).

To analyse the overall stability of system (5), when coupling between subsystems is taken into account, we introduce Lyapunov functions of subsystems $v_i = v_i(t, x_i)$ i=1,...,s and define regions

$$\bar{D}_i = \{x_i\colon v_i(t, x_i) < v_i^o\} \quad i=1,\ldots,s \qquad (10)$$

The constants v_i^o are chosen so as to provide that $D \in \bar{D} = \bar{D}_1 \times \cdots \times \bar{D}_s$. Then, an aggregate system

$$\dot{v} = Wv \qquad (11)$$

where $v = [v_1,\ldots,v_s]^T$, $W = [w_{ij}]_{s\times s}$ is a Metzler matrix, which satisfies the condition

$$(\text{grad } v_i)^T(f_i(x)+b_i(t, x)u) < \sum_{j=1}^{s} w_{ij}v_j \quad i=1,\ldots,s \qquad (12)$$

for $\forall x \in \bar{D}$, is determined. If the condition

$$Wv^o < 0 \qquad (13)$$

where $v^o = [v_1^o \cdots v_s^o]^T$, holds, the system (5) is stable in $\bar{D}$. If not, one can adjust local control (8) or introduce global control $u^g(t, x)$, i.e. set

$$u_i = u_i^\ell(t, x_i) + u_i^g(t, x) \quad i=1,\ldots,s \qquad (14)$$

to compensate for the coupling effects between subsystems. This means minimization of numbers w_{ij}. Both u^ℓ and u^g can be introduced gradually and in a selective way, as a result of trade-aft between system performances and control law complexity.

Since the synthesis method is based on sufficient stability conditions, it is possible to achieve desired system performances without strictly satisfying aggregate stability conditions (13). The system performances can be checked by simulation. Since the procedure gives a quantitative measure of subsystem stability and interactions, its applicability is greatly widened in computer-aided design enviroment because it allows, feedbacks to be introduced step by step and various factors to be compensated.

NUMERICAL ASPECTS OF THE SYNTHESIS PROCEDURE

A six-degree-of-freedom aircraft simulation model for the moving base flight simulator is chosen as the complete model for actual application. The simulation model is usually avoilable in algorithmic form, i.e. as an actual computer implementation. A large number of data, fabulated functions, logical branching and conditions make it very difficult to use such a model in a classical manner-analyzing complex analytical expressions, making differentiations and approximations. For this reason the assumption has been made that the complete model is available only in a form of algorithmic mapping (1):

$$(t, X, U) \xrightarrow{F,\ B} \dot{X} \qquad (15)$$

by which triplet (t, X, U) maps into X. The software designed with these principles allows access to object dynamics only through a single dynamic module (15) and does not require user manipulations with the complete model after modeling is completed.

To speed up computations and provide users with simpler equations for analysis and synthesis, the approximative model, valid in restricted regions $D_x \subset R^n$, $D_u \subset R^n$ of state and input space is needed. This tork is easily solved if we choose the structure of approximative model of the form

$$\dot{x} = \xi^o(t, x, u) + P^T\xi(t, x, u) \qquad (16)$$

where P is a vector of unknown parameters and ξ^o, ξ are chosen functions. To determine P for the best approximative model in region $D_x \times D_u$ in the least square sense, the set of points $\{(x_j, u_j)\colon x_j \in D_x, u_j \in D_u\}_{j=1}^N$, for N chosen, is generated, and $\{\dot{x}_j\}_{j=1}^N$ is calculated at points (x_j, u_j) using relation (5). Then P is given by

$$P^T = Y_N Z_N^{-1} \qquad (17)$$

$$Y_N = \frac{1}{N}\sum_{j=1}^{N}(\dot{x}_j - \xi^o(t, x_j, u_j))\xi^T(t, x_j, u_j) \qquad (18)$$

$$Z_N = \frac{1}{N}\sum_{j=1}^{N}\xi(t, x_j, u_j)\xi^T(t, x_j, u_j) \qquad (19)$$

For the linear approximative model,

$$\dot{x} = Ax + Bu = [A \quad B]\left[\frac{x}{u}\right] \qquad (20)$$

we obtain

$$[A \quad B] = \left(\frac{1}{N}\sum_{j=1}^{N}\dot{x}_j\begin{bmatrix}x_j\\u_j\end{bmatrix}^T\right)\left(\frac{1}{N}\sum_{j=1}^{N}\begin{bmatrix}x_j\\u_j\end{bmatrix}\begin{bmatrix}x_j\\u_j\end{bmatrix}^T\right)^{-1} \qquad (21)$$

In a special case if one wants the linear model to be valid in the vicinity of the point x=0, then, taking $x_j = [0,\ldots,0, \varepsilon_j, 0,\ldots,0]$ j = 1,...,n, where ε_j are chosen small constants, the column vectors A_j of matrix A are given by

$$A_j = f(t, x_j)/\varepsilon_j \qquad (22)$$

and

$$B = b(t, 0) \qquad (23)$$

The nominal control and state space trajectory are calculated applying a numerical procedure based on solving the inverse dynamics problem (Stojić, 1985).

System decomposition into subsystems is, in fact, a partitioning of X, U and Y into subvectors, which is usually proceeded by coordinate permutation to group subsystem coordinates.

Local control synthesis can be done at subsystem level by well known classical or modern methods. Since subsystems are ot low order, this task can be successfully solved by various state space or frequency techniques.

The key step in nonlinear system stability analysis is to construct the Lyapunov function in such a way as to get better estimation and avoid potentional conservatism. The best-choice are Lyapunov functions $v_i(t, x_i)$ which exactly estimate subsystems exponential stability degree (Šiljak, 1978). We chose $v_i(t, x_i)$ in the class of functions

$$v_i(t, x) = (x_i^T P_i x_i)^r \quad r>0 \tag{24}$$

where P is a positive definite matrix and r>0, real number. To obtain the estimate

$$\dot{v}_i(t, x_i) < \alpha_i v_i(t, x_i) \tag{25}$$

with a minimal α, the following optimization problem is solved:

$$\max_P \min_{x \in D_x} (-\dot{v}_i(t, x)/v_i(t, x_i)) \tag{26}$$

If v_i is chosen on the basis of linearized model, the procedure given in (Šilja, 1978; Vukobratović and Stokić, 1980) can be applied. For the nonlinear complete or approximate model the optimization problem (26) should be solved.

Since it is difficult to calculate "by hand" the partial derivatives of $\dot{v}_i$ especially for the complete model, the optimization procedure without calculating the derivatives is used (Bazaraa M. and Shetty C., 1979). Numerical differentiation is avoided because it is time consuming and causes numerical errors ($\dot{v}_i$ is often nondifferentiable in D_x).

To determine the aggregate matrix W, a special procedure has been developed. Since the choice of w_{ij} satisfying (13) is not unique, we search for such w_{ij} which gives the best estimate of subsystem exponential stability degrees. This task is realized by solving the following optimization problems

$$\min_W \sum_{i=1}^{s} w_{ij} \tag{27}$$

with constraints

$$\dot{v}_i(x) < \sum_{j=1}^{s} w_{ij} v_j(x_j), \quad x \in D_x \tag{28}$$

We have a linear cost function with an infinite number of linear constraints (i.e. for every $x \in D_x$).

To solve this problem a secant-type algorithm is applied.

SOFTWARE IMPLEMENTATION

The goal during software design was to automatize control system design, allowing flexibility in use, easy software development and its extension. An assumption was made that a designer directly interacts with a computer and is familiar with existing file servicing routines for creating, deleting or editing files. The design procedure, whoose global flow chart is given in Fig. 1, is implemented on VAX-11/780 computer. For user communication, the keyboard for input and alphanumerical or graphical terminal for output are used.

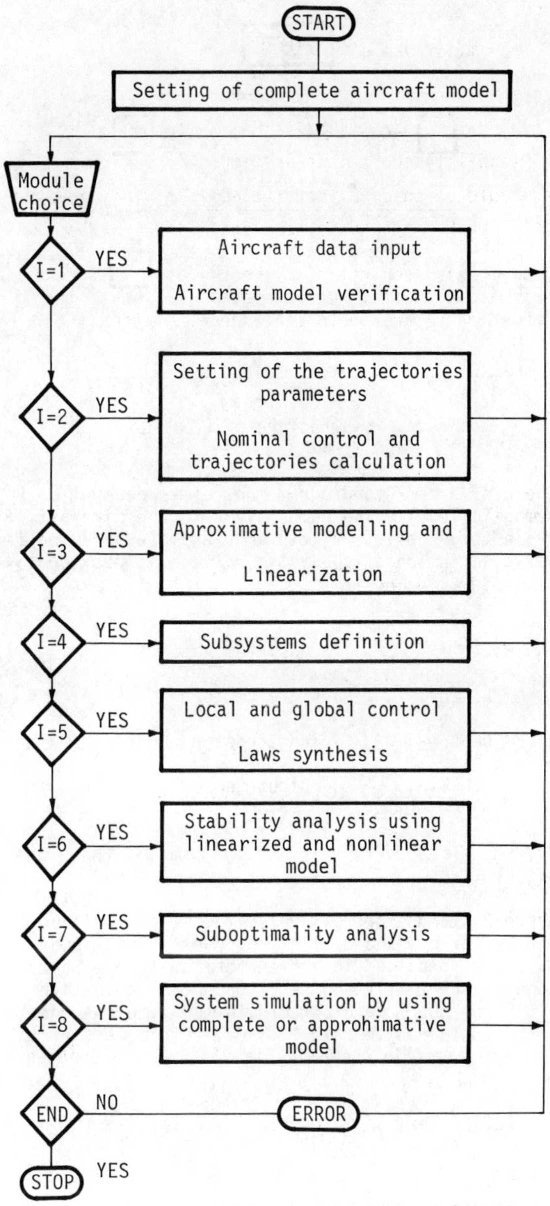

Fig. 1. Global flow chart of software support organization

The package, called FCS CAD (Flight Control Systems CAD), is organized on the same principles as the earlier version for linear control system design (Stojić and Mitrović, 1983) FCS CAD has a two-level structure, as shown in Fig. 2.

The first level is command decoding performed by procedure INTERP which also assigns input and output files and devices. A command oriented interaction was chosen for user communication with the computer. The form of command is

CNAME OF1,...,OFK = IF1,...,IFL

where CNAME is command (module) name OF1,...,OFK are output file names and IF1,...,IFL are input file names (command arguments). Some of the input arguments are optional and they are usually specified if a user wonts to avoid default control parameters and constants in algorithms. If other arguments are missing, the INTERP prompts for them. All output arguments are optional.

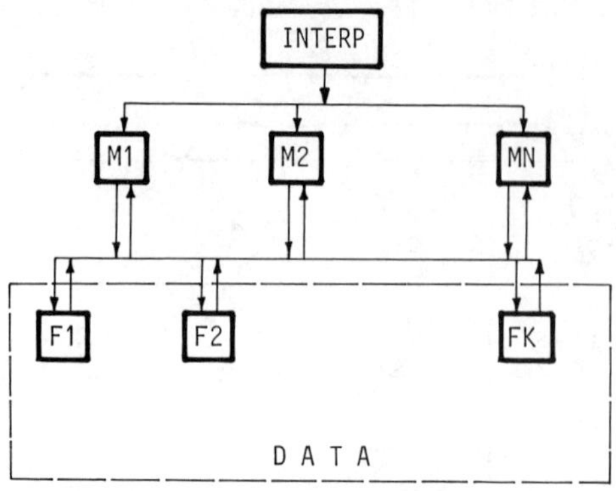

Fig. 2. Structure of FCS CAD

The user access to program control is done at a symbolic level using names given to the aircraft, trajectory subsystems, etc. For example, a symbolical description of subsystem rolling is provided by the file shown in Fig. 3a

ROLL	subsystem name
P FI	state vector
DL	input vector
P FI	output vector

Fig. 3a. File for subsystem definition

where P stands for roll velocity, FI for roll angle and DL for command deflection. User edited input data are specified in a free format. If one wants to plot the time response of P contained in file A1.TRJ on the same diagram with the time response of FI contained in file F2.TRJ, it is necessary to specify the file shown in Fig. 3b where T denotes

```
PLOT  "Text shown on plot"

A4  2  1

DIJAGRAM   "Text on diagram"

PAR  A1.TRJ  T  P

PAR  F2.TRJ  T  FI
```

Fig. 3b. Graphical output control file

time. The statement "A4 2 1" specifies that the format of plot will be A4 with 2 by 1 diagram on each page.

The design process is decomposed into a number of typical tasks which may be solved by the computer without user intervention during execution. Such tasks are called modules. They form the second level and perform assigned operation on input data and give output data. Every module is separately executable and knows nothing about the others. This feature is especially suitable for developing and adding new modules to the package.

One can distinguish modules for nonlinear aircraft models and for linear models. In the nonlinear modules the aircraft dynamics is represented by a subroutine realizing the mapping (15). The nonlinear control law and structure of approximative models are set by user defined subroutines. A correction in mathematical model requires subroutine changing. In the linear modules, which are mainly intended for local control synthesis, the model is represented parametrically. Thus, they are designed to be applicable for general linear systems of order lower than 20 (Stojić and Mitrović, 1983). It is assumed that the linear system can be represented by:

a) (A, B, C) - a triplet of matrices for state space representation

b) $\frac{1}{d(s)} G(s)$ - matrix transfer function

c) $\{t_i, x(t_i), u(t_i)\}_{i=1}^{N}$ - time response

d) $\{s_i, \frac{1}{d(s_i)} G(s_i)\}_{i=1}^{N}$ - frequency response

e) (A) - linear static system $y = Au$

Standard internal computer representation (disc file) is associated with each type of representation, a) to e). The modules are constrained to have input and output in a form of standard representation, enabling design to be done by a sequence of commands without intermediate data preparation. Other module inputs are optional, and contain control parameters and constants, if the user wants to avoid default values.

The principle of using the standard form of input and output files has been retained for nonlinear modules, except for aircraft data which are used to initiate flight dynamic subroutine.

Procedure INTERP is realized on VAX-VMS operating system as using Digital Command language. Execution of every module is done as a subprocess. This permits any sequence of modules to be defined as a new module using command files. Adding a new module to the package requires no change of INTERP. As a module is indenpendent of the others, and itself represents a single executable program, it is sufficient to complete its testing. For HELP purposes, the INTERP has to be supplied by a file containing short information about modules. The modules are mainly written in FORTRAN and some of them in PASCAL.

AN EXAMPLE OF CONTROL SYNTHESIS

The procedure is illustrated by automatic flight control synthesis of a combat aircraft. Aircraft data from the example are used for forming a six-degree-of-freedom flight simulator. The desired maneuver is an asymmetric pull-up maneuver with a normal acceleration of about 4.5g, shown in Fig. 4.

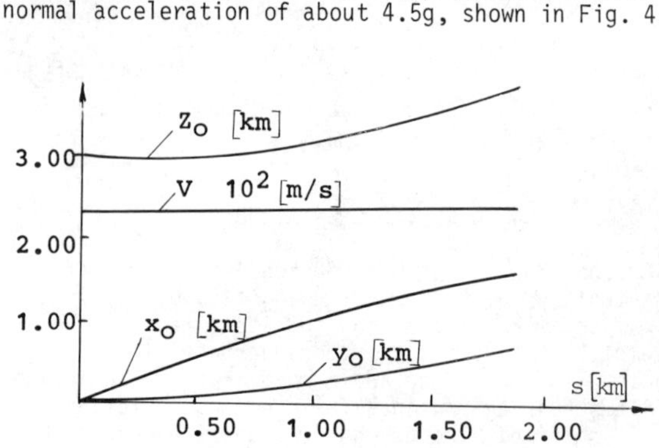

Fig. 4. Desired aircraft maneuver

Synthesized nominal control U^0 and state space trajectory X^0 are shown in Figs. 5 and 6.

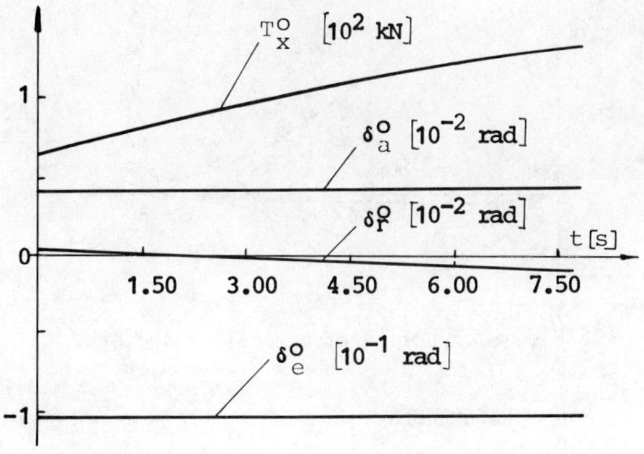

Fig. 5. Nominal control

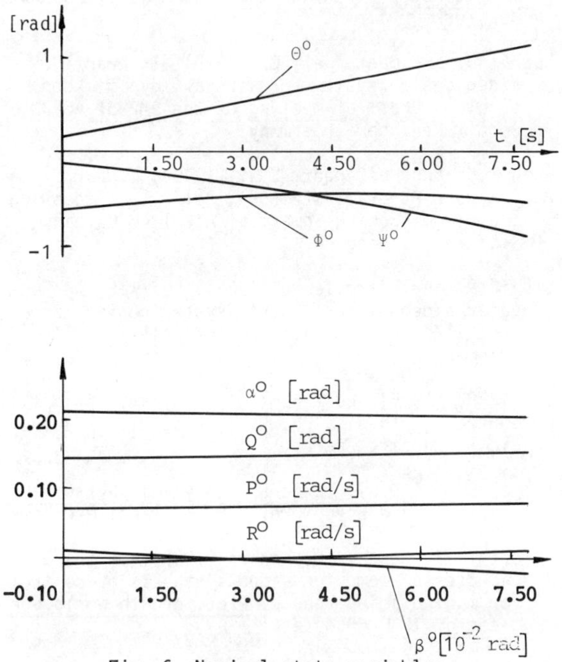

Fig. 6. Nominal state variables

In Figs. 4, 5, 6 the following notation is used: x_0, y_0, z_0 are the coordinates of the aircraft center of mass in the inertial system; $U = [T_x, \delta_e, \delta_a, \delta_r]$ is the thrust, the elevator deflection, rudder deflection and aileron deflection respectively; P, Q, R are the projections of the aircraft angular velocity on the body axes system; Φ, Θ, Ψ are Euler's angles; v, β, α are the flight path velocity, the sideslip angle and the angle of attack, respectively.

As the third step of the procedure the linearized model (19), describing small perturbations around the initial point of nominal state space trajectory is formed.

For the purpose of decentralized control systhesis, we adopt the subsystems according to Fig. 7, where "Δ" denote increments in the previously defined variables with respect to their nominal values.

The local control is chosen in the form

$$u_i^\ell = -K_i^L x_i \qquad i=1,\dots,4 \tag{29}$$

to minimize the local quadratic criterion

$$J_i(x_i(0)) = \int_0^\infty e^{2\sigma_i t}(x_i^T Q_i x_i + u_i^T R_i u_i)dt \tag{30}$$

with respect to the linearized subsystem equations

$$\dot{x}_i = A_i x_i + B_i u_i \qquad i=1,\dots,4 \tag{31}$$

where σ_i is the prescribed exponential stability degree of subsystem, Q_i, R_i are the positive semi-definite weighting matrices.

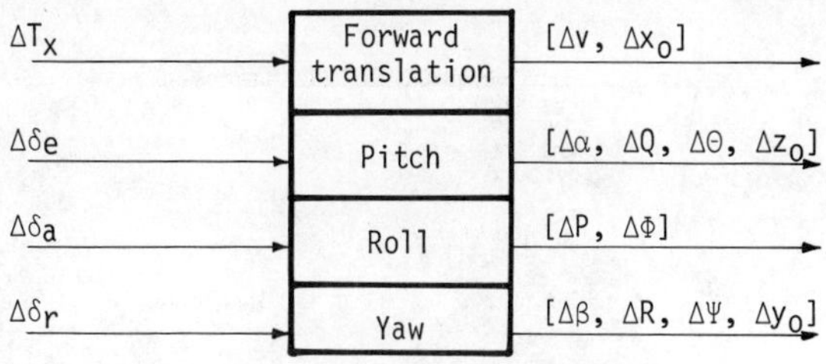

Fig. 7. System decomposition

The parameters of local control in the example are shown in Table 1.

TABLE 1 Parameters of local control

i	Q_i	R_i	σ_i	K_i^L
1	diag(10^{-7}, 10^{-2})	10^{-7}	0.5	[$0.15\cdot10^5$, 7938]
2	diag(10^{-4}, 10, 0.02, 0.05)	$0.12\cdot10^5$	0.3	[0.75, -0.076, -0.98, $0.31\cdot10^{-2}$]
3	diag(0.05, 5.)	120	0.	[0.039, 0.204]
4	diag(10^{-5}, 100., 10^{-3}, 10^{-2})	5.10^3	0.32	[-6.806, -0.668, -7.451, $-0.93\cdot10^{-2}$]

The subsystem Lyapunov functions are chosen in the form (24) on the basis of linearized subsystem models.

Aggregate matrix W_L in (11) for the overall linearized system (19) with local control

$$W_L = \begin{bmatrix} -0.9705 & 597.7 & 68.27 & 183.0 \\ 0.1036E-01 & -1.231 & 0.3785 & 4.994 \\ 0.8274E-03 & 0.1122E-01 & -3.527 & 55.95 \\ 0.4935E-02 & 0.3189 & 34.95 & -0.6621 \end{bmatrix}$$

shows significant interaction between subsystems. Moreover, eigenvalues of the A-BK^L matrix show that the fourth subsystem is unstable. Thus, we decided to apply global control in the form

$$u^g = -K^G x \tag{32}$$

To reduce interaction factors W_{12}, W_{34}, W_{43} in the matrix W_L appropriate gains are determined for linear global control such that

$$K^G = \begin{bmatrix} 0. & 0. & -0.8109E+6 & 0.300 & 0.3619E+6 & -1402. \\ 0. & 0. & 0. & 0. & 0. & 0. \\ 0. & 0. & 0. & 0. & 0. & 0. \\ 0. & 0. & 0. & 0. & 0. & 0. \end{bmatrix}$$

$$\begin{matrix} 0. & 0. & 0. & 0. & 0. & 0. \\ 0. & 0. & 0. & 0. & 0. & 0. \\ 0. & 0. & 2.188 & 0.3151 & 3.386 & 0.4241E-2 \\ -0.3114 & 0.1637 & 0. & 0. & 0. & 0. \end{matrix} \Bigg]$$

Closing the global loop gains reduced the couplings between subsystems, as can be seen from aggregate matrix W_{L+G} for the overall linearized system

$$W_{L+G} = \begin{bmatrix} -0.9705 & 15.01 & 68.27 & 32.98 \\ 0.8141E-02 & -1.333 & 0.3235 & 0.6384 \\ 0.8268E-03 & 0.3499E-01 & -3.526 & 0.1110 \\ 0.7246E-02 & 1.777 & 37.50 & -0.5771 \end{bmatrix}$$

Simulation results of tracking for disturbances of initial conditions type using the complete model are shown in Fig. 8.

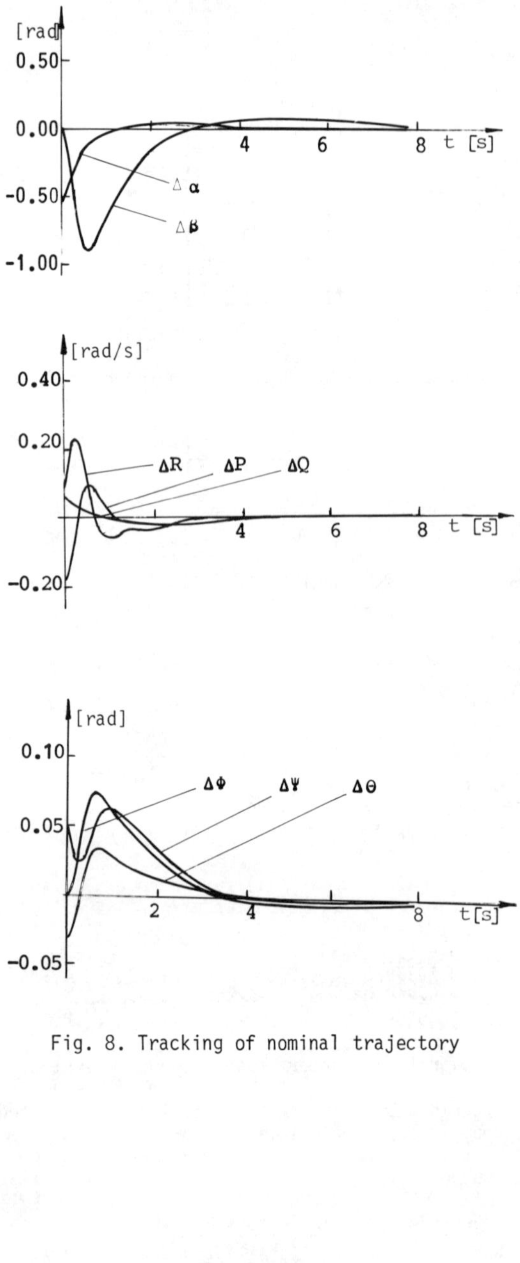

Fig. 8. Tracking of nominal trajectory

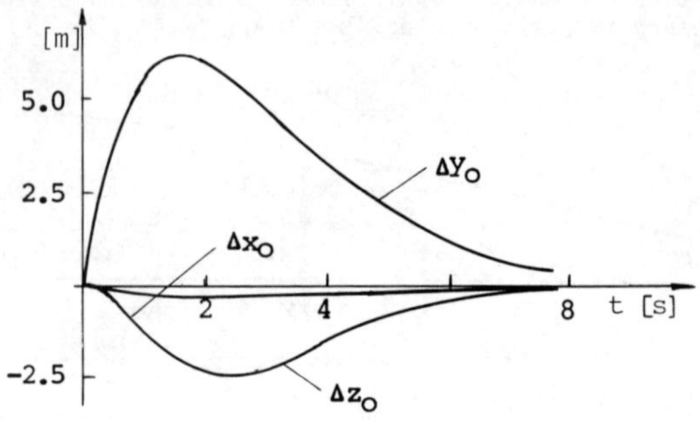

Fig. 8. Tracking of naminal trajectory

REFERENCES

Bazaraa M. and Shetty C. (1979). Nonlinear Programming: Theory and Algorithms. John Wiley and Sons, New York

Kubbat W. and Oesterhelt G. (1979). Interactive aided design system for aircraft dynamic control problem. Proc. of meeting of the SMP of AGARD, Sept., Neubiberg, Germany.

Shah P.C. and Heidergott K.W. (1979). Linear System Analysis Program, L224(QR), Vol. 1. Engineering and Usage, Boeing Comercial Airplane Company, Seattle, Washington.

Stojić R. and Mitrović V. (1983). Software for computer-aided linear control system design. Proc. of IASTED Symp. on Applied Informatics, Lile, France.

Stojić R. (1985). Contribution to dynamic flight control synthesis. Ph. D. Thesis. Military Technical Academy, Zagreb.

Šiljak D.D. (1978). Large Scale Dynamic Systems: Stability and Structure. Nort-Holland, New York.

Vukobratović M. and co-warkers (1983). Software package for computer-aided synthesis of control for manipulation robots. Proc. of 9th world congress of IFAC, Budapest.

Vukobratović M. and Stokić D. (1980). Contribution to the decoupled control of large-scale mechanical systems. Automatica, Vol. 16, pp. 9-22.

Vukobratović M., Stokić D. and Stojić R. (1982). One new method of dynamic flight control. Proc. of joint IFAC/ESA Symp. on automatic control in space, Holland.

COMPUTER AIDED DESIGN OF A HYDROFOIL CONTROL SYSTEM

R. Whalley* and P. C. Gregory**

**School of Mechanical Engineering, University of Bradford, West Yorkshire, UK*
***Department of Control Engineering, Royal Naval Engineering College, Plymouth, Devon, UK*

Abstract In this paper the regulation of the heave, pitch and roll motion of a 31 metre surface piercing hydrofoil model will be investigated using computer aided design software. The structure of the suite employed, the options available and the scope of the software, will be outlined together with the evolutionary changes in theoretical development which promoted the distinguishing features of the package. In describing the basis of the numerical procedures, emphasis will be placed upon the flexibility afforded by the approach adopted which has enabled continuous alteration and amendment to be accommodated for over a decade. By way of demonstration, optimal control theory will be used to synthesise controllers which minimise a quadratic performance index comprising the weighted sum of the squares of the state and input vectors. Variations on the closed loop system eigenvalues have been computed for changes in the penalty functions to give a family of curves relating the speed of response of the system to the performance index. Comparisons of the results obtained will be made with earlier work, based on non-interacting control theory, by selecting coincident closed loop eigenvalues and steady state characteristics. Frequency response representations of these configurations will be made and the relative performance of the optimal and non-interacting designs will be commented upon.

Keywords Computer aided system design; decoupling; multivariable systems; optimal control; dynamic response; ships.

INTRODUCTION

Almost twenty years ago work began on a computer aided design suite for control systems at the Royal Naval Engineering College, Manadon, Plymouth. The intention at that time was to build modular routines which could be run on small machines having 16k core with hardcopy plotting facilities. As the complexity of the task grew it became necessary to link modules via data files and common storage enabling series or cyclic connections to be made at the exit from the mainline routines. It was agreed that program construction should be in the form of a sequence of subroutine calls where possible incorporating the IBM scientific subroutines and in-house library routines which would be designed along similar lines. In keeping with the trend at that time Fortran 4 was selected as the programming language in the hope that it would be supported in accordance with IBM's guarantee of compatability with later versions of this language. Finally the starting date of the project determined that the programs embodied transfer function rather than state space concepts as time domain procedures were then still under development.

With the benefit of hindsight it is easy to see that a delay of one or two years would have altered certain aspects of this software study profoundly. In particular, the adoption of a computationally difficult polynomial approach in contrast to state space methods of analysis and design altered the future course of events significantly. Though managable for single input - single output systems the polynomial operations required in the scheme became increasingly difficult to sustain as large multi input - multi output descriptions were considered whereas in state space terms input and output scalars simply become vectors in the multivariable case. Thereafter calculations could be performed on arrays using all the manipulative power invoked by matrix algebra. In contrast the equivalent operations using polynomials were beset with dimensionality and efficiency problems which were inflated by coding difficulties.

For a short period of time the philosophical basis of the package appeared to be suspect as controller designs based on the minimisation of performance indices, known as optimal control, and then the procedures leading to arbitrary pole assignment were introduced. Lamentably there were few controller design techniques based on the input - output description of multivariable systems although it was appreciated that the strength of the approach depended on the introduction of procedures using the terminal characteristics of power and process plant.

Essentially the system equivalence afforded by software based on polynomials remained the focus of attention of the engineers constructing the package who investigated alternative approaches with great diligence. These studies showed that packages employing state space methods for design and simulation suffered from utilisation deficiences in that it was difficult to access intermediate variables such as controller inputs and outputs, inner loop feedback signals and actuator signals as a consequence of the system transformations required for computational convenience.

The inhibiting factor inherent in controller design suites based on state space methods arose from the minimum coordinate representation and hence the minimal storage requirements for system models together with the algorithmic efficiency associated with array operations. These "attributes" rendered such packages inflexible when attempts to expand the software to include design methods based on the input - output map of a system became increasingly popular in the seventies.

Fortunately, embedding state space design methods into software constructed to accommodate polynomial system descriptions and design procedure was, on the other hand, relatively easy. This aspect of the configuration together with the careful standardisation, documentation and testing of the routines used guaranteed the longevity of the suite which appears to have considerable potential for development today.

In the following section a description of the suite the structure and links between the principal modules and the facilities available for linear system design, analysis and simulation, are presented. The structure of the software has proved to be effective enabling a continuous program of alteration and addition to be absorbed since inception. Currently the package is being extended to include minimum variance and none linear design procedures.

SOFTWARE SUITE

The suite of programs for control system design, analysis and simulation known as CONPACK will accommodate system models having up to six input and six outputs with transfer functions of up to nineteenth order. State space descriptions may have up to 120 states although most of these dimensions can be lifted by appropriate changes in storage vector lengths. The multivariable section of the package which is approximately 200k is shown diagrammatically in Figure 1.

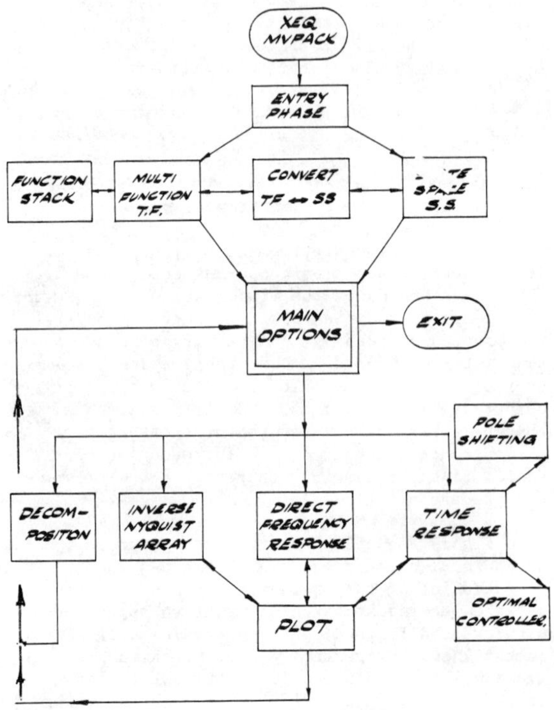

Fig. 1. Schematic of CAD Package.

HYDROFOIL MODEL

In this example the model of a 31 ft surface piercing hydrofoil will be considered. Optimal control and decoupling theory will be used in the feedback design which is given in detail in Bono et al (1983), Whalley, R., Gregory, P. C. (1984) and Whalley, R. (1978). The model of the craft was determined from sea trials and parameter estimates using maximum likelihood techniques.

State space representation of the hydrofoil is given in Gregory, P. C. and Whalley, R. (1984) from which the equivalent transfer function matrix description is

$$y(s) = G(s)\,u(s)$$

where the elements of G(s) are:

$$g_{11}(s) = (-.32810 - .36858s - .59678s^2 -.21364s^3 -.16600s^4)/\Delta(s)$$

$$g_{12}(s) = (-.27431 - .15600s - .27903s^2 - .71880E\text{-}01s^3 - .60000E\text{-}01s^4)/\Delta(s)$$

$$g_{21}(s) = (.18975 +.21315s + .34513s^2 + .12355s^3 + .96000E\text{-}01s^4)/\Delta(s)$$

$$g_{22}(s) = (.64006 + .36399s + .65108s^2 + .16772s^3 + .14000s^4)/\Delta(s)$$

$$g_{23}(s) = (.659327 + .81642s + 1.3263s^2 + .57247s^3 + .40400s^4)/\Delta(s)$$

$$g_{31}(s) = (.18975 + .21315s + .34513s^2 + .12355s^3 + .96000E\text{-}01s^4)/\Delta(s)$$

$$g_{32}(s) = (.64006 + .36399s + .65108s^2 + .16772s^3 + .14000s^4)/\Delta(s)$$

$$g_{33}(s) = (.65927 - .81642s - 1.3263s^2 - .57247s^3 - .40400s^4)/\Delta(s)$$

$$g_{13} = 0$$

$$\Delta(s) = 3.8404 + 5.6266s + 10.436s^2 + 7.1081s^3 + 6.3927s^4 + 1.9510s^5 + 1.0000s^6$$

$$y(s) = (\text{pitch, heave, roll})^t$$

$$u(s) = (\text{aft flap angle change, frd port flap angle change, frd stbd flap angle change})^t$$

GRAPHICAL ANALYSIS

The simulated open loop response of the hydrofoil following step changes on frd port and stbd flaps and the aft flaps are shown in Figures 2, 3 and 4, respectively.

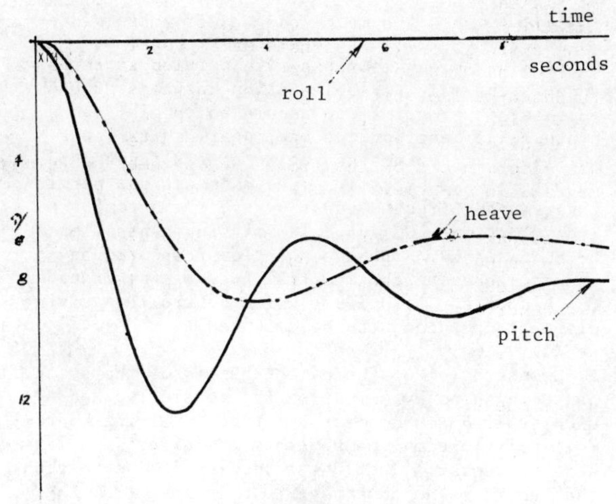

Fig. 2. Transient response of open loop system following a 10% step change in aft flaps.

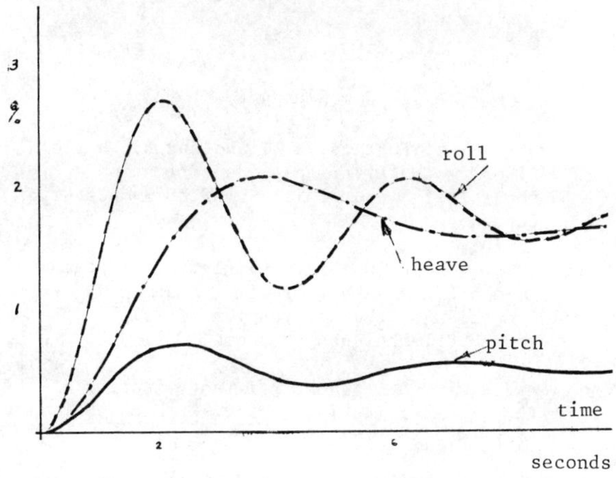

Fig. 3. Transient response of open loop system following a 10% step change in fwd port flap.

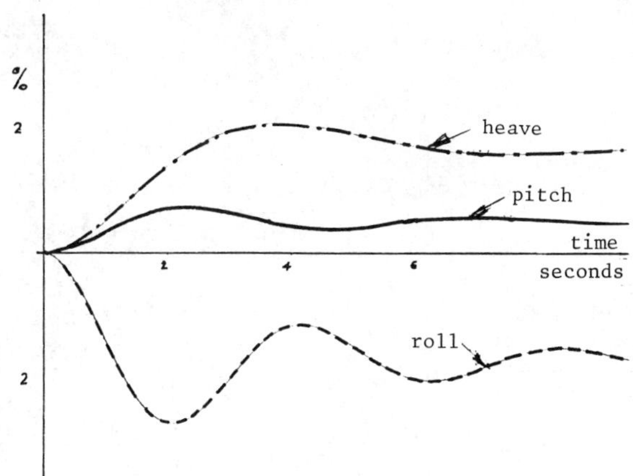

Fig. 4. Transient response of open loop system following a 10% step change in fwd stbd flap.

Moreover, the Nyquist array indicates the high degree of interaction as shown in Figure 5 below where each element has large Gershgorin circles arising from the magnitude of the off diagonal elements. In Bono et al (1983) and Whalley and Gregory (1984) the decoupling and LQG controllers are derived using the decomposition phase and the optimal controller module in MV PACK. By varying the performance index the closed loop eigenvalues of the decoupled and optimal control strategies were arranged to be coincident and using the divided gain procedure in the Pole-Shifting module steady state equivalence was achieved thereafter.

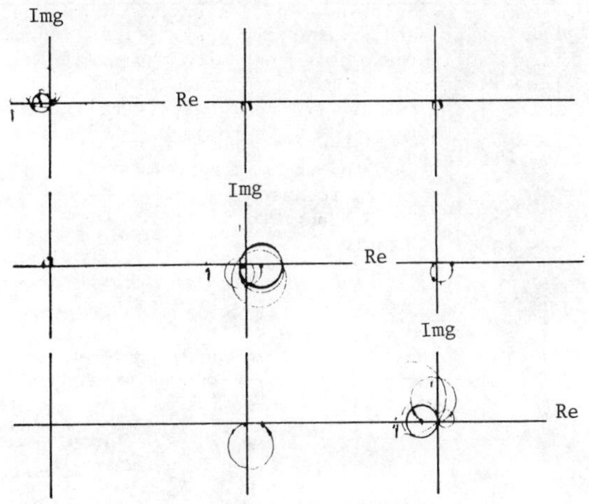

Fig. 5. DNA of system with G Bands superimposed from .01 to 1.0 R/sec.

In Figures 6, 7 and 8 the closed loop optimal and none-optimal (decoupled) response characteristics are shown on the same axis following 10% step changes on each desired value, in turn.

Naturally the optimal and none-optimal (decoupled) response characteristics are similar in each case. However there is significant interaction following a 10% step change in the desired value of heave where the **pitch** transient gives almost 40% coupling between these variables.

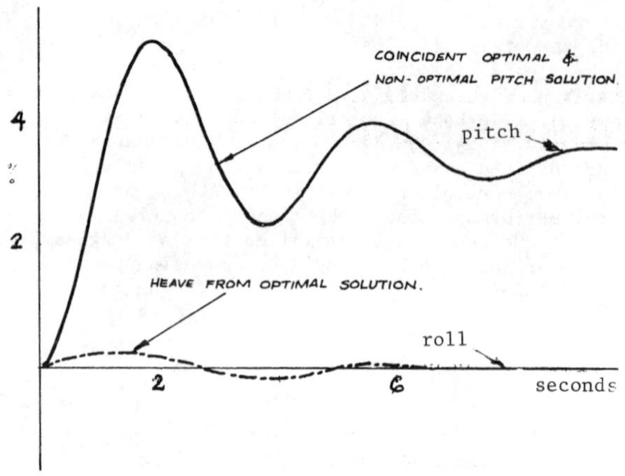

Fig. 6. Transient response of optimal and non-optimal systems with unity feedback following a 10% step change to desired value of pitch.

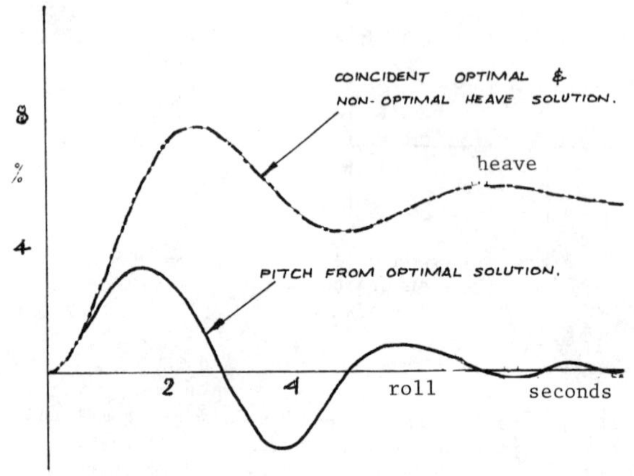

Fig. 7. Transient response of optimal and non-optimal systems with unity feedback following a 10% step change to desired value of heave.

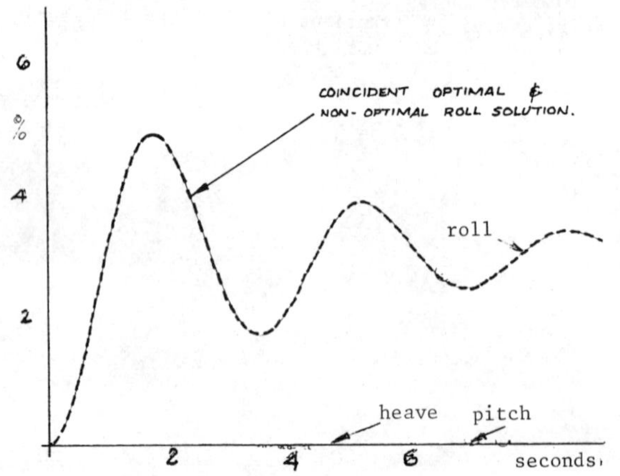

Fig. 8. Transient response of optimal and non-optimal systems with unity feedback following a 10% step change to desired value of roll.

CONCLUSION

An interconnected suite of software known as CONPACK has been described with particular reference to the multivariable option known as MV PACK. This set of programs can handle polynomial and state space descriptions of linear, multi-input, multi-output systems having up to six inputs and outputs and 120 states. The main options permit inverse and direct Nyquist array, characteristic locus, Optimal LQG and Pole-Shifting controllers to be designed. A decomposition phase can be used to generate the spectral form of the transfer function description from which controllers giving diagonal dominance can be computed.

A demonstration of the effectiveness of the suite has been shown by computing the controlled and uncontrolled response of a surface piercing hydrofoil using optimal control and the direct Nyquist array methods. From these results it can be seen that there is no advantage to be gained from the use of optimal control technique which gives a poorer result than the non-interacting controller even though a more complicated feedback strategy was used.

REFERENCES

Bono, R., R. L. Fuzzard, A. Tiano and R. Whalley (1983). Multivariable identification and control of a hydrofoil. IASTED Symposium, Copenhagen.

Gregory, P. C. and Whalley, R. (1984). Optimal regulation of a hydrofoil. Inst. M.C. Symposium on the Application of Multivariable Techniques, Manadon, Plymouth.

Whalley, R. (1978). Spectral factorisation of the transfer matrix. Proc. I. Mech. E., Vol. 192, No. 29.

Whalley, R. and Gregory, P. C. (1984). Multivariable control of a hydrofoil. Seventh Ship Control Conference, Bath.

ON-LINE DECISION AIDS IN COMPUTER INTEGRATED MANUFACTURING

C. G. Cassandras

Department of Electrical and Computer Engineering, University of Massachusetts, Amherst, MA 01003, USA

Abstract. Control and performance optimization of modern manufacturing systems is characterized by an increasing degree of complexity that accompanies automation and flexibility. Because the functional relationship between typical performance measures and parameters of interest generally cannot be determined, analytic (queueing) models are only of limited help. Detailed analysis requires extensive simulation, an approach limited by its high, sometimes prohibitive, computational cost and infeasible when real-time decisions must be made. An alternative is offered by Perturbation Analysis, a recently developed technique which efficiently estimates performance gradients with respect to parameters from a single interval of observing real on-line data. This technique can be used to exploit the data collection capabilities of computer integrated manufacturing systems and automatically provide valuable "sensitivity" information. It essentially creates a decision aid whose uniqueness lies in supplying on-line answers to "what if..." questions. An application of perturbation analysis is discussed, where emphasis is placed on optimizing control strategies for dynamic (i.e. state dependent) scheduling and routing of individual parts in a computer integrated manufacturing system.

Keywords. Perturbation techniques, discrete event systems, routing strategies, material handling.

INTRODUCTION

A typical automated manufacturing process consists of interdependent operations linked with a transportation (material handling) system to form a network through which a variety of production parts can circulate. Recently, the term 'flexible manufacturing system' (FMS) has emerged to describe such processes. Another popular term is 'computer integrated manufacturing' (CIM). The latter captures the fact that the design and real-time operation of such an environment is heavily dependent on the use of computers for performance evaluation, scheduling, monitoring and control.

Viewed as dynamic systems, automated manufacturing processes are characterized by a high level of complexity, compounded by the presence of uncertainties such as random mechanical failures. Given a performance measure J and a set of decision parameters $\theta_1,\dots,\theta_N$, it is important to be able to evaluate 'sensitivities' of the form $\Delta J/\Delta\theta_i$, i=1,...,N, over one or more stochastic realizations of the system. At the design stage, this contributes to the selection of the right components. In normal operation, these sensitivites are valuable tools in answering a variety of "what if..." questions for fine tuning as well as short-term scheduling purposes.

Queueing models of manufacturing processes generated by stochastic analysis (Solberg, 1977; Buzacott and Shanthikumar, 1980) are limited by a considerable number of restrictive assumptions. For most real systems of interest they provide approximations only useful in first-cut feasibility studies. Since the function $J(\theta_1,\dots,\theta_N)$ is generally unknown, simulation is a widely accepted tool in analyzing the complicated dynamics of these systems (Fishman, 1978) and studying the effect of a parametric perturbation $\Delta\theta_i$ on some performance measure J. This approach, however, is limited by its high, sometimes prohibitive, computational cost. For this reason, it is also infeasible as an on-line decision tool.

A new approach that bypasses the limitations of both queueing theory and simulation has recently been developed for efficiently estimating sensitivities of the form $\Delta J/\Delta\theta_i$, i=1...,N (Ho and Cassandras, 1983; Ho and co-workers, 1984). This approach is based on the availability of real data processed on-line and has been termed Perturbation Analysis (PA). Exploiting the fast monitoring capabilities of automated manufacturing systems, this technique adds a new dimension to complicated decision making problems in a computer integrated environment. Our purpose here is to describe the fundamental results of PA and consider its applicability to computer integrated manufacturing. A specific scheduling problem is also discussed as an example of this technique's real-time potential.

PERTURBATION ANALYSIS OF DISCRETE EVENT SYSTEMS

A discrete event dynamic system (DEDS) is characterized by state transitions completely controlled by the occurrences of discrete events over time. This is the case in most manufacturing processes, where the arrival (departure) of production parts at machines initiates (terminates) specific activities, generally stochastic in nature.

Using models developed by classical queueing theory (for example, Jackson, 1963), a DEDS consists of interconnected servers and queues of some finite capacity (infinite capacity being just a special case). Through this network circulate customers, generally belonging to different classes. For simplicity, we initially assume a single customer class is present. Two key observation can immediately be made:

1. Servers are always in one of three states:

(i) Busy (ii) Idle (iii) Blocked (by one or more downstream queues whose capacities have been reached.

2. Only 'idling' (also termed NO Input (NI)) and 'blocking' (also termed Full Output (FO)) are interesting phenomena from a performance optimization point of view (if all servers were always busy, then there would be nothing to improve upon).

In systems and control theory, it is often possible to observe the state trajectory of a complex nonlinear system and then attempt to linearize around this nominal trajectory. A similar approach is followed by PA for a stochastic DEDS: in the course of a particular stochastic realization, performance gradients are computed with respect to various decision parameters. The discrete event nature of these systems gives rise to simple state perturbation dynamics. Furthermore, as an estimation procedure, it has recently been pointed out by Cao (1984) and Kumar (1984) that PA is often characterized by attractive statistical properties.

Viewed as a dynamic system evolving over a particular stochastic realization, a DEDS can be described by state equations with two categories of state variables: (i) Queue lengths, q_i, i=1,...,N, (ii) Times of occurrence of specified events, e_i, i=1,...,N, such as service completions. Another important realization discussed by Cassandras and Ho (1984) is that the discrete event nature of such systems allows one to replace the time domain by a more natural event domain. In other words, time as the independent variable in the state equations is conveniently replaced by an event counter k=0,1,... The resulting state equations are of the form:

$$q(k+1) = q(k) + \sigma[q(k), e(k), \omega(k;\theta)]$$

$$e(k+1) = e(k) + \tau[q(k), e(k), \omega(k;\theta)]$$

where $[q(k)\ e(k)]^T$ is the state vector when the kth event occurs, $\sigma[\cdot]$ and $\tau[\cdot]$ are transition functions for $q(k)$, $e(k)$ respectively and $\omega(k;\theta)$ is the stochastic input parametrized by θ, which can be viewed as simply a stream of random variables. Details of this formulation are provided elsewhere (Cassandras and Ho, 1984).

Note that $e_i(k)$ is a real-valued variable, whereas $q_i(k)$ is integer-valued. Since we would first like to focus on "infinitesimal" perturbations, we study only the dynamics of $\delta e_i(k)$, i=1,...,N, which could be made arbitrarily small, while the smallest possible perturbation in queue length would yield $\delta q_i(k) = +1$ or -1. Nonetheless, "infinitesimal" perturbations in DEDS should really be interpreted as those causing no change in event ordering (Cassandras and Ho, 1984), even though this property can be satisfied by $\delta e_i(k)$ of finite value.

In Pa three steps are involved:

1. Perturbation Generation: the process through which a parameter perturbation $\delta\theta$ is transformed into some state perturbation $\delta e_i(k)$.

2. Perturbation Propagation: for "sufficiently small" $\delta e_i(k)$ (rigorously defined by Cassandras and Ho, 1984), the process by which these perturbations propagate through the network can be described as follows:

   'a perturbation δe_i at server i propagates to a neighboring server j whenever an idle or a blocked state at j is terminated by i' (PP)

Note that this process is universal, i.e. it is independent of the original parameter perturbation $\delta\theta$. A formal proof and discussion of this result can be found elsewhere (Cassandras and Ho, 1984).

3. Effect on Performance Measure: for most performance measures of interest (e.g. throughput), state perturbations $\delta e_i(k)$ are directly translated into perturbations δJ.

Example: Consider a manufacturing process consisting of two operations performed serially. Let the parameter of interest θ be the mean service time of the first operation (modelled as server 1), regardless of the actual service time distribution, and let $\delta\theta$ be a given perturbation. Of interest are the state variables $e_i(k)$, i=1,2, representing the times when service completions occur at server i; k=0,1,... counts occurrences of these events at either server. Finally, let J_K be the throughput time for K customers, i.e. the time when K production parts complete the second operation (server 2) of the process.

The perturbation $\delta\theta$ is interpreted as a fixed change to every service time actually observed at server 1 in the course of an observation interval (Suri, 1983). It is then easy to see what the three steps of PA are for this example:

1. Perturbation Generation: for all service completions at server 1:

$$\delta e_1(k+1) = \delta e_1(k) + \delta\theta, \quad \text{if kth event occurs at server 1} \tag{1}$$

2. Perturbation Propagation: according to (PP) we have:

$$\delta e_2(k+1) = \begin{cases} \delta e_1(k), & \text{if kth event occurs at server 1 and terminates NI at server 2} \\ \delta e_2(k), & \text{otherwise} \end{cases} \tag{2}$$

3. Effect on Performance Measure: since the throughput time for k events is given be $e_2(k)$, we can predict that for k < K:

$$\delta J_K(k;\delta\theta) = \delta e_2(k) \tag{3}$$

Thus, our estimate of the gradient $\frac{\partial J}{\partial\theta}$ is updated as follows:

$$\frac{\delta J_K(k)}{\delta\theta} = \frac{\delta J_K(k)}{\delta e_2(k)} \frac{\delta e_2(k)}{\delta e_1(k)} \frac{\delta e_1(k)}{\delta\theta}$$

where each part of the right-hand-side is determined from (1) - (3).

The Real-Time Aspect of Perturbation Analysis

Consider a manufacturing process viewed as a DEDS and some performance measure J believed to depend on two parameters θ_1, θ_2. To determine the sensitivity of J to θ_1, θ_2, we need to estimate the gradient of J with respect to these parameters. The 'brute-force' approach, applied either to the actual system or a simulation model, requires three separate experiments as shown in Fig. 1a: the first is a nominal run to evaluate $J(\theta_1, \theta_2)$ and the other two are perturbed runs to evaluate $J(\theta_1+\Delta\theta_1, \theta_2)$ and $J(\theta_1, \theta_2+\Delta\theta_2)$ respectively, or equivalently: $\Delta J/\Delta\theta_1$, $\Delta J/\Delta\theta_2$.

Now imagine a 'block box' connected to the system as shown in Fig. 1b. This is an on-line processor of data collected through automated monitoring. Its function is to observe the actual state, evaluate the perturbed states resulting from $\Delta\theta_1$, $\Delta\theta_2$ and ultimately provide the perturbed performance measure values. Using the framework described in the previous section, PA can accomplish this function with minimum additional computational cost. Note that if N parameters were involved, this approach eliminates the need for (N+1) separate experiments.

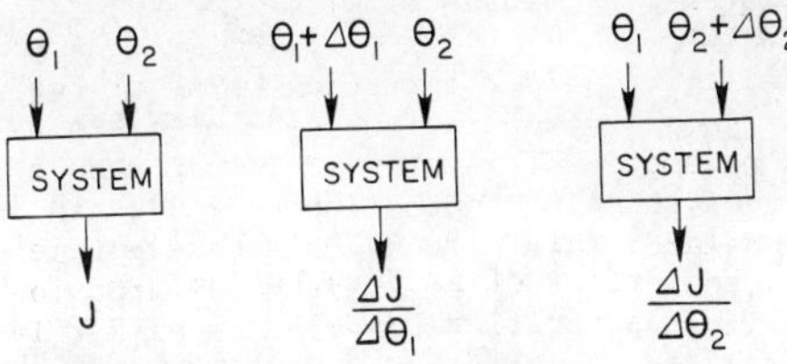

Fig. 1a. 'Brute-Force' Sensitivity Analysis

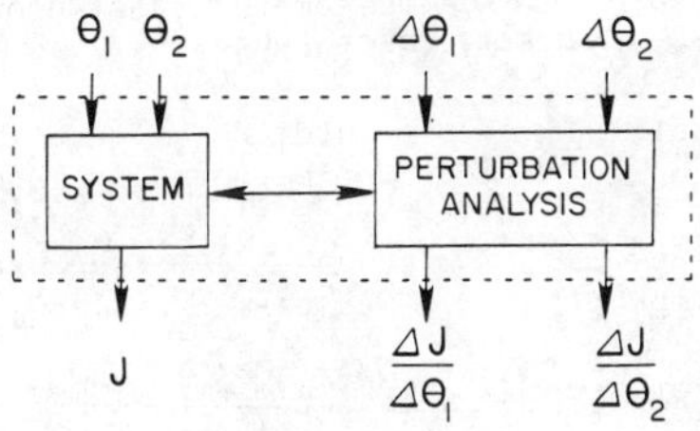

Fig. 1b. Evaluating Performance Gradient Through Perturbation Analysis

A LOW LEVEL SCHEDULING PROBLEM

A computer integrated manufacturing environment represents a complex resource allocation problem. 'Resources' - such as production parts, tools, pallets, fixtures and material handling carriers - are dispatched to servers on the basis of certain control and scheduling strategies. At a high level, long-term scheduling algorithms are needed to control inventory and plan production in order to meet forecasted demand. At a lower level, short-term scheduling strategies are used to achieve efficient resource allocation and respond to disturbances such as equipment failures and temporary traffic congestion.

Focusing on the lower level scheduling problem, we model an automated material handling system (MHS) as a network through which resources circulate and are routed to various servers. A typical example is shown in Fig. 2, where resources are dispatched from a central storage area (warehouse) to one or more servers, and eventually returned to it in the form of finished or intermediate products. This resource allocation problem becomes particularly interesting when: (1) Some sharable resources exist, e.g. some particular unitized load in the MHS can be processed by either server 1 or server 3. (2) The service times are short compared to the transfer times from the storage area. In this case, resource flows are set up so that the MHS is efficiently used to buffer unitized loads. Furthermore, routing strategies must be determined in order to send sharable resources to those servers most urgently needing them instead of blindly executing deterministic transfers. There are two major advantages to such a 'dynamic routing' approach: (1) Traffic is kept low. (2) The system is more reliable, in the sense that it can continue to function (in a 'degraded mode') even after a failure occurs.

A manufacturing system is usually driven by the peak service rates of operations. One can therefore set up flow-balancing equations, so that to any pair of nodes (i,j) there corresponds a desired

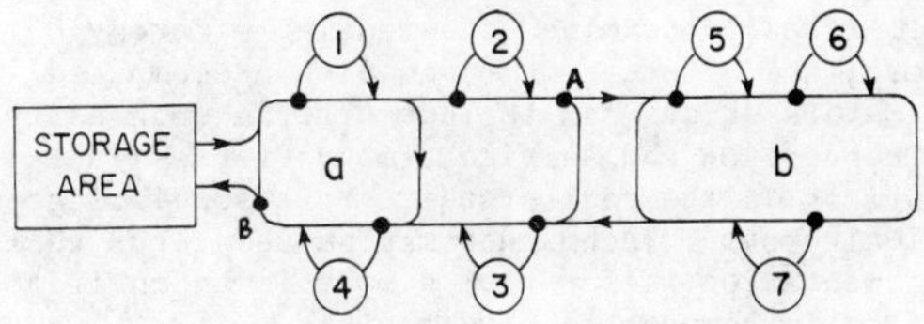

Fig. 2. Material Handling System Example

flow u_{ij}^r for a resource of type r, r=1,...,R. If t_{ij} is the average travel time from node i to node j, the corresponding desired resource level, s_{ij}^r (also referred to as 'set-point'), is given by:

$$s_{ij}^r = u_{ij}^r t_{ij} \tag{4}$$

We are implicitly assuming here that travel times are independent of resource type.

Now consider a unitized load of type r arriving at some branching node i. Because of equipment cost limitations, only two branches are usually present, so that a binary decision must be made. Let j and k be two possible destination nodes accessed from i, with set-points s_{ij}^r, s_{ik}^r respectively. Suppose that x_{ij}^r, x_{ik}^r are the actual numbers of resources (in unitized loads) for (i,j) and (i,k) respectively, representing state information available to a local controller at node i. If node j is favored over node k, then the routing strategy $D_{ij}^r(t)$ at time t is as follows:

$$D_i^r(t) = \begin{cases} j, & x_{ij}^r(t) < s_{ij}^r \\ \\ k, & \text{otherwise} \end{cases} \tag{5}$$

where node k is always the default destination. This type of routing strategy is discussed by Tanenbaum (1981). Simple means for selecting whether to favor node j or k in the decision are described by Cassandras (1984).

In practice, it is not realistic to expect making decisions for every pair of neighboring nodes (i,j). Instead, 'aggregate' set-points can be used, since a factory usually consists of distinct areas with similar types of operations performed in each area (for example, a 'processing' area and a 'packaging' area). This is illustrated in Fig. 2, where two areas are identified with one material handling loop serving area a and the other serving area b. For some resource type r, we may therefore define set-points s_a^r and s_b^r. A routing strategy at node A for instance is based on whether area b has currently a low resource level, i.e. $x_b^r(t) < s_b^r$. Similarly, at node B the decision may be to return to the warehouse unless the global resource level is low, i.e. $x_a^r(t) + x_b^r(t) < s_a^r + s_b^r$.

The routing strategy (5) is parametrized by s_{ij}^r. The determination of these parameters is made at a higher level and passed down to the local controller at node i, which executes the strategy. For systems subject to disturbances and variations in travel times t_{ij} due to traffic interaction, a simple flow analysis may be quite inadequate, except as a first-cut approach. Using actual data from the system, however, it is possible to evaluate sensitivities of the form $\frac{\Delta J}{\Delta s_{ij}^r}$, where J is some performance measure, and therefore periodically adjust the values of s_{ij}^r. Typical performance measures include throughput, server utilization and total traffic on the MHS. It is worthwhile pointing out that this approach 'closes the loop' between the actual system and the coordinating controller that determines the routing strategy parameters; the alternative - based on precomputed parameters - does not include effects such as traffic congestion and service time variations actually occurring in the real system. More sophisticated optimal control techniques may be used, but their implementation is far from simple in an environment limited in computational capacity.

Perturbation Analysis in a State-Dependent Routing Environment

As the previous discussion suggests, the 'building block' shown in Fig. 3 is of particular interest in MHS. In addition to the state variables $q_i(k)$, $e_i(k)$, i=1,2,3, defined earlier, we will also use the following notation:

$s(k)$ = server index where kth event (service completion) occurs

$S_i(k)$ = service time at server i when kth event occurs

$D_i(k)$ = server index representing the destination of a customer leaving server i when kth event occurs

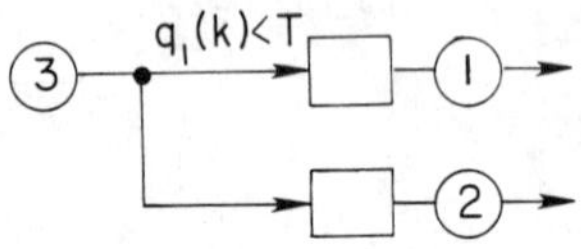

Fig. 3. Threshold Dependent Routing Strategy

A routing strategy similar to (5) has the form:

$$D_3(k) = \begin{cases} 1, & q_1(k) < T \\ 2, & \text{otherwise} \end{cases} \tag{6}$$

where T is the 'threshold' parameter value (the set-point defined in the previous section).

Infinitesimal PA is inadequate in this situation. The reason is that queue length perturbations may now be created and in turn result in large event occurrence time perturbations. Suppose, for example, that the threshold parameter T is perturbed by δT. For simplicity, assume that $\delta T = 1$. The first opportunity to exploit this change occurs when we observe $q_1(k) = T$ and $s(k) = 3$, i.e an arrival takes place while queue 1 is at the nominal threshold level. At that time, if $\delta T = 1$ were implemented, the arriving customer would have entered queue 1 and generated a queue length perturbation $\delta q_1(k+1) = 1$. Since that customer would normally enter queue 2, we also have $\delta q_2(k+1) = -1$. This information must be stored and subsequently used to generate event time perturbations at servers 1 and 2.

A perturbation analytic approach to the system of Fig. 3 is described by Cassandras (1984b). Our purpose here is to provide a simpler descriptive view leading to an algorithm for updating state perturbations. For simplicity, we shall focus on server 1 and set T = 1. This is equivalent to a G/G/1/1 queueing system, where all arrivals occurring while the queue length is 1 are lost. A simple state space model for this system is given by:

$$q_1(k+1) = \begin{cases} q_1(k) + 1 & s(k) = 3,\ q_1(k) = 0 \\ q_1(k) - 1 & s(k) = 1 \\ q_1(k) & \text{otherwise} \end{cases} \tag{7}$$

$$e_1(k+1) = \begin{cases} e_3(k) + S_1(k) & s(k) = 3,\ q_1(k) = 0 \\ e_1(k) & \text{otherwise} \end{cases} \tag{8}$$

Suppose the parameter of interest here is the threshold T, initially set at T = 1. In order to study the effect of a perturbation $\delta T = 1$, we can identify four critical states which are discussed below. Fig. 4 is a representative sample path of the system for the two cases T = 1 and T = 2. Dotted lines represent idle periods, arrival times are denoted by a_i and departure times by d_i. Note that in the nominal sample path (T=1) only arrivals at a_1, a_4, a_6 and a_7 enter the queue.

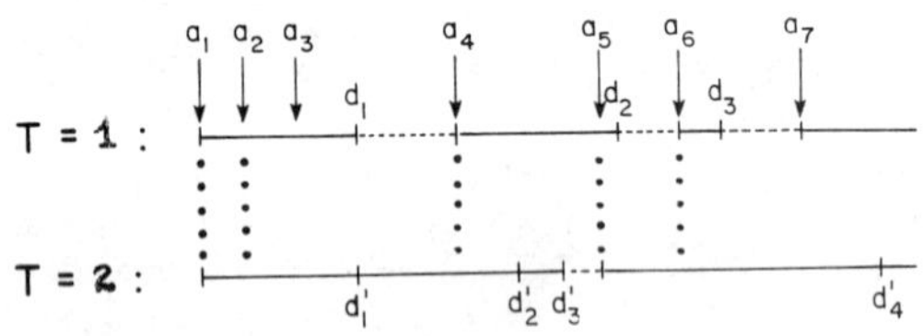

Fig. 4. Unperturbed and Perturbed Sample Path Example.

1. Opportunity to gain an arrival: when $q_1(k) = 1$ and an arrival occurs, this arrival is lost. If $\delta T = 1$, however, it is possible to gain the arrival provided $\delta q_1(k) = 0$. In Fig. 4 this happens at a_2. It does not happen at a_3, since when this event occurs we already have $\delta q_1(k)=1$.

2. Opportunity to lose an arrival: when $q_1(k) = 0$ and an arrival occurs, this arrival terminates an idle period (a_4 in Fig. 4). It is possible, however, that $\delta q_1(k) = 2$, in which case the arrival is actually lost in the perturbed path. This is illustrated with a_7 in Fig. 4, an arrival not lost in the nominal path. In the perturbed path, a_5 and a_6 both occur during a single busy interval, so that a_7 finds the queue at its perturbed threshold value.

3. Opportunity to gain in event time: as in the previous case, when $q_1(k) = 0$ and an arrival occurs, this arrival terminates an idle period. If $\delta q_1(k) = 1$ when this arrival occurs, we can infer that the idle period is eliminated in the perturbed path, since one more customer would be present for processing. In Fig. 4, this is illustrated with the gained arrival a_2: when d_1' occurs in the perturbed path, the server enters a busy instead of an idle period. In fact, the entire busy interval (a_4,d_2) is pushed forward in time by the nominal idle interval (d_1,a_4)

4. Opportunity to lose in event time: In case 3, event time is gained because of an idle period elimination. Subsequently, however, when a departure occurs, it is possible that an idle period is present in the perturbed path and terminated by a different arrival. In Fig. 4, the idle period following d_3 is terminated by a_7; in the perturbed path, there is an idle period following d_3' which is terminated by a_5. Therefore, at a_7 we predict an event time gain of magnitude (d_3,a_7), but at d_3 we should infer that this gain is to be reduced by (d_3',a_5).

An important advantage of infinitesimal PA is that no memory is needed in updating $\delta e_i(k)$ and queue length perturbations are never involved. This is generally not the case in studying even the simple G/G/1/1 system, where $\delta q_i(k)$ plays a crucial role as arrivals are sometimes gained and sometimes lost when perturbations are present. Furthermore, unless service times are fixed, we must also maintain some memory of state information. In particular, it is convenient to define a <u>pending queue</u>, $P(k) = \{p_1,\ldots,p_n\}$, where p_j, $j = 1,\ldots,n$, is the arrival time of a customer as yet unprocessed in the perturbed path at the time the kth event is observed. The following conceptual algorithm describes how $\delta q(k)$, $\delta e(k)$, and $P(k)$ are updated for a G/G/1/1 system observed, assuming the parametric perturbation $\delta T = 1$ is of interest:

1. Fix N. Initialize: $q = 0$, $\delta e = 0$, $P = \{ 0 \}$.

2. Observe kth event occurrence, $k = 1,2,\ldots$ Stop if $k = N$.

   Let a_i be the time of the ith ARRIVAL, d_i the time of the ith DEPARTURE.

   2.1. If kth event is ARRIVAL, then:

   if $q = 0$ and $p_1 \neq 0$, then: delete p_1 from P
   add a_i to P
   $\delta e \leftarrow \delta e + (a_i - e)$

   else if $q > 0$, then: add a_i to P

   2.2 If kth event is DEPARTURE and $p_1 \neq 0$, then

   if $e + \delta e < p_1$, then: $\delta e \leftarrow \delta e - (p_1 - e - \delta e)$
   $\delta q \leftarrow \delta q - 1$
   delete p_1 from P

   else do for all $p_j \varepsilon P$:

   if $e + \delta e > p_j$ and $\delta q = 0$, then:
   $\delta q \leftarrow \delta q + 1$

   else if $e + \delta e > p_j$ and $\delta q = 1$, then:
   delete p_j form P

   else if $e + \delta e < p_j$, then:
   stop, go to 2

This algorithm stores in the queue P all arrival times observed over a busy interval (step 2.1). At the end of such an interval (step 2.2), the queue is scanned, in order to identify 'critical states' 1,2. Critical state 4 occurs if $e + \delta e < p_1$ in step 2.2, reflecting the fact that an idle period would be present in the perturbed path. Critical state 3 is identified in step 2.1 when an arrival terminates an idle interval and the queue P is non-empty.

As already mentioned, PA in this non-infinitesimal case is considerably more involved and its practicality has yet to be put to the test of sufficient experimentation through simulation. This approach, nonetheless, is indicative of the structure of DEDS, which generally allows the development of conceptually simple procedures for prediciting perturbed behavior.

CONCLUSIONS

The complexity of CIM systems requires fast and efficient decision aids to enable reliable operation under varying conditions. Perturbation Analysis provides such an on-line decision support tool, by evaluating parametric sensitivities while simply observing an actual system. Fig. 5 illustrates the generation of on-line reports from actual system data, in addition to conventional monitoring reports.

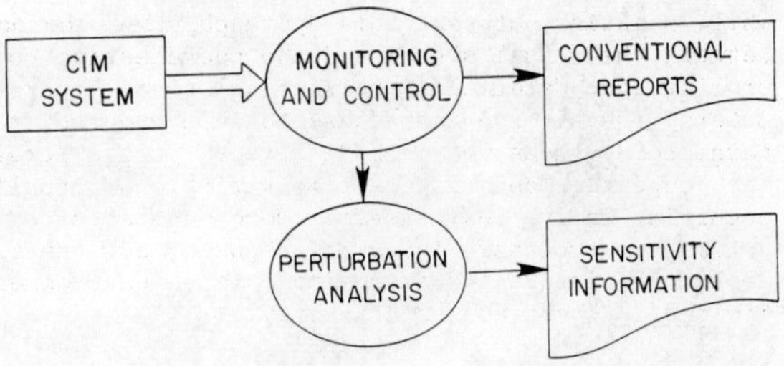

Fig. 5. On-Line Sensitivity Information Collection.

As is often emphasized (Buzacott, 1982), automated material handling is the crucial feature of CIM. It is for this reason that a significant effort is placed on enhancing control strategies in material handling. Perturbation Analysis promises to be an important technique towards developing and optimizing such strategies.

REFERENCES

Buzacott, J.A., and Shanthikumar, J.G. (1980). Models for Understanding Flexible Manufacturing Systems, <u>AIIE Trans.</u>, 12,4, 339-350.

Buzacott, J.A. (1982). 'Optimal' operating rules for automated manufacturing systems. <u>IEEE Trans. on Automatic Control</u>, <u>27</u>, <u>1</u>, 80-86.

Cao, X. (1984). Convergence of parameter sensitivity estimates in a stochastic experiment. Subm. to <u>IEEE Trans. on Automatic Control</u>.

Cassandras, C.G., and Ho, Y.C. (1984). An event domain formalism for sample path perturbation analysis of discrete event dynamic systems. To appear in <u>IEEE Trans. on Automatic Control</u>.

Cassandras, C.G. (1984). A hierarchical routing control scheme for material handling systems. <u>Proc. First ORSA/TIMS Conf. on Flexible Manufacturing Systems</u>.

Cassandras, C.G. (1984b). Efficient simulation techniques for performance evaluation of queueing networks. <u>Proc. IASTED Intl. Symp. Applied Identification, Modeling and Simulation</u>.

Fishman, G.S. (1978). <u>Principles of Discrete Event</u> Simulation, Wiley.

Ho, Y.C., and Cassandras, C.G. (1983). A new approach to the analysis of discrete event dynamic systems. <u>Automatica</u>, <u>19</u>, <u>2</u>, 149-167.

Ho, Y.C. (ed) (1984). SPEEDS: a new technique for the analysis and optimization of queueing networks. <u>Technical Report No. 675</u>, Division of Applied Sciences, Harvard University.

Jackson, J.R. (1963). Jobshop-like queueing systems. <u>Management Science</u>, <u>10</u>, <u>1</u>, 131-142.

Kumar, P.R. (1984). On estimating derivatives of cost functions from simulated or real data. <u>Proc. 23rd IEEE Conf. Decision and Control</u>.

So lberg, J.J. (1977). A mathematical model of computerized manufacturing systems. <u>Proc. 4th Intl. Conf. Prod. Res.</u>

Suri, R. (1983). Implementation of sensitivity calculation on a Monte Carlo experiment. <u>J. Opt. Th. and Applic.</u>, <u>40</u>, <u>4</u>, 625-630.

Tanenbaum, A.S. (1981). <u>Computer Networks</u>, Prentice-Hall.

SELECTION AND EVALUATION OF CAD/CAM SYSTEMS FOR THE MECHANICAL INDUSTRY

T. T. El-Midany and A. K. Abd El Latif

Department of Production Engineering and Mechanical System Design, PO Box 9027, King Abdulaziz University, Jeddah 21413, Saudi Arabia

Abstract. The application of computers to design and manufacturing in engineering, CAD/CAM is not new. However, it is only relatively recently that falling costs and technology advances in microelectronics have made CAD/CAM systems widely available to the mechanical industry. Automobile and aeronautical industries were among the pioneersim their use for the CAD/CAM. The primary advantages in adopting CAD/CAM are well known and include faster tendering response with improved documentation, improved accuracy and quality, reduced lead times this is often the key factor; the possibility of higher productivity ; and cost savings; or more commonly, greater many making ability. The greatest benefits are obtained wherever products are complex such as the design of aircraft, cars and computers. CAD/CAM technology is moving very quickly on both the software and hardware fronts and CAD/CAM systems can be very quickly outdated unless their design and structure is flexible enough to enable then to incorporate the latest technology. In correct selection of CAD/CAM system has sometimes resulted in the in position of unreasonable constraints upon the freedom of design and manufacture decision making and reduced the confidence in the use of CAD/CAM systems.

In this paper we attempt to develop the guide lines and criteria for selection a CAD/CAM to meat the mechanical industry requirements, including the interactive graphic system and associated software for design and software packages for manufacturing. It does not describe in detail a formal procedure to be followed, but gives general approach.

Keywords. CAD/CAM for Mechanical Industry, Selection of CAD/CAM, Evaluation of CAD/CAM, CAD/CAM systems.

INTRODUCTION

CAD/CAM systems available today provide either, an aid to the engineer which enables him interactively to combine his skill with the speed, memory and computational ability of the computer, or a computerised and automated batch processing facility which from a relatively small amount of input data will complete a sequence of tasks previously undertaken by the engineer/planner, El-Midany (1982). In fact the CAPP systems have been developing way of automating this procedure and have using many approaches, the aim of the entire interlinking and integration efforts is the reduction of the input effort by multiple use of stored workpiece information, Fig.1 show the various scopes of tasks that can be covered by individual integrated CAM/CAD system, Spur and Grottke (1984).

Mechanical industry recognize now the vital role that CAD/CAM will play in our future and both lay stress on the need to invest more and to expand the existing CAD/CAM systems. Implementing a CAD/CAM systems for the mechanical industry are an important investment. However, the experience of many companies has shown that the potential benefits for exceed those of other investments competing for limited investment dollars. In most companies, the areas that provide the greatest potential return to justify the investment are; reduction in times, increasing plant productivity and improving the manufacturing procedure by updating the methods with the more advanced technology.

CAD/CAM is the catalyst to bring about a new spirit of professional achievement in industry. CAD/CAM is an integrated part, they share a common data base, Fig.2, to apply in one without the other is wasting half of the available resources and data. CAD/CAM technology responded to mechanical industry needs for sophisticated interactive graphics, computer controlled machine tools, intellegent robots, improved inspection techniques and a most of other innovation to do manufacturing better Groover and Zimmers (1984).

NES (1984) reported that, CAD system is used to create a new data base. Investigation of companies who use CAD will readily show that a company has been very successful indeed if 50% of the data base resides on the CAD computer with a period of five years. It will not be surprising to learn that during the first year, hardly any of the new data base is created because of the long learing time required for the CAD system.

Many companies continue to use the traditional means of creating new drawing because the most investment in CAD will only allow a small number of draughtsmen or designers to transfer from the drawing board of the design system. It is seen from Fig.3 that during the first five years of the life of a CAD/CAM system, approximately 80% of the data base will still come from the traditional drawing board source. It would reasonable to conclude that the manufacturing engineer requires a system that will work today and integrate tomorrow.

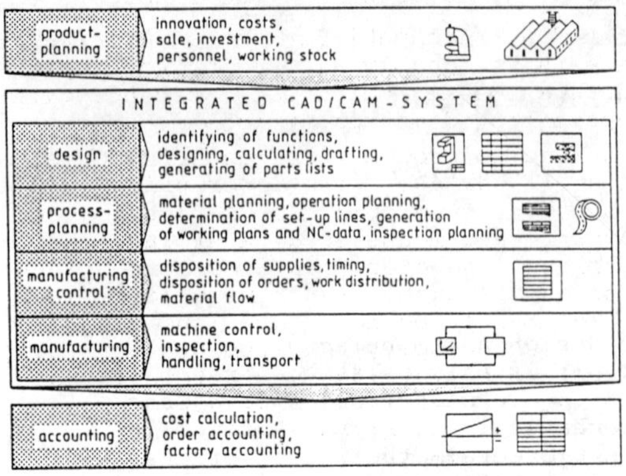

Fig.1. Scope of tasks of an Integrated CAD/CAM system.

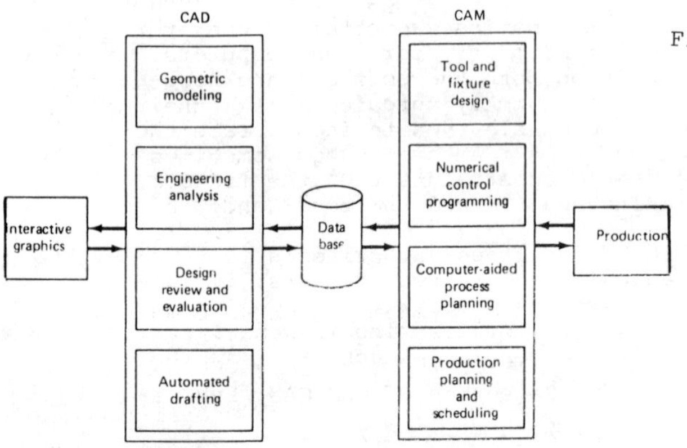

Fig.2. Relationship of CAD/CAM database to CAD and CAM.

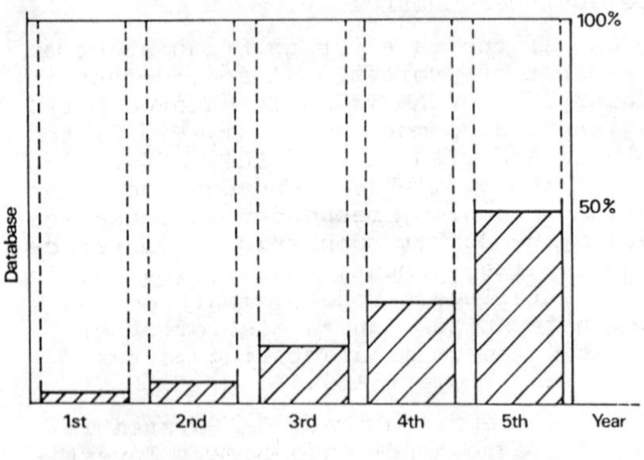

Fig.3. Rate of groth of CAD database.

CAD/CAM systems involve activities which make use of computer systems to assist in the creation, modification, analysis, optimization of a design, documentation of the product design, process planning, scheduling and to monitor and control the manufacturing operations. It consist of the hardware which includes, the computer, one or more graphic display terminals, key boards and other peripheral equipment. The software consists of the computer programs to implement computer graphics on the system plus application programs to facilitate the functions of the user company Fig.4.

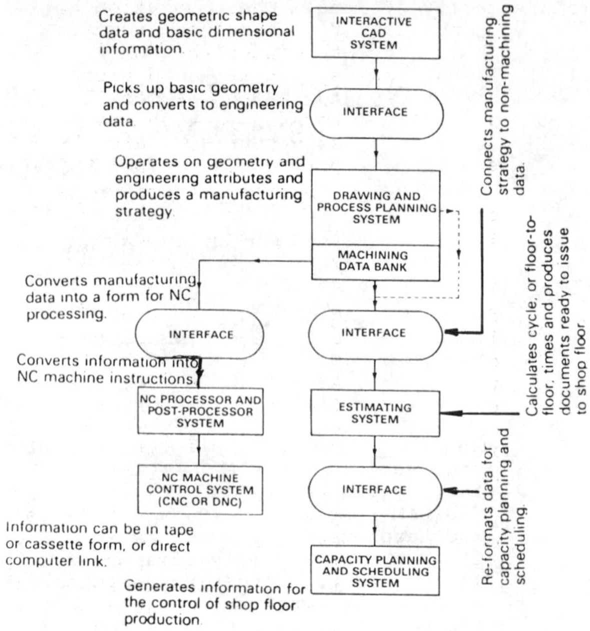

Fig.4. The software requirements of a CAD/CAM linked manufacturing system

There are different types for the computer aided systems which are required for the different companies. Their classification is depending on such factors as: availability of computing facilities, Jobbing on batch production, size of organization, component variety, outputs required and links with other systems. Most of the following systems, Computer Aided Design and Drafting, process planning, NC tape preparation, Scheduling Quality Control and Estimating , required a considerable amount of conceptual thinking and decision making. Interactive systems specially for CAD are now available in a whole range of packages varying from relatively low-cost 2D systems to expensive and extremely powerful 3D system.

In the case of the inter-active approach the systems tend to be adaptable to a very wide range of tasks on the basis that much of the critical decision taking process is carried out by the operation. Whereas in the case of the data-input systems the very fact that the computer system has to make on often complex series of decisions tends to limit their application to specific types of activity.

Bumberg and Gerwin (1981) found that firms in the UK, West Germany and US all had considerable problems in evaluating CAD/CAM systems, and discounted cash flow was of very limited use. Insufficient evidence was available to estimate future net returns, especially if it was proposed to use the new equipment to produce a new product. A critical issue was how to quantify the benefits likely to accrue from enhanced flexibility. In the light of their conclusions about the difficulties involved in evaluating sposticated systems and of the difficulties in implementing them, they concluded that new manufacturing technology was outstripping firm's capabilities for coping with it. The skills, systems and procedures available to firms were inadequate to enable them to support the sophisticated new manufacturing systems. Their solution was for firms not to strive for the automatic factory but rather to concentrate on simpler technologies such as stand alone CNC machine tools and they suggested that the ideal of the automated factory may be illusory. But, if some countries are making substantial progress towards it and, thereby , enhancing their competitiveness, it may be necessary for other countries to follow suit. If a technology is being applied successfully in other developed countries, it may have to be applied in all.

The products offered on the market and generally called CAD/CAM systems are very manifold and very different in their capacity performance. Also the level of the communication interface is an important criterion for the user. The only way to discover about CAD/CAM is to use it. Many cases are on record of a firm buying a turnkey system only to discover that it was the wrong one and to through it out in favour of another better adopted to the firms specific problems. This is the sort of expense that a small firm in particular cannot afford. The important thing for a firm of whatever size is to buy experience as cheaply as possible so as to determine what it is that it wants. This is often best done on a time-sharing bureau for small capital outlay Duncan and Emerton (1983).

Potential users of CAD/CAM can become very confused when trying to decide between the rival claims of different vendors of this type of equipment. The literature available from the vendors concentrates on the hardware offered and the software available, Harrow (1983). The main benefits which attract potential users are the prospects of producing drawings more quickly, able to respond to design changes without long delays and a reduction in the number of scorce and expensive staff. The selection of a CAD/CAM system and its successful implementation is a highly complex task requiring skills which span several disciplines which normally have little point of contact. The first constraint on any proposed system is the amount of many available system can vary in piece from a few thousand pounds/dollars to a few millian, dependent upon the number of work stations required and the central processor used , e.g. a system based on a general-purpose mini computer with plotter and three workstations together with a graphics data base and general purpose graphics software could be expected to cost about £ 250000 at 1983 price as reported by Harrow. The best way for many firms in mechanical industry is to get into CAD/CAM by buying or renting a terminal and using a time-sharing computer bureau. This reduces initial capital commitments and allows experience to be gained before making expensive final and vital long term decisions.

CAD/CAM SYSTEMS

The CAD/CAM systems consist of the hardware which includes, the computer, graphic display terminals and other peripheral equipments. The software consists of the computer programs. If CAD/CAM is to be effective tool, it is cru cial to require in each case which type of device is appropriate. The choice is confused by the rapid development of technology; The features dependent on main frame computers five years ago are becoming available on intelligent terminals. The choice of a system needs to balance today's situation with future evaluation both of the system and of the user. However it should not be attempted without an in depth of feasibility study to determine the concept.

Groth in the capability of computer hardware is reasonably assured by the rapid advances of microelectronic technology. Sophisticated manufacturing techniques continue to produce more powerful computers in small packages with lower price tags concurrent development of more efficient software and means of data storage and retrieval, however, requires more attention. Data bases are growing larger and are contantly being updated and altered. Access and display requirements generally vary. Voice-interactive systems for data storage and retrieval have advanced in recent years as a replacement for key-board storage systems.

The CAD/CAM systems throught the world have acquired capabilities making them suitable for use in the mechanical industry. The maxi-main frame systems are quite suitable for large companies, however the problems of making remote access by graphic terminals to the central processor is a worthwhile task to be solved. The system overhead supporting remote entry graphic and graphic data handling on these machine tends to impair response times at the terminal to an unacceptable degree. Local plotting is not usually possible with maxi-installations.

The turnkey system is bought complete with central processor, graphics terminals, plotter...etc, and can be brought into operation by simply "turn a key". Several years ago such systems offered little more than high speed draghting facilities. Now, however, the better known systems have a considerable three dimensional capabilities including from surface design, finite element model creation, calculation of section and volume properties and solid model generation; process planning, scheduling and to monitor and control the manufacturing operations, Fig.5 it is more superior to the remote access maxi machine, Abd El Latif and El-Midany (1984).

The general prupose mini computer is usually superior to the turnkey system interms of flexibility and potential for expansion, particularly with the advent of networking or linking processors. The new generation of mini and super mini computers have a tremendous role to play, particularly in process control, data logging and providing terminals with intelligence of the pre-programmed variety.

The CAD/CAM hardware would be useless without the software to support it. It includes programs for implementing certain specialized functions as design analysis programs, finite element analysis, modeling, etc. The advent of solid modellers "as romalus Fig.6" have a dramatic effect in providing a large amount of dedicated power to each user. CAD/CAM technology is moving very quickly on both the software and hardware fronts and CAD/CAM systems can be very quickly outdated unless their design and structure is flexible enough to enable them to incorporate the latest technology, Cox (1983).

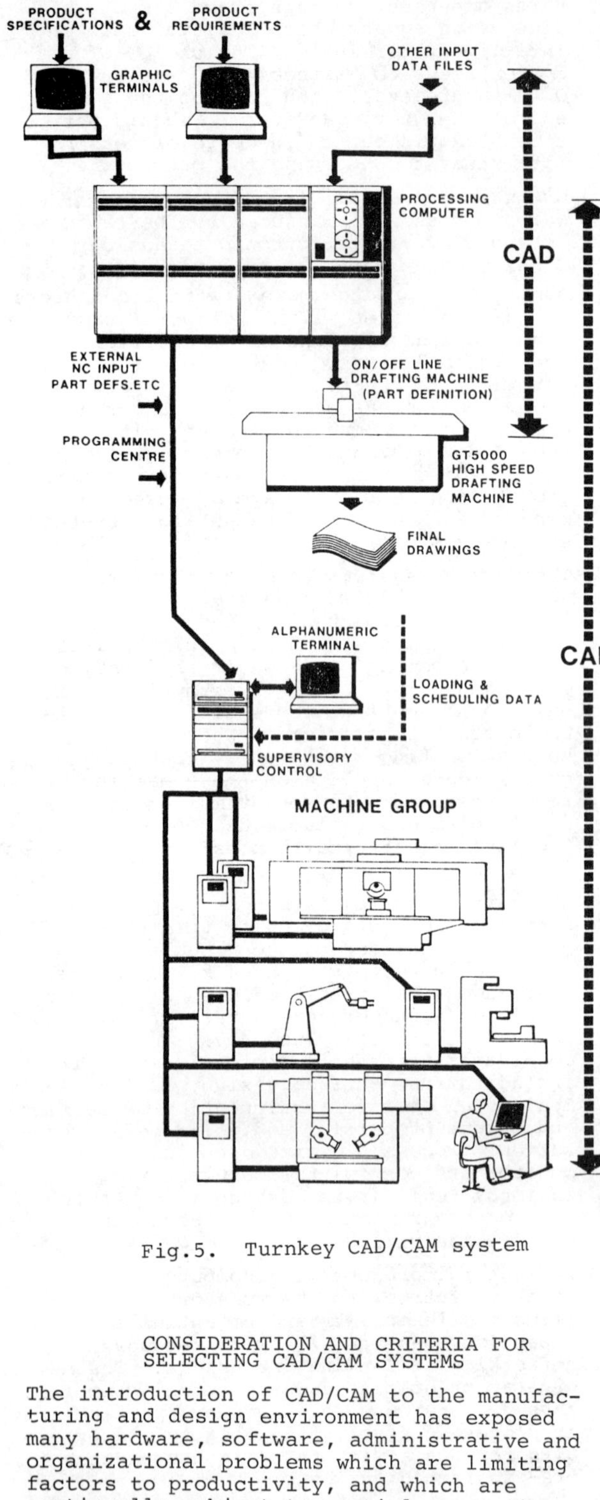

Fig.5. Turnkey CAD/CAM system

The benchmark component modelled with Romulus

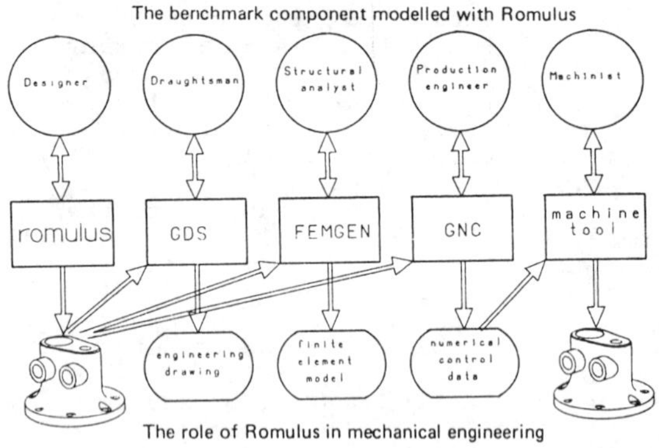

The role of Romulus in mechanical engineering

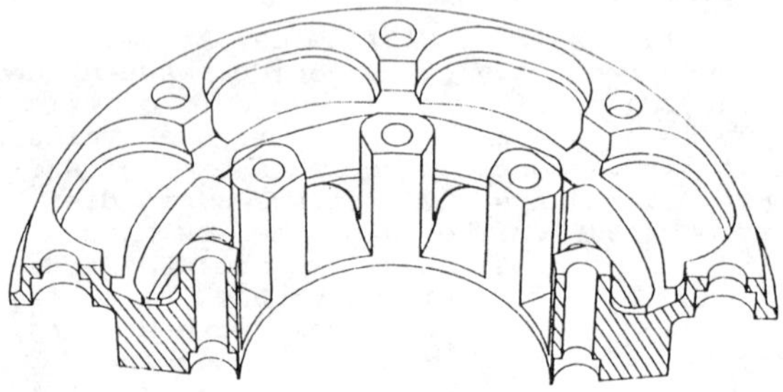

The ICMB component – Romulus model sectioned

Fig.6. The role of Romulus in mechanical engineering.

CONSIDERATION AND CRITERIA FOR SELECTING CAD/CAM SYSTEMS

The introduction of CAD/CAM to the manufacturing and design environment has exposed many hardware, software, administrative and organizational problems which are limiting factors to productivity, and which are continually subject to partial corrective actions. In many cases the initial aim in introducing CAD/CAM is to reduce the quantity of subcontracting and to make better use of skilled staff who are in short supply. If speed of production is the only criterian then simple drawings and schematic diagrams will show the greatest improvement, about 5 or 8:1 compared with manual methods.

The overall speed of the system will be dependent upon a number of factors, most of which are not under the control of the user, and among these are the power of the computer used, the design of the graphics database, the design of the workstations, the speed of transfer from the discs, the speed of the communication links and the mix of work being performed at one time. Only the last can be effectively controlled by the user.

It is up to the user to decide the relative importance of each criteria based upon his needs because the correct investment in manufacturing systems will probably be the most important decision made during the following years. The selection of CAD/CAM systems are a complex task (that requires a multidisciplinary approach on the part of the user organisation. The system chosen should be sufficiently span-ended to allow expansion into areas of interest that become apparent with experience.

The implementation of CAD/CAM systems will consider the feasibility of purchase the system from one of the vendors. The choice of the most suitable system is one of the most difficult the user has to face, one of the complications being the very large number of factors which have a bearing on the choice. The considerations and criteria for selecting the appropriate system, Table 1, is considered as the most useful. It may be used to evaluate the alternative systems, and make the recommendations for the prefered system to be selected.

The first step for the prospective user company is to assign a weight to the most important factors Table 1 , a weight of 450 is assigned to the general considerations and mechanical applications and 100 for the other applications each is subdivided to the various subdivisions,n,. Each company emphasize upon the most important factors considered by giving it a higher point weight. The indicated weight is based on limited industrial experience. Then each system is evaluated with repeat to the considered factors by giving a grade point,x, from the following point gradation:

Very suitable	8 to 10	points
Quite suitable	6 to 7	points
Suitable	4 to 5	points
Limited Suitability	1 to 3	points
Unsuitable	0	points

Then multiply n by x give a value, which reflect the importance of that general factor to the user company. Summing up $\Sigma n.x$ for each candidate give a comparison value which differentiate between them, higher values indicate more suitable system. In this process, candidate has the most favourable greater rate is being used.

Selecting a CAD/CAM system based upon the benefit/Cost ratio give more emphasizes upon the cost which is becoming less importance due to the availability of very many systems with reasonable comparable costs. Also the cost factor in the general considerations reflects the importance and weight of that factor to most of the user companies.

TABLE 1 Selecting a CAD/CAM System

Important Factors Considered	Relative importance factor (n)		Grading out of 10 for diff.candidates. A	B	C	D	Total points of diff. candidates nA	nB
A-General Considerations	450							
1. Cost	150							
Hardware								
CPU		20	x_1	x_2	x_3	x_4		
Added Stations		15						
Peripherals		10						
Software								
Turnkey basic package		15						
Added speciality packages		10						
Hardware maintenance		10						
Software maintenance		10						
Specials		10						
Spares		10						
Documentation		10						
Training		5						
Transportation		5						
Facilities		5						
Supplies		5						
Field support		5						
Support personnel		5						
2. Quality	100							
User group existence and support								
Corporate quality								
Responsiveness								
Financial stability								
Number of installations								
Growth								
Reliability								
Simultaneous operation								
Crash recovery								
Power loss recovery								
Environmental sensitivity								
Human factors considerations								
Hardware								
Software								
Response time								
Output device speed								
Interfaces								
Communication								
Plotters								
Product documentation								
Training								
3. Programming	50							
High level vendor language								
Standard language								
Assembly language								
Source availability								
Clear documentation								
User protection								
Diagnostic aids								
Data base access								
Clear interfaces								
4. Compatibility	50							
Compatibility with existing equipment								
Graphic terminal control								
Operator input methods prompting								
Plotting control								
5. Service	50							
Contract								
Parts location								
Turnaround								
Warrantee and discontinuance clause								
Software bug service								
Software service								
6. System management	30							
High speed peripherals								
User diagnostics								
File management software								
Expense reports and logging								
7. Delivery and logistics support	20							
Staging in-plant benchmark								
Packing								
Installation aid								
Pre-delivery inspection								
Installation guide								
Receipt acceptance								
Supplies								
Revisions								
Biling								
Proprietary agreement								
On-site debugging								
B-Mechanical Applications 450								
1. Mechanical design	150							
Logic diagrams								
Process diagrams								
Schematics								
Architectural drawing								
Plant layout								
Structural steel design								
Piping								
From/To lists								
Mapping								
Sheet metal design								
Two-dimensional mechanical design and drafting								
Point functions								
Line functions								
Arc functions								
Conic functions								
Dimensioning								
Line type								
Text.arrows								
Witness line suppression								
Character height								
Flag note								
Three-dimensional mechanical design and drafting								
Wire-frame or solid modeling								
Hidden line removal capabilities								
Surfacea and planes								
Curves and curved sections								
Tool design								
Jig and fixture design								
Nameplates								
Dimensionless drawings								
2. Mechanical interfaces	100							
Stress analysis								
Finite element modeling (FEM)								
Modal analysis								
Flow analysis								
Mechanism analysis								

mass properties determination Geometric properties Spline analysis									
3. Manufacturing Planning packages		100							
Computer-aided process planning capabilities Metal forming NC programming Nesting of sheet metal parts for efficient cutting Machining NC programming Automatic NC functions Other NC functions NC default values NC interfaces									
C-Other Applications.	100								
1. Electrical design applications		50							
2. Building design applications		30							
3. Architecture design applications		20							
Σn Total points of each candidates.								ΣnA	ΣnB

CONCLUSION

The selection of a suitable CAD/CAM system is a highly complex task involving a multi-discrplinary approach. Whichever system is choosen a great deal of work must be done by the potential user. In most cases the user will not fully appreciated the potential which these systems can offer. It is therefore essential that any system choosen is sufficiently open-ended to allow expansion into areas of interest which will be discovered with experience. Evaluation of each system is based on considering the most important factors for the user. Each factor is then graded and the system with the high total points should be selected.

REFERENCES

AbdElLatif, A.K. and T.T. El-Midany (1984). Some aspects of CAD in design of Mech. Components and system. 1st Intr. conf. on advanced manuf.(IFS), Singapore.

Blumberg, M. and D. Gerwin (1981). C-ping with advanced manufacturing technology. Intr. Inst. of Manag. Labour Market Policy , Discussion Paper, Berlin.

Cox, D. (1983), Formula one shapes up, Effective CAD/CAM, I. Mech. Eng. pp 117-120,UK.

Duncan, M.E. and G.F. Emerton (1983)- CAD/CAM for mechanical parts in the electronic industry, Effective CADCAM I. Mech. Eng. Pp. 53-60, UK.

El-Midany, T.T. and A.M. El-Taimi (1982) Computer Aided Production Planning (CAPP) systems and approach, 2nd IFAC Symposium on CAD of Multivariable Tech. System. pp 537-554, USA.

Harrow, P. (1983), Factors in Selecting a Computer-Aided Design System, Computer-Aided Eng. Jr. pp 16-24, UK

NES Oct. (1984) Computer graphics in manufacturing The Jr. of the NE Society pp, 22-24,UK.

Spur, G. and F.L. Krause (1984)-CAD-Technik, C.H. Verlog, Munchen, Wien.

THE SPECIFICATION OF A COMPUTER AIDED SYSTEM FOR THE DESIGN AND MANUFACTURE OF LIGHTING COLUMNS

P. F. McGoldrick*, C. O'Brien*, N. A. Rusby* and J. B. Lightbody**

**Department of Production Engineering and Production Management, University of Nottingham, Nottingham NG7 2RD, UK*
***Abacus Municipal Limited, Sutton-in-Ashfield, Nottingham NG17 5FT, UK*

Abstract. This paper considers the problem of specifying special purpose CAD/CAM systems by presenting a case study which discusses the requirements for software that will be used to aid the design and manufacture of lighting columns. Whilst many of the details of the programs will be unique to this particular set of problems, much of what is described may have wider applicability particularly, perhaps, for the smaller company or for an organisation with limited knowledge and experience of this technology.

Keywords. Computer-aided design; computer-aided manufacture; computer software; manufacturing processes; columns.

INTRODUCTION

Specification is the key to good design; this is true whether one is dealing with a very tangible piece of hardware or, as in this case, with a computer program which is intended to aid the design and manufacture of a group of products.

In particular, this paper presents a case study to illustrate the specification of computer software for use in the design and manufacture of lighting columns.

The work is being undertaken under the CASE (Collaborative Awards in Science and Engineering) award scheme organised under the auspices of the SERC (Science and Engineering Research Council). In this scheme a research student is nominated to work on a project of joint interest to a university department and to the collaborating company.

Abacus Municipal Limited is a small manufacturing company which makes a variety of products but principally a range of steel lighting columns. The chief component in the price of these columns is their material content. Since the market for these columns is highly competitive it is absolutely essential that efficient use is made of that material content.

In most cases columns are designed and manufactured to a variety of different national and international Standards and Codes of Practice. As well as these conditions, others which may be imposed by a particular client as part of an individual contract, and then there is the range of grades of steel or other materials, cross-section shapes, sizes and thicknesses. All this means that only a computer is capable of coping with the enormous variety of alternatives which need to be evaluated in order to arrive at anything approaching an optimal design.

Business is of course not just about design, and from the outset it was clear that any software had to enable consideration to be made of manufacture as well. There is no point in designing a product which cannot be made by the company's existing or projected future plant, nor indeed one which ignores the variety of raw material needed and the cost of holding stocks of that material.

Whilst many of the details of the structural design software are unique to this particular problem and of little interest outside, it was felt, by both the company and the University, that the problems considered in this paper could be of interest to those who might be thinking of investing in CADCAM software.

THE SERC CASE SCHEME

The Science Research Council (subsequently renamed the Science and Engineering Research Council - SERC) began a scheme called CAPS (Co-operative Awards in Pure Science) in 1967. In 1972 this scheme was broadened to include university departments, and in 1975, following a one-year pilot study, the scheme - now with the CASE acronym - was expanded to include engineering projects.

Throughout this period of expansion and change the objectives of the scheme remained broadly unaltered and, with a now mature scheme, that is still the case. It provides a postgraduate training which enables a student to register for a university research degree (almost invariably a PhD) whilst working on a project which has very real links with the co-operating body. It is usually the case that the student spends at least three months of his three years working in and with the parent organisation, which is required to nominate an industrial supervisor who will advise the student on his work. Thus, in addition to providing an opportunity for graduates to gain firsthand experience of work outside the academic laboratory, the scheme encourages the development and strengthening of contacts between academic institutions and outside bodies.

All British organisations, both public and private, are eligible to co-operate with academic institutions in the CASE Scheme, provided that a satisfactory training for a research student can be arranged. The Council of SERC decides each year how many CASE awards will be made and these are allocated to specific projects which have previously been judged as suitable for the training of a research student. Projects submitted for the allocation of an award are judged in competition and assessed by the relevant SERC subject committee on their scientific merit, the value of the training offered, the degree of collaboration involved and the contribution of the co-operating body. Once an award has been allocated to a project a student is nominated by the university in consultation with the co-operating organisation.

There are special arrangements under this scheme about the exploitation of any commercially valuable results. In recognition of the co-operating body's contributions to the work, the SERC waives its usual rule that the results should be exploited for the public benefit.

THE COMPANY

The Abacus group of companies is centred on Sutton-in-Ashfield - a small town in the centre of a coal-mining region about twenty five kilometres north of the city of Nottingham. The largest volume of trading in the group is undertaken by Abacus Municipal Limited, which is a wholly owned subsidiary of the parent company Abacus Holdings.

Abacus Municipal manufacture a range of street furniture including bus shelters, litter bins, covered walkways, high masts and most importantly lighting columns. The selling, design, manufacture and erection of these columns is a major contributor to turnover and employment among the one hundred and eighty workforce.

Contacts between the company and the University have been strong. Both the first two authors have acted as consultants and advisors to the company whilst being full-time employees of the University; the fourth author is Works Director of Abacus Municipal and has responsibility for both design and manufacture of the company's products.

PREVIOUS COMPANY EXPERIENCE WITH CAD

The company has had limited but successful experience of CAD. In late 1979 Abacus commissioned software to assist in the positioning of lighting masts used to floodlight large areas such as goods yards, airport marshalling aprons and sports fields.

The calculation of ground light intensity when using various lantern types at various orientations and heights on columns at various locations is a straightforward but extremely time-consuming task, and hence one which was highly suited to computerisation.

The software, written by consultants, runs on a Tektronix T4052 microcomputer which is equipped with two magnetic tape cartridge drives for data storage and retrieval as well as an A4 pen plotter for hardcopy. Currently (November 1984) the system is being reconfigured to run on a Hewlett Packard HP 9836 micro with twin floppy disc drives and an A1 pen plotter. This revised system will also incorporate an A0 digitiser, which will greatly aid the input to the machine's memory of the particular ground layout specified on the client's blueprint.

The second area in which the company has had experience is computer assistance with the structural design of rigid lighting columns to British Standard BS 5649 (1978). This program calculates the bending moments, torsional forces, and deflections of a design of column as specified by the design engineer; the design methodology specified by the Standard (see later) takes into account wind force loading and gravitational forces. This software was written "in house" by the design engineer concerned and runs on the HP 9836 machine.

OTHER COMPANY INVOLVEMENT WITH COMPUTING

A management information and control system has been installed and implemented over the past two years. This system, consisting of IBM System 34 hardware running the MAPICS software, makes available on-line management information which includes marketing, production control and inventory levels.

The system is not used to produce market forecasts; this exercise would be difficult because of the variability in demand for the small quantities of non-standard columns which tend to dominate current sales when compared to the rather lower volume of standard columns currently being sold. The stored data is however used to give up-to-date information on product costs and to monitor performance of the company's marketing strategies.

The production control data base includes full Bill of Materials and product routings for the orders on hand. In conjunction with the inventory level records, this enables a weekly net requirement plan to aid production scheduling to be produced. No automatic monitoring of shop floor progress is available currently and sequencing is, in the absence of manual intervention, sequential.

Inventory levels are updated in a batch mode from requisition orders and delivery notes. Commonly used stock items have a safety level indicated; when the number held goes below this level the company buyer is warned that the item should be re-ordered.

THE PROBLEM IN OUTLINE

As so often, the problem can be expressed as one of efficiency and productivity. It has already been noted that the market into which lighting columns are sold is highly competitive and that a large proportion of the finished cost of the product is represented by the cost of the raw material. Whilst this cost can be reduced by judicious and timely purchasing, it is vital that the material is efficiently used - in particular that the design criteria are met but not grossly exceeded - so that every piece of expensive material in the finished product is absolutely essential.

Value is added to the raw material during column manufacture, and again it is crucial that the minimum amount of work is done and that it be done as cost-effectively as possible. Paradoxically it is possible to use material more efficiently by adding more manufacturing operations, and the major reason for looking to CADCAM for help is that the number of alternatives which the designer can try is vastly increased, and so design and thus manufacture can be truly optimised.

STANDARDS FOR DESIGN

Basically the company designs and manufactures to the standard specified by the customer. In general for the United Kingdom this will be one of three British Standards; frequently these standards will be adapted for other countries by taking into account the different topological and climatic conditions which will prevail and which affect design.

The situation in regard to British Standards is changing. The first relevant Standard, BS 1840 (1960), is now obsolescent and rapidly becoming obsolete. It makes general recommendations on construction and installation; in addition it specifies mandatory levels on maximum allowable stress as well as deflection under a given distributed wind load. Several modifications to this Standard have been introduced over the years; the most significant of these is BE 4/72 (1972). This reduced the scope of the Standard - by limiting it to columns between eight and fifteen metres high and with up to three metres lantern outreach - and more significantly increased the wind loading.

For higher masts or lighting columns the company works to BS CP3 (1972), a code of practice which indicates more specific information on wind speeds and hence loading across the country; site surroundings and local topography are also taken into account by this document which really allows for the greater care needed when dealing with large structures.

A new British Standard BS 5649 (1978) is being introduced in stages. This much more stringent Standard specifies dynamic design criteria for the first time, and these loadings, as well as the static ones, take account of geography, topography, height of the column and the column profile (for example whether it is circular or hexagonal). This standard applies to columns up to eighteen metres in height.

DESIGN AND CONSTRUCTION

Across the industry, many different materials and overall configurations of design are used; in Abacus's case however all the columns are made from steel with two alternative hollow cross-sections. They are either manufactured from constant diameter circular tube or from a tapered, octagonal tubing which is folded from sheet steel with the seam continuously welded.

Manufacture from circular tube places two main constraints on the designer. Firstly the suppliers stipulate a minimum order quantity for non-standard size materials, and so, when net batch requirements planning and inventory level limitations do not permit a special raw materials order, standard stocked sizes must be used. Similarly, the material sizes currently stocked by the company may be specified in order to minimise lead time with the additional benefit of increasing stock turnover.

The second constraint relates to the production of columns using stepped sizes of tubes; this solution to the design problem is commonly used since the cross-section needed reduces with increasing distance up the column, and so changing the size periodically gives a step-wise approximation to the tapering section which might otherwise be ideal. Each step of course involves a joint and so the problem is to weigh up the cost of this joint against the material saving which is made by using it more efficiently. Each joint potentially involves the swaging of the larger diameter tube, welding (which may have to be both internal and external), weld reinforcement, and weld defect detection (using penetrant dyes), and thus can be expensive.

The octagonal, tapered sections which are bought in are press fitted together and hence the constraints mentioned do not apply.

Base sections for the larger columns are treated as standard sub-assemblies which are manufactured as stock items in order to reduce lead time; this tends to mean that, at present, the bases are not redesigned for individual columns though this situation is to be reviewed.

To protect against environmental corrosion - and to improve appearance - the finished columns are surface coated, either by galvanising or by metal spraying with zinc or aluminium.

EXTENT AND SCOPE OF CAD

The CAD software is seen as having two main objectives. The simpler function of the two is to act as a computational aid to calculate the performance factors of a suggested design.

These performance criteria typically include: the strength, in both bending and torsion of the various sections of the column; the maximum deflections of the column in two planes; and the natural frequency of the column structure giving consideration to the elasticity and damping coefficient of its foundations. These criteria are themselves determined by a number of variables such as: site or location conditions; overall dimensions and wall thickness of the various sections; cross-sectional profile; second moment of area; and material properties such as stiffness and elastic moduli of the material.

Although all these functions will be a useful part of the CAD package they would hardly constitute a major task for the computer; they would not realise the machine's full potential. It has long been recognised that the design stage largely dictates the manufacturing process (and vice versa of course) and thus, in the design of the software, the manufacturing as well as the structural and marketing aspects have to be borne in mind.

It is intended that the software will be capable of producing its own structurally optimum design given the overall configuration of the column as well as any other specified or necessary conditions. From this structural basis, manufacturing constraints can be imposed:

(i) Raw material availability to produce acceptable lead times and hence delivery dates; this availability can either be from stock or on an appropriate delivery timescale. This use of non-optimum material cross-sections and lengths will usually incur cost penalties which the system can quantify;

(ii) the batching of many small orders which use similar material sizes in order to produce gross and net requirement plans. Simulation techniques which could reflect the current shop floor loading might be most appropriate for determining the optimum batch size or batching arrangement. Too large a batch could increase work in progress and lead times for other pieceparts, but there would be savings in pre-process planning - such things as bills of materials, material flow charts and documentation - as well as in machine set-up costs and in the familiarisation of the workforce with that particular type of job. Flexibility to allow for customer preferences would have to be offset against the cost of production;

(iii) a nesting algorithm used in conjunction with the design algorithm to minimise the material wasted when column piecepart lengths are cut from stock tube. Tube which is delivered cut to the correct length is more expensive;

and (iv), as discussed earlier, the balance between material costs and the costs of joints.

LINKS BETWEEN LIGHT INTENSITY AND CAD PACKAGES

In many cases the customer's contract will specify the lantern type, the overall column height and outreach (the offset of the centre line of the lantern from the centre line of the column); in these circumstances the lighting intensity program mentioned earlier would serve no useful purpose and would not be used. Where flexibility exists however the situation is different, in that many of the variables examined in the light intensity program interact with those of the structural design part of the CAD package and thus an optimal design will need to bear in mind both parts of the overall design.

The lantern type has a particular lighting "footprint" and this is carefully examined in the intensity program and is central to its success. The lantern also affects the structure in two ways; its weight is a load at the end of the column arm, and its cross sectional area projected to the wind gives rise to a torsional load and a bending moment which act on the column.

The outreach (as well as the detailed design of the bracket arm) is a factor in the intensity program from the point of view of focus and projection, but it also applies further dead weight and wind-pressure-induced loading to the structure.

The overall column height governs the area which can be lit with an acceptable intensity and of course is a major factor in governing the column design and controlling its cost.

LIMITATIONS OF CAD

The authors firmly believe in the <u>AIDED</u> aspect of CADCAM. It is very much a principle which will be borne in mind when writing the software that the design decisions must be made by the designer. The computer, in addition to being a supremely quick calculating machine, enables more alternatives to be examined; it will enable "what-if" questions to be asked and answered. If this were not the case then precious design skills and experience would fall into disrepair or be lost forever; the company could be in danger of becoming totally dependent on a "black box" which would become moribund with the passage of time.

Cost penalties are another factor which should be mentioned. It is clearly the case that it is more expensive to develop software for a specific purpose than it would be to buy a more standard package where the development costs can be amortised over a number of customers. If the software does not exist though then there is no real alternative; the costs must be accepted.

LINKS INTO CAM

In order to achieve the maximum benefit from this system it will have to link with the management database contained in the IBM system; equally it should have the ability to produce data which will drive any automatic manufacturing plant which the company has or envisages acquiring in the foreseeable future.

In order that the optimum designs can be produced, the CAD system will interrogate the MAPICS database to recover current costs of all the manufacturing operations involved in the production of a joint in a column incorporating stepped diameter circular tubes.

No major plant purchases are envisaged immediately, but clearly the availability of data suitable for automatic machines will ensure that the acquisition of such equipment will be an easier decision to make. The conversion of an existing flame cutting machine, which contours by following an ink line on a drawing, to numerical control is being pursued.

CAD/CAM INTERACTIONS

These interactions are displayed diagramatically in Figure 1. At Level 1 the interaction takes place between the two programs which were discussed earlier - the structural design program and the light intensity package.

The outcome of these two packages interacts with three further areas at Level 2. The BATCHING program has already been mentioned; this will group together similar products. In the EVALUATION OF DESIGN sector, cost assessments will be made and manufacturing lead times estimated. The reduction or increase in costs as a result of the previously suggested batching will be indicated. The initial link into the IBM system will be made in the STOCK AND SHOP FLOOR INTERROGATION routines. These will have access to current inventory levels - to data on stocks of raw materials, and finished parts and sub-assemblies which may be needed. Data on shop floor loading with an indication of potential future capacity limitations will also be needed. Bottlenecks and delays could then be alleviated or eliminated by indicating that a modified design which would utilise less time on the manufacturing plant is possible with what could be adjudged to be an acceptable cost penalty.

At Level 3, three further links into the IBM system will take place. The DATA UPDATE area carries out the function suggested by its title. As well as the feedback of information to MAPICS of data which will interact with inventory levels and re-order quantities, work in progress levels will be predicted and product lead times, together with required customer delivery dates, can be estimated. In the PRE-PROCESS PLANNING sector bills of materials and requisition orders will be generated and parts routings and the net requirements plan suggested. If future investment encompasses NC or CNC plant then the requisite data will be generated here. The COSTINGS link is intended for any update of information which may be indicated.

CONCLUDING REMARKS

Many companies - perhaps particularly the smaller ones - are reluctant to embrace CADCAM and related technologies. Often this is because senior management do not fully understand the technology. The benefits that it will bring are often difficult enough even to enumerate let alone to evaluate financially; the penalties of disruption and dependence can be very severe; the risks are high.

The situation can be exacerbated by the hard-sell approach adopted by many companies involved in computing and the innate conservatism of many of the individuals involved in the customer's decision-making process.

The authors hope that this case study will encourage other organisations to sit down and analyse their own company structure and methods to see if CAD/CAM could help.

ACKNOWLEDGEMENTS

The third-named author is in receipt of a postgraduate studentship from the Science and Engineering Research Council, to whom thanks are due.

The authors also wish to gratefully acknowledge the financial support given to the project by Abacus Municipal Limited.

REFERENCES

BE 4/72 (1972). Technical Memorandum. Department of the Environment, London.

BS 1840 (1960). Specification for Steel Columns for Street Lighting, First Revision. British Standards Institution, London.

BS 5649 (1978). Specification for Lighting Columns. British Standards Institution, London.

BS CP3 (1972). Code of Basic Data for the Design of Buildings. Chapter V. Loading. British Standards Institution, London.

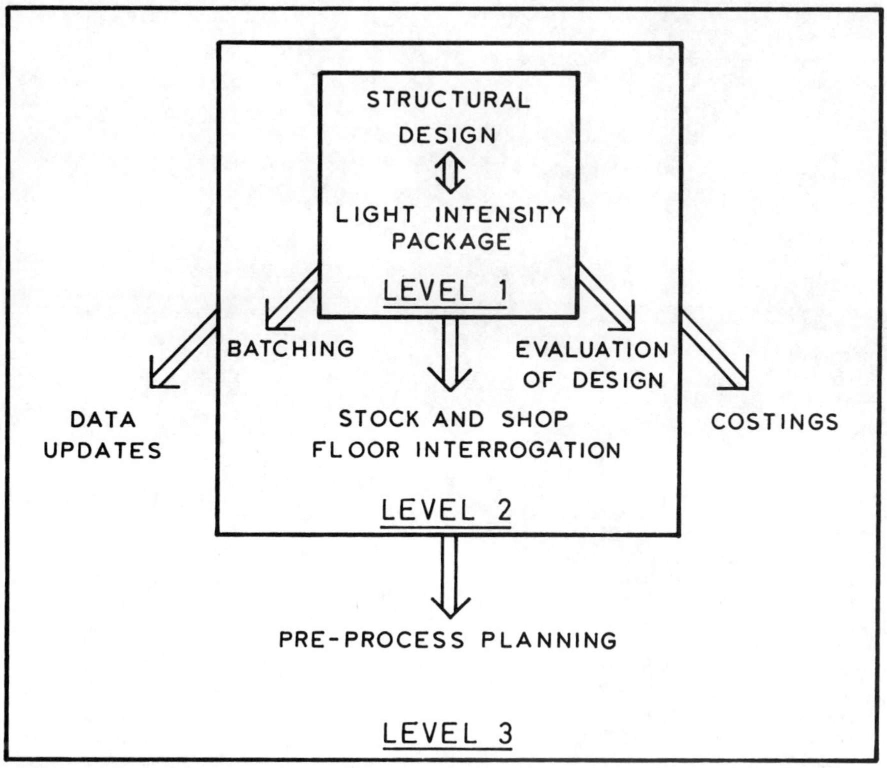

FIGURE 1. CAD/CAM INTERACTIONS

A NUMERICALLY RELIABLE ALGORITHM FOR FITTING A SUM OF EXPONENTIALS OR SINUSOIDS TO NOISY DATA

B. De Moor and J. Vandewalle

ESAT Laboratory, Department of Electrical Engineering, Katholieke Universiteit Leuven, Heverlee, Belgium

Abstract. Many problems of practical CAD importance in systems, modelling and control can be formulated mathematically as a problem in signal decomposition where the signal is modelled as a finite number of independent elementary signals which are additively combined and corrupted by an additive zero mean white noise source. In this paper the parameters of a sequence of real valued and equidistant data of the form

$$y_k = \sum_{i=1}^{n} a_i e^{\beta_i k} e^{j(\gamma_i k + \delta_i)} + \eta_k \qquad k=0,N$$

where $\{\eta_k\}_{0,N}$ is a white zero mean noise sequence, are estimated using a realization technique, based upon the singular value decomposition of certain matrices. First, several possible practical combinations of the parameters α_i, β_i, γ_i and δ_i are considered with respect to their practical application. Secondly, the signal is modelled using a time-invariant autonomous state space model. Furthermore, some theoretical results are established which lead to a state space realization procedure that estimates the unknown parameters in a numerically reliable way. When the data are noise corrupted, it is shown that the techniques used still deliver remarkably reliable results, that however can be shown to be nearly optimal in a least squares sense. Furthermore, some numerical simulations reveal the reliability and possible interactive development of the algorithm.

Keywords. Computational methods, Harmonic analysis, Numerical methods, Parameter estimation, Signal processing, Singular value decomposition, Spectral analysis, State space method, Sum of exponentials.

INTRODUCTION

Let's consider a sequence of equidistant, real valued data, generated by the following expression

$$y_k = \sum_{i=1}^{n} \alpha_i e^{\beta_i k} e^{j(\gamma_i k + \delta_i)} \qquad (1)$$

$$k = 0, N$$

where α_i, β_i, γ_i, δ_i, $\in \mathbb{R}$

The n terms of the sum can be considered as "elementary signals", each of them is characterized by four parameters : the amplitude α_i, a damping factor β_i, a pulsation γ_i that corresponds with an oscillation frequency $\delta_i = \gamma_i/2\pi$ and a phase δ_i (in radians).

Then the following special cases can occur

I. $\gamma_i = \delta_i = o$ for $i = 1,\ldots,n$ then

$$y_k = \sum_{i=1}^{n} \alpha_i e^{\beta_i k} \qquad (2)$$

is a sum of n real exponentials.

II. $\gamma_i \neq o$ for $i = 1, n$; the n is even for real data.

Since the data y_k are assumed to be real, the parameters $\alpha_i, \beta_i, \gamma_i, \delta_i$, must occur in conjugated pairs, i. e. for each quadruplet (α_i, β_i, γ_i, δ_i,) there is a corresponding one $(\alpha_i, \beta_i, -\gamma_i, -\delta_i)$

For l= 1.. n/2 we put

$$\begin{aligned} a_\ell &= 2\alpha_{2\ell} = 2\alpha_{2\ell-1} \\ b_\ell &= \beta_{2\ell} = \beta_{2\ell-1} \\ \omega_\ell &= -\gamma_{2\ell} = \gamma_{2\ell-1} \\ \psi_\ell &= -\delta_{2\ell} = \delta_{2\ell-1} \end{aligned} \qquad (3)$$

Then using Euler's formula the summation (1) is written as a sum of cosinoids with varying amplitudes.

$$y_k = \sum_{\ell=1}^{n/2} a_\ell e^{b_\ell k} \cos(\omega_\ell k + y_\ell) \qquad (4)$$

When $b_\ell < o$, the corresponding elementary signal is damped. When $b_\ell = o$, the elementary signal is a pure cosine.

III. Not all $\gamma_i \neq o$ for $i = 1,n$ i.e. there are some i for which $\gamma_i = \delta_i = o$. The general signal (1) now becomes a mixed sum of real exponentials and cosines.

A compact formulation of the problem is the following .

* Sponsored by a grant from the Belgian IWONL.

** This work was partially supported as a project of the Belgian FKFO.

For a given n, minimize over all $\alpha_i, \beta_i, \gamma_i, \delta_i \in IR$

$$\sum_{k=o}^{N} w_k [(\sum_{i=1}^{n} \alpha_i e^{\beta_i k} e^{j(\gamma_i k + \delta_i)} - y_k]^2 \quad (5)$$

for a given set of N +1 real valued and equidistant y_k and for a known set of weights w_k. The determination of the optimal "n" is considered further on. Signals of the general form (1) occur frequently in various applications. In fact, the general solution to any phenomenon described by a system of linear differential equations with constant coefficients is of the form (1) .

For the discussion of the statitistical question, whether it is justified to approximate a given data series by an exponential sum of the form (1), we refer to the literature [Ruhe, 1980] .

Although this contribution presents a unifying approach to the several 'special cases' I, II, III, generally, they are treated separately. Here we also stress the numerical aspects and the noise rejection qualities of the algorithm . The first solution to the problem of fitting experimental data by a sum of real exponentials is due to de Prony [1795] . Variations on his method are described in several books on numerical analysis [Lanczos, 1957] ('problem of "weighted moments" '). The problem occurs in fitting radioactive decay measurements (see ref. in [Ruhe, 1980]), transmission function analysis in atmospheric research [Wiscombe, 1977] compartmental analysis [Jacquez, 1972], electrical network analysis (Heaviside's expansion theorem [Lanczos, 1957], analysis of human lung exhalation and chemical rate constant estimation (see ref in [Ruhe, 1980] and [Wiscombe 1977]).

In many of these applications, the amplitudes a_i are expected to be positive. In [Wiscombe, 1977] a survey of various methods is given. The fundamental theorems guaranteeing existence and uniqueness of positive sums of a 'best fit' are treated by Cantor & Evans (1970). Algorithms based upon these results are described in [Wiscombe, 1977] and [Evans, Gragg, Leveque, 1980] . In [Ruhe, 1980] a non linear weighted least squares approach is studied while in [Ruhe, Wedin 1977] the problem is considered as a separable non-linear least squares problem since it is linear in the amplitudes a_i once the exponents β_i are known. A Remes-type algorithm is developped in [Braess, 1970] while a Padé-approximant approach together with conditions for existence and uniqueness are discussed in [Sidi, 1982].

The case of fitting data by a sum of cosines is handled in [Lanczos, 1957] as the famous 'search for hidden periodicities'. However, more generally, the problem belongs to the domain of spectral analysis [Kay, Marple, 1981], where so called 'non traditional approaches' can perform much better then the Fast Fourier transforms in high resolution spectrum estimation with application to beamforming, direction finding, array signal processing and other areas. An excellent survey of various spectral estimation methods together with some 300 references, is given in [Kay, Marple, 1981]. The singular value decomposition as a key tool is used in [Tufts, Kumaresan, 1982] to obtain least squares approximations of limited rank of the signal correlation matrix estimate. In [Kung, 1981] a Toeplitz approximation method is presented which inspired this work.

The case of exponentially damped sinusoïds is considered in [Kumaresan, Tufts, 1982] where also the singular value decomposition of a matrix shows up.

Special care has to be taken with respect to the sometimes ill-conditioned nature of the 'sum of real exponentials'-problem. The parameters that describe the signal, are badly determined by the data. If these are perturbed by a small amount, there may be quite a large charge in the best-fit parameters. In [Lanczos, 1957] it is shown how several different exponential sums can approximate the same data series equally well. The fact that large variations are produced in the results by small variations in the data, is found in all methods and is an intrinsic property of the problem itself. It is customary to blame this ill-conditioning on the non orthogonality of the set of exponential functions [Lanczos, 1957]. However, a detailed sensivity analysis in [Ruhe, 1980] and [Wiscombe, 1977] illuminates the nature of the ill-conditioning more precisely.

This work fits in a general trend towards numerically reliable software for systems and control.

STATE SPACE MODEL AND SOME USEFUL PROPERTIES.

The problem of determining the parameters of the signal (4) , can be reformulated mathematically as a linear system realization problem. The signal of the general form (1) can be modelled by an autonomous linear state space model of dimension n

$$x_{k+1} = A . x_k \qquad y_k = C . x_k \quad (6)$$

Where

$$A = \begin{bmatrix} e^{\beta_1 + j\gamma_1} & 0 \cdots\cdots 0 \\ 0 & \ddots & \vdots \\ \vdots & \ddots & \vdots \\ 0 \cdots\cdots & & e^{\beta_n + j\gamma_n} \end{bmatrix} \quad (7)$$

$$C = [\alpha_1 e^{j\delta_1} \;\; \cdots\cdot \;\; \alpha_n e^{j\delta_n}] \quad (8)$$

$$X_o = [1 \; 1 \; \ldots \; 1]^t \quad (9)$$

Proof :obvious by repeated substitution.

The key observation to understand the solution presented further, is to recognize that the general signal (1) is the homogeneous solution to a constant coefficient linear difference equation,

$$y_k = \sum_{i=1}^{n} a_i \, y_{k-i} \; , \qquad k= n,N \quad (10)$$

For a derivation of this expression we refer to [Kay, Marple, 1981]

The coefficients a_i are related to the elementary signals by Vieta's polynomial root theorem. From (10), it directly follows that the pxq Hankelmatrix with p, q> n, consisting of the p + q - 1 first data y_k, must have a rank equal to n.

$$H_{pq} = \begin{bmatrix} y_o \; y_1 \; y_2 \; \cdots\cdot \; y_{q-1} \\ y_1 \; y_2 \; \cdots\cdots \; y_q \\ y_2 \; \cdots\cdots\cdot \\ \cdot \\ \cdot \\ \cdot \\ y_{p-1} \; \cdots\cdots\cdot \; y_{p+q-2} \end{bmatrix} \quad (11)$$

and

$$\text{rank } [H_{pq}] = n \quad (12)$$

Proof : Any m-th column (m > n) can be written as a linear combination of the n preceding columns using (10) .

Further more, from the state space model (6) it follows that

$$y_k = C.A^{k-1} . X_0 \quad (13)$$

This leads to a decomposition of the Hankelmatrix H_{pq} :

$$H_{pq} = \begin{bmatrix} y_o \cdots & y_{q-1} \\ \vdots & \vdots \\ y_{p-1} \cdots & y_{p+q-2} \end{bmatrix} = \begin{bmatrix} Cx_o & CAx_o & \cdots & CA^{q-1}x_o \\ CAx_o & \cdots & & \\ CA^{p-1}x_o & \cdots & & CA^{p+q-2}x_o \end{bmatrix}$$

$$= \begin{bmatrix} C \\ CA \\ \cdot \\ \cdot \\ \cdot \\ CA^{p-1} \end{bmatrix} [\, x_o \; A.x_o \ldots A^{q-1}x_o \,] = \Gamma\cdot\Delta \qquad (14)$$

defining the matrices Γ(pxn) and Δ(nxp) that are both obviously of rank n.

$$\text{rank}\,(H_{pq}) = \text{rank}\,\Gamma = \text{rank}\,\Delta = n \qquad (15)$$

Now a new Hankelmatrix $\vec{H}_{pq}$ is defined which is derived from the previous one by shifting it over one colum to the right, while inserting one new date sample y_{p+q-1}

$$\vec{H}_{pq} = \begin{bmatrix} y_1 & y_2 & \cdots & y_q \\ y_2 & & & \\ \vdots & & & \vdots \\ y_p & \cdots & & y_{p+q-1} \end{bmatrix} \qquad (16)$$

where again due to (10)

$$\text{rank}\,(\vec{H}_{pq}) = n \qquad (17)$$

Using (12) and (13), it's easy to prove that

$$\vec{H}_{pq} = \Gamma\cdot A\cdot\Delta \qquad (18)$$

and the least squares estimate of A is computed as

$$A = \Gamma^{+}.\,\vec{H}_{pq}\,.\,\Delta^{+} \qquad (19)$$

where Γ^{+} and Δ^{+} denote the pseudo-inverse .

DERIVATION OF THE ALGORITHM.

Following [Zeiger, Mc Ewen, 1974] and [Kung,1978] the singular valuedecomposition of the data Hankelmatrix H_{pq} is used [Klema, Laub, 1980]

- to estimate the (numerical) rank and then the number of elementary signals to model.
- to obtain numerically reliable estimation of the matrices Γ, Δ, A and C.

Suppose the (partial) SVD of H_{pq} is given by

$$H_{pq} = \underset{pxn}{U}\;.\;\underset{nxn}{\Sigma}.\;\underset{nxq}{V^t} \qquad (20)$$

then the 'balanced realization technique' computes Γ and Δ up to an nxn similarity transform T, which is to be determined furtheron.

$$\Gamma\cdot T = U.\,\Sigma^{1/2} \qquad (21)$$

and

$$T^{-1}.\Delta = \Sigma^{1/2}.\,V^t \qquad (22)$$

The least-squares estimate of A, up to the similarity transformation T, is computed using (19) as

$$T^{-1}.A.T. = (\Gamma.T)^{+}.\,\vec{H}_{pq}\,.\,(T^{-1}.\Delta)^{+} = \Sigma^{-1/2}.U^t.\,\vec{H}_{pq}\,.\,V.\,\Sigma^{-1/2} \qquad (23)$$

Now the similarity transformation matrix T is computed by taking into consideration that

1) the systemmatrix A of the state space model(6) is diagonal and
2) the first column of the matrix Δ (14) is given by (9)

Since A and $T^{-1}.A.T$ are similar,they share the same eigenvalues. So condition 1) is fulfilled by computing the eigenvalues and -vectors of $T^{-1}.A.T.$

Let X be the nxn matrix containing as columvectors the eigenvectors of $T^{-1}.A.T.$ and Λ be the diagonal nxn matrix with the eigenvalues, then clearly

$$T^{-1}.A.T. = X.\Lambda.X^{-1} \qquad (24)$$

and using (18)

$$\begin{aligned}\vec{H}_{pq} &= (\Gamma.T)\,.\,(T^{-1}.A.T.).(T^{-1}.\Delta) \\ &= U.\Sigma^{1/2}.X.\Lambda.X^{-1}.\Sigma^{1/2}.V^t \\ &= (U.\Sigma^{1/2}.X).\,\Lambda\,(X^{-1}.\Sigma^{1/2}.V^t)\end{aligned} \qquad (25)$$

Now let t denote the (complex) first column of $X^{-1}.\Sigma^{1/2}.V^t$

$$t = \begin{bmatrix} t_1 \\ \cdot \\ \cdot \\ \cdot \\ t_n \end{bmatrix} = X^{-1}.\Sigma^{1/2}.V^t \begin{bmatrix} 1 \\ 0 \\ \cdot \\ \cdot \\ 0 \end{bmatrix} \qquad (26)$$

Define by D^{-1} the complex diagonal nxn matrix containing the inverse entries of t

$$D^{-1} = \begin{bmatrix} \frac{1}{t_1} & 0 & \cdots & 0 \\ \vdots & \ddots & & \vdots \\ 0 & \cdots & & \frac{1}{t_n} \end{bmatrix} \qquad (27)$$

Then it is obvious that

$$D^{-1}.\,t = x_o = \begin{bmatrix} 1 \\ 1 \\ \cdot \\ \cdot \\ 1 \\ 1 \end{bmatrix} \qquad (28)$$

Hence

$$D^{-1}.X^{-1}.\Sigma^{1/2}.V^t.\begin{bmatrix} 1 \\ 0 \\ \cdot \\ \cdot \\ 0 \end{bmatrix} = \begin{bmatrix} 1 \\ 1 \\ \cdot \\ \cdot \\ 1 \end{bmatrix} \qquad (29)$$

and (24) is to be modified as

$$\vec{H}_{pq} = (U.\Sigma^{1/2}.X.D)\,.\Lambda.(D^{-1}.X^{-1}.\Sigma^{1/2}.V^t) \qquad (30)$$

which finally results in the parameter estimation

$$A = \Lambda = \begin{bmatrix} e^{\beta_1 + j\gamma_1} & & \\ & \ddots & \\ & & e^{\beta_n + j\gamma_n} \end{bmatrix} \qquad (31)$$

and C as the first row of $U.\Sigma^{1/2}.X.D$

$$C = [1\;0\;\ldots 0].U.\Sigma^{1/2}.X.D = [\alpha_1 e^{j\delta_1} \ldots \alpha_n e^{j\delta_n}] \qquad (32)$$

In conclusion of this section we summarize the algorithm

Step 1 : Construct the matrices H_{pq} and $\vec{H}_{pq}$ (eq. 11,16)

Step 2 : Compute SVD of H_{pq} (eq. 20)

Step 3 : Compute $X.\Lambda$ from (23), (24) via an eigenvalue decomposition.

Step 4 : Determine all parameters from (31), (32).

COMPUTATIONAL CONSIDERATIONS

We now consider some computational refinements. The singular value decomposition (20) and the complex eigenvalue decomposition (24) can be computed using fully tested standard subroutines.
In the special case of real exponentials and when square Hankelmatrices are used (p=q), the realized matrix $T^{-1}.A.T$ can be proven to be symmetric and definite positive, making it's eigenvalue decomposition well conditioned [Wilkinson, 1965]

In the other cases and using square Hankelmatrices, the realized matrix $T^{-1}.A.T$ is symmetric when the absolute value of its entries are considered.
The diagonal matrix (27) is computed as the complex solution to the set of equations defined by (29)

$$X.t = \Sigma^{1/2}.V^t . \begin{bmatrix} 1 \\ 0 \\ . \\ . \\ . \\ 0 \end{bmatrix} \quad (33)$$

Equating real and imaginary parts of (33)

$$X = X_r + jX_i$$

$$t = t_r + jt_i$$

$$\begin{bmatrix} X_r & -X_i \\ X_i & X_r \end{bmatrix} . \begin{bmatrix} t_r \\ t_i \end{bmatrix} = \Sigma^{1/2}.V.t. \begin{bmatrix} 1 \\ 0 \\ \vdots \\ 0 \end{bmatrix} \quad (34)$$

This set can be solved by another SVD. Despite the additional computational requirements, the condition of (34) can be monitored.
Having calculated the matrices A (31) and C (32) the parameters $\alpha_i, \beta_i, \gamma_i, \delta_i$, are computed as follows

$$A = \text{diag}\ (\lambda_i) \text{ then } \beta_i = |\lambda_i| \quad (35)$$

$$\delta_i = \arg\ (\lambda_i)$$

$$C = [\alpha_1 e^{j\delta_1} \dots \alpha_n e^{j\delta_n}] \text{ then } \alpha_i = |\alpha_i e^{j\delta_i}|$$

$$\delta_i = \arg\ (\alpha_i e^{j\delta_i}) \quad (36)$$

INFLUENCE OF THE NOISE : SUBOPTIMALITY

So far, we only considered the case in which

I . the data were known to be exactly representable by a finite sum of elementary signals (1)

and II. no noise corrupted the data.

However, when the data are additively corrupted by noise with variance σ^2, several effects that are difficult to separate, cause the method presented here to be suboptimal in the sense defined by (5).
-If n elementary signals are involved causing the noiseless Hankelmatrix H_{pq} to have n singular values to be non-equal to zero

$$\sigma_1 \geqslant \sigma_2 \geqslant \dots \geqslant \sigma_n > 0$$

and (min (p,q) - n) to equal zero, the noise corrupted version of this Hankelmatrix will have all its singular values σ_i different from zero. It is proven in [Anderson, 1963] that, asymptotically, as $N \to \infty$, the eigenvalues of a perturbed covariance matrix behave like

$$\text{for } i = 1, n \quad \tilde{\lambda}_i = \lambda_i + \sigma^2$$
$$\text{and for } i = n+1, \min(p,q) \quad \tilde{\lambda}_i = \sigma^2 \quad (37)$$

and also some likelihood ratio criterions are proposed in order to obtain the best rank estimation. A similar approach is followed in [Kung, 1981] where, due to the same criterion (37) the ambient space is divided into a n-dimensional signal subspace and a [min (p,q) - n] - dimensional noise subspace. Another reliable rank criterium is derived [De Moor, 1984] and [Staar, 1982]. Only those singular values are retained that satisfy

$$\tilde{\sigma}_i > \sqrt{(R_o+1) \max\ (p,q)} . \sigma \quad (38)$$

where R_o is a user defined maximal minimal signal to noise ratio (SNR).
The corresponding singular vectors form an estimate of the signal subspace.
It is obvious that if one of the signal components falls below the level of accuracy of the date points it is completely unrecoverable. This is an ever present caveat in the problem of separating elementary signals.
-If the numerical rank n is estimated, using one of the preceding criterions, the noise corrupted Hankelmatrix of full rank is approximated in a least squares sense by a matrix computed from the partial SVD (20), according to the theory developped in [Eckart, Young, 1936]. However, this matrix generically shows no longer the Hankelstructure and hence does not represent a n-th order linear system (6). Although this problem is not yet solved for the least squares case with Frobenius-norm, the approximation, obtained using the method here presented, can be shown to be close to optimal [Kung, 1978,1981]. This is due to the well known insensitivity of the singular vector subspaces to perturbations in the data [Wilkinson, 1965].
-Although the singular value decomposition is well conditioned , the singular vectors and values differ from the original ones due to the noise.
Usually, the effect on the singular vectors is neglected, not only because of the insensitivity of the signal subspace but also because the effect of the perturbation is rather complicated, as is shown by a first order perturbation analysis of the SVD in [De Moor, 1984].
Concerning the perturbations of the singular values practical tests have shown that an improvement in the results (reduction in bias) is to be expected, when the partial SVD (20) is replaced by

$$\tilde{H}_{pq} = U.[\Sigma - S\ ].V^t \quad (39)$$

where S is a diagonal matrix.

$$S = \text{diag}\ (\hat{\sigma})$$

Here $\hat{\sigma}$ is an estimate of the supposed isotropic noise singular values.
(maximum likehood estimate [Anderson, 1963], arithmetic mean [Kumaresan, Tufts, 1982]).
We conclude that the noise causes the algorithm to be suboptimal but on the other hand, the partial SVD increases the signal to noise ratio dramatically because the date are projected into the signal subspace. The noise compensation procedure produces a reduction in bias of the estimates.

NUMERICAL RESULTS

In this section we present some striking numerical results that illustrate the power of the algorithm.

I. Fitting a sum of exponentials.

In [Wiscombe, Evans, 1977] it is reported that all known methods experience great difficulty in reconstructing a particular exponential sum, given data sampled from that sum. The ability of our algorithm is however excellent in doing this job, even if the data are only accurate to 2 decimal digits (rounded). 97 data points of the sum

$$y_k = 0.1\ e^{-0.01k} + 0.3\ e^{-0.1k} + 0.6\ e^{-k}$$
$$= a_1.b_1^k + a_2.b_2^k + a_3\ b_3^k$$

were generated for k=0,96 and stored in 48x48 Hankelmatrices.
5 runs with the algorithm using unrounded data and data rounded to 5,4,3 and 2 decimal digits were performed on an IBM 3033 machine with 24 bit mantissa and 7 bit exponent. Following quantities were computed : coefficients and exponents $\hat{x}$, relative deviation defined as 100. $(\hat{x}_{estimate} - x_{exact}) / x_{exact}$, and the error measures

$$e_{max} = \max_{o<k<96} |\ y_k - \hat{y}_k\ |$$

$$e_{rms} = \left[\sum_{k=1}^{96} [y_k - \hat{y}_k\ ]^2 \right]^{1/2} \Big/ 96$$

$$e_{\%} = (100/96) \sum_{i=1}^{96} |y_k - \hat{y}_k| / |y_k|$$

where $\hat{y}_k$ denotes the estimate of the k-th value of the sum.
The results are listed in table 1-3

TABLE 1 ACCURACY OF ESTIMATED PARAMETERS IN %

	6D	5D	4D	3D	2D
a_1	-0.01	-0.02	-0.3	-0.4	0.6
b_1	-0.0001	0.0003	0.0004	-0.001	-0.008
a_2	0.006	0.01	0.06	-0.09	0.1
b_2	0.0004	0.0006	0.002	0.01	-0.07
a_3	-0.0006	0.001	0.002	0.07	-0.1
b_3	-0.0006	-0.0002	0.0003	0.1	-0.3

TABLE 2 ERROR MEASURES

	6D	5D	4D	3D	2D
e_{rms}	0.21E-06	0.23E-06	0.47E-06	0.39E-05	0.35E-04
$e_{\%}$	0.24E-02	0.33E-02	0.41E-02	0.12E-01	0.95E-01
e_{max}	0.36E-05	0.38E-05	0.12E-04	0.17E-03	0.26E-02

TABLE 3 SINGULAR VALUES OF THE 48x48 HANKELMATRICES

	6D	5D	4D	3D	2D
σ_1	4.319	4.319	4.319	4.319	4.315
σ_2	0.890	0.890	0.890	0.890	0.889
σ_3	0.256	0.256	0.256	0.256	0.257
σ_4	$\sim 10^{-5}$	$\sim 10^{-4}$	$\sim 10^{-4}$	$\sim 10^{-3}$	$\sim 10^{-2}$

Comparing the results listed in table 1 and 2 to those reported in [Wiscombe, Evans, 1977], it can be stated that the algorithm, we proposed, obtains the same level of accuracy. Table 3 confirms the fact that the singular values are quite insensitive to perturbation in the data.
It's interesting to note that the choice of the Hankelmatrix dimensions is quite arbitrary as long as the minimal dimension is larger than the number n of elementary signals. Several tests on matrices with large varying dimensions, could detect no significant difference in the accuracy of the estimate.

II. Sum of (damped) cosines.
In order to illustrate the reliability, 91 data were sampled from the four term damped cosine series

$$y_k = 21.(0.9)^k.\cos(0.2k+0.7)+7.(0.8)^k.\cos(0.1k+0.4)$$
$$+38.(0.4)^k.\cos(0.45k+0.8)+20.(0.9)^k.\cos(0.3k+0.9)$$
$$= \sum_{i=1}^{4} \alpha_i \beta_i^k \cos(\gamma_i k + \delta_i)$$

The data were stored in a 45X45 Hankelmatrix with singular spectrum

$\sigma_1=99.68$; $\sigma_2=81.04$; $\sigma_3=9.881$; $\sigma_4=4.328$; $\sigma_5=3.377$; $\sigma_6=0.2933$; $\sigma_7=0.1941E-01$; $\sigma_8=0.1857E-01$

and the remaining ones all smaller then 0.15E-03. This leads to the obvious decision that the Hankelmatrix rank must be 8.
Full precision accuracy results are listed in Table 4 and 5 .

TABLE 4 ACCURACY OF ESTIMATED PARAMETERS IN %

i	1	2	3	4
α_i	0.09	0.3	0.03	0.1
β_i	-0.003	-0.08	0.005	-0.003
γ_i	0.03	-0.2	0.04	-0.01
δ_i	-0.4	1.1	0.05	0.05

The error measures are :

$e_{rms} = 0.208E-04$

$e_{\%} = 0.397$

$e_{max} = 0.643E-03$

These results are excellent, taking into consideration that machineprecision corresponds to 6 à 7 decimal digits.

III. Even the identification of so-called "unstable" signals causes no problem as long as overflow is avoided.
Sampled data from the sum :

$$y_k = 10.(1.1)^k.\cos(0.4k + 0.6)$$
$$+ 7.(0.9)^k \cos(0.2k + 0.4)$$
$$= \sum_{i=1}^{2} \alpha_i \beta_i^k \cos(\gamma_i k+\delta_i)$$

were additively corrupted with a pseudorandom white, zero mean, Gaussian noise sequence with variance $\sigma^2=1$.
51 data samples were used and the Hankelmatrix dimensions were 25x25.
The singular values are $\sigma_1=3434.5$; $\sigma_2=2108.5$; $\sigma_3=18.511$; $\sigma_4=15.214$; $\sigma_5=6.8939$; $\sigma_6=6.1657$; $\sigma_7=\ldots$
From σ_5 on, the singular values reach a saturation level, indicating that the noise level is reached. The accuracy of the estimates, taking a rank estimate of n=4, is listed in Table 5.

TABLE 5 ACCURACY OF ESTIMATED PARAMETERS IN %

i	1	2
α_i	0.005	0.6
β_i	-0.7	2.3
γ_i	0.005	-6.7
δ_i	-1.59	33.0

Remark that the errors for the second signal are larger than that of the first because the first signal contribution is larger than the second and this is less perturbed by the noise.
The error measures are :

$e_{rms} = 0.814$

$e_{\%} = 5.178$

$e_{max} = 25.1$

The last error being caused by the amplification of the misfitting due to the instability.

IV. In the following example the effect of additive white noise of increasing variance is investigated, using the sum

$$y_k = 10.(0.95)^k\cos(0.2k + 0.8)$$
$$+ 12(0.85)^k\cos(0.4k + 0.3)$$

41 data were generated and 20x20 Hankelmatrices were used.
The signal-to-noise ratio SNR is defines as

$$SNR = 10 \log \left[\left(\sum_{k=0}^{40} y_k^2 \right) / 41\sigma^2 \right] \text{ [dB]}$$

The error measures are listed in Table 6.

TABLE 6 ERRORS AS A FUNCTION OF THE NOISE LEVEL

	σ^2	SNR(dB)	e_{rms}	$e_{\%}$	e_{max}
0	0	-	0.45E-05	0.35E-02	0.57E-04
1	0.0001	54	0.68E-03	0.51E+00	0.82E-02
2	0.0005	47	0.15E-02	1.14	0.183-01
3	0.001	44	0.22E-02	1.6	0.26E-01
4	0.005	37	0.47E-02	3.6	0.57E-01
5	0.01	34	0.67E-02	5.1	0.80E-01
6	0.05	27	0.15E-01	11.6	0.18
7	0.1	24	0.20E-01	16.7	0.26
8	0.5	17	0.49E-01	46.5	0.67
9	1.0	14	0.72E-01	71.2	0.98
10	2.0	11	0.84E-01	78.7	1.36
11	5.0	7	0.12	80.0	2.54

To conclude this section, we now mention some of the software aspects.
The program is written in FORTRAN. It's up to the user to choose between a fitting of the data I) by a sum of real exponentials II) by real cosines or III) a combination of the two. A list of the computed singular values is shown to let the user decide whether the numerical rank is to be determined automatically (using one of the rank criterions) or not. We want to stress here that this is the only decision step in the algorithm in contrast with e.g. convergence and coalesce decision to be made in the method of [Wiscombe, Evans, 1977]. At a simple request intermediate results are printed together with all kinds of a posteriori indicators of the numerical reliability (condition numbers, scaling factors, first-order perturbation analysis using the estimated Jacobian etc..)

CONCLUSIONS

A numerically reliable algorithm for the estimation of the parameters of a complex sum of exponentials is derived. The data must be equidistant and may be corrupted by noise. The method is based upon the balaced realization technique of a linear state space model.
The key tool is the singular value decomposition. It guarantees a reliable estimation of the number of elementary signals and of the parameters. Although the noise causes the algorithm to be suboptimal, some practical examples prove the method to have a more than sufficient degree of accuracy. Further research will concentrate upon a detailed sensitivity analysis, extending the algorithm for complex valued data and taking into consideration a possible adaptive scheme based upon adaptive singular value decomposition techniques.

REFERENCES

Anderson, T.W. (1963). Asymptotic theory for principal component analysis. Ann. Math. Statist., 34, 122-148

Braess, D. (1970). Die Konstruktion der Tschebyscheff Approximierenden bei der Anpassung mit exponentialsummen. Journal of approximation theory, 3, 261-273

Cantor, D.G. and J.W. Evans (1970). On approximations by positive sums of powers. Siam J. Appl. Math., 18, 380.

De Moor, B. (1984). First order perturbation analysis of the singular value decomposition. Int. report, ESAT lab., K.U. Leuven

De Moor, B., J. Vandewalle and J. Vanderschoot (1984). A filtering principle for vector signals based on the singular value decomposition. Proc. Europ. Conf. on Circuit Theory and Design (ECCTD), 1985, Prague. Submitted.

De Prony, R. (1795). Essay expérimentale et analytique. J. EC. Polytech. Paris 1, 24-76

Dongarra, J.J., J.R. Bunch, C.B. Moler, and G.W. Stewart (1979). Linpack User's guide. SIAM Philadelphia.

Eckart, C. and G. Young (1936). The approximation of a matrix by another of lower rank. Psychometrika, vol 1, 211-218.

Evans, J.W, W.B. Gragg, and R.J. Leveque (1980) On least squares exponential sum approximation with positive coefficients. Math. Comp., 34, 203-212.

Jacquez, J.A. (1972). Compartmental analysis in biology and medecine. Elsevier Publishing Company, Amsterdam

Kay, S.M. and S.L. Marple (1981). Spectrum analysis, A modern perspective. Proc. IEEE, Vol. 69, no 11.

Klema, V.C. and A.J. Laub (1980). The singular value decomposition : its computation and some applications. IEEE Trans. Aut. Contr., Vol AC-25,2, 164-176.

Kumaresan, R. and D.W. Tufts (1982). Estimating the parameters of exponentially damped sinusoïds and pole-zero modelling in noise. IEEE. Trans. on acoustics, speech and signal processing, Vol ASSP-30,6.

Kung, S.Y. (1978). A new identification and model reduction algorithm via singular value decomposition. Proc. 12th Asilomar Conf. Circuits, Syst. Computer., Pacific Grove, Calif., 705-714.

Kung, S.Y. (1981). A Toeplitz approximation method and some applications. Int. Symp. on math. theory of networks and systems, Vol. 4, Santa Monica, California.

Lanczos, C. (1956). Applied Analysis. Prentice Hall, Inc. Englewood Cliffs, N.J.

Ruhe, A. and Per Ake Wedin (1980a) Algorithms for separable nonlinear least squares problems SIAM review , Vol. 22, 3.

Sidi Avram (1982). Interpolation at equidistant points by a sum of exponential functions. Journal of approximation theory, 34, 194-210

Staar, J. (1982). Concepts for reliable modelling of linear systems with application to online identification of multivariable state space descriptions. PhD thesis, ESAT Lab, K.U. Leuven.

Tufts, D.W. and R. Kumaresan (1982). Estimation of frequencies of multiple sinusoïds: Making linear prediction perform like maximum likehood. Proc. IEEE, Vol. 70, 9.

Vandewalle, J., J. Staar, B. De Moor, J. Lauwers (1984). An adaptive singular value decomposition algorithm and its use in adaptive realization. 6th. int. conf. anal. optim. of systems, June 1984, Nice.

Wilkinson, J.H (1965). The algebraic eigenvalue problem. London, Oxford Univ. Press.

Wiscombe, W.J. and J.W. Evans (1977) Exponential sum fitting of radiative transmission functions. Journal of Comput. Physics, 24, 416-444.

Zeiger, H.P. and A.J. Mc Ewen (1974) Approximate linear realization of given dimension via Ho's algorithm. IEEE Trans. Aut. Control, Vol. AC-19, 153.

COMPUTER AIDED DESIGN OF DIGITAL CONTROLLERS FOR MULTIVARIABLE PROCESSES BASED ON SERIES EXPANSIONS APPROACH

G. Obinata*, T. Nakamura* and H. Inooka**

**Department of Mechanical Engineering for Production, Akita University, Akita, Japan*
***Department of Mechanical Engineering, Tohoku University, Sendai, Japan*

Abstract. This paper considers a computer aided design of a digital controller for multivariable system which realizes approximately an arbitrarily specified closed-loop response. The algorithm is derived on the relations of series expansions of pulse transfer function matrices; a linear equation is obtained for parameters of a controller. A computer program is developed based on the algorithm so as to allow a designer to assess quickly the effect of specifications on a resultant closed-loop response. For example, one can understand that a large control input is needed when a short settling time is specified. Some numerical examples are given.

Keywords. Digital control; discrete time systems; control system synthesis; computer aided design; least squares approximations.

Introduction

Discrete control has advantages of flexibility over continuous control in implementing various algorithms for controlling a plant in a desired manner: one paticular advantage is the achievement of finite time settling control (Ragazzini and Franclin, 1958). The authors (Inooka, Obinata and Takeshima, 1983) proposed a digital control scheme for single-input and single-output linear processes based on series expansions of pulse transfer functions. Using the least square method, a linear equation is obtained for unknown parameters of the reduced-order controller. The digital controller realizes approximately any closed-loop response. This paper generalizes the results to a class of multi-variable systems and developes a computer program for computer aided design based on the proposed method.

Merits of introducing computer aided design are as follows. First, one can see how the closed-loop response approximates a specified response; the response may not resemble the specified response when a short settling time response is specified. It is particularly undesirable in multivariable control systems that the resultant feedback system cannot be decoupled even if decoupling control scheme is given as the closed-loop specification. Second, one can assess easily the effect of the order of the controller on the response of the resultant closed-loop system. In general, a severe specification may be realized when the order of the controller is high. Third, one can see the behavior of the control input and the disturbance effect; it is frequently required for practical applications to avoid a larger control input and to reject the effect of a disturbance. Also, the coupling between the control loops can be checked easily.

Mathematical operations included in the design procedure are simple and not iterative. These are several matrix additions and multiplications, and one matrix inversion. Because of the simplicity of the algorithm, the computational time required in a trial is very short and necessary memory size is small. We used a personal computer with a two-dimensional digitizer and a graphical display. Two examples are given to illustrate the design procedure and to examine the closed-loop system behavior under the finite time settling and decoupling control scheme.

Relations of series expansions of pulse transfer function matrices

Consider the output feedback system with a digital controller shown in Fig.1. It is assumed that the plant is stable and the input u and the output y are both r-dimensional vectors. Let us denote the pulse transfer function matrices of a process and a controller G(z) and C(z), respectively. These transfer function matrices can be expanded in negative power of z as follows:

$$G(z) = \sum_{i=0}^{\infty} G_i z^{-i} \tag{1}$$

$$C(z) = \sum_{i=0}^{\infty} C_i z^{-i} \tag{2}$$

where G_i and C_i are both $r \times r$ coefficient matrices. In many industrial processes G_0 may be equal to zero matrix; therefore, we assume that $G_0 = 0$ in what follows. The open-loop pulse transfer function matrix E(z) can be expanded as follows:

$$E(z) \triangleq G(z)C(z) = \sum_{i=0}^{\infty} E_i z^{-i} \tag{3}$$

From equations (1), (2) and (3), the coefficient matrix: E_i can be expressed by the matrices: G_i

and C_i; that is,

$$E_0 = 0$$

$$E_i = \sum_{j=1}^{i} G_j C_{i-j}, \quad i = 1, 2, 3, \ldots \tag{4}$$

Expand the closed-loop pulse transfer function matrix as follows:

$$W(z) \triangleq E(z)\{ I + E(z)\}^{-1}$$

$$= \sum_{i=0}^{\infty} W_i z^{-i} \tag{5}$$

By the use of equation (3), the coefficient matrix: W_i is determined by the following iteration formula:

$$W_0 = 0$$

$$W_1 = E_1$$

$$W_i = E_i - \sum_{j=1}^{i-1} W_{i-j} E_j, \quad i = 2, 3, \ldots \tag{6}$$

Thus the coefficient matrices of the series expansions of the closed-loop pulse transfer function are expressed in terms of the coefficient matrices corresponding to open-loop elements.

Remark 1: It should be noted that k-th coefficient : W_k of the closed-loop system is determined only by E_i ($i \leq k$); in other words, W_k is determined by G_i and C_i ($i \leq k$). The same relations hold for the time moments (Inooka, 1976; Inooka and Obinata, 1982; Obinata, 1978).

Design of a control system

Let matrix sequence : $\{ W_i ;\ i=1, 2, 3, \ldots \}$ represent some desired unit pulse response. We consider the design problem that the matrix sequence of closed-loop system is matched approximately as long as required with the desired matrix sequence : $\{ W_i \}$. We formulate the following procedure for the computer aided design system.

Step 1: Specify the desirable pulse response : $\{ W_i ;\ i=1, 2, 3, \ldots, m\}$. If finite time settling and decoupling control scheme is adopted, the desired matrix sequence : $\{ W_i ;\ i=1, 2, 3, \ldots m\}$ must satisfy the following condition:

$$W_i = \begin{bmatrix} w_{11}(i) & 0 & \cdots & 0 \\ 0 & w_{22}(i) & & 0 \\ \vdots & & \ddots & \vdots \\ 0 & \cdots & 0 \cdots & w_{rr}(i) \end{bmatrix} \quad i = 1, 2, \ldots m \tag{7}$$

where the scalar sequence : $\{ w_{jj}(i);\ i=1, 2, \ldots, m\}$ satisfies the condition: $\sum_{k=1}^{m} w_{jj}(k) = 1.0$ ($j= 1, 2, \ldots, r$). Also specify the order of the controller : n ($n < m-1$).

Step 2: Solve the equation (6) for E_i ($i=1, 2, \ldots, m$). The results are expressed by the following iteration formula:

$$E_1 = W_1$$

$$E_i = W_i + \sum_{j=1}^{i-1} W_{i-j} E_j, \quad i = 2, 3, \ldots, m \tag{8}$$

Step 3: Substitute E_i ($i=1, 2, \ldots, m$) into equation (4) and construct an equation for C_i under the condition that

$$C_n = \ldots = C_m \triangleq C \tag{9}$$

Then the equation for C_i is reduced to

$$\overline{G}\,\overline{C} = \overline{E} \tag{10}$$

where

$$\overline{C} = [\, C_0 \ \ C_1 \ \ldots \ C_n \,]'$$

$$\overline{E} = [\, E_1 \ \ E_2 \ \ldots \ E_m \,]'$$

$$\overline{G} = \begin{bmatrix} G_1 & 0 & \cdots & 0 & 0 \\ G_2 & G_1 & 0 \cdots & 0 & 0 \\ G_3 & G_2 & G_1 & \cdots & 0 \\ \vdots & \vdots & & & \vdots \\ G_{n+1} & G_n & \cdots & G_2 & \tilde{G}_1 \\ G_{n+2} & G_{n+1} & \cdots & G_3 & \tilde{G}_2 \\ \vdots & \vdots & & \vdots & \vdots \\ G_m & G_{m-1} & \cdots & G_{m-n+1} & \tilde{G}_{m-n} \end{bmatrix}$$

$$\tilde{G}_j = \sum_{k=1}^{j} G_k$$

Step 4: Solve the equation (10) by the method of least squares. The least squares solution $\overline{C}^* = [\, C_0^* \ C_1^* \ldots C_n^* \,]'$ is given by

$$\overline{C}^* = (\, \overline{G}'\overline{G} \,)^{-1} \overline{G}'\overline{E} \tag{11}$$

Step 5: The controller is expressed as follows:

$$C(z) = C_0^* + C_1^* z^{-1} + \ldots + C_{n-1}^* z^{-n+1} + C_n^* z^{-n}$$

$$+ C z^{-n-1} + C z^{-n-2} + \ldots$$

$$= (\, \tilde{C}_0 z^n + \ldots + \tilde{C}_n)/(\, z^{n-1}(z - 1)) \tag{12}$$

where

$$\tilde{C}_0 = C_0^*$$

$$\tilde{C}_i = C_i^* - C_{i-1}^*, \quad i = 1, 2, \ldots, n$$

Step 6: Substitute equation (12) into equation (4), and then calculate $\{ W_i^* ;\ i=1, 2, \ldots, m \}$ using equation (6) and the control inputs. If the matrix sequence : $\{ W_i^* \}$ is not close sufficiently to the specified sequence : $\{ W_i \}$, then increase the order of the controller and go back to Step 2. Also, if the amplitude of the control input exceeds the allowable range, then modify the specified pulse response to more generous one and go back to Step 1.

Step 7: Calculate the continuous unit step response of the resultant closed-loop system with a hold on the control input, and examine the variation of the continuous output on the interval between each sampling instance. If the amplitude of the variation is larger than allowable one, then decrease the sampling time and go back to Step 1.

Remark 2: In step 1, W_i should be specified long enough so as to have a stable closed-loop response even if $W_i = 0$ for large i.

Remark 3: The constraint (9) is introduced in order to add the integral action to the controller. The steady state error of the closed-loop system for the step reference input is zero under the integral action.

Computer aided design system

The computer aided design system based on the proposed design procedure is implemented into a personal computer (NEC PC-9801). The computer system has a random access memory unit of 224K bytes capacity and a CPU unit with a 16-bit microprocessor (Intel 8086). The peripheral equipments of the computer are a two-dimentional digitizer, a graphical display, and a plotter. The BASIC language is used to write the program for the computer aided design. Concerning memory requirements, the program area needs 15K bytes and the variable area is proportional to $r^2 \times m$. In Step 1 of the design procedure, a designer enters the desirable specified response :$\{W_i\}$, the order of the controller :n, and the number of the specified response :m using the keyboad of the console or the digitizer. The arithmetic computations in Step 2~Step 5 are done automatically, and then the informations of the resultant closed-loop system are shown on the graphical display or are plotted out on the sheet. These informations are the sequence :$\{W_i^*\}$, the discrete and continuous response for the step reference input, the control input, and the response for the step disturbance. Based on these informations, the designer judges whether a modification of the specified response or the order of the controller is needed.

Examples

In this section two examples of finite time settling and decoupling control are given.

Example 1: Tomizuka (1979) has presented a example of level control problem with three tanks. The linearized state $[x_1, x_2, x_3]'$ around the equilibrium are modeled by the state equation:

$$\begin{bmatrix} \dot{x}_1 \\ \dot{x}_2 \\ \dot{x}_3 \end{bmatrix} = \begin{bmatrix} -0.375 & 0.125 & 0 \\ 0.25 & -1.0 & 0.25 \\ 0 & 0.5 & -1.5 \end{bmatrix} \begin{bmatrix} x_1 \\ x_2 \\ x_3 \end{bmatrix} + \begin{bmatrix} 0.25 & 0 \\ 0 & 0.5 \\ 0 & 1 \end{bmatrix} \begin{bmatrix} u_1 \\ u_2 \end{bmatrix}$$

$$\begin{bmatrix} y_1 \\ y_2 \end{bmatrix} = \begin{bmatrix} 0 & 1 & 0 \\ 0 & 0 & 1 \end{bmatrix} \begin{bmatrix} x_1 \\ x_2 \\ x_3 \end{bmatrix} \tag{13}$$

The unit step responses are plotted in Fig.2. In figures, the (i,j) means the i-th variable when only the j-th reference input is set to a unit step function. In design (Step 1~ Step 6), the plant was represented by a discrete time system which was obtained from equation (13) and an added zero-hold on the input; thus, the output of the continuous plant exactly matches the output of the discrete model under digital control at each sampling time. Let the sampling time be set to 1 second, and the desirable pulse response of the closed-system be:

$$\begin{bmatrix} 0.6 & 0 \\ 0 & 0.6 \end{bmatrix}, \begin{bmatrix} 0.4 & 0 \\ 0 & 0.4 \end{bmatrix}, \begin{bmatrix} 0 & 0 \\ 0 & 0 \end{bmatrix}, \ldots \tag{14}$$

The matrix sequence gives a decoupling response with a settling time of 2 seconds. In order to examine the effect of the order n of the controller while m is set to a constant value, we use the follwing index as the measure of approximation.

$$J_{ij} = \frac{1}{m+1} \sum_{k=0}^{m} (w_{ij}^*(k) - w_{ij}(k))^2 \tag{15}$$

We may observe from Fig.3 that the response becomes more close to specified one as n is increased and the approximation is almost perfect for $n \geq 4$. As n=6, the continuous outputs of plant and control inputs for unit step inputs are shown in Fig.4. The outputs show almost exactly the decoupling response with a settling time of 2 seconds and show the small variations on the interval between sampling instances. But, the control input (1, 1) has a large deviation from the steady state value. If the deviation is larger than the allowable value, it is necessary to change the specified pulse response.

Example 2: The system to be considered here consists of the automotive gas-turbin, the fuel system, and the nozzle actuator (Patel and Munro, 1982). The system was modeled by the transfer function matrix:

$$G(s) = \begin{bmatrix} \dfrac{0.806s + 0.264}{s^2 + 1.15s + 0.202} & \dfrac{-(15.0s+1.42)}{s^3 + 12.8s^2 + 13.6s + 2.36} \\ \dfrac{1.95s^2 + 2.12s + 0.490}{s^3 + 9.15s^2 + 9.39s + 1.62} & \dfrac{7.14s^2 + 25.8s + 9.35}{s^4 + 20.8s^3 + 116.4s^2 + 111.6s + 18.8} \end{bmatrix} \tag{16}$$

The unit step responsess are plotted in Fig.5. The system with a zero-hold on the input was represented by the discrete time system in design procedures Step 1 ~ Step 6. We set the sampling time to 1 second and specify the desirable pulse response of the closed-loop system as follows:

$$\begin{bmatrix} 1.0 & 0 \\ 0 & 1.0 \end{bmatrix}, \begin{bmatrix} 0 & 0 \\ 0 & 0 \end{bmatrix}, \begin{bmatrix} 0 & 0 \\ 0 & 0 \end{bmatrix}, \ldots \tag{17}$$

The matrix sequence gives a decoupling response with a settling time of 1 second. The discrete responses of the closed-loop system for unit step inputs and the control inputs are shown in Fig.6 (m=16, n=10). It is observed that the approximation is almost perfect. Under the same conditions, the continuous response of the closed-loop system for unit step input are shown in Fig.7. We may observe some undesirable variations on the interval between sampling instances. In order to reduce this variations, we change the sampling time to 0.5 second and the desirable pulse response to

$$\begin{bmatrix} 0.5 & 0 \\ 0 & 0.5 \end{bmatrix}, \begin{bmatrix} 0.5 & 0 \\ 0 & 0.5 \end{bmatrix}, \begin{bmatrix} 0 & 0 \\ 0 & 0 \end{bmatrix}, \ldots \tag{18}$$

The sequence gives the same finite time settling response as the former. The resultant responses are shown in Fig.8. These are almost satisfactory for the finite time settling and decoupling control scheme.

Conclusions

The method of designing a digital controller for a multivariable process has been considered which realizes approximately a desired closed-loop response. The unit pulse response of a controller is determined from a pulse response of a plant and a specified pulse response of a closed-loop system by use of a relation of series expansions. Using the least square method, a linear equation is obtained for unknown parameters of the reduced-order controller. Because the approximation is used in the design procedure, we cannot ascertain in advance whether the rsultant feedback system achieve the desired response; therefore, a number of trials are necessary in general for the choice of the order of the controller and for the modification of the closed-loop specifications. A computer aided design system has been developed suitable for such a trial. Because of the simplicity of the proposed algorithm, the computational time required in a trial is very short. Using the computer aided design system, a designer can select appropreate design parameters in a short time.

REFERENCES

Inooka, H. (1976). Moments of an output feedback system. Electronics Letters, 12, 588-589.

Inooka, H., and Obinata, G. (1982). Relations of moments in composite systems. IEEE Trans. Automatic Control, 27, 967-969.

Inooka, H., and Obinata, G., and Takeshima, M. (1983). Design of a digital controllers based on series expansions of pulse transfer functions, ASME Journal of Dynamic Systems, Measurements, and Control. 105, 204-206.

Obinata, G. (1978). Remarks on the time moments of composite systems. Electronics Letters, 14, 509-511.

Patel, R. V., and Munro, N. (1982). Multivariable system theory and design. Pergamon Press, 249.

Ragazzini, J. R., and Franclin, G. F. (1958). Sampled-data control systems, McGraw-Hill, New York City.

Tomizuka, M. (1979). A simple digital control scheme for a class of multi-input, multi-output industrial processes. ASME Journal of Dynamic Systems, Measurements, and Control, 101, 339-344.

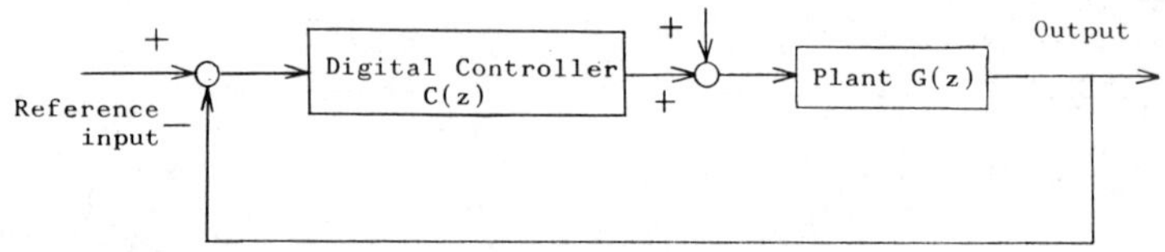

Fig.1 Output feedback control system.

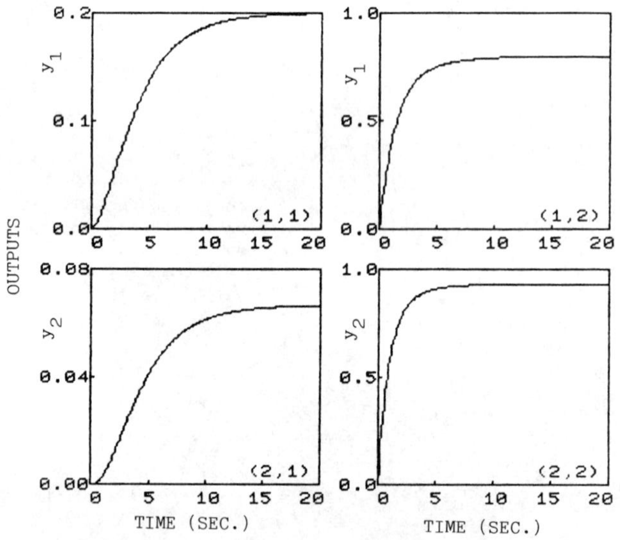

Fig.2 Unit step responses of three tank system.

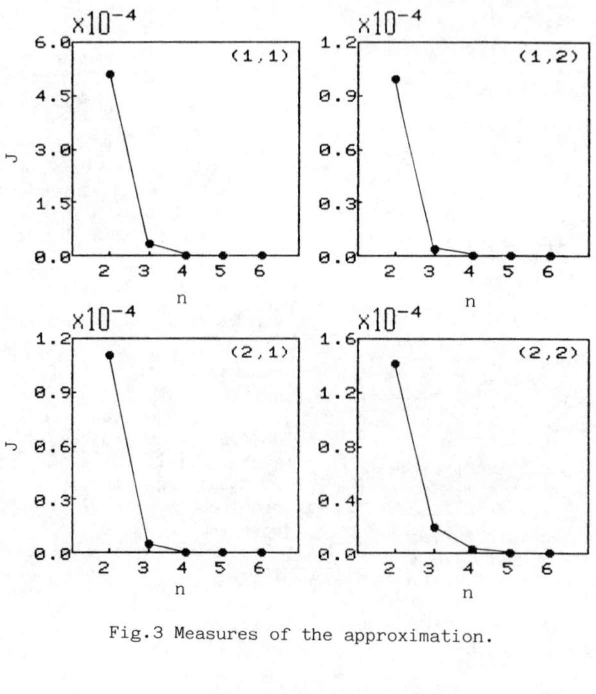

Fig.3 Measures of the approximation.

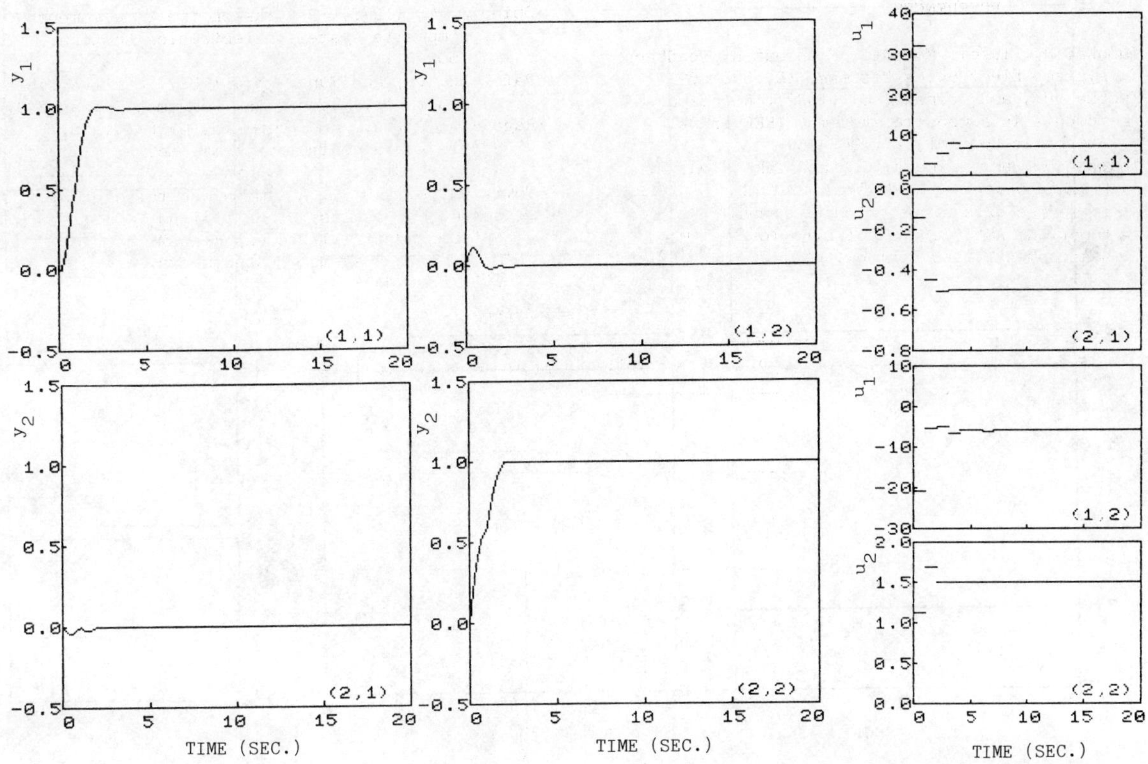

Fig.4 Outputs of continuous plant and control inputs for a unit step reference input (n=6, m=16).

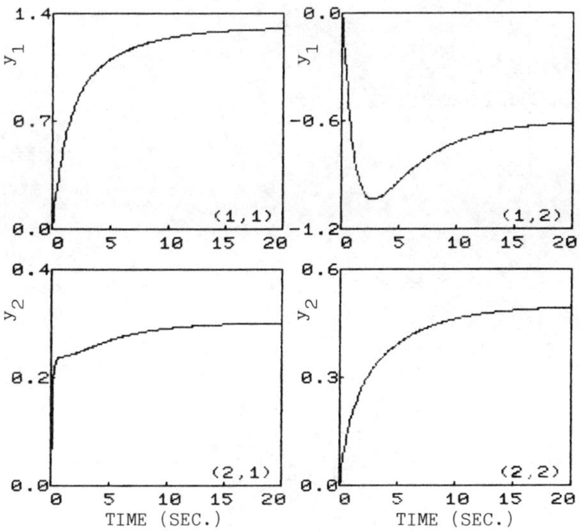

Fig.5 Unit step responses of automotive gas-turbine.

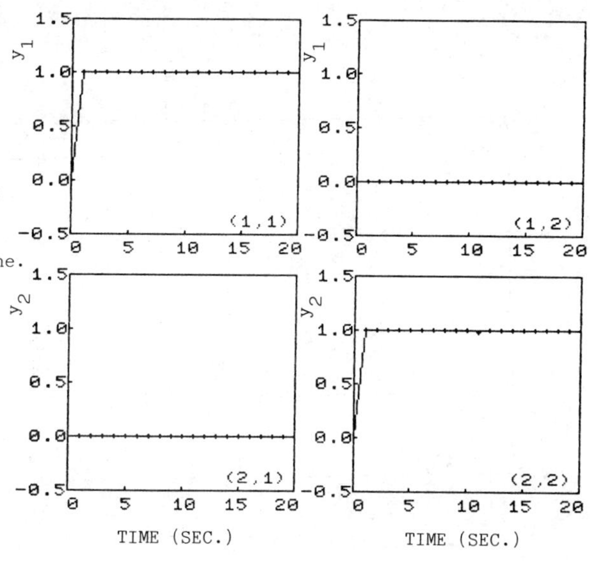

Fig.6 Outputs for a unit step reference input (n=10, m=16).

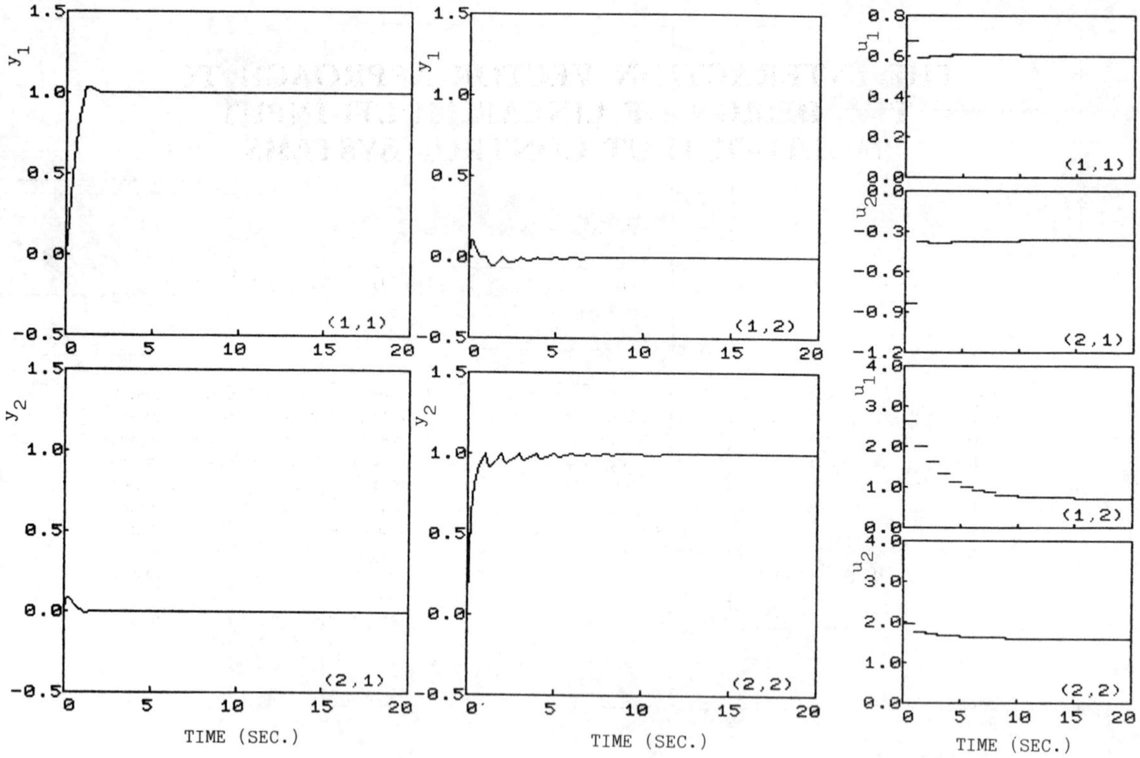

Fig.7 Outputs of continuous plant and control inputs for a unit step reference input (ΔT=1 sec.).

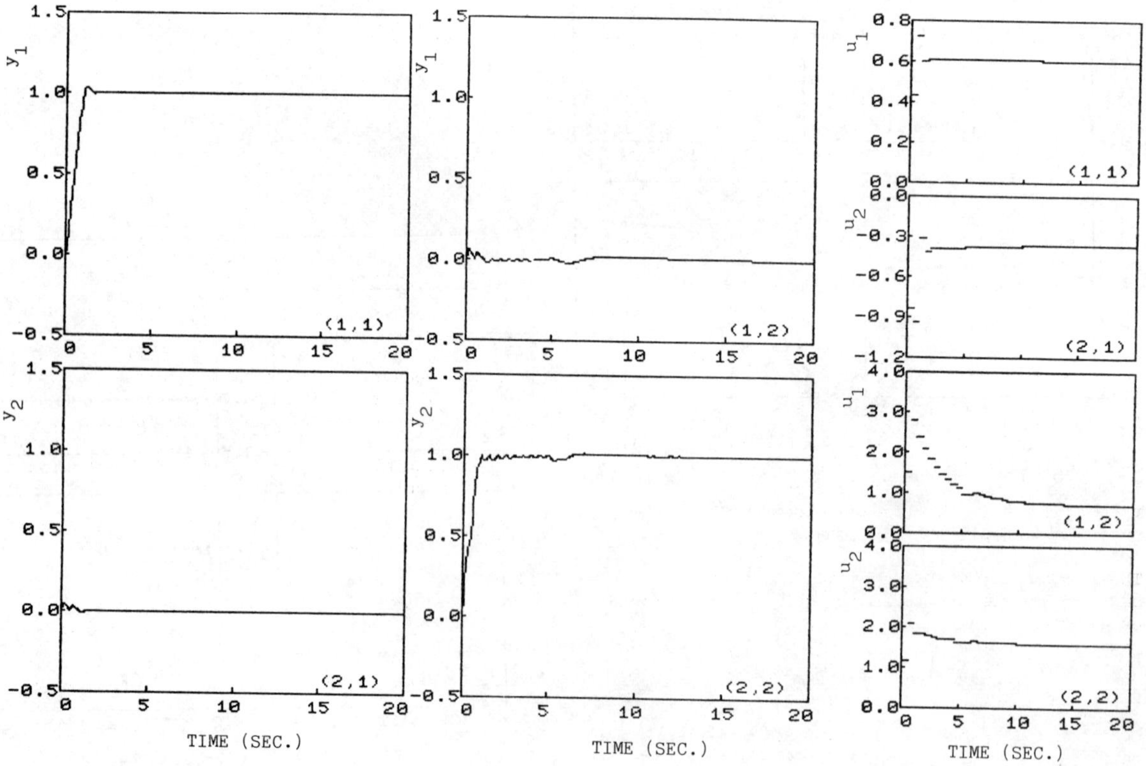

Fig.8 Outputs of continuous plant and control inputs for a unit step reference input (ΔT=0.5 sec.).

THE INTERACTION VECTOR APPROACH TO THE DESIGN OF LINEAR MULTI-INPUT MULTI-OUTPUT CONTROL SYSTEMS

P. M. Taylor

Department of Electronic Engineering, University of Hull, Hull HU6 7RX, UK

Abstract. Interaction vector displays were first presented as a means of indicating frequency dependent interactions in linear multivariable systems by the superposition of interaction vectors onto an inverse Nyquist array. The vectors can also be defined for a Nyquist array and show how each element $q_{ii}(j\omega)$ of the open loop frequency response is modified to $h_i(j\omega)$ when all loops are closed excepting loop i. It has been shown how the interaction vectors, or approximations to them, are modified by compensation. Uncertainties in the frequency response of the system model are considered by surrounding each frequency evaluation of the uncompensated $g_{ik}(j\omega)$ by an uncertainty circle in which the true $g_{ik}(j\omega)$ must be. It is shown how each approximation interaction vector can then be represented by a nominal vector surrounded by an offset circular bound. The designer can thereby assess the effects of uncertainties at all stages of the interactive design of compensation. The approach is illustrated with examples from the literature.

Keywords. Computer-aided design, control system analysis, control system design, frequency response, multivariable control systems.

INTRODUCTION

The frequency domain techniques of Rosenbrock (1974) using the inverse Nyquist array with superimposed Gershgorin or Ostrowski bands retain several desirable properties for the designer of linear multivariable feedback systems. Firstly the displays used are merely an aid, the designer is given complete freedom over choice of compensation. A simple framework is provided whereby the designer may assess stability and estimate the effects on closed loop performance of the injection of compensation into the open loop. Secondly, particularly if Nyquist arrays are used, the effects of uncertainties in the compensated open loop frequency responses may be treated by simple enlargement of the Gershgorin discs. As these discs are based on only the moduli of off diagonal elements the designs will be inherently tolerant to phase variations in the off diagonal elements of the open loop compensated system. However, the requirement for diagonal dominance can be an unnecessarily restrictive condition, especially for systems having many inputs and outputs. The neglect of phase information means that any design must assume the worst case, that the presence of the off diagonal coupling terms has a destabilising influence. Clearly this need not be so. Also, as discussed by Doyle and Stein (1971), the resultant designs may well be unacceptably sensitive to uncertainties in the frequency responses of the uncompensated plant.

The interaction vector approach which may be based on Nyquist or inverse Nyquist arrays provides a more precise design aid by fully utilising phase information to locate the $h_i(j\omega)$ loci, the open loop frequency response loci of loop i with all other loops closed, or the $\hat{h}_i(j\omega)$ loci respectively. The approach is outlined briefly in the next section, followed by details of the treatment of uncertainties. Programming aspects are described followed by illustrative examples and conclusions.

INTERACTION VECTOR PLOTS

Interaction vector plots were first introduced by Gray and Taylor (1979) to show the precise results on the open loop frequency response of any loop of adding in the interaction terms linking the loop to all others (closed). The system configuration is shown in Fig. 1. In the inverse case

$$\hat{h}_i = \hat{q}_{ii} + \hat{V}_{ii} \tag{1}$$

where $$\hat{V}_{ii} = \sum_{\substack{k=1,m \\ k\neq i}} \hat{V}_{ik} \tag{2}$$

and $$\hat{V}_{ik} = \hat{q}_{ik}\, y_k / y_i \quad k\neq i \tag{3}$$

The terms $\hat{V}_{ii}$ and $\hat{V}_{ik}$ are termed the inverse interaction vectors and are displayed on the basic inverse Nyquist array as shown in Fig. 2 which also shows the Gershgorin bands for comparison. Three possible measures of interaction were suggested by Gray and Taylor (1979) each one of which has a graphical interpretation, as shown in Fig. 3.

Similar displays and functionals may be obtained for the Nyquist arrays with

$$h_i = q_{ii} + V_{ii} \tag{4}$$

where

$$V_{ii} = \sum_{\substack{k=i,m \\ k\neq i}} V_{ik} \tag{5}$$

and

$$V_{ik} = q_{ik}\, e_k/e_i \quad k\neq i \tag{6}$$

It has been shown by Taylor and Hayton (1981) how the designer can estimate the effects of compensation on the interaction vectors simultaneously with estimation of the effects on the basic elements of the displayed array. It was shown that for ease of estimation a convenient approximation

$$\hat{v}_{ik} \simeq \hat{q}_{ik}\,\hat{q}_{ki} \;/\; (1 = \hat{q}_{kk}) \qquad (7)$$

may be used for the inverse Nyquist displays, giving

$$\hat{v}_{ii} \simeq \sum_{\substack{k=1,m\\ k\neq i}} q_{ik}\, q_{kc} \;/\; (1 + \hat{q}_{kk}) \qquad (8)$$

Note that no approximation is involved for two input two output systems. Exact and approximate interaction vectors are computed and may be displayed separately or simultaneously as desired. It has been found in practice so far that the approximation improves as the compensation proceeds and the second order interaction terms reduce. Similarly in the direct Nyquist case the following approximation may be made

$$v_{ii} \simeq \sum_{\substack{k=1,m\\ k\neq i}} q_{ik}\, q_{ki} \;/\; (1 + q_{kk}) \qquad (9)$$

Note that each element of Q appears only once in the expansion, making compensation estimation easier and, as will be seen, simplifying sensitivity analysis.

UNCERTAINTY BOUNDS ON THE INTERACTION VECTORS

One of the primary uses of feedback ought to be to diminish effects of plant uncertainties. As Horowitz (1982) has highlighted it is only recently that plant uncertainties have been considered in a quantified way for multi-input multi-output systems. It is often not clear how measures of sensitivity actually relate to any physically measureable effects. The interaction vectors and functionals are measureable experimentally and become more attractive if some means can be found of computing their bounds when the plant model is uncertain. It will be assumed here that the plant uncertainties can be characterised by circular uncertainty bounds around each complex value of $g_{ik}(j\omega)$ for the set of ω under consideration, as shown in Fig. 4, Since h_i may be approximated by

$$h_i \simeq q_{ii} + \sum_{\substack{k=1,m\\ k\neq i}} q_{ik}\, q_{ki} / (1 + q_{kk}) \qquad (10)$$

a method is needed for finding bounds on the uncertainties for sums, products and quotients of complex numbers with circular uncertainty bounds. The problem is then essentially one of complex interval arithmetic with circular intervals as discussed in Alefeld and Herzberger (1983). The uncertainties in each element of Q will be treated as being independent of each other. This may not be so in practice because of the particular structure of the plant and the technique will then give overly conservative bounds. When compensation is present the uncertainties in G will be reflected into uncertainties in Q and conservatism may be introduced because the uncertainties in each element of Q will no longer be independent of each other if the compensators are not diagonal. The method now follows.

Let uncertain complex numbers C1 and C2 lie anywhere within circles centred on C1N and C2N and with radii U1|C1N| and U2|C2N| respectively. C1N and C2N are the nominal values of C1 and C2 and U1 and U2 are their fractional uncertainties. It is simple to show that the sum C1 + C2 will lie somewhere within the bounds given by a circle radius U1|C1N| +U2|C2N| centred on C1N + C2N. Multiplication is a more complex problem.

Consider C1 and C2 normalised to lie inside circles centred on (1, j0) and of radii U1 and U2 respectively. The product of C1 and C2 then lies within a cardioid formed by the outer values of the products of all points on the circumference of the C1 bounds with all points on the circumference of the C2 bounds as shown in Fig. 5. This cardioid must now be bounded by a circle ready for further operations. A program has therefore been written which finds the smallest circle enclosing the cardioid for a range of discrete values of U1 and U2 lying between 0 and 1. This program, run once only, stores the results in a data file. The interaction vector program in the CAD suite, reads in this data file as a lookup table. When complex interval multiplication is required the two complex numbers are normalised, the result (new centre and radius) is interpolated from the lookup table and then denormalised.

If the quotient C1/C2 is required it is merely necessary to normalise C1 and C2, invert C2 and multiply as above. Inversion of the circle bounding C2 results in an offset circle bounding 1/C2. If U2 > 1 a warning is given as points inside the C2 circle are then mapped outside the 1/C2 circle and the uncertainty area can then cover most of the complex plane.

PROGRAMMING ASPECTS

The interaction vector technique is computationally simple and rapid. In the inverse case with all loops closed

$$\tilde{y} = A\tilde{r} \qquad (11)$$

$$\text{where} \quad A = [\hat{Q} + I]^{-1} \qquad (12)$$

if say $r_i = 1$ and $r_k = o$ for all $k = i$

$$\text{then} \quad y_k/y_i = a_{ki} \;/\; a_{ii} \qquad (13)$$

This ratio will also hold when all loops are closed excepting loop i. The exact inverse interaction vectors can now be calculated from equations 2 and 3. The approximate vectors merely require knowledge of the elements of Q (jω). In the direct Nyquist case if $A = [\,Q + I\,]^{-1}$ then $e_k/e_i = a_{ki}/a_{ii}$. Thus the same software can be used in both cases by interchanging $\hat{Q}$ with Q as required. When compensation is present $\hat{G}$ is interchanged with G, $\hat{K}$ with L and $\hat{L}$ with K. The largest computational burden is the complex matrix inversion in equation 12 which is performed for each frequency value used.

The CAD programs are written in FORTRAN IV as part of the Hull Computer Aided Control System Design Suite (Taylor, 1984), which runs on a PDP11/34 running under RSXM V3.2. The interaction vectors are requested by the user as alternatives to Gershgorin or Ostrowski bands after display of the basic Nyquist or inverse Nyquist array. Colour graphics has proven useful particularly when the interaction vectors are closely aligned with the basic array curves.

EXAMPLES

Figure 2 shows the inverse Nyquist array of the illustrative example of Gray and Taylor (1979) with interaction vectors and Gershgorin bands superimposed. In this case

$$G(s) = \begin{bmatrix} \frac{20}{s(s+2)} & \frac{5}{s(s+1)} \\ \frac{10}{s(s+2)} & \frac{20}{s(s+1)} \end{bmatrix} \quad (14)$$

The stabilising effect of the interaction is clearly seen. Figure 6 illustrates the effects of uncertainties in $G(j\omega)$ with uncertainty fractions $u_{ij} = 0.1$ for all i and j. The closed loop system will remain stable, although with poor phase margins.

The example shown in Fig. 7 is due to Rekasius (1979) and has

$$G(s) = \begin{bmatrix} \frac{9}{s+1} & \frac{-10}{s+1} \\ \frac{-8}{s+2} & \frac{9}{s+2} \end{bmatrix} \quad (15)$$

This is a highly interactive system, nominally completely decoupled by a precompensator

$$K(s) = \begin{bmatrix} 9 & 10 \\ 8 & 9 \end{bmatrix} \quad (16)$$

However this compensator is effectively cancelling out one interaction signal by deliberately adding another out of phase. It is not surprising that the resultant system is extremely sensitive to any variations in $G(j\omega)$ about the nominal values. The interaction vector plot of Fig. 8 shows vectors of zero length but very large uncertainty circles.

CONCLUSIONS

The interaction vector displays give the designer quantitative and visual information about frequency dependent loop interactions at all stages during interactive design of compensation based on a Nyquist or inverse Nyquist array. The use of complex interval arithmatic allows the vectors to be bounded when the elements of the model frequency response $G(j\omega)$ have unstructured uncertainties defined as lying within discs surrounding each $g_{ik}(j\omega)$ frequency point. The designer can thereby detect when compensation will result in a closed loop system sensitive to uncertainties in the system model. The size of the displayed bounds is expressible in terms of bounds of measureable signals. The use of approximate interaction vectors allows the technique to be extended easily to systems with more than two inputs and outputs although further study is needed to determine more rigorously the resultant errors in the uncertainty analysis.

The second example given in the paper shows an extreme case of bad compensation using cancellation of large interacting signals which results in severe sensitivity problems. Such cancellation is in essence done by all decoupling or partial decoupling methods and the designer may be unaware of robustness problems unless the sensitivities are checked by techniques such as those above.

REFERENCES

Alefeld, G., and J. Herzberger (1983). *Introduction to Interval Computations*. Academic Press, London

Doyle, J.C., and G. Stein (1981). Multivariable feedback design: concepts for a classical/modern synthesis. *IEEE Trans.* AC-26 No. 1 pp 4-16.

Gray, J.O., and P.M. Taylor (1979) Frequency domain functionals for the assessment of interaction effects in multivariable feedback systems. *Proc IFAC Symp. on Computer Aided Design of Control Systems*, Zurich. pp 81-86.

Horowitz, I. (1982). Quantitative feedback theory. *IEE Proc.* Vol 129, Pt.D No.6, pp 215-226

Rosenbrock, H.H. (1974). *Computer Aided Control System Design* Academic Press, London

Taylor, P.M. (1984). The Hull control system design suite. *Proc. IMC Workshop on Computer Aided Control System Design*, Brighton, pp 73-78

Taylor, P.M., and G. Hayton (1981). The manipulation of interaction effects in multivariable feedback systems. *Proc. IFAC 81, 3rd World Congress*, Kyoto, Vol. 4, pp 6-11

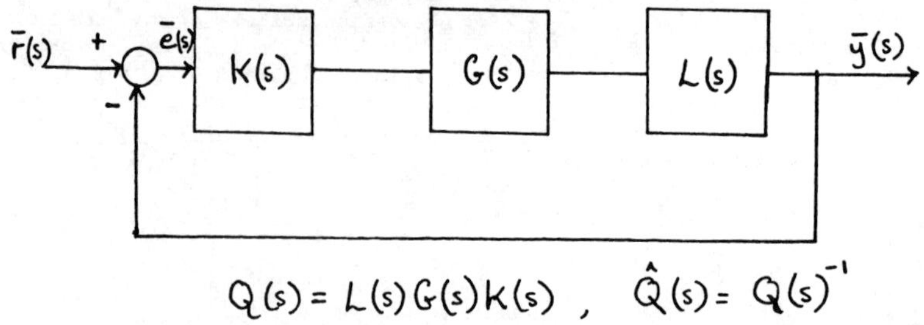

Fig. 1. System configuration

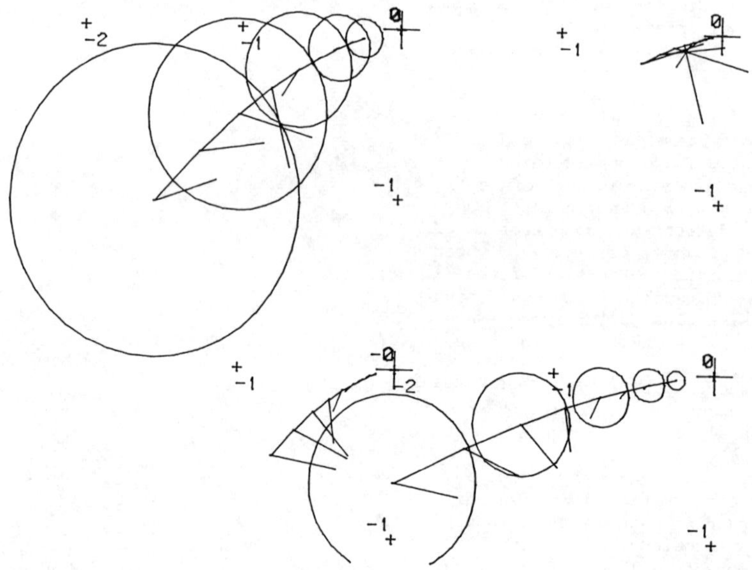

Fig. 2. Nyquist Array with Gershgorin bands and interaction vectors for the example of Gray and Taylor (1974)

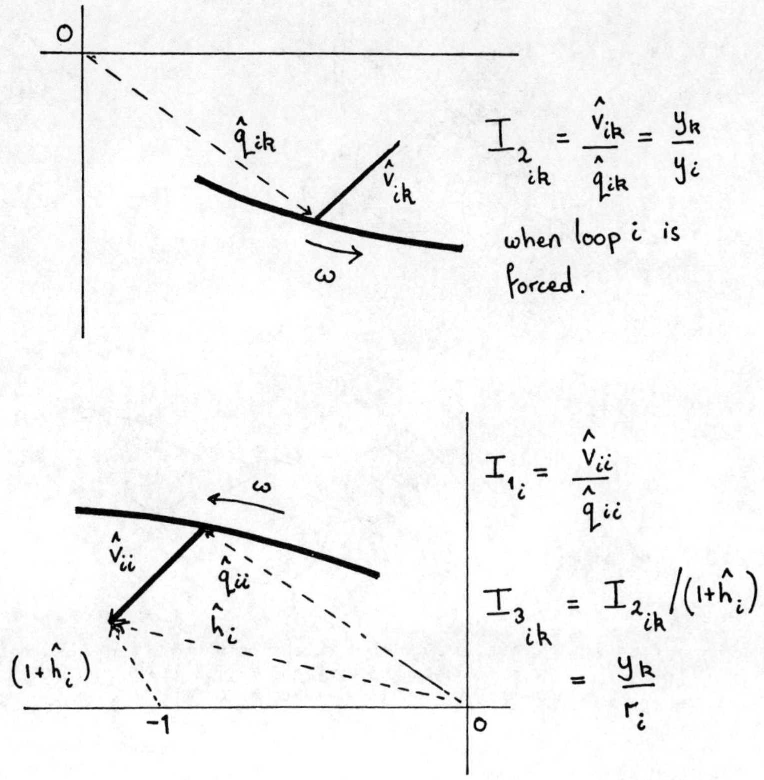

Fig. 3. Graphical interpretation of interaction functionals

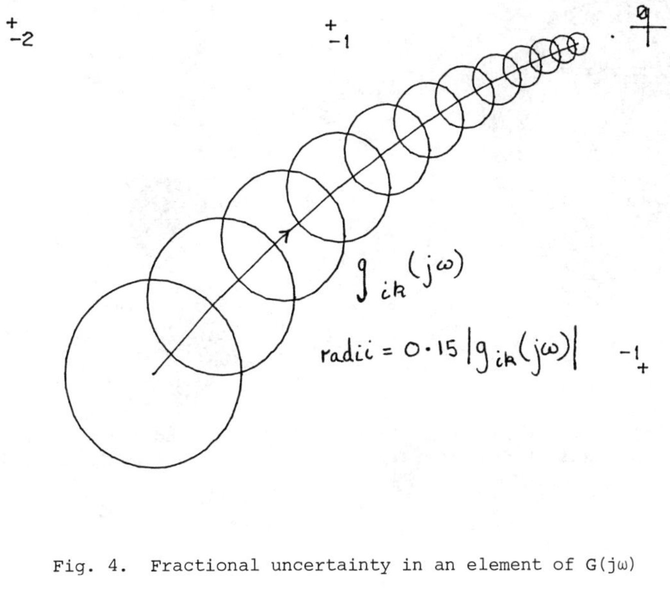

Fig. 4. Fractional uncertainty in an element of G(jω)

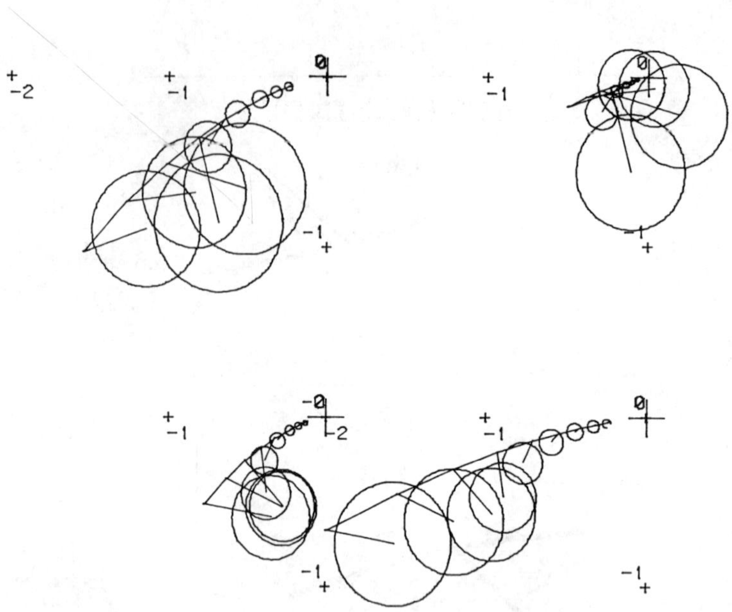

Fig. 5. Nyquist array with interaction vectors and uncertainty circles for the example of Gray and Taylor (1979) with 0.1 fractional uncertainty in all elements of $G(j\omega)$

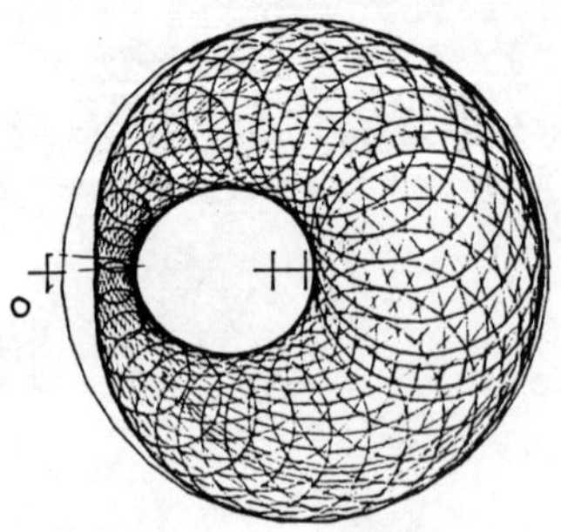

Fig. 6. Smallest circle enclosing cardoid formed by multiplying two normalised complex numbers with fractional uncertainties 0.3 and 0.7

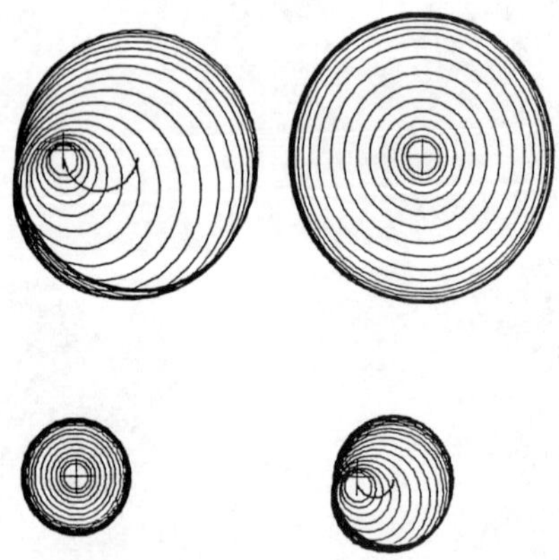

Fig. 7. Nyquist array with interaction vectors and uncertainty circles for the example of Rekasius (1979)

AUTHOR INDEX

SUBJECT INDEX